Polymeric Carriers for Delivery Systems in Biomedical Applications – in Memory of Professor Andrzej Dworak

Polymeric Carriers for Delivery Systems in Biomedical Applications – in Memory of Professor Andrzej Dworak

Marek M. Kowalczuk
Iza Radecka
Barbara Trzebicka

MDPI

Basel • Beijing • Wuhan • Barcelona • Belgrade • Novi Sad • Cluj • Manchester

Editors
Marek M. Kowalczuk
Centre of Polymer and
Carbon Materials
Polish Academy of Sciences
Zabrze
Poland

Iza Radecka
Faculty of Science and
Engineering
University of Wolverhampton
Wolverhampton
United Kingdom

Barbara Trzebicka
Centre of Polymer and
Carbon Materials
Polish Academy of Sciences
Zabrze
Poland

Editorial Office
MDPI AG
Grosspeteranlage 5
4052 Basel, Switzerland

This is a reprint of articles from the Special Issue published online in the open access journal *Polymers* (ISSN 2073-4360) (available at: www.mdpi.com/journal/polymers/special_issues/polym_carr_deliv_biomed_appli_mem_prof_Dworak).

For citation purposes, cite each article independently as indicated on the article page online and as indicated below:

Lastname, A.A.; Lastname, B.B. Article Title. *Journal Name* **Year**, *Volume Number*, Page Range.

ISBN 978-3-7258-2212-6 (Hbk)
ISBN 978-3-7258-2211-9 (PDF)
doi.org/10.3390/books978-3-7258-2211-9

Contents

About the Editors . . . . **vii**

Preface . . . . **ix**

Alicja Utrata-Wesołek, Barbara Trzebicka, Jerzy Polaczek, Iza Radecka and Marek Kowalczuk
Polymeric Carriers for Delivery Systems in Biomedical Applications—In Memory of Professor Andrzej Dworak
Reprinted from: *Polymers* **2023**, *15*, 1810, doi:10.3390/polym15081810 . . . . **1**

Stanislaw Slomkowski, Teresa Basinska, Mariusz Gadzinowski and Damian Mickiewicz
Polyesters and Polyester Nano- and Microcarriers for Drug Delivery
Reprinted from: *Polymers* **2024**, *16*, 2503, doi:10.3390/polym16172503 . . . . **6**

Kriti Kapil, Shirley Xu, Inseon Lee, Hironobu Murata, Seok-Joon Kwon and Jonathan S. Dordick et al.
Highly Sensitive Detection of Bacteria by Binder-Coupled Multifunctional Polymeric Dyes
Reprinted from: *Polymers* **2023**, *15*, 2723, doi:10.3390/polym15122723 . . . . **42**

Pascal Bevan, Maria Vicenta Pastor, María Pilar Almajano and Idoia Codina-Torrella
Antioxidant and Antiradical Activities of *Hibiscus sabdariffa* L. Extracts Encapsulated in Calcium Alginate Spheres
Reprinted from: *Polymers* **2023**, *15*, 1740, doi:10.3390/polym15071740 . . . . **58**

Matylda Szewczyk-Łagodzińska, Andrzej Plichta, Maciej Dębowski, Sebastian Kowalczyk, Anna Iuliano and Zbigniew Florjańczyk
Recent Advances in the Application of ATRP in the Synthesis of Drug Delivery Systems
Reprinted from: *Polymers* **2023**, *15*, 1234, doi:10.3390/polym15051234 . . . . **77**

Subramaniyan Ramasundaram, Sivasangu Sobha, Gurusamy Saravanakumar and Tae Hwan Oh
Recent Advances in Biomedical Applications of Polymeric Nanoplatform Assisted with Two-Photon Absorption Process
Reprinted from: *Polymers* **2022**, *14*, 5134, doi:10.3390/polym14235134 . . . . **128**

Varvara Chrysostomou, Aleksander Foryś, Barbara Trzebicka, Costas Demetzos and Stergios Pispas
Amphiphilic Copolymer-Lipid Chimeric Nanosystems as DNA Vectors
Reprinted from: *Polymers* **2022**, *14*, 4901, doi:10.3390/polym14224901 . . . . **148**

Katya Kamenova, Lyubomira Radeva, Krassimira Yoncheva, Filip Ublekov, Martin A. Ravutsov and Maya K. Marinova et al.
Functional Nanogel from Natural Substances for Delivery of Doxorubicin †
Reprinted from: *Polymers* **2022**, *14*, 3694, doi:10.3390/polym14173694 . . . . **172**

Mattia Parati, Louisa Clarke, Paul Anderson, Robert Hill, Ibrahim Khalil and Fideline Tchuenbou-Magaia et al.
Microbial Poly-γ-Glutamic Acid (γ-PGA) as an Effective Tooth Enamel Protectant
Reprinted from: *Polymers* **2022**, *14*, 2937, doi:10.3390/polym14142937 . . . . **185**

Judit E. Puskas, Gayatri Shrikhande, Eniko Krisch and Kristof Molnar
Multifunctional PEG Carrier by Chemoenzymatic Synthesis for Drug Delivery Systems: In Memory of Professor Andrzej Dworak
Reprinted from: *Polymers* **2022**, *14*, 2900, doi:10.3390/polym14142900 . . . . **203**

Paria Pouyan, Mariam Cherri and Rainer Haag
Polyglycerols as Multi-Functional Platforms: Synthesis and Biomedical Applications †
Reprinted from: *Polymers* **2022**, *14*, 2684, doi:10.3390/polym14132684 . . . . **216**

Andra-Cristina Humelnicu, Petrișor Samoilă, Corneliu Cojocaru, Raluca Dumitriu, Andra-Cristina Bostănaru and Mihai Mareș et al.
Chitosan-Based Therapeutic Systems for Superficial Candidiasis Treatment. Synergetic Activity of Nystatin and Propolis
Reprinted from: *Polymers* **2022**, *14*, 689, doi:10.3390/polym14040689 . . . . **234**

Chen Jiao, Franziska Obst, Martin Geisler, Yunjiao Che, Andreas Richter and Dietmar Appelhans et al.
Reversible Protein Capture and Release by Redox-Responsive Hydrogel in Microfluidics
Reprinted from: *Polymers* **2022**, *14*, 267, doi:10.3390/polym14020267 . . . . **253**

Abeer Aljubailah, Wafa Nazzal Odis Alharbi, Ahmed S. Haidyrah, Tahani Saad Al-Garni, Waseem Sharaf Saeed and Abdelhabib Semlali et al.
Copolymer Involving 2-Hydroxyethyl Methacrylate and 2-Chloroquinyl Methacrylate: Synthesis, Characterization and In Vitro 2-Hydroxychloroquine Delivery Application
Reprinted from: *Polymers* **2021**, *13*, 4072, doi:10.3390/polym13234072 . . . . **269**

Aleksander Forys, Maria Chountoulesi, Barbara Mendrek, Tomasz Konieczny, Theodore Sentoukas and Marcin Godzierz et al.
The Influence of Hydrophobic Blocks of PEO-Containing Copolymers on Glyceryl Monooleate Lyotropic Liquid Crystalline Nanoparticles for Drug Delivery
Reprinted from: *Polymers* **2021**, *13*, 2607, doi:10.3390/polym13162607 . . . . **301**

Martin Studenovský, Anna Rumlerová, Libor Kostka and Tomáš Etrych
HPMA-Based Polymer Conjugates for Repurposed Drug Mebendazole and Other Imidazole-Based Therapeutics
Reprinted from: *Polymers* **2021**, *13*, 2530, doi:10.3390/polym13152530 . . . . **319**

Athanasios Skandalis, Dimitrios Selianitis and Stergios Pispas
PnBA-b-PNIPAM-b-PDMAEA Thermo-Responsive Triblock Terpolymers and Their Quaternized Analogs as Gene and Drug Delivery Vectors
Reprinted from: *Polymers* **2021**, *13*, 2361, doi:10.3390/polym13142361 . . . . **333**

Ken Cham-Fai Leung, Kathy W. Y. Sham, Josie M. Y. Lai, Yi-Xiang J. Wang, Chi-Hin Wong and Christopher H. K. Cheng
Citrate-Coated Magnetic Polyethyleneimine Composites for Plasmid DNA Delivery into Glioblastoma †
Reprinted from: *Polymers* **2021**, *13*, 2228, doi:10.3390/polym13142228 . . . . **348**

About the Editors

Marek M. Kowalczuk

Marek M. Kowalczuk is a Professor at the Centre of Polymer and Carbon Materials at the Polish Academy of Sciences, Zabrze, Poland, and Emeritus Professor in synthetic/polymer chemistry at the University of Wolverhampton, UK. He received his PhD in 1984 from the Faculty of Chemistry, Silesian University of Technology, Gliwice, Poland, and his DSc in 1994 from the same university. Since 2010, he has been a Professor of Chemistry, nominated by the President of Poland. He has been a Visiting Professor at Ohio State University, Columbus, USA; a Visiting Lecturer at the University of Massachusetts, Amherst, USA; and a Marie Curie fellow at the University of Bologna, Italy. He was an elected member of the Chemistry Committee of the Polish Academy of Sciences. He is an author and co-author of over 200 scientific papers and several patents. His main scientific interests are biodegradable and functional polymer biomaterials, novel initiators and mechanisms of anionic polymerisation related to the synthesis of biodegradable polymers possessing the desired architecture, the biodegradation of polymers, polymer mass spectrometry, and forensic engineering of advanced polymeric materials.

Iza Radecka

Iza Radecka is a professor at the Faculty of Science and Engineering at the University of Wolverhampton, UK. Iza received an MSc in biological sciences from the University of Silesia, Poland, in 1990. Within biosciences, her research first focused on anatomy and histology (interactions between heavy metals and enzymatic activity in brain tissue). In 1991, she took a position as a junior researcher at the Institute of Polymer Chemistry at the Polish Academy of Sciences in Zabrze, Silesia, with a special interest in the biodegradability of different synthetic polymer blends. After one year there, she decided to take a PhD researcher/lecturer post at the University of Silesia, Katowice, Poland, with a special interest in microbial biotechnology concerning the production of biodegradable polymers by bacteria under different environmental conditions. She completed her PhD at the Department of Biochemistry, University of Silesia, Katowice, Poland. After that, in 2000, she joined the School of Applied Sciences at the University of Wolverhampton, United Kingdom. Since then, her research has focused on the cost-effective synthesis of new biomaterials using bacterial biopolymers from eco-sustainable feedstock and their chemical derivatization, which can transform crude polymers into a range of highly valuable products. Iza has published numerous research papers in highly ranked scientific journals and authored several chapters in biotechnological books. Iza has also given a broad number of invited lectures at international conferences. Iza teaches a wide variety of microbiology and biotechnology courses, both undergraduate and postgraduate level where she puts her knowledge and experience to good use. Iza is a Fellow of the Royal Society of Biology.

Barbara Trzebicka

Barbara Trzebicka is currently the director of the Centre of Polymer and Carbon Materials Polish Academy of Sciences and a professor in the Laboratory of nano- and microstructural materials in the Centre. Barbara Trzebicka received an MSc in physics from the University of Silesia, Poland in 1978. Within physics, her research was focused on the application of infrared and Raman spectroscopy to the study of silane molecules. Then, she took a position as a junior researcher at the Institute of Polymer at the Polish Academy of Sciences in Zabrze, with a special interest in the synthesis of polyepichlorohydrin and its characterization in a solid state. She completed her PhD in 1987 at the Faculty of Chemistry, Silesian University of Technology in Gliwice.

In 1988, she started working as an assistant professor at the Institute of Coal Chemistry of the Polish Academy of Sciences in Gliwice, focusing her research on the physicochemical characterization of various forms of hard coal and then returning to polymer research. She participated in the study of polyoxazolines and polyethers, as well as amphiphilic copolymers and their organization in water. From 2007, she continued her scientific work at the Centre of Polymer and Carbon Materials Polish Academy of Sciences in Zabrze. She obtained her habilitation in chemistry at the Faculty of Chemistry, Warsaw University of Technology, in 2011, and the academic title of professor in chemistry in 2016. Since then, her research has focused on polymers with branched structures, thermoresponsive polymers, polymer–peptide conjugates, lipid–polymer systems, polymer nanoparticles for encapsulation of biologically active substances, aggregation processes of polymers in solutions, polymer surfaces for cell culture, and use of dynamic and static light scattering in polymer studies. Prof. Trzebicka is a Fellow of the Polish Chemical Society and a member of the European Chemical Federation Board and the Polish Academy of Arts and Sciences.

Preface

Polymeric drug delivery enables the introduction of a therapeutic substance into the body. A successful therapeutic outcome requires the appropriate consideration of pharmacodynamics and pharmacokinetics. Among polymeric materials, both natural and synthetic polymers can be applied for controlled drug delivery under the condition that they possess the ability to be in contact with a living system without producing any adverse effects. This is especially so with biodegradable polymers, which are a suitable choice for numerous new drug-delivery systems. This reprint, dedicated to the memory of the late Professor Andrzej Dworak, scientist and friend, covers recent advances in natural and synthetic polymeric materials with the desired physical, chemical, biological, biomechanical, and degradation properties to match the various requirements of controlled drug delivery.

Marek M. Kowalczuk, Iza Radecka, and Barbara Trzebicka
Editors

polymers MDPI

Editorial

Polymeric Carriers for Delivery Systems in Biomedical Applications—In Memory of Professor Andrzej Dworak

Alicja Utrata-Wesołek [1,*], Barbara Trzebicka [1], Jerzy Polaczek [2], Iza Radecka [3] and Marek Kowalczuk [1,*]

1 Centre of Polymer and Carbon Materials Polish Academy of Sciences, M. Curie-Sklodowskiej St. 34, 41-800 Zabrze, Poland; btrzebicka@cmpw-pan.pl
2 SIGMA-NOT Eds., Ratuszowa St. 11, 03-450 Warszawa, Poland
3 Faculty of Science and Engineering, University of Wolverhampton, Wolverhampton WV1 1 LY, UK; i.radecka@wlv.ac.uk
* Correspondence: autrata@cmpw-pan.pl (A.U.-W.); mkowalczuk@cmpw-pan.pl (M.K.); Tel.: +48-32-271-077 (ext. 261) (A.U.-W.)

check for updates

Citation: Utrata-Wesołek, A.; Trzebicka, B.; Polaczek, J.; Radecka, I.; Kowalczuk, M. Polymeric Carriers for Delivery Systems in Biomedical Applications—In Memory of Professor Andrzej Dworak. *Polymers* **2023**, *15*, 1810. https://doi.org/10.3390/polym15081810

Received: 4 April 2023
Accepted: 5 April 2023
Published: 7 April 2023

Since the appearance of the first civilizations, various substances have been used to improve human health. Although nature is undoubtedly the source of many medicinal substances, since the early 19th century there has been an enormous development of pharmaceutical products aimed at treating or improving human health. Over time, the emergence of new generations of therapeutic agents has been observed, including not only small molecules, but also proteins and peptides, monoclonal antibodies, nucleic acids, and living cells. This progress has opened up new possibilities for the use of these therapeutic agents in the treatment of specific diseases, while at the same time causing enormous challenges in delivering them into the body while maintaining their therapeutic usefulness. For the effective action of these agents, it is necessary, among other things, for them to have proper solubility, stability, target location, off-target non-toxicity, controlled pharmacokinetics, possibility of bypassing biological barriers (cell membranes, nucleus) or, in the case of cells, maintaining cell viability and phenotypes [1]. These challenges have compelled scientists to develop a variety of advanced drug delivery systems (DDSs). DDSs are formulations or devices used to transport and release (preferably in a controlled manner) therapeutic agents to their destinations in the body, and to minimize off-target drug accumulation. Thanks to such control of the rate, time, and localization of release of these drugs in the body, the DDSs enable us to achieve maximum therapeutic efficacy. The developed systems for delivering the bioactive agents include various forms, from the conventional (e.g., tablets, capsules, drops, sprays, injections, or creams) to controlled ones (smart, modulated, and targeted delivery systems in the form of micro- and nanoparticles, micelles, liposomes, polymersomes, conjugates, gels, or implants) [2,3].

Polymers, along with metals and ceramics, constitute the most common platform used for DDSs [4,5]. They have many advantages over the other classes, and many of them are well tolerated during contact with a living system without producing any adverse effects. They are non-toxic and biologically inert. Both natural and synthetic polymers can be applied for controlled drug delivery, although the use of synthetic ones is preferable. This is due to the fact that, thanks to the controlled processes of polymerization, it is possible to obtain materials with a highly reproducible structure–function relationship. Some of the polymers are biodegradable, so the problem of their accumulation within the body is eliminated, providing that their degradation by-products are non-toxic, that they do not produce any immune response, or lie below the renal threshold level. Moreover, polymers can be fabricated into complex structures and shapes (e.g., homo- and copolymers, star polymers, dendrimers, branched polymers, micelles, nanoparticles, vesicles, gels, surfaces) leading to a wide range of physicochemical properties. Last but not least, polymers have tunable chemistries, including controllable and responsive properties (e.g., stimuli responsive) and countless possibilities for modifications to achieve desired properties and mimic biological systems. All these beneficial properties of polymers have enabled huge progress in the development of drug delivery technologies, allowing for transportation of not only small drugs, both hydrophilic and hydrophobic, but also proteins and nucleic acids (Figure 1). This achievement, by improving drug safety and efficacy, may unquestionably facilitate better patient comfort and quality of life.

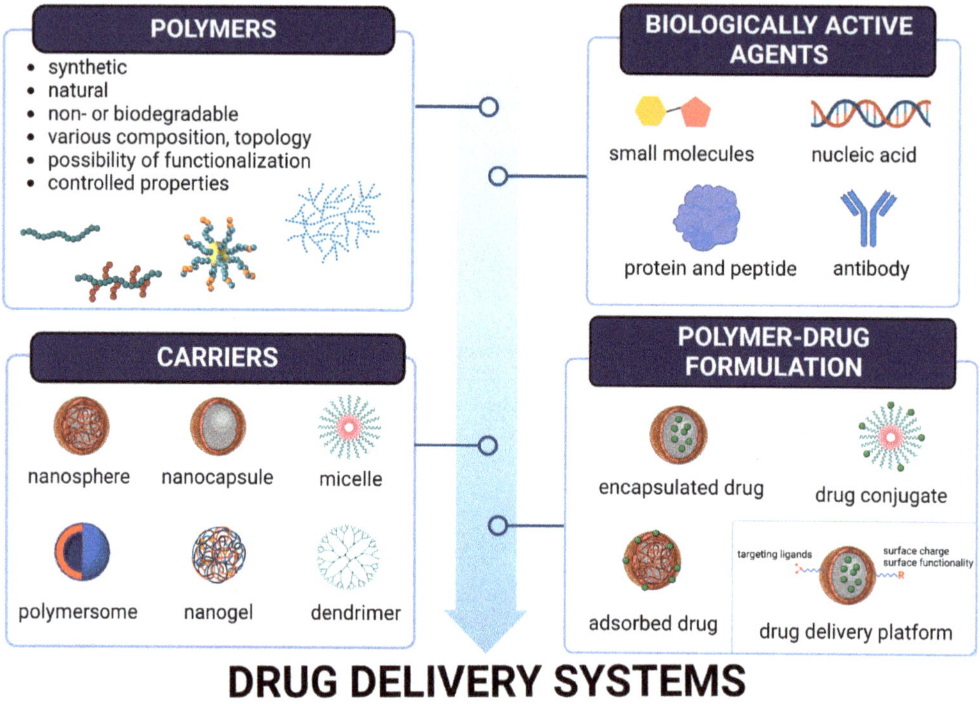

Figure 1. Schematic representation of the formulation of polymeric carriers for delivery systems in biomedical applications.

Over the past 70 years, DDS polymer materials and their designs have progressed from external devices and simple off-the-shelf macroscopic supplies through to microscopic particles and ultimately to complex and rationally designed nanocarriers. The application of DDSs has improved the clinical usefulness of many drugs and enabled new therapeutics, such as anti-cancer and siRNA therapies. However, there are still remaining challenges that drug delivery systems have to overcome to be clinically viable. The next generation of DDSs will need to be able to overcome the biological barriers which limit the delivery of complex therapeutic molecules, and utilize less invasive systems which secrete biomolecules into specific tissues, at specific times and concentrations, sometimes for a prolonged period of time. Therefore, the innovations in the field of polymer materials will be the driving force in overcoming existing boundaries, while the polymer DDSs will still be considered the most active field of biomedical research in pharmaceutical industries and academic laboratories.

This book comprises articles concerning recent advances in natural and synthetic polymeric materials with the desired physical, chemical, biological, and biomechanical properties to match the various requirements of controlled delivery of various therapeutic agents. The book contains reviews and original articles of eminent scientists and friends of Professor Dworak, each of which was published previously as original contributions to the *Polymers* Special Issue dedicated to the memory of late Professor Andrzej

Dworak https://www.mdpi.com/journal/polymers/special_issues/polym_carr_deliv_biomed_appli_mem_prof_Dworak (accessed on 23 February 2021)

Prof. Dworak was an outstanding specialist with experience in the studies of the mechanisms of oxirane and cyclic imines polymerization and controlled radical polymerization of various types of monomers in order to obtain macromolecules with carefully planned structures and properties, including polymers sensitive to stimuli. In his research, he successfully translated knowledge in the field of polymer chemistry and polymer materials into their potential use in medicine and pharmacy. Prof. Dworak initiated many national and European projects in which he carried out studies in the field of basic research, including the preparation and therapeutic use of various types of nanoparticles and polymer nanocontainers carrying different types of active substances. He also managed projects related to the development of polymer supports for the culture and transfer of cell sheets, which significantly accelerated the healing process of burn wounds. The results of Prof. Dworak's research has been published in nearly 150 articles in international journals, and he was the co-author of several patents.

On 17 March 2022, the Silesian Meeting on Polymer Materials Conference (Polymat 2022) was organized and dedicated to the memory of Prof. Andrzej Dworak (Figure 2). The conference was held by the Centre of Polymer and Carbon Materials of the Polish Academy of Sciences (CMPW PAN) at Zabrze on the first anniversary of Prof. Dworak's death. The participants were welcomed by Prof. Barbara Trzebicka, Director of the Centre, and Mr. Krzysztof Lewandowski, Vice President of the City of Zabrze. After a brief opening speech delivered by the Chairman of the Committee of Chemistry of the Polish Academy of Sciences, Professor Janusz Jurczak, Professor Zbigniew Florjańczyk, Warsaw University of Technology, and Professor Stanisław Penczek, Centre of Molecular and Macromolecular Studies of Polish Academy of Sciences, Łódź, presented the work and scientific activity of Professor A. Dworak. His medicine-oriented studies on polymeric materials were emphasized in particular.

Figure 2. POLYMAT 2022 Conference in memory of Prof. Andrzej Dworak (photo by CMPW PAN). Top row: Prof. Andrzej Dworak; Prof. Axel Müller giving a lecture. Bottom row: Mattia Parati, University of Wolverhampton, United Kingdom, receiving the award for the best poster.

According to the conference program, Prof. Axel Müller (Figure 2), Johannes Gutenberg University Mainz, Germany, delivered a lecture on tapered block and multiblock copolymers made via statistical anionic copolymerization; Prof. Rainer Haag, Free University Berlin, Germany, lectured on the synthesis and biomedical applications of multifunctional polyglycerols; and Prof. Stergios Pispas, National Hellenic Research Foundation, Athens, Greece, presented thermoresponsive linear and hyperbranched copolymers synthesized by reversible addition-fragmentation chain-transfer (RAFT) polymerization. After a lunch break, Prof. Petar Petrov, Institute of Polymers, Bulgarian Academy of Sciences, Sofia, reviewed multifunctional block copolymer nanocarriers; and Prof. Brigitte Voit, Leibniz Institute of Polymer Research, Dresden, Germany, gave a lecture on responsive nanocapsules and multicompartments used as cellular mimics. Prof. Marek Kowalczuk (Centre of Polymer and Carbon Materials of Polish Academy of Sciences, Zabrze), Prof. Stanisław Słomkowski (Centre of Molecular and Macromolecular Studies of Polish Academy of Sciences, Łódź), Prof. Christo Tsvetanov (Institute of Polymers of Bulgarian Academy of Sciences), and Prof. Hans-Jürgen Adler (Dresden University of Technology, Germany), performed the roles of chairpersons very efficiently.

During poster sessions, more than 130 scientific communications were presented and discussed (according to the rule "one participant—one poster"). The posters were available also on the Internet. Many of them were of high practical importance. The posters were evaluated by a special jury (headed by Prof. Neli Koseva, Institute of Polymers, Bulgarian Academy of Sciences). The three best communications were selected and awarded prizes of EUR 450 each, funded by the Swiss journal *Polymers*, published by the Multidisciplinary Digital Publishing Institute. In the category of advanced synthesis and study of polymeric materials, Aneta Medaj, Jagiellonian University, Cracow, Poland, was awarded for her presentation entitled "Polymer nanocapsules templated on liquid cores as model photoreactors". In the category of interdisciplinary and international studies, Erik Dimitrov of the Institute of Polymers, Bulgarian Academy of Sciences, Sofia, Sofia University, and the Centre of Polymer and Carbon Materials, Zabrze, Poland, was awarded for the presentation "Nucleolipid vesicles. Supramolecular structures resembling spherical nucleic acids". In the category of advanced biopolymer materials, Mattia Parati (Figure 2), University of Wolverhampton, United Kingdom, was distinguished with an award for the paper "Algal and yeast waste fraction valorization for the biosynthesis of poly(γ-glutamic acid). A versatile biomaterial".

The conference was held under the auspices of the European Federation of Polymers, the Committee of Chemistry of the Polish Academy of Sciences, and the Polish Chemical Society. The media patronage over the conference was taken by *Polymers*, MDPI, and the *Chemical Industry*.

The conference was accompanied with a small exhibition of research instruments (polymer analysis and testing). The Scientific Committee, headed by Prof. Zbigniew Florjańczyk, was responsible for the high scientific level of the presented papers, and the Organizing Committee, headed by Prof. Alicja Utrata-Wesołek, orchestrated the smooth running of the event. The conference was attended by about 170 scientists from Poland and abroad. After the closing remarks given by Prof. Barbara Trzebicka, chairwoman of the Conference, the participants proposed a farewell toast: to meet again at Zabrze in the near future! An extended report on the conference was also published in Polish [6].

Author Contributions: Conceptualization and writing original draft, A.U.-W. and J.P.; writing—review and editing, B.T. and I.R.; supervision, M.K. All authors have read and agreed to the published version of the manuscript.

Conflicts of Interest: The authors declare no conflict of interest.

References

1. Vargason, A.M.; Anselmo, A.C.; Mitragotri, S. The evolution of commercial drug delivery technologies. *Nat. Biomed. Eng.* **2021**, *5*, 951–967. [CrossRef] [PubMed]
2. Benoit, D.S.W.; Overby, C.T.; Sims, K.R., Jr.; Ackun-Farmmer, M.A. Drug Delivery Systems. In *Biomaterials Science*, 4th ed.; Wagner, W.R., Sakiyama-Elbert, S.E., Zhang, G., Yaszemski, M.J., Eds.; Elsevier: Amsterdam, The Netherlands; Academic Press: Cambridge, MA, USA, 2020; pp. 1237–1266. ISBN 9780128161371.

3. Adepu, S.; Ramakrishna, S. Controlled Drug Delivery Systems: Current Status and Future Directions. *Molecules* **2021**, *26*, 5905. [CrossRef] [PubMed]
4. Sung, Y.K.; Kim, S.W. Recent advances in polymeric drug delivery systems. *Biomater. Res.* **2020**, *24*, 12. [CrossRef] [PubMed]
5. Englert, C.; Brendel, J.C.; Majdanski, T.C.; Yildirim, T.; Schubert, S.; Gottschaldt, M.; Windhab, N.; Schubert, U.S. Pharmapolymers in the 21st century: Synthetic polymers in drug delivery applications. *Prog. Polym. Sci.* **2018**, *87*, 107–164. [CrossRef]
6. Polaczek, J. Polymat2022. *Przem. Chem.* **2022**, *101*, 225–228.

polymers

MDPI

Review

Polyesters and Polyester Nano- and Microcarriers for Drug Delivery

Stanislaw Slomkowski *, Teresa Basinska, Mariusz Gadzinowski and Damian Mickiewicz

Division of Functional Polymers and Polymer Materials, Centre of Molecular and Macromolecular Studies, Polish Academy of Sciences, H. Sienkiewicza 112, 90-363 Lodz, Poland; teresa.basinska@cbmm.lodz.pl (T.B.); mariusz.gadzinowski@cbmm.lodz.pl (M.G.); damian.mickiewicz@cbmm.lodz.pl (D.M.)
* Correspondence: stanislaw.slomkowski@cbmm.lodz.pl

Abstract: Many therapies require the transport of therapeutic compounds or substances encapsulated in carriers that reduce or, if possible, eliminate their direct contact with healthy tissue and components of the immune system, which may react to them as something foreign and dangerous to the patient's body. To date, inorganic nanoparticles, solid lipids, micelles and micellar aggregates, liposomes, polymeric micelles, and other polymer assemblies were tested as drug carriers. Specifically, using polymers creates a variety of options to prepare nanocarriers tailored to the chosen needs. Among polymers, aliphatic polyesters are a particularly important group. The review discusses controlled synthesis of poly(β-butyrolactone)s, polylactides, polyglycolide, poly(ε-caprolactone), and copolymers containing polymacrolactone units with double bonds suitable for preparation of functionalized nanoparticles. Discussed are syntheses of aliphatic polymers with controlled molar masses ranging from a few thousand to 10^6 and, in the case of polyesters with chiral centers in the chains, with controlled microstructure. The review presents also a collection of methods useful for the preparation of the drug-loaded nanocarriers: classical, developed and mastered more recently (e.g., nanoprecipitation), and forgotten but still with great potential (by the direct synthesis of the drug-loaded nanoparticles in the process comprising monomer and drug). The article describes also in-vitro and model in-vivo studies for the brain-targeted drugs based on polyester-containing nanocarriers and presents a brief update on the clinical studies and the polyester nanocarrier formulation approved for application in the clinics in South Korea for the treatment of breast, lung, and ovarian cancers.

Keywords: poly(β-butyrolactone); polylactide; poly(ε caprolactone); macrolactone; drug carrier; nanoparticle preparation

check for updates

Citation: Slomkowski, S.; Basinska, T.; Gadzinowski, M.; Mickiewicz, D. Polyesters and Polyester Nano- and Microcarriers for Drug Delivery. *Polymers* **2024**, *16*, 2503. https://doi.org/10.3390/polym16172503

Academic Editor: Anton Blencowe

Received: 5 February 2024
Revised: 26 August 2024
Accepted: 29 August 2024
Published: 3 September 2024

1. Introduction

There are many reasons for using bioactive substances hidden in carriers. Several classes of bioactive compounds, for example, ifosfamide, cisplatin, or doxorubicin, designed for use as cytostatic or cytotoxic components of anticancer drugs, should be encapsulated in nanocarriers, shielding healthy cells from the drug's action and thus making it more selective when targeted to cancer cells [1–3].

The bioactive compounds should be entrapped in the nanocarriers also when the patient's organism recognizes them as a "foreign" material that should be destroyed or other way eliminated by the immune system. This is rather common when nucleic acids, proteins, medium oligonucleotides, or oligopeptides are used as drugs [4–6].

Nanocarriers have been specifically designed for transport across various barriers in the body, such as the very tight blood-brain [7,8], blood-spinal cord [9], and blood-peripheral nerve barriers [10] or membranes that separate cells from their surroundings [11,12].

Of particular interest are carriers equipped with molecular structures on their surface that bind to receptors specific to cancer cells [13–17].

Importantly, there are so-called "stealth" nanocarriers, i.e., nanocarriers that are not recognized by the patients' immune system and therefore remain in the blood circulation for a longer time [18–20]. To this list should be added carriers that disintegrate, releasing their contents under the influence of external or internal stimuli, such as local temperature change, pH, glucose, glutathione, enzymes, light, ultrasound, and magnetic field [21–24]. Also of interest are superparamagnetic nanoparticles (SPION), which can be easily "immobilized" in a selected tissue (such as a tumor). Then, when subjected to an alternating electromagnetic field, they increase their temperature and, by hyperthermia, destroy the surrounding cancer cells [25].

After fulfilling their function of transporting bioactive compounds and releasing them at the required location, the carries should be decomposed and eliminated to avoid the undesirable accumulation of their residues. Most often, drug carriers are prepared from synthetic or natural compounds of high molar mass, which, with time, are suitable for hydrolytic degradation. It is worth noting that chemical bonds can be ordered concerning their increased susceptibility to hydrolysis as follows: carbon-carbon bond in aliphatic hydrocarbons $<$ amide $<$ carbonate $\leq$ ester $<$ anhydride. bonds Aliphatic hydrocarbons and polyamides under normal conditions (human body temperature, aqueous environment) are resistant to hydrolysis and therefore are not suitable for the preparation of drug carriers. Of all other synthetic polymers, aliphatic polyesters are the most promising. They represent the largest and most diverse group, including polymers approved by the US Food and Drug Administration for medical applications.

Susceptibility to hydrolysis is not the only requirement that polymer carriers should meet. They should also be mechanically and thermally stable to ensure convenient processing under various conditions. Moreover, the polymers used for the fabrication of drug carriers should be free of any by-products and traces of unreacted monomers. Scientists are also concerned about the possible distant harmful effects of remaining catalyst and initiator fragments (free or chemically bound) in polymeric materials used to make drug carriers.

According to the Web of Science database, out of thousands of papers published in the past ten years on the polymer drug nanocarriers, 35% were made of aliphatic polyesters. This review focuses on the synthesis and properties of aliphatic polyesters that are used or have the potential to be used in the production of drug carriers. Advances in the development of methods for preparing drug carrier systems that are suitable for encapsulating drugs with various properties are discussed. The review ends with a brief presentation of examples of carriers for targeted drug delivery and conclusions.

2. Polyesters for Fabrication of Drug Carriers

Structures of polyester blocks in homo- and copolymers traditionally used for the preparation of nano- and microparticulate carriers of bioactive substances are shown in Figure 1.

It is worth noting that each constitutional repeating unit (CRU) of poly(β-butyrolactone) and polylactide contains a chiral carbon atom (with *R* or *S* configuration), which broadens possible variation of chain microstructure affecting polymer properties such as the ability to crystallize and degrade as well as their thermo-mechanical behavior and, as a result, a field of their applicability.

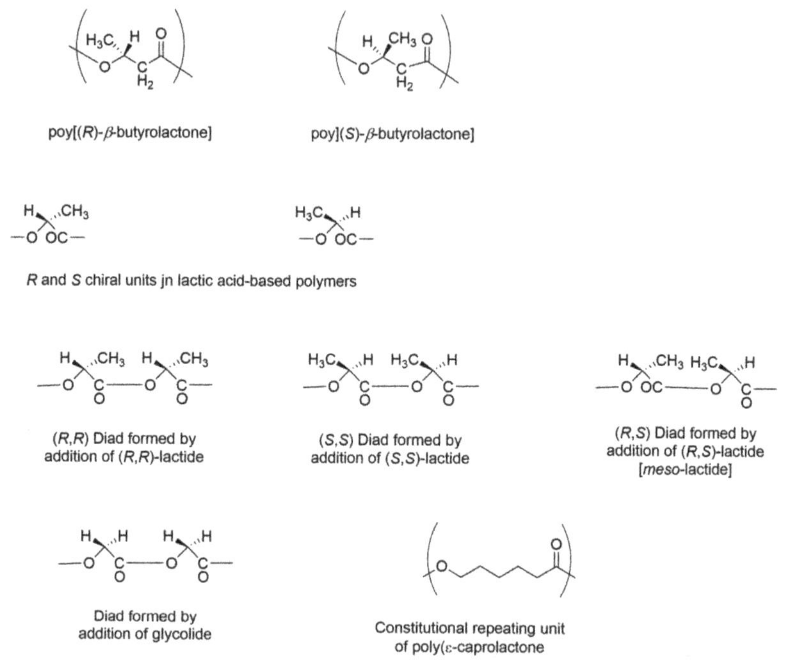

Figure 1. Structures of polymers, which are most often used for the preparation of carriers of drugs and other bioactive compounds.

2.1. *Synthesis of Poly(β-Butyrolactone) (Other Name Poly(Hydroxybutyrate))*

There are three major routes for the synthesis of aliphatic polyesters: ring-opening polymerization of cyclic esters (lactones and lactides), polycondensation of hydroxy-carboxylic acids, and bio-related methods. Poly(β-butyrolactone) (often referred to as poly(hydroxybutyrate)—PHB) is produced by over twenty various strains of microbes [26–30]. Later, in the review, the acronym PHB will be used for bacterial and plant-produced polymers, whereas PBL (poly(β-butyrolactone)) for polymers produced by invitro methods.

The polymer synthesized inside of bacteria aggregates into granules, accounting for up to 90% (wt) of cell dry weight [31]. Molar masses of bacterial PHB are usually in the range of 9×10^3 to 2.4×10^6 g/mol [32]. Polymers with much lower molar masses, often oligomers, are usually needed to produce drug delivery carriers. Therefore, it is worth noting that high-molar-mass polyesters, which like PHB belong to the polyhydroxyalkanoate family, can be enzymatically (e.g., by *Pseudomonas putida (oleovorans)*) hydrolyzed or catalytically (using ethylene glycol/dibutyltin dilaurate) transesterified to low-molar-mass products [33,34].

PHB granules were also produced by transgenic plants created by the introduction of bacterial genes. For example, by alfalfa modified by introduction of genes (*phb*A, *phb*B, *phb*C) from *Ralstonia eutropha* [35], sugar beet clone 93161p with introduced plasmid pRi15834 from *Agrobacterium tumefaciens* [36], flax with cDNA encoding the β-ketothiolase or phbA, phbB, phbC genes [37], oil palm by introduction of phbA, phbB, phbC genes [38] and by transgenic sugarcane (*Saccharum* sp. cv Q117) and switchgrass (*Panicum virgatum* L.) plants [39]. However, PHB content in harvested biomass was usually below 4% dry weight content of harvested biomass. Only quite recently some transgenic *Camelina sativa* plants (known also as false flax) with seeds containing up to 15.2% (*wt*/*wt*) of PHB were obtained [40]. However, this value is still very much lower than the aforementioned 90% (wt) produced by bacteria.

It is worth mentioning that the vast majority of bacteria and plants produce PHB only with the main chain containing only *R* stereocenters and therefore pure isotactic. Synthetic methods make it possible to obtain *R* stereoisomers and a variety of *R/S* polymers with different microstructures.

The polymerization of racemic β-butyrolactone carried in bulk and initiated with Et_2Zn-H_2O was found to lead to a product that consists of two fractions, the one soluble in chloroform and the other insoluble in this solvent [41]. The soluble fraction turned out to be non-crystalline and optically inactive. This means that the said PBL is racemic and atactic. The chloroform-insoluble fraction was also optically inactive, but its X-ray spectra were essentially identical to those of the bacterial PHB, which contains only *R* stereocenters. The conclusion was that this fraction consists only of isotactic chains (or chains with very long isotactic blocks) with about the same content of *R* and *S* stereo-centers.

Shelton et al. reported on the synthesis of β-butyrolactone enriched in *R* isomer (about 73% content) and showed that in polymerization initiated with the Et_3Al-H_2O system, this monomer can be converted to an optically active polymer similar to natural (i.e., bacterial) PHB [42].

Comprehensive studies of the polymerization of racemic β-butyrolactone and other β-alkyl-β-propiolactones initiated with Et_3Al-H_2O and Et_2Zn-H_2O initiators with various proportions of (metal alkyl)-water components have greatly improved the understanding of the mechanism of polymerization but have not led to the required level of stereospecificity of this process [43,44].

Le Borgne and Spassky reported the results of their studies on the stereoelective polymerization of racemic β-butyrolactone initiated with products of reactions of $ZnEt_2$, $AlEt_3$, and $CdMe_2$ with the *R* enantiomer of 3,3 dimethyl-1,2 butanediol (*R*(-)-DMBD) [45]. It was found that all polymerizations were stereoelective but differed concerning the efficiency and character of stereoelectivity. The most efficient was the $ZnEt_2/R$(-)-DMBD system leading to homosteric polymerization (polymerization with asymmetric carbon atoms in the initiator and in the preferentially incorporated lactone isomer being the same) and with the stereoelectivity ratio (a measure of preferential consumption of a given enantiomer) $r_R = 1.7$. Polymerization initiated with the $AlEt_3/R$(-)-DMBD system was also homochiral but its stereoelectivity was lower ($r_R = 1.1$). It should be noted that the $CdMe_2/R$(-)-DMBD system initiates the antisteric polymerization. (Configurations of chiral carbon atoms in initiator and preferentially incorporated monomer were opposite); however, its stereoelectivity was very low ($r_R = 1.01$).

It has been noticed that alkoxides and carboxylic acid salts with sodium or potassium cations do not initiate anionic polymerization of β-butyrolactone [46]. However, complexing the aforementioned cations by crown ethers with the proper size of cavity is known to convert tight ion pairs to the loose ones, facilitating ion pair dissociation to free ions, making these compounds efficient initiators of the β-butyrolactone polymerization. The polymerization initiated with $CH_3O^-K^+$'18crown6 ($CH_3O^-K^+$'18CR6) is a good example. In this case, although the initiation is relatively complicated (see Scheme 1), the propagation consists of the simple addition of monomer molecules to the carboxylate active centers, proceeding with cleavage of the alkyl-oxygen bond [47]. The propagation is shown in Scheme 2. The resulting macromolecules have HO- or crotonate ester end-groups at one and carboxylic acid groups at the other end.

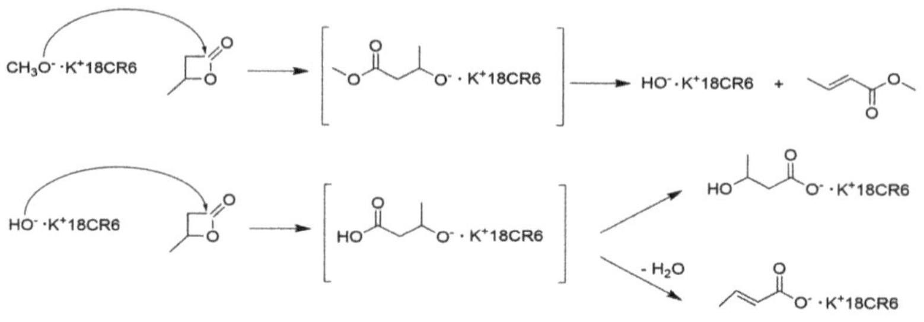

Scheme 1. Polymerization of *β*-butyrolactone—mechanism of initiation with $CH_3O^{-}K^{+}$'18CR6.

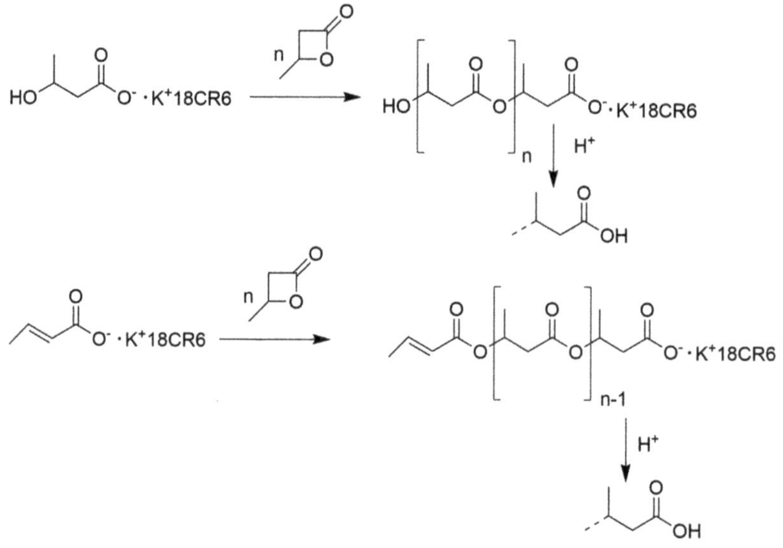

Scheme 2. Polymerization of *β*-butyrolactone—propagation.

Several attempts were undertaken to obtain synthetic PBL [poly(*β*-butyrolactone)] with exclusively (or mainly) ® stereo-centers, i.e., the polymers resembling natural PHB. Very successful in this research were Lenz et al., who used Et_3Al-H_2O and Et_2Zn-H_2O as initiators polymerized (*S*)-*β*-butyrolactone. In these processes the propagation occurred with alkyl-oxygen bond scission, with inversion of configuration, yielding eventually ®-PBL [48]. In the same paper, authors noted that in polymerization of (*S*)-*β*-butyrolactone catalyzed with ethylaluminoxane, the monomer is added via the acyl-oxygen bond scission route with retention of configuration [(*S*)-PBL was produced].

It should be mentioned that although polymerizations of (*S*)-*β*-butyrolactone initiated with alkali metal alkoxides yield polymers almost exclusively with *R* stereocenters, the products still may differ from the natural ones. Whereas the natural PHB contains only HO-. . . . . .-COOH groups, the synthetic one may also contain a fraction of macromolecules with crotonate ester end-groups (CH_sCH=CHCOO-. . . . . .-COOH; see Scheme 2). However, it was found that polymerization of (*S*)-*β*-butyrolactone initiated with ®-3-hydroxybutyric acid sodium' (crown ether) salt gave polymer containing only end-groups, which are present in natural PHB [21].

Bacterial and plant polyhydroxybutyrates with pure isotactic structures have good mechanical properties, enabling their application as degradable replacements of conventional commodity thermoplastics such as polyethylene and polypropylene and as specialty polymers for usage in medicine [49]. However, these polymers also have some undesirable properties. One of them is brittleness [50,51]. Moreover, their melting temperature (about 180 °C) is very close to the temperature at which thermal degradation becomes important, which makes the processing from melt difficult [32,50,52]. The problem was usually solved by using additives, decreasing polymer melting temperature and/or increasing temperature, at which polymer begins to degrade. However, in the case of polymers for medical applications, such an approach would require the selection of not only biocompatible polymers but biocompatible additives as well. Thus, an interesting alternative was the development of the synthesis of preferentially syndiotactic poly(β-butyrolactone) with low melting temperatures (usually below 120 °C) by using racemic (*R*, *S*)-β-butyrolactone as monomer and distannoxane derivatives as catalysts [53].

Recently, important progress in the synthesis of poly(β-butyrolactone)s with various stereo configurations was achieved by Chen et al. [54]. These studies included the synthesis of a mixture of cyclic dimers of β-butyrolactone (4,8-dimethyldioxocane-2,6-dione) with structures shown in Figure 2.

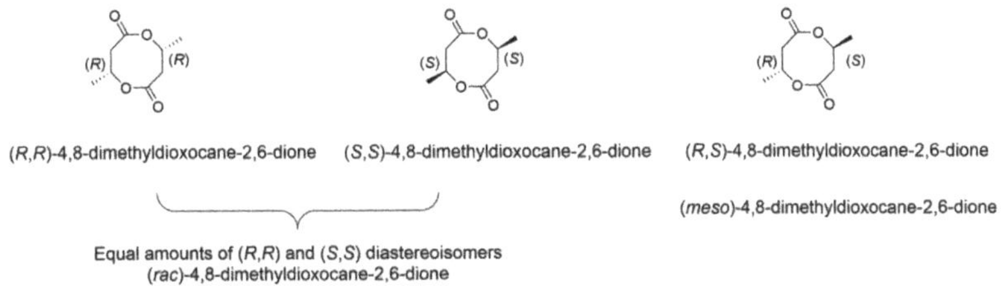

Figure 2. Structures of the cyclic dimers of β-butyrolactone.

From the mixture, they isolated by crystallization pure racemate [(*R*,*R*)-4,8-dimethyldioxocane-2,6-dione + (*S*,*S*)-4,8-dimethyldioxocane-2,6-dione] and by column chromatography and crystallization pure (*R*,*S*)-4,8-dimethyldioxocane-2,6-dione *meso*-diastereoisomer. By using complex initiators inducing stereoselective polymerization of (*rac*)-4,8-dimethyldioxocane-2,6-dione, they obtained a racemic mixture of crystalline isotactic poly[®-β-butyrolactone] and poly[(*S*)-β-butyrolactone]. Polymerization of (*meso*)-4,8-dimethyldioxocane-2,6-dione yielded an amorphous pure syndiotactic polymer. Diastereoselective copolymerization of (*rac*)-4,8-dimethyldioxocane-2,6-dione with various amounts of added at start (*meso*)-4,8-dimethyldioxocane-2,6-dione enabled synthesis of stereosequenced poly(β-butyrolactones) with modified mechanical and thermal properties.

2.2. *Polylactides*

Aliphatic polyesters derived from lactic acid enantiomers, or their cyclic dimers are an important class of polymers used in medicine. The chains of these polymers contain the same *R* and *S* stereorepeating units, regardless of whether they were obtained from lactic acids or lactides. In both cases, the term polylactides is commonly used.

Oligolactides were first obtained by Gay-Lussac and Pelouze in the 19th century as by-products formed during the drying of lactic acid by distillation [55]. However, studies directed toward determining the mechanism of polymerization of lactones and lactides were not conducted until about a century later by Carothers et al. [56]. Unfortunately, the polyesters obtained in these studies had low molar masses and were susceptible to hydrolysis. Both of these properties were considered highly undesirable, and ultimately polylactides did not attract much attention at the time. Even later, during two decades from 1945–1964, only four papers mentioning polylactides have been published. The great interest in polylactides started in the last two decades of the 20th century, during which, according to the Web of Science database, the number of publications on polylactides reached about 3500. The field is still very hot, with the average number of papers published per week approaching 80.

Among the approximately forty thousand journal publications and book chapters on PLA that have appeared in the last twenty years, there have been several very comprehensive reviews [57–61]. It is also worth mentioning the recent IUPAC recommendations for nomenclature and terminology for linear lactic acid-based polymers [38].

In this review, we do not intend to discuss in detail the existing knowledge on the synthesis and properties of PLA, but we briefly summarize the most important information relevant to the application of polylactides as matrices for drug carriers.

The main raw material for the synthesis of polylactides is lactic acid. There are two stereoisomers of this compound (see Figure 3).

(*R*)-lactic acid
IUPAC approved name

(*S*)-lactic acid
IUPAC approved name

(*D*)-lactic acid
other often used name

(*L*)-lactic acid
other often used name

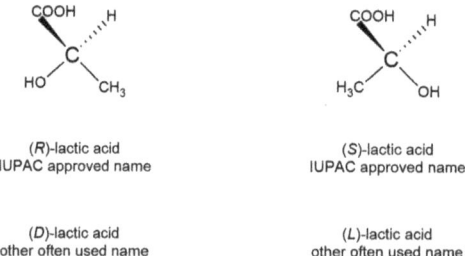

Figure 3. Structures lactic acid stereoisomers. Names according to Ref. [62].

Lactic acid is produced on a mass scale by biotechnological processes (lactic acid fermentation). The following bacteria strains were used for the synthesis of ®-lactic acid: *Lactobacillus delbrueckii* LD0025, *Lactobacillus delbrueckii bulgaricus* (LB12), *Lactobacillus coryniformis subsp. torquens* (DSM20004), *Sporolactobacillus inulinus* (NBRC13595), *Lactobacillus coryniformis* (NCDC369), Sporolactobacillus inulinus SI0073, metabolically engineered *Lactobacillus plantarum*, and *Lactobacillus coryniformis subsp. torquens* (DSM20004) [63–67].

There are also strains of bacteria that produce (*S*)-lactic acid: *Alkaliphilic Bacillus* sp. WL-S20, Bacillus coaggulaans, Bacillus subtilis MUR1, Candida sonorensis, Lactococcus lactis LL0018, *Lactobacillus casei* sp., and recombinant *Escherichia coli* [63,68,69].

Recent advances in microbial lactic acid synthesis, purification process, and purity of final products are significant. Namely, the efficiency of sugar conversion to *R*-lactic acid approached 90%, and the optical purity of the resulting product was 99.9% [67].

However, it is also worth mentioning the chemical synthesis of racemic lactic acid from ethylene in a process consisting of three steps: oxidation of ethylene to acetaldehyde in a process catalyzed with $PdCl_2$, catalytic conversion of acetaldehyde to lactonitrile in reaction with hydrogen cyanide, and hydrolysis of lactonitrile using sulfuric acid as a catalyst [70].

Seemingly the simplest way to synthesize polylactides should be polycondensation of lactic acids. However, this process faces a significant difficulty. In the polycondensation of hydroxyacids, every additional step is accompanied by the formation of a water molecule. Since polycondensation of hydroxyacids is reversible, without the removal of water, the process would reach equilibrium, at which only short polymer chains would be present in the system. However, when the polycondensations were carried out in the presence of catalysts in a high boiling solvent (diphenyl ether) at 130 °C with the removal of water by azeotropic distillation and turning back the dried solvent, polylactides were formed. If after 24 h of the above process, the molar mass of the polylactides was still low (M_w about 5000 g/mol), the polycondensation was continued for an additional 16 h at 160 °C, yielding polylactides with M_w ranging usually from 10^4 to 10^5 g/mol, depending on the catalyst [71]. Many compounds have been tested as catalysts for this solution polycondensation, including protonic acids (H_3PO_4, H_2SO_4, CH_3SO_3H, p-toluenesulfonic acid, Nafion-H®), metals/metalloids (Al, Mg, Sn, Ti, Zn), metal/metalloid derivatives such as oxides (GeO_2, Sb_2O_3, SnO, SnO_2, TiO_2, ZnO, ZrO_2); chlorides ($SnCl_2$, $SnCl_4$, $ZnCl_2$); salts of metals/metalloins and organic acids such as acetic acid (AcO), lactic acid (LA), oleic acid (OA) e.g., $Co(AcO)_2$, $Cu(OA)_2$, $Fe_2(LA)_3$, $Mn(AcO)_2$, $Ni(AcO)_2$, $Ti(acac)_2$, $Y(OA)_3$,

$Zn(LA)_2$; and alkoxides $Al(i\text{-}PrO)_3$, $Ti(BuO)_4$ and also $(Bu)_2SnO$ [71]. The polylactides with the highest molar masses (M_w from 230,000 to 240,000 g/mol) were obtained using Sn, SnO, and $SnCl_2$ catalysts.

It should be mentioned that in 1995 the high-temperature solution condensation process was commercialized by Mitsui Chemicals Co. (Tokyo, Japan). However, production was suspended a few years later.

In 2001, Kimura et al. reported on the development of a melt/solid polycondensation of the (*S*)-lactic acid yielding (*S*)-PLA with M_w reaching 600,000 g/mol (determined by GPC using polystyrene standards for calibration) [72,73]. The process begins with the classical thermal dehydration of monomers with the formation of the oligomers with a degree of polymerization of about 8. To this mixture were added $SnCl_2 \cdot 2H_2O$) and *p*-toluenesulfonic acid. Then, the temperature was increased to 180 °C, and the polycondensation was carried out in the molten state at pressure that decreased gradually to 10 Torr with stirring for 5 h. After this time, the M_w of the polymer was 13,000 g/mol, and the polymerization mixture was cooled to room temperature. The prepolymer was then heated to 105 °C and kept at this temperature for 1 or 2 h. At this stage, the prepolymer crystalized. During crystallization, the polymer chains segregate. The amorphous phase should be more abundant in shorter chains and the crystalline phase in longer chains. As a result, the local concentration of the hydroxyl and carboxyl end-groups should be high in the amorphous phase. Moreover, the end-groups of the crystalline phase chains should be located preferentially in the interfacial areas. In this way, the aforementioned segregation of chains should facilitate polycondensation. After the crystallization step, the polycondensation of the prepolymer was carried out in the solid state at 150 °C for up to 30 h. The highest molar mass (600,000 g/mol) was observed for a crystallization time lasting 2 h and polycondensation lasting 20 h. After a longer (30 h) time, some degradation was observed, and M_w decreased to about 200,000 g/mol. However, despite evident advantages, this process is not used for the industrial synthesis of (*S*)-polylactide [73].

Currently, the most advanced technology for the synthesis of PLAs is based on ring-opening polymerization of lactides. This process involves lactic acid oligo condensation (initiated, for example, by ZnO or $Sn(Oct)_2$ (tin 2-ethylhexanoate)), oligo condensation of lactic acid carried out at 140 °C, catalytic (using ZnO as a catalyst) decomposition of linear PLA oligomers to the cyclic dimers of lactic acids (i.e., lactides), their thorough purification, and their subsequent use as monomers in ring-opening polymerization [61,74,75].

The technology involving the ring-opening polymerization of lactides is the most attractive because it is less demanding and relatively simple. Both polycondensation and ring-opening polymerization are preceded by oligocondensation of lactic acid. However, it is worth noting that oligocondensation does not require a dry monomer. Very good results have been obtained using an aqueous solution with a lactic acid content of only about 40% [50]. In the polycondensation process, the next step is a tedious and energy-demanding, much more advanced dehydration. In the ring-opening process, the second step involves the depolymerization of oligopeptides into lactides, which are easily purified and dried by distillation, sublimation, and crystallization from ethyl acetate and subsequently from ethyl methyl ketone and washing with ice-cold dry ethyl ether [74,75]. The lactides can be distilled and sublimed at temperatures slightly below the thermal decomposition becomes significant, i.e., about 200 °C [76]. The distillation is carried out at a pressure of 0.1–15 Torr and a temperature ranging from 180 to 215 °C, while sublimation is carried out at a pressure of 0.1 Torr and temperatures from 60 to 100 °C [75].

Syntheses of lactides from pure *R* and pure *S* lactic acid stereoisomers yield (*R*,*R*)-lactide and (*S*,*S*)-lactide, respectively. However, similar syntheses from mixtures of *R* and *S* lactic acids in addition to a mixture of (*R*,*R*)-lactide and (*S*,*S*)-lactide diastereoisomers also yield meso-lactide ((*R*,*S*)-lactide). Structures of the aforementioned diastereoisomers are shown in Figure 4. Their proportions depend on the stereoenantiomeric composition of the initial reaction mixture.

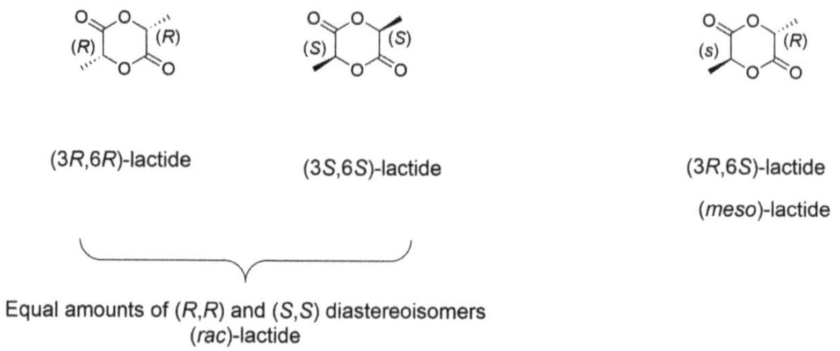

Figure 4. Structures of lactide stereoisomers.

Several classes of compounds were positively verified as initiators and initiators/catalysts. They include: for anionic and coordination ring-opening polymerization metal alkoxides, inorganic salts (salts of organic acids)/alcohol systems, metal oxides, so-called organic superbases, and enzymes; for cationic polymerization protic acids. A large number of metal and metalloid alkoxide initiators was discussed in a review by Slomkowski, Penczek, and Duda [60]. In addition to aluminum tris-isopropoxide ($Al(O\text{-}i\text{-}Pr)_3$) and tin di-butoxide ($Sn(OBu)_2$) to this group belong also many other alkoxides of lithium, sodium, potassium, calcium, magnesium, scandium, scandium, yttrium, lanthanum, titanium hafnium, and zirconium. A characteristic feature of metal alkoxides (both initiators and propagating species) is their tendency to form aggregates that differ in reactivity. For example, $Al(O\text{-}i\text{-}Pr)_3$ may exist as a mixture of reactive trimer and much less reactive tetramer [77]. Thus, for controlled polymerization, aluminum tris-isopropoxide with a known proportion of trimer and tetramer should be used. Methods have been developed to obtain both trimer and tetramer as almost pure individual compounds [78]. However, it is also worth mentioning that aggregation is often reduced by preparing alkoxides complexed by compounds with bulky organic groups [60].

Tin, zinc, calcium, and magnesium salts of 2-ethylhexanoic acid belong to an important group of compounds used to initiate polymerization of lactides [60,79,80]. The most common is the polymerization initiated by tin (II) bis(2-ethylhexanoate) (often called tin (II) octanoate, abbreviated $Sn(Oct)_2$). Several, sometimes contradictory, models of the polymerization mechanism based on the use of the said compound have been proposed. Finally, a complete mechanism consistent with all experimental observations was established by Penczek and Duda [81,82]. They noted that for initiation and propagation, tin (II) octanoate alone is not sufficient, but the polymerization mixture must also contain a certain, usually small, amount of alcohol (ROH) or water. Moreover, it has been established that neither tin (II) octanoate nor alcohol alone can initiate the polymerization of lactides. However, the reaction of these compounds produces in situ the true initiating species, tin (II) mono- and di-alkoxides. These compounds (-Sn-OR and RO-Sn-OR) react with the lactide to form the propagating species responsible for chain growth. All reactions involved in the polymerization process are reversible. The rate of polymerization is proportional to the product of the actual lactide concentration and concentration of the tin alkoxide moieties. The concentration of polymer chains is equal to the initial concentration of alcohol molecules. The whole process involving also chain transfer is shown in Figure 5 and Schemes 3–5.

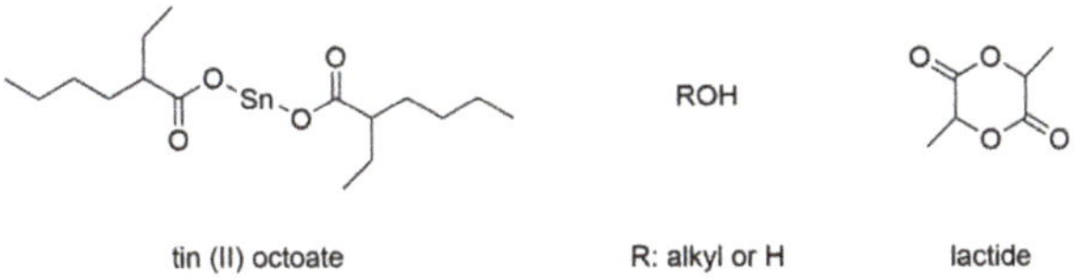

Figure 5. Initial composition of the polymerization mixture.

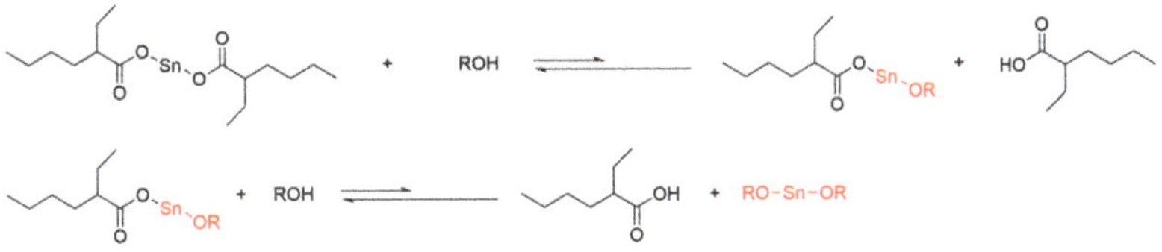

Scheme 3. Formation of initiators (red marked are groups participating in initiation).

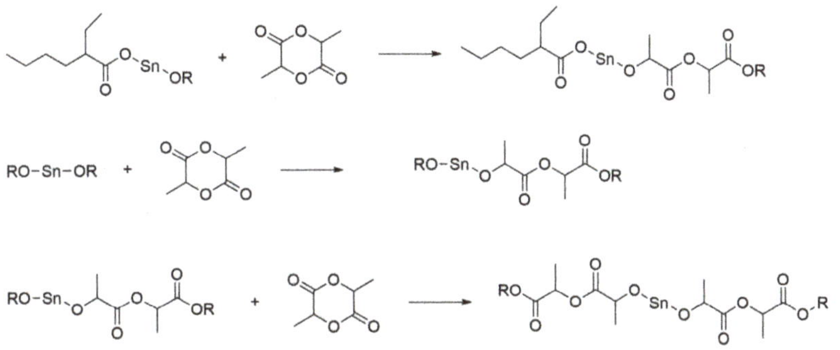

Scheme 4. Initiation of lactide polymerization by an initiator produced in situ from the $Sn(Oct)_2$˙ROH.

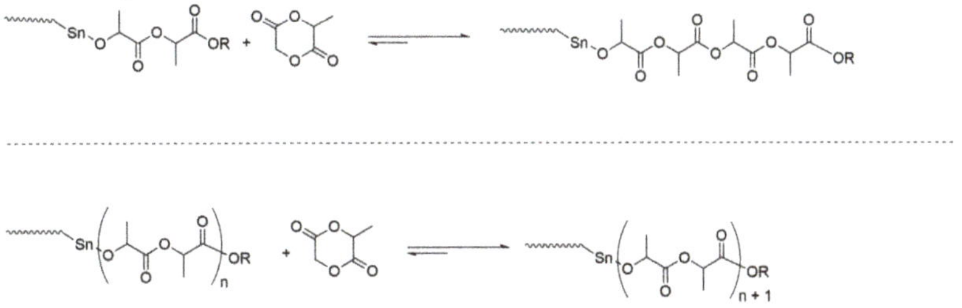

Scheme 5. Propagation in lactide polymerization by an initiator produced in situ from the $Sn(Oct)_2$˙.ROH.

It has been established that propagation in the polymerization of (*S*,*S*)-lactide is a reversible reaction, and at equilibrium, a certain amount of monomer remains in the system [83]. For example, for polymerization carried out in 1,4-dioxane at temperatures ranging from 80 °C to 135 °C, the equilibrium monomer concentration increases from 0.05 to 0.15 mol/L (i.e., from 5 to 15% of the initial lactide concentration). Such a high content of unreacted monomer is usually unacceptable for polymers used for medical purposes because the presence of the residual lactide has the effect of causing uncontrolled degradation of the polymer and may also affect its interactions with biological systems. Two polymerization methods were developed to obtain poly[(*S*,*S*)-lactide] free of traces of unreacted monomer.

The first method consists of the polymerization initiated using $Sn(Oct)_2$ and carried out under two temperature regimes [84]. The polymerization starts at 140 °C in molten monomer. With monomer conversion, the polymerizing mixture solidifies into a supercooled amorphous substance, which, when remained even up to 10 h under the aforementioned conditions, contained 5.6% of unreacted monomer. However, when after the first hour of polymerization at 140 °C the temperature is lowered to 120 °C the crystalline phase is formed and separates from the amorphous one. Finally, after 9 h, the monomer conversion approaches 100%. This occurs because the whole amount of unreacted monomer remains in the amorphous phase and its local concentration exceeds the equilibrium concentration close to the complete monomer conversion. It is even more convenient to perform the polymerization at a higher temperature, i.e., the first step at 170 °C and the second one at 140 °C, because at higher temperatures the [(S,S)-lactide] polymerization proceeds faster.

The second method is based also on a two-step process. In the first step, a simple homopolymerization of (S,S)-lactide is carried out until the monomer concentration approaches the equilibrium concentration. At this stage, the required amount of a second monomer, ε-caprolactone, is added, starting the copolymerization of (*S*,*S*)-lactide and ε-caprolactone [60]. To limit transesterification, the polymerization should be initiated with an initiator that selectively favors propagation over transesterification, e.g., with the reaction product of the trimer of $Al[OCH(CH_3)_2]_3$ with the (S)-(+)-2,2′-[1,1′-binaphthyl-2,2′-diylbis(nitrylomethilidyne)]diphenol [(S)-SB(OH)2]. It has been shown that when the initial concentration of added ε-caprolactone sufficiently exceeds the equilibrium concentration of lactide in homopolymerization, the copolymerization yields products practically free from unreacted lactide, containing chains composed mainly from poly[(S,S)-lactide] with short end-segments rich in poly(ε-caprolactone) [85].

Polylactides belong to polyesters, which were synthesized in the broadest range of molar masses. Almost all the above-mentioned metal alkoxide-based and metal octoate-based initiators were suitable for the synthesis of polymers with M_n up to 40,000. There are many reports on polylactides with M_n exceeding 100,000 that were synthesized using the $Sn(Oct)_2$-based initiators. Polylactides with M_n up to 100,000 were obtained using $Al(O^iPr)_3$ [86]. Polymer with the highest M_n approaching 10^6 was obtained in the polymerization initiated with $Sn(OBu)_2$ [87].

Many researchers were concerned about the possibility of the harmful effects of organometallic compounds on the human body. Therefore, there was concern about the use of polymers obtained with initiators containing metals, especially heavy metal derivatives, in medical applications. In general, this topic has not been well studied, except for tin. After extensive research, the US Food and Drug Administration (FDA) approved using for medical purposes polylactides containing up to 20 ppm of tin [60,80].

Concern over the use of polylactides containing heavy metal compounds has led to an interest in the polymerization of lactides with metal-free initiators. One type of these initiators are compounds formed in situ in reactions of alcohols with bases. The term superbase has been adopted for some very strong bases used for this purpose. The first review papers on the metal-free polymerization systems were published about two decades ago [88,89]. Such compounds were used as strong bases as: 4-pyrrolidinopyridine (PPY) [88–90], 4-dimethylaminopyridine (DMAP) [88–90], 1,8-diazabicycloundec-7-ene (DBU) [91–93], 1,5,7-triazabicyclo [4.4.0]dec-5-ene (TBD) [90], N-methyl-1,5,7-triazabicyclo [4.4.0]dec-5-ene (MTBD) [94], phosphines (PBu_3, $PPhMe_2$, PPh2Me, PPh_3, and chiral phosphines: 1,1′-bis(di-i-propylphosphino) ferrocene, ®-[(*S*)-2-(dicyclohexylphosphino)-ferrocynyl] ethyldicyclohexylphosphine, (−)-1,1′-bis(2*S*,4*S*)-2,4-(diethylphospholano) ferrocene, and (−)-1,2-bis(2*R*,5*R*)-2,5-(dimethylphospholano)-benzene) [95], carbenes [96,97], and thiourea-amine [98].

Another group of initiating systems consists of a combination of alcohols, which act as initiators, and acids, which function as catalysts leading to polymerization according to the activated monomer mechanism [99–102]. In polymerization by activated monomer mechanism, the active centers are usually protonated monomer molecules, and propagation involves their addition to the hydroxyl end-groups of the chains. Therefore, in the metal-free propagating systems, neither basic nor acidic catalysts are covalently bonded to macromolecules and therefore can diffuse out from the matrices of polylactide drug carriers and interact in a harmful way with patients' organisms. Recently, Penczek showed that this problem can be solved by using the initiating systems containing both initiating and catalytic fragments in one molecule [103,104]. The authors use the acronym CINICAT for this class of initiators. Polymerization was initiated using 5-ethyl-2-hydroxy-5-hydroxymethyl-1,3,2-dioxaphosphorinane-2-oxide (GM) and hydroxymethyl phosphonic acid (HMPA), in the case of *S,S*-lactide and ε-caprolactone, respectively. The CINICAT moieties were shown to be covalently bound to the synthesized polyester chains as their end groups and did not diffuse out of the polymers. Examples of the structures of base components of the metal-free initiating systems and of CINICAT compounds are shown in Figures 6 and 7, respectively.

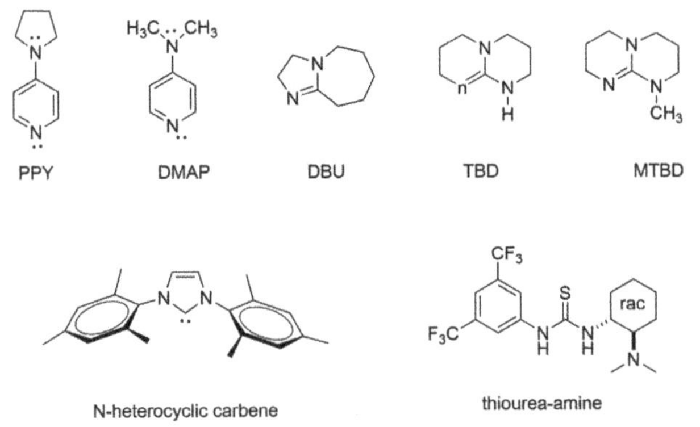

Figure 6. Base components of metal-free initiating systems of the polymerization of lactides: PPY (4-pyrrolidinopyridine), DMAP (4-dimethylaminopyridine), DBU (1,8-diazabicycloundec-7-ene), TBD (1,5,7-triazabicyclo[4.4.0]dec-5-ene), MTBD (N-methyl-1,5,7-triazabicyclo[4.4.0]dec-5-ene), *N*-heterocyclic carbene, thiourea-amine.

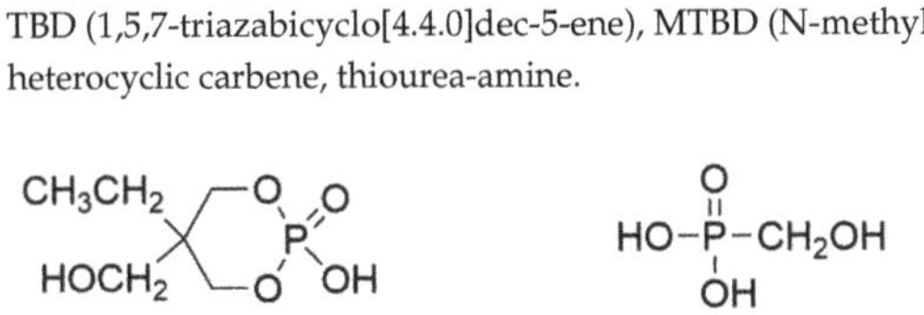

Figure 7. Basic components of the metal-free initiating systems of the polymerization of lactides (GA) and ε-caprolactone HMPA.

Since lactides contain chiral centers in their molecules, some researchers investigated their stereospecific polymerization. While the conversion of pure lactide enantiomers into stereoregular polymers is rather trivial, the attention was mainly concentrated on the polymerization of the racemic mixtures of the (R,R)- and (S,S)-lactides yielding racemic mixtures of the enantiomerically pure homopolymers and/or multiblock copolymers with long stereo sequences. There were also investigated stereospecific polymerization processes, which preferentially yielded polymers with a particular enantiomer chosen from the initially racemic monomer mixture. The first report on the aforementioned stereoelective polymerization of lactide has been published by N. Spassky et al. [105]. The authors obtained a chiral initiator by a three-step process. First synthesized was a chiral Shiff base ligand, ®-2,2′-[1,1′-binaphthyl-2,2′-diylbis(nitrilomethylidyne)]diphenol, denoted as ®-SALBinapht, in the reaction of ®-1,1′-binaphthyl-2,2′-diamine with salicylaldehyde. Then, in reaction with $AlEt_3$, the synthesized ®-SALBinapht base was converted into ®-SALBinaphtAlEt.

The last step consisted of a reaction of ®-SALBinaphtAlEt with CH_3OH, producing the ®-SALBinaphtAlOCH_3 initiator. The described synthetic process is illustrated in Scheme 6.

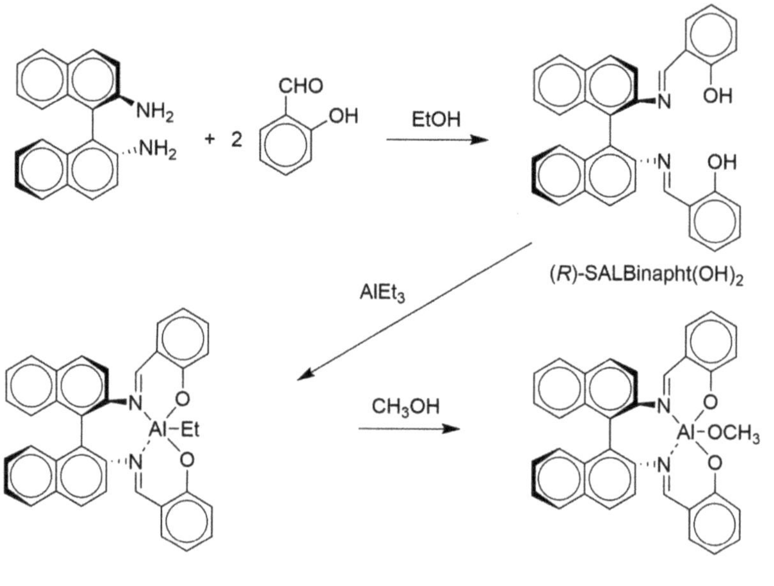

(*R*)-SALBinaphtAlEt (*R*)-SALBinaphtAlOCH$_3$

Scheme 6. A pathway for the synthesis of ®-SALBinaphtAlOCH_3, an initiator for the stereoelective polymerization of racemic lactide.

The stereoselectivity of the polymerization of a racemic mixture of (*R*,*R*)-lactide and (*S*,*S*)-lactide initiated with ®-SALBinaphtAlOCH_3 was high, with up to 88% enrichment in *R* constitutional units at 19% of monomer conversion.

In the following years, ways were developed to synthesize many other initiators containing Schiff base-type moieties and aluminum, zinc, magnesium, titanium, or zirconium alkoxides, useful for inducing stereospecific (stereoselective and stereoelective) polymerization of racemic lactide [106–117]. These polymerizations made it possible to obtain polymers with controlled microstructure and thus with controlled degradability required for drug carriers. Examples of these initiators are shown in Figure 8.

Systems that allow ligand exchange during polymerization provide interesting opportunities for the synthesis of enantio-different di-block copolymers from the *rac*-lactide by the one-pot consecutive stereoelective polymerization. An interesting example was presented by A. Duda et al. [118]. In this process, the polymerization of *rac*-lactide was initiated with (*S*)-SALBinaphtAlO^{i}Pr (produced in situ in the reaction of (*S*)-SALBinapht$(OH)_2$ with Al(O^{i}Pr)$_3$ trimer), and propagation was carried on until the monomer conversion approached 50%. At this moment, when almost the whole amount of (*S*,*S*)-lactide has been polymerized, an equimolar amount of ®-SALBinapht$(OH)_2$ was added to the mixture. It is worth noting that the exchange does not have to be quantitative, but because it is reversible, all chains have a chance to be capped with the ®-SALBinapht ligand for some time and thus enable the addition of the remaining (*R*,*R*)-lactide. As a result, the poly[(*S*,*S*)-lactide]-poly[(*S*,*S*)/(*R*,*R*)-grad-lactide]-poly[(*R*,*R*)-lactide] copolymer has been produced. The reaction path is shown in Scheme 7. It should be mentioned that despite low molar masses (a few thousand), the melting temperature was very high (210 °C) due to the formation of stereocomplexes between the poly[(*R*,*R*)-lactide] and poly[(*S*,*S*)-lactide] chains.

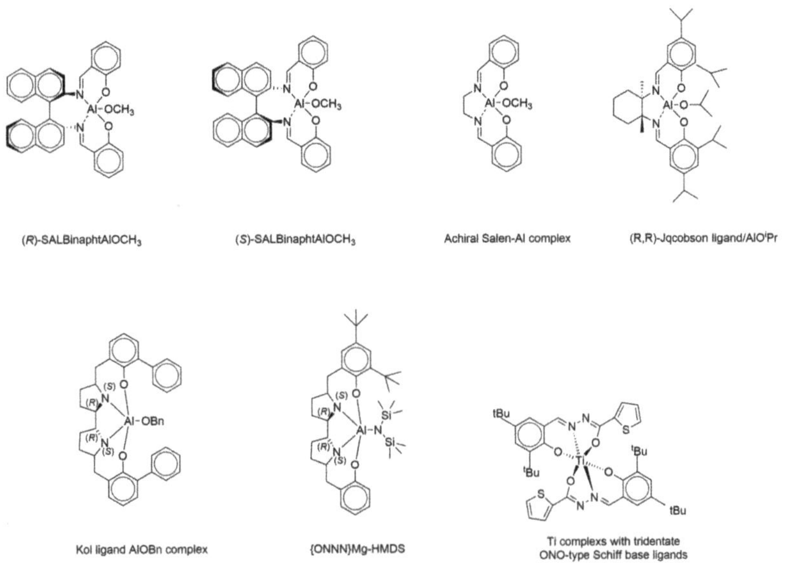

Figure 8. Initiators used for stereospecific polymerization of lactides.

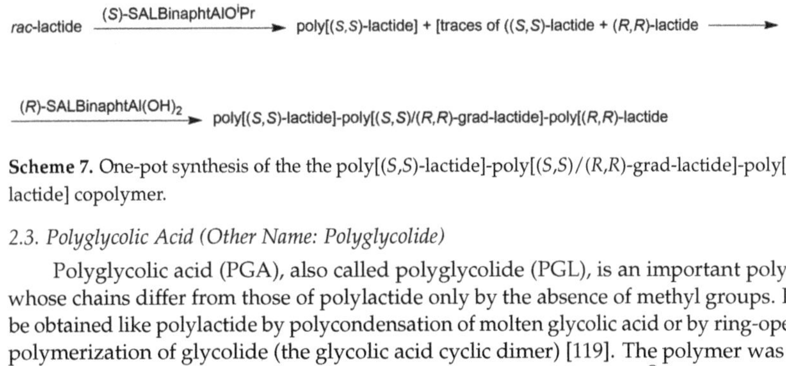

Scheme 7. One-pot synthesis of the the poly[(*S*,*S*)-lactide]-poly[(*S*,*S*)/(*R*,*R*)-grad-lactide]-poly[(*R*,*R*)-lactide] copolymer.

2.3. *Polyglycolic Acid (Other Name: Polyglycolide)*

Polyglycolic acid (PGA), also called polyglycolide (PGL), is an important polyester whose chains differ from those of polylactide only by the absence of methyl groups. It can be obtained like polylactide by polycondensation of molten glycolic acid or by ring-opening polymerization of glycolide (the glycolic acid cyclic dimer) [119]. The polymer was commercialized in 1970 by DuPont for the production of degradable Dexon® sutures. It is worth mentioning that the USA Drug and Food Administration has approved its use for medical purposes, including drug delivery systems. The glass transition of PGA ranges from 35 to 40 °C, which is close to the temperature of the human body, and melting temperature is in the range 200–225 °C. The PGA degrades relatively quickly. Samples of nearly amorphous PGA (about 5% degree of crystallinity) with 36 mg mass and initial thickness 0.5 mm required about a month for almost complete degradation in phosphate-buffered saline with an initial pH = 7.4 and ionic strength of 0.01 M [120,121]. As degradation progressed, the pH of the liquid phase decreased due to the increased concentration of carboxylic end-groups produced during hydrolysis [120,121].

Except for HFIP (hexafluoroisopropanol), PGA is insoluble in organic solvents. This is a major drawback since all methods used to produce nano- and microcarriers require polymers in solution. This problem has been solved by replacing PGA with glycolide-lactide copolymers (PLGA). There are hundreds of publications on PLGA carriers, mainly on drug-loaded nanofibers [122–124]. PLGA has also been used to prepare drug-loaded nanoparticles [125–127].

2.4. Poly(ε-Caprolactone) and Macrolactones

Poly(ε-caprolactone) (PCL) is a polyester often considered a promising candidate for the preparation of polymeric nano- and microparticle drug delivery systems. Of the polyesters commonly used to make drug carriers, poly(ε-caprolactone) chains contain the longest flexible aliphatic segments separating degradable ester groups. The glass transition temperature (T_g) of PCL reported in the literature ranged from −60 to −40 °C and the melting temperature (T_m) from 58 to 64 °C [128–130]. Therefore, particles made of PCL can be manufactured at mild temperatures, eliminating the danger of destroying the encapsulated bioactive compounds. When choosing PCL as a drug carrier, it is important to remember that this polymer degrades slowly and should therefore be used mainly in continuous drug delivery systems. In the human body, PCL is degraded by hydrolysis of the ester groups. The rate of degradation depends not only on the molar mass of PCL but also on the shape and size of the objects being degraded and the local pH. For macroscopic objects, it may need even more than 24 months [131]. However, PCL nanoparticles can degrade faster.

ε-Caprolactone was polymerized by cationic, anionic, and coordination processes. The cationic polymerization was carried out according to the activated monomer mechanism [99,132]. In this process, the polymerizing mixture, in addition to the monomer, contains protic acid and growing chains with hydroxyl end groups. At the beginning of the polymerization, the mixture must also contain an alcohol that functions as an initiator. The first step involves protonation of the ester group of the monomer. Then a protonated monomer is added to the initiator (alcohol) or the hydroxyl chain-ends and produces chain-ends with protonated hydroxyls (...-$CH_2OH_2^+$). Migration of a proton from the charged end-group completes the reaction cycle, extending the polymer chain by one constitutional unit. PCL with M_n up to 140,000 was obtained by this method [132].

Studies of anionic polymerization initiated by lithium t-butoxide and sodium trimethylsiloxide were carried out in the late 1970s and early 1980s and showed that the process is inevitably accompanied by the formation of cyclic oligomers, whose concentration for polymerization in solution carried out at 0 °C at equilibrium was 0.25 mol/L [133–135]. It was assumed that the anionic forms of alkoxides are too reactive and react with ester bonds not only in the monomer but also in the polymer chains (via back-biting). Therefore, less reactive covalent aluminum alkoxides, tin (II) alkoxides, tin(II) bis(2-ethyl hexanoate) ($Sn(Oct)_2$) derivatives ($Sn(Oct)_2/ROH$ and $SnOct)_2/RNH_2$) were used in later studies. First, it was shown that the polymerization of ε-caprolactone initiated with $CH_3OAl(CH_2CH_3)_2$ is effectively suppressed [136]. Later, it was observed that the reactivity-selectivity principle is general in the polymerization of ε-caprolactone and that ...-$OC(O)(CH_2)_5OAl(^iC_4H_9)_2$, albeit much less reactive than ...-$OC(O)(CH_2)_5O$-Na+ are more reactive in propagation than in back-biting (ratio of propagation to the back-biting rate constant $k_p/k_b = 7.7 \times 10^4$ L/mol), practically eliminating cyclic oligomers from the synthesized PCL [137]. The polymerization of ε-caprolactone initiated with $^iPrOAl(OCH_3)_2$ yielded polymer with M_n, 430,000 [138].

Polyesters commonly used to make drug carriers typically lack groups that would allow them to be functionalized on demand and in a controlled manner. Therefore, recently macrolactones containing double bonds have attracted the attention of researchers who have used them as comonomers in the copolymerization of ε-caprolactone [139–141]. Examples of these macrolactones are shown in Figure 9.

It should be noted that macrolactones containing a double bond have been effectively incorporated into PCL chains. The copolymers can be used as an additive that, when dissolved together with standard PCL, can be used to prepare carriers equipped with double bonds suitable for functionalizing them in the thiol-ene reaction [139,140].

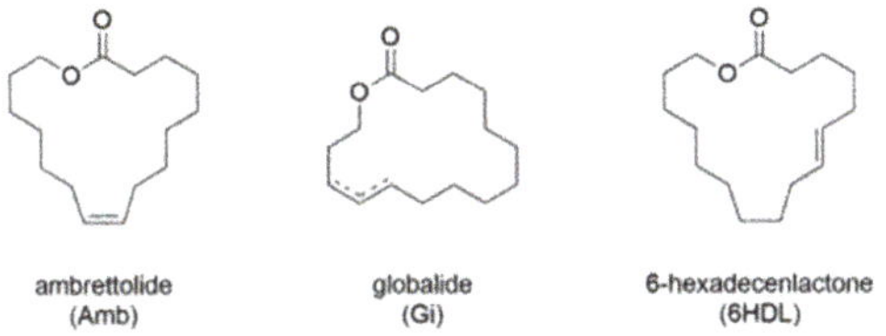

Figure 9. Macrolaactones containing double bonds in the ring.

3. Preparation of Polyester Drug Carriers

There are two major strategies and several methods for the preparation of polyester carriers of drugs and other therapeutic compounds (for example, compounds used for diagnostic purposes). These strategies are schematically presented in Figure 10.

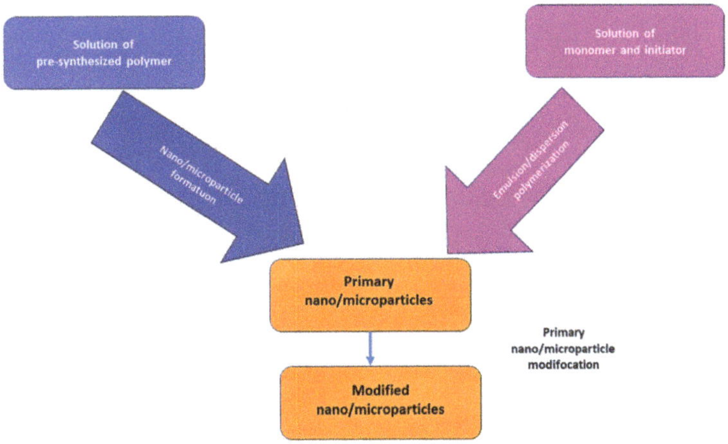

Figure 10. Main strategies for obtaining polyester drug carriers: from previously synthesized polymers and directly from monomers by polymerization.

Some strategies and methods are general and have been used to prepare nano- and microcarriers from many types of polymers. However, some are specific to polyester carriers. However, they all have one thing in common—they are solution-based processes.

The most common strategy is based on preparing drug carriers from previously synthesized polymers. It involves a series of methods, sometimes quite complex, that ultimately transform polymer pieces into nano- or microparticles of the required size. Since ways to mechanically disintegrate solid polymer into nano- and microparticles of the required and uniform size are unknown, simpler methods involving dispersing a polymer solution in the gas or liquid phase and then solidifying the formed particles have been used. Polymer solutions containing the drug allow the preparation of polymer particles loaded with the drug.

The second strategy involves dispersion or emulsion polymerization of suitable monomers. Very often, the original particles require further modification before they can be used for drug formulation.

Methods of preparing nano- and microcarriers and their advantages and disadvantages are discussed below.

3.1. Drug Carriers by Oil-in-Water (O/W) Method

A method called "oil in water" is used to prepare particles containing hydrophobic bioactive compounds. This is a general method that is not specific only to polyester particles but is also suitable for other particles made of hydrophobic polymers. First, the polymer and drug are dissolved in an organic solvent with a low boiling point and miscible with water to a few percent. The polymer and drug solution is gradually added to water, which may contain a surfactant and may contain a hydrophilic polymer (such as polyvinyl alcohol (PVA)). The mixture is effectively homogenized or otherwise blended. Methylene dichloride is often used as an "oil." Dispersion produces nano- or micro-droplets of "oil" (organic solvent) containing the polymer and drug. Migration of the organic solvent through the water-based solution toward the air and its eventual evaporation transforms the droplets into solid particles suspended in a water-based continuous phase. The crude product is often purified by isolating it several times by centrifugation-resuspension in an aqueous continuous phase. The O/W method has been successfully used to prepare drug-containing particles from polyhydroxybutyrate (PHB), polylactides (PLA), poly(lactide/glycolide) (PLGA), poly(ε-caprolactone) (PCL), and copolymers containing blocks of the above-mentioned polyesters. The O/W method is probably the oldest used to obtain polymeric drug carriers. Therefore, it may be surprising how many interesting and important improvements have been made over the past fifteen years. As a result, the O/W method has been adapted to produce polyester particles with average diameters ranging from 60 nm to ≈50 μm [142–150]. It is worth mentioning that particle diameters are of great importance for drug carriers, as they affect their distribution in the body and uptake by cells [126,151]. Some studies have yielded unexpected results. For example, it was shown that appropriate adjustment of the viscosity of the "oil" and "water" (containing polyvinyl alcohol) phases, the mixing rate, the molar mass, and the structure of the end groups of the copolymer constituting the drug carrier (PLGA) made it possible to obtain drug-free and drug-loaded (paclitaxel) spheroidal PLGA particles [152]. The results of studies on obtaining PLGA particles using vitamin E as an emulsifier also have great potential [147]. The advantage of vitamin E is that it is a natural substance (containing tocopherols and tocotrienols) that can replace synthetic emulsifiers. In the European Union, vitamin E is approved as a common food additive. Its use has made it possible to obtain PLGA nanoparticles up to about 60 nm in diameter.

Recently, a simple O/W method was developed to prepare indented PLGA microparticles containing encapsulated budesonide, a drug used in the treatment of chronic asthma and other obstructive pulmonary diseases [150]. The final product was obtained by lyophilizing a suspension of the frozen particles and applying it using a standard powder inhaler. The advantage of the method mentioned above is the simplicity of the required particle preparation equipment and better aerodynamic performance than with smooth surface carriers.

3.2. Drug Carriers by Water-in-Oil-in-Water (W_1/O/W_2) Method

The water-in-oil-in-water (W_1/O/W_2) method, which is an extension of the O/W procedure, is designed to prepare water-soluble carriers for bioactive compounds, particularly nucleic acids, oligonucleotides, proteins, and oligopeptides. "Water" refers to a water-based ingredient (often a buffer) that is a solvent for the bioactive compound but is not a solvent for the polymer that makes up the carrier body. The subscripts "1" and "2" indicate that the water-based ingredients can be different. "Oil" refers to an easily evaporated organic liquid that mixes slightly with water and is a good solvent for the polymer but not for the drug. The two phases may contain various surfactants, and the "water" may contain a small amount of hydrophilic polymer (such as PVA). The entire process begins with dissolving the drug in the "water" and dispersing this solution in the "oil" (e.g., in dichloromethane). The resulting W/O mixture consists of a continuous "oil" phase containing the dissolved polymer (e.g., PLGA) and very small droplets of "water" containing the drug. In the next step, the mixture is dispersed in "water". The product is a mixture of an aqueous continuous phase containing microdroplets of "oil" with nanodroplets of an aqueous phase containing the dissolved drug. The low boiling "oil" component diffuses through the aqueous continuous phase and evaporates. Its elimination transforms microdroplets into drug-containing microparticles. The W/O/W method has been used to obtain nano- and/or microparticles from PLGA [153,154] and PCL [155,156] polymers. Some interesting examples are presented below.

G. Golomb et al. described the preparation of PLGA particles containing an encapsulated human placental alkaline phosphatase gene cloned in the pcDNA3 vector [153]. The particles were produced by a standard W/O/W procedure, using buffers as the "water" phase and chloroform as the "oil." This work is worth mentioning because it represents the first case of sustained gene delivery allowing pDNA release for about a month.

Antigens used in vaccine production are often water/buffer-soluble proteins or oligopeptides that can be conveniently encapsulated by the $W_1/O/W_2$ method. Therefore, the development of a process suitable for modifying the aforementioned molecules in a way that allows their uptake by dendritic cells and macrophages has been much appreciated. Such a method recently developed for PLGA particles involved conjugation of PLGA particles coated with polydopamine to exosomes and ovalbumin [129]. It is worth mentioning that PLGA nanoparticles conjugated with exosomes were more efficiently taken up by antigen-presenting cells.

Most of the polyester nano- and microparticles produced by the $W_1/O/W_2$ double emulsion method were obtained from linear homo- and copolymers, and the interest in the influence of the copolymer architecture on their assembly and drug loading is understandable. Therefore, it is worth mentioning the study of particles obtained from six-arm copolymers with poly(ε-caprolactone)-poly(ethylene oxide) arm structure (6S-PCL-PEO) [155]. This paper presents the preparation and characterization of model particles loaded with ovalbumin (OVA). The average diameter of the obtained unloaded nanoparticles was about 191 nm. The diameter of nanoparticles with encapsulated OVA was slightly larger (233 nm). Nanoparticles with such small diameters are well "visible" to macrophages. The authors of the above study have provided a model system that, when compared with nanoparticles from linear copolymers with similar molar mass ratios of PCL and PEO blocks, could reveal the influence of the polymer architecture on the nanoparticle properties. Recently, R.D. Gökberk et al. developed a pulmonary drug delivery method based on PCL particles produced by a $W_1/O/W_2$ double emulsion process [156]. The preparation of PCL particles containing lysozyme, an antimicrobial peptide, was optimized to provide high encapsulation efficiency (EE 65.15%). Their median aerodynamic parameter was 5.44 ± 0.19 μm, and the fine particle delivery fraction (fraction of particles with less than 5 μm in diameter) was $51.0 \pm 2.9\%$. The long-term release of lysozyme from particles lasting up to 35 days is very important. The above formulation can be considered as an alternative to antibiotic-based systems.

A disadvantage of hydrophilic drug particles prepared by the $W_1/O/W_2$ method is the often-observed initial rapid release of the drug. Recently, this disadvantage has been alleviated by converting a hydrophilic drug (e.g., bovine serum albumin—BSA—used as a hydrophilic drug model) into a hydrophobic drug [157]. The "hydrophilic-hydrophobic" conversion was carried out by mixing an aqueous solution of BSA with a tert-butanol solution of lectin; the mixture was frozen and lyophilized. The resulting BSA-lectin (S) nanoparticles were easily dispersed in dichloromethane "oil" containing dissolved PLGA. This dispersion was transformed into BSA-loaded PLGA particles, similar to the second O/W_2 step of a typical $W_1/O/W_2$ process. The entire modified process, which can be referred to as $W_1/S/O/W_2$, yielded nanoparticles with a significantly reduced initial burst.

3.3. *Drug Carriers by Nanoprecipitation*

Nanoprecipitation is a process involving two mutually miscible liquids (at least in certain proportions). The first liquid should be a solvent for the polymer and the drug (e.g., acetone). The second liquid (e.g., water) should be a non-solvent for the polymer and preferably also a non-solvent for the drug, but the latter requirement is not crucial. Both liquids can also contain surfactants. Mixing these liquids results in the formation of nanodroplets of the first liquid containing the polymer and the drug, with accompanying deposition of the polymer at the interface. The solvent rapidly diffuses into the non-solvent phase, and this displacement results in the almost immediate formation of drug-loaded polymer nanoparticles. This method was developed by Fessi et al. and used to prepare indomethacin-loaded poly(*R*,*S*-lactide) nanoparticles [158]. It should be noted that there are not only reports describing cases in which the polymer and drug solution were added to the insolvent but also many cases in which the non-solvent was added to the polymer and drug solution [158,159]. It is noteworthy that the first system (acetone/water) mentioned earlier has almost become the standard and is still widely used after many years. In the review, however, we draw readers' attention to its recent modifications.

For the preparation of nanoparticle drug carriers by nanoprecipitation, all kinds of typical aliphatic polyesters were used, including poly(*R*,*S*)-lactide-*co*-glycolide, poly(*S*,*S*)-lactide-*co*-glycolide, poly(S,S-lactide), poly(ε-caprolactone), poly(3-hydroxybutyrate), and poly(3-hydroxybutyrate-*co*-3-hydroxyvalerate) (PHBV). However, while the most comprehensive studies have been conducted for polylactides and related polymers, the least studied (only one paper [160]) have been nanoparticles containing butyric acid ester segments. Here are some examples indicating progress in the controlled preparation of nanoparticles by nanoprecipitation.

In some cases, combination treatment is required with two active compounds delivered in the required proportion to the tissue in question. Such treatment is often desirable for patients suffering from certain cancers. The solution is to prepare double-loaded nanoparticles. Bandyopadhyaya et al. developed the preparation of PLGA nanoparticles with encapsulated curcumin (CUR) and niclosamide (NIC) and compared their properties with those of similar but single-drug-loaded nanoparticles. The nanoparticles were produced by the classical nanoprecipitation method, using acetone as the organic solvent for PLGA, CUR, and NIC and PVA-containing water as the nonsolvent [160]. The average diameters of the doubly charged nanoparticles were 257 nm. The diameters of not loaded nanoparticles only-CUR and only-NIC loaded were slightly smaller (207, 214, 216, respectively). The loading efficiency for doubly drug-loaded nanoparticles was higher than for single drug loading. The authors noted that the advantage of the doubly drug-loaded PLGA-CUR-NIC nanoparticles was their more efficient uptake by cancer cells.

Dual loading is also beneficial for the oral use of drugs sensitive to low pH. Loading the nanoparticles with the drug and a pH stabilizer (such as magnesium oxide used as an alkalizer) can protect the drug from degradation in the stomach. In this way, PLGA nanoparticles loaded with lansoprazole were prepared for prolonged release after oral administration [161].

Classical nanoprecipitation was developed and adapted for the encapsulation of hydrophobic drugs, which for many years was an obvious limitation of this method. Recently, a simple modification has been introduced to enable the use of nanoprecipitation of hydrophilic bioactive compounds. The modified nanoprecipitation has been verified for the encapsulation of water-soluble ciprofloxacin (a drug used in ophthalmic therapy) in poly[(*R*,*S*)-lactide]-dextran (PLA-DEX) or poly(lactide-*co*-glycolide)-polyethylene glycol (PLGA-PEG) [162]. DMSO was used as a solvent for the aforementioned hydrophilic polymers, and HCl/water was used as a solvent for ciprofloxacin and a non-solvent for PLA-DEX and PLGA-PEG. Nanoparticles were obtained by adding a DMSO solution of the polymers to an aqueous HCl solution of ciprofloxacin at a controlled rate. The diameters of the particles obtained ranged from 82 to 205 nm, depending on the polymer used and the loading of the nanoparticles with the drug. The maximum mass loading of the drug reached 27.24%.

3.4. Preparation of Drug Carriers Using Spray-Drying Equipment and Chips Used in Microfluidic Techniques

The above subsections describe the physicochemical processes used to produce polymeric drug nanocarriers. This one describes the use of commercially available equipment for this purpose.

Some diseases require direct treatment of the lung tissue. In such cases, the method of choice is treatment with dry powder preparations administered by inhalation. Drug preparations in dry powder form are usually prepared using spray dryers of a design similar to equipment commonly used in the food industry to produce powdered milk or instant coffee. For laboratory purposes, equipment is sold that allows the preparation of inhalation formulations on a scale of a few grams. Each spray dryer consists of the following parts: an atomizer powered by a polymer and drug dissolved preferably in a volatile solvent; a drying chamber powered by the fine droplets produced by the atomizer and connected to a source of hot air or nitrogen that transforms the droplets into particles and transports them to a cyclone where a vortex of gas directs them to a collector [163]. Traces of the remaining particles suspended in air or nitrogen flow toward the filters and are eventually expelled. The main advantages of spray-drying nanoparticle manufacturing are its versatility, tunability, simplicity, and scalability. Disadvantages include exposure of bioactive components to hot air, the high pumping capacity required, and the large volume of liquid (often organic) to be evaporated and recycled. Moreover, it is worth noting that the diameters of polyester particles produced by spray drying typically exceed the nanometer range. For example, the diameters of PLGA particles containing Artemisone (an antimalarial drug) range from 1 to 1.7 μm [164].

Microfluidic techniques are based on microreactors consisting of various microchannels and reaction chambers of fixed or variable arrangement. They differ from the usual laboratory glassware used to produce nanoparticles (such as beakers, flasks, stirrers, rotary evaporators, etc.) by their working volume, which is much smaller. The parameters of particles produced in microfluidic reactors depend not only on the design of these devices. For example, when particles are formed by injecting a solution of a polymer and any other bioactive component dissolved in an organic solvent (e.g., in dichloromethane) into a containing water microreactor, the fine droplets are formed and solidify into microparticles. The diameters of particles depend on the flow rate of the organic and aqueous phases and the inner diameter of the injection needle. A device with an inner needle diameter of 1.3 mm yielded microparticles with diameters ranging from 130 to 700 μm. Large microparticles were suitable for the embolization of capillary blood vessels in the treatment of early-stage aneurysms [165].

Drug nanocarriers are known to be produced by combining two or even three methods in a single process. The above approach was used to prepare PLGA particles loaded with niclosamide and equipped with a hydrophilic coating of poly(vinyl alcohol) PVA and hyaluronic acid (HA) [166]. Niclosamide is a multifunctional drug used to treat cancer, metabolic disorders, and viral infections. It is insoluble in water and, as a solid, is highly hydrophobic. As such, it must be encapsulated in nanoparticles coated with hydrophilic compounds to be dispersed conveniently. The first step of the process involves dissolving equal amounts of PLGA and niclosamide in an appropriate volume of ethyl acetate ("oil" phase). In parallel, a known volume of aqueous PVA solution (aqueous phase) is prepared. The two phases are mixed until a homogeneous mixture is obtained. Then a certain amount of hyaluronic acid is added to the mixture, which is injected into the microfluidic system in the next step and further mixed slowly. The mixture is then subjected to ultrasound and stirred until the ethyl acetate evaporates, accompanied by particle formation. In summary, the entire process combined the oil-in-water method using classical mixing and stirring in a microfluidic system, with a final treatment involving ultrasound, dialysis, and lyophilization.

Recently, Schneider et al. described an example of multidrug-carrying nanoparticles prepared by elegantly combining nanoprecipitation in a microfluidic microreactor with a spray-drying technique [167]. Primary PLGA nanoparticles containing curcumin were obtained by nanoprecipitation, injecting a solution of the drug and PLGA dissolved in acetonitrile and an aqueous solution of Pluronic F68 into a microfluidic reactor. The optimized process yielded nanoparticles with a diameter of 105 nm. Spray drying of combined PLGA-curcumin nanoparticles and antibiotics (tobramycin, ciprofloxacin, or azithromycin) and N-acetylcysteine (NAC) yielded inhalable formulations with a median mass aerodynamic parameter (MMAD) ranging from 2.16 to 2.63 μm and a fine particle fraction (FPF; fraction of particles with MMAD < 5 μm) ranging from 61.6 to 97.5%.

3.5. Direct Synthesis of Polyester Nano- and Microparticles by Ring-Opening Dispersion Polymerization of Lactides and ε-Caprolactone

Direct synthesis of particles has the following advantages: easy scalability, similar to dispersion polymerization of vinyl monomers; easy adjustment of particle diameters and molar masses of polymers to requirements; narrow dispersion of diameters and molar masses; and control of interfacial properties of particles. The first syntheses of nano- and microparticles of polylactide and poly(ε-caprolactone) were carried out in 1994 by Slomkowski et al. [168]. Dispersion polymerization of ε-caprolactone was initiated using diethylaluminum ethanolate and carried out in the presence of poly(dodecyl acrylate)-g-poly(ε-caprolactone) surfactant in a mixed 1,4-dioxane-heptane continuous phase (1:8 *v*/*v*). Poly[(*R*,*S*) lactide] microspheres were prepared by a process with tin (II) octanoate used as catalyst/initiator and poly(dodecyl acrylate)-*g*-poly(ε-caprolactone) added as surfactant. The continuous phase was a mixture of 1,4-dioxane and heptane in a ratio of 1:4 *v*/*v*. The average diameters (D_n) and diameter dispersities ($Đ_D = D_v/D_n$) were 0.628 μm and 1.38, and 1.25 μm and 1.15 for poly(ε-caprolactone) and polylactide particles, respectively. Subsequent publications provided comprehensive information on the kinetics and mechanism of dispersion polymerization used to better control particle properties. It was noted that the dispersity of directly synthesized poly[(*S*,*S*)-lactide] microspheres strongly depended on the concentration and composition of the surfactant poly(dodecyl acrylate)-g-poly(ε-caprolactone) [169,170]. The best results (the dispersion with the smallest diameter, $\chi = 1/(Đ_D\text{-}1) = 33$; $Đ_D = 1.03$) were obtained for the ratio of M_n of poly(ε-caprolactone) strains and M_n of the whole surfactant macromolecule equal to 0.23 [145]. For poly(dodecyl acrylate)-*g*-poly(ε-caprolactone) with an average number of strains of 1.3 ($M_n = 4700$) $Đ = 1.08$ [169]. A summary of the most important results gives a very simple picture of the dispersive ring-opening polymerization of ε-caprolactone and lactides [170–173]. At the very beginning of polymerization, the system is homogeneous and contains 1,4-dioxane and heptane in the appropriate proportion, a monomer (ε-caprolactone or lactide), a surfactant [poly(dodecyl acrylate)-*g*-poly(ε-caprolactone)], and an initiator (diethylaluminum ethanolate or tin (II) octanoate). Initiation takes place in solution, and the chains begin to grow in solution as well. When the growing chains reach a certain critical length, they undergo a chain-globule transition and aggregate into particle nuclei. In ε-caprolactone polymerization, this process takes about 100 s. After this initial period, all active centers participate in propagation inside the particles, which are supplied with monomers from the continuous phase. It has also been noted that in the polymerization of ε-caprolactone and any of the lactides, the number of nanoparticles initially formed during nucleation remains constant at later stages, i.e., neither new particles are formed nor do they aggregate [170–172]. As a result, in the ring-opening dispersion, polymerization particles should be treated as nano or microreactors containing all active propagating species and a significant part of the ε-caprolactone or lactide monomers. Because the volume of particles is much smaller (in the majority of experiments, ca. eight times) than the whole volume of the polymerizing mixture (the volume of particles + the volume of continuous phase), the local concentration of monomers and propagating species is higher, and the rate of polymerization should be higher than the otherwise analogous polymerization in solution. Indeed, it was noticed that rates of polymerization in the dispersed systems (i.e., in 1,4-doxane/heptane mixtures) were about ten times higher than rates of polymerization in solution (in THF) [170,172]. At the end of polymerization, the propagating centers were deactivated by the addition of a small amount of acetic acid. The particles were purified by the several times repeated cycle of sedimentation-isolation-resuspension in fresh portions of heptane. Heptane was also a medium for their storage.

In later years, several new surfactants were synthesized: poly(ε-caprolactone)-*b*-poly(dodecyl acrylate), poly(ε-caprolactone)-*b*-poly(octadecyl methacrylate)-*b*-poly(dimethylaminoethyl methacrylate), poly{(dodecyl methacrylate)-*co*-poly{α-methacryloxyethoxy-poly{(*S*,*S*)-lactide}}}, poly{(dodecyl methacrylate)-*co*-poly{α-methacryloxyethoxy-poly{(*R*,*S*)-lactide}}}, poly[(dodecyl methacrylate)-*co*-(hydroxyethylmethacrylate)], poly[(dodecyl methacrylate)-*g*-poly[(*R*,*S*)-lactide] [126,174–176]. Their use has not led to the development of a fundamentally new mechanism for the dispersion polymerization of cyclic esters. However, it is worth noting that the use of poly[(dodecyl methacrylate)-*co*-(hydroxyethyl methacrylate)], which undergoes in situ transformation into poly[(dodecyl methacrylate)-*co*-(hydroxyethyl methacrylate)-*g*-poly[(*R*,*S*)-lactide], simplified the synthesis [175]. Poly(ε-caprolactone) and polylactide are synthesized by dispersion polymerization and form stable suspensions in organic media. However, for use as drug carriers, they should form stable suspensions in water, buffers, and other water-based media. The method developed consisted of replacing the suspending medium (heptane) with an ethanol-KOH solution containing surfactants. Controlled hydrolysis time allowed to obtain particles with hydroxyl and carboxyl groups in the interfacial layer, providing additional electrostatic stabilization for stabilization by surfactants. The range of pH and ionic strength at which suspensions of poly(ε-caprolactone) and poly[(S,S)-lactide] are colloidally stable has been determined [177].

3.6. *Preparation of the Directly Synthesized Polyester Nano- and Microparticles Loaded with Drugs or Drug Models*

The use of the strategies listed below to prepare directly synthesized drug-loaded particles has been verified for the following methods:

Adsorption or covalent immobilization of selected compounds on the particle surface. Examples include the adsorption of human serum albumin (HSA) and γ globulin (γG) on poly(ε-caprolactone) and poly[(*R*,*S*) lactide] particles [168]. In the case of poly(ε-caprolactone), the maximum surface concentration of adsorbed protein (at saturation) was 1.0 mg/m^2 and 2.6 mg/m^2 for HSA and γG, respectively. For adsorption on poly[(*R*,*S*) lactide] molecules, the maximum surface concentration was 0.9 mg/m^2 and 1.7 mg/m^2 for HSA and γG, respectively. Another example is the covalent immobilization of 6 amino-quinoline (6AQ) on the surface of poly[(*S*,*S*) lactide] microspheres. The maximum surface area of covalently immobilized 6AQ was 0.27 mg/m^2 [177].

Very simple is loading polyester particles with a hydrophobic, liquid drug able to swell the carriers. For example, the poly(ε-caprolactone) particles with diameters D_n = 620 nm were suspended in a 70%/30% *v*/*v* ethanol/water mixture. To portions of this suspension, a varied amount of ethyl salicylate was added. The concentration of nanoparticles was 5.4 mg/mL in all samples. The concentration of ethyl salicylate ranged from 1.7 mg/mL to 56.6 mg/mL. The samples were shaken at room temperature for 48 h. The particles were separated by centrifugation, and the content of the encapsulated drug was calculated as the difference between its concentration in the continuous before and after incubation. The particles with a degree of loading in the range from 9.7 to 37.0 wt% were obtained [177]. Higher loading was also possible, but such particles were converted to droplets of viscous liquid.

The use of bioactive compounds containing hydroxyl groups as transfer agents in the dispersion polymerization of cyclic esters has proven suitable for the direct synthesis of drug-loaded nanoparticles. For example, the polymerization of ε-caprolactone ($[\varepsilon\text{-caprolactone}]_0 = 3.8 \times 10^{-1}$ mol/L) initiated by $(CH_3)_3SiONa$ ($[(CH_3)_3SiONa]_0 = 1.76 \times 10^{-3}$ mol/L) and carried out in a 1,4-dioxane/heptane 1:9 *v/v* mixture in the presence of *N*,*N*-bis(2-hydroxyethyl)isonicotinamide (transfer agent; $[N,N\text{-bis(2-hydroxyethyl)isonicotinamide}]_0 = 2.2 \times 10^{-2}$ mol/L) and poly(dodecyl acrylate)-*g*-poly(ε-caprolactone) (surfactant; 1.6 g/L) yielded poly(ε-caprolactone)/*N*,*N*-bis[poly(ε-caprolactone)]isonicotinamide nanoparticles with a drug content of 6.4 wt%.

Direct entrainment of 5-methoxy-2-{[(4-methoxy-3,5-dimethyl-2-pyridyl)-methyl 1]-sulfinyl}-1H-benzimidazole (omeprazole) during the synthesis of poly[(*S*,*S*)-lactide] nanoparticles by ring-opening dispersion polymerization, it was used as a simple method to obtain polyester drug carriers [178]. The polymerization was carried out as described in Section 3.5 with one small modification. Since the nucleation of the particles was completed, omeprazole (an inhibitor of (H^*-K^+)ATPase, the "pH pump" responsible for the secretion of HCl in the stomach) was gradually added. The following initial reagent concentrations were used in the synthesis: $[(S,S)\text{-lactide}]_0 = 2.5 \times 10^{-1}$ mol/L, $[\text{tin (II) 2-ethylhexanoate}]_0 = 4.6 \times 10^{-3}$ mol/L, $[\text{poly(dodecyl acrylate)-}g\text{-poly(}\varepsilon\text{-caprolactone)}]_0 = 1.6 \times 10^{-1}$ g/L. After adding the appropriate amount of omeprazole, the total concentration of the drug in the polymerization mixture was 5×10^{-3} mol/L. The percentage of omeprazole in the produced particles was 11% [179].

4. Targeted Drug Delivery to the Brain Based on Polyester Nanoparticles

There is a long list of active substances that, as drugs, can be administered orally in tablets or loaded into macroscopic capsules, for which it is only required to pass from the gastrointestinal tract into adjacent blood vessels and be distributed throughout the body with the blood. Typical examples include soya lipids to support liver function, progesterone (lutein) often used by patients with vision problems, and even such simple compounds as KCl prescribed to many cardiac patients. However, for many drugs, simple systemic administration of untargeted drugs is undesirable. This is the case, for example, when systemic administration causes undesirable side effects by placing the drug primarily in an organ other than where it should act or does not allow the drug to reach its chosen site of action. This section presents the main types of polyester carriers for targeted drug delivery to selected parts of the brain or to particular types of brain cells.

The brain is a very delicate organ that requires effective mechanical protection. Moreover, it can function properly only in a suitable, chemically stable environment and should not be directly exposed to uncontrolled electromagnetic stimuli. The skull usually provides sufficient mechanical protection. Since bones are not permeable by water-based liquids, the skull also protects the brain from uncontrolled interactions with body fluids. However, the brain must maintain some contact with the environment, from which nutrients are supplied and into which unwanted metabolites are removed. The brain must maintain the information exchange channels provided by nerves. The nutrients are transported in the blood vessels crossing the scull. However, the endothelium of these vessels is so tight that it creates the blood-brain barrier, which only some compounds can overcome [7,8]. The nerves used for an exchange of information transport electric signals and as a route for the transportation of medically active compounds can be used only in very special instances such as nasal drug administration. This is because the nasal cavity contains the only location (the so-called olfactory region), where the nerves of the central neural system present in the mucosa are exposed to external contacts. Unfortunately, the surface of the olfactory structures in humans is quite small and approaches 0.3 m^2 (taking into account microvilli) [180], which limits the nasal transport potential. However, the research on using the aforementioned methods continues.

There are some life-threatening illnesses, which should be considered manifestations of structural and functional disorders of parts or of the whole brain. The hopes of their cure or at least alleviation of symptoms depend on finding the appropriate drug and its targeting carrier to the needed site.

As the first example, there are presented carriers for treatment of the brain-degenerating Parkinson's disease. Parkinson's disease is manifested by such symptoms as tremors, rigidity, bradykinesia, and postural instability. All of them are caused by low dopamine levels in the striatum that control movement and muscle function [181,182]. Dopamine is a product of dopaminergic neurons in the part of the brain called the substantia nigra pars compacta, and regardless of the reason for their lesion, the production of dopamine is decreased. It should be noted that the metabolism of dopamine also includes a tract involving its oxidation by monoamine oxidase type B, known for producing hydroxyl radicals harmful to dopaminergic neurons. Therefore, for the Parkinson's disease-ill patients, it was desirable to block this tract. A few years ago, N. Ahmad investigated the possibility of using rasagiline, a compound known for the deactivation of monoamine oxidase type B [182]. The rasagiline containing PLGA nanoparticles was designed for nasal delivery. They were prepared by the double emulsification-solvent evaporation method. The particles were coated with chitosan to make them mucosa adhesive. The bare PLGA-rasagiline and chitosan-coated (chitosan-PLGA-rasagiline) nanoparticles were comprehensively characterized by providing their diameters and diameter distributions, zeta potential, drug loading, and in vitro drug release. Ex vivo nasal mucosa permeation and brain pharmacokinetics studies for the rat model revealed that the chitosan-PLGA-rasagiline nanoparticles can be effectively used for rasagiline delivery to the brain using the olfactory pathway. Recently, S. Lee et al. reported on using the resveratrol-loaded PLGA nanoparticles for the protection of the dopaminergic neurons of the substantia nigra pars compacta [183]. The lactoferrin-conjugated PLGA-resveratrol nanoparticles were used for this purpose since lactoferrin conjugation enhances the internalization of nanoparticles into the brain microvascular endothelial cells and makes the crossing of the blood-brain barrier more effective. Indeed, the bioluminescent imaging analyses revealed the accumulation of the lactoferrin-PLGA-resveratrol nanoparticles in the brain higher than the accumulation of the bare PLGA-resveratrol nanoparticles.

There are two potentially complementary approaches for the treatment of patients suffering from Parkinson's disease. The one described above focused on the delivery of drugs protecting the dopaminergic neurons from degradation, and the second one was based on the delivery of the dopamine to the brain, aiming at alleviating the symptoms related to its insufficient level. Nanoparticles used to deliver dopamine to the brain were similar to nanoparticles used to deliver compounds that protect dopaminergic neurons. However, it is worth mentioning the research of R. Pahuja et al., who developed PLGA-dopamine nanoparticles that, despite not containing any targeting molecules, were able to deliver dopamine effectively to the brain [184]. The nanoparticles were prepared by the double emulsification solvent evaporation method (W/O/W). First, dopamine in water was emulsified in dichloromethane. Then this emulsion was emulsified in water containing poly(vinyl alcohol) (PVA), which was used as a stabilizer of nanoparticles' suspension. Interactions of the dopamine-loaded nanoparticles with the model SH-SY5Y cells (in vitro studies) and with brain cells (in vivo studies using the Parkinsonian rat model). The studies revealed that the dopamine-loader nanoparticles administered intravenously are effectively delivered to the rat's brain and that this treatment reversed neurochemical and neurobehavioral symptoms in the Parkinsonian rats. Positive results were obtained also using the dopamine-loaded PLGA nanoparticles bearing the lactoferrin and borneol functions [185] and a much more complex albumin/PLGA-dopamine nanosystem [186].

Another illness related to the degeneration of the brain is Alzheimer's disease. Its early symptoms showing some problems with memory are usually not alarming. Rare difficulties with recalling names of people or places occur almost to everybody. However, in the case of Alzheimer's disease, the difficulties increase with time, and suffering people do not recognize their friends and family or are disoriented about time and places. Eventually the patients have problems with speech, changing positions, eating, and swallowing. The structure of their brains is very different from that of healthy people. They contain amyloid beta (Aβ) aggregates, which are toxic to neurons. Alzheimer's disease is incurable, and its etiology is not known. The only available treatment is based on protection of neurons against their degradation. The only drug approved for this purpose, both in the US and Europe, is memantine. However, the effectiveness of its direct use has been quite low. Several years ago, E. Sánchez-López et al. developed PLGA-based nanocarriers with the hope of improving the transport of memantine across the blood-brain barrier [187]. The memantine-loaded PLGA-PEO nanoparticles were prepared by the double emulsification W/O/W method. During the first step, the memantine was dissolved in deionized water. Then, this solution was emulsified in ethyl acetate containing dissolved PLGA-PEO copolymer. Eventually, the primary emulsion was emulsified in water containing the poly(vinyl alcohol) stabilizer. Evaporation of the liquid organic phase (ethyl acetate) produced a stable suspension of the memantine-loaded nanoparticles dispersed in the water continuous phase. The control memantine-free nanoparticles were produced similarly but without memantine addition. The fluorescent nanoparticles containing rhodamine were also prepared for the biodistribution studies. The research included comprehensive characterization of nanoparticles, in vitro studies of nanoparticles' interactions with cells (mouse microvascular endothelial cells (bEnd.3) and astrocytes from brain rat cortex, and in vivo studies using the mouse model. It should be noted that nanoparticles used in animal studies did not contain any special tags facilitating their crossing through the blood-brain barrier. Any traces of PVA, which were weakly adsorbed onto the particles, were removed by the repeated sequence of the centrifugation-isolation-resuspension steps. The most important results of the aforementioned studies are as follows: The orally administered memantine-loaded nanoparticles with diameters lower than 200 nm were transported from the gastrointestinal tract to the blood. Circulating with blood, they crossed the blood-brain barrier and, according to histological analysis, accumulated in brain tissue. The behavioral tests revealed that animals treated with the memantine-loaded nanoparticles had better learning capability than mice treated with unencapsulated memantine.

Recently, Q. Wu et al. reported interesting studies on treating the primary mouse cortical cultured neurons with PLGA nanoparticles (632 nm diameter) [188]. The suspension of nanoparticles in the phosphate-buffered saline (0.01 M, pH 7.4) was prepared by dispersing the commercial product of Phosphorex (Hopkinton, MA, USA). For details of experiments, see Ref. [188]. The most important findings revealed that uptake of pure PLGA by neurons not only protects them against amyloid-β (Aβ) toxicity but reduces the conversion of the amyloid-β-protein precursor (APP) to Aβ. However, the aforementioned studies are at an early stage, and the mechanism of PLGA action on neurons requires more research.

Nowadays, chemotherapy is a standard procedure consisting of using drugs that, selectively and irreversibly internalized in the cancer cells, cause cell death or stop their proliferation (cytotoxic and cytostatic effects, respectively). In the case of brain cancer, these requirements are especially important because problems with crossing the BBB barrier hinder the delivery of drugs to the brain, and insufficient selectivity may be detrimental to the neurons, resulting in serious side effects. Some examples of recent research based on using polyester nanoparticle carriers are listed below. All these carriers were made from PLGA and its derivatives; all were designed for the preparation of drugs against glioma, the most often encountered primary brain cancer; however, they differed in detail to test various administration strategies.

Recently, Ye et al. proposed a new strategy based on targeting the epidermal growth factor receptors (EGFRs) [189]. These receptors are overexpressed on glioma cells, and their mutations and amplification result in the growth and proliferation of gliomas. Growth of glioma is also facilitated by Golgi phosphoprotein 3 (GPH3). Since Gefitinib (an anti-tyrosine kinase inhibitor) hampers EGFR production and GOLPH3 siRNA blocks the production of GPH3, authors prepared the angiopep-2 (A2)-modified cationic lipid-poly (lactic-co-glycolic acid) (PLGA) nanoparticles loaded with the aforementioned active substances. The in vitro and in vivo studies revealed the anti-glioma effects, justifying further studies.

In recent years, researchers turned their attention to hybrid nanoparticles of biological-synthetic origin with the hope of better tailoring their blood-brain barrier (BBB) and blood-brain tumor barrier (BBTB). As an example, may be mentioned PLGA nanoparticles coated with erythrocyte membrane and modified with two peptides from bacteria, known to enhance transport across BBB [190]. The synthesized nanoparticles loaded with Euphorbia factor L1 (EFL1) extracted from Euphorbia semen (model drug) were able to cross the BBB and BBTB and kill the glioma cells.

It is worth mentioning that still there is some space for the preparation of new drugs using the more traditional formulation methods, such as the preparation of PLGA/PLGA-PEO nanoparticles loaded with paclitaxel and R-flurbiprofen [191]. The nanoparticles were prepared by the nanoprecipitation method and by coating them with cationogenic chitosan. The amphiphilic PLGA-PEO copolymer was used to maintain the colloidal stability of nanoparticle suspension and chitosan coating to facilitate their uptake across the negatively charged membrane of glioma cells. The nanoparticles were administered by intraperitoneal route to rats with implanted tumor cells. It was noticed that animals treated with nanoparticles loaded with paclitaxel and R-flurbiprofen exerted anti-tumor activity.

Improvement in selectivity in targeting glioma cells required also basic research. Therefore, it is worth mentioning the work by S. Acharya et al. who investigated the effect of the folic acid moieties introduced into the interfacial layer of the prednisolone-loaded PLGA nanoparticles [192]. The studies proved that the presence of the folic acid labels strongly enhances the uptake of the nanoparticles by the cancer cells and that the anti-tumor activity of the nanoparticles is long-lasting.

The last example in this subsection describes the preparation and efficacy of the temozolomide-loaded PLGA nanoparticles, which were coated with fused stroma and glioma cell membrane fragments [193]. The nanoparticles were designed to target the glioma tumor microenvironment (TME), a structure containing glioma and stroma cells. Determination of the efficacy of the said nanoparticles as an anti-tumor agent was the main objective of the study. The drug-loaded PLGA nanoparticles were prepared by the oil-in-water emulsification solvent evaporation method. The membranes were obtained by lysis of the glioma and stroma cells, purification of the cell fragments, and their fusion. The nanoparticle coating was performed by several times repeated extrusion of the nanoparticles/membranes suspension through the 200 nm Whatman membrane filter. The details of preparation are described in Ref. [193]. It has been found that the treatment of mice with implanted gliomas using temozolomide-loaded PLGA nanoparticles targeting the gliomas TME extended animals' lives.

5. Preclinical and Clinical Studies of Polyester Nanocarriers

Hundreds of research papers have been published on model drug polyester carrier systems. Far fewer are reports of preclinical studies using cell cultures, animals, or both. Recently, H.O. Alsaab et al. prepared a list of selected examples of preclinical studies based on nanocarriers made of PLGA polymers, or copolymers containing PLGA blocks [194]. These preclinical studies were related to the treatment of the following health problems (in parentheses are indicated the encapsulated drugs): lung cancer (platinum-curcumin or sorafenib), colon cancer (afatinib or 5-fluorouracil-chrisin), ovarian cancer (curcumin), and prostate cancer (uncaria tomentosa extract). The aforementioned publication (Ref. [184]) provides also information on the clinical trials of some PLGA-based formulations (data from the clinicaltrials.gov website of the US National Institute of Health). The following list contains names of the drugs and, in parentheses, indications for applications of drugs: Arestin (periodontal disease), Atridox® (chronic adult periodontitis), Eligard (advanced prostatic cancer), Decapepty (prostate cancer), Lupron Depot (prostate cancer), Nutropin Depot® (growth deficiency), Pamorelin® (prostate cancer), Risperidal® Consta (antipsychotic), Sandostatin® LAR (acromegaly and carcinoid), Somatuline® LA (acromegaly), Suprecur® MP (prostate cancer), TrelstarTM Depot (advanced prostatic cancer), Zoladex® (breast cancer and prostate cancer).

Four years ago, M. do Carmo Pereira published an interesting editorial paper discussing problems on the way from in vitro through preclinical and clinical studies to the pharmaceutical market [195]. Discussed were the immune nanocarriers for the targeted brain delivery. These nanocarriers are decorated with covalently immobilized antibodies specific to the receptors present mainly on the cancer cells. The most significant progress based on the aforementioned strategy was made for the liposomes and solid lipid nanoparticles. Some formulations of this kind (produced by Gilead Sciences, Foster City, CA,

USA, and Cephalon/TEVA Pharmaceutical Industries, Fairfield, NJ, USA) are already on the market. Available are also some formulations based on the PLGA copolymers: Nutropin Depot® (Genentech, South San Francisco, CA, USA) and Trelstar® (Pfizer, New York, NY, USA) [195]. M. do Carmo Pereira et al. discussed also the strategy utilizing the receptor-ligand-mediated transport of nanoparticles across the blood-brain barrier. The aforementioned approach included the identification of receptors highly expressed on the surface endothelial cells of the capillary blood vessels, preparation of monoclonal antibodies against these receptors, and covalent binding of the antibodies to the surface of the drug-loaded nanoparticles. Such nanoparticles bind to receptors on the surface of endothelium cells of capillary blood vessels of the brain and enter into them by endocytosis, initiating the process of transport across the blood-brain barrier. Among the receptors are not only receptors for the most abundant transferrin and insulin but also receptors of folic acid, lipoprortein, lactoferrin, and some others. However, despite positive results in the in vitro studies, none of them was yet approved for use in clinics [195]. More successful were attempts at the development of polyester nanocarriers targeting other organs than the brain. Samyang Biopharmaceuticals (Seongnam-si, Republic of Korea) elaborated on the PEO-PLA polymeric micelle paclitaxel-loaded formulation. In 2007, this formulation under the name Genexol-PM was approved in South Korea for clinical application for the treatment of breast, lung, and ovarian cancers (information from the US National Institute of Health website updated in 2023 [196]. In 2018, S.-W. Lee et al. published an open-access article describing the results of the phase II trial to evaluate the efficacy and safety of a Cremophor-free polymeric micelle formulation of paclitaxel as a first-line treatment for ovarian cancer [197]. However, until 2023, no information is available about approval by the US Food and Drug Administration (FDA) or European Medicines Agency (EMA) for using Genexol-PM for anticancer therapy.

Since Genexol-PM micelles are stabilized with copolymers containing PEO blocks, it is worth mentioning a very important problem, which should attract the attention of the scientific community involved in studies in the field of drug delivery. The problem is related to the immunogenic properties of poly(ethylene oxide) (PEO), known also under the name poly(ethylene glycol) (PEG). In 2007, the paper entitled "PEGylated liposomes elicit an anti-PEG IgM response in a T cell-independent manner" [198] was published. The studies described in this publication proved that the intravenous injection of PEGylated liposomes, which did not contain any protein or peptide, results in weak IgG and intensive IgM responses. Later studies revealed that siRNA lipoplexes with surfaces containing bonded PEG are quickly cleared from the circulating blood when they are injected again after the first injection [199]. Further studies showed the that injection of siRNA-containing PEGylated lipoplex in mice results in PEG interaction with B cell-intrinsic toll-like receptor 7, leading to an immune response manifested by enhanced production of anti-PEG IgM [200]. As a result of the aforementioned research, it became clear that PEO should not be used for nanoparticle coating enhancing their circulation in blood but should be replaced by another non-immunogenic polymer. Probably, the immunogenicity of PEO/PEG developed with time due to the long-lasting exposure of humans to these water-soluble synthetic polymers commonly used in pharmacy, cosmetics, and even as food additives. However, the replacement of PEO or PEG with another synthetic polymer may be only a temporary solution because the problem may repeat.

6. Conclusions

A review of previous and recent publications discussing the synthesis of aliphatic polyesters of carboxylic acids shows that to date the researchers have at their disposal a rich library of homo- and copolymers with basic properties suggesting their suitability for using them for the preparation of drug nanocarriers. The simplest are the semi-crystalline poly(ε-caprolactone) (PCL) and highly crystalline polyglycolide (PGL). Another source-based name for PGL is poly(glycolic acid) (PGA), depending on whether the polymer is made by ring-opening polymerization from glycolide (the cyclic dimer of glycolic acid) or by polycondensation of the glycolic acid. Both PCL and PGA have only the methylene groups in the aliphatic partis, separating the ester groups. However, their thermal properties, solubility in organic solvents, and degradability differ significantly due to the very different microstructure. In PCL, the flexible aliphatic segment is long, consisting of five methylene groups. In PGA, only one methylene group separates two consecutive ester groups, making the PGA chains more rigid. As a result, for PCL Tg and melting temperatures are low, −60 and 60 °C, respectively, whereas for PGA Tg is in a range of 35 to 40 °C, and melting temperature ranges from 200 to 225 °C. Moreover, PCL is easily soluble in chloroform, dichloromethane, dimethyl sulfide, acetyl chloride, tetrahydrofuran, and furan that can be used in the processes of nanocarriers' preparation [201], whereas PGA is soluble only in hexafluoroisopropanol, which makes the preparation of nanocarriers from PGA homopolymers very difficult. The problem was solved by replacing PGA with copolymers of glycolic acid and lactic acid (PLGA). Thus, most of the research on polyester drug carriers has been completed for polylactide/polyglycolide nano- and microparticles. It should also be stressed that the final formulation of the nanocarriers is free from the residual traces of the organic solvents. The current state of knowledge makes it possible to synthesize polylactides with the required stereoregularity and controlled crystallinity. It is possible to synthesize polylactides with the required molar masses in the range of a few thousand to about 6×10^5 g/mol and with a narrow molar mass dispersion. Proper selection of methods used for the preparation of drug carriers allows for obtaining them with diameters from a few dozen nanometers to microns. The time of the drug release from polylactide and poly(lactide/glycolide) particles can be controlled to some extent from days to weeks. A significant advance in the synthesis of polylactides is the development of its metal-free synthesis. The process yields polymers that do not contain heavy metals and, at least by some researchers, are considered safer for medical applications. Noteworthy is the polymerization initiated by systems with initiator and catalytic groups combined into a single molecule, so that they later remain permanently bound to the polymer chain and thus cannot migrate outside the drug carrier particle.

A very special case is poly(β butyrolactone) (PBL) and poly(hydroxybutyrate) (PHB) (two source names are used for a polymer with the same chain structure, depending on whether the polymer was derived from β-butyrolactone or hydroxybutyric acid). Their synthesis and biosynthesis have been comprehensively studied, and their overall structure-property relationships suggest their usefulness for medical applications, but what is surprising is that research in this area is still scarce and the field requires more extensive exploration. Since PLAs and PBLs contain chiral carbon atoms for the preparation of nanocarrier polymers with well-defined content of the stereoisomeric units and with the known microstructure, they should be used. There is great potential in using macrolactones with double bonds to synthesize functional nanoparticles for targeted drug delivery. However, so far no reports on the preparation of such nanoparticles are available. However, there are articles on the copolymerization of ε-caprolactone and macrolactones with double bonds and their subsequent functionalization in the click thiol-ene reaction. The relationship between drug release and nanoparticle degradation is not well understood. The need for multiple-drug-loaded nanoparticles and sequential (or triggered by external stimuli) drug release from them is well recognized. It should be noted that there were obtained PLGA nanocarriers that, even without any targeting moieties, can efficiently cross the blood-brain barrier, opening the way for the simple synthesis of drugs for the patients suffering from Alzheimer's and Parkinson's diseases. Thus, there is still a lot of space for further research in the field of polyesters tailored for drug carriers.

Funding: This research received no external funding.

Acknowledgments: Authors acknowledge support from the CMMS PAS Statutory Fund.

Conflicts of Interest: The authors declare no conflicts of interest.

References

1. Chen, B.; Yang, J.-Z.; Wang, L.-F.; Zhang, Y.-J.; Lin, X.-J. Ifosfamide-loaded poly (lactic-co-glycolic acid) PLGA-dextran polymeric nanoparticles to improve the antitumor efficacy in osteosarcoma. *BMC Cancer* **2015**, *15*, 752. [CrossRef] [PubMed]
2. Lilienthal, I.; Herold, N. Targeting molecular mechanisms underlying treatment efficacy and resistance in osteosarcoma: A review of current and future strategies. *Int. J. Mol. Sci.* **2020**, *21*, 6885. [CrossRef]
3. Si, M.; Xia, Y.; Cong, M.; Wang, D.; Hou, Y.; Ma, H. In situ co-delivery of doxorubicin and cisplatin by injectable thermosensitive hydrogels for enhanced osteosarcoma treatment. *Int. J. Nanomed.* **2022**, *17*, 1309–1322. [CrossRef] [PubMed]
4. Ding, D.; Zhu, Q. Recent advances of PLGA micro/nanoparticles for the delivery of biomacromolecular therapeutics. *Mater. Sci. Eng. C* **2018**, *92*, 1041–1060. [CrossRef] [PubMed]
5. Piperno, A.; Sciortino, M.T.; Giusto, E.; Montesi, M.; Panseri, S.; Scala, A. Recent advances and challenges in gene delivery mediated by polyester-based nanoparticles. *Int. J. Nanomed.* **2021**, *16*, 5981–6002. [CrossRef]
6. Conte, C.; Monteiro, P.F.; Gurnani, P.; Stolnik, S.; Ungaro, F.; Quaglia, F.; Clarke, P.; Grabowska, A.; Kavallaris, M.; Alexander, C. Multi-component bioresponsive nanoparticles for synchronous delivery of docetaxel and TUBB3 siRNA to lung cancer cells. *Nanoscale* **2021**, *13*, 11414–11426. [CrossRef]
7. Thangudu, S.; Cheng, F.-Y.; Su, C.-H. Advancements in the blood-brain barrier penetrating nanoplatforms for brain related disease diagnostics and therapeutic applications. *Polymers* **2020**, *12*, 3055. [CrossRef]
8. Maher, R.; Moreno-Borrallo, A.; Jindal, D.; Mai, B.T.; Ruiz-Hernandez, E.; Harkin, A. Intranasal polymeric and lipid-based nanocarriers for CNS drug delivery. *Pharmaceutics* **2023**, *15*, 746. [CrossRef]
9. Shen, K.; Sun, G.; Chan, L.; He, L.; Li, X.; Yang, S.; Wang, B.; Zhang, H.; Huang, J.; Chang, M.; et al. Anti-inflammatory nanotherapeutics by targeting matrix metalloproteinases for immunotherapy spinal cord injury. *Small* **2021**, *17*, 2102102. [CrossRef]
10. Sun, Y.; Zabihi, M.; Li, Q.; Li, X.; Kim, B.J.; Ubogu, E.E.; Raja, S.N.; Wesselmann, U.; Zhao, C. Drug permeability: From the blood–brain barrier to the peripheral nerve barriers. *Adv. Therap.* **2023**, *6*, 2200150. [CrossRef]
11. Zhao, M.; Zhu, T.; Chen, J.; Cui, Y.; Zhang, X.; Lee, R.J.; Sun, F.; Li, Y.; Teng, L. PLGA/PCADK composite microspheres containing hyaluronic acid–chitosan siRNA nanoparticles: A rational design for rheumatoid arthritis therapy. *Int. J. Pharm.* **2021**, *596*, 120204. [CrossRef] [PubMed]
12. Sunoqrot, S.; Niazi, M.; Al-Natour, M.A.; Jaber, M.; Abu-Qatouseh, L. Loading of coal tar in polymeric nanoparticles as a potential therapeutic modality for psoriasis. *ACS Omega* **2022**, *7*, 7333–7340. [CrossRef] [PubMed]
13. Powell, D.; Chandra, S.; Dodson, K.; Shaheen, F.; Wiltz, K.; Ireland, S.; Syed, M.; Dash, S.; Wiese, T.; Mandal, T.; et al. Aptamer-functionalized hybrid nanoparticle for the treatment of breast cancer. *Europ. J. Pharm. Biopharm.* **2017**, *114*, 108–118. [CrossRef] [PubMed]
14. Martínez-Jothar, L.; Sofia Doulkeridou, S.; Raymond, M.; Schiffelers, R.M.; Javier Sastre Torano, J.S.; Oliveira, S.; van Nostrum, C.E.; Hennink, W.E. Insights into maleimide-thiol conjugation chemistry: Conditions for efficient surface functionalization of nanoparticles for receptor targeting. *J. Control. Release* **2018**, *282*, 101–109. [CrossRef]
15. Farran, B.; Montenegro, R.C.; Kasa, P.; Pavitra, E.; Yun Suk Huh, Y.S.; Han, Y.-K.; Kamal, M.A.; Nagarajua, G.P.; Rajue, G.S.R. Folate-conjugated nanovehicles: Strategies for cancer therapy. *Mater. Sci. Eng. C* **2020**, *107*, 110341. [CrossRef]
16. Lee, S.E.; Lee, C.M.; Won, J.F.; Jang, G.-Y.; Lee, J.H.; Park, S.H.; Kang, T.H.; Han, H.D.; Park, Y.-M. Enhancement of anticancer immunity by immunomodulation of apoptotic tumor cells using annexin A5 protein-labeled nanocarrier system. *Biomaterials* **2022**, *288*, 121677. [CrossRef]
17. Sanjanwala, D.; Vandana Patravale, V. Aptamers and nanobodies as alternatives to antibodies for ligand-targeted drug delivery in cancer. *Drug Discov. Today* **2023**, *28*, 103550. [CrossRef]
18. Pelosi, C.; Tinè, M.R.; Wurm, F.R. Main-chain water-soluble polyphosphoesters: Multi-functional polymers as degradable PEG-alternatives for biomedical applications. *Eur. Polym. J.* **2020**, *141*, 110079. [CrossRef]
19. Panyue Wen, P.; Ke, W.; Dirisala, A.; Toh, K.; Tanaka, M.; Li, J. Stealth and pseudo-stealth nanocarriers. *Adv. Drug Deliv. Rev.* **2023**, *198*, 114895. [CrossRef]
20. Bona, B.L.; Lagarrigue, P.; Chirizzi, C.; Espinoza, M.I.M.; Pipino, C.; Metrangolo, P.; Cellesi, F.; Bombelli, F.B. Design of fluorinated stealth poly(ε-caprolactone) nanocarriers. *Colloids Surf. B Biointerfaces* **2024**, *234*, 113730. [CrossRef]
21. Chountoulesi, M.; Naziris, N.; Pippa, N.; Pispas, S.; Demetzos, D. Stimuli-responsive nanocarriers for drug delivery. In *Nanomaterials for Clinical Applications, Case Studies in Nanomedicines a Volume in Micro and Nano Technologies*; Pippa, N., Demetzos, C., Eds.; Elsevier: Amsterdam, The Netherlands, 2020. [CrossRef]
22. Guo, Q.; Liu, J.; Yang, H.; Lei, Z. Synthesis of photo, oxidation, reduction triple-stimuli-responsive interface-cross-linked polymer micelles as nanocarriers for controlled release. *Macromol. Chem. Phys.* **2021**, *222*, 2000365. [CrossRef]
23. Sun, Y.; Davis, E. Nanoplatforms for targeted stimuli-responsive drug delivery: A review of platform materials and stimuli-responsive release and targeting mechanisms. *Nanomaterials* **2021**, *11*, 746. [CrossRef] [PubMed]
24. Singh, D.; Sharma, Y.; Dheer, D. Stimuli responsiveness of recent biomacromolecular systems (concept to market): A review. *Int. J. Biol. Macromol.* **2024**, *261*, 129901. [CrossRef]

25. Kiamohammadi, L.; Asadi, L.; Shirvalilou, S.; Khoei, S.; Khoee, S.; Soleymani, M.; Minaei, S.E. Physical and biological properties of 5-fluorouracil polymer-coated magnetite nanographene oxide as a new thermosensitizer for alternative magnetic hyperthermia and a magnetic resonance imaging contrast agent: In vitro and in vivo study. *ACS Omega* **2021**, *6*, 20192–20204. [CrossRef] [PubMed]
26. Trotsenko, Y.A.; Belova, L.L. Biosynthesis of poly(3-hydroxybutyrate) and poly(3-hydroxybutyrate-co-3-hydroxyvalerate) and its regulation in bacteria. *Microbiology* **2000**, *6*, 635–645. [CrossRef]
27. Verlinden, R.A.J.; Hill, D.J.; Kenward, M.A.; Williams, C.D.; Radecka, I. Bacterial synthesis of biodegradablepolyhydroxyalkanoates. *J. Appl. Microbiol.* **2007**, *102*, 1437–1449. [CrossRef] [PubMed]
28. Jendrossek, D.; Pfeiffer, D. New insights in the formation of polyhydroxyalkanoate granules (carbonosomes) and novel functions of poly(3-hydroxybutyrate). *Environment. Microbiol.* **2014**, *16*, 2357–2373. [CrossRef]
29. Nagarajan, D.; Aristya, G.R.; Lin, Y.-J.; Chang, J.-J.; Yen, H.-W.; Chang, J.-S. Microbial cell factories for the production of polyhydroxyalkanoates. *Essays Biochem.* **2021**, *65*, 337–353. [CrossRef]
30. Zhou, W.; Bergsma, S.; Colpa, D.I.; Euverink, G.-J.W.; Krooneman, J. Polyhydroxyalkanoates (PHAs) synthesis and degradation by microbes and applications towards a circular economy. *J. Environ. Manag.* **2023**, *341*, 118033. [CrossRef]
31. Koller, M. A review on stability replacing shed and emerging fermentation schemes for microbial production of polyhydroxyalkanoate (PHA) biopolyesters. *Fermentation* **2018**, *4*, 30. [CrossRef]
32. Lee, S.Y. Bacterial polyhydroxyalkanoates. *Biotechnol. Bioeng.* **1996**, *49*, 1–14. [CrossRef]
33. Hirt, T.D.; Neuenschwander, P.; Suter, U.W. Telechelic diols from poly[®-3-hydroxybutyric acid] and poly([®-3-hydroxybutyric acid]-co-[®-3-hydroxyvaleric acid]. *Macromol. Chem. Phys.* **1996**, *197*, 1609–1614. [CrossRef]
34. Andrade, A.P.; Witholt, B.; Hany, R.; Egli, T.; Li, Z. Preparation and characterization of enantiomerically pure telechelic diols from mcl-Poly[®-3-hydroxyalkanoates]. *Macromolecules* **2002**, *35*, 684–689. [CrossRef]
35. Saruul, P.; Srienc, F.; Somers, D.A.; Samac, D.A. Production of a biodegradable plastic polymer, poly-β-hydroxybutyrate, in transgenic alfalfa. *Crop Sci.* **2002**, *42*, 919–927. [CrossRef]
36. Menzel, G.; Harloff, H.-J.; Jung, C. Expression of bacterial poly(3-hydroxybutyrate) synthesis genes in hairy roots of sugar beet (*Beta vulgaris* L.). *Appl. Microbiol. Biotechnol.* **2003**, *60*, 571–576. [CrossRef]
37. Wróbel, M.; Zebrowski, J.; Szopa, J. Polyhydroxybutyrate synthesis in transgenic flax. *J. Biotechnol.* **2004**, *107*, 41–54. [CrossRef]
38. Parveez, G.K.A.; Bahariah, B.; Ayub, M.H.; Masani, M.Y.A.; Rasid, O.A.; Tarmizi, A.H.; Ishak, Z. Production of polyhydroxybutyrate in oil palm (*Elaeis guineensis* Jacq.) mediated, by microprojectile bombardment of PHB biosynthesis genes into embryogenic calli. *Front. Plant Sci.* **2015**, *6*, 598. [CrossRef]
39. McQualter, R.B.; Somleva, M.N.; Gebbie, L.K.; Li, X.; Petrasovits, L.A.; Snell, K.D.; Nielsen, L.K.; Brumbley, S.M. Factors affecting polyhydroxybutyrate accumulation in mesophyll cells of sugarcane and switchgrass. *BMC Biotechnol.* **2014**, *14*, 83. [CrossRef]
40. Malik, M.R.; Yang, W.; Patterson, N.; Tang, J.; Wellinghoff, R.L.; Preuss, M.L.; Burkitt, C.; Sharma, N.; Ji, Y.; Jez, J.M.; et al. Production of high levels of poly-3-hydroxybutyrate in plastids of *Camelina sativa* seeds. *Plant Biotechnol. J.* **2015**, *13*, 675–688. [CrossRef]
41. Agostini, D.E.; Lando, J.B.; Shelton, J.R.J. Synthesis and characterization of poly-β-Hydroxybutyrate. I. Synthesis of crystalline DL-poly-β-hydroxybutyrate from DL-β-Butyrolactone. *J. Polym. Sci. A-1* **1971**, *9*, 2775–2787. [CrossRef]
42. Shelton, J.R.; Agostini, D.E.; Lando, J.B. Synthesis and characterization of poly-β-hydroxybutyrate. II. Synthesis of D-poly-β-hydroxybutyrate and the mechanism of ring-opening polymerization of β-butyrolactone. *J. Polym. Sci. A-1* **1971**, *9*, 2789–2799. [CrossRef]
43. Teranishi, K.; Iida, M.; Araki, T.; Yamashita, S.; Tani, H. Stereospecific polymerization of β-alkyl-β-propiolactone. *Macromolecules* **1974**, *7*, 421. [CrossRef]
44. Iida, M.; Araki, T.; Teranishi, K.; Tani, H. Effect of substituents on stereospecific polymerization of /β-alkyl- and β-chloroalkyl-β-propiolactones. *Macromolecules* **1977**, *10*, 275–284. [CrossRef]
45. Le Borgne, A.; Spassky, N. Stereoelective polymerization of β-butyrolactone. *Polymer* **1989**, *30*, 2312–2319. [CrossRef]
46. Jedlinski, Z.; Kurcok, P.; Lenz, R.W. First facile synthesis of biomimetic poly-®-3-hydroxybutyrate via regioselective anionic polymerization of (*S*)-β-butyrolactone. *Macromolecules* **1998**, *31*, 6718–6720. [CrossRef]
47. Kurcok, P.; Kowalczuk, M.; Hennek, K.; Jedlinski, Z. Anionic polymerization of beta-lactones initiated with alkali-metal alkoxides: Reinvestigation of the polymerization mechanism. *Macromolecules* **1992**, *25*, 2017–2020. [CrossRef]
48. Zhang, Y.; Gross, R.A.; Lenz, R.W. Stereochemistry of the ring-opening polymerization of (S)-S-Butyrolactone. *Macromolecules* **1990**, *23*, 3206–3212. [CrossRef]
49. Philip, S.; Keshavarz, T.; Roy, I. Polyhydroxyalkanoates: Biodegradable polymers with a range of applications. *J. Chem. Technol. Biotechnol.* **2007**, *82*, 233–247. [CrossRef]
50. Khanna, S.; Srivastava, A. Recent advances in microbial polyhydroxyalkanoates. *Process Biochem.* **2005**, *40*, 607–619. [CrossRef]
51. Domínguez-Díaz, M.; Meneses-Acosta, A.; Romo-Uribe, A.; Peña, C.; Segura, D.; Espin, G. Thermo-mechanical properties, microstructure and biocompatibility in poly-b-hydroxybutyrates (PHB) produced by OP and OPN strains of *Azotobacter vinelandii*. *Eur. Polym. J.* **2015**, *63*, 101–112. [CrossRef]
52. Hong, S.-G.; Gau, T.-K.; Huang, S.-C. Enhancement of the crystallization and thermal stability of polyhydroxybutyrate by polymeric additives. *J. Therm. Anal. Calorim.* **2011**, *103*, 967–975. [CrossRef]

53. Arcana, M.; Giani-Beaune, O.; Schue, F.; Amass, W.; Amass, A. Structure and morphology of poly(b-hydroxybutyrate) synthesized by ring-opening polymerization of racemic (R,S)-b-butyrolactone with distannoxane derivatives. *Polym. Int.* **2000**, *49*, 1348–1355. [CrossRef]
54. Tang, X.; Westlie, A.H.; Watson, E.M.; Chen, E.Y.-X. Stereosequenced crystalline polyhydroxyalkanoates from diastereomeric monomer mixtures. *Science* **2019**, *366*, 754–758. [CrossRef] [PubMed]
55. Gay-Lussac, H.J.; Pelouze, H. Über die Milchsäure. *Ann. Phys.* **1833**, *105*, 108. [CrossRef]
56. Carothers, W.H.; Dorough, G.L.; Van Natta, F.J. Studies of polymerization and ring formation. x. the reversible polymerization of cyclic esters. *J. Am. Chem. Soc.* **1932**, *54*, 761–772. [CrossRef]
57. Garlotta, D. A literature review of poly(lactic acid). *J. Polym. Environ.* **2001**, *9*, 63–83. [CrossRef]
58. Duda, A. ROP of Cyclic Esters. Mechanisms of Ionic and Coordination Processes. *Polym. Sci.* **2012**, *IV*, 213–246. [CrossRef]
59. Penczek, S.; Cypryk, M.; Duda, A.; Kubisa, P.; Slomkowski, S. Living ring-opening polymerizations of heterocyclic monomers. *Prog. Polym. Sci.* **2007**, *32*, 247–282. [CrossRef]
60. Slomkowski, S.; Penczek, S.; Duda, A. Polylactides-an overview. *Polym. Adv. Technol.* **2014**, *25*, 436–447. [CrossRef]
61. Pretula, J.; Slomkowski, S.; Penczek, S. Polylactides—Methods of synthesis and characterization. *Adv. Drug Deliv. Rev.* **2016**, *107*, 3–16. [CrossRef]
62. Vert, M.; Chen, J.; Hellwich, K.-H.; Hodge, P.; Nakano, T.; Scholz, C.; Slomkowski, S.; Vohlidal, J. Nomenclature and terminology for linear lactic acid-based polymers (IUPAC Recommendations 2019). *Pure Appl. Chem.* **2020**, *92*, 193–211. [CrossRef]
63. Matsutani, K.; Kimura, Y. PLA synthesis. Rrom the monomer to the polymer. In *Poly(Lactic Acid) Science and Technology: Processing, Properties, Additives and Applications*; Jiménez, A., Peltzer, M., Ruseckaite, R., Eds.; The Royal Society of Chemistry: Cambridge, UK, 2015.
64. Okano, K.; Hama, S.; Kihara, M.; Noda, H.; Tsutomu Tanaka, T.; Kondo, A. Production of optically pure D-lactic acid from brown rice using metabolically engineered *Lactobacillus plantarum*. *Appl. Microbiol. Biotechnol.* **2017**, *101*, 1869–1875. [CrossRef] [PubMed]
65. Balakrishnan, R.; Tadi, S.R.R.; Sivaprakasam, S.; Rajaram, S. Optimization of acid and enzymatic hydrolysis of kodo millet (Paspalum scrobiculatum) bran residue to obtain fermentable sugars for the production of optically pure D (−) lactic acid. *Ind. Crops Prod.* **2018**, *111*, 731–742. [CrossRef]
66. Zaini, N.A.M.; Chatzifragkou, A.; Tverezovskiy, A.; Charalampopoulos, D. Purification and polymerisation of microbial D-lactic acid from DDGS hydrolysates fermentation. *Biochem. Eng. J.* **2019**, *150*, 107265. [CrossRef]
67. Din, N.A.S.; Lim, S.J.; Maskat, M.Y.; Zaini, N.A.M. Microbial D-lactic acid production, *In Situ* separation and recovery from mature and young coconut husk hydrolysate fermentation broth. *Biochem. Eng. J.* **2022**, *188*, 108680. [CrossRef]
68. Abedi, E.; Hashemi, S.M.B. Lactic acid production—Producing microorganisms and substrates sources-state of art. *Heliyon* **2020**, *6*, e04974. [CrossRef]
69. Huang, Y.; Wang, Y.; Shang, N.; Li, P. Microbial fermentation processes of lactic acid: Challenges, solutions, and future prospects. *Foods* **2023**, *12*, 2311. [CrossRef]
70. Yankov, D. Fermentative lactic acid production from lignocellulosic feedstocks: From source to purified product. *Front. Chem.* **2022**, *10*, 823005. [CrossRef]
71. Ajioka, M.; Enomoto, K.; Suzuki, K.; Yamaguchi, A. Basic properties of polylactic acid produced by the direct condensation polymerization of lactic-acid. *Bull. Chem. Soc. Jpn.* **1995**, *68*, 2125–2131. [CrossRef]
72. Moon, S.-I.; Taniguchi, I.; Miyamoto, M.; Kimura, Y.; Lee, C.-W. Synthesis and properties of high-molecular-weight poly(L-lactic acid) by melt/solid polycondensation under different reaction conditions. *High Perform. Polym.* **2001**, *13*, S189–S196. [CrossRef]
73. Moon, S.-I.; Lee, C.-W.; Taniguchi, I.; Miyamoto, M.; Kimura, Y. Melt/solid polycondensation of l-lactic acid: An alternative route to poly(l-lactic acid) with high molecular weight. *Polymer* **2001**, *42*, 5059–5062. [CrossRef]
74. Kulkarni, R.K.; Pani, K.; Neuman, C.; Leonard, F. Lactic acid for surgical implants. *Arch. Surg.* **1966**, *93*, 839–843. [CrossRef] [PubMed]
75. Kohn, F.E.; van Den Berg, J.W.A.; van de Ridder, G.; Feijen, J. The ring-opening polymerization of D,L-lactide in the melt initiated with tetraphenyltin. *J. Appl. Polym. Sci.* **1984**, *29*, 4265–4277. [CrossRef]
76. Lin, Z.; Guo, X.; He, Z.; Liang, X.; Wang, M.; Jin, G. Thermal degradation kinetics study of molten polylactide based on Raman spectroscopy. *Polym. Eng. Sci.* **2021**, *61*, 201–210. [CrossRef]
77. Duda, A.; Penczek, S. Polymerization of ε-caprolactone initiated by aluminum isopropoxide trimer and/or tetramer. *Macromolecules* **1995**, *28*, 5981–5992. [CrossRef]
78. Duda, A.; Penczek, S. On the difference of reactivities of various aggregated forms of aluminium triisopropoxide in initiating ring-opening polymerizations. *Macromol. Rapid Commun.* **1995**, *16*, 67–76. [CrossRef]
79. Kowalski, A.; Libiszowski, J.; Majerska, K.; Duda, A.; Penczek, S. Kinetics and mechanism of ε-caprolactone and L,L-lactide polymerization coinitiated with zinc octoate or aluminum acetylacetonate: The next proofs for the general alkoxide mechanism and synthetic applications. *Polymer* **2007**, *48*, 3952–3960. [CrossRef]
80. Gadomska-Gajadhur, A.; Ruśkowski, P. Biocompatible catalysts for lactide polymerization—Catalyst activity, racemization effect, and optimization of the polymerization based on design of experiments. *Org. Process Res. Dev.* **2020**, *24*, 1435–1442. [CrossRef]
81. Kowalski, A.; Duda, A.; Penczek, S. Kinetics and mechanism of cyclic esters polymerization initiated with Tin(II) octoate. 3. Polymerization of L,L-dilactide. *Macromolecules* **2000**, *33*, 7359–7370. [CrossRef]

82. Majerska, K.; Duda, A.; Penczek, S. Kinetics and mechanism of cyclic esters polymerization initiated with tin(II) octoate, 4—Influence of proton trapping agents on the kinetics of epsilon-caprolactone and L,L-dilactide polymerization. *Macromol. Rapid Commun.* **2000**, *21*, 1327–1332. [CrossRef]
83. Duda, A.; Penczek, S. Thermodynamics of L-Lactide polymerization. equilibrium monomer concentration. *Macromolecules* **1990**, *23*, 1636–1639. [CrossRef]
84. Shinno, K.; Miyamoto, M.; Kimura, Y.; Hirai, Y.; Yoshitome, H. Solid-state postpolymerization of l-lactide promoted by crystallization of product polymer: An effective method for reduction of remaining monomer. *Macromolecules* **1997**, *30*, 6438–6444. [CrossRef]
85. Mosnacek, J.; Duda, A.; Libiszowski, J.; Penczek, S. Copolymerization of LL-lactide at its living polymer-monomer equilibrium with ε-caprolactone as comonomer. *Macromolecules* **2005**, *38*, 2027–2029. [CrossRef]
86. Degee, P.; Dubois, P.; Jerome, R. Bulk polymerization of lactides initiated by aluminium isopropoxide.3. Thermal stability and viscoelastic properties. *Macromol. Chem. Phys.* **1997**, *198*, 1973–1984. [CrossRef]
87. Kowalski, A.; Libiszowski, J.; Duda, A.; Penczek, S. Polymerization of L,L-dilactide initiated by Tin(II) butoxide. *Macromolecules* **2000**, *33*, 1964–1971. [CrossRef]
88. Bourissou, D.; Moebs-Sanchez, S.; Martín-Vaca, B. Recent advances in the controlled preparation of poly(α-hydroxy acids): Metal-free catalysts and new monomers. *C. R. Chimie* **2007**, *10*, 775–794. [CrossRef]
89. Kamber, N.E.; Jeong, W.; Waymouth, R.M.; Pratt, R.C.; Lohmeijer, B.G.G.; Hedrick, J.L. Organocatalytic ring-opening polymerization. *Chem. Rev.* **2007**, *107*, 5813–5840. [CrossRef]
90. Nederberg, F.; Connor, E.F.; Möller, M.; Glauser, T.; Hedrick, J.L. New paradigms for organic catalysts: The first organocatalytic living polymerization. *Angew. Chem. Int. Ed.* **2001**, *40*, 2712–2715. [CrossRef]
91. Zhang, X.; Waymouth, R.M. Zwitterionic Ring opening polymerization with isothioureas. *ACS Macro Lett.* **2014**, *3*, 1024–1028. [CrossRef]
92. Alba, A.; Thillaye du Boullay, O.; Martin-Vaca, B.; Bourissou, D. Direct ring-opening of lactide with amines: Application to the organo-catalyzed preparation of amide end-capped PLA and to the removal of residual lactide from PLA samples. *Polym. Chem.* **2015**, *6*, 989–997. [CrossRef]
93. Lee, G.S.; Moon, B.R.; Jeong, H.; Shin, J.; Kim, J.G. Mechanochemical synthesis of poly(lactic acid) block copolymers: Overcoming the miscibility of the macroinitiator, monomer and catalyst undersolvent-free conditions. *Polym. Chem.* **2019**, *10*, 539–545. [CrossRef]
94. Lohmeijer, R.G.G.; Pratt, R.C.; Leibfarth, F.; Logan, J.W.; Long, D.A.; Dove, A.P.; Nederberg, F.; Choi, J.; Wade, C.; Waymouth, R.M.; et al. Guanidine and amidine organocatalysts for ring-opening polymerization of cyclic esters. *Macromolecules* **2006**, *39*, 8574–8583. [CrossRef]
95. Myers, M.; Connor, E.F.; Glausser, T.; Moeck, A.; Nyce, G.W.; Hedrick, J.L.J. Phosphines: Nucleophilic organic catalysts for the controlled ring-opening polymerization of lactides. *J. Polym. Sci. Part A Polym. Chem.* **2002**, *40*, 844. [CrossRef]
96. Connor, E.F.; Nyce, G.W.; Myers, M.; Moeck, A.; Hedrick, J.L. First example of N-heterocyclic carbenes as catalysts for living polymerization:: Organocatalytic ring-opening polymerization of cyclic esters. *J. Am. Chem. Soc.* **2002**, *124*, 914–915. [CrossRef]
97. Coulembier, O.; Dove, A.P.; Pratt, R.C.; Sentman, A.C.; Culkin, D.A.; Mespouille, L.; Dubois, P.; Waymouth, R.M.; Hedrick, J.L. Latent, thermally activated organic catalysts for the on-demand living polymerization of lactide. *Angew. Chem. Int. Ed.* **2005**, *44*, 4964–4968. [CrossRef]
98. Dove, A.P.; Pratt, R.C.; Lohmeijer, B.G.G.; Waymouth, R.M.; Hedrick, J.L. Thiourea-based bifunctional organocatalysis: Supramolecular recognition for living polymerization. *J. Am. Chem. Soc.* **2005**, *127*, 13798–13799. [CrossRef] [PubMed]
99. Basko, M.; Kubisa, P. Cationic copolymerization of ε-caprolactone and L,L-lactide by an activated monomer mechanism. *J. Polym. Sci. A Polym. Chem.* **2006**, *44*, 7071–7081. [CrossRef]
100. Basko, M.; Kubisa, P. Mechanism of propagation in the cationic polymerization of L,L-lactide. *J. Polym. Sci. A Polym. Chem.* **2008**, *46*, 7919–7923. [CrossRef]
101. Basko, M.; Kubisa, P. Cationic polymerization of L,L-lactide. *J. Polym. Sci. A Polym. Chem.* **2010**, *48*, 2650–2658. [CrossRef]
102. Basko, M. Activated monomer mechanism in the cationic polymerization of L,L-lactide. *Pure Appl. Chem.* **2012**, *84*, 2081–2088. [CrossRef]
103. Lewinski, P.; Kaluzynski, K.; Pretula, J.; Mielniczak, G.; Penczek, S. Catalysis in polymerization of cyclic esters. Catalyst and initiator in one molecule. Polymerization of lactide. *J. Catal.* **2022**, *405*, 249–264. [CrossRef]
104. Kaluzynski, K.; Pretula, J.; Lewinski, P.; Kazmierski, S.; Penczek, S. Catalysis in polymerization of cyclic esters. *Catalyst and initiator in one molecule. Polymerization of ε-caprolactone, J. Catal.* **2020**, *392*, 97–107. [CrossRef]
105. Spassky, N.; Wisniewski, M.; Pluta, C.; LeBorgne, A. Highly stereoelective polymerization of rac-(D,L)-lactide with a chiral Schiff's base/aluminium alkoxide initiator. *Macromol. Chem. Phys.* **1996**, *197*, 2627–2637. [CrossRef]
106. Nomura, N.; Ishii, R.; Akakura, M.; Aoi, K. Stereoselective ring-opening polymerization of racemic lactide using aluminum-achiral ligand complexes: Exploration of a chain-end control mechanism. *J. Am. Chem. Soc.* **2002**, *124*, 5938–5939. [CrossRef] [PubMed]
107. Tang, Z.H.; Yang, Y.K.; Pang, X.; Hu, J.L.; Chen, X.S.; Hu, N.H.; Jing, X.B. Controlled and stereospecific polymerization of rac-lactide with a single-site ethyl aluminum and alcohol initiating system. *J. Appl. Polym. Sci.* **2005**, *98*, 102–108. [CrossRef]
108. Nomura, N.; Ishii, R.; Yamamoto, Y.; Kondo, T. Stereoselective ring-opening polymerization of a racemic lactide by using achiral salen- and homosalen-aluminum complexes. *Chem.-A Eur. J.* **2007**, *13*, 4433–4451. [CrossRef]

109. Radano, C.P.; Baker, G.L.; Smith, M.R. Stereoselective polymerization of a racemic monomer with a racemic catalyst: Direct preparation of the polylactic acid stereocomplex from racemic lactide. *J. Am. Chem. Soc.* **2000**, *122*, 1552–1553. [CrossRef]
110. Ovitt, T.M.; Coates, G.W. Stereoselective ring-opening polymerization of rac-lactide with a single-site, racemic aluminum alkoxide catalyst: Synthesis of stereoblock poly(lactic acid). *J. Polym. Sci. Part A Polym. Chem.* **2000**, *38*, 4686–4692. [CrossRef]
111. Ovitt, T.M.; Coates, G.W. Stereochemistry of lactide polymerization with chiral catalysts: New opportunities for stereocontrol using polymer exchange mechanisms. *J. Am. Chem. Soc.* **2002**, *124*, 1316–1326. [CrossRef]
112. Zhong, Z.Y.; Dijkstra, P.J.; Feijen, J. [(salen)Al]-mediated, controlled and stereoselective ring-opening polymerization of lactide in solution and without solvent: Synthesis of highly isotactic polylactide stereocopolymers from racemic D,L-lactide. *Angew. Chem. Int. Ed.* **2002**, *41*, 4510–4513. [CrossRef]
113. Dijkstra, P.J.; Du, H. Feijen, Single site catalysts for stereoselective ring-opening polymerization of lactides. *J. Polym. Chem.* **2011**, *2*, 520–527. [CrossRef]
114. Tsuji, H. Poly(lactide) stereocomplexes: Formation, structure, properties, degradation, and applications. *Macromol. Biosci.* **2005**, *5*, 569–597. [CrossRef]
115. Hador, R.; Botta, A.; Venditto, V.; Lipstman, S.; Goldberg, I.; Kol, M. The dual-stereocontrol mechanism: Heteroselective polymerization of *rac*-lactide and syndioselective polymerization of meso-lactide by chiral aluminum salan catalysts. *Angew. Chem. Int. Ed.* **2019**, *58*, 14679–14685. [CrossRef]
116. Rosen, T.; Rajpurohit, J.; Lipstman, S.; Venditto, V.; Kol, M. Isoselective polymerization of *rac*-lactide by highly active sequential {ONNN} magnesium complexes. *Chem. Eur. J.* **2020**, *26*, 17183–17189. [CrossRef] [PubMed]
117. Roymuhury, S.K.; Mandal, M.; Chakraborty, D.; Ramkumar, V. Homoleptic titanium and zirconium complexes exhibiting unusual O_{iminol}–metal coordination: Application in stereoselective ring-opening polymerization of lactide. *Polym. Chem.* **2021**, *12*, 3953–3967. [CrossRef]
118. Majerska, K.; Duda, A. Stereocontrolled polymerization of racemic lactide with chiral initiator: Combining stereoelection and chiral ligand-exchange mechanism. *J. Am. Chem. Soc.* **2004**, *126*, 1026–1027. [CrossRef]
119. Sanko, V.; Sahin, I.; Sezer, U.A.; Sezer, S. A versatile method for the synthesis of poly(glycolic acid): High solubility and tunable molecular weights. *Polym. J.* **2019**, *51*, 637–647. [CrossRef]
120. Hurrell, S.; Cameron, R.E. Polyglycolide: Degradation and drug release. Part I: Changes in morphology during degradation. *J. Mater. Sci. Mater. Med.* **2001**, *12*, 811–816. [CrossRef]
121. Hurrell, S.; Cameron, R.E. Polyglycolide: Degradation and drug release. Part II: Drug release. *Mater. Sci. Mater. Med.* **2001**, *12*, 817–820. [CrossRef]
122. Goh, Y.-F.; Shakir, I.; Hussain, R. Electrospun fibers for tissue engineering, drug delivery, and wound dressing. *J. Mater. Sci.* **2013**, *48*, 3027–3054. [CrossRef]
123. Contreras-Cáceres, R.; Cabeza, L.; Perazzoli, G.; Díaz, A.; López-Romero, J.M.; Melguizo, C.; Prados, P. Electrospun nanofibers: Recent applications in drug delivery and cancer therapy. *Nanomaterials* **2019**, *9*, 656. [CrossRef] [PubMed]
124. Li, J.; Liu, Y.; Hend, E.; Abdelhakim, H. Drug delivery applications of coaxial electrospun nanofibres in cancer therapy. *Molecules* **2022**, *27*, 1803. [CrossRef] [PubMed]
125. Snoddy, B.; Jayasuriya, A.C. The Use of nanomaterials to treat bone infections. *Mater. Sci. Eng. C-Mater. Biol. Appl.* **2016**, *67*, 822–833. [CrossRef]
126. Reddy, P.G.; Domb, A.J. Formation of micro/nanoparticles and microspheres from polyesters by dispersion ring-opening polymerization. *Polym. Adv. Technol.* **2021**, *32*, 3835–3856. [CrossRef]
127. Procopio, A.; Lagreca, E.; Jamaledin, R.; La Manna, S.; Corrado, B.; Di Natale, C.; Onesto, V. Recent fabrication methods to produce polymer-based drug delivery matrices (experimental and in silico approaches). *Pharmaceutics* **2022**, *14*, 872. [CrossRef]
128. Rusa, C.C.; Tonelli, A.E. Polymer/polymer inclusion compounds as a novel approach to obtaining a PLLA/PCL intimately compatible blend. *Macromolecules* **2000**, *33*, 5321–5324. [CrossRef]
129. Douglas, P.; Andrews, G.; Jones, D.; Walker, G. Analysis of in vitro drug dissolution from PCL melt extrusion. *Chem. Eng. J.* **2010**, *164*, 359–370. [CrossRef]
130. Douglas, P.; Albadarin, A.B.; Al-Muhtaseb, A.H.; Mangwandi, C.; Walker, G.M. Thermo-mechanical propertiesofpolyε-caprolactone/poly L-lactic acid blends: Additionofnalidixicacid and polyethyleneglycoladditives. *J. Mechan. Behav. Biomed. Maater.* **2015**, *45*, 154–165. [CrossRef]
131. Woodruff, M.A.; Hutmacher, D.W. The return of a forgott in the 21st century. *Prog. Polym. Sci.* **2010**, *35*, 1217–1256. [CrossRef]
132. Lewinski, P.; Pretula, P.; Kaluzynski, K.; Kazmierski, S.; Penczek, S. ε-Caprolactone: Activated monomer polymerization; controversy over the mechanism of polymerization catalyzed by phosphorus acids (diarylhydrogen phosphates). Do acids also act as initiators? *J. Catal.* **2019**, *371*, 305–312. [CrossRef]
133. Ito, K.; Hashizuka, Y.; Yamashita, Y. Equilibrium cyclic oligomer formation in the anionic polymerization of ε-caprolactone. *Macromolecules* **1977**, *10*, 821–824. [CrossRef]
134. Ito, K.; Yamashita, Y. Propagation and depropagation rates in the anionic polymerization of ε-caprolactone cyclic oligomers. *Macromolecules* **1978**, *11*, 68–72. [CrossRef]
135. Sosnowski, S.; Slomkowski, S.; Penczek, S.; Reibel, L. Kinetic-caprolactone polymerization and formation of cyclic oligomers. *Makromol. Chem.* **1983**, *1984*, 2159–2179. [CrossRef]

136. Hofman, A.; Slomkowski, S.; Penczek, S. Polymerization of ε-caprolactones with kinetic suppression of macrocycles. *Makromol. Chem. Rapid Commun.* **1987**, *8*, 387–391. [CrossRef]
137. Penczek, S.; Duda, A.; Slomkowski, S. The reactivity-selectivity principle in polymerization. The case of polymerization of ε-caprolactone. *Makromol. Chem. Macromol. Symp.* **1992**, *54–55*, 31–40. [CrossRef]
138. Biela, T.; Kowalski, A.; Libiszowski, J.; Duda, A.; Penczek, S. Progress in polymerization of cyclic esters: Mechanisms and synthetic applications. *Macromol. Symp.* **2006**, *240*, 47–55. [CrossRef]
139. van der Meulen, I.; Li, Y.; Deumens, R.; Joosten, E.A.J.; Koning, C.E.; Heise, A. Copolymers from unsaturated macrolactones: Toward the design of cross-linked biodegradable polyesters. *Biomacromolecules* **2011**, *12*, 837–843. [CrossRef]
140. Claudino, M.; van der Meulen, I.; Trey, S.; Jonsson, M.; Heise, A.; Johansson, M. Photoinduced thiol-ene crosslinking of globalide/ε-caprolactone copolymers: Curing performance and resulting thermoset properties. *J. Polym. Sci. A Polym. Chem.* **2012**, *50*, 16–24. [CrossRef]
141. Wilson, J.A.; Ates, Z.; Pflughaupt, R.L.; Dove, A.P.; Heise, A. Polymers from macrolactones: From pheromones to functional materials. *Prog. Polym. Sci.* **2019**, *91*, 29–50. [CrossRef]
142. Lee, J.; Jung, S.G.; Cheon-Seok Park, C.S.; Kim, H.-Y.; Batt, C.A.; Kim, Y.-R. Tumor-specific hybrid polyhydroxybutyrate nanoparticle: Surface modification of nanoparticle by enzymatically synthesized functional block copolymer. *Bioorg. Med. Chem. Lett.* **2011**, *21*, 2941–2944. [CrossRef]
143. Kajjari, P.B.; Manjeshwar, L.S.; Aminabhavi, T.M. Novel blend microspheres of poly(3-hydroxybutyrate) and pluronic f68/127 for controlled release of 6-mercaptopurine. *J. Appl. Polym. Sci.* **2014**, *131*, 40196. [CrossRef]
144. Lee, J.; Saparbayeva, A.; Hlaing, S.P.; Kwak, D.; Kim, H.; Kim, J.; Lee, E.H.; Yoo, J.-W. Cupriavidus necator-produced polyhydroxybutyrate/EudragitFS hybrid nanoparticles mitigates ulcerative colitis via colon-targeted delivery of cyclosporine A. *Pharmaceutics* **2022**, *14*, 2811. [CrossRef]
145. Pan, C.-T.; Hwang, Y.-M.; Lin, Y.-M.; Zeng, S.-W.; Wang, S.-Y.; Kuo, S.-W.; Ju, S.-P.; Liang, S.-S.; Liu, Z.-H.; Yen, C.-K. Development of polycaprolactone microspheres with controllable and uniform particle size by uniform design experiment in emulsion progress. *Sens. Mater.* **2019**, *31*, 311–318. [CrossRef]
146. Ponjavić, M.; Nikolić, M.S.; Jevtić, S.; Jeremić, S.; Dokić, L.; Donlagić, J. Star-shaped poly(ε-caprolactones) with well-defined architecture as potential drug carriers. *J. Serb. Chem. Soc.* **2022**, *87*, 1075–1090. [CrossRef]
147. Mozafari, M. Synthesis and characterisation of poly(lactide-co-glycolide) nanospheres using vitamin E emulsifier prepared through one-step oil-in-water emulsion and solvent evaporation techniques. *IET Nanobiotechnol.* **2014**, *8*, 257–262. [CrossRef]
148. Harguindey, A.; Domaille, D.W.; Fairbanks, B.D.; Wagner, J.; Bowman, C.N.; Cha, J.N. Synthesis and assembly of click-nucleic-acid-containin PEG–PLGA nanoparticles for DNA delivery. *Adv. Mater.* **2017**, *29*, 1700743. [CrossRef]
149. Takeuchi, I.; Kimura, Y.; Makino, K. Effect of the conformation of poly(L-lactide-co-glycolide) molecules in organic solvents onnanoparticle. size. *J. Oleo Sci.* **2020**, *69*, 1125–1132. [CrossRef]
150. Han, C.-S.; Kang, J.-H.; Kim, Y.-J.; Kim, D.-W.; Park, C.W. Inhalable nano-dimpled microspheres containing Budesonide-PLGA for improved aerodynamic performance. *Int. J. Nanomed.* **2022**, *17*, 3405–3419. [CrossRef]
151. Slomkowski, S. Polyester nano- and microparticles by polymerization and self-assembly of macromolecules. In *Nanoparticles for Pharmaceutical*; Domb, A.J., Tabata, Y., Kumar, M.N.V.R., Farber, S., Eds.; American Scientific Publishers: California, CA, USA, 2007.
152. Heslinga, M.J.; Mastria, E.M.; Eniola-Adefeso, O. Fabrication of biodegradable spheroidal microparticles for drug delivery applications. *J. Control. Release* **2009**, *138*, 235–242. [CrossRef]
153. Cohen, H.; Levy, R.J.; Gao, J.; Fishbein, I.; Kousaev, V.; Sosnowski, S.; Slomkowski, S.; Golomb, G. Sustained delivery and expression of DNA encapsulated in polymeric nanoparticles. *Gene Ther.* **2000**, *7*, 1896–1905. [CrossRef]
154. You, G.; Kim, Y.; Lee, J.H.; Song, J.; Mok, H. Exosome-modified PLGA microspheres for improve internalization into dendritic cells and macrophages. *Biotechnol. Bioproc. Eng.* **2020**, *25*, 521–527. [CrossRef]
155. Duan, J.; Liu, C.; Liang, X.; Li, X.; Chen, Y.; Chen, Z.; Wang, X.; Kong, D.; Li, Y.; Yang, J. Protein delivery nanosystem of six-arm copolymer poly(ε-caprolactone)–poly(ethylene glycol) for long-term sustained release. *Int. J. Nanomed.* **2018**, *13*, 2743–2754. [CrossRef]
156. Gökberk, B.D.; Erdinç, N. Design, Optimization, and characterization of lysozyme-loaded poly(ε-caprolactone) microparticles for pulmonary delivery. *J. Pharm. Innov.* **2023**, *18*, 325–338. [CrossRef]
157. Chen, L.; Mei, L.; Feng, D.; Huang, D.; Tong, X.; Pan, X.; Zhu, C.; Wu, C. Anhydrous reverse micelle lecithin nanoparticles/PLGA compositemicrospheres for long-term protein delivery with reduced initial burst. *Colloids Surf. B Biointerfaces* **2018**, *163*, 146–154. [CrossRef]
158. Fessi, H.; Puisieux, F.; Devissaguet, J.P.; Ammoury, N.; Benita, S. Nanocapsule formation by interfacial polymer deposition following solvent displacement. *Int. J. Pharm.* **1989**, *55*, R1–R4. [CrossRef]
159. Rivas, C.J.M.; Tarhini, M.; Badri, W.; Miladi, K.; Greige-Gerges, H.; Nazari, Q.A.; Rodríguez, S.A.G.; Román, R.A.; Fessi, H.; Elaissari, A. Nanoprecipitation process: From encapsulation to drug delivery. *Int. J. Pharm.* **2017**, *532*, 66–81. [CrossRef] [PubMed]
160. Prabhuraj, R.S.; Bomb, K.; Srivastava, R.; Bandyopadhyaya, R. Dual drug delivery of curcumin and niclosamide using PLGA nanoparticles for improved therapeutic effect on breast cancer cells. *J. Polym. Res.* **2020**, *27*, 133. [CrossRef]
161. Alsulays, B.B.; Anwer, M.K.; Aldawsari, M.F.; Aodah, A.; Adam, E.; Alshehri, S.; Abdel-Kader, M.S. Preparation and evaluation of a stable and sustained releaseof lansoprazole-loaded poly(d,l-lactide-*co*-glycolide) polymeric nanoparticles. *J. Polym. Eng.* **2019**, *39*, 822–829. [CrossRef]

162. Xu, J.; Chen, Y.; Jiang, X.; Gui, Z.; Zhang, L. Development of hydrophilic drug encapsulation and controlled release using a modified nanoprecipitation method. *Processes* **2019**, *7*, 331. [CrossRef]
163. Marante, T.; Viegas, C.; Duarte, I.; Macedo, A.S.; Fonte, P. An overview on spray-drying of protein-loaded polymeric nanoparticles for dry powder inhalation. *Pharmaceutics* **2020**, *12*, 1032. [CrossRef]
164. Heidari, M.; Golenser, J.; Greiner, A. Meeting the needs of a potent carrier for malaria treatment: Encapsulation of Artemisone in poly(lactide-coglycolide). *Part. Part. Syst. Charact.* **2022**, *39*, 2100152. [CrossRef]
165. Nosrati, Z.; Li, N.; Michaud, F.; Ranamukhaarachchi, S.; Karagiozov, S.; Soulez, G.; Martel, S.; Saatchi, K.; Häfeli, U.O. Development of a coflowing device for the size-controlled preparation of magnetic-polymeric microspheres as embolization agents in magnetic resonance navigation technology. *ACS Biomater. Sci. Eng.* **2018**, *4*, 1092–1102. [CrossRef] [PubMed]
166. Tai, Y.; Tian, M.; Chen, Y.; You, P.; Song, X.; Xu, B.; Duan, C.; Jin, D. Preparation of PLGA microspheres loaded with niclosamide via microfluidic technology and their inhibition of Caco-2 cell activity in vitro. *Front. Chem.* **2023**, *11*, 1249293. [CrossRef] [PubMed]
167. Lababidi, N.; Montefusco-Pereira, C.V.; Carvalho-Wodarz, C.S.; Lehr, C.-M.; Schneider, M. Spray-dried multidrug particles for pulmonary co-delivery of antibiotics with *N*-acetylcysteine and curcumin-loaded PLGA-nanoparticles. *Eur. J. Pharm. Biopharm.* **2020**, *157*, 200–210. [CrossRef] [PubMed]
168. Sosnowski, S.; Gadzinowski, M.; Slomkowski, S.; Penczek, S. Synthesis of bioerodible poly(ε-caprolactone) latexes and poly(D,L-lactide) microspheres by ring-opening polymerization. *J. Bioact. Compat. Polym.* **1994**, *9*, 345–366. [CrossRef]
169. Sosnowski, S.; Gadzinowski, M.; Slomkowski, S. Poly(L,L-lactide) microspheres by ring-opening polymerization. *Macromolecules* **1996**, *29*, 4556–4564. [CrossRef]
170. Slomkowski, S.; Sosnowski, S.; Gadzinowski, M. Polyesters from lactides and caprolactone. Dispersion polymerization versus polymerization in solution. *Polym. Degrad. Stab.* **1998**, *59*, 153–160. [CrossRef]
171. Gadzinowski, M.; Sosnowski, S.; Slomkowski, S. Kinetics of the dispersion ring-opening polymerization of ε-caprolactone initiated with diethylaluminum ethoxide. *Macromolecules* **1996**, *29*, 6404–6407. [CrossRef]
172. Slomkowski, S.; Gadzinowski, M.; Sosnowski, S. Mechanism of particle formation and kinetics of the dispersion polymerization of cyclic esters. *Macromol. Symp.* **1998**, *132*, 451–462. [CrossRef]
173. Slomkowski, S.; Sosnowski, S.; Gadzinowski, M.; Pichot, C.; Eaissari, A. Tailored synthesis of polyesters by dispersion ring opening polymerization of ε-caprolactone and lactides. *Macromol. Symp.* **2000**, *150*, 259–268. [CrossRef]
174. Muranaka, M.; Kitamura, Y.; Yoshizawa, H. Preparation of biodegradablemicrospheres by anionic dispersion polymerization with PLA copolymeric dispersion stabilizer. *Colloid Polym. Sci.* **2007**, *285*, 1441–1448. [CrossRef]
175. Muranaka, M.; Ono, T. Preparation of monodisperse polylactide microspheresby dispersion polymerization using a polymeric stabilizer with hydroxy groups. *Macromol. Rapid Commun.* **2009**, *30*, 152–156. [CrossRef] [PubMed]
176. Muranaka, M.; Yoshizawa, H.; Ono, T. Design of polylactide-grafted copolymeric stabilizer for dispersion polymerization of D,L-lactide. *Colloid Polym. Sci.* **2009**, *287*, 525–532. [CrossRef]
177. Gadzinowski, M.; Slomkowski, S.; Elaïssari, A.; Pichot, C. Phase transfer and characterization of poly(ε-caprolactone) and poly(L-lactide) microspheres. *J. Biomater. Sci. Polym. Ed.* **2000**, *11*, 459–480. [CrossRef] [PubMed]
178. Slomkowski, S. Preparation of biodegradable particles by polymerization processes. In *Colloidal Biomolecules, Biomaterials and Biomedical Applications*; Elaissari, A., Ed.; Surfactant Science Series; Marcel Deker: New York, NY, USA, 2003; Volume 16.
179. Slomkowski, S.; Sosnowski, S.; Gadzinowski, M.; Pichot, C.; Elaissari, A. Direct synthesis of polyester microspheres, potential carriers of bioactive compounds. *ACS Symp. Ser.* **1998**, *709*, 143–153. [CrossRef]
180. Gizurarson, S. Anatomical and histological factors affecting intranasal drug and vaccine delivery. *Curr. Drug Deliv.* **2012**, *9*, 566–582. [CrossRef]
181. Taylor, J.P.; Hardy, J.; Fischbeck, K.H. Toxic proteins in neurodegenerative disease. *Science* **2002**, *296*, 1991–1995. [CrossRef]
182. Ahmad, N. Rasagiline-encapsulated chitosan-coated PLGA nanoparticles targeted to the brain in the treatment of Parkinson's disease. *J. Liquid Chrom. Related Technol.* **2017**, *40*, 677–690. [CrossRef]
183. Katila, N.; Duwa, R.; Bhurtel, S.; Khanal, S.; Maharjan, S.; Jee-Heon Jeong, J.-H.; Lee, S.; Choi, D.-Y.; Yook, S. Enhancement of blood–brain barrier penetration and the neuroprotective effect of resveratrol. *J. Control. Release* **2022**, *346*, 1–19. [CrossRef]
184. Pahuja, R.; Seth, K.; Shukla, A.; Shukla, R.K.; Bhatnagar, P.; Chauhan, L.K.S.; Saxena, P.N.; Arun, J.; Chaudhari, B.P.; Patel, D.K.; et al. Trans-blood brain barrier delivery of dopamine-loaded nanoparticles reverses functional deficits in Parkinsonian rats. *ACS Nano* **2015**, *9*, 4850–4871. [CrossRef]
185. Tang, S.; Wang, A.; Yan, X.; Chu, L.; Yang, X.; Song, Y.; Sun, K.; Yu, X.; Liu, R.; Wu, Z.; et al. Brain-targeted intranasal delivery of dopamine with borneol and lactoferrin co-modified nanoparticles for treating Parkinson's disease. *Drug Deliv.* **2019**, *26*, 700–707. [CrossRef]
186. Monge-Fuentes, V.; Mayer, A.B.; Lima, M.R.; Geraldes, L.R.; Zanotto, L.N.; Moreir, K.G.; Martins, O.P.; Piva, H.L.; Felipe, M.S.S.; Amaral, A.C.; et al. Dopamine-loaded nanoparticle systems circumvent the blood-brain barrier restoring motor function in mouse model. *Sci. Rep.* **2021**, *11*, 15185. [CrossRef]
187. Sánchez-López, E.; Ettcheto, M.; Egea, M.A.; Espina, M.; Cano, A.; Calpena, A.C.; Camins, A.; Carmona, N.; Silva, A.M.; Souto, E.B.; et al. Memantine loaded PLGA PEGylated nanoparticles for Alzheimer's disease: In vitro and in vivo characterization. *J. Nanobiotechnol.* **2018**, *16*, 32. [CrossRef]

188. Wu, Q.; Karthivashan, G.; Nakhaei-Nejad, M.; Anand, B.G.; Giuliani, F.; Kar, S. Native PLGA nanoparticles regulate APP metabolism and protect neurons against β-amyloid toxicity: Potential significance in Alzheimer'sdisease pathology. *Int. J. Biol. Macromol.* **2022**, *219*, 180–1196. [CrossRef] [PubMed]
189. Ye, C.; Pan, B.; Xu, H.; Zhao, Z.; Shen, J.; Lu, J.; Yu, R.; Liu, H. Co-delivery of GOLPH3 siRNA and gefitinib by cationic lipid-PLGA nanoparticles improves EGFR-targeted therapy for glioma. *J. Mol. Med.* **2019**, *97*, 1575–1588. [CrossRef]
190. Cui, Y.; Sun, J.; Hao, W.; Chen, M.; Wang, Y.; Xu, F.; Gao, C. Dual-target peptide-modified erythrocyte membrane-enveloped PLGA nanoparticles for the treatment of glioma. *Front. Oncol.* **2020**, *10*, 563938. [CrossRef]
191. Caban-Toktas, S.; Sahin, A.; Lulei, S.; Esendagli, G.; Vural, I.; Oguz, K.K.; Soylemezoglu, F.; Mut, M.; Dalkara, T.; Mansoor Khan, M.; et al. Combination of paclitaxel and R-flurbiprofen loaded PLGA nanoparticles suppresses glioblastoma growth on systemic administration. *Int. J. Pharm.* **2020**, *578*, 119076. [CrossRef]
192. Acharya, S.; Praveena, P.; Raja Guru, B.R. In vitro studies of prednisolone loaded PLGA nanoparticles-surface functionalized with folic acid on glioma and macrophage cell lines. *Pharm. Sci.* **2021**, *27*, 407–417. [CrossRef]
193. Ma, J.; Dai, L.; Yu, J.; Cao, H.; Bao, Y.; Hu, J.; Zhou, L.; Yang, J.; Sofia, A.; Chen, H.; et al. Tumor microenvironment targeting system for glioma treatment via fusion cell membrane coating nanotechnology. *Biomaterials* **2023**, *295*, 122026. [CrossRef] [PubMed]
194. Alsaab, H.O.; Alharbi, F.D.; Alhibs, A.S.; Alanazi, N.B.; Alshehri, B.Y.; Saleh, M.A.; Alshehri, F.S.; Algarni, M.A.; Almugaiteeb, T.; Uddin, M.N.; et al. PLGA-based nanomedicine: History of advancement and development in clinical applications of multiple diseases. *Pharmaceutics* **2022**, *14*, 2728. [CrossRef]
195. Loureiro, J.A.; Ramalho, M.U.; do Carmo Pereira, M. Immuno-nanocarriers for brain delivery: Limitations from in vitro to preclinical and clinical studies. *Nanomedicine* **2020**, *15*, 543–545. [CrossRef]
196. Available online: https://www.cancer.gov/nano/cancer-nanotechnology/current-treatments (accessed on 4 February 2024).
197. Lee, S.-W.; Kim, Y.-M.; Cho, C.H.; Kim, Y.T.; Kim, S.M.; Hur, S.Y.; Kim, J.-H.; Kim, B.-G.; Kim, S.-C.; Ryu, H.-S.; et al. An open-label, randomized, parallel, phase II trial to evaluate the efficacy and safety of a cremophor-free polymeric micelle formulation of paclitaxel as first-line treatment for ovarian cancer: A Korean Gynecologic Oncology Group Study (KGO G-3021). *Cancer Res. Treat.* **2018**, *50*, 195–203. [CrossRef] [PubMed]
198. Ishida, T.; Wang, X.-Y.; Shimizu, T.; Nawata, K.; Kiwada, H. PEGylated liposomes elicit an anti-PEG IgM response in a T cell-independent manner. *J. Control. Release* **2007**, *122*, 349–355. [CrossRef] [PubMed]
199. Tagami, T.; Nakamura, K.; Shimizu, T.; Ishida, T.; Kiwada, H. Effect of siRNA in PEG-coated siRNA-lipoplex on anti-PEG IgM production. *J. Control. Release* **2009**, *137*, 234–240. [CrossRef]
200. Hashimoto, Y.; Abu Lila, A.S.; Shimizu, T.; Ishida, T.; Kiwada, H.J. B cell-intrinsic toll-like receptor 7 is responsible for the enhanced anti-PEG IgM production following injection of siRNA-containing PEGylated lipoplex in mice. *J. Control. Release* **2014**, *184*, 1–8. [CrossRef]
201. Bordes, C.; Fréville, V.; Ruffin, E.; Marote, P.; Gauvrit, J.Y.; Briançon, S.; Lantéri, P. Determination of poly(ε-caprolactone) solubility parameters: Application to solvent substitution in a microencapsulation process. *Int. J. Pharm.* **2010**, *383*, 236–243. [CrossRef] [PubMed]

polymers

MDPI

Article

Highly Sensitive Detection of Bacteria by Binder-Coupled Multifunctional Polymeric Dyes

Kriti Kapil [1,†], Shirley Xu [2,†], Inseon Lee [2], Hironobu Murata [1], Seok-Joon Kwon [2], Jonathan S. Dordick [2,*] and Krzysztof Matyjaszewski [1,*]

1 Department of Chemistry, Carnegie Mellon University, 4400 Fifth Avenue, Pittsburgh, PA 15213, USA; kkapil@andrew.cmu.edu (K.K.); hiromura@andrew.cmu.edu (H.M.)
2 Department of Chemical and Biological Engineering, Center for Biotechnology & Interdisciplinary Studies, Rensselaer Polytechnic Institute, Troy, NY 12180, USA; xus11@rpi.edu (S.X.); leei5@rpi.edu (I.L.); kwons2@rpi.edu (S.-J.K.)
* Correspondence: dordick@rpi.edu (J.S.D.); km3b@andrew.cmu.edu (K.M.)
† These authors contributed equally to this work.

Abstract: Infectious diseases caused by pathogens are a health burden, but traditional pathogen identification methods are complex and time-consuming. In this work, we have developed well-defined, multifunctional copolymers with rhodamine B dye synthesized by atom transfer radical polymerization (ATRP) using fully oxygen-tolerant photoredox/copper dual catalysis. ATRP enabled the efficient synthesis of copolymers with multiple fluorescent dyes from a biotin-functionalized initiator. Biotinylated dye copolymers were conjugated to antibody (Ab) or cell-wall binding domain (CBD), resulting in a highly fluorescent polymeric dye-binder complex. We showed that the unique combination of multifunctional polymeric dyes and strain-specific Ab or CBD exhibited both enhanced fluorescence and target selectivity for bioimaging of *Staphylococcus aureus* by flow cytometry and confocal microscopy. The ATRP-derived polymeric dyes have the potential as biosensors for the detection of target DNA, protein, or bacteria, as well as bioimaging.

Keywords: pathogen identification; bioimaging; fluorescence; copolymer; ATRP; flow cytometry; confocal imaging

check for updates

Citation: Kapil, K.; Xu, S.; Lee, I.; Murata, H.; Kwon, S.-J.; Dordick, J.S.; Matyjaszewski, K. Highly Sensitive Detection of Bacteria by Binder-Coupled Multifunctional Polymeric Dyes. *Polymers* **2023**, *15*, 2723. https://doi.org/10.3390/polym15122723

Academic Editors: Marek Kowalczuk, Iza Radecka and Barbara Trzebicka

Received: 23 May 2023
Revised: 15 June 2023
Accepted: 16 June 2023
Published: 18 June 2023

1. Introduction

Infectious diseases caused by pathogens such as bacteria, viruses, and fungi remain a great burden on humanity [1]. As an example, antimicrobial resistance (AMR) is a major threat to human health. AMR-related infections have killed as many people as AIDS (acquired immunodeficiency syndrome) or malaria [2]. Under these circumstances, target-specific identification of pathogens is critical for effective medical intervention and decontamination of the infected areas [3]. Colony counting, immunological, and polymerase chain reaction (PCR) techniques are the traditional methods for pathogen identification [4–7]. However, these methods require time-consuming and complicated procedures such as cell culture, antigen/antibody treatment, and cell lysis/DNA amplification [5–7]. In this context, fluorescent labeling and detection have emerged as promising tools for pathogen visualization and identification, due to their simple labeling procedure, sensitivity, and stability [3,8–10]. Moreover, conjugation of fluorescent materials with a biological binder such as an antibody [11,12], an aptamer [13,14], or the cell-wall binding domain of a lytic enzyme (CBD) [15–17] allows for the targeting of a specific pathogen.

Various classes of fluorescent materials such as small-molecule organic dyes [18–21], fluorescent proteins [22–24], self-fluorescent polymers [10,25–27], dye-labeled polymeric nanoparticles [28–34], and quantum dots [35–38] have been explored. Among these, fluorescent dyes have gained popularity due to their commercial availability, ease of operation, and high resolution [39]. This has paved the way for the development of fluorescent dye copolymers, which combine various physicochemical properties with optical emission. The copolymers are less prone to sequestration from cells and tissues and typically exhibit lower toxicity and better photostability than low-molecular-weight dyes [40], hence offering a simple yet effective approach for low-sensitivity, high-contrast imaging of bacterial cells.

The fluorescent labeling of block copolymers has been achieved through various methods, such as noncovalent encapsulation, direct labeling with fluorescent initiators or monomers, or post-polymerization modification [41]. A variety of dyes were used for fluorescent labeling, including carbocyanine dye (e.g., Cy5, Cy5.5, Cy7-azide), benzopyrylium dyes (e.g., DY-676 and DY-700), push–pull dyes (e.g., coumarins, Nile red), and xanthene dyes (e.g., rhodamine, fluorescein). Xanthene dyes are particularly attractive due to their brightness, high extinction coefficients, quantum yield, and exceptional chemical stability. Noncovalent encapsulation may lead to leakage of dye in biological media, resulting in high background signals and cytotoxicity [42,43]. Direct labeling enables the incorporation of fluorescent dye during the polymerization step [44]. The fluorescent initiator-bearing hydroxyl groups were used to initiate the ring-opening polymerization [45,46]. This approach limits the number of fluorescent markers per polymer chain [47,48]. Alternatively, the use of a fluorescent monomer allows for control over the dye content by controlling the number of fluorescent monomers incorporated during the polymerization. Before its incorporation, the fluorophore is converted to a monomer by functionalization with a polymerizable vinyl group [49]. Techniques such as free-radical polymerization [50–56], reversible addition–fragmentation chain transfer (RAFT) [57–62], atom transfer radical polymerization (ATRP) [63–74], and ring-opening metathesis polymerization (ROMP) [75,76] have been used to incorporate fluorophores into block copolymers.

Despite the long history of employing polymers for bioanalytical applications, high-dispersity polymers generally offer limited control over functionality and topology [77–79]. ATRP has emerged as one of the most versatile and powerful reversible deactivation radical polymerization (RDRP) techniques, offering precise control over molecular weight, molecular weight distribution, functionality, architecture as well as tolerance to most functional groups [80–94]. ATRP is characterized by an equilibrium established through an inner-sphere electron-transfer process mediated by a transition metal complex, usually the activator $[Cu(I)\text{-}L]^+$ (L typically being a polydentate amine ligand), which reacts with an alkyl halide initiator (R-X), leading to the formation of a $[X\text{–}Cu(II)/L]^+$ deactivator and a propagating radical (R*). Radical propagation occurs until the radical chain ends are deactivated by $[X\text{–}Cu\ (II)/L]^+$, forming X-capped dormant species and regenerating $[Cu(I)\text{-}L]^+$. ATRP equilibrium is shifted toward dormant species since the rate constant of activation of dormant species is typically much smaller than the rate constant of radical deactivation, i.e., $k_{act} \ll k_{deact}$. Thus, the key aspect of the ATRP mechanism is a low concentration of active propagating species and a larger number of dormant chains [81,95–98]. Over the years, the scope of ATRP has been expanded to various solvents and reaction conditions, including water at room temperature using a low concentration of copper catalyst and no protective atmosphere of inert gas [99–104]. By optimizing the polymerization conditions and parameters, the copolymerization kinetics can be controlled.

Over the last decade, photoinduced ATRP techniques have been developed to harness the power of light to generate radicals [85,105,106]. Recently, photoinduced ATRP using copper complexes to achieve controlled radical propagation and photocatalyst to trigger and drive polymerization has been explored [107]. Our group reported green-light-induced ATRP with dual catalysis, using eosin Y (EYH_2) in combination with a copper complex as a highly efficient method for rapid and well-controlled polymerization of oligo (ethylene oxide) methyl ether methacrylate [108] and oligo (ethylene oxide) methyl ether acrylate [109] in water under ambient conditions without the need for deoxygenation. The scope of the technique has been demonstrated by controlled polymerization of a variety of monomers, hyperbranched polymers with a tunable degree of branching [110] and grafting from the surface of biomolecules to synthesize well-defined protein–polymer [97,105,111–115] and DNA–polymer bioconjugates [97,116,117].

Herein, we report the synthesis and characterization of a polymeric dye complex containing a binder such as an antibody or the cell-wall binding domain (CBD) from a lytic enzyme for a highly sensitive bioimaging technique for pathogen identification. Putative autolysin from *Staphylococcus aureus* (SA1) contains a cysteine, histidine-dependent amidohydrolases/peptidases (CHAP) domain and a putative CBD [118]. The CBD from SA1 was successfully expressed and exhibited high selectivity to *Staphylococcus aureus*, similar to the CBD of lysostaphin, a lytic enzyme known for its specific targeting of *Staphylococcus aureus* [119]. The approach involves the coupling of *Staphylococcus aureus* targeting polyclonal antibody or CBD of SA1 to rhodamine dye-labeled copolymers comprised of oligo (ethylene oxide) methyl ether methacrylate or carboxy betaine methacrylate (CBMA). The polymers were grafted from a biotin-functionalized ATRP initiator under blue-light irradiation using Eosin-Y/Cu-mediated fully oxygen-tolerant ATRP technique. The nondestructive binding properties of the copolymeric dye complex were tested on the target bacterial cells and were applied to the bioimaging of target bacteria using fluorescence detection and confocal microscopy analysis.

2. Materials and Methods

2.1. Materials

All chemicals were purchased from commercial sources and used as received unless otherwise noted. Tris(2-pyridylmethyl) amine (TPMA, 99%) was purchased from AmBeed (Arlington Heights, IL, USA). Methacryloxyethyl thiocarbamoyl rhodamine B (RDMA) was purchased from Polysciences (Warrington, PA, USA). 3-[[2-(Methacryloyloxy)ethyl] dimethylammonio] propionate (CBMA) was purchased from TCI (Tokyo, Japan). Water (HPLC grade), dimethylformamide (DMF, ≥99.8%), and dimethyl sulfoxide (DMSO, ≥99.7%) were purchased from Fisher (Waltham, MA, USA). *Staphylococcus aureus* (ATCC 6538) (*S. aureus*) and *Bacillus anthracis* Sterne (*B. anthracis*) were purchased from the American Type Culture Collection (ATCC) (Manassas, VA, USA). Polyclonal antibody against *Staphylococcus aureus* was purchased from Invitrogen (Waltham, MA, USA). Eosin Y (EYH_2, 99%), copper (II) bromide ($CuBr_2$, 99.99%), triethanolamine (TEOA, ≥99.0%), and NeutrAvidin were purchased from Thermo Fisher Scientific (Waltham, MA, USA). Biotinylated rhodamine B was purchased from Nanocs (New York, NY, USA). Luria–Bertani (LB) medium and agar were purchased from Becton Dickinson (Franklin Lakes, NJ, USA). BL21(DE3) competent cells and restriction enzymes such as NdeI and XhoI were purchased from New England Biolabs (NEB) (Ipswich, MA, USA). Oligo (ethylene oxide) methyl ether methacrylate (average M_n = 500, $OEOMA_{500}$), 1,4-bis(3-isocyanopropyl) piperazine (QA), ampicillin, isopropyl-β-d-thiogalactoside (IPTG), deoxyribonuclease (Dnase) I from bovine pancreas, phenylmethanesulfonylfluoride (PMSF), Tween20, glycerol, imidazole, phosphate-buffered saline (PBS), D-biotin, paraformaldehyde (PFA), sodium phosphate, sodium hydroxide (NaOH), and sodium chloride (NaCl) were purchased from Sigma Aldrich (St. Louis, MO, USA). Lysozyme and nickel-NTA agarose beads were purchased from Gold Biotechnology (St. Louis, MO, USA). All solutions were prepared with purified water by a Milli-Q purification system from Millipore (Burlington, MA, USA). The biotinylated ATRP initiator (Biotin-I) was synthesized according to a previously reported procedure [110].

2.2. Synthesis of Biotinylated Dye Copolymers

$OEOMA_{500}$ was passed through a column of basic alumina to remove the inhibitor. The stock solutions of RDMA (20 mg in 2.0 mL of DMSO), biotin-I (20 mg in 1.0 mL of DMSO), $CuBr_2$ (33.5 mg in 20.0 mL of DMSO), TPMA (13.06 mg in 1.0 mL of DMSO), and EYH_2 (0.97 mg in 1.0 mL of DMSO) were prepared prior to polymerization.

BT-p(CBMA-RDMA (2)): In a 5 mL volumetric flask, 344 mg (1.5 mmol) of CBMA were weighed. $CuBr_2$ stock (200 μL), TPMA stock (100 μL), biotin-I (200 μL), EYH_2 stock (50 μL), RDMA stock (1 mL), DMF (50 μL), and 10X PBS solution (500 μL) were added. Finally, HPLC water was added to reach a final volume of 5 mL, and the reaction mixture was stirred on a vortex. The final concentrations were CBMA (300 mM), RDMA (3 mM), biotin-I (1.5 mM), EYH_2 (15 μM), $CuBr_2$ (0.3 mM), TPMA (0.9 mM), and DMSO (30% v/v). ([CBMA]/[RDMA]/[biotin-I]/[EYH_2]/[$CuBr_2$]/[TPMA] = 200/2/1/0.02/0.4/1.2). Then, 4.4 mL of the reaction "cocktail" were added to a 1-dram (12/96 mm) vial equipped with a magnetic stirring bar. The polymerization mixture was stirred in an open vial at 500 rpm for 60 min under blue LEDs (450 nm, 25.0 mW/cm^2).

BT-p($OEOMA_{500}$-RDMA (2/4): In a 5 mL volumetric flask, 750 mg (1.5 mmol) of $OEOMA_{500}$ were weighed. $CuBr_2$ stock (200 μL), TPMA stock (100 μL), biotin-I (100 μL), EYH_2 stock (50 μL), RDMA stock (500 μL), DMF (50 μL), and 10X PBS solution (500 μL) were added. Finally, HPLC water was added to reach a final volume of 5 mL, and the reaction mixture was stirred on a vortex. The final concentrations were $OEOMA_{500}$ (300 mM), RDMA (1.5 mM), biotin-I (0.75 mM), EYH_2 (15 μM), $CuBr_2$ (0.3 mM), TPMA (0.9 mM), and DMSO (30% v/v). ([$OEOMA_{500}$]/[RDMA]/[biotin-I]/[EYH_2]/[$CuBr_2$]/[TPMA] = 400/2/1/0.02/0.4/1.2). For the synthesis of BT-p($OEOMA_{500}$-RDMA (4)), RDMA stock (1.0 mL) was added, resulting in the final concentration of RDMA (3 mM) in the polymerization mixture. ([$OEOMA_{500}$]/[RDMA]/[biotin-I]/[EYH_2]/[$CuBr_2$]/[TPMA] = 400/4/1/0.02/0.4/1.2). Then, 4.4 mL of the CRBP cocktail were added to a 1-dram (12/96 mm) vial equipped with a magnetic stirring bar. The polymerization mixture was stirred in an open vial at 500 rpm for 60 min under blue LEDs (450 nm, 25.0 mW/cm^2).

2.3. Characterization of Biotinylated Copolymeric Rhodamine B by ^1H NMR Spectroscopy and Size Exclusion Chromatography with Multi-Angle Light Scattering (SEC–MALS) Detectors

Before analysis, the synthesized biotinylated copolymeric dyes were purified by dialysis in deionized water using SpectraPor® 10 kDA cutoff dialysis membrane for 48 h and then lyophilized. ^1H NMR spectra were recorded on Bruker Avance III 500 MHz spectrometers with D_2O as the solvent. SEC–MALS measurements were performed using the Agilent SEC system (Agilent, 1260 Infinity II with UV detector) coupled with MALS, DLS, Viscometer, and RI detectors (Wyatt Technology Corporation, Santa Barbara, CA, USA). Measurements were performed using the Waters Ultrahydrogel Linear column with 1X DPBS as an eluent at room temperature and a flow rate of 0.5 mL/min.

2.4. Kinetics of Photoinduced ATRP

The ATRP reaction mixture (5 mL) was prepared according to the general procedure described above for the synthesis of BT-p($OEOMA_{500}$-RDMA (4)). The final concentrations were $OEOMA_{500}$ (300 mM), RDMA (3.0 mM), biotin-I (0.75 mM), EYH_2 (15 μM), $CuBr_2$ (0.3 mM), TPMA (0.9 mM), and DMSO (30% v/v). ([$OEOMA_{500}$]/[RDMA]/[biotin-I]/[EYH_2]/[$CuBr_2$]/[TPMA] = 400/4/1/0.02/0.4/1.2). Then, 4.4 mL of the ATRP cocktail ([$OEOMA_{500}$]/[RDMA]/[biotin-I]/[EYH_2]/[$CuBr_2$]/[TPMA] = 400/4/1/0.02/0.4/1.2) were added to a 1-dram (12/96 mm) vial equipped with a magnetic stir bar. The polymerization mixture in an uncapped vial was stirred at 500 rpm for 40 min under green LEDs (520 nm, 9.0 mW/cm^2). The samples were taken at regular intervals, quenched with 1,4-bis(3-isocyanopropyl) piperazine [120], and then analyzed by ^1H NMR and SEC.

2.5. Photostability Assessment of Biotinylated Copolymeric Rhodamine B

The photostability of the biotinylated copolymeric dyes was accessed by measuring the fluorescence readout every 30 min for 18 h. The experiment was carried out overnight using a BioTek Synergy H1 microplate reader. The copolymer dye was dissolved in 1X PBS buffer to reach a final concentration of 60 μM and placed in a 96-well polystyrene black-bottom plate. Triplicate samples of the same concentration were measured with a control sample containing only the buffer. Mineral oil (50 μL) was added to each well to prevent evaporation. Samples were incubated in a plate reader (xenon flash lamp, high energy) at 37 °C in the measurement chamber. The fluorescence intensity was measured by top optics using the monochromator filter set: excitation at 540 nm and emission at 570 nm, every 30 min.

2.6. Preparation of Cell-Wall Binding Domain (CBD)

The DNA sequence of the cell-wall binding domain of SA1 [119] including an avi-tag and *BirA* (biotin ligase) was subcloned into a pGS-21a and pCDF-duet vector between NdeI and XhoI, respectively. BL21(DE3) competent cells were then cotransformed with pGS-21a with SA1BD gene and pCDF-duet with *birA* gene. One milliliter of the saturated overnight culture was inoculated into 100 mL of fresh LB media containing ampicillin (100 μg/mL) and grown until the absorbance at 600 nm reaches approximately 0.4. IPTG and D-biotin were added to a final concentration of 1 mM and 50 μM, respectively, and cells were cultured at 16 °C and 150 rpm overnight. Afterward, cells were pelleted by centrifuging (4000 rpm) at 4 °C for 15 min and resuspended in 10 mL of cell lysis buffer in native purification buffer (NPB, 20 mM sodium phosphate, and 500 mM NaCl, pH 8.0) containing PMSF (1 mM), lysozyme (100 μg/mL), bovine pancrease Dnase I (100 μg/mL), and glycerol (5%, *v*/*v*). The cell suspensions were sonicated in ice using Misonix Sonicator® 3000 (Farmingdale, NY, USA) for 30 min with 1 s pulses and then centrifuged at 4000 rpm for 15 min to collect the supernatant. The His-tagged protein in the supernatant was purified using nickel nitrilotriacetic acid (Ni-NTA) affinity chromatography. The bound protein was washed once with NPB containing PMSF (1 mM) and five times with NPB containing imidazole (20 mM). The protein was eluted with an elution buffer (NPB containing imidazole (200 mM)) and was dialyzed against PBS at pH 7.4 using an 8 kDa molecular weight cutoff membrane (SpectrumLabs, Arden, NC, USA). Protein concentrations were determined spectrophotometrically at 280 nm using a NanoDrop ND-1000 (ThermoFisher, Waltham, MA, USA).

2.7. Construction and Characterization of Antibody/CBD-Copolymeric Rhodamine B Complex

To construct the antibody/CBD complexes with rhodamine B dye, biotinylated polyclonal antibody or CBD solution was first mixed with NeutrAvidin in phosphate-buffered saline (PBS, pH 7.4) and incubated at room temperature for 30 min. Biotinylated monomeric or copolymeric rhodamine B solution was then added to the mixture, followed by incubation at room temperature for 30 min. The molar ratio of biotinylated antibody/CBD, NeutrAvidin, and monomeric/copolymeric rhodamine B was 1:1:1.

Fluorescence from the prepared complexes in the presence of target bacteria were measured to determine the binding. Briefly, 30 μL of the saturated overnight culture were inoculated into 3 mL of fresh LB media and grown until the absorbance at 600 nm reached approximately 0.4. Afterward, cells were centrifuged and washed three times with PBS (pH 7.4). Antibody/CBD complex fusion proteins were added to the target cells (5×10^8 cells/mL) and incubated at room temperature for 15 min. After incubation, the resulting mixtures were washed three times with PBS with 0.2% Tween20 to remove the unbound complex and adjusted to 5×10^8 cells/mL. The fluorescence intensity of the mixture was then measured using a microplate reader (SpectraMax M5, San Jose, CA, USA) (λ_{ex} = 546 nm and λ_{em} = 580 nm). The fluorescence from complexes at the surface of bacteria was also measured using flow cytometry. Mixtures were diluted 10-fold in PBS to 10^7 cells/mL prior to flow cytometry. Flow cytometry was performed using the BD LSRII flow cytometer (BD Biosciences, Franklin Lake, NJ, USA) with 20,000 events collected for each sample. Gating and further flow cytometry analysis were performed using FlowJo.

2.8. Confocal Laser Scanning Microscopy (CLSM)

Antibody/CBD complex fusion proteins were added to bacterial cells (5×10^8 cells/mL) and incubated at room temperature for 15 min. After incubation, the resulting mixtures were washed three times with PBS with 0.2% Tween20 to remove the unbound complex and adjusted to 5×10^8 cells/mL. A PFA solution (4%) was then added to the resulting mixtures to fix the bacteria cells. After incubation on ice for 30 min, the mixture was washed twice with PBS. Each 10 μL aliquot of the prepared mixed suspensions was added to clean glass slides and lightly covered using coverslips. The samples were excited at 546 nm, and the emission was recorded between 556 and 632 nm. The samples were examined by confocal laser scanning microscopy (LSM780) with a 93× glycerol immersion objective lens using a 546 nm laser (Carl Zeiss A.G., Oberkochen, Germany). Microscopy images were prepared and analyzed using ImageJ (Version 1.53k, National Institutes of Health, Bethesda, MD, USA). Briefly, raw TIF images for each microscopy sample were imported into ImageJ. For each image, the fluorescence was measured for 25 cells, and 5 background spots were measured for correction. The size of measurement for each cell and background was kept consistent within images. The corrected total cell fluorescence (CTCF) for each cell was calculated. Confocal microscopy was also used to examine the specificity of the antibody/CBD complex fusion proteins by using a mixture of both *S. aureus* and *B. anthracis* cells. The same procedure as above was used where the only modification is using a 1:1 mixture of *S. aureus* and *B. anthracis* cells. Confocal microscopy was performed using the same parameters above. Brightfield images for bacterial mixtures were also taken.

3. Results and Discussion

3.1. Synthesis and Characterization of Biotinylated Copolymeric Rhodamine B

The synthesis of biotinylated multifunctional dye copolymers was performed using a recently developed fully oxygen-tolerant, photoinduced atom transfer radical polymerization (ATRP) [108]. The copolymerization of zwitterionic and hydrophilic CBMA monomer and dye-labeled rhodamine-B-methacrylate (RDMA) monomer was performed in an aqueous medium under blue-light irradiation (λ_{max} = 450 nm, 25.0 mW/cm^2), using biotin-I as the initiator, Eosin Y (EYH_2) as the organic photoredox catalyst, and $CuBr_2$/TPMA as the deactivator (Figure 1).

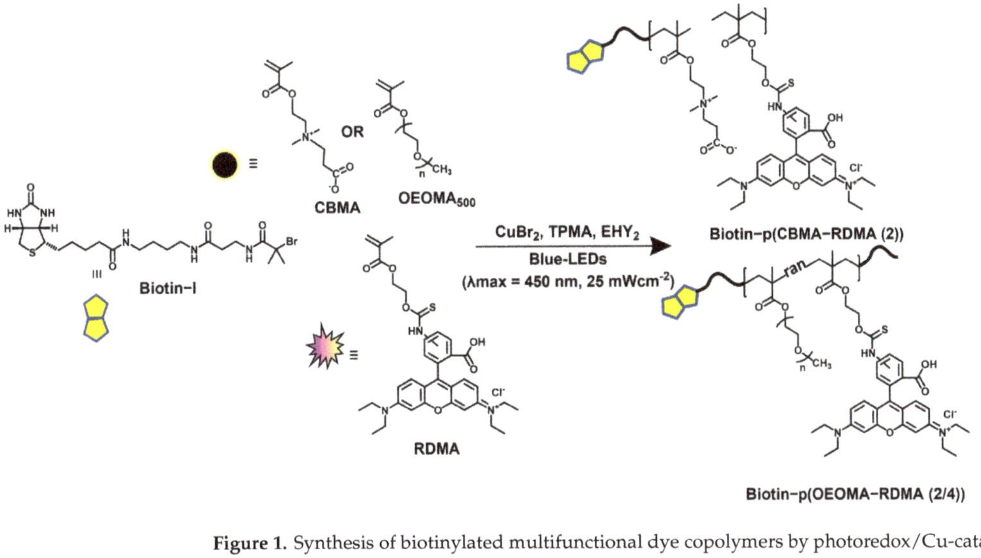

Figure 1. Synthesis of biotinylated multifunctional dye copolymers by photoredox/Cu-catalyzed, oxygen-tolerant ATRP.

Similarly, amphiphilic copolymers comprised of PEG backbone were also synthesized by copolymerization of oligo (ethylene oxide) methyl ether methacrylate ($OEOMA_{500}$) monomer with RDMA. ^{1}H NMR was analyzed to compute conversion of monomers; the peak integral from 4.26 to 4.32 ppm corresponding to the monomer was set at 100 at t = 0, and the broad peak between 4.05 and 4.20 ppm corresponding to the polymer was observed at t = 60 min. Within 60 min, the CBMA and $OEOMA_{500}$, and RDMA reached a high conversion (≈80%), which was used to compute theoretical molecular weight ($M_{n,th}$). ^{1}H NMR spectra of the purified polymer samples was also recorded (Figures S1 and S2). The purified polymer samples were then analyzed by the SEC–MALS technique (Table 1), where a good agreement between $M_{n,th}$ and the observed absolute molecular weight ($M_{n, abs}$) revealed well-controlled polymerizations (Figure 2a).

Table 1. Synthesis of biotinylated multifunctional dye copolymers in water [a].

Entry	Sample Name	[M]/[Dye-M]/[I]	αM [b] (%)	Mn,th [c]	Mn, abs [d]	Đ
1	Biotin-p(CBMA-RDMA (2))	200/2/1	80%	36 600	40 200	1.25
2	Biotin-p($OEOMA_{500}$-RDMA (2))	400/2/1	84%	168 000	150 850	1.17
3	Biotin-p($OEOMA_{500}$-RDMA (4))	400/4/1	78%	158 600	149 200	1.23

[a] Reactions conditions: [M] = 300 mM, [Dye-M] = [RDMA] = 1.5–3.0 mM, [I] = [Biotin-I] = 0.75–1.5 mM, [EYH_2] = 15 M, [$CuBr_2$] = 0.3 mM, [L] = [TPMA] = 0.9 mM in PBS with DMSO (30% v/v), irradiated for 60 min under blue LEDs (450 nm, 25.0 mW cm^2), in an open vial with stirring at 500 rpm. Reaction volume 4.4 mL. [b] Monomer conversion was determined by using ^{1}H NMR spectroscopy. [c] Theoretical molecular weight ($M_{n,th}$) was calculated using the equation $M_{n,th}$ = [M] × MW_M × α + $MW_{\text{-Biotin-I}}$. [d] Absolute molecular weight ($M_{n, abs}$) and Đ were determined by using SEC–MALS.

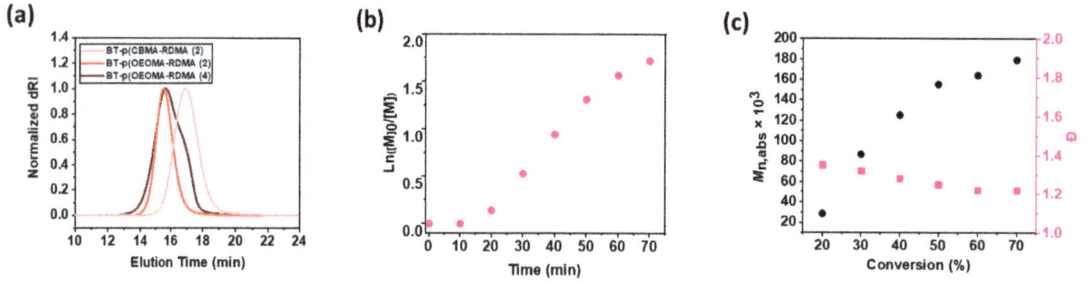

Figure 2. (**a**) SEC analysis of the biotinylated copolymer rhodamine dye in Table 1 after 60 min. Copolymerization kinetics of the optimized EYH_2/Cu-catalyzed ATRP for the synthesis of BT-p($OEOMA_{500}$-RDMA (4)). (**b**) First-order kinetic plot. (**c**) Evolution of molecular weight and molecular weight distribution with conversion (black dots represent molecular weight and pink dots represent molecular weight distribution).

The copolymerization kinetics of the photoinduced ATRP were performed under the optimized conditions ([$OEOMA_{500}$]/[RDMA]/[biotin-I]/[EYH_2]/[$CuBr_2$]/[TPMA] = 400/4/1/0.02/0.4/1.2). The samples were taken at regular intervals, quenched with 1,4-bis(3-isocyanopropyl) piperazine, and monitored by ^{1}H NMR in D_2O. A short induction period (10 min) was followed by rapid polymerization, reaching 78% monomer conversion within 60 min (Figures 2b and S3), and exhibited first-order kinetics. The monomer conversion determined via ^{1}H NMR revealed statistical incorporation of the fluorescent monomer (RDMA) within the polymer chain. A good agreement between theoretical and experimental molecular weights was observed. In addition, SEC traces revealed that molecular weights increased as a function of monomer conversion, and dispersity values remained low $Đ < 1.3$). The ATRP technique enabled efficient and rapid synthesis of copolymer chains without the need for deoxygenation, with the desired degree of polymerization, desired molar ratio of RDMA, predictable molecular weights, narrow molecular weight distribution, and homogenously distributed fluorescent dye monomers (Figure 2c).

3.2. Characterization of Biotinylated Polymeric Dyes and Their Complexes with Selective Binders

We complexed the biotinylated polymeric dyes with NeutrAvidin and a cell binder such as biotinylated *S. aureus* polyclonal antibody or biotinylated CBD using biotin/NeutrAvidin interactions and tested their binding toward the target *S. aureus* cells using fluorescence detection (Figure 3). For fluorescence measurements of antibody/CBD-polymeric dye complexes on the surface of *S. aureus*, we used an excitation and emission at 546 nm and 580 nm, respectively, based on the fluorescent spectra of p(OEOMA500-RDMA (4)) (Figure S4). In both antibody and CBD cases, the p(OEOMA$_{500}$-RDMA (4)) complex showed the best performance without any background fluorescence (Figure 4a) compared to p(OEOMA$_{500}$-RDMA (2)) and p(CBMA-RDMA (2)) complexes. The result implies that a higher molar ratio of rhodamine B dye was incorporated in the p(OEOMA$_{500}$-RDMA (4)) backbone than in the p(OEOMA$_{500}$-RDMA (2)) backbone. Also, in the case of zwitterionic BT-p(CBMA-RDMA (2)), we observed a background fluorescence, which can be attributed to the distinct characteristics of the copolymer backbone, contributing to its zwitterionic and hydrophilic nature. This unique feature of the copolymer backbone may lead to nonspecific interactions to zwitterionic teichoic acid polymers located within the Gram-positive cell wall. [121,122], [1] (resulting in high background fluorescence. In contrast, both p(OEOMA$_{500}$-RDMA (2)) and p(OEOMA$_{500}$-RDMA (4)) complexes with PEGylated side chains can effectively minimize the nonspecific interactions with the cell surface. In all cases, the antibody-induced fluorescence was higher than the CBD-induced fluorescence. This may be because the number of binding sites for the polyclonal antibody is higher than that of CBD at the upper cell wall structure of the Gram-positive *S. aureus*. The concentration of p(OEOMA$_{500}$-RDMA (4)) species in the complexation process was optimized and fixed at 6 μM (Figure S5).

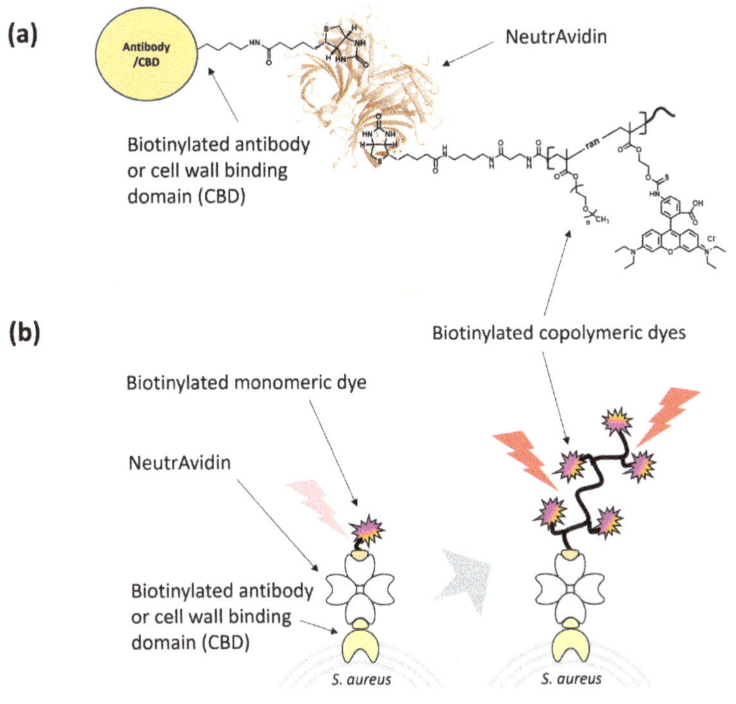

Figure 3. Schematic of (**a**) antibody/CBD-copolymeric rhodamine B complex, and (**b**) signal generation of antibody/CBD-monomeric and -copolymeric rhodamine B complexes on the target *S. aureus*.

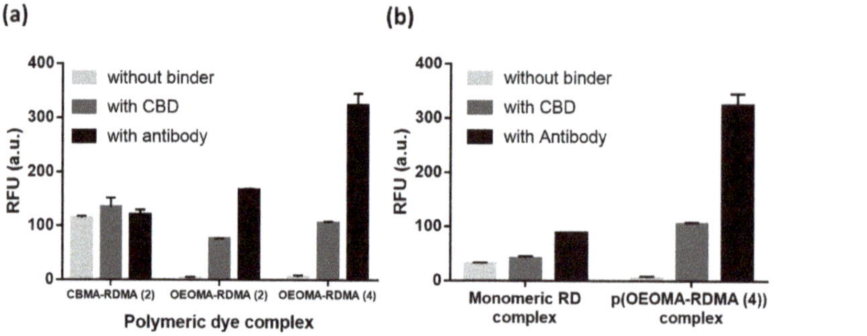

Figure 4. Comparison of fluorescence from CBD/antibody complex with (**a**) BT-p(CBMA-RDMA (2)), BT-p(OEOMA-RDMA (2)), and BT-p(OEOMA-RDMA (4)), and (**b**) BT-RD and BT-p($OEOMA_{500}$-RDMA (4)) at the surface of target bacteria.

To assess the stability of biotinylated copolymer rhodamine B, the fluorescence intensity of the synthesized copolymers was monitored overnight ([BT-p(OEOMA-RDMA (4)] = 60 μM). The polymer sample was incubated in 1X PBS buffer in a plate reader at 37 °C in the measurement chamber for 18 h. The negligible change in the fluorescence intensity of the polymer samples during the experiment indicated the high stability of the polymeric dyes under these experimental conditions (Figure S6).

Next, the fluorescence of antibody/CBD-p($OEOMA_{500}$-RDMA (4)) complexes was measured and compared with biotinylated monomeric rhodamine B complexes with NeutrAvidin and antibody/CBD (antibody/CBD-RD complexes) (Figure 3). The fluorescence of the antibody- and CBD-p($OEOMA_{500}$-RDMA (4)) complexes were 2.6 and 3.7 times higher than that of the antibody- and CBD-RD complexes (Figure 4b), respectively, suggesting that the signal of each binder-p($OEOMA_{500}$-RDMA (4)) complex was improved by increasing the number of rhodamine B dyes on the copolymer without background fluorescence from nonspecific binding of the copolymer.

We then assessed binding onto single bacterial cells using flow cytometry. The addition of each binder-p($OEOMA_{500}$-RDMA (4)) complex generated a clear change in the fluorescence intensity compared with the target *S. aureus* cells alone (Figure 5), suggesting that the fluorescence of each binder-p($OEOMA_{500}$-RDMA (4)) complex comes from binding to the surface of *S. aureus* cells. Furthermore, we confirmed the specificity of these complexes using flow cytometry. In both binder cases, the complex did not bind to *B. anthracis* Sterne cells, as no fluorescence was detected. These results suggest that we have generated complexes that are target-specific and can bind to the target bacterium, induced by antibody or CBD.

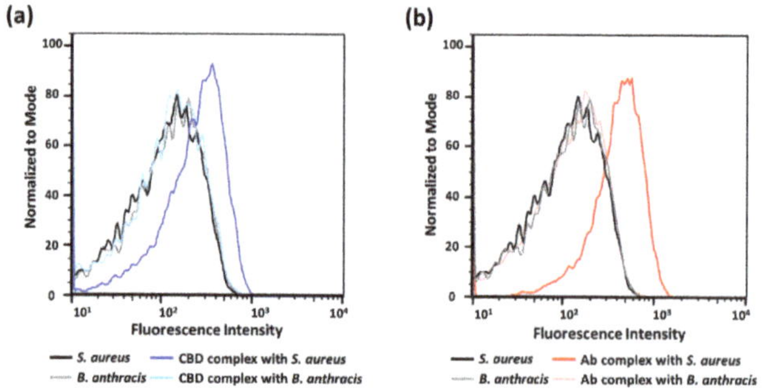

Figure 5. Flow cytometry analysis for (**a**) CBD-p($OEOMA_{500}$-RDMA (4)) complex and (**b**) antibody-p($OEOMA_{500}$-RDMA (4)) complex at the surface of bacteria.

3.3. Antibody/CBD-Polymeric Dyes Complex for Bioimaging Application

We applied the signal-enhancing property of each binder-p(OEOMA$_{500}$-RDMA (4)) complex for the bioimaging of target bacteria. When we prepared *S. aureus* cells with binder-p(OEOMA$_{500}$-RDMA (4)) complex, all the target bacterial cells with binder-p(OEOMA$_{500}$-RDMA (4)) complex showed red emission in each image (Figure 6a–d). In addition, the fluorescence images showed the same trends that we had previously obtained using fluorescence detection on a plate reader. Further analysis of these images was performed to quantify the fluorescence from the single cell by corrected total cell fluorescence (CTCF) analysis using the ImageJ program. The CTCF of antibody- and CBD-p(OEOMA$_{500}$-RDMA (4)) complexes were 1.6 and 3.6 times higher than that of antibody- and CBD-RD complexes, respectively (Figure 6e). Additionally, in a mixture of cells, under confocal microscopy, we specifically distinguished the presence of the target bacteria from the nontarget using these complexes (Figure 6f,g). These results suggest that these binder-p(OEOMA$_{500}$-RDMA (4)) complexes can be applied for bioimaging to visualize specific bacteria.

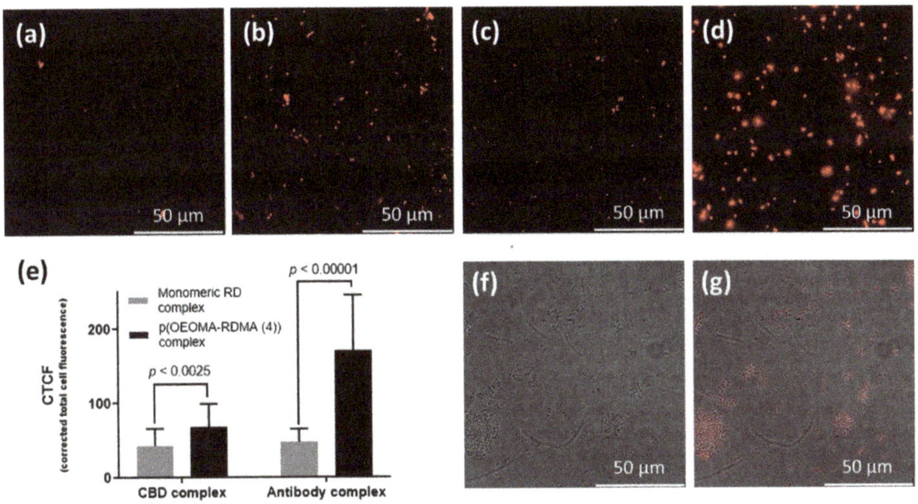

Figure 6. Confocal images of (**a**) CBD-monomeric RD and (**b**) CBD-p(OEOMA$_{500}$-RDMA (4)), (**c**) antibody-monomeric RD, and (**d**) antibody-p(OEOMA$_{500}$-RDMA (4)) complexes with *S. aureus* cells. (**e**) Comparison of corrected total cell fluorescence (CTCF) normalized by bacterial cell size in each confocal image (n = 15*). Confocal images for (**f**) fluorescent field, and (**g**) merged bright and fluorescent fields of antibody-p(OEOMA$_{500}$-RDMA (4)) complex in the mixture of sphere-shaped *S. aureus* and the rod-shaped *B. anthracis* cells. The scale bar for all confocal images represents 10 µm.

4. Conclusions

We have developed ATRP-derived copolymeric multifunctional rhodamine B dyes and attached them to binders such as an antibody or CBD for selective binding of target bacterial cells. The photoredox/Cu-catalyzed ATRP technique enabled the efficient and rapid synthesis of well-defined copolymers with multiple fluorescent dyes. Antibody/CBD-polymeric dye complex showed both enhanced fluorescence and target selectivity for bioimaging. This is due to the special structural property of this complex, consisting of multiple fluorescent dyes and a single binding molecule. The combination of this unique property of the polymeric dye and binder-induced targeting can also be applied to conjugate multiple signaling molecules such as quantum dots, DNA, and enzymes, followed by the potential applications in pathogen detection, and selective microbial decontamination, as well as bioimaging. The present work has also opened the potential application of ATRP-derived polymeric dyes in biosensors for detection of the DNA or protein biomarkers.

Supplementary Materials: The following supporting information can be downloaded at: https://www.mdpi.com/article/10.3390/polym15122723/s1, Figure S1: ^{1}H NMR spectrum of BT-p(CBMA-RDMA) in D_2O; Figure S2: ^{1}H NMR spectrum of BT-p(OEOMA-RDMA) in D_2O; Figure S3: Overlapped ^{1}H NMR spectra showing kinetics of EYH_2/Cu-catalyzed blue-light-induced ATRP for copolymerization of $OEOMA_{500}$ and RDMA; Figure S4: Fluorescence spectra of BT-p(OEOMA500-RDMA (4) (5 mg/mL) in PBS; Figure S5: Binding test using *S. aureus* cells with various concentrations of antibody/CBD-p($OEOMA_{500}$-RDMA (4)) complex; Figure S6: Evolution of the fluorescence intensity of BT-p($OEOMA_{500}$-RDMA (4)) (60 μM) overnight (37 °C, in PBS Buffer).

Author Contributions: All authors contributed to writing this manuscript. Conceptualization: K.K., S.X., I.L., H.M., S.-J.K., J.S.D. and K.M.; methodology, K.K., S.X. and I.L.; resources, K.K.; investigation, K.K., S.X. and I.L.; formal analysis, K.K., S.X. and I.L.; writing—original draft preparation, K.K., S.X. and I.L.; writing—review and editing, H.M., S.-J.K., J.S.D. and K.M.; visualization, H.M., S.-J.K., J.S.D. and K.M.; supervision, J.S.D. and K.M.; project administration, J.S.D. and K.M.; funding acquisition, J.S.D. and K.M. All authors have read and agreed to the published version of the manuscript.

Funding: Financial support from the DTRA (HDTRA-1-20-1-0014) and NSF (DMR 2202747) is acknowledged.

Institutional Review Board Statement: Not applicable.

Data Availability Statement: Not applicable.

Acknowledgments: Dedicated to the memory of the late Andrzej Dworak—a great scientist and a longtime friend.

Conflicts of Interest: The authors declare no conflict of interest.

References

1. Holmes, K.K.; Bertozzi, S.; Bloom, B.R.; Jha, P.; Gelband, H.; DeMaria, L.M.; Horton, S. Major Infectious Diseases: Key Messages from Disease Control Priorities, Third Edition. In *Disease Control Priorities, Third Edition (Volume 6): Major Infectious Diseases*; World Bank: Washington, DC, USA, 2017; pp. 1–27.
2. Antimicrobial Resistance Collaborators. Global burden of bacterial antimicrobial resistance in 2019: A systematic analysis. *Lancet* **2022**, *399*, 629–655. [CrossRef] [PubMed]
3. Yoon, S.A.; Park, S.Y.; Cha, Y.J.; Gopala, L.; Lee, M.H. Strategies of Detecting Bacteria Using Fluorescence-Based Dyes. *Front. Chem.* **2021**, *9*, 743923. [CrossRef] [PubMed]
4. Yang, S.; Rothman, R.E. PCR-based diagnostics for infectious diseases: Uses, limitations, and future applications in acute-care settings. *Lancet Infect. Dis.* **2004**, *4*, 337–348. [CrossRef] [PubMed]
5. Váradi, L.; Luo, J.L.; Hibbs, D.E.; Perry, J.D.; Anderson, R.J.; Orenga, S.; Groundwater, P.W. Methods for the detection and identification of pathogenic bacteria: Past, present, and future. *Chem. Soc. Rev.* **2017**, *46*, 4818–4832. [CrossRef] [PubMed]
6. Hameed, S.; Xie, L.; Ying, Y. Conventional and emerging detection techniques for pathogenic bacteria in food science: A review. *Trends Food Sci. Technol.* **2018**, *81*, 61–73. [CrossRef]
7. Franco-Duarte, R.; Černáková, L.; Kadam, S.; Kaushik, K.S.; Salehi, B.; Bevilacqua, A.; Corbo, M.R.; Antolak, H.; Dybka-Stępień, K.; Leszczewicz, M.; et al. Advances in Chemical and Biological Methods to Identify Microorganisms—From Past to Present. *Microorganisms* **2019**, *7*, 130. [CrossRef]
8. Yao, Z.; Carballido-López, R. Fluorescence Imaging for Bacterial Cell Biology: From Localization to Dynamics, From Ensembles to Single Molecules. *Annu. Rev. Microbiol.* **2014**, *68*, 459–476. [CrossRef]
9. Guo, Z.; Zeng, J.; Liu, W.; Chen, Y.; Jiang, H.; Weizmann, Y.; Wang, X. Formation of bio-responsive nanocomposites for targeted bacterial bioimaging and disinfection. *Chem. Eng. J.* **2021**, *426*, 130726. [CrossRef]
10. Si, Y.; Grazon, C.; Clavier, G.; Rieger, J.; Tian, Y.Y.; Audibert, J.F.; Sclavi, B.; Meallet-Renault, R. Fluorescent Copolymers for Bacterial Bioimaging and Viability Detection. *ACS Sens.* **2020**, *5*, 2843–2851. [CrossRef]
11. Lu, L.L.; Suscovich, T.J.; Fortune, S.M.; Alter, G. Beyond binding: Antibody effector functions in infectious diseases. *Nat. Rev. Immunol.* **2018**, *18*, 46–61. [CrossRef]
12. Dammes, N.; Peer, D. Monoclonal antibody-based molecular imaging strategies and theranostic opportunities. *Theranostics* **2020**, *10*, 938–955. [CrossRef] [PubMed]
13. Wan, Q.; Liu, X.; Zu, Y. Oligonucleotide aptamers for pathogen detection and infectious disease control. *Theranostics* **2021**, *11*, 9133–9161, Review. [CrossRef]
14. Davydova, A.; Vorobjeva, M.; Pyshnyi, D.; Altman, S.; Vlassov, V.; Venyaminova, A. Aptamers against pathogenic microorganisms. *Crit. Rev. Microbiol.* **2016**, *42*, 847–865. [CrossRef]
15. Fischetti, V.A. Bacteriophage lysins as effective antibacterials. *Curr. Opin. Microbiol.* **2008**, *11*, 393–400. [CrossRef]

16. Wu, X.; Kwon, S.-J.; Kim, J.; Kane, R.S.; Dordick, J.S. Biocatalytic Nanocomposites for Combating Bacterial Pathogens. *Annu. Rev. Chem. Biomol. Eng.* **2017**, *8*, 87–113. [CrossRef]
17. Bhagwat, A.; Mixon, M.; Collins, C.H.; Dordick, J.S. Opportunities for broadening the application of cell wall lytic enzymes. *Appl. Microbiol. Biotechnol.* **2020**, *104*, 9019–9040. [CrossRef] [PubMed]
18. Li, B.H.; Lu, L.F.; Zhao, M.Y.; Lei, Z.H.; Zhang, F. An Efficient 1064 nm NIR-II Excitation Fluorescent Molecular Dye for Deep-Tissue High-Resolution Dynamic Bioimaging. *Angew. Chem. -Int. Ed.* **2018**, *57*, 7483–7487. [CrossRef] [PubMed]
19. Liu, C.C.; Li, M.F.; Ma, H.L.; Hu, Z.B.; Wang, X.Y.; Ma, R.; Jiang, Y.Y.; Sun, H.T.; Zhu, S.J.; Liang, Y.Y. Furan Donor for NIR-II Molecular Fluorophores with Enhanced Bioimaging Performance. *Research* **2023**, *2023*, 0039. [CrossRef]
20. Chang, P.; Han, C.M.; Xu, H. Research progress of near infrared organic small-molecule electroluminescent materials. *Chin. J. Liq. Cryst. Disp.* **2021**, *36*, 62–77. [CrossRef]
21. Hama, H.; Kurokawa, H.; Kawano, H.; Ando, R.; Shimogori, T.; Noda, H.; Fukami, K.; Sakaue-Sawano, A.; Miyawaki, A. Scale: A chemical approach for fluorescence imaging and reconstruction of transparent mouse brain. *Nat. Neurosci.* **2011**, *14*, 1481–1488. [CrossRef]
22. Romei, M.G.; Boxer, S.G. Split Green Fluorescent Proteins: Scope, Limitations, and Outlook. *Annu. Rev. Biophys.* **2019**, *48*, 19–44. [CrossRef] [PubMed]
23. Zimmer, M. Green fluorescent protein (GFP): Applications, structure, and related photophysical behavior. *Chem. Rev.* **2002**, *102*, 759–781. [CrossRef] [PubMed]
24. Vetschera, P.; Mishra, K.; Fuenzalida-Werner, J.P.; Chmyrov, A.; Ntziachristos, V.; Stiel, A.C. Characterization of Reversibly Switchable Fluorescent Proteins in Optoacoustic Imaging. *Anal. Chem.* **2018**, *90*, 10527–10535. [CrossRef] [PubMed]
25. Ban, Q.F.; Li, Y.; Wu, S. Self-fluorescent polymers for bioimaging. *View* **2022**, *3*, 20200135. [CrossRef]
26. Bentolila, A.; Totre, J.; Zozulia, I.; Levin-Elad, M.; Domb, A.J. Fluorescent Cyanoacrylate Monomers and Polymers for Fingermark Development. *Macromolecules* **2013**, *46*, 4822–4828. [CrossRef]
27. Deng, H.P.; Su, Y.; Hu, M.X.; Jin, X.; He, L.; Pang, Y.; Dong, R.J.; Zhu, X.Y. Multicolor Fluorescent Polymers Inspired from Green Fluorescent Protein. *Macromolecules* **2015**, *48*, 5969–5979. [CrossRef]
28. Adjili, S.; Favier, A.; Massin, J.; Bretonniere, Y.; Lacour, W.; Lin, Y.C.; Chatre, E.; Place, C.; Favard, C.; Muriaux, D.; et al. Synthesis of multifunctional lipid-polymer conjugates: Application to the elaboration of bright far-red fluorescent lipid probes. *RSC Adv.* **2014**, *4*, 15569–15578. [CrossRef]
29. Duret, D.; Haftek-Terreau, Z.; Carretier, M.; Berki, T.; Ladaviere, C.; Monier, K.; Bouvet, P.; Marvel, J.; Leverrier, Y.; Charreyre, M.T.; et al. Labeling of native proteins with fluorescent RAFT polymer probes: Application to the detection of a cell surface protein using flow cytometry. *Polym. Chem.* **2018**, *9*, 1857–1868. [CrossRef]
30. Jiang, R.M.; Huang, L.; Liu, M.Y.; Deng, F.J.; Huang, H.Y.; Tian, J.W.; Wen, Y.Q.; Cao, Q.Y.; Zhang, X.Y.; Wei, Y. Ultrafast microwave-assisted multicomponent tandem polymerization for rapid fabrication of AIE-active fluorescent polymeric nanoparticles and their potential utilization for biological imaging. *Mater. Sci. Eng. C-Mater. Biol. Appl.* **2018**, *83*, 115–120. [CrossRef]
31. Huang, H.Y.; Jiang, R.M.; Ma, H.J.; Li, Y.S.; Zeng, Y.; Zhou, N.G.; Liu, L.J.; Zhang, X.Y.; Wei, Y. Fabrication of claviform fluorescent polymeric nanomaterials containing disulfide bond through an efficient and facile four-component Ugi reaction. *Mater. Sci. Eng. C-Mater. Biol. Appl.* **2021**, *118*, 111437. [CrossRef]
32. Dong, J.D.; Jiang, R.M.; Huang, H.Y.; Chen, J.Y.; Tian, J.W.; Deng, F.J.; Dai, Y.F.; Wen, Y.Q.; Zhang, X.Y.; Wei, Y. Facile preparation of fluorescent nanodiamond based polymer nanoparticles via ring-opening polymerization and their biological imaging. *Mater. Sci. Eng. C-Mater. Biol. Appl.* **2020**, *106*, 110297. [CrossRef]
33. Reisch, A.; Klymchenko, A.S. Fluorescent Polymer Nanoparticles Based on Dyes: Seeking Brighter Tools for Bioimaging. *Small* **2016**, *12*, 1968–1992. [CrossRef] [PubMed]
34. Thapaliya, E.K.; Zhang, Y.; Dhakal, P.; Brown, A.S.; Wilson, J.N.; Collins, K.M.; Raymo, F.M. Bioimaging with Macromolecular Probes Incorporating Multiple BODIPY Fluorophores. *Bioconjugate Chem.* **2017**, *28*, 1519–1528. [CrossRef] [PubMed]
35. Kloepfer, J.A.; Mielke, R.E.; Wong, M.S.; Nealson, K.H.; Stucky, G.; Nadeau, J.L. Quantum dots as strain- and metabolism-specific microbiological labels. *Appl. Environ. Microbiol.* **2003**, *69*, 4205–4213. [CrossRef] [PubMed]
36. Chalmers, N.I.; Palmer, R.J.; Du-Thumm, L.; Sullivan, R.; Shi, W.Y.; Kolenbrander, P.E. Use of quantum dot luminescent probes to achieve single-cell resolution of human oral bacteria in biofilms. *Appl. Environ. Microbiol.* **2007**, *73*, 630–636. [CrossRef]
37. Gazouli, M.; Liandris, E.; Andreadou, M.; Sechi, L.A.; Masala, S.; Paccagnini, D.; Ikonomopoulos, J. Specific Detection of Unamplified Mycobacterial DNA by Use of Fluorescent Semiconductor Quantum Dots and Magnetic Beads. *J. Clin. Microbiol.* **2010**, *48*, 2830–2835. [CrossRef]
38. Xue, X.H.; Pan, J.; Xie, H.M.; Wang, J.H.; Zhang, S. Fluorescence detection of total count of Escherichia coli and Staphylococcus aureus on water-soluble CdSe quantum dots coupled with bacteria. *Talanta* **2009**, *77*, 1808–1813. [CrossRef] [PubMed]
39. Resch-Genger, U.; Grabolle, M.; Cavaliere-Jaricot, S.; Nitschke, R.; Nann, T. Quantum dots versus organic dyes as fluorescent labels. *Nat. Methods* **2008**, *5*, 763–775. [CrossRef]
40. Wolfbeis, O.S. An overview of nanoparticles commonly used in fluorescent bioimaging. *Chem. Soc. Rev.* **2015**, *44*, 4743–4768. [CrossRef]
41. Bou, S.; Klymchenko, A.S.; Collot, M. Fluorescent labeling of biocompatible block copolymers: Synthetic strategies and applications in bioimaging. *Mater. Adv.* **2021**, *2*, 3213–3233. [CrossRef]

42. Zhang, L.E.; Zhang, Z.K.; Liu, C.R.; Zhang, X.K.; Fan, Q.L.; Wu, W.; Jiangsu, X.Q. NIR-II Dye-Labeled Cylindrical Polymer Brushes for in Vivo Imaging. *Acs Macro Lett.* **2019**, *8*, 1623–1628. [CrossRef] [PubMed]
43. Trofymchuk, K.; Valanciunaite, J.; Andreiuk, B.; Reisch, A.; Collot, M.; Klymchenko, A.S. BODIPY-loaded polymer nanoparticles: Chemical structure of cargo defines leakage from nanocarrier in living cells. *J. Mater. Chem. B* **2019**, *7*, 5199–5210. [CrossRef] [PubMed]
44. Li, G.; Zhu, X.L.; Cheng, Z.P.; Zhang, W.; Sun, B. Synthesis of poly(methyl methacrylate) labeled with fluorescein moieties via atom transfer radical polymerization. *J. Macromol. Sci. Part A-Pure Appl. Chem.* **2008**, *45*, 495–501. [CrossRef]
45. Chaney, E.J.; Tang, L.; Tong, R.; Cheng, J.J.; Boppart, S.A. Lymphatic Biodistribution of Polylactide Nanoparticles. *Mol. Imaging* **2010**, *9*, 153–162. [CrossRef] [PubMed]
46. Zhang, Y.; Chen, Y.J.; Li, X.; Zhang, J.B.; Chen, J.L.; Xu, B.; Fu, X.Q.; Tian, W.J. Folic acid-functionalized AIE Pdots based on amphiphilic PCL-b-PEG for targeted cell imaging. *Polym. Chem.* **2014**, *5*, 3824–3830. [CrossRef]
47. Wang, K.; Luo, Y.M.; Huang, S.; Yang, H.B.; Liu, B.; Wang, M.F. Highly Fluorescent Polycaprolactones Decorated with Di(thiophene-2-yl)-diketopyrrolopyrrole: A Covalent Strategy of Tuning Fluorescence Properties in Solid States. *J. Polym. Sci. Part A-Polym. Chem.* **2015**, *53*, 1032–1042. [CrossRef]
48. Lu, X.J.; Zhang, L.F.; Meng, L.Z.; Liu, Y.H. Synthesis of poly(N-isopropylacrylamide) by ATRP using a fluorescein-based initiator. *Polym. Bull.* **2007**, *59*, 195–206. [CrossRef]
49. Breul, A.M.; Hager, M.D.; Schubert, U.S. Fluorescent monomers as building blocks for dye labeled polymers: Synthesis and application in energy conversion, biolabeling and sensors. *Chem. Soc. Rev.* **2013**, *42*, 5366–5407. [CrossRef]
50. Wan, Q.; Liu, M.Y.; Mao, L.C.; Jiang, R.M.; Xu, D.Z.; Huang, H.Y.; Dai, Y.F.; Deng, F.J.; Zhang, X.Y.; Wei, Y. Preparation of PEGylated polymeric nanoprobes with aggregation-induced emission feature through the combination of chain transfer free radical polymerization and multicomponent reaction: Self-assembly, characterization and biological imaging applications. *Mater. Sci. Eng. C-Mater. Biol. Appl.* **2017**, *72*, 352–358. [CrossRef]
51. Li, F.J.; Zhu, A.P.; Song, X.L.; Ji, L.J.; Wang, J. The internalization of fluorescence-labeled PLA nanoparticles by macrophages. *Int. J. Pharm.* **2013**, *453*, 506–513. [CrossRef]
52. Guan, X.L.; Lai, S.J.; Su, Z.X. Facile Preparation and Potential Application of Water-Soluble Polymeric Temperature/pH Probes Bearing Fluorescein. *J. Appl. Polym. Sci.* **2011**, *122*, 1968–1975. [CrossRef]
53. Kim, C.; Wallace, J.U.; Chen, S.H.; Merkel, P.B. Effects of dilution, polarization ratio, and energy transfer on photoalignment of liquid crystals using coumarin-containing polymer films. *Macromolecules* **2008**, *41*, 3075–3080. [CrossRef]
54. Manickasundaram, S.; Kannan, P.; Kumaran, R.; Velu, R.; Ramamurthy, P.; Hassan, Q.M.A.; Palanisamy, P.K.; Senthil, S.; Narayanan, S.S. Holographic grating studies in pendant xanthene dyes containing poly(alkyloxymethacrylate)s. *J. Mater. Sci. -Mater. Electron.* **2011**, *22*, 25–34. [CrossRef]
55. Berger, S.; Synytska, A.; Ionov, L.; Eichhorn, K.J.; Stamm, M. Stimuli-Responsive Bicomponent Polymer Janus Particles by "Grafting from"/"Grafting to" Approaches. *Macromolecules* **2008**, *41*, 9669–9676. [CrossRef]
56. Li, G.; Bai, L.P.; Tao, F.R.; Deng, A.X.; Wang, L.P. A dual chemosensor for Cu2+ and Hg2+ based on a rhodamine-terminated water-soluble polymer in 100% aqueous solution. *Analyst* **2018**, *143*, 5395–5403. [CrossRef] [PubMed]
57. Hu, J.M.; Dai, L.; Liu, S.Y. Analyte-Reactive Amphiphilic Thermoresponsive Diblock Copolymer Micelles-Based Multifunctional Ratiometric Fluorescent Chemosensors. *Macromolecules* **2011**, *44*, 4699–4710. [CrossRef]
58. Hu, J.M.; Zhang, X.Z.; Wang, D.; Hu, X.L.; Liu, T.; Zhang, G.Y.; Liu, S.Y. Ultrasensitive ratiometric fluorescent pH and temperature probes constructed from dye-labeled thermoresponsive double hydrophilic block copolymers. *J. Mater. Chem.* **2011**, *21*, 19030–19038. [CrossRef]
59. Ma, C.P.; Xie, G.Y.; Zhang, X.Q.; Yang, L.T.; Li, Y.; Liu, H.L.; Wang, K.; Wei, Y. Biocompatible fluorescent polymers from PEGylation of an aggregation-induced emission dye. *Dye. Pigment.* **2017**, *139*, 672–680. [CrossRef]
60. Nicolas, J.; San Miguel, V.; Mantovani, G.; Haddleton, D.M. Fluorescently tagged polymer bioconjugates from protein derived macroinitiators. *Chem. Commun.* **2006**, *45*, 4697–4699. [CrossRef]
61. Madsen, J.; Canton, I.; Warren, N.J.; Themistou, E.; Blanazs, A.; Ustbas, B.; Tian, X.H.; Pearson, R.; Battaglia, G.; Lewis, A.L.; et al. Nile Blue-Based Nanosized pH Sensors for Simultaneous Far-Red and Near-Infrared Live Bioimaging. *J. Am. Chem. Soc.* **2013**, *135*, 14863–14870. [CrossRef]
62. Truong, N.P.; Jones, G.R.; Bradford, K.G.E.; Konkolewicz, D.; Anastasaki, A. A comparison of RAFT and ATRP methods for controlled radical polymerization. *Nat. Rev. Chem.* **2021**, *5*, 859–869. [CrossRef] [PubMed]
63. You, J.; Yoon, J.A.; Kim, J.; Huang, C.F.; Matyjaszewski, K.; Kim, E. Excimer Emission from Self-Assembly of Fluorescent Diblock Copolymer Prepared by Atom Transfer Radical Polymerization. *Chem. Mater.* **2010**, *22*, 4426–4434. [CrossRef]
64. Neugebauer, D.; Charasim, D.; Swinarew, A.; Stolarzewicz, A.; Krompiec, M.; Janeczek, H.; Simokaitiene, J.; Grazulevicius, J.V. Polymethacrylates with anthryl and carbazolyl groups prepared by atom transfer radical polymerization. *Polym. J.* **2011**, *43*, 448–454. [CrossRef]
65. Spiniello, M.; Blencowe, A.; Qiao, G.G. Synthesis and characterization of fluorescently labeled core cross-linked star polymers. *J. Polym. Sci. Part A-Polym. Chem.* **2008**, *46*, 2422–2432. [CrossRef]
66. Madsen, J.; Warren, N.J.; Armes, S.P.; Lewis, A.L. Synthesis of Rhodamine 6G-Based Compounds for the ATRP Synthesis of Fluorescently Labeled Biocompatible Polymers. *Biomacromolecules* **2011**, *12*, 2225–2234. [CrossRef]

67. Yang, Q.A.; Jin, H.; Xu, Y.D.; Shen, Z.H.; Fan, X.H.; Zou, D.C.; Zhou, Q.F. Electroluminescent Block Copolymers Containing Oxadiazole and Thiophene via ATRP. *J. Polym. Sci. Part A-Polym. Chem.* **2010**, *48*, 5670–5678. [CrossRef]
68. Yang, Q.; Xu, Y.D.; Jin, H.; Shen, Z.H.; Chen, X.F.; Zou, D.C.; Fan, X.H.; Zhou, Q.F. A Novel Mesogen-Jacketed Liquid Crystalline Electroluminescent Polymer with Both Thiophene and Oxadiazole in Conjugated Side Chain. *J. Polym. Sci. Part A-Polym. Chem.* **2010**, *48*, 1502–1515. [CrossRef]
69. Trzebicka, B.; Szweda, R.; Kosowski, D.; Szweda, D.; Otulakowski, L.; Haladjova, E.; Dworak, A. Thermoresponsive polymer-peptide/protein conjugates. *Prog. Polym. Sci.* **2017**, *68*, 35–76. [CrossRef]
70. Dimitrov, I.; Trzebicka, B.; Muller, A.H.E.; Dworak, A.; Tsvetanov, C.B. Thermosensitive water-soluble copolymers with doubly responsive reversibly interacting entities. *Prog. Polym. Sci.* **2007**, *32*, 1275–1343. [CrossRef]
71. Yamamoto, S.; Pietrasik, J.; Matyjaszewski, K. Temperature- and pH-responsive dense copolymer brushes prepared by ATRP. *Macromolecules* **2008**, *41*, 7013–7020. [CrossRef]
72. Nese, A.; Lebedeva, N.V.; Sherwood, G.; Averick, S.; Li, Y.C.; Gao, H.F.; Peteanu, L.; Sheiko, S.S.; Matyjaszewski, K. pH-Responsive Fluorescent Molecular Bottlebrushes Prepared by Atom Transfer Radical Polymerization. *Macromolecules* **2011**, *44*, 5905–5910. [CrossRef]
73. Mielanczyk, A.; Skonieczna, M.; Bernaczek, K.; Neugebauer, D. Fluorescein nanocarriers based on cationic star copolymers with acetal linked sugar cores. Synthesis and biochemical characterization. *Rsc Adv.* **2014**, *4*, 31904–31913. [CrossRef]
74. Zhao, K.; Cheng, Z.P.; Zhang, Z.B.; Zhu, J.; Zhu, X.L. Synthesis of fluorescent poly(methyl methacrylate) via AGET ATRP. *Polym. Bull.* **2009**, *63*, 355–364. [CrossRef]
75. Li, Y.; Bai, Y.G.; Zheng, N.; Liu, Y.; Vincil, G.A.; Pedretti, B.J.; Cheng, J.J.; Zimmerman, S.C. Crosslinked dendronized polyols as a general approach to brighter and more stable fluorophores. *Chem. Commun.* **2016**, *52*, 3781–3784. [CrossRef] [PubMed]
76. Chien, M.P.; Carlini, A.S.; Hu, D.H.; Barback, C.V.; Rush, A.M.; Hall, D.J.; Orr, G.; Gianneschi, N.C. Enzyme-Directed Assembly of Nanoparticles in Tumors Monitored by in Vivo Whole Animal Imaging and ex Vivo Super-Resolution Fluorescence Imaging. *J. Am. Chem. Soc.* **2013**, *135*, 18710–18713. [CrossRef]
77. Yu, K.K.; Li, K.; Hou, J.T.; Yang, J.; Xie, Y.M.; Yu, X.Q. Rhodamine based pH-sensitive "intelligent" polymers as lysosome targeting probes and their imaging applications in vivo. *Polym. Chem.* **2014**, *5*, 5804–5812. [CrossRef]
78. Hench, L.L.; Polak, J.M. Third-generation biomedical materials. *Science* **2002**, *295*, 1014–1017. [CrossRef]
79. Anderson, D.G.; Burdick, J.A.; Langer, R. Materials science—Smart biomaterials. *Science* **2004**, *305*, 1923–1924. [CrossRef]
80. Wang, J.S.; Matyjaszewski, K. Controlled/"Living" Radical Polymerization. Halogen Atom Transfer Radical Polymerization Promoted by a Cu(I)/Cu(II) Redox Process. *Macromolecules* **1995**, *28*, 7901–7910. [CrossRef]
81. Patten, T.E.; Xia, J.; Abernathy, T.; Matyjaszewski, K. Polymers with very low polydispersities from atom transfer radical polymerization. *Science* **1996**, *272*, 866–868. [CrossRef]
82. Matyjaszewski, K.; Xia, J.H. Atom transfer radical polymerization. *Chem. Rev.* **2001**, *101*, 2921–2990. [CrossRef]
83. Corrigan, N.; Jung, K.; Moad, G.; Hawker, C.J.; Matyjaszewski, K.; Boyer, C. Reversible-deactivation radical polymerization (Controlled/living radical polymerization): From discovery to materials design and applications. *Prog. Polym. Sci.* **2020**, *111*, 101311. [CrossRef]
84. Chmielarz, P.; Fantin, M.; Park, S.; Isse, A.A.; Gennaro, A.; Magenau, A.J.D.; Sobkowiak, A.; Matyjaszewski, K. Electrochemically mediated atom transfer radical polymerization (eATRP). *Prog. Polym. Sci.* **2017**, *69*, 47–78. [CrossRef]
85. Pan, X.C.; Tasdelen, M.A.; Laun, J.; Junkers, T.; Yagci, Y.; Matyjaszewski, K. Photomediated controlled radical polymerization. *Prog. Polym. Sci.* **2016**, *62*, 73–125. [CrossRef]
86. Matyjaszewski, K.; Tsarevsky, N.V. Macromolecular Engineering by Atom Transfer Radical Polymerization. *J. Am. Chem. Soc.* **2014**, *136*, 6513–6533. [CrossRef] [PubMed]
87. Matyjaszewski, K.; Tsarevsky, N.V. Nanostructured functional materials prepared by atom transfer radical polymerization. *Nat. Chem.* **2009**, *1*, 276–288. [CrossRef]
88. Ribelli, T.G.; Lorandi, F.; Fantin, M.; Matyjaszewski, K. Atom Transfer Radical Polymerization: Billion Times More Active Catalysts and New Initiation Systems. *Macromol. Rapid Commun.* **2019**, *40*, 1800616. [CrossRef]
89. Enciso, A.E.; Fu, L.; Russell, A.J.; Matyjaszewski, K. A Breathing Atom-Transfer Radical Polymerization: Fully Oxygen-Tolerant Polymerization Inspired by Aerobic Respiration of Cells. *Angew. Chem. Int. Ed.* **2018**, *57*, 933–936. [CrossRef]
90. Oh, J.K.; Min, K.; Matyjaszewski, K. Preparation of poly(oligo(ethylene glycol) monomethyl ether methacrylate) by homogeneous aqueous AGET ATRP. *Macromolecules* **2006**, *39*, 3161–3167. [CrossRef]
91. Baker, S.L.; Kaupbayeva, B.; Lathwal, S.; Das, S.R.; Russell, A.J.; Matyjaszewski, K. Atom Transfer Radical Polymerization for Biorelated Hybrid Materials. *Biomacromolecules* **2019**, *20*, 4272–4298. [CrossRef]
92. Jakubowski, W.; Min, K.; Matyjaszewski, K. Activators Regenerated by Electron Transfer for Atom Transfer Radical Polymerization of Styrene. *Macromolecules* **2006**, *39*, 39–45. [CrossRef]
93. Magenau, A.J.D.; Strandwitz, N.C.; Gennaro, A.; Matyjaszewski, K. Electrochemically Mediated Atom Transfer Radical Polymerization. *Science* **2011**, *332*, 81–84. [CrossRef] [PubMed]
94. Matyjaszewski, K. Atom Transfer Radical Polymerization (ATRP): Current status and future perspectives. *Macromolecules* **2012**, *45*, 4015–4039. [CrossRef]
95. Fung, A.K.K.; Coote, M.L. A mechanistic perspective on atom transfer radical polymerization. *Polym. Int.* **2021**, *70*, 918–926. [CrossRef]

96. Lorandi, F.; Fantin, M.; Matyjaszewski, K. Atom Transfer Radical Polymerization: A Mechanistic Perspective. *J. Am. Chem. Soc.* **2022**, *144*, 15413–15430. [CrossRef] [PubMed]
97. Jin-Shan Wang and Krzysztof, M. Controlled/"living" radical polymerization. atom transfer radical polymerization in the presence of transition-metal complexes. *J. Am. Chem. Soc.* **1995**, *117*, 5614–5615.
98. Tsarevsky, N.V.; Pintauer, T.; Matyjaszewski, K. The rate of deactivation in atom transfer radical polymerization in protic and aqueous media. *Polym. Prepr. (Am. Chem. Soc. Div. Polym. Chem.)* **2004**, *45*, 1067–1068.
99. Simakova, A.; Averick, S.E.; Konkolewicz, D.; Matyjaszewski, K. Aqueous ARGET ATRP. *Macromolecules* **2012**, *45*, 6371–6379. [CrossRef]
100. Tsarevsky, N.V.; Matyjaszewski, K. "Green" atom transfer radical polymerization: From process design to preparation of well-defined environmentally friendly polymeric materials. *Chem. Rev.* **2007**, *107*, 2270–2299. [CrossRef]
101. Ouchi, M.; Terashima, T.; Sawamoto, M. Transition Metal-Catalyzed Living Radical Polymerization: Toward Perfection in Catalysis and Precision Polymer Synthesis. *Chem. Rev.* **2009**, *109*, 4963–5050. [CrossRef]
102. Pintauer, T.; Matyjaszewski, K. Atom transfer radical addition and polymerization reactions catalyzed by ppm amounts of copper complexes. *Chem. Soc. Rev.* **2008**, *37*, 1087–1097. [CrossRef] [PubMed]
103. Konkolewicz, D.; Magenau, A.J.D.; Averick, S.E.; Simakova, A.; He, H.; Matyjaszewski, K. ICAR ATRP with ppm Cu catalyst in water. *Macromolecules* **2012**, *45*, 4461–4468. [CrossRef]
104. Szczepaniak, G.; Fu, L.Y.; Jafari, H.; Kapil, K.; Matyjaszewski, K. Making ATRP More Practical: Oxygen Tolerance. *Acc. Chem. Res.* **2021**, *54*, 1779–1790. [CrossRef]
105. Fu, L.; Wang, Z.; Lathwal, S.; Enciso, A.E.; Simakova, A.; Das, S.R.; Russell, A.J.; Matyjaszewski, K. Synthesis of Polymer Bioconjugates via Photoinduced Atom Transfer Radical Polymerization under Blue Light Irradiation. *ACS Macro Lett.* **2018**, *7*, 1248–1253. [CrossRef] [PubMed]
106. Szczepaniak, G.; Łagodzińska, M.; Dadashi-Silab, S.; Gorczyński, A.; Matyjaszewski, K. Fully oxygen-tolerant atom transfer radical polymerization triggered by sodium pyruvate. *Chem. Sci.* **2020**, *11*, 8809–8816. [CrossRef] [PubMed]
107. Sun, M.K.; Szczepaniak, G.; Dadashi-Silab, S.; Lin, T.C.; Kowalewski, T.; Matyjaszewski, K. Cu-Catalyzed Atom Transfer Radical Polymerization: The Effect of Cocatalysts. *Macromol. Chem. Phys.* **2023**, *224*, 2200347. [CrossRef]
108. Szczepaniak, G.; Jeong, J.; Kapil, K.; Dadashi-Silab, S.; Yerneni, S.S.; Ratajczyk, P.; Lathwal, S.; Schild, D.J.; Das, S.R.; Matyjaszewski, K. Open-air green-light-driven ATRP enabled by dual photoredox/copper catalysis. *Chem. Sci.* **2022**, *13*, 11540–11550. [CrossRef]
109. Kapil, K.; Jazani, A.M.; Szczepaniak, G.; Murata, H.; Olszewski, M.; Matyjaszewski, K. Fully Oxygen-Tolerant Visible-Light-Induced ATRP of Acrylates in Water: Toward Synthesis of Protein-Polymer Hybrids. *Macromolecules* **2023**, *56*, 2017–2026. [CrossRef]
110. Kapil, K.; Szczepaniak, G.; Martinez, M.R.; Murata, H.; Jazani, A.M.; Jeong, J.; Das, S.R.; Matyjaszewski, K. Visible-Light-Mediated Controlled Radical Branching Polymerization in Water. *Angew. Chem. Int. Ed.* **2023**, *62*, e202217658. [CrossRef]
111. Cummings, C.; Murata, H.; Koepsel, R.; Russell, A.J. Tailoring enzyme activity and stability using polymer-based protein engineering. *Biomaterials* **2013**, *34*, 7437–7443. [CrossRef]
112. Pan, X.; Lathwal, S.; Mack, S.; Yan, J.; Das, S.R.; Matyjaszewski, K. Automated Synthesis of Well-Defined Polymers and Biohybrids by Atom Transfer Radical Polymerization Using a DNA Synthesizer. *Angew. Chem. Int. Ed.* **2017**, *56*, 2740–2743. [CrossRef] [PubMed]
113. Li, F.; Cao, M.; Feng, Y.; Liang, R.; Fu, X.; Zhong, M. Site-Specifically Initiated Controlled/Living Branching Radical Polymerization: A Synthetic Route toward Hierarchically Branched Architectures. *J. Am. Chem. Soc.* **2019**, *141*, 794–799. [CrossRef] [PubMed]
114. Kaupbayeva, B.; Murata, H.; Lucas, A.; Matyjaszewski, K.; Minden, J.S.; Russell, A.J. Molecular Sieving on the Surface of a Nano-Armored Protein. *Biomacromolecules* **2019**, *20*, 1235–1245. [CrossRef] [PubMed]
115. Kaupbayeva, B.; Boye, S.; Munasinghe, A.; Murata, H.; Matyjaszewski, K.; Lederer, A.; Colina, C.M.; Russell, A.J. Molecular Dynamics-Guided Design of a Functional Protein-ATRP Conjugate That Eliminates Protein-Protein Interactions. *Bioconjugate Chem.* **2021**, *32*, 821–832. [CrossRef] [PubMed]
116. Averick, S.E.; Dey, S.K.; Grahacharya, D.; Matyjaszewski, K.; Das, S.R. Solid-phase incorporation of an ATRP initiator for polymer-DNA biohybrids. *Angew. Chem. Int. Ed.* **2014**, *53*, 2739–2744. [CrossRef] [PubMed]
117. Lathwal, S.; Yerneni, S.S.; Boye, S.; Muza, U.L.; Takahashi, S.; Sugimoto, N.; Lederer, A.; Das, S.R.; Campbell, P.G.; Matyjaszewski, K. Engineering exosome polymer hybrids by atom transfer radical polymerization. *Proc. Natl. Acad. Sci. USA* **2021**, *118*, e2020241118. [CrossRef]
118. Bateman, A.; Rawlings, N.D. The CHAP domain: A large family of amidases including GSP amidase and peptidoglycan hydrolases. *Trends Biochem. Sci.* **2003**, *28*, 234–237. [CrossRef]
119. Kim, D.; Kwon, S.-J.; Sauve, J.; Fraser, K.; Kemp, L.; Lee, I.; Nam, J.; Kim, J.; Dordick, J.S. Modular Assembly of Unique Chimeric Lytic Enzymes on a Protein Scaffold Possessing Anti-Staphylococcal Activity. *Biomacromolecules* **2019**, *20*, 4035–4043. [CrossRef]
120. Szczepaniak, G.; Piatkowski, J.; Nogas, W.; Lorandi, F.; Yerneni, S.S.; Fantin, M.; Ruszczynska, A.; Enciso, A.E.; Bulska, E.; Grela, K.; et al. An isocyanide ligand for the rapid quenching and efficient removal of copper residues after Cu/TEMPO-catalyzed aerobic alcohol oxidation and atom transfer radical polymerization. *Chem. Sci.* **2020**, *11*, 4251–4262. [CrossRef]

121. Erfani, A.; Seaberg, J.; Aichele, C.P.; Ramsey, J.D. Interactions between Biomolecules and Zwitterionic Moieties: A Review. *Biomacromolecules* **2020**, *21*, 2557–2573. [CrossRef]
122. Brown, S.; Maria, J.P.S.; Walker, S. Wall Teichoic Acids of Gram-Positive Bacteria. *Annu. Rev. Microbiol.* **2013**, *67*, 313–336. [CrossRef] [PubMed]

polymers

MDPI

Article

Antioxidant and Antiradical Activities of *Hibiscus sabdariffa* L. Extracts Encapsulated in Calcium Alginate Spheres

Pascal Bevan [1], Maria Vicenta Pastor [1], María Pilar Almajano [1,*] and Idoia Codina-Torrella [2,*]

[1] Chemical Engineering Department, Universitat Politècnica de Catalunya, Av. Diagonal 647, 08028 Barcelona, Spain
[2] Agri-Food Engineering and Biotechnology Department, Universitat Politècnica de Catalunya, Esteve Terrades 8, 08860 Castelldefels, Spain
* Correspondence: m.pilar.almajano@upc.edu (M.P.A.); idoia.codina@upc.edu (I.C.-T.)

Abstract: The interest in natural sources with high antioxidant powder has recently increased in several sectors. Ionic gelation methods could be used to protect bioactive substances to control the kinetics and release of these ingredients to the food matrix. This study dealt with the evaluation of the antioxidant capacity and scavenging activity of extracts of *Hibiscus Sabdariffa* L. (HSL) (with 50% ethanol) encapsulated in calcium alginate spheres as a new source for preserving food against oxidative damage. Their antioxidant activity was measured in different o/w emulsions in which HSL spheres reduced the formation of hydroperoxides (~80%) and thiobarbituric-acid-reactive substance products (~20%). The scavenging activity of HSL extracts was measured in different food simulants (water, water acidified with 3% acetic acid, ethanol at 50%, and pure ethanol), and corresponded to 0.20–0.43, 0.31–0.62, and 11.13–23.82 mmol Trolox/mL extract for Trolox equivalent antioxidant capacity (TEAC), 2,2-diphenylpicrylhydrazyl (DPPH), and oxygen radical absorbance capacity (ORAC) assays, respectively. In general, the best antiradical activity was observed in the ethanolic and acidified mediums, in which the highest concentration of released polyphenols ranged from 0.068 to 0.079 mg GAE/mL. This work indicates the potential of alginate spheres for encapsulating antioxidant compounds as an innovative strategy for several industrial applications.

Keywords: spherification; alginate; *Hibiscus sabdariffa* L.; encapsulation; oxidation; antiradical

Citation: Bevan, P.; Pastor, M.V.; Almajano, M.P.; Codina-Torrella, I. Antioxidant and Antiradical Activities of *Hibiscus sabdariffa* L. Extracts Encapsulated in Calcium Alginate Spheres. *Polymers* **2023**, *15*, 1740. https://doi.org/10.3390/polym15071740

Academic Editors: Marek Kowalczuk, Iza Radecka and Barbara Trzebicka

Received: 25 February 2023
Revised: 26 March 2023
Accepted: 27 March 2023
Published: 31 March 2023

1. Introduction

The oxidation of nutritional components (such as proteins, vitamins, minerals, etc.) determines the overall quality of food and directly impacts the consumer's acceptance of the product. There are different strategies to delay the oxidation of these components during the shelf life of food, of which the incorporation of synthetic and natural antioxidants is one of the most widely used. These substances counteract oxidation by acting through different strategies: as reducing agents, quenchers of radical species, free radical scavengers, or by inactivating prooxidants [1]. Nowadays, the interest in natural sources with high antioxidant power has recently increased in several sectors to replace synthetic antioxidants in industrial applications [2]. Regarding this, the incorporation of plant extracts with a high antioxidant capacity is a current focus of interest for developing new functional products. Apart from that, the economic impact of using these extracts from by-products or under-exploited plant species also has increased the interest in these natural sources [2]. Regrettably, the incorporation of these extracts makes necessary their protection against environmental and processing conditions to preserve their stable properties over time [3]. Different techniques have been proposed, of which encapsulation is one of the most widely used [4]. This technique could use ionic gelation methods, which lead to gel formation [5]. This molecular process protects the bioactive components with a hydrogel formed with specific structural and mechanical characteristics [5]. In general, different polysaccharides and calcium salts are used to obtain these spheres [5,6], which determines their physicochemical characteristics (viscosity, porosity, shape, etc.). Alginate is one of the most widely used polymers, which is used by the industry as a thickener or stabilizer. Alginate is formed by a chain of B-D-mannuronic (M units) acid and alpha-L-guluronic acid (G units) linked with 1–4 glycoside linkage. The gelation of alginate depends on the polymer's chemical structure (molecular weight and G/M unit ratio, among others) [5]. This polysaccharide has a high affinity to form hydrogels through ionic cross-linking with divalent or trivalent cations, such as Ca^{2+}. When calcium ions replace Na^{+} to cross-link different chains of alginate, the network which is formed is commonly described as an "eggbox model". Two main gelation methods are traditionally used (external and internal gelation). In the internal or indirect gelation process (which has been used in the current study), the gel formation is due to the calcium release into the alginate solution and produces a homogeneous gel structure with pores. The bioactive compounds are mixed with the solution of calcium, and then the mixture is added dropwise into an alginate bath. This method allows better control of the kinetics, and it is commonly used for delivery systems of bioactive compounds in food [5]. A disadvantage of this method of protecting the antioxidant compounds is that encapsulation hydrophilic of low-molecular substances presents an easy diffusion and fast release through the ionic gel network depending on the value of different physicochemical parameters of the medium (pH, temperature, etc.).

In the literature, different authors have encapsulated plant extracts to protect bioactive compounds against their degradation [3,5,6]. *Hibiscus sabdariffa* L. (HS) (roselle; Malvaceae) is an autogenous plant with around 200 varieties, which is cultivated in different tropical or subtropical areas worldwide [7–9]. This edible plant is an important source of dietary nutrients and it is traditionally consumed for its health and nutritional benefits [9]. Frequently, in the food industry, this plant has also been used in herbal beverages and as a flavoring agent [10]. Roselle composition depends on the flowering stage, but also their different varieties, the part of the plant, genetics, environmental conditions, and ecology. Carbohydrates are the main component (considering the whole plant) (~12–54%) [11], followed by protein (3–27%), fat (0.1–13%), and crude fiber (2.3–27%) [11,12]. Moreover, Roselle is an appreciable source of vitamins, minerals, and bioactive components [13], such as the organic acids or phenolic components, which make this edible plant exploitable in the treatment of various chronic diseases (such as cardiovascular disease, cancer, diabetes, or neurological disorders). Their content in phenolic compounds, which are mainly located in the flower's calyx [14], is principally composed of anthocyanins and flavonoids [12]. De Moura et al. (2018) [6] encapsulated extract from *Hibiscus sabdariffa* L. by the ionic gelation method and evaluated the release of anthocyanins under simulated gastrointestinal conditions. The results of their studies demonstrated that microcapsules protected the bioactive compounds of HS and suggested that this application could be technically feasible.

The aim of this study was to determine the antioxidant capacity and scavenging activity of extracts of *Hibiscus sabdariffa* L. encapsulated in small calcium alginate spheres as a new potential natural source for preserving food products against oxidative damage. In this context, primary and secondary oxidation activity was measured in o/w emulsions in which these spheres were previously incorporated. The radical scavenging activity of these samples was also evaluated by using different methods (TPC, TEAC, DPPH, and ORAC, in four food simulants (water, water with 3% acetic acid, ethanol 50% and, ethanol 100%). The release of the main polyphenolic bioactive compounds of *Hibiscus sabdariffa* L. extract (chlorogenic acid, quercetin and, kaempferol) during the storage of these samples was also quantified. This study responds to the evaluation of the potential viability of these HSL gelling spheres as new ingredients for industrial applications.

2. Materials and Methods

2.1. Materials

2.1.1. Reagents

All reagents were purchased from Sigma Aldrich (Barcelona, Spain) except for HPLC-MS solvents, which were supplied by Panreac (Barcelona, Spain) and alginate (AlgogelTM 3001, INS 401), which was supplied by Algaia (Paris, France).

2.1.2. Plant Material and Preparation of *Hibiscus sabdariffa* L. Extracts

Hibiscus sabdariffa L. plants were provided by a local market in Barcelona. Flowers (from red calyces) were accurately separated from the rest of the plant and dried at ~35 °C for ~1 week, until constant weight. Then, dried flowers were ground with an electric blender (Krups F203, Barcelona, Spain) and sieved. Plant powder (HSP) was then packaged in different amber glass bottles and stored in a desiccator to be preserved from light until the performance of the analyses.

Plant extracts were obtained with ethanol/water (50%, v/v). In of all them, 9.1 g of HSP was extracted at a 1:10 ratio (powder/solvent, w/v), in a dark flask glass, and was stirred using a multiposition magnetic stirrer (Ovan, MM90E, Barcelona, Spain) for 6 h at room temperature (~20 °C); then, after leaving for 30 min, the supernatant was accurately separated after centrifugation (Orto Alresa Mod. Consul, Ajalvir, Madrid, Spain) at 2500 rpm for 15 min, extracts were vacuum filtered, and, subsequently, the supernatant was accurately collected. All supernatants were put in small vials and stored at −80 °C until performing the corresponding analyses.

2.2. *Sphere Preparation and Encapsulation of HSL Extract*

In the current study, two types of spheres were prepared: (type 1) spheres with HSL extract in 50% EtOH and (type 2) spheres with ethanol 50% ethanol and 0.1% blue colorant (to make them visible), without extract, as control. The type of sphere used (with or without HSL extract) depended on the objective of the study. A solution of sodium alginate (S1) was firstly prepared (5 g alginate/L of deionized water), which was left at rest overnight, in order to make the air bubbles disappear. The day after, other homogeneous solutions (S2) were prepared, which consisted of 24.8% of each type of initial solution ((1) or (2)), xanthan gum (0.5%, w/v), calcium chloride (0.25%, w/v), saccharose (9.95%, w/v), and the rest was deionized water (64.5%, w/v). The solution that contained the HSL extract (50% EtOH) was then beaded dropwise onto the alginate solution (S1). For this purpose, we used a device designed for Mym Group (Barcelona) (Figure 1). As shown in Figure 1, after absorbing the solution with alginate (S1), this solution was added dropwise in a tube that contained S2 solution, and the gelation of the S1 rapidly occurred. Spheres were finally washed in a distilled water bath, and they were immediately collected and superficially dried through a cellulose filter at room temperature (~20 °C). Spheres with EtOH 50% (EtOH:H_2O), without HSL extract, were prepared following the same methodology. All spheres were stored at room temperature until use.

2.3. *Stabilization Studies of Spheres*

2.3.1. Physical Characterization of Spheres

The size, morphology, and color of type 1 and type 2 spheres were determined. Visual morphology and color of spheres were estimated by direct observation. After performing the diffusion assay with HSL spheres (type 1), spheres were carefully separated and superficially dried (filter paper) from their corresponding simulants. The weight and size of spheres were measured over the storage time. To determine the size of spheres, the longest and shortest diameters of ten randomly selected units were determined. The diameter was measured by using a caliper, and it was expressed by averaging the longest and shortest diameters. The weight of spheres was determined by means of ten spheres by using an analytical balance.

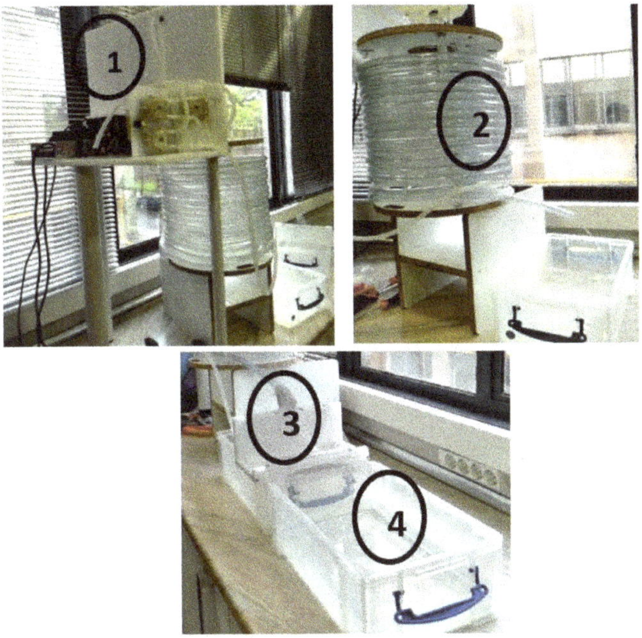

Figure 1. Device for sherification, developed by Mym Group (Barcelona). (**1**) Feeding tube; (**2**) sodium alginate tube (5 m, 3 min); (**3**) collector of sodium alginate solution; and (**4**) bath of distilled water.

2.3.2. Diffusivity Assay

A diffusivity assay of antioxidant bioactive components from HSL spheres was performed by using spheres with HSL extract (50% EtOH) and four different food systems (water, water with acetic acid (3%), pure EtOH, and EtOH 50% (EtOH:H_2O)). This selection was based on the agreement with the Regulation of the European Commission (EU) No.10/2011 on plastic materials and articles intended to come into contact with food. Spheres were immersed with the selected solvent (2 spheres in 1 mL) for a period of time of 72 h and samples were collected every 5 min (the first hour), every hour (until 6 h), and later every 24 h. The food systems (without spheres) were stored at −80 °C until their subsequent analyses. The antiradical activity of solvents was then evaluated according to those exposed in Section 2.3.3. The overall assay was performed in triplicate.

2.3.3. Radical Scavenging Activity

Radical scavenging activity of food simulants was found by determining the total phenolic compound content, Trolox equivalent antioxidant capacity, 2,2-Difenil-1-picrilhidrazil radical scavenging activity, and oxygen radical antioxidant capacity. All analyses were performed in triplicate.

Content of Total Phenolic Compounds

Total phenolic compound (TPC) content was determined by colorimetric spectrophotometry according to the Folin–Ciocalteu (FC) method and considering some modifications [15]. Simulant solvent (7.6%, v/v) was mixed with a FC-reactive (30.8, v/v) sodium carbonate solution at 20% (30.8% v/v) and Milli-Q water (30.8%, v/v). Mixtures were allowed to stand for 1 h at room temperature and preserved from light. The absorbance was measured at 765 nm, at room temperature, with a plate reader UV spectrophotometer (FLUOstar OMEGA, Perkin-Elmer, París, France). TPC was calculated by interpolating the absorbance of the sample against a calibration curve made with a standard solution of gallic acid at fifteen different concentrations (0 to 18 μg/mL) (R^2 = 0.999). The results were expressed as mg gallic acid equivalents (GAE) per mL of solvent (w.s.).

Trolox Equivalent Antioxidant Capacity

The Trolox equivalent antioxidant capacity (TEAC) assay was performed as described by Gallego et al. [16] using a micro-plate reader. Extracts were evaluated by 2,2′-azinobis(3-ethylbenzthiazoline-6-sulfonate) (ABTS) (7 mM). The absorbance of samples was measured by means of a plate reader UV–vis spectrophotometer (FLUOstar OMEGA, Perkin-Elmer, París, France) at 734 nm for 20 min and at 30 °C. TEAC values were obtained according to a calibration curve made with Trolox at eight different concentrations, which ranged from 0.1 to 0.5 mmol Trolox/mL (R^2 = 0.996). The results were expressed as μmol of Trolox equivalents per mL of extract (w.s.).

2,2-Difenil-1-Picrilhidrazil (DPPH) Radical Scavenging Activity Assay

The DPPH radical scavenging activity assay was performed according to Villasante et al. (2020) [17]. Solutions were prepared according to 9.1% (v/v) of the extract with 90.9% (v/v) of 2,2-Diphenyl-1-picrylhydrazyl solution. The absorbance was measured by using a UV–vis spectrophotometer (FLUOstar OMEGA, Perkin-Elmer, Paris, France) at 517 nm by using a micro-plate reader, over 75 min, measuring every 15 min. DPPH values were determined according to a calibration curve, which ranged from 0.02 to 0.3 mmol Trolox/mL (R^2 = 0.993). The results were expressed as mmol of Trolox equivalents per mL of extract (w.s.).

Oxygen Radical Antioxidant Capacity (ORAC)

The ORAC method was adapted from [15]. Samples were previously diluted with PBS at the proportion of 1:50. The assay was performed with an automated fluorescence microplate reader equipped with a temperature-controlled incubation chamber (Fluostar Omega, BMG, Ortenberg, Germany) and incubated at 37 °C. In each microplate, samples were added at the concentration corresponding to 18.2% (w/v), and then fluorescein was added at a concentration corresponding to 63.6% (w/v). AAPH (0.3 M) was finally added at the final proportion of 18.2%, and fluorescein was measured at 2-min intervals for 2 h. The ORAC value was calculated using a regression equation relating to Trolox concentration between 20 and 350 mmol Trolox/mL (R^2 = 0.994). The net area under the fluorescence decay curve in each concentration was also calculated. The results were expressed as mM of Trolox per mL of extract (w.s.).

2.3.4. Characterization of Phenolic Compounds by HPLC

Three different compounds (chlorogenic acid, quercitin, and kaemferol) were analyzed in the four previously mentioned food systems by HPLC-DAD [14] at the moment in which the maximum antiradical activity was reached (once the diffusivity stabilization occurred, at 6 h of storage). An HPLC system (Spectra-Physics 8810, Thermo Electron Corporation, Asheville, NC, USA) which was equipped with a diode array detector (DAD SP 8490 UV-Visible, Thermo Electron Corporation, Asheville, NC, USA) was used. A volume of 75 μL of the extract was filtered (0.22 μm) and injected into an analytical C18 column (Mediterranean 15 cm, Teknokroma, S.C.L. Spain) at 25 °C. The mobile phase was composed of 0.5% formic acid (v/v) in MilliQ water (eluent A) and 0.5% formic acid in acetonitrile (eluent B). The elution gradient corresponded to 95% A and 5% B, under isocratic conditions. The run time was 25 min and the solvent flow rate corresponded to 0.8 mL/min. The detector wavelength was set at 280 nm and 325 nm in order to quantify the corresponding polyphenols, the concentration of which was calculated using specific calibration curves. The results were expressed as ppm of the corresponding phenolic compound. Characterization of these compounds was performed in triplicate.

2.4. *Antioxidant Activity of in o/w Emulsions*

2.4.1. Oil-in-Water Emulsion Preparation

Oil-in-water (o/w) emulsion preparation was as follows: Tween-20 (1%, w/w), Milli-Q water (89%, w/w), and methyl linoleate (MeLo, 10%, w/w) oil were used. Tween-20 and Milli-Q water were previously diluted by using a magnetic stirrer to disperse them. In this mixture, the MeLo oil was added dropwise with continuous sonication, for 10 min, using an ultrasonic homogenizer (Hielscher, UP200S, Teltow, Germany) in an ice bath. In order to evaluate the antioxidant capacity of HSL extract on the emulsions, two different batches of samples were prepared. Spheres (10 or 20 units) (sphere type 1, see Section 2.2) were introduced into their respective plastic containers, and then, 10 g of the emulsion was carefully incorporated (Figure 2). All plastic containers were then allowed to oxidize in an oven, at the same time, and were incubated at 33 °C for 10 days in darkness and with constant slow agitation. A third batch of samples (control samples, sphere type 2, see Section 2.2) was also prepared, in which 20 units of spheres with 50% ethanol were added to 10 g of the emulsion, following the same procedure.

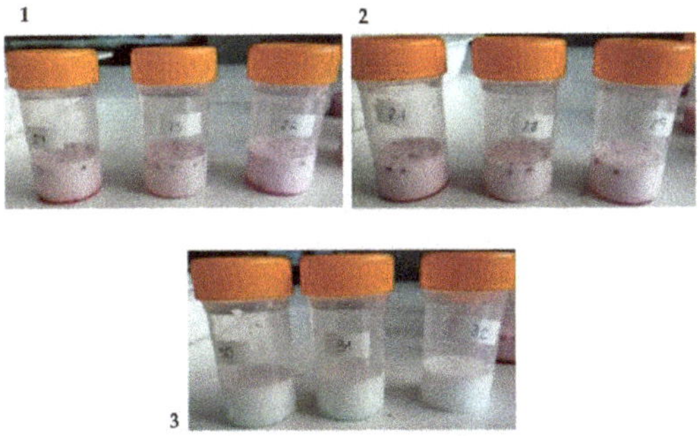

Figure 2. Oil-in-water emulsions with different type of spheres. (**1**) Emulsions with 20 spheres of HSL extract (HSL20), (**2**) emulsions with 10 spheres of HSL extract (HSL10), and (**3**) emulsions with 20 spheres of 50% ethanol (EtOH). Note: HSL: *Hibiscus sabdariffa* L.

2.4.2. Primary Oxidation Measures (Peroxide Value)

The primary oxidation of emulsions was determined by the peroxide value (PV) method, using ferric cyanide [17]. Emulsion drops (of ~19–25 mg of weight) were mixed with 1 mL of absolute EtOH, and the mixture was then completely homogenized with a vortex. In a plastic cuvette, a specific volume of this solution was a mixture with the necessary amount of absolute EtOH to have a final volume of 4 mL, and then 1.8% (v/v) $FeCl_2$ (in 37% HCl) and 1.8% (v/v) of ammonium thiocyanate were added. The blank contained 4 mL of absolute EtOH and 1.8% (v/v) of each reactant. PV values were determined from a calibration curve obtained with the official method to determine the hydroperoxides in pure oil (R^2 = 0.998). The absorbance measure was determined during storage time (10 days), each day, by using a UV–vis spectrophotometer (Zuzi spectrophotometer 4201/20, Auxilab, Navarra, Spain) at 500 nm, and the results are expressed as milli-equivalents of hydroperoxides per kilogram of emulsion (meq hydroperoxides/kg emulsion). Measures were taken in triplicate.

2.4.3. Secondary Oxidation Reactions (Thiobarbituric Acid Reactive Substances, TBARS)

Malondialdehyde (MDA) content was determined according to [18]. A percentage of 8.3% (v/v) of each emulsion was mixed with 8.3% (v/v) of BHA and 83.4% (v/v) of thiobarbituric acid (TBA) solution. Samples were then incubated in a water bath (which was at 95 °C) for 10 min, and then, they were centrifuged (Orto Alresa Mod. Consul, Ajlvir, Madrid, Spain) at 2500 rpm for 5 min. The supernatant of each tube was accurately removed, and the absorbance of this fraction was measured at 531 nm by using a UV–vis spectrophotometer (Zuzi spectrophotometer 4201/20, Auxilab, Navarra, Spain). Results were expressed as milligrams of malondialdehyde per kilogram of emulsion (mg MDA/kg emulsion). Measures were taken in triplicate.

2.4.4. pH Value

Measures of pH were obtained by using a pH meter (GLP21, Criston Instruments, Barcelona, Spain). Measures were performed in triplicate.

2.5. Statistical Analysis

The overall study was performed in triplicate. Statistical analysis was performed with MINITAB 17 software (Minitab, Inc., State College, PA, USA). Data were analyzed using one-way analysis of variance (ANOVA), and Tukey's multiple comparison test was used to determine significant differences between samples, with a value of $p < 0.05$ being considered significant.

3. Results and Discussion

3.1. Morphology, Size and Color of HSL Spheres

Figure 3 shows the images of the spheres obtained in this study, produced with and without HSL extract (50% EtOH). The diameter of all spheres ranged from 2 to 3 mm. No differences were observed among spheres with HSL extract and spheres without the extract. The color of spheres depended on the pigments contained in their corresponding compositions. Spheres elaborated with HSL extract (Figure 3A,B) were red-colored due to their content of the most important pigment of this plant, the anthocyanins [13]. On the contrary, spheres produced only 50% EtOH (Figure 3C,D) were blue-colored because of the blue colorant which was incorporated in their formula to improve their visibility.

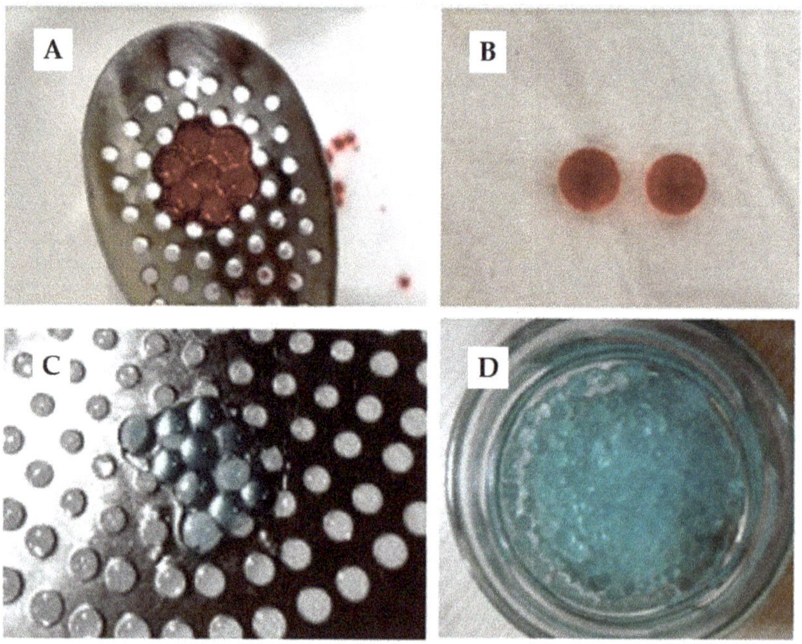

Figure 3. Morphology and size of alginate spheres obtained with (**A**,**B**) HSL extract (50% EtOH) and (**C**,**D**) with blue colorant. Note: HSL: *Hibiscus sabdariffa* L.

As observed in Figure 3, the spheres obtained were morphologically similar, and they presented a homogeneous spherical shape. The high viscosity of the alginate solution probably explains why these particles became spherical. It has been described that the gelation process can be impacted by the concentration of alginate solution [4,19] such that a higher concentration of this biopolymer (1–4%, w/w) molecules increases the number of cross-linking interactions with the bivalent ions and reinforces the structure of the matrix. Similar results were obtained by other authors. Machado et al. (2022) [19] studied the encapsulation of phenolic extracts from spirulina sp. with alginate and reported spheres became more spherical at higher concentrations of sodium alginate. De Moura et al. (2019) [6] also encapsulated bioactive compounds from Hibiscus extract by internal gelation and obtained similar particle-sized spheres (~2 mm). Other physicochemical parameters also impact the gel formation process and determine the characteristics of the sphere, such as the concentration of the calcium ions, the characteristics of the alginate (molecular weight or ratio of M/G units), the processing parameters (pH, time of contact the temperature of processing), and the composition of alginate gels (mixtures of alginate with other polymers or ingredients) [5].

3.2. *Difussivity Assay and Scavenging Activity*

The release of total phenolic compounds from the alginate spheres with HSL extract was evaluated in different mediums: water (W), acidified water with 3% acetic acid (v/v) (AA), pure ethanol (100EtOH), and EtOH 50% (v/v) (50EtOH). Different mediums were set with the objective of simulating some of the most representative food systems. Figure 4 shows the release behavior of TPC in spheres contained in these simulants. As observed, the maximum release of the polyphenolic compounds occurred during the first 6 h. From that moment, this content remained similar in all samples. In the W medium, the maximum value of TPC corresponded to 0.056 mg GAE/mL, which was lower ($p < 0.05$) than the one observed in the other samples. Simulants with ethanol (100EtOH and 50EtOH) presented a similar released TPC content ($p > 0.05$), which corresponded to 0.068 and 0.070 mg GAE/mL, respectively. On the contrary, the AA matrix presented the highest content in TPC (0.080 mg GAE/mL) ($p < 0.05$), which demonstrated that the low pH of the medium improved the release of these bioactive substances from the gelling structure. The diffusion of the polyphenolic substances from the sphere to the external medium could be explained by different release mechanisms, which are highly affected by environmental conditions. Among others, the pH determines the percentage of release in substances encapsulated by alginate gels, which is attributed to the behavior in solubility of the alginate constituents (β-D-mannuronic and α-L-guluronic acids) at different pH values. When the pH value of the external medium is above the pKa value of both acids (3.38 and 3.65, for β-D-mannuronic and α-L-guluronic acids, respectively), the release of internal compounds occurs because of the increase in solubility of the polymeric gel [20].

Concerning the regression results of TPC release in these four mediums (Figure 4), all samples presented a model which was characterized by a calibration line with three different sections: a first section that ranged from 0 to 50 min, a second section that ranged from 50 min to 6 h, and a last section that ranged from 6 to 72 h. In all samples, the first stage was characterized by the highest velocity of the diffusivity phenomena. Data obtained from the regression equations showed that the velocity of the release of TPC (slope of the line) corresponded to water (0.020) < EtOH50% (0.030) < pure ethanol (0.040) ≈ acid water (0.040). The efficiency of ethanol solvents for the selective extraction of polyphenols probably could explain these results. In the second part of the regression line (from 50 min to 6 h), the diffusivity of TPC decreased by ~90% in all mediums, probably because of the saturation of the external medium (Table 1). Finally, from 6 to 72 h of storage (third step), the diffusion of total polyphenols remained constant in all samples, which demonstrated that from 6 h onwards, no further diffusion of compounds occurred. Luong et al. [21] reported similar results in tea (*Camellia chrysantha*) extracts encapsulated in alginate and chitosan nanoparticles. These authors studied the release kinetics of polyphenols from above these nanoparticles depending on the pH of the solution and the different compositions of the hydrogel membrane. In all cases, during the first hour,

50% of total polyphenols were released. A continuous release stage was described during the following 10 h, in which over 90% of total polyphenol content was released from those nanoparticles. From 10 to 30 h, TPC release practically remained stable. These authors also described that TPC released from the nanoparticles in the acidic environment was lower than that observed in neutral solutions. Koksal E. et al. [22] studied the controlled release of vitamin U (S-mehylmethionine) from microencapsulated Brassica oleracea L. extracts with gelatin/sodium alginate (3:5:1) polymers at different pH values. These authors also reported that during the first hour, 50% of total vitamin U was released. However, in that case, it took practically 24 h to reach the maximum value, which indicates the slow release of vitamin U, probably due to the composition of the hydrogel. Maximum release was observed at neutral pH, whereas at acid pH, the diffusion values corresponded to about one-fifth of that value.

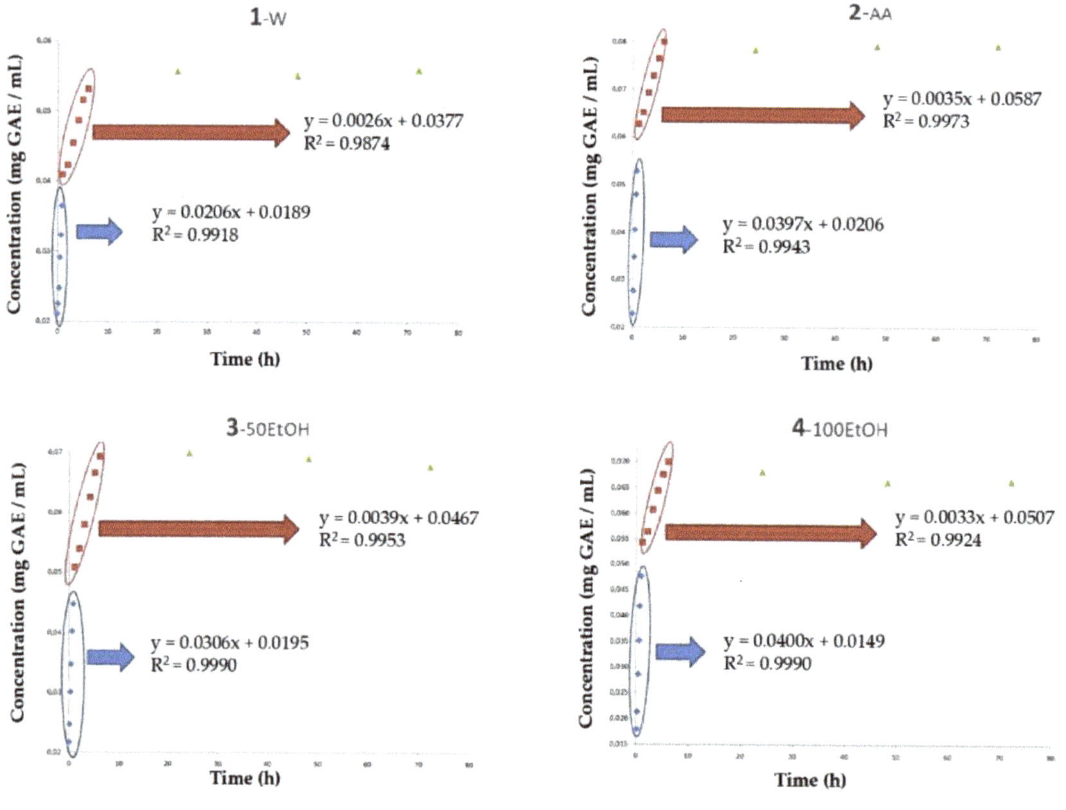

Figure 4. Evolution of total phenolic compounds (TPC) of samples during a storage time of 72 h. Simulants: (**1**) W: water medium, (**2**) AA: acidified water with 3% aceic acid (v/v), (**3**) 50EtOH: ethanol/water 50:50 (v/v), (**4**) 100EtOH: pure ethanol; Concentration: mg gallic acid equivalents (GAE) per mL of extract (w.s.). Time: hours. HSL: *Hibiscus sabdariffa* L.

Table 1. Regression kinetics of the evolution of total phenolic compound (TPC) content of the samples.

Simulants [2]	TPC [1]				
	Section 1		Section 2		Section 3
	Slope [3]	Intercept	Slope	Intercept	Maximum Value [mg GAE/mL] [4]
W	0.0206	0.0189	0.0026	0.0377	0.0560 [c]
AA	0.0397	0.0206	0.0035	0.0587	0.0793 [a]
50EtOH	0.0306	0.0195	0.0039	0.0467	0.0699 [b]
100EtOH	0.0400	0.0149	0.0033	0.0507	0.0682 [b]

[1] TPC: Total phenolic compounds; [2] Simulants: W: water medium, AA: Acidified water with 3% acetic acid (v/v), 50EtOH: ethanol/water 50:50 (v/v), 100EtOH: pure ethanol; [3] Slope: mg GAE/mL × h; [4] Concentration: mg gallic acid equivalents (GAE) per mL of extract (w.s.). [a–c] Values with the same superscript did not present significant differences ($p < 0.05$).

All samples were also evaluated for their scavenging activities by DPPH free radical scavenging, TEAC, and ORAC assays. The results obtained were in accordance with those observed in diffused TPC. Although TEAC and DPPH methods are similar (both evaluate the ability of the antioxidants to neutralize free radicals, using the radical cation $ABTS^{\bullet+}$ in the TEAC method and the radical $DPPH^{\bullet}$ in the DPH method), some differences were observed among samples. Figures 5 and 6 show the evolution of oxidative stability (TEAC and DPPH assays) of all simulants during 75 h of storage. Concerning the TEAC method (Figure 5), along the first section, the slope obtained in AA sample was ~50% lower than the observed in the other simulants, which indicated that in this medium existed a lower antiradical capacity to inhibit the $ABTS^{\bullet+}$ radicals. On the contrary, the simulant that presented the best antiradical capacity corresponded to 50EtOH. From 50 min to 6 h of storage, the AA simulant showed the best radical scavenging capacity. In Section 3, it was observed that no differences were observed among these samples at the end of the storage time (Table 2). A hypothesis that could explain these results in the acidified medium is that they are not due so much to the antiradical capacity as to the fact that the acidic medium prevents an adequate development of the method.

Regarding the DPPH method, during the first section of the regression line, results obtained showed that the 100EtOH simulant presented the highest slope if compared with the other simulants, which could be attributed to the highest antiradical efficiency of this sample with $DPPH^{\bullet}$ radical. In the second section, simulants that presented the best antiradical capacity corresponded to 50EtOH, followed by AA and 100EtOH simulants, which exhibited similar results (Figure 6). As observed in Figure 6 and Table 3, W simulant exhibited the worst antiradical capacity. Concentration values of DPPH reached once the radical scavenging products in the simulants were saturated demonstrated that the released in water was lower and, for this reason, the W simulant exhibited the worst antiradical capacity. While, on the contrary, the AA, 50EtOH, and 100EtOH simulants exhibited similar results ($p < 0.05$). These samples reached concentrations of Trolox ~ 31% higher than those observed in W.

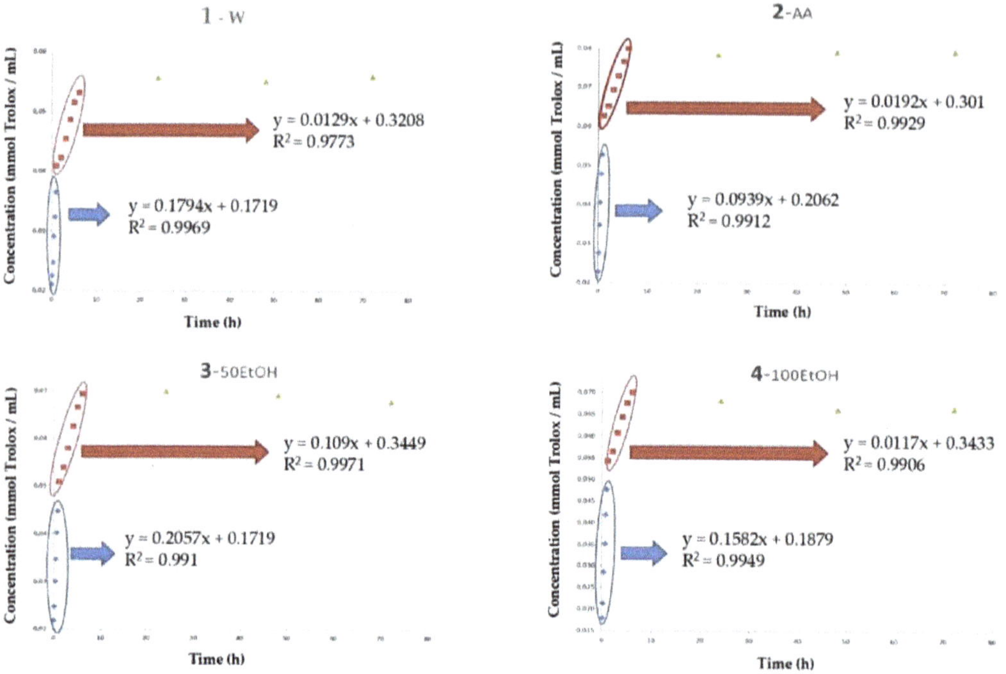

Figure 5. Evolution of Trolox equivalent antioxidant capacity (TEAC) of samples during a storage time of 72 h. Simulants: (**1**) W: water medium, (**2**) AA: acidified water with 3% acetic acid (v/v), (**3**) 50EtOH: ethanol/water 50:50 (v/v), (**4**) 100EtOH: pure ethanol; Concentration: mmol Trolox per mL of extract (w.s.); Time: hours; HSL: *Hibiscus sabdariffa* L.

Table 2. Regression kinetics of the evolution of Trolox equivalent antioxidant capacity (TEAC) of the samples.

Simulants [2]	TEAC [1]				
	Section 1		Section 2		Section 3
	Slope [3]	Intercept	Slope [3]	Intercept	Maximum Value [mmol Trolox/mL] [4]
W	0.1794	0.1719	0.0129	0.3208	0.4177 [a]
AA	0.0939	0.2062	0.0192	0.301	0.4370 [a]
50EtOH	0.2057	0.1719	0.0109	0.3449	0.4222 [a]
100EtOH	0.1582	0.1879	0.0117	0.3433	0.4164 [a]

[1] TEAC: Trolox equivalent antioxidant capacity; [2] Simulants = W: water medium, AA: acidified water with 3% acetic acid (v/v), 100EtOH: pure ethanol, 50EtOH: ethanol/water 50:50 (v/v); [3] Slope: mmol Trolox/mL × h; [4] Concentration: mmol of Trolox/mL of extract (w.s.). [a] Values with the same superscript did not present significant differences ($p < 0.05$).

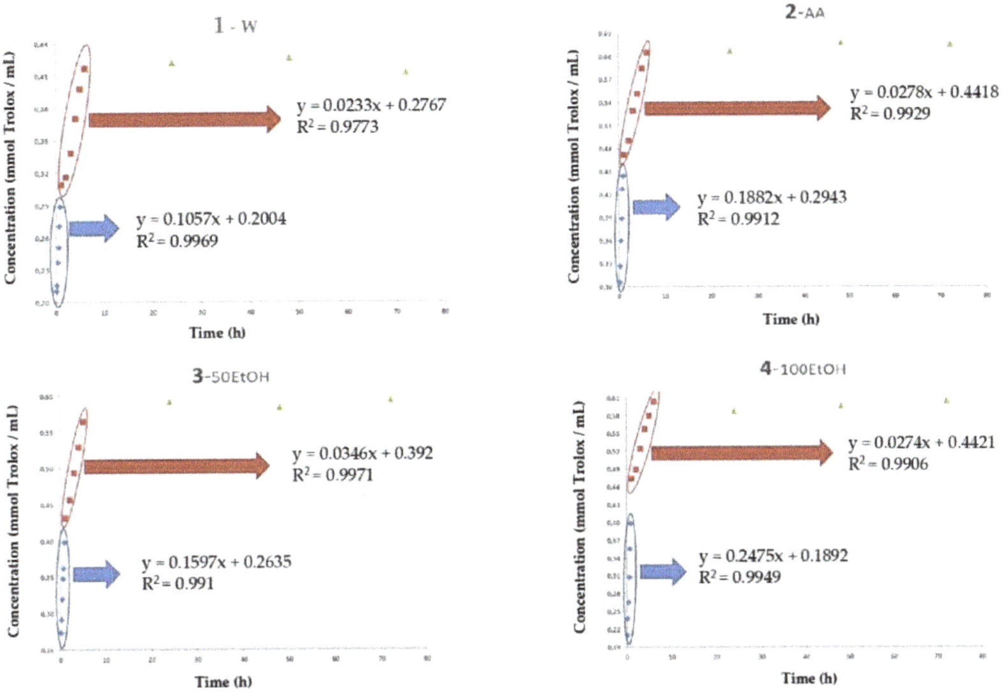

Figure 6. Evolution of 2,2-difenil-1-picrilhidracilo (DPPH) of samples during a storage time of 72 h. Simulants: (**1**) W: water medium, (**2**) AA: acidified water with 3% aceic acid (v/v), (**3**) 50EtOH: ethanol/water 50:50 (v/v), (**4**) 100EtOH: pure ethanol; Concentration: mmol Trolox per mL of extract (w.s.); Time: hours; HSL: *Hibiscus sabdariffa* L.

The ORAC assay measures the global capacity of antioxidants to neutralize peroxide radicals. In this case, since this test is highly sensitive to pH, the ORAC method was not carried out on the samples that contained acidified water. Results obtained with this method also presented some differences in the sections of the regression model. Concerning the first section, the solvent that presented the best scavenging activity corresponded to the W, followed by 50EtOH and 100EtOH (Figure 7 and Table 4). Constrariously, in the second and third sections, 100EtOH medium exhibited the best results.

Table 3. Regression kinetics of the evolution of Trolox equivalent antioxidant capacity (DPPH) of the samples.

Simulants [2]	DPPH [1]				
	Section 1		Section 2		Section 3
	Slope [3]	Intercept	Slope [3]	Intercept	Maximum Value [mmol Trolox/mL] [4]
W	0.1057	0.2004	0.0233	0.2767	0.4268 [b]
AA	0.1882	0.2943	0.0278	0.4418	0.6176 [a]
50EtOH	0.1597	0.2635	0.0346	0.392	0.5936 [a]
100EtOH	0.2475	0.1892	0.0274	0.4421	0.6050 [a]

[1] DPPH: 2,2-difenil-1-picrilhidracilo; [2] Simulants = W: water medium, AA: acidified water with 3% acetic acid (v/v), 100EtOH: pure ethanol, 50EtOH: ethanol/water 50:50 (v/v); [3] Slope: mmol Trolox/mL × h; [4] Concentration: mmol of Trolox/mL of extract (w.s.). [a,b] Values with the same superscript did not present significant differences ($p < 0.05$).

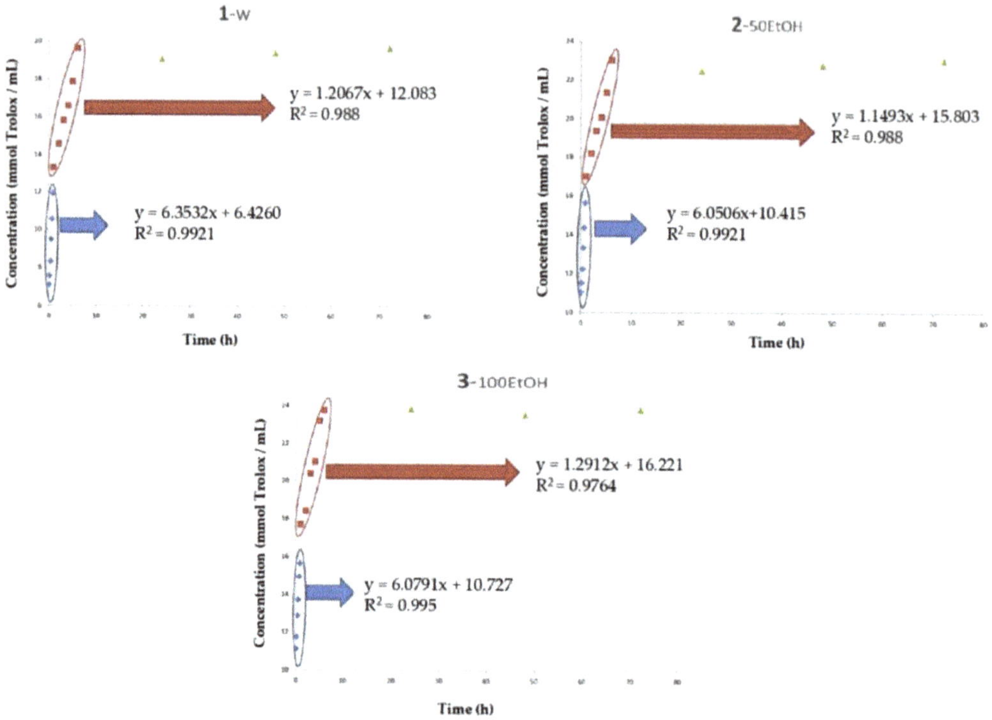

Figure 7. Evolution of the oxygen radical absorption capacity (ORAC) of samples during a storage time of 72 h. Simulants: (**1**) W: water medium, (**2**) AA: acidified water with 3% acetic acid (v/v), (**3**) 100EtOH: pure ethanol, 50EtOH: ethanol/water 50:50 (v/v); Concentration: mmol Trolox per mL of extract (w.s.); Time: hours; HSL: *Hibiscus sabdariffa* L.

Considering the overall results, the anti-radical capacity of the simulant that contained ethanol exhibited the best results. On the contrary, the medium with water was presented as the worst. Polyphenolic compounds seem to have better antiradical activity in polar mediums because of their possible dissociation [15]. Alkaline conditions also may enhance their power as radical scavengers. The best scavenging activity was observed in all simulants after 6 h of storage, which remained constant in all samples until the end of their storage time, probably because of the stabilization in the diffusion of TPC after 6 h.

Table 4. Regression kinetics of the evolution of oxygen radical absorption capacity (ORAC) of the samples.

Simulants [2]	ORAC [1]				
	Section 1		Section 2		Section 3
	Slope [3]	Intercept	Slope [3]	Intercept	Maximum Value [mmol Trolox/mL] [4]
W	6.3532	6.426	1.2067	12.083	19.6686 [b]
50EtOH	6.0506	10.415	1.1493	15.803	23.0273 [a]
100EtOH	6.0791	10.727	1.2912	16.221	23.8159 [a]

[1] ORAC: oxygen radical absorption capacity; [2] Simulants = W: water medium, AA: acidified water with 3% acetic acid (v/v), 100EtOH: pure ethanol, 50EtOH: ethanol/water 50:50 (v/v); [3] Slope: mmol Trolox/mL × h; [4] Concentration: mmol of Trolox/mL of extract (w.s.). [a,b] Values with the same superscript did not present significant differences ($p < 0.05$).

In order to calculate the migration through the hydrogel of the individual compounds, the concentration of three individual compounds (chlorogenic acid, quercetin, and kaempferol) in the simulant was determined by HPLC-DAD. The moment in which, based on the results obtained previously, the maximum had been reached, close to saturation, was chosen. As observed (Table 5), chlorogenic acid was the most abundant phenolic compound (initial HSL extract and in all simulants), followed by quercetin and kaempferol. In the literature, it has been described that more than 95 phenolic compounds have been identified in this plant, of which quercetin, kaempferol, and chlorogenic acid are some of the most frequently reported [6,10,23]. Among others, the presence of these substances has been strongly correlated with the nutraceutical value and medical properties of these extracts [12,24]. Spheres immersed in 100EtOH simulant exhibited the highest release of chlorogenic acid ($p < 0.05$), probably because of the best extraction affinity of phenolic acids for ethanol [23]. Similar results were observed in the chlorogenic acid release in W and AA mediums (Table 5). Obouayeba et al. (2013) [25] also observed the major extraction yields of flavonoids of HSL in mediums with methanol. On the contrary, the amounts of quercetin and kaempferol increased in spheres immersed in water, probably because of the high solubility of both components in this medium. The results showed that quercetin and kaempferol losses decreased ($p < 0.05$) according to W > AA > 100EtOH > EtOH:W.

Table 5. Concentration of HSL extract and percentage of loss (from the interior to the simulant, in brackets) of chlorogenic acid, quercetin, and kaempferol in the spheres after 6 h of contact. The percentage was calculated taking into account the initial concentration (inside the sphere) and the final concentration (in the simulant) and the volume of simulant.

Reference [1]	Concentration [2] (% Loss [3])		
	Chlorogenic Acid	Quercetin	Kaempferol
HSL extract	73.911	13.583	1.332
HSL in W	0.919 (37.6)	0.083 (69.2)	0.006 (75.6)
HSL in AA	1.049 (28.8)	0.109 (59.6)	0.010 (63.8)
HSL in EtOH:W	0.766 (48.0)	0.123 (54.7)	0.013 (51.7)
HSL in 100EtOH	0.291 (80.2)	0.116 (57.0)	0.011 (57.9)

[1] W: water medium, AA: acidified water with 3% acetic acid (v/v), 100EtOH: pure ethanol, EtOH:W: ethanol/water 50:50 (v/v); HSL: *Hibiscus sabdariffa* L.; [2] Concentration: ppm.; [3] % loss: percentage of loss.

3.3. Changes of Diameter and Weight of HSL Spheres Contained in Different Simulants with Storage

The percentage of variation in diameter and weight of spheres is shown in Figures 8 and 9. As observed, variation of both parameters depended on the time that spheres remained in contact with each simulant, and also on the composition of the simulant (water, acidified water, ethanol at 50%, or pure ethanol).

As observed in Figures 8 and 9, the highest differences in variation of weight of spheres corresponded to those particles contained in mediums with ethanol (50EtOH and 100EtOH), probably because of the higher volatility of ethanol in comparison with the other solvents. Spheres immersed in 100EtOH medium showed the biggest changes. On the contrary, spheres immersed in water and acidified water did not present significant ($p < 0.05$) differences if comparing them. Although from 6 to 72 h the diameter and the weight of spheres were still changing, no diffusion of their internal liquid was observed in any simulant.

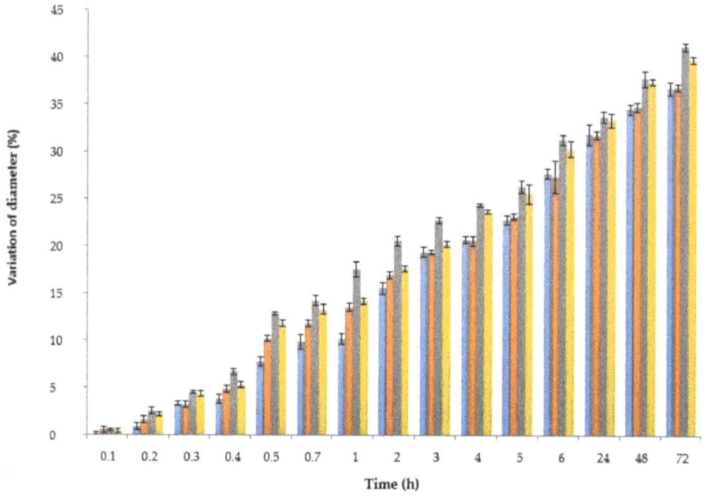

Figure 8. Variation of diameter (%) of spheres with HSL extract during their storage. Water (•), acidified water 3% (•), ethanol 50% (•), and ethanol 100% (•). Values expressed corresponded to the average values and vertical bats to their corresponding standard deviation (n = 3).

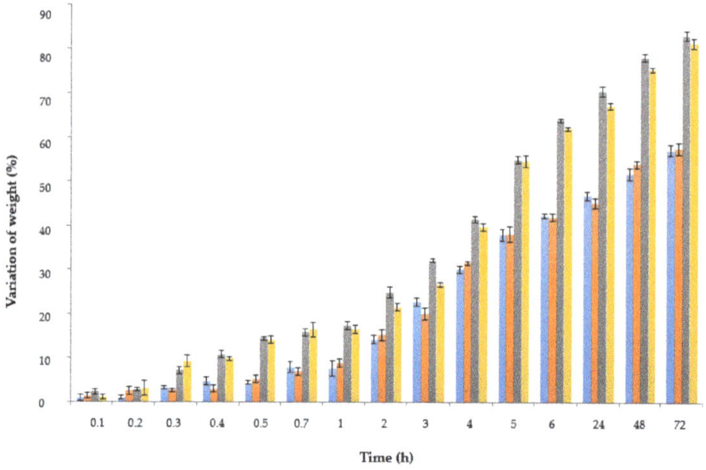

Figure 9. Variation of weight (%) of spheres with HSL during their storage. Water (•), acidified water 3% (•), ethanol 50% (•), and ethanol 100% (•). Values expressed corresponded to the average values and vertical bats to their corresponding standard deviation (n = 3).

3.4. *Antioxidant Activity of Spheres in o/w Emulsions*

The evaluation of primary lipid oxidative reactions was carried out through a peroxide value assay, which was performed during a storage time of 240 h. In the beginning, all emulsions presented a low content of peroxides (<5 meq hydroperoxide/kg emulsion), which did not differ ($p > 0.05$) among them. After 60 h, results showed that samples with higher values of primary oxidation products corresponded to the control (sample without HSL extract) and the emulsion with spheres of 50% ethanol (100EtOH). In these samples, PV increased according to the storage time, which corresponded to 236 ± 8.6 and 224 ± 20.4 meq. hydroperoxide/kg emulsion, respectively, at the end of the storage time. On the contrary, both emulsions that contained spheres of HSL extract (10HSL and 20HSL) showed the best oxidative stability. After ~154 h of storage, 10HSL and 20HSL samples remained stable against oxidation (Figure 10), although, from this point, these emulsions exhibited an exponential increase in their corresponding PV. No differences ($p > 0.05$) were observed between 10HSL and 20HSL samples after 240 h of storage (51.9 ± 0.6 and 52.0 ± 0.4 meq. hydoperoxides/kg emulsion, respectively). As observed in Figure 10, HSL spheres (20HSL and 10HSL) reduced the formation of hydroperoxides by ~80% if compared with the control emulsion (without spheres and 100EtOH).

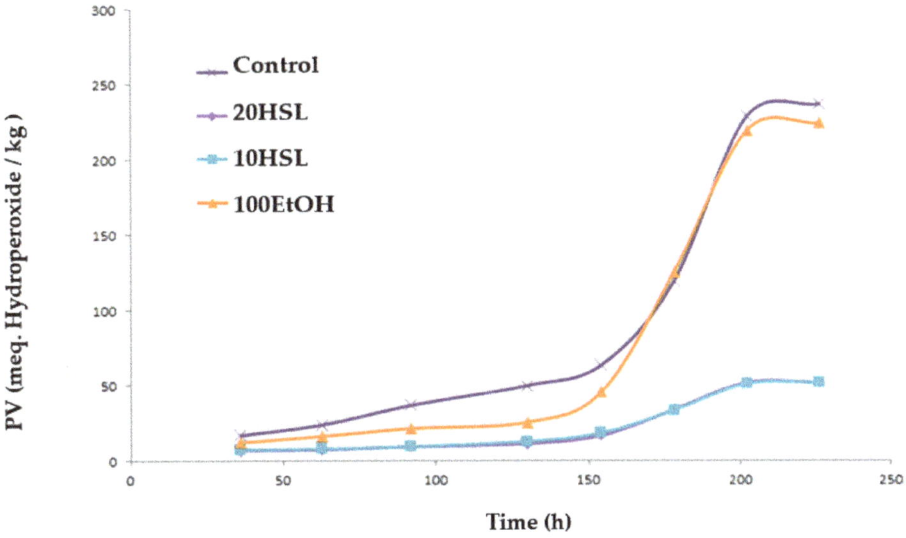

Figure 10. Evolution of peroxide value of emulsions during their storage at 33 °C. Control: emulsion without anything; 100EtOH: emulsion with spheres of pure 50% ethanol; 10HSL: emulsions with 10 spheres of HSL extract; 20HSL: emulsions with 20 spheres of HSL extract. HSL: *Hibiscus sabdariffa* L. Emulsion: Tween-20 (1%, w/w), Milli-Q water (89%, w/w) and methyl linoleate (MeLo, 10%, w/w) oil. PV: Peroxide value, meq. Hydroperoxide/kg emulsion.

Secondary oxidative products were monitored through TBARS assay at the initial time and after 96 h of the preparation of the samples. In the beginning, all samples presented similar contents of thiobarbituric-acid-reactive substances, although emulsions with spheres with HSL extract (10HSL and 20HSL) showed the lowest values. During the storage, the control exhibited a remarkable increase of TBARS, while these compounds showed a slower increase in emulsions with spheres (HSL or EtOH). After ~300 h of storage, 20HSL and 10HSL samples exhibited the highest inhibitory activity towards the formation of TBARS, followed by 10HSL > EtOH > Control (Figure 11). These results could also be attributed to the least hydroperoxide accumulation in 20HSL and 10HSL samples. Phenolic compounds contained in HSL extracts (such as chlorogenic acid, quercetin, and kaempferol) could interfere with the oxidative cycle of these emulsions [20,26]. These natural antioxidants probably played an important role in inhibiting the breakdown of hydroperoxides to TBARS, and supporting the previous findings concerning the potential antioxidant activity of the extracts of these plants against food systems [10,26]. Figure 11 shows that the formation of TBARS products was reduced by ~20% when HSL spheres were included in o/w emulsions if compared with the control.

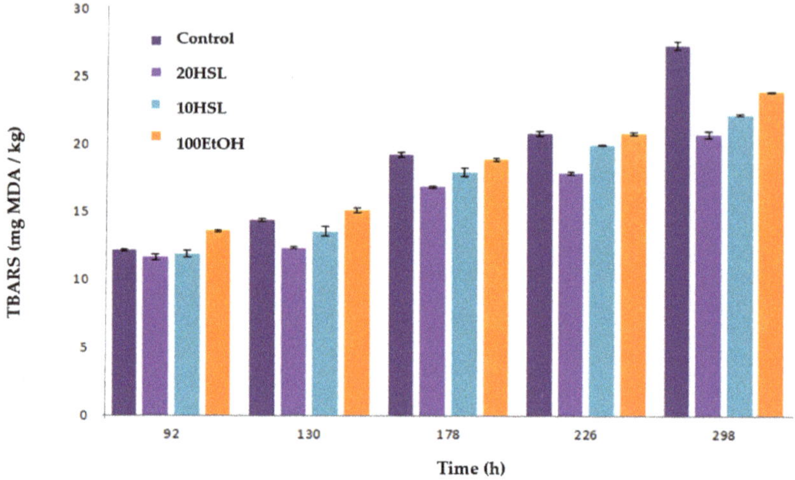

Figure 11. Evolution of thiobarbituric-acid-reactive substances (TBARS) of emulsions during their storage at 33 °C. Control: emulsion without anything; 100EtOH: emulsion with spheres of 50% ethanol; 10HSL: emulsions with 10 spheres of HSL extract, 20HSL: emulsions with 20 spheres of HSL extract. HSL: *Hibiscus sabdariffa* L. Emulsion: Tween-20 (1%, w/w), Milli-Q water (89%, w/w) and methyl linoleate (MeLo, 10%, w/w) oil. mg MDA/kg: mg of malondialdehyde/kg emulsion; error bars represent the standard deviation (n = 3).

The evolution of the pH value of emulsions with HSL spheres exhibited a similar and slight decrease during their storage time (from 2.57 ± 0.30 to 2.11 ± 0.09 in 10HSL and from 2.60 ± 0.13 to 2.06 ± 0.06 in 20HSL). On the contrary, a greater decrease in pH values of the control (from 2.02 ± 0.21 to 1.61 ± 0.16) and the emulsions with EtOH spheres (from 2.38 ± 0.20 to 1.76 ± 0.26) was observed, probably because of the new by-products generated as a consequence of the propagation of oxidation reactions. At lower pH values, the velocity of lipid oxidative reactions tends to accelerate, since the antioxidants could lose their effectiveness [27,28].

4. Conclusions

This study demonstrated the antioxidant capacity and scavenging activity of extracts of *Hibiscus Sabdariffa* L. (HSL) (with 50% ethanol) encapsulated in calcium alginate spheres. These spheres improved the oxidative stability of o/w emulsions (by reducing the produced substances of primary and secondary oxidation reactions), which directly depended on the number of HSL spheres included in these emulsions. The composition of the external medium was demonstrated to be an important factor to determine the physical stability of the spheres (diameter and weight) and also the release of the phenolic compounds contained in these alginate spheres. The scavenging capacity of simulants with spheres that contain HSL extracts was also demonstrated, which directly depends on the composition of the medium in which the spheres are dispersed. This work also indicated the potential of alginate spheres for encapsulating antioxidant compounds as a disruptive strategy of innovation for industrial applications. Nowadays, the encapsulation of bioactive compounds from plants is a potential tendency of industry in order to protect these substances through innovative delivery systems. This study sought to complement the current state of the art in order to overcome the above-mentioned challenges.

Author Contributions: Conceptualization, P.B. and M.P.A.; data curation, M.V.P.; formal analysis, P.B. and M.V.P.; funding acquisition, P.B.; investigation, M.P.A. and I.C.-T.; methodology, P.B.; project administration, M.P.A.; resources, P.B. and I.C.-T.; supervision, M.P.A.; validation, P.B., and I.C.-T.;

writing—original draft, P.B., I.C.-T.; writing—review and editing, M.P.A. and I.C.-T. All authors have read and agreed to the published version of the manuscript.

Funding: This research received no external funding.

Institutional Review Board Statement: Not applicable.

Informed Consent Statement: Not applicable.

Data Availability Statement: The data presented in this study are available on request from the corresponding author.

Acknowledgments: To Selene Rosado for her indispensable work in the laboratory.

Conflicts of Interest: The authors declare no conflict of interest.

References

1. Loi, M.; Paciolla, C. Plant antioxidants for food safety and quality: Exploring new trends of research. *Antioxidants* **2021**, *10*, 972. [CrossRef]
2. Lourenço, S.C.; Moldão-Martins, M.; Alves, V.D. Antioxidants of natural plant origins: From sources to food industry applications. *Molecules* **2019**, *24*, 4132. [CrossRef] [PubMed]
3. Nedovic, V.; Kalusevic, A.; Manojlovic, V.; Levic, S.; Bugarski, B. An overview of encapsulation technologies for food applications. *Procedia Food Sci.* **2011**, *1*, 1806–1815. [CrossRef]
4. Bennacef, C.; Desobry-Banon, S.; Probst, L.; Desobry, S. Advances on alginate use for spherification to encapsulate biomolecules. *Food Hydrocoll.* **2021**, *118*, 106782. [CrossRef]
5. Mohammadalinejhad, S.; Kurek, M.A. Microencapsulation of Antocyanins. Critical review of techniques and wall materials. *Appl. Sci.* **2021**, *11*, 3936. [CrossRef]
6. De Moura, S.C.S.R.; Berling, C.L.; Germer, S.P.M.; Alvim, I.D.; Hubinger, M.D. Encapsulating anthocyanins from *Hibiscus sabdariffa* L. calyces by ionic gelation: Pigment stability during storage of microparticles. *Food Chem.* **2018**, *241*, 317–327. [CrossRef]
7. Owoade, O.; Adetutu, A.; Olorunnisola, O. A review of chemical constituents and pharmacological properties of *Hibiscus sabdariffa* L. *Int. J. Curr. Res. Biosci. Plant Biol.* **2019**, *6*, 42–51. [CrossRef]
8. Amer, S.A.; Al-Khalaifah, H.; Gouda, A.; Osman, A.; Goda, N.I.A.; Mohammed, H.A. Potential Effects of Anthocyanin-Rich Roselle (*Hibiscus sabdariffa* L.). Extract on the Growth, Intestinal Histomorphology, Blood Biochemical Parameters, and the Immune Status of Broiler Chickens. *Antioxidants* **2022**, *11*, 544. [CrossRef]
9. Salami, S.O.; Afolayan, A.J. Evaluation of nutritional and elemental compositions of green and red cultivars of roselle: *Hibiscus sabdariffa* L. *Sci. Rep.* **2021**, *11*, 1030. [CrossRef]
10. Da-Costa-Rocha, I.; Bonnlaendr, B.; Sievers, H.; Pischel, I.; Heinrich, M. *Hibiscus sabdariffa* L.—A phytochemical and pharmacological review. *Food Chem.* **2014**, *165*, 424–443. [CrossRef]
11. Yuniati, Y.; Elim, P.E.; Alfanaar, R.; Kusuma, H.S.; Mahfud. Extraction of anthocyanin pigment from *Hibiscus Sabdariffa* L. by ultrasonic-assisted extraction. In *IOP Conference Series: Materials Science and Engineering*; IOP Publishing: Bristol, UK, 2021; Volume 1010.
12. Peredo Pozos, G.I.; Ruiz-López, M.A.; Zamora-Nátera, J.F.; Álvarez-Moya, C.; Barrientos Ramírez, L.; Reynoso Silva, M. Antioxidant Capacity and Antigenotoxic Effect of *Hibiscus sabdariffa* L. Extracts Obtained with Ultrasound-Assisted Extraction Process. *Appl. Sci.* **2020**, *10*, 560. [CrossRef]
13. Hiromi, Y.; Goro Kajimoto, S.E. Antioxidant effects of d-tocopherols at different concentrations in oils during microwave heating. *J. Am. Oil Chem. Soc.* **1993**, *70*, 989–995.
14. Wang, J.; Cao, X.; Qi, Y.; Ferchaud, V.; Chin, K.L.; Tang, F. High-performance thin-layer chromatographic method for screening antioxidant compounds and discrimination of *Hibiscus sabdariffa* L. by principal component analysis. *JPC–J. Planar Chromatogr. –Mod. TLC* **2015**, *28*, 274–279. [CrossRef]
15. Ouerfelli, M.; Bettaieb, B.K.; Almajano, M.P. Radical Scavenging and Antioxidant Activity of Anthyllis Vulneraria Leaves and Flowers. *Molecules* **2018**, *23*, 1657. [CrossRef]
16. Gallego, M.G.; Gordon, M.H.; Segovia, F.J.; Skowyra, M.; Almajano, M.P. Antioxidant Properties of Three Aromatic Herbs (Rosemary, Thyme and Lavender) in Oil-in-Water Emulsions. *J. Am. Oil Chem.* **2013**, *90*, 1559–1568. [CrossRef]
17. Villasante, J.; Codina, E.; Hidalgo, G.I.; de Ilarduya, A.M.; Muñoz-Guerra, S.; Almajano, M.P. Poly (α-dodecyl γ-glutamate) (PAAG-12) and polylactic acid films charged with α-tocopherol and their antioxidant capacity in food models. *Antioxidants* **2019**, *8*, 284. [CrossRef]
18. Grau, A.; Guardiola, F.; Boatella, J.; Barroeta, A.; Codony, R. Measurement of 2-thiobarbituric acid values in dark chicken meat through derivative spectrophotometry: Influence of various parameters. *J. Agric. Food Chem.* **2000**, *48*, 1155–1159. [CrossRef]
19. Machado, A.R.; Silva, P.M.P.; Vicente, A.A.; Souza-Soares, L.A.; Pinheiro, A.C.; Cerqueira, M.A. Alginate Particles for Encapsulation of Phenolic Extract from Spirulina sp. LEB-18: Physicochemical Characterization and Assessment of In Vitro Gastrointestinal Behavior. *Polymers* **2022**, *14*, 4759. [CrossRef]

20. Jabeur, I.; Pereira, E.; Caleja, C.; Calhelha, R.C.; Soković, M.; Catarino, L. Exploring the chemical and bioactive properties of: *Hibiscus sabdariffa* L. calyces from Guinea-Bissau (West Africa). *Food Funct.* **2019**, *10*, 2234–2243. [CrossRef]
21. Luong, P.H.; Nguyen, T.C.; Pham, T.D.; Tran, D.M.T.; Ly, T.N.L.; Vu, Q.T.; Tran, T.K.N.; Thai, H. Preparation and assessment of some characteristics of nanoparticles based on sodium alginatet, chitosan and *Camellia Chrysantha* polyphenols. *Int. J. Polym. Sci.* **2021**, *2021*, 5581177. [CrossRef]
22. Koksal, E.; Gode, F.; Ozaltin, K.; Karakurt, I.; Suly, P.; Saha, P. Controlled release of vitamin U from microencapsulated brassica oleracea L. var. capitata extract for peptic ulcer treatment. *Food Bioproc. Tech.* **2023**, *16*, 677–689. [CrossRef]
23. Piovesana, A.; Rodrigues, E.; Zapata Noreña, C.P. Composition analysis of carotenoids and phenolic compounds and antioxidant activity from hibiscus calyces (*Hibiscus sabdariffa* L.) by HPLC-DAD-MS/MS. *Phytochem. Anal.* **2019**, *30*, 208–2017. [CrossRef] [PubMed]
24. Hapsari, B.W.; Manikharda Setyaningsih, W. Methodologies in the analysis of phenolic compounds in roselle (*Hibiscus sabdariffa* L.): Composition, biological activity, and beneficial effects on human health. *Horticulturae* **2021**, *7*, 35. [CrossRef]
25. Obouayeba, A.P.; Djyh, N.B.; Sekou, D.; Djaman, A.J.; N'guessan, J.D.; Kone, M.; Kouakou, T.H.P. No Titl Phytochemical and Antioxidant Activity of Roselle (*Hibiscus sabdariffa* L.) Petal Extracts. *Res. J. Pharm. Biol. Chem. Sci.* **2013**, *4*, 1694–1720.
26. Neacsu, M.; Vaughan, N.; Raikos, V.; Multari, S.; Duncan, G.J.; Duthie, G.G. Phytochemical profile of commercially available food plant powders: Their potential role in healthier food reformulations. *Food Chem.* **2015**, *179*, 159–169. [CrossRef]
27. Gustav, K.M.; Haroon, E.T.; Osei-Kwarteng, M.; Abdalbasit, A.M.; Onyago Gweyi, J. Food use of whole and extracts of *Hibiscus sabdariffa*. In *Roselle (Hibiscus sabdariffa)*; Academic Press: Cambridge, MA, USA, 2021; pp. 123–136.
28. Edwin, N.F.; Shu-Wen, H.; Aeschbach, R.; Prior, E. Antioxidant Activity of a Rosemary Extract and Its Constituents, Carnosic Acid, Carnosol, and Rosmarinic Acid, in Bulk Oil and Oil-in-Water Emulsion. *J. Agric. Food Chem.* **1999**, *44*, 131–135.

polymers MDPI

Review

Recent Advances in the Application of ATRP in the Synthesis of Drug Delivery Systems

Matylda Szewczyk-Łagodzińska *, Andrzej Plichta, Maciej Dębowski, Sebastian Kowalczyk, Anna Iuliano and Zbigniew Florjańczyk *

Faculty of Chemistry, Warsaw University of Technology, Noakowskiego 3, 00-664 Warsaw, Poland
* Correspondence: matylda.lagodzinska.dokt@pw.edu.pl (M.S.-Ł.); zbigniew.florjanczyk@pw.edu.pl (Z.F.)

Abstract: Advances in atom transfer radical polymerization (ATRP) have enabled the precise design and preparation of nanostructured polymeric materials for a variety of biomedical applications. This paper briefly summarizes recent developments in the synthesis of bio-therapeutics for drug delivery based on linear and branched block copolymers and bioconjugates using ATRP, which have been tested in drug delivery systems (DDSs) over the past decade. An important trend is the rapid development of a number of smart DDSs that can release bioactive materials in response to certain external stimuli, either physical (e.g., light, ultrasound, or temperature) or chemical factors (e.g., changes in pH values and/or environmental redox potential). The use of ATRPs in the synthesis of polymeric bioconjugates containing drugs, proteins, and nucleic acids, as well as systems applied in combination therapies, has also received considerable attention.

Keywords: drug delivery system; drug nanocarrier; micelle; ATRP; block copolymer; polymer–drug conjugate; polymersome; polyplexes; self-assembly; branched copolymer

check for updates

Citation: Szewczyk-Łagodzińska, M.; Plichta, A.; Dębowski, M.; Kowalczyk, S.; Iuliano, A.; Florjańczyk, Z. Recent Advances in the Application of ATRP in the Synthesis of Drug Delivery Systems. *Polymers* **2023**, *15*, 1234. https://doi.org/10.3390/polym15051234

Academic Editors: Marek Kowalczuk, Iza Radecka and Barbara Trzebicka

Received: 6 February 2023
Revised: 26 February 2023
Accepted: 27 February 2023
Published: 28 February 2023

1. Introduction

Polymeric systems for controlled drug release have been the subject of intensive academic and industrial research for more than half a century, with the aim of extending the period of therapeutic action, enabling drug delivery to a specific site in the body, and reducing the negative side effects induced by bioactive substances. The mechanism of action of these systems is very diverse, ranging from the protection of active pharmaceutical ingredients from an aqueous living environment for a programmed time period to the targeting and formation of conjugates with polymers covalently attached to drugs to increase their stability and immunogenicity [1–7]. Most of the recent studies have focused on biocompatible nanocarriers, such as micelles, vesicles (polymersomes), nanogels, dendrimers, and hybrid nanoparticles with porous inorganic cores [8–21]. The reduction in polymeric containers to submicron sizes (typically 10 to 200 nm) allows for the delivery of drugs in the form of a stable colloidal dispersion, hence promoting more efficient absorption of therapeutic loads when compared with larger carriers. In addition, the large surface area of nanoparticles provides space for functionalization, targeting, and bioconjugation. Therefore, nanocarriers can also exhibit other desirable properties, such as biodegradability, the ability to change size and permeability under the influence of external stimuli (e.g., temperature, pH), or the formation of multicompartment containers, in which the simultaneous loading and release of different drugs are possible [10]. The simplest and most common method of assembling polymers into well-defined nanoparticles is the self-organization of amphiphilic block copolymers in water or water–oil systems. By adjusting the macromolecular properties of the copolymers and the self-organization conditions, various nanostructures can be formed, including separated micelles or polymersomes, and cluster aggregates. Amphiphilic block copolymers are also often used in resorbable implantable plates, containing a drug with targeted cytotoxicity that can be inserted directly into the area altered by pathological cells [7].

Modern polymer chemistry offers many synthetic strategies that can be used to prepare block copolymers with specific macromolecular architecture, composition, functionality, and low-molecular-weight dispersity. These include various types of living and controlled polymerization, often combined with post-polymerization selective chemical modification of the terminal functional groups [21]. In the past two decades, controlled radical polymerization (CRP) methods, particularly reversible addition–fragmentation chain transfer (RAFT) polymerization and atom transfer radical polymerization (ATRP) [22–33], have become some of the leading tools in the synthesis of block copolymers. They can be carried out in a variety of solvents, including water, and are tolerant of most functional groups. The CRP method is typically used to obtain segments derived from active vinyl monomers (e.g., (meth)acrylates, (meth)acrylamides, styrene derivatives), which are combined with other synthetic or natural polymers employing end groups that can act as CRP initiators or reactants in "click" reactions. This strategy has been successfully applied to build multi-block linear chains and to construct macromolecules with a more complex topology, such as star-like polymers, comb-like polymers, hyperbranched structures, networks, and hybrids, with a covalent bond attachment of well-defined functional polymers to therapeutics or other polymer-grafted materials. Many attempts have been made to use these materials in drug delivery systems and early-stage results have been described in comprehensive articles published in 2009 and 2012 [34,35]. Nowadays, the field is growing rapidly and many exciting new results have been published in the last decade.

In this paper, we would like to present some new developments in the ATRP-utilized design of novel polymeric structures for drug delivery systems (DDSs). For the purpose of this review, DDSs obtained via ATRP are divided into the four most recently studied types: micelles, polymersomes, and polyplexes formed by linear block copolymers; carriers formed by branched copolymers; hybrid nanoparticles; and bioconjugates.

We dedicate this article to the late Professor Andrzej Dworak, who made tremendous contributions to the development of research on the self-organization processes of amphiphilic polymers and stimuli-responsive materials and their practical use in modern medicine and pharmacy. He was the author of many original papers as well as fundamental review papers, which provide valuable information and inspiration for future generations of chemists undertaking research in this fascinating area of science.

2. The Principles of ATRP

ATRP is a robust polymerization method that was developed and presented by Professor Matyjaszewski's group in 1995 [36,37]. It was inspired by atom transfer radical addition, which was successfully used in the synthesis of low-molecular-weight compounds [38]. ATRP, next to nitroxide-mediated polymerization or RAFT [39,40], is a method of CRP [41].

The idea of ATRP is based on reducing the concentration of radicals in the polymerization system through reversible reactions of the activation and deactivation of the active center as a result of halogen atom transfer, based on a dynamic equilibrium between working radical centers and dormant organic halides, with a relatively low homolytic dissociation activation energy of C-X bonds (Figure 1) [42,43]. Therefore, this equilibrium must be strongly shifted to the left (towards the organic halides R-X). The transfer of the halogen atom (X) from the initiator molecule or the growing polymer chain (R-X) takes place to the catalyst molecule, which is an inorganic salt, i.e., a transition metal halide with two oxidation states differing by one electron ($Mt^{(n+)}X_n$), complexed with a ligand (L). For accepting a halogen, the catalyst is in a reduced form and acts as an activator ($Mt^{n+}X_n$), increasing its degree of oxidation, and the process itself is an activation reaction (k_a). As a result of activation, a radical ($R^{\bullet}$) is formed capable of attaching to monomer (M) molecules (initiation or propagation (k_p) stage) or other reactions typical of radical polymerization. However, the participation of termination (reaction with the other radical, k_t) is drastically diminished by reducing the concentration of radicals to a level several orders of magnitude lower than that of typical radical polymerization. During the operation of the radical, an appropriate number of monomer molecules are attached to the growing chain. In contrast, the number of attached monomers per one act of activation depends on many rate constants, including those that affect the ATRP equilibrium (k_a, k_d). The latter, in turn, depends on the structure of the organic halide and the type of catalyst, and in particular on the type of ligand whose task is to create a complex with an inorganic salt that is soluble in organic media and to give the catalyst an appropriate reduction potential [44]. It is worth noting that most often the role of ATRP ligands is played by polydentate nitrogen compounds, including aliphatic, cycloaliphatic, or aromatic amines [45,46]. However, other systems, for example based on phosphorus compounds, are also used [47,48]. In the next stage, the oxidized catalyst (deactivator, $Mt^{[(n+1)+]}X_{n+1}/L$), having an additional halogen atom, reacts with one of the radicals present in the system, transferring a halide to it, which results in the formation of a dormant form of the polymerization center and the reduction of the catalyst. Due to the statistical nature of the activation and deactivation reactions, random organic halides and radicals undergo it, respectively, which, with a sufficiently fast deactivation, ensures a relatively uniform growth in all the polymer chains present in the system (linear increase in average molar mass with the degree of monomer conversion and a small dispersity in the molar masses of the obtained polymer). Copper salts were one of the first and the best-known catalytic systems used in ATRP; numerous studies have confirmed the possibility of using the halides of other metals, i.e., Fe [47–50], Ru [51,52], Ga, or Ir [53,54]. The required amount of catalyst in a normal ATRP mechanism is relatively large, as, typically, one molecule of catalyst is used per one initiation site. This has been a huge disadvantage of the method and constitutes a limitation for implementing it into industrial practice.

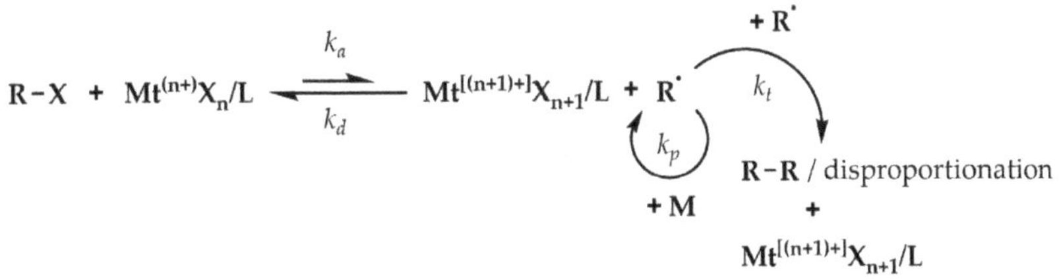

Figure 1. General scheme of an ATRP mechanism.

In the last 15 years, numerous related methods have been developed that use much smaller amounts of catalysts. Such systems, however, require more active ligands and additional agents whose task is to regenerate the activator molecules that were irreversibly formed in the system due to some termination processes and/or the contamination of the system with oxygen. Such agents might be glucose, ascorbic acid, hydrazine, tin(II) compounds (in activators regenerated by electron transfer variant, ARGET) [55], typical radical initiators as sources of radicals (in initiators for continuous activator regeneration variant, ICAR) [29,51,56,57], zero covalent metals, e.g., Cu^0 (in supplemental activation reducing agent variant, SARA) [58], or external stimuli such as electrical current/potential (in eATRP) [59], and ultrasound in the presence or absence of piezoelectric materials (in mechano/sonoATRP) [60,61]. The other and "greener" type of ATRP is photoinduced organocatalyzed polymerization (O-ATRP), which does not involve any metal catalyst [22,62]. These developments changed ATRP into a technique well suited to the principles of green chemistry [63].

Activated organic halides are used as initiators of ATRP. They can be simple compounds, such as the commonly used 2-bromobutyric acid ethyl ester, or functional initiators, i.e., those that allow the introduction of an end-group capable of chemical reactions or interactions, e.g., dye moiety [64,65], drug or proteins molecules [66–68], stimuli-sensitive molecules [69] or groups able to chemically react with other molecules or chains [70], etc. In some cases, the initiator can be covalently bonded to some surface or is able to influence the topology of macromolecules [71,72]. Therefore, ATRP allows the synthesis of linear, cyclic, and branched polymers, including star-shaped, comb, bottle-brush, and hyperbranched structures, nanogels, and polymer–drug conjugates (Figure 2).

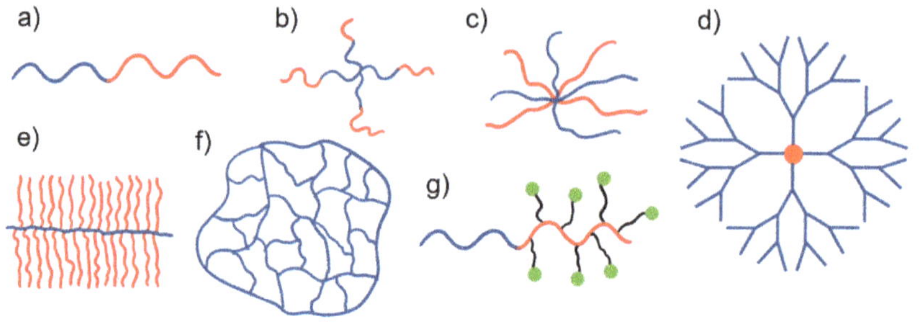

Figure 2. Representative molecular architectures synthesized via ATRP and applied in the formation of polymeric DDSs: (**a**) linear block copolymers, (**b**) regular star copolymers, (**c**) miktoarm star (co)polymers, (**d**) polymer brushes, (**e**) dendrimers, (**f**) nanogels, and (**g**) polymer–drug conjugate.

In the case of polymer stars, there are several synthetic strategies possible; they can be obtained via "core-first", "arm-first", and "coupling-onto" methods utilizing ATRP. While the "core-first" method uses a multifunctional initiator with a specified number of initiating species, which defines the number of arms, the "arm-first" method relies on ready, halogen functionalized polymer chains that are activated in the presence of the crosslinking agent, so core formation takes place in situ. It does not allow us to control the number of arms in a very good manner, however; it enables the synthesis of stars with structurally different arms (miktoarm). Polymer grafts (combs and bottle-brushes) might be synthesized by the polymerization of macromonomers (obtained by ATRP or in the other way), by the initiation of ATRP with a formed polymer chain comprising initiating species along it, or by coupling the side chains to the main chain consisting of appropriate chemical groups in monomeric units, when the side and main chains may be obtained via ATRP as well [73]. There have been successful attempts at star-brush molecule ("hairy" stars) synthesis as well [74].

It is essential that the halogen atom is formally present at the other end of the polymer chain after polymerization. This provides a vast opportunity to exchange the atom into the other, more reactive group, enabling further reactions, i.e., linking specific molecules to that chain-end or coupling the macromolecule with another one. The example of a halogen-to-azide group exchange best demonstrates this, which then opens the way to a variety of coupling possibilities with alkyne moieties, called "click" chemistry [75].

On the other hand, the residual halogen atom at the end of the macromolecule can be used as a macroinitiator for the synthesis of block copolymers, thus depriving ionic polymerizations of the monopoly for obtaining well-defined segment systems [76,77]. This, in turn, enables the design of materials with specific morphology obtained due to the self-assembly of block copolymers [78]. These include hydrophobic copolymers capable of forming ordered nanostructures from a polymer melt, e.g., lamellae, gyroids, hexagonally packed cylinders, or spheres. It is worth noting that the type of nanostructure can be influenced by the basic structural parameters included in the phase diagrams, i.e., the interaction parameter, degree of polymerization, and volume fractions of components as a predominance factor. However, ATRP allows us to control other subtle parameters, e.g., the topology or dispersion of blocks [79,80], which can change phase boundaries and stabilize phases considered to be thermodynamically metastable, e.g., hexagonal perforated lamellae [81].

ATRP also enables the synthesis of hydrophobic segments for amphiphilic block copolymers or complete amphiphilic block copolymers and double hydrophilic block copolymers [82–85]. Amphiphilic materials can self-assemble in aqueous systems to form vesicles (liposomes) or micelles (spherical, cylindrical, hexagonally packed, or lamellar "neat" micelles in parallel arrangement) [86], which are often used in DDSs, for instance. In some cases of doubly hydrophilic copolymers, one of the segments is a polyelectrolyte (often a polyanion), which can interact with particles of inorganic salts, being the basis for artificial biomineralization processes. Among the monomers giving hydrophilic segments in ATRP, acrylamide, 2-hydroxyethyl methacrylate (HEMA), or poly(ethylene glycol) methacrylates should be mentioned [87]. However, polyelectrolytes usually have to be created using hydrophobic precursors, e.g., tert-butyl acrylate (tBuA), which, after polymerization, are hydrolyzed to poly(acrylic acid) (PAA) [88]. The direct use of acrylic acid in polymerization causes its reaction with a deactivator, i.e., copper(II) salts. This shows that although ATRP is quite robust to reaction conditions, this method has some limitations. The main challenge of this method is the limited range of monomers that can be used compared to that in simple radical polymerization. It is required that these monomers have high resonance stabilization. Hence styrenes, acrylates, methacrylates, or acrylonitrile are most often used. At the same time, the well-controlled ATRP of ethylene, butadiene, or vinyl acetate is practically impossible. It should be noted, however, that ATRP allows us to obtain the large family of amphiphilic copolymers by using hydrophilic macroinitiators, such as poly(ethylene glycol) (PEG), and naturally occurring polysaccharides. The typical synthetic approach involves the transformation of the hydroxyl end-group of the hydrophilic reagent with 2-bromoisobutyryl bromide (BIBB) to form an ATRP macroinitiator (Figure 3) [82].

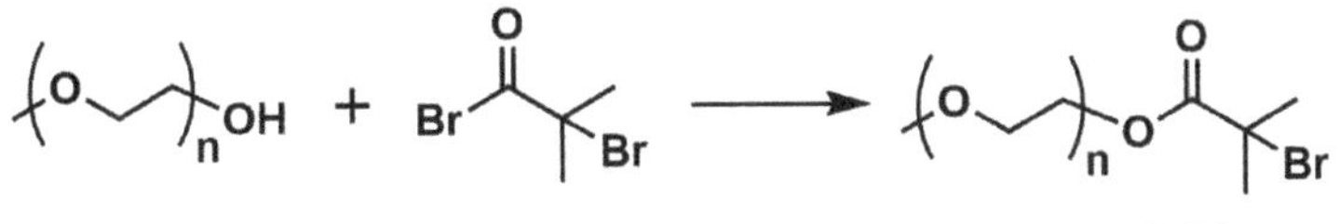

Figure 3. A typical synthesis of PEG macroinitiator of ATRP for amphiphilic block copolymers.

3. DDSs Based on Linear Block Copolymers

Due to their ability to self-assemble into nanostructures in aqueous media, linear block copolymers bearing both hydrophilic and hydrophobic blocks have gained much attention as potential drug nanocarriers. Although there are several factors that impede the behavior of linear block copolymers (LBCPs) in aqueous dispersions (e.g., the molecular weight of polymeric chains, as well as the number, chemical composition, and length of their constituting macroblocks, temperature, pH, and ionic strength of the environment) [78,89], one can categorize the resulting nano-sized particles (usually with diameters up to 200 nm) into two main groups: micelles and polymersomes (polymeric vesicles) [78,89,90]. The former are small aggregates exhibiting a core-shell structure, in which the hydrophobic blocks of amphiphilic macromolecules are stacked together inside, whereas the hydrophilic parts of LBCPs are directed outside (Figure 4a). On the other hand, the structure of polymersomes resembles that of liposomes: they exist as hollow spheroids, in which the LBCPs' chains are located at the surface and form an amphiphilic double layer with their hydrophilic blocks sticking out on both sides (Figure 4b). It should be noted that each part of such nanoassemblies makes a distinctive and significant contribution to the overall successful performance of a DDS: the hydrophobic core of the micelle, or the interior of the amphiphilic double layer in a polymersome, is responsible mainly for the drug loading capacity (LC) and its controlled release, whereas hydrophilic corona increase the biocompatibility of the whole system and protect drug molecules from any unfavorable destructive interaction with enzymes, serum proteins, and other constituents present in the bloodstream before a DDS reaches its target site [91,92].

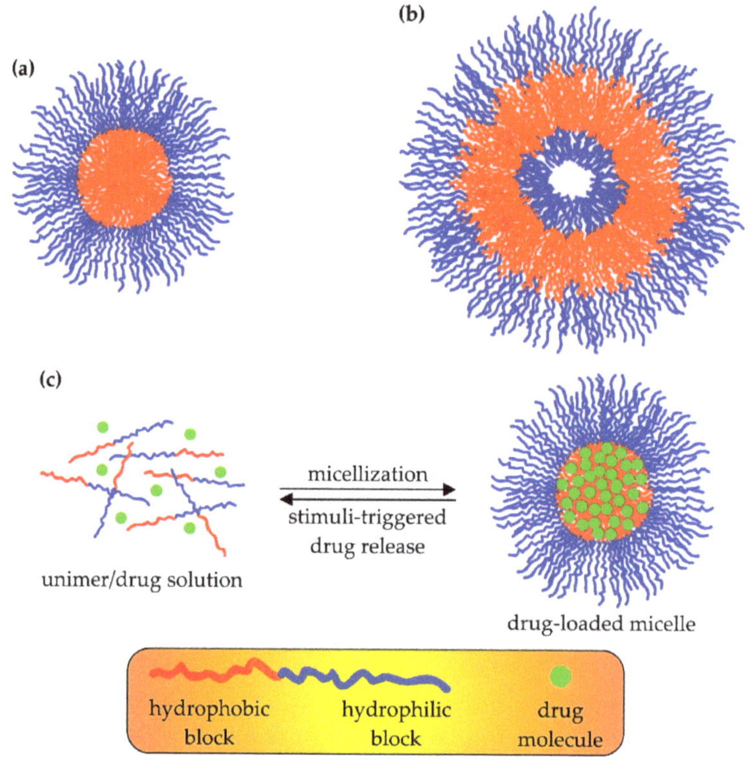

Figure 4. Schematic representations of (**a**) micelle and (**b**) polymersome formed by LBCPs, as well as (**c**) general synthetic strategy applied for the formulation of micelle-based DDSs.

Since the loading of polymeric micelles with drugs is performed via a physical entrapment of the latter within the hydrophobic core of the micelle, one can accomplish it by self-assembling LBCPs in the presence of drug molecules (a bottom-up approach, Figure 4c). This strategy is characterized by simplicity and ease of performance, although several important processing conditions (e.g., temperature, pH, the concentration of reagents) must be carefully controlled. It involves the dissolving of amphiphilic LBCPs and hydrophobic drugs in an appropriate organic solvent, followed by a drop-wise mixing of the resulting solution with the aqueous phase. Usually, water-miscible organic solvents are utilized in this method (e.g., THF [93,94], DMF [95], or DMSO [92,96,97]), and they have to be subsequently separated from the drug-containing micelles, together with the unentrapped drug molecules, by means of dialysis [94–98].

Micelles must be biocompatible and sufficiently resistant to the internal environment of the living organism to transport the chosen drug to its destined location (e.g., tumor cells or pathologically altered tissues). Thus, there has been continuous interest in the development of methodologies for the stabilization, cross-linking, and functionalization of polymer micelles. However, their stability cannot be too high since at the targeted site a DDS has to liberate its therapeutic payload in a strictly controlled manner, thus solely affecting the targeted cells as well as maintaining the concentration of the delivered drug within the optimal therapeutic limits for an appropriate amount of time [91,99]. Moreover, in order to minimize the negative side effects occurring during the application of many drugs (especially those used in cancer treatment), or to protect the therapeutic agents that are otherwise easily destroyed in the bloodstream (e.g., nucleic acids used in gene therapy), efforts have recently been made for the synthesis of DDSs that change their physicochemical and/or structural properties in response to some external stimuli, either physical (e.g., light, ultrasound, or temperature) or chemical factors (e.g., changes in pH values and/or redox potential of the environment) [100].

3.1. *Smart DDSs Based on Micelles*

Generally, there are two types of endogenous stimuli (factors independent of human influence and related to the functioning of a living organism) that have been used to trigger a drug release from smart DDSs in cancer cells or tissues—both of them are connected to the altered metabolism of the target site. For instance, it is well known that the increased glycolytic metabolism occurring in tumor cells (the Warburg effect) produces a larger amount of lactic acid and CO_2 than that observed during the standard metabolism of healthy cells. Since both compounds are expelled from the cells, their increased concentration leads to the acidification of the extracellular microenvironment of the tumor cells, which exhibits a slightly lower pH value (usually below 6.5) than that observed in blood or healthy tissues (ca. 7.4) [100,101]. Moreover, some organelles inside cells (e.g., endosomes or lysosomes) exhibit even lower pH values (usually 4.5–5.5) [100,102]. Taking this into account, pH is one of the most intensively exploited natural triggers for drug liberation from DDSs.

On the other hand, all living cells combat destructive reactive oxygen species (radicals) by utilizing several natural antioxidants, among which glutathione (GSH) is the most abundant [103,104]. Because of this, the concentration of GSH inside the healthy cells (up to 10 mM) is about 1000 times higher than that in the bloodstream, and in the case of tumor cells it is even higher (at least a few times higher). The obvious increase in the reductive conditions of the intracellular environment can be a trigger for safe drug release directly inside a tumor cell [104].

In order to introduce pH responsiveness into micelle-based DDSs, two main strategies can be applied, both of which are focused on the introduction of a physicochemical imbalance into drug-loaded LBCPs' micelles leading to their destruction and the release of their therapeutic content in the acidic environment of the tumor cells. This can be performed either via the protonation of functional groups incorporated into LBCP chains, or the acid-catalyzed hydrolytic cleavage of the chemical bonds present in the polymer backbone or pendant groups.

3.1.1. Drug Release Induced by Protonation

This approach has been exploited by several research groups; however, a common feature of their works has been the utilization of tertiary amine groups as proton acceptors and PEG as a hydrophilic block of LBCPs. Wang and Zhang synthesized a series of double hydrophobic triblock copolymers, poly(2-(N,N′-diethylamino)ethyl methacrylate)-b-poly(ethylene glycol)-b-poly(2-(N,N′-diethylamino)ethyl methacrylate) (PDEAEMA-*b*-PEG-*b*-PDEAMA) [93], by conducting the ATRP of 2-(N,N′-diethylamino)ethyl methacrylate (DEAEMA) in the presence of 2-bromoisobutyrate-terminated PEG (Br–PEG–Br, a product of PEG esterification with BIBB) as an ATRP macroinitiator. From this work, it is evident that the elongation of hydrophobic blocks does not increase the cytotoxicity of the blank micelles (even at their highest concentration, in which the viability of the tested cell cultures was above 80%), whereas the critical micelle concentration (CMC) decreases. Moreover, by increasing the number of DEAEMA units, one can increase both the hydrodynamic diameter of the blank or drug-loaded micelles (within the 40–180 nm or 50–220 nm ranges, at the physiological pH of 7.4, respectively), as well as the doxorubicin (DOX) loading contents and entrapment efficiencies (up to ca. 8.1% and 89%, respectively, for the copolymer with the highest molar weight). The pH responsiveness of the investigated DDSs was proven by their faster DOX release observed at the endo-/lysosomal pH conditions (between 65 and 90% at pH = 5.4 in comparison to 25–35% at pH = 7.4). In another study [94], Wang, Zhang, and coworkers showed that the length of the hydrophilic block in PDEAEMA also impedes its micellar properties and DOX uptake: the increase in the PEG block length leads to smaller micelles that exhibit lower DOX loading contents and efficiencies (8.0–6.4% and 86–68%, respectively), as well as cumulative drug release (a decrease from ca. 90% to 70% at the endo-/lysosomal pH).

In order to make PDEAEMA micelles more thermodynamically stable, Chen and coworkers proposed a symmetrical elongation of the hydrophobic block by copolymerizing it with the second hydrophobic monomer. For that purpose, they conducted a sequential ARGET ATRP of DEAEMA with either methyl methacrylate (MMA) [105] or HEMA [106]. This method resulted in a symmetrical, double hydrophobic pentablock copolymer, although in the case of the HEMA-based system, its hydrophilic 2-hydroxyethyl groups required hydrophobization via an amidation reaction with the amine groups of folic acid [106]. The obtained micelles showed comparatively low CMC values (especially the system containing MMA units, 2.4–2.8 mg/L) indicative of their potentially better stability in the bloodstream [105], as well as a larger encapsulation efficiency (e.g., 20–35% for the MMA-based system [105] and 45–48% in the HEMA-based copolymer) [106], although the DOX loading contents were substantially lower (below 25%) in comparison to the previously investigated triblock copolymer PDEAEMA. It should be noted that, at the same length of the second monomer, the copolymers with longer pH-sensitive PDEAEMA blocks exhibited higher values of the DOX loading content and efficiency and formed slightly larger micelles while their CMC values decreased. More importantly, in both cases, the DOX-loaded micelles rapidly released their content at an acidic pH and were characterized by cytotoxicity against tumor cells (a tumor-suppressing effect) close to that of free DOX—for example, after 48 h of incubation with DOX-loaded HEMA-based micelles, the viability of HepG2 tumor cells was reduced to ca. 20%.

An interesting option for the synthesis of DDSs based on PDEAEMA-containing micelles is the process of co-micellization, in which drug-encapsulating nano-assemblies are formed due to the entanglement of two different copolymers. It is especially helpful if the desired building blocks of LBCPs have to be synthesized according to different polymerization mechanisms—in many cases, the separation of such polymerization procedures makes laboratory work less tedious and quicker. An exemplification of this strategy is the research work carried out by Yang and coworkers [98,107], in which the authors described mixed micelles formulated from amphiphilic copolymers obtained via the ARGET ATRP of DEAEMA and ring-opening polymerization (ROP) of ε-caprolactone (ε-CL). In reference [98], DOX-loaded mixed micelles consisting of the mixture of amphiphilic diblock copolymers poly(2-(N,N′-diethylamino)ethyl methacrylate)-*b*-poly(poly(ethylene glycol) methyl ether methacrylate) (PDEAEMA-*b*-PPEGMA) and poly(ε-caprolactone)-*b*-poly (poly(ethylene glycol) methyl ether methacrylate) (PCL-*b*-PPEGMA) were investigated. Both the experimental studies on drug release and its computational simulations employing dissipative particle dynamics (DPD) indicated that mixed micelles exhibited good pH-responsivity: they were very stable at the physiological pH, showing very limited leakage of DOX, whereas, at pH = 5.0, they were characterized by an accelerated release of DOX. Moreover, these co-micelles were completely biocompatible, with no cytotoxicity detected during in vitro tests, and their capability for DOX up-take was high: the estimated DOX loading contents and DOX encapsulation efficiencies were in the 23–31% and 60–91% ranges, respectively, depending on their composition [98]. Another study simplified the topology of the copolymers by utilizing PEG monomethyl ether (MPEG) as a macroinitiator in ε-CL ROP and ARGET ATRP of DEAEMA (after prior esterification of MPEG hydroxyl terminal with BIBB) [107]. The DOX loading contents of these nanosystems (hydrodynamic diameters in the 200–300 nm range) were on a comparable level to their poly(poly(ethylene glycol) methyl ether methacrylate (PPEGMA)-based analogs, whereas they exhibited slightly lower DOX encapsulating efficiencies (values between 21% and 63%). By combining experimental work with DPD simulations, the authors showed that DOX tended to distribute in the mixed inner core formed by poly(ε-caprolactone) (PCL) and pH-sensitive PDEAEMA chains, owing to the hydrophobic interactions; however, as the PCL/PDEAEMA chains of the polymers increased in length, the ability of the micelles for DOX loading decreased, suggesting that protective effect of the hydrophilic corona (shell) also plays a role in ensuring an appropriate LC of this type of DDSs [107].

3.1.2. Drug Release Induced by Acid Bond Cleavage

Alternatively to the protonation/deprotonation mechanism, pH-responsive polymeric micelles can be disintegrated via the acid-promoted cleavage of the chemical bonds incorporated into their structure with ester, ortho ester, β-thiopropionate ester, hydrazone, imine (e.g., benzoic imine), acetal, ketal, oxime, vinyl ether, or amide groups [108–111]. Based on the location of these acid-labile linkages within block copolymer chains, the latter (as well as nano-assemblies formed by them in aqueous solutions) can be categorized into three main groups (see Figure 5): backbone acid-cleavable (i.e., those characterized by bond breaking between monomeric units of the hydrophobic block), pendant acid-cleavable (i.e., LBCPs whose hydrophobic blocks are fabricated from monomers bearing acid-labile substituents), and shell acid-cleavable (shell-sheddable, i.e., those characterized by bond breaking at the junction between their hydrophilic and hydrophobic blocks) [109].

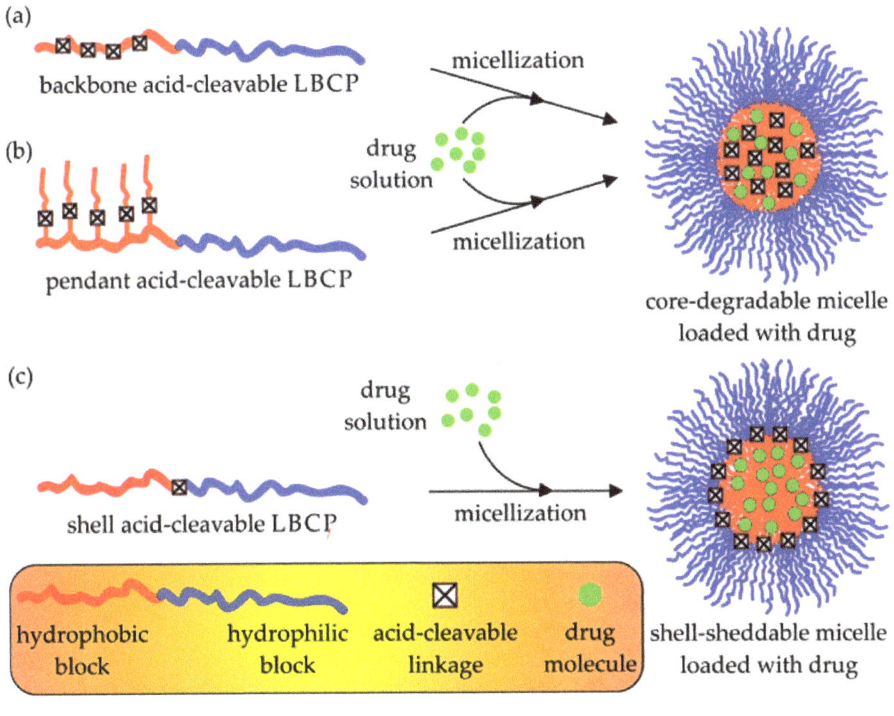

Figure 5. Possible locations of acid-labile linkages within amphiphilic LBCPs before and after micellization: (**a**) backbone acid-cleavable, (**b**) pendant acid-cleavable, and (**c**) shell acid-cleavable nanoassemblies.

Since all ATRP techniques produce new hydrolytically stable C–C bonds between monomeric units, they have no use in the formulation of backbone acid-cleavable micelles. Only an extension of ATRP, namely atom transfer radical polyaddition (ATRPA), has been able to show some results in this field. Li and coworkers proved that biodegradable pH-responsive polyester showing a low critical solution temperature (around 37 °C) could be obtained via the ATRPA of bis(styrenic)- and bis(bromoisobutyrate)-type monomers. As expected, the synthesized copolymer underwent depolymerization at a pH of around 5.5 (due to the acidolysis of ketal bonds incorporated via bis(bromoisobutyrate)-type monomer); however, no drug release tests have been conducted for this system [112].

The concept of a controlled drug release in the case of block copolymers having pendant groups prone to acid-triggered bond breaking is based on the fact that, by changing their chemical composition (due to the cleavage of the side groups), one can disrupt a delicate balance in the hydrophilic/hydrophobic interactions responsible for keeping the micelles intact, thus leading to the destabilization and disassembling of the latter. Such a process is especially promoted if bond breaking increases hydrophilicity or results in the formation of electrically charged moieties within the hydrophobic core of the micelle. Two main synthetic strategies can be applied for the synthesis of this type of copolymers: a direct polymerization or a chemical modification of the reactive side groups present in the already synthesized polymeric chains [109,110]. The latter strategy is mostly used for the attachment of drugs to polymeric carriers having different molecular structures, resulting in the formation of polymer–drug conjugates; thus, it will be discussed in detail in a separate section of this review. The direct polymerization approach usually requires the design and subsequent synthesis of the appropriate monomers bearing the acid-labile functional groups and exhibiting a sufficiently high reactivity in the chosen type of polymerization reactions. For that purpose, as far as ATRP techniques are considered, three synthetic pathways can be utilized: (a) the polymerization of a functionalized, hydrophobic, unsaturated monomer in the presence of a hydrophilic ATRP macroinitiator; (b) the polymerization of a functionalized, hydrophilic, unsaturated monomer in the presence of a hydrophobic ATRP macroinitiator, or the copolymerization of the appropriately functionalized hydrophilic and hydrophobic unsaturated monomers. It should be noted that only the first of these leads to micelles being formed by amphiphilic LBCPs, as shown by the results obtained by Li and coworkers, who used 2-chloropropionate-ended MPEG chains to initiate the ATRP of N-substituted acrylamide (trans-N-(2-ethoxy-1,3-dioxan-5-yl) acrylamide, tNEA) and synthesized a diblock copolymer-containing ortho ester moiety in the pendant substituents [113]. The authors proved that by changing the degree of polymerization within the hydrophobic poly(trans-N-(2-ethoxy-1,3-dioxan-5-yl) acrylamide) (PtNEA) block, one can easily alter the self-assembling properties of the resulting copolymer and obtain nanostructures with morphologies ranging from spherical micelles (for the shortest PtNEA blocks) and rod-like clusters to polymersomes (for the longest PtNEA blocks). Interestingly, regardless of their morphological features, these nanoassemblies were stable at physiological pH but underwent disintegration at the endo-/lysosomal mildly acidic environment (pH 4.6–5.0), with micelles being the most prone to hydrolysis. Moreover, they all were capable of DOX loading and displayed pH-dependent drug release profiles, as well as concentration-dependent cytotoxicity against HepG2 tumor cells [113].

The other two ATRP-based strategies leading to copolymers with labile pendant groups were explored by Wei and coworkers [114], as well as Oh and Khorsand [115]; however, they both resulted in graft copolymers, which subsequently self-assembled into drug-loading micelles. The first group polymerized a newly synthesized hydrophilic methacrylate monomer (a derivative of HEMA acetalized with MPEG oligomers) using a hydrophobic bromine-terminated PCL, a product of the ring-opening polymerization of ε-CL initiated with 2-hydroxyethyl 2′-bromoisobutyrate as the ATRP macroinitiator [114], whereas the second group sequentially copolymerized hydrophilic poly(ethylene glycol) methyl ether methacrylate (PEGMA) with pH-responsive hydrophobic tert-butyl methacrylate using a small-molecule ATRP initiator (2-hydroxyethyl 2′-bromoisobutyrate) [115].

In the last decade, new concepts regarding the utilization of ATRP in the synthesis of copolymers with pH-cleavable pendant substituents have been proposed. Although some of them do not relate to amphiphilic LBCPs (e.g., the usage of deactivation-enhanced ATRP conditions and the acetal-containing diacrylate monomer for the in situ synthesis of an amphiphilic hyperbranched copolymer capable of DOX loading and subsequent releasing at the endosomal pH) [116], the others are directly concerned with such copolymers. For example, Li and coworkers observed that ATRP of a mixture of HEMA and DEAEMA, followed by the reaction between the hydroxyl groups of poly(2-hydroxyethyl methacrylate) (PHEMA) blocks with 2-ethylidene-4-methyl-1,3-dioxolane, resulted in a triblock LBCP, in which at a pH of ca. 5, the protonation of the tertiary amine groups from the PDEAEMA units promoted the hydrolysis of the surrounding pendant ortho ester moieties. Depending on the content of PDEAEMA, their hydrolysis half-times ranged from hundreds of minutes at a pH of 5.4 to several days at physiological pH, but the copolymer with the highest amount of the tertiary amine units was characterized by the most accelerated loss of ortho ester groups [117]. It should be noted that a similar synergistic behavior, including an accelerated DOX release at an acidic medium, was also observed by Dong and coworkers in the case of the related copolymer exhibiting a more complicated, multi-grafted molecular architecture, which was obtained via ATRP of a mixture of 2-(N,N′-dimethylamino)ethyl methacrylate and (2,2′-dimethyl-1,3-dioxolane-4-yl)methyl methacrylate carried out in the presence of a polymeric macroinitiator bearing pendant 2-bromoisobutyrate groups (a triblock ATRP macroinitiator obtained in the course of the ROP of ε-CL and γ-(2-bromo-2-methylpropionate)-ε-caprolactone initiated with MPEG) [118].

Shell-sheddable micelles formulated by a third group of the amphiphilic LBCPs prone to pH-triggered bond cleaving have been developed in response to the so-called "PEG dilemma" encountered during the in vivo application of PEGylated drug nanocarriers: their low cellular uptake and an increased possibility of the production of anti-PEG antibodies in the case of the prolonged presence of PEG in the bloodstream [119,120]. In the mildly acidic microenvironment of the tumor cells and tissues, these micelles lose their PEG coronas due to the cleavage of pH-labile bonds, linking hydrophobic and hydrophilic blocks of the amphiphilic LBCPs, whereas their hydrophobic cores undergo aggregation at the cellular walls. Although many pH-breakable bonds can be utilized for that purpose, in the case of the ATRP-derived shell-sheddable amphiphilic LBCPs, only two types have been recently used: acetal and imine linkages [121,122]. They can be incorporated into the copolymer chain structure by the chemical conjugation of two different homopolymers bearing reactive chain-ends [121,122] or the direct polymerization of the unsaturated monomer started with the appropriately functionalized ATRP macroinitiator [123]. A good example of the first approach is the work of Patil and Wandgaonkar [121], who reported the synthesis and self-assembling properties of an acetal-linked diblock copolymer of ε-CL and N-isopropylacrylamide (NIPAM) PCL-*b*-PNIPAM. This copolymer was formed through the alkyne-azide click reaction between the propargyl-terminated PCL and azide-ended poly(N-isopropylacrylamide) (PNIPAM) chains. The acetal linkers were introduced to the PNIPAM blocks via the ATRP of NIPAM carried out with the usage of a newly designed ATRP initiator, namely 2-(1-(2-azidoethoxy)ethoxy)ethyl 2-bromo-2-methylpropanoate. At room temperature, the resulting copolymer self-assembled in an aqueous solution into micelles (ca. 74 nm in diameter) that could encapsulate rhodamine B (as a model drug) at a pH of 7.4 and subsequently release it at the endosomal pH or upon heating up to 40 °C. A very similar approach was tested by Ni and coworkers [124], who conjugated the acetal-containing coumarin- and azide-terminated hydrophobic chains of PCL, with the hydrophilic monoalykynyl-terminated graft block copolymer of PEGMA and dimethylaminoethyl methacrylate (DMAEMA), synthesized via ATRP in the presence of a propargyl 2-bromoisobutyrate initiator. This linear-graft triblock copolymer showed good DOX loading content (5.3%), DOX loading efficiency (28.5%), and increased DOX release at a pH of 5.0; additionally, it was capable of simultaneous nucleic acid loading and subsequent delivery into HeLa tumor cells.

An interesting option for the application of click chemistry in the synthesis of the shell-sheddable amphiphilic LBCPs is the in situ formation of the pH-labile junction during the final step of conjugation of the copolymer building blocks. However, this approach requires the presence of the appropriately selected reactive groups at the terminals of (co)polymers subjected to conjugation. Dimitrov and coworkers tested it on the shell-sheddable micelles formulated by a copolymer containing a benzyl imine moiety, as well as PEG, PDMAEMA, and polylactide (PLA) blocks [122]. The simultaneous formation of both the imine linker and final copolymer chains proceeded in the mixture of monoamine-terminated DMAEMA/lactide (LA) diblock copolymer and PEG homopolymer terminated with benzaldehyde group, without the use of any catalyst. In this study, the ATRP of DMAEMA was conducted on the monoalkynyl-terminated brominated polyester macroinitiator (A-PLA-PDMAEMA) resulting from the ROP of LA (initiated with propargyl alcohol) and subsequent esterification with BIBB. Interestingly, in earlier work, Dimitrov and coworkers also showed that A-PLA-PDMAEMA itself formed micelles in aqueous media that were capable of the delivery and controlled release of curcumin inside acute promyelocyte leukemia-derived HL-60 cells [125]. It is worth noting that the research works of Dimitrov's group cited above are indicative of a new trend emerging in the field of the micelle-based DDSs [122,125], namely, the incorporation of some precisely designed subcellular targeting ligands into amphiphilic LBCP chains. Dimitrov and coworkers explored this by introducing (via a post-polymerization reaction with (4-bromobutyl)triphenylphosphonium bromide) pendant triphenylphosphonium and quaternary ammonium cations, which facilitated the transportation of the dePEGylated copolymer through the phospholipid barrier of the cellular and lysosomal walls.

Recently, several research groups have investigated an interesting option of combining in one DDS two different mechanisms of pH-dependent micelle destabilization and/or drug release. Oh and coworkers utilized a newly designed ATRP macroinitiator (a product of ethylene glycol vinyl ether (EGVE) esterification with BIBB followed by acetalization with a hydroxyl-terminated MPEG) and a methacrylate monomer bearing a pendant acetaldehyde acetal linkage (a product of EGVE esterification with acetyl chloride followed by acetalization with HEMA) for the synthesis of a dual location acid-cleavable amphiphilic LBCP (Figure 6) [123]. This copolymer was characterized by the presence of pH-labile acetal linkers both in the pendant groups of the hydrophobic block, as well as at the hydrophobic/hydrophilic block junction. Therefore, its nano-assemblies in aqueous media were prone to both corona detachment (upon cleavage of the acetal block junction) and core destruction (via the cleavage of the pendant acetal moieties). It is believed that such a combination of these two micelle destabilization mechanisms may allow us to overcome some of their limitations, e.g., the sluggish degradation of the core-degradable systems and undesired aggregation of dePEGylated cores for shell-sheddable systems. Moreover, the authors further enhanced the DOX loading capability of this system as well as the acid-catalyzed hydrolysis of the acetal moieties by the copolymerization of additional monomer-bearing acid-ionizable imidazole groups [123].

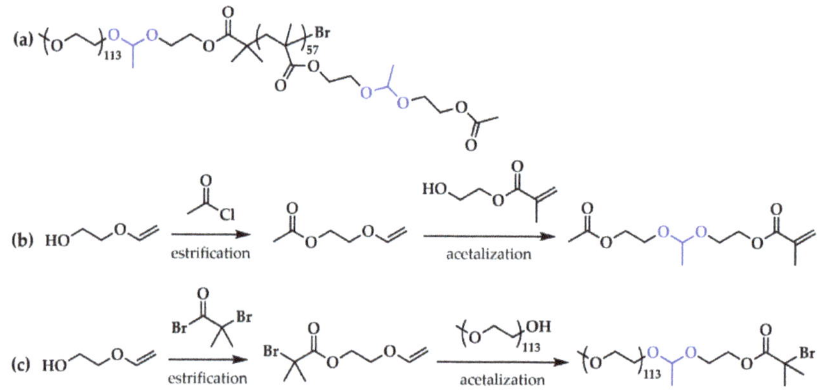

Figure 6. (**a**) Structure of a dual location acid-cleavable amphiphilic LBCP containing pH-labile acetal groups, as well as synthetic pathways, resulting in the preparation of (**b**) ATRP monomer and (**c**) ATRP macroinitiator used for its synthesis [123].

Zhang and coworkers explored the concept of a dual mechanism of drug encapsulation in the case of micelles formulated from a double hydrophilic triblock glycopolymer containing MPEG and poly(2-gluconamidoethyl methacrylate) (PGAMA) as the hydrophilic end-blocks, separated by a hydrophobic PDEAEMA block [97]. For that purpose, a sequential ATRP of DEAEMA (first step) and 2-gluconamidoethyl methacrylate (GAMA) (second step) was initiated with the BIBB-modified MPEG macroinitiator. For micellization carried out in the presence of boron-containing anticancer drug bortezomib (BTZ), they obtained BTZ-loaded micelles with hydrodynamic diameters of ca. 80 nm and a rather high value of CMC (30 mg/L), in which BTZ entrapment was achieved by both hydrophobic interactions with the PDEAEMA core and the covalent complexation (conjugation) of BTZ boronic acid functionality with glucose groups of PGAMA. The achieved BTZ loading content and entrapment efficiency were estimated to be 7.6% and 72%, respectively. At physiological pH, the BTZ release from these micelles was very slow (ca. 20% after 10 h), whereas it increased substantially (up to ca. 60% after 10 h) when the endo-/lysosomal pH conditions were applied. In addition, this DDS showed an appreciable prolonged release profile, since even after 60 h, the cumulative release of BTZ did not exceed 70%.

Zeng and coworkers verified the applicability of the pH-dependent double-triggered drug release strategy during combination anticancer therapy [96]. Starting from the Br-containing ATRP macroinitiator (synthesized from MPEG, 4-formylbenzoic acid, and BIBB), they polymerized the conjugate of HEMA and ibuprofen into a hydrophobic block of amphiphilic LBCP. In an aqueous environment and at physiological pH, the obtained diblock copolymer easily formed spherical micelles, with mean hydrodynamic diameters around 200 nm and a small CMC value of 2.5 mg/L, which could be filled with DOX molecules and exhibited a LC and encapsulation efficiency of ca. 10% and 33%, respectively. Under endo-/lyposomal acidic conditions, these micelles disintegrated due to the cleavage of the benzoic imine bonds linking their hydrophobic cores with MPEG hydrophilic shells (thus they belong to the group of shell-sheddable nano-assemblies), which was accompanied by the hydrolysis of the ester bonds connecting the ibuprofen to the HEMA-derived monomeric units. Although both drugs (DOX and ibuprofen) were released from micelles at sufficiently high rates, each process was controlled by different factors: the DOX release depended solely on micelle collapse, while the ibuprofen release was additionally controlled by ester bond hydrolysis. It should be noted that the synthesized DDS exhibited an anti-tumor behavior against B16 murine melanoma cells similar to the free DOX hydrochloride, as evidenced by both in vitro and in vivo tests.

3.1.3. Redox and Dual Redox/pH-Sensitive Systems

Changes in the redox conditions in the microenvironment of tumor cells are another important factor that is used as endogenous stimuli for targeted drug delivery. Up to now, several types of oxidation- and/or reduction-sensitive chemical functionalities have been proposed for that purpose [126–128], among which two have gained particular popularity in the case of ATRP-derived polymeric micelles: arylboronate moiety (for an oxidation-responsive DDS) and disulfide (SS) bond (for a reduction-responsive DDS).

Boronate-Bearing Oxidation-Responsive Systems

Li and coworkers utilized an ATRP macroinitiator, MPEG esterified with a proper acyl halide (2-chloropropionate chloride or BIBB) [129,130], to copolymerize a phenylboronic pinacol ester-containing acrylate monomer (4-(4,4,5,5-tetramethyl-1,3,2-dioxaborolan-2-yl)benzyl acrylate) with NIPAM [129], or fluorescent acrylic acid ester bearing 1,8-naphthalimide groups [130]. In both cases, an amphiphilic LBCP was produced, which at 37 °C easily formed micellar systems in aqueous media (a double hydrophilic copolymer containing NIPAM exhibited a critical aggregation temperature between ca. 10 °C and 20 °C, which depended on its molecular weight) [129,130]. The subsequent loading with a hydrophobic DOX resulted in the micellar DDSs, which were susceptible to a well-known arylboronate oxidation mechanism (Figure 7) and underwent destabilization under the influence of H_2O_2 or the intracellular reactive oxygen species (ROS) [131].

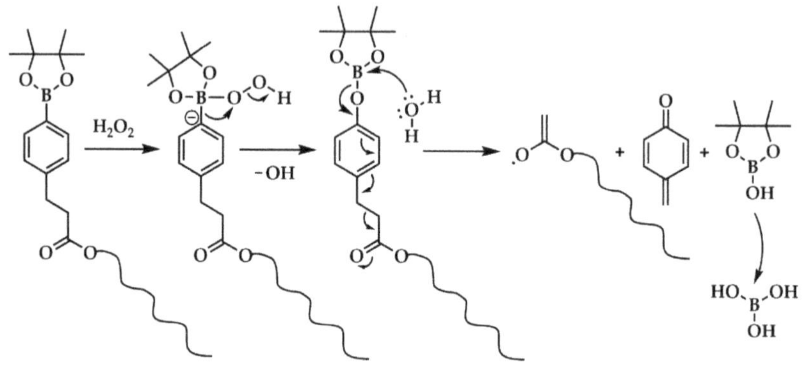

Figure 7. Mechanism of arylboronate oxidation under the influence of hydrogen peroxide.

In another study, Li's group showed that arylboronate oxidation chemistry could be effectively combined with the pH-induced destruction of LBCP micelles. Starting with the BIBB-esterified MPEG, they carried out a simultaneous ATRP, obtaining an amphiphilic triblock copolymer, in which the hydrophobic block contained monomeric units of the aforementioned oxidation-sensitive arylboronate monomer and another acrylate bearing the pH-labile ortho ester substituents [132]. The results of the pH- and/or H_2O_2-triggered degradation of the Nile red (NR)-loaded micelles showed that the oxidation of the phenylboronic ester moieties promoted the subsequent hydrolysis of the ortho ester pendant groups (due to a catalytic effect of the newly formed pendant carboxylic groups). Moreover, the kinetics of both these reactions could be easily tuned by changing the copolymer composition and oxidant concentration, as well as the pH of the environment.

Disulfide-Bearing Reduction-Responsive Systems

The principle of operation of the SS-containing reduction-responsive LBCPs is based on a well-known two-step thiol-disulfide exchange reaction proceeding under the influence of a suitable reducing agent bearing thiol groups (e.g., intracellular glutathione). First, one molecule of the reducing agent cleaves the SS bond, resulting in the formation of a mixed SS moiety and the liberation of a free sulfhydryl group. In the next step, another molecule of reductant breaks the mixed SS bond, releasing the second sulfhydryl group, while dimerizing to the oxidized form of the reducing agent (Figure 8) [133].

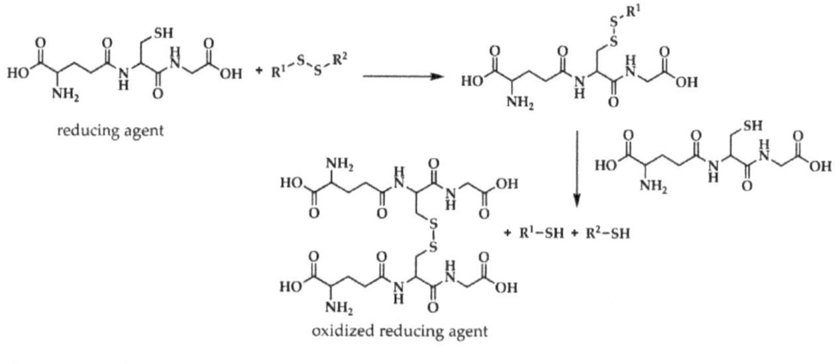

Figure 8. Mechanism of the reductive cleavage of disulfide bonds under the impact of the thiol-containing glutathione.

Similarly to the acid-cleavable groups of the pH-responsive amphiphilic copolymers, the reduction-cleavable SS linkages can be incorporated into three different locations within LBCP macromolecules, i.e., in the hydrophobic block backbone (resulting in the core-cleavable micelles), at the junction between the hydrophobic and hydrophilic blocks (leading to the shell-sheddable micelles), or as components of the pendant groups attached to the hydrophobic block [134]. However, in the case of the ATRP-derived LBCP micellar DDSs, only the last two of these have any significance. In fact, the backbone SS multi-cleavable drug-delivering nanoassemblies are usually produced via step-growth (e.g., hydroxyl-carboxylic or disulfide-thiol polycondensation, and a combination of Michael-type polyaddition with the "click" chemistry-based post-polymerization modification) or chain-growth polymerization techniques other than ATRP (e.g., ROP and RAFT) [134], whereas the usage of ATRP is limited to the processes in which the SS-containing ATRP macroinitiator is utilized for the polymerization of PEGMA, thus resulting in a copolymer with a non-linear topology [134–136].

LBCPs with SS linkage between their blocks can be synthesized in the course of ATRP started by a properly designed SS-containing hydrophobic or hydrophilic ATRP macroinitiator. Wang and coworkers utilized bis(2-hydroxyethyl) disulfide (HO-SS-OH) as an initiator for the ROP of ε-CL and esterified the resulting SS-containing hydroxyl-terminated PCL with BIBB, thus obtaining an SS-containing hydrophobic macroinitiator [137]. The latter was applied in the ATRP of *tert*-butyl methacrylate (tBuMA), producing a symmetrical hydrophobic tetrablock copolymer, which, upon the subsequent hydrolysis of the *tert*-butyl ester groups with trifluoroacetic acid, formed a desired amphiphilic LBCP composed of two symmetrical PCL-*b*-poly(methacrylic acid) blocks separated via a single SS linkage. Interestingly, Wang's group also demonstrated that a simple change in the order of the ROP and BIBB esterification steps could cause LBCP to exhibit a different topology (i.e., a classic amphiphilic diblock copolymer), even when using the same reagents (HO-SS-OH, ε-CL, and tBuMA). Via a reaction of HO-SS-OH with an equimolar amount of BIBB, they synthesized a bifunctional initiator active both in the ATRP and ROP of cyclic ester, namely 2-hydroxyethyl-2′-(bromoisobutyryl)ethyl disulfide (HO-SS-Br), and then separately polymerized ε-CL and tBuMA [138]. The micelles formulated from both these copolymers showed very good paclitaxel-loading properties (exhibited by the hydrophobic PCL cores) [137,138], whereas diblock copolymer-containing micelles could also encapsulate amine-containing cisplatin via electrostatic interactions with its carboxylic groups located in the hydrophilic shell [138]. Drug release profiles proved that in a reductive microenvironment, these DDSs showed an accelerated release of drug molecules, while in vitro tests also indicated their increased cellular up-take and cytotoxicity to a non-small-cell lung cancer CRL-5802 cell line, even when compared to the free drugs [137,138].

Separately, Huang's group and Oh's group synthesized a PLA-based hydrophobic ATRP macroinitiator (PLA-Br) via the ROP of racemic D,L-LA initiated by HO-SS-Br [139,140]. Subsequently, PLA-Br was utilized in the ATRP of methacrylate monomers bearing pH-sensitive amine groups: 2-aminoethyl methacrylate (AEMA) [139] or DMAEMA [140]. In both cases, the final step of amphiphilic LBCP synthesis included a post-polymerization chemical modification of the pendant amine groups aimed at the formation of cations (via the quaternization of DMAEMA units with methyl iodide) [140] or anions (via the amidation of AEMA units with dicarboxylic acid cyclic anhydrides) [139]. On one hand, the incorporation of ionic moieties within the micelles' shells made them more hydrophilic, but also endowed them with some additional features: increased stability and resistance to protein-fouling during circulation in the bloodstream (a case of nano-assemblies with polyanionic shells) [139] or the capability to form polyplexes with oligonucleotides used in the gene therapy (a case of the polycationic nanocarriers) [140], without compromising the ability of the micelles' hydrophobic cores to DOX encapsulation. It is worth noting that a shell charge of the micelles studied by Huang and coworkers showed an interesting dynamic dependence on the pH of the microenvironment: though negative at physiological pH, it became positive at the more acidic tumor site and thus improved the up-take of the drug-loaded micelles by tumor cells (cationic species are more prone to endocytosis) [139].

The use of the PEG-based hydrophilic ATRP macroinitiator in the synthesis of reduction-susceptible shell-sheddable LBCP micelles was reported by Oh's group [141,142]. MPEG activated with 1,1′-carbonyldiimidazole was esterified with HO-SS-Br via a carbonate linkage and then served as a macroinitiator in the ATRP of hydrophobic unsaturated monomers [141]. Model studies carried out for styrene ATRP showed that the resulting amphiphilic LBCPs self-assembled in aqueous media into nanostructures with different morphologies (micelles, vesicles, or rod-like particles), which depended on both the structure of copolymer chains and processing conditions [141]. In another study, the same general procedure was applied for the preparation of DOX-loaded micelles, albeit styrene was replaced with a more hydrophobic methacrylate monomer derived from rosin, namely dehydroabietic ethyl methacrylate [142]. In vitro tests showed that this type of DDS was characterized by a good stability toward proteins, increased cellular uptake, and promptly released DOX in contact with reductive GSH upon internalization into HeLa tumor cells (a cumulative DOX release after 24 h of incubation changed from ca. 10% to almost 50% when the concentration of GSH changed from 0% to 10 mM) [142].

The third class of SS-cleavable LBCP micellar DDSs comprises copolymers, in which hydrophobic polymeric blocks contain pendant substituents bearing SS bonds. Similarly to their pH-sensitive analogs, this type of drug-loaded micelles delivers their therapeutic cargo as a result of the disturbance of their hydrophilic/hydrophobic balance occurring upon a reductive breaking of SS bonds at the target site. This type of LBCPs is easily produced by a direct ATRP approach using MPEG-Br, as a hydrophilic macroinitiator, and the SS-containing methacrylate monomer. One of the most commonly used monomers for that purpose is the product of a two-step Steglich esterification of 3,3′-dithiodipropionic acid with HEMA and ethanol (HMSSEt) [143,144]. Oh and coworkers used this approach for the preparation of the micelle-forming diblock copolymer MPEG-*b*-PHMSSEt and reported its good encapsulation and drug-releasing properties toward NR (as a simple drug-modeling molecule) [143], as well as antitumor DOX [144]. A DOX-loaded system also exhibited desirable cytotoxicity toward HeLa cancer cells, comparable with free DOX molecules [144].

Although LBCPs with SS-containing pendant groups can themselves be used in the formulation of the micellar DDSs, another attractive possibility of their application has been also investigated in recent years—the synthesis of micelles with cross-linked cores. Such systems have been proposed as a way to circumvent the problems associated with CMC values that are too high, exhibited by many classical (non-cross-linked) micellar DDSs, which experience instability upon dilution (i.e., during circulation in the bloodstream). Oh and coworkers explored this idea in the case of a symmetrical double hydrophobic triblock copolymer consisting of the PEG hydrophilic central block and two terminal hydrophobic PHMSSEt blocks [95]. The authors observed that by subjecting PHMSSEt-*b*-PEG-*b*-PHMSSEt micelles to a catalytic amount of a reductant (e.g., D,L-dithiothreitol, DTT), one could cleave only a small number of SS moieties—too few to lead to the disassembling of the micelles. Instead, the newly formed sulfhydryl groups attached to one copolymer chain acted as reductants in the thiol–disulfide exchange reaction with the SS groups of the adjacent chains, thereby resulting in the formation of the covalent cross-links between those chains (see Figure 9a). Differential light scattering (DLS) measurements showed that when subjected to an excess of solvent, these cross-linked micellar systems increased their sizes (an effect of core swelling), contrary to the non-cross-linked analogs experiencing disassembling. It should be noted that both in vitro and in vivo tests proved the complete degradation of the SS-cross-linked cores of the DOX-loaded micelles upon contact with intracellular concentrations of GSH, although they released their payloads a little bit slower than their non-cross-linked counterparts [95].

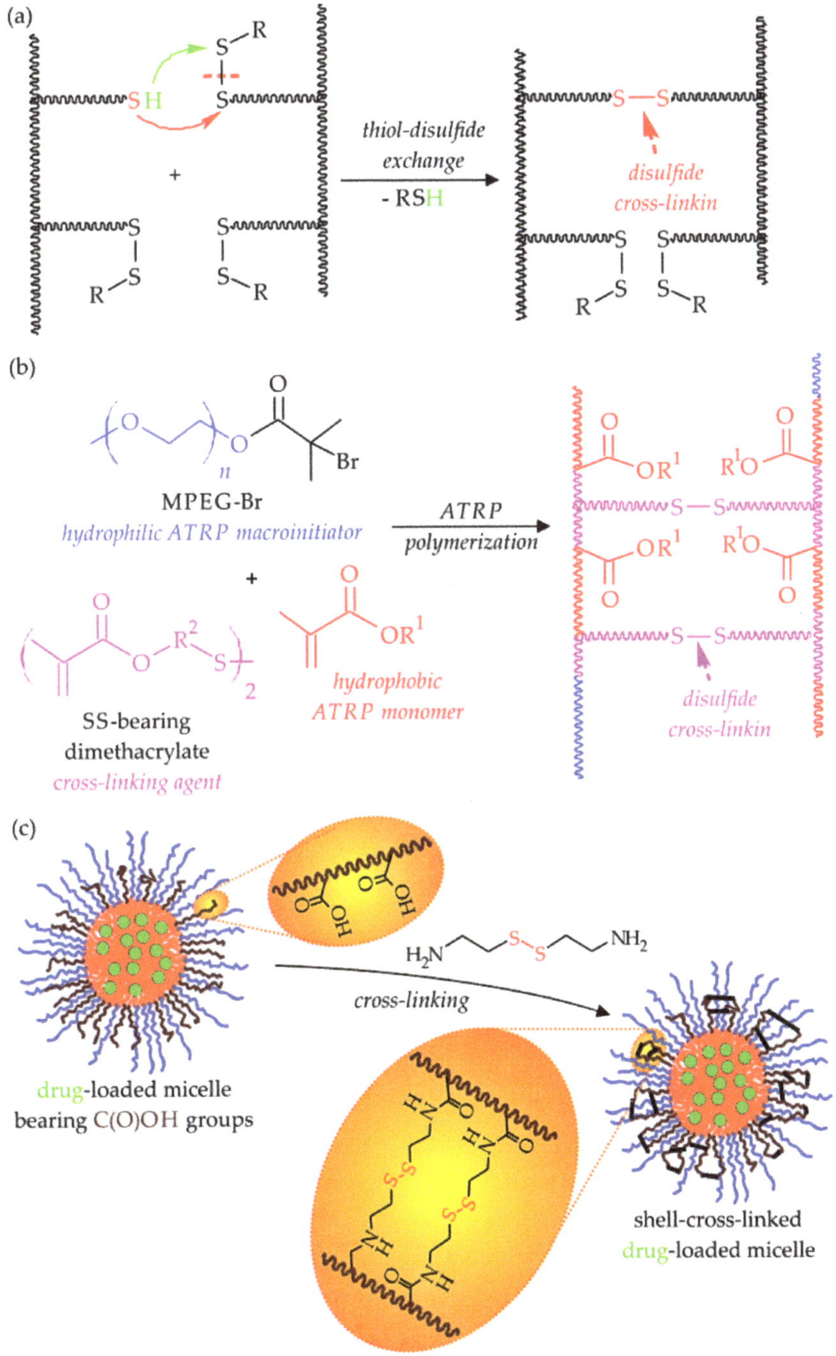

Figure 9. Different synthetic strategies of the SS-based chemical cross-linking of the micellar DDSs: (**a**) reductive cross-linking of the pendant SS-bearing groups, (**b**) cross-linking via a direct ATRP (co)polymerization, (**c**) cross-linking via a post-polymerization modification.

Interestingly, the cross-linking of the ATRP-derived LBCP micelles via reduction-degradable SS bonds can be achieved by methods other than the one described above. For example, Chhikara and coworkers utilized the inverse miniemulsion AGET ATRP technique to copolymerize DMAEMA with a newly synthesized SS cross-linker (a symmetrical dimethacrylate-containing short PEG blocks linked via SS bonds), in the presence of an MPEG-Br initiator (Figure 9b). After loading with DOX, they obtained nanoassemblies responding to both the changes in the concentration of GSH and the pH of tumor tissues, which were also cytotoxic toward HeLa cells [145]. A similar strategy, utilizing direct polymerization for core-cross-linking of the micelles, was described by Liu and coworkers [146]. First, they synthesized an amphiphilic, bromine-terminated diblock copolymer (via ATRP of tBuA initiated with MPEG-Br) and then used it as an ATRP macroinitiator to copolymerize tBuA and SS-bearing diacrylate cross-linking agent, namely *N,N'*-bis(acryloyl)cystamine. The resulting core-cross-linked micelles, after additional hydrophilization (i.e., acidolysis of *tert*-butyl side groups), could easily encapsulate hydrophobic DOX molecules and subsequently release them in a controlled manner when triggered by the acidification of the environment and an increase in the concentration of GSH. Their cytotoxicity toward HeLa tumor cells was on a comparable level (e.g., ca. 40% cell viability at a dosage of 10 μg/mL) with that of a free DOX [146].

A different approach was proposed by Petrov and coworkers, who investigated the post-polymerization cross-linking of mixed co-micelles composed of two double hydrophilic symmetrical triblock copolymers containing hydrophobic PCL as their central block [126]. One of these copolymers contained long hydrophilic PEG blocks (PEG-*b*-PCL-*b*-PEG), whereas the other one, PAA-*b*-PCL-*b*-PAA, contained much shorter hydrophilic segments of PAA and was obtained via the ATRP of tBuA initiated on both chain-ends of the brominated PCL, followed by the acidolysis of the ester moieties to free carboxylic groups. The co-micellization of these two copolymers resulted in nanoassemblies composed of a common PCL core and a mixed, bilayer, hydrophilic shell containing PEG and PAA blocks of different lengths. After loading with caffeic acid phenethyl ester (CAPE), the hydrophilic coronas of these micelles were cross-linked via the introduction of cystamine dichloride and its reaction (i.e., amidation) with the carboxylic groups of PAA segments (Figure 9c). Because of the difference in the length of PEG and PAA segments, the cross-linking reaction took place in the inner layer of the micelles' hydrophilic coronas and stabilized the hydrophobic core without comprising its ability to encapsulate CAPE. In fact, under simulated physiological conditions, no CAPE leakage was detected; however, at mildly acidic pH or in a reductive environment, these micellar DDSs quickly released their therapeutic cargo [126].

Since the quick and efficient release of drug molecules from the drug-loaded nanoassemblies internalized by tumor cells is a key factor in the case of smart micellar DDSs, the design and synthesis of micelles that can be disassembled via two (or more) different mechanisms operating simultaneously have seen growing interest in the last decade. This concept was also investigated in the case of reduction-susceptible DDSs. For example, Oh and coworkers reported dual-site redox-responsive micelles formed by a linear diblock copolymer containing MPEG and PHMSSEt blocks [147]. These micelles were susceptible to disintegration not only via the changes in hydrophobicity of the PHMSSEt core (resulting from a GSH-induced cleavage of SS bonds in its pendant groups), but also due to a detachment of the MPEG hydrophilic corona caused by the breaking of the SS bond at the block junction. In order to synthesize this new type of shell-sheddable/core-degradable LBCP, the authors utilized the ARGET ATRP technique and the SS-containing bromine-ended MPEG macroinitiator (a product of MPEG esterification with HO-SS-Br) to copolymerize HMSSEt. The authors also showed that these micelles loaded earlier with DOX experienced no drug leakage at the non-reductive conditions, whereas in the presence of any reductant, they very quickly released their payload (e.g., up to 70% of the encapsulated NR indicator after ca. 5 h of incubation in the reductive environment). During in vitro tests, they also exhibited a similar profile of cytotoxicity toward HeLa tumor cells as the free DOX [147]. Recently, the same research group has optimized the processing of the abovementioned micellar DDS by introducing an interesting concept: the so-called "lab-on-chip" flow synthesis of micelles in a two-phase microfluidic reactor [148,149].

The concept of an enhanced drug release through the synergistic effect of two micelle destabilization mechanisms was further extended to the joint action of two endogenous stimuli. The correctness of this strategy is proven in the study which Liu's group reported, that in the presence of GSH, the cystamine-cross-linked micelles containing pH-sensitive PAA segments showed two to three times more accelerated DOX release at a pH of 5.0 compared to non-reducing conditions [146]. Several other research groups have applied a direct polymerization approach in the synthesis of this type of pH/redox dual stimuli-responsive micellar systems. Oh and Jazani obtained acid shell-sheddable pendant SS-cleavable micelles via a multistep synthetic procedure: starting from MPEG and ethanolamine, they synthesized ketal-containing amine-capped MPEG, which, upon amidization with BIBB, resulted in the macroinitiator utilized in a subsequent step of the ATRP of HMSSEt. Thus, a diblock copolymer was obtained, consisting of a hydrophilic MPEG segment linked through a pH-labile ketal moiety to a hydrophobic polymethacrylate block having multiple redox-responsive SS pendant groups [150]. Oh's research group also reported on a more sophisticated redox-responsive amphiphilic LBCP additionally equipped with two types of pH-sensitive sites: an acid-cleavable acetal group located at the junction of their hydrophilic and hydrophobic blocks as well as the DMAEMA units, which are easily protonated in acidic media [151]. This copolymer was produced via the ARGET ATRP of HMSSEt, starting from an acetal- and bromine-bearing bifunctional MPEG macroinitiator, and formed colloidally stable nano-sized micelles (ca. 83 nm in diameter), whose hydrophobic cores were additionally cross-linked under reducing conditions. Interestingly, due to the presence of dimethylamine moieties, the obtained dual-location dual pH/reduction-degradable micelles were capable of pH-reversible nucleic acid complexing, potentially useful in a gene therapy [151]. This type of gene-delivering system, namely polyplexes, will be discussed in a separate section of this review.

3.1.4. Micelles Responsive to External Stimuli

In order to increase the efficiency of drug release and to accelerate polymer degradation, dual- and multi-responsive DDSs, utilizing external stimuli, have been repeatedly highlighted. Of the possible stimuli, thermosensitive systems are advantageous for clinical applications, as several spatial heating systems, such as focused high-intensity ultrasound, are already used to treat tumors. The temperature-induced change in micellar function can be designed using hydrophilic blocks that become hydrophobic above the lower critical solution temperature (LCST), tuned to the local body temperature. Below the LCST, these blocks are located in the micelle corona, while above this temperature, they move to the hydrophobic core, which promotes the release of the drug molecules stored there. NIPAM copolymers are most commonly used to prepare thermosensitive blocks. Still, the additional hydrophilic segments must be introduced into the micelle corona to suppress the formation of large intercellular aggregates above the LSCT. For the synthesis of such micelles, the RAFT method is mainly used [152], but in the following section, we will show examples of carriers obtained using ATRP. For example, it was proven that micelles formed by the copolymer poly(ethylene glycol)-*ss*-(poly(dimethylaminoethyl methacrylate)-*co*-poly(2-nitrobenzyl methacrylate)) (PEG-*ss*-(PDMAEMA-*co*-PNBM)) could respond to various stimuli, such as pH, dithiothreitol (DTT), temperature, and UV light irradiation [153]. In the presented study, NR was used as a hydrophobic model drug. Using a single stimuli trigger, the NR release was 28% when the temperature was increased to 50 °C, 40% after 10 mM of DTT addition, 80% at basic conditions (pH = 11), and 89% in 30 min when the UV light irradiation was applied. However, the highest cumulative release was achieved when the combination of triggers was used, that is, UV irradiation under pH = 11. In that case, the NR release increased up to 93% in 60 min. Other combinations of the triggers, that is, UV irradiation under pH = 11 with a reductant and pH trigger with a reductant, did not result in an increase in the release rate. The combination of UV irradiation with 10 mM DDT was suitable for core-crosslinked micelles prepared from amphiphilic block copolymer methoxy poly(ethylene glycol)-*b*-poly(3-azido-2-hydroxy-propyl methacrylate-*co*-*o*-nitrobenzyl methacrylate) (mPEG-*b*-P(GMA-N_3-*co*-NBM)) and alkyne-functionalized crosslinking agent containing a disulfide bond in the structure [154]. In that case, the UV irradiation was accelerating the cleavage of disulfide crosslinkers, increasing the release rate after 360 min from 56.9%, which was achieved in the reductive environment, to 73.8%. Similar results were obtained when light irradiation was combined with oxidation. The amount of NR released increased from 53.3% to 76.7%. An interesting example of UV light-breakable and thermosensitive block copolymer poly(2-nitrobenzyl methacrylate)-*b*-poly(2-(2-methoxyethoxy)ethyl methacrylate-*co*-oligo(ethylene glycol) methacrylate) (PNBM-*b*-P(MEO_2MA-*co*-OEGMA)) was proposed by Yuan and Guo [155]. Under UV irradiation, hydrophobic poly(2-nitrobenzyl methacrylate) (PNBM) was converted into hydrophilic poly(methacrylic acid) (PMA) and the micelles were dissociated. When the solutions were heated, poly(methacrylic acid)-*b*-poly(2-(2-methoxyethoxy)ethyl methacrylate-*co*-oligo(ethylene glycol) methacrylate) (PMA-*b*-P(MEO_2MA-*co*-OEGMA)) copolymers re-self-assembled into micelles with poly(2-(2-methoxyethoxy)ethyl methacrylate-*co*-oligo(ethylene glycol) methacrylate) (P(MEO_2MA-*co*-OEGMA)) core and PMA shell. Smart block copolymers have also been synthesized by Jazani and coworkers [156]. A triple stimuli-responsive copolymer exhibited responses to acid, reduction, and light. The preparation of block copolymers included the ATRP of carbonyl imidazole methacrylate in the presence of a difunctional initiator with disulfide bonds and two acetal linkages, followed by the postpolymerization reaction of carbonyl imidazole with an *o*-nitrobenzyl amine. It was proven that UV irradiation caused NR release up to 70% in 10 h; meanwhile, the diffusion of NR was enhanced when both stimuli, 10 mM GSH and pH = 4.2, were simultaneously applied. In that case, the NR release rose from 20% to 90%, compared with that in the experiment when the single stimuli was used.

3.2. *Smart DDSs Based on Polymersomes*

Polymersomes are artificial vesicles made from amphiphilic copolymers, which are more stable than liposomes and show less toxicity in vivo [157,158]. If polymersomes are assembled from polymers capable of carrying and releasing drugs, then they can be used as DDSs. Recent studies have demonstrated that synthesizing polymers and copolymers using ATRP techniques results in better drug delivery performance, such as enhanced colloidal dispersion stability, raised swelling ratios, and responsiveness to a pH change, when compared to polymers synthesized by traditional radical polymerization techniques [159,160]. Polymersomes have many advantages, such as the ability to encapsulate both hydrophobic and hydrophilic substances, good biocompatibility, physical and chemical robustness, high LC, and high colloidal stability, so they are commonly researched as potential DDSs [157,161]. Moreover, they can release drugs in different target sites, depending on environmental conditions.

3.2.1. pH-Triggered Drug Release

Polymersomes release drugs while subjected to external stimuli or a change in the environment, such as a pH change (Figure 10) [162]. Lorella Izzo et al. researched pH-responsive polymersomes that could swell without disaggregation, which significantly lowered their cytotoxicity [163]. They synthesized a three-component amphiphilic copolymer utilizing Br-terminated MPEG as a copolymerization macroinitiator and MMA and DMAEMA as ATRP monomers. The MPEG formed a hydrophilic block, MMA provided hydrophobicity to the poly(methyl methacrylate)-*ran*-poly(dimethylaminoethyl methacrylate) (PMMA-*ran*-PDMAEMA) block (*ran* stands for random distribution of monomers within this block), and DMAEMA was used to trigger the pH-dependent size-change of the polymersomes. Moreover, DMAEMA can form strong hydrogen bonds, hence acting as a non-covalent cross-linker between different polymers forming the vesicle. Both linear and branched copolymers were synthesized, differing in mol% of DMAEMA (22 to 62 mol%) and M_n of the product (7–51 kDa). Linear polymers with 22–28 mol% of DMAEMA provided the best results. At a pH of 7.4, they were able to form polymersomes with a monomodal size distribution, which suggests that at this pH, no release of copolymers took place, while after the pH was reduced to 4.4, the vesicles increased 10 times in size. These vehicles were loaded with paclitaxel (PTX), releasing only 5–7% of the drug in 48 h under neutral and slightly acidic conditions and 52–41% under acidic conditions in just 2 h. Therefore, it was stated that the systems developed are able to release PTX at lysosomal pH.

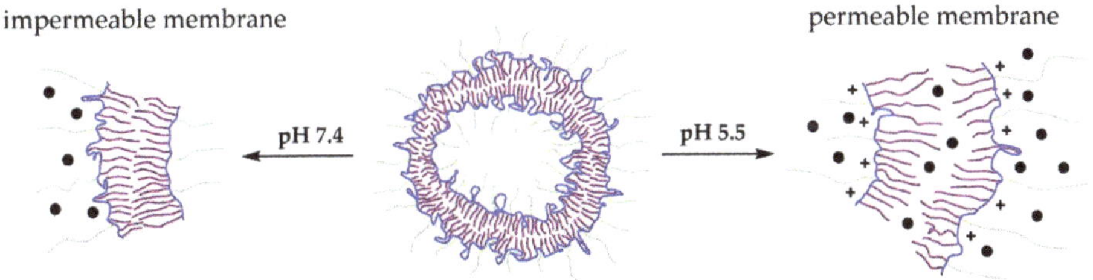

Figure 10. Schematic overview of pH-responsive polymersomes' membrane behavior in neutral and acidic environments.

3.2.2. Miscellaneous Systems

Polymersomes do not have to be only pH-sensitive. Jianzhong Du et al. developed both a pH- and ultrasound-responsive system, utilizing Br-terminated poly(ethylene oxide) PEO as a macroinitiator and DEAEMA and methoxyethyl methacrylate (MEMA) as ATRP monomers [164]. PEO was chosen as a hydrophilic block due to its biocompatibility and prolonged in vivo circulation time, MEMA to provide hydrophobicity to the poly(2-(N, N′-diethylamino)ethyl methacrylate)-*stat*-methoxyethyl methacrylate) (P(DEAEMA-*stat*-MEMA)) block (*stat* stands for the statistical distribution of monomers within this block) and ultrasound responsiveness, while DEAEMA was chosen for its pH responsiveness. Polymersomes, assembled from those polymers, decrease by 40% in size when sonicated with 40 W power and disassemble at a pH of 5.83. Polymersomes loaded with DOX hydrochloride (DOX·HCl), a chemotherapeutic agent, were tested both in vitro and in vivo. The results demonstrated that ultrasound, a non-invasive stimulus, is a valid drug-release switch, that polymersomes can successfully escape endo-/lysosomes, and that this polymersome drug system can significantly inhibit tumor growth (95% reduction in tumor mass in mice).

Other promising ATRP-synthesized polymersomes, which could serve as stimuli-responsive drug-releasing systems, are being researched. ARGET ATRP was utilized to synthesize giant, hybrid lipid vesicles from a MPEG-based macroinitiator as well as MMA and DMAEMA (monomers) [165]. 1-palmitoyl-2-oleoyl-*sn*-glycero-3-phosphocholine was used as a lipid fraction. Alan B. Gamble et al. developed polymersome systems by the ATRP reaction of 4-azidobenzyloxycarbonylaminoethyl methacrylate (ABOC) or 4-fluorobenzyloxycarbonylaminoethyl methacrylate (FBOC) (monomers), utilizing an MPEG-based macroinitiator [166]. This system proved to be pH-sensitive: it did not show particle distribution at neutral conditions, while at a pH of 4.5, the vesicles were distributed. However, to the best of the authors' knowledge, both of these systems have not yet been tested as DDSs. Table 1 shows a literature review of polymersome-forming polymers, which were synthesized utilizing ATRP methods, which includes both the already tested DDSs and those which are not fully developed yet.

Table 1. A review of the literature on the synthesis of modern polymersomes using ATRP techniques.

Encapsulated Drug	Monomers Used	The Variant of the ATRP Technique	Applications	Ref
PTX	MMA, DMAEMA	normal	pH-triggered drug release	[163]
DOX·HCl	MEMA, DEAEMA	normal	Ultrasound-triggered drug release	[164]
-	MMA, DMAEMA	ARGET	Gateway to stimuli-responsive giant hybrid vesicle DDS	[165]
-	ABOC, FBOC	normal	Gateway for pH-triggered DDS	[166]

3.3. Polyplexes

Polyplexes are artificial vesicles composed of interpolyelectrolyte complexes, which are typically made from two oppositely charged polyelectrolytes: positively charged polymers and negatively charged nucleic acids [167,168]. These opposite charges are responsible for the self-assembly of polyplexes through electrostatic condensation [169]. They can encapsulate drug molecules without chemically binding them, thus delivering them to target sites without any chemical modification and with unaffected intermolecular drug activity [170]. Moreover, positively charged polyplexes can destabilize the endosomal membrane of targeted cells as they cause the inflow of anionic molecules, creating osmotic pressure, which causes the disruption of the cellular membrane, thus leading to the internalization of the polyplexes [171,172].

Since polyplexes are capable of high-density payload condensation, they can penetrate the cell membrane, protect its contents (nucleic acids) from enzymatic degradation, and release it at the target site (for example, the tumor site). Therefore, they are widely researched as pDNA and mRNA delivery systems, especially as the injection of naked nucleic acids provides efficient protein expression in only very limited cases [173]. Moreover, polyplexes can be designed in such a way as to specifically recognize target cells. To achieve this, the polymers constituting polyplexes need to have well-defined properties, and controlled polymerization methods allow for such polymer synthesis [174,175]. ATRP is one of these methods, as it allows for great control of the polymerization or copolymerization of several monomers with low $Đ_M$ and it allows for the incorporation of functionalized side-chains in the construction of block, alternate, and grafted copolymers. Polyplexes can be divided into vesicles serving solely nucleic acid-delivery functions and both drug- and nucleic acid-delivery functions.

3.3.1. Nucleic Acid Delivery

T. Vermonden et al. recently designed a polyplex system utilizing two polymerization techniques. First, an NIPAM-based thermosensitive copolymer, PNIPAM-PEG-PEG-PNIPAM, was synthesized through traditional ATRP, using a hydrophilic PEG-based macroinitiator [176]. Then, it was polymerized with DMAEMA through free radical polymerization, leading to cationic block formation and a cloud point of 34 °C. The copolymers obtained formed polyplexes with pDNA under physiologically relevant conditions. The group compared this polyplex system with non-thermoresponsive polyplexes, assembled from PEG-based macroinitiators and DMAEMA only. They showed that the chain length of the copolymer determines the polyplex stability and that the NIPAM introduction to the polymer backbone through ATRP enables the formation of polyplexes with improved cytocompatibility, which could be caused by higher surface charge shielding. Transfection experiments revealed that thermosensitive polyplex systems could deliver nucleic acids to HeLa cancer cells, even in the presence of serum proteins. In a later study, the group showed that the thermosensitive polyplex systems obtained could be anchored in a thermosensitive hydrogel, which allowed for more controlled and sustained siRNA delivery when compared to free siRNA-hydrogel systems, leading to potential localized tumor treatment applications [177].

3.3.2. Simultaneous Nucleic Acid and Drug Delivery

Polyplexes can be multifunctional—they can be designed to simultaneously deliver nucleic acids to a target site and additionally release drugs. Such a system was proposed by P. Ni et al. [178]. They reported a reduction- and pH-triggered dual-responsive triblock copolymer galactosamine-poly(ethylethylene phosphate)-*a*-poly(ε-caprolactone)-*ss*-poly(2-(dimethylamino)ethyl methacrylate) (Gal-PEEP-*a*-PCL-*ss*-PDMAEMA), prepared via the multi-step synthetic pathway, which included the ATRP of DMAEMA. These triblock copolymers could self-assemble into micelles and form polyplexes with green fluorescence protein-encoded DNA, were biodegradable, possessed low cytotoxicity, had a decent drug (DOX) loading capability, and could release the drug load in cancer cells in a fast manner. Moreover, they targeted HepG2 cells over HeLa cells as the former were overexpressing asialoglycoprotein receptors, which interacted with the galactosamine (Gal) ligand of the copolymers. This group's research showed that their system is a promising dual-responsive DDS for simultaneous nucleic acid delivery and drug release.

Another simultaneous drug and nucleic acid delivery system was proposed by Y. Wu et al. [179]. Their biodegradable copolymer was based on a poly(3-hydroxybutyrate) (PHB)-based macroinitiator, with PHB obtained from renewable resources, and an ATRP-synthesized PDMAEMA block. The PHB block was introduced to counteract the cationic poly(dimethylaminoethyl methacrylate) (PDMAEMA) toxicity. The copolymer could form polyplexes with pRL-Relina plasmid DNA (pDNA). Tests showed that the proposed system had a better transfection efficiency than PEI (gold standard gene carrier) and that the nucleic acid and PTX drug co-delivery resulted in the death of increased drug-resistant cancer cells with a high expression of antiapoptosis Bcl-2 protein. Therefore, the PHB-PDMAEMA copolymers could be used for chemotherapy to effectively inhibit drug-resistant cancer cell growth. Table 2 shows a literature review of polyplex-forming polymers utilizing ATRP as their synthesis method.

Table 2. A review of the literature on the synthesis of modern polyplexes using normal ATRP technique.

Encapsulated Drug	Nucleic Acid Used	Monomers Used	Applications	Ref
-	pDNA	NIPAM	Thermosensitive nucleic acid delivery	[176]
-	siRNA	NIPAM	Hydrogel-aided nucleic acid delivery	[177]
DOX	DNA	DMAEMA	Redox-triggered drug and nucleic acid release	[178]
PTX	pDNA	DMAEMA	Bcl-2 targeted drug and nucleic acid delivery	[179]

4. Branched Copolymers in DDSs

4.1. Polymer Stars

The main advantage of using star copolymers in DDSs is their small hydrodynamic radius, which makes them easy to clear from in vivo systems. Compared with their linear analogs, amphiphilic star block copolymers can form aggregates in an aqueous solution with a high thermodynamic stability, leading to a relatively low critical aggregation concentration, which is very important for drug delivery carriers [180]. Additionally, if the controlled degradation of the star is applied, the system can be used to control the drug release rate. Star copolymers can be synthesized by core-first, arm-first, or coupling-onto methods [181]; however, the majority of the examples found in the literature in the last decade describe the use of the core-first strategy. This process relies on controlled polymerization in the presence of a well-defined initiator with a known number of initiating groups. The star copolymer is created in a one-step process; however, the number of arms is rather limited due to the small core molecules which are usually applied [182]. Star copolymers used for pH-responsive, thermo-sensitive drug delivery, and the delivery of nucleic acid-based drugs have been prepared by several research groups. Chmielarz et al. described the synthesis of six-armed copolymers with meso-inositol as the core and hydrophilic poly(di(ethylene glycol) methyl ether methacrylate) (PDEGMA) and amphiphilic poly(di(ethylene glycol) methyl ether methacrylate)-*b*-poly(methyl methacrylate) (PDEGMA-PMMA) as the arms [183]. These vitamin-based star polymers, produced by low-ppm ATRP, potentially can work as thermo-sensitive DDSs. In another study, a six-armed star triblock copolymer poly(2-(diethylamino)ethyl methacrylate)-*b*-poly(methyl methacrylate)-*b*-poly(poly(ethylene glycol) methyl ether methacrylate) (*s*-$(PDEA_{62}$-*b*-$PMMA_{195}$-*b*-$PPEGMA_{47})_6$) was tested as a potential pH-responsive delivery carrier [184]. The results showed that the LC of the star copolymer was 33–35 wt% (relative to the polymer) at a pH of 7.4, 26–28 wt% at a pH of 10.5, and 10–15 wt% at a pH of 2.0. A bit lower LC was exhibited by the miktoarm star block copolymer, MPEG-*b*-P(MMA-*co*-$MAA)_2$ [185]. The maximum values of the LC and drug encapsulation efficiency were 10.3% and 48.7% for MPEG-*b*-P(MMA_9-*co*-$MAA_{35})_2$ micelles and 16.5% and 82.3% for MPEG-*b*-P(MMA_{24}-*co*-$MAA_{25})_2$ micelles, respectively. Between 50% and 90% efficiencies of indomethacin encapsulation were also obtained for the four-arm star copolymers containing methyl (meth)acrylate and (meth)acrylic acid units [186]. The largest amount of drug (85%) was released within 96 h from micelles based on MA/MAA stars containing 24% of the hydrophobic fraction. An example of UV-cleavable unimolecular micelles was described by Liu [187]. Star-PMMA-PPEGMA, synthesized with photolabile *o*-nitrobenzyl groups at the cyclotriphosphazene core, turned out to have a great tendency to dissociate and release an encapsulated drug on dilution under physiological conditions.

In recent decades, cationic polymers have shown great competence in medical applications, including drug delivery. One of the most popular cationic polymers synthesized by ATRP is water-soluble pH-sensitive PDMAEMA. Due to the tertiary amine groups at the surface of the polymer, which become partially protonated at the physiological solution, DMAEMA possesses cationic charges [188]. An interesting example of an eight-armed star, positively charged copolymer was proposed by Zheng et al. [189]. In this case, star PDMAEMA was synthesized using a calix [4]-resorcinarene initiator and in the next step, hydrophobic blocks of poly(methyl methacrylate) or poly(butyl acrylate) were incorporated via the "one-pot" method. Star polymers with a narrow molecular weight distribution and particle size in the range of 20.3–36.6 nm were successfully obtained. A similar copolymer structure was proposed by Dworak [190]. In that case, star block copolymers were created from 28-arm poly(arylene oxindole) core, cationic DMAEMA, and nonionic (ethylene glycol) methyl ether methacrylate (DEGMA). The introduction of DEGMA segments into the star arms allowed for lower cytotoxicity in comparison to homopolymer PDMAEMA. These systems are dedicated to the delivery of plasmid DNA in gene therapy. In another study, Cho et al. designed PEG-based star polymers with a cationic core and evaluated their feasibility for nucleic acid delivery [191]. The star polymers were synthesized by the ATRP of DMAEMA and ethylene glycol dimethacrylate (EGDMA). The obtained polyplexes exhibited a high efficiency in nucleic acid delivery, particularly at relatively low star polymer weights or molar ratios.

Degradation is important in drug delivery to reduce the accumulation of polymeric materials in the body. Smaller fragments can be easily metabolized and subsequently excreted out of the body [192]. A combination of ATRP and ROP can bring interesting star copolymers with a well-defined molecular weight, architecture, functionality, and biodegradability [193]. PCL is used as a biodegradable block in the majority of cases; however, some examples of PLA and polyglycolide also can be found [194]. Biodegradable polyesters, synthesized by the ring-opening polymerization of cyclic esters in the presence of tris(hydroxymethyl)ethane [195,196], pentaerythritol (redox and lower–upper critical solution temperature (LCST-UCST) thermoresponsive transition) [197,198], 2-azidoethyl D-gluconamide [199], hexakis[p-(hydroxymethyl)phenoxy]cyclotriphosphazen [200], and β-cyclodextrin core, can act as hydrophobic macroinitiators in ATRP reactions [180]. For example, the PCL-based core was modified to yield halogen-terminated, three-arm or six-arm star-shaped PCL-*b*-PHEMA macroinitiators for ATRP, from which self-assembling noncytotoxic micelles were formed [196]. The LC and drug encapsulation efficiency were higher for the six-arm structure than for the three-arm structure and reached 9.16 and 69.8%, respectively. The highest drug release cumulant of 6sPCL-*b*-PHEMA micelles could reach a high level of 75%. Thermosensitive and highly drug-loaded micelles were also prepared using the three-arm PLA macroinitiator (3-arm PLA-*b*-PNIPAM) [195]. The obtained system offers a stable and effective platform for cancer chemotherapy with camptothecin (CPT). In some cases, not only the macroinitiator was degradable. Recently, Teng and co-workers prepared a biobased miktoarm star copolymer from soybean oil, isosorbide, and caprolactone [201]. In their studies, they used 1,4: 3,6-Dianhydro-D-glucitol 2-acrylate 5-acetate monomer as the substitute for styrene.

4.2. Polymer Combs and Brushes

Polymer brushes and combs are long-chain polymers (backbones) or surfaces, to which linear polymers (side-chains) are attached [202,203]. In brushes, the distance between grafting points is smaller than the side-chains' end-to-end distance, while in combs this distance is larger [204]. Both of these polymer classes can be stimuli-responsive, form vesicles, and be tailored to target specific cells; hence, they can be effectively utilized as DDSs. They can be synthesized by "grafting-to" (a chemical reaction between reactive groups of side-chains and backbone) and "grafting-from" (monomer polymerization from backbone active sites) approaches. The technique, which is nowadays most commonly used to synthesize polymer brushes by the "grafting from" approach, is ATRP [205].

To encapsulate drugs, polymer brushes and combs can form micelles. F. Cellesi et al. synthesized a series of comb and brush block PCL and PEG copolymers, with PEGMA being utilized as an ATRP monomer [206]. The copolymers' self-assembly and dexamethasone (DEX) drug encapsulation capabilities were based on the PCL/PEG ratio and molecular weight. The best copolymer for drug delivery application was a brush grafted from a four-arm star-shaped backbone. Another interesting system was proposed by H. Wei et al. [207]. They synthesized a reduction-sensitive amphiphilic cyclic brush PHEMA-*g*-PCL-disulfide link-poly(oligo(ethyleneglycol) methacrylate) with an ATRP monomer oligo(ethyleneglycol) methacrylate (OEGMA). The copolymer self-assembles into micelles with enhanced stability, which could be destabilized by the reducing environment, such as the one in tumor cells. It can also encapsulate DOX, an anti-tumor drug. Therefore, this system could be useful in chemotherapy.

Polymer brushes can also form stimuli-responsive polymersomes. As cancer cells have different redox potentials than normal cells and the extracellular matrix, redox-sensitive polymersomes could be suitable for cancer therapy [208]. Veena Koul et al. tested this idea by utilizing a PLA-based macroinitiator for PEGMA ATRP [209]. The polymers formed were biocompatible, biodegradable, hemocompatible, and conjugated with folic acid, hence they were also redox-sensitive, pH-sensitive, easily self-assembled into polymersomes, and possessed a disulfide bridge in their polymer backbone, which prevents rapid drug release in cancer cells. Moreover, hydrophilic, polymeric chains of PEGMA monomeric units prevent proteins from being adsorbed on the vesicle surface, thus preventing the immune system response. Polymersomes loaded with DOX have shown different drug-releasing behavior in different pH and GSH concentrations (the substance responsible for different redox potential of cancer cells). In vivo studies have shown that polymersomes loaded with DOX lead to a 96% decrease in tumor volume in mice; hence, they are greatly superior to the free drug (25% decrease in tumor volume), but also to the marketed drug DOXIL (PEG-modified liposomal DOX, 70% decrease in tumor volume). Moreover, they did not display significant toxicity to the organism [210]. Yue Zhang et al. also proposed a stimuli-responsive polymersome system composed of brush copolymers. They polymerized 2-((adamantan-1-yl)amino)-1-(4-((2-bromo-2-methylpropanoyl)oxy)phenyl)-2-oxoethyl methacrylate (ABMA) by ATRP, which, after further reactions, formed P(OEGMA-

co-ABMA)-*g*-PDEGMA graft copolymer and P(OEGMA-*co*-ABMA)-*g*-PDEGMA/β-CD-SG [211]. Both copolymers could self-assemble into polymersomes and proved to be thermo-responsive, as the PDEGMA chains collapse at 37 °C. Hence, these copolymers could be utilized as thermos-responsive DDSs; however, they have not yet been tested as such.

Polymer brushes and combs have also been widely utilized to form polyplexes. R. P. Vieira et al. in 2023 presented a deactivation-enhanced atom transfer radical polymerization (DE-ATRP)-synthesized copolymer, which could form polyplexes and be used for targeted nucleic acid delivery [212]. The DE-ATRP method was used as it allows for greater kinetic control compared to traditional ATRP [213]. The monomers used were EGDMA, DMAEMA, and plant-based β-pinene. PDMAEMA was chosen for its high gene compatibility and buffering capacity, EGDMA for its cross-linking ability and vinyl groups, allowing for post-polymerization functionalization reactions, and β-pinene to provide solution stability to polyplexes, as well as for its antibiotic resistance modulation and anticoagulant, antitumor, antimicrobial, antioxidant, anti-inflammatory, and cytoprotective properties [214,215]. The copolymer formed in a one-pot DE-ATRP reaction was a nanometric, hyperbranched amphiphilic material, which formed polyplexes with gWiz-GFP plasmid DNA (pDNA), with encapsulation values up to 75.1%. The β-pinene monomeric unit proved to provide the material with an excellent solution stability and high positive charge, allowing for smooth cellular membrane penetration. Polyplexes showed different transfection efficiency with different cell lines; hence, after further research, they can be potentially used as organ-targeted cell vectors for gene therapy. Another nucleic acid delivery polyplex system synthesized with the aid of ATRP was proposed by S. Averick et al. in 2017 [216]. They prepared fentanyl-chain-ended polymers for targeted delivery to neurons, or more specifically, to Mu opioid receptor (MOR) expressing cells. They used glycidyl methacrylate (GMA) and oligo((ethylene oxide)methacrylate) (OEOMA) as ATRP monomers and fentanyl species (Fen-Acry-EtBPA) as an ATRP initiator to produce diblock copolymers, which had a high affinity to MOR. The OEOMA was chosen for its hydrophilic properties and biocompatibility, while the GMA was chosen to provide post-polymerization functionalization opportunities [217,218]. At this step, the fentanyl conjugate was obtained; therefore, it will be discussed from this point of view in Section 6.2. The chain-end fentanyl polymers formed were fully biocompatible, formed polyplexes with siRNA, and could be bound and internalized by SH-SY5Y cells, which express MOR endogenously. The siRNA binding properties were proven to be correlated to polymer length and charge, with longer polymer chains with a higher cationic charge binding siRNA more efficiently. This allows for tuning the copolymer and hence the polyplex properties; however, further studies are required to improve and optimize nucleic acid delivery. In 2017, an acid-sensitive polyplex gene vector system was proposed by X. Jiang et al., which was achieved through ATRP and ring-opening reactions [219]. GMA was used as an ATRP monomer and modified poly(β-cyclodextrin) as an initiator. The copolymer formed was later modified with ethanolamine to form brush-shaped, pH-sensitive, cationic host modules. The polyplexes were assembled with pcDNA3-Luc pDNA and then modified with adamantly based guest molecules. This not only provided the polyplexes with a stealth effect, which improved nanoparticle stability, but also allowed for targeted nucleic acid delivery, as the polyplexes targeted cells that were over-expressing folate receptors, just like cancer cells. This system also possessed a high gene condensation capability, low cytotoxicity, and high transfection efficiency. V. Koul et al. also proposed a polyplex DDS in 2017 [220]. They used ATRP to synthesize redox-sensitive polymer PPEGMA-*s*-*s*-PCL, with PEGMA as an ATRP monomer. The PEGMA side chains were short, which led to a negligible chance of evoking an immune system response and a stealth effect. The copolymers could self-assemble with pololike kinase 1 siRNA to form polyplexes, which could be loaded with DOX. The drug release could be regulated by low pH and redox conditions and the simultaneous nucleic acid and drug release has led to tumor growth inhibition during tests, making these polyplex systems suitable for tumor-specific delivery.

Table 3 shows a literature review of polymer combs and brushes synthesized utilizing ATRP, which includes both the already tested drug-releasing systems and those which are promising but not fully developed yet.

Table 3. A review of the literature on the synthesis of modern polymer combs and brushes using ATRP techniques.

Encapsulated Drug/Nucleic Acid	Monomers Used	The Variant of the ATRP Technique	Applications	Ref
DEX	PEGMA	normal	Tuning molecular architecture for tailoring drug-releasing properties	[206]
DOX	OEGMA	normal	Redox-triggered drug release	[207]
DOX	PEGMA	normal	Redox-triggered drug release	[209,210]
-	ABMA	normal	Gateway for thermos-responsive drug release	[211]
pDNA	EGDMA, DMAEMA, β-Pinene	DE/AGET	Nucleic acid delivery	[212]
siRNA	OEOMA, GMA	ARGET	MOR-targeted nucleic acid delivery	[216]
pDNA	GMA	normal	pH-triggered nucleic acid delivery	[219]
DOX, siRNA	PEGMA	normal	Redox-triggered drug and nucleic acid release	[220]

5. Smart DDSs Based on Nanoparticles Coated with Polymers Obtained via ATRP

Recently, several research groups proposed a novel strategy for utilizing ATRP in smart DDS synthesis. It envisages drug delivery via hybrid nanoparticles (HNPs) exhibiting a core-shell structure—these hybrid nanocarriers consist of an inorganic core made of metal or metal oxide nanoparticles, whose surface is modified with polymeric chains made by ATRP. A substantial advantage of this type of DDS is its obvious resistance to the destruction and premature release of a therapeutic payload (due to the dilution in the bloodstream) in comparison to micelle-based analogs, which are always characterized by the specific values of CMC.

5.1. Metal Oxide-Based Nanocarriers

Ensafi and coworkers utilized 2-bromopropionyl bromide for the modification of the hydroxyl-containing surface of the hydrothermally synthesized nanoparticles of ZnO or TiO_2, thus obtaining macroinitiator nanoparticles from which the ATRP of DEAEMA was started [221]. Blank HNPs grafted with PDEAEMA chains had hydrodynamic diameters of ca. 55 nm, which were increased to 75–85 nm after loading with the anticancer drug flutamide. The drug was loaded into the PDEAEMA shell in situ (during ATRP in the presence of the dissolved flutamide) and kept there via hydrogen bonding between its amine groups and carbonyl functionalities of the grafted polymer. HNPs with flutamide percentages 2–10% (in relation to the content of PDEAEMA) were tested showing an accelerated drug release at acidic conditions (pH = 5) [221].

A different type of HNP was investigated by Alswieleh and coworkers, who modified mesoporous silica nanoparticles with amphiphilic (co)polymers containing PDEAEMA and poly(oligo(ethylene glycol) methyl ether methacrylate) (POEGMA) or poly(2-(*tert*-butylamino)ethyl methacrylate) (PTBAEMA) and PEGMA as hydrophobic and hydrophilic blocks, respectively [222,223]. In both cases, before the ATRP procedure, the surface hydroxyl groups of the SiO_2 nanoparticles were subjected to silanization with (3-aminopropyl)triethoxysilane (APTES) and then amidation with BIBB. In the case of the PDEAEMA-based system, hydrophobic and hydrophilic blocks were linked together via succinic acid and cysteine linkages [222], whereas the PTBAEMA block was directly copolymerized (by ATRP) with hydrophilic PEGMA [223]. Therefore, the synthesized HNPs were utilized as nanocarriers for water-soluble drugs, such as anticancer DOX (in its hydrochloride form, DOX·HCl) [222] or doxycycline (a tetracycline antibiotic) [223], exhibiting drug loading efficiencies of 69% or 38–44%, respectively. Interestingly, the mechanism of drug loading in these polymer-modified HNPs is based on the physical entrapment of drug molecules inside pores present in the silica core, although it can be additionally supported by interactions with drug-complexing functional moieties (e.g., amine groups) introduced to the surface of silica pores [223]. Moreover, in this type of DDSs, the ATRP-derived chains of hydrophobic polymers, containing tertiary amine units, function as the pH-responsive gatekeepers. Due to electrostatic repulsion forces, caused by protonation at acidic pH, hydrophobic polymers stretch out from the surface of HNPs opening pores for the diffusion of drug molecules, then at physiological pH, they collapse onto the surface of HNPs, closing the pores and entrapping the drug within them, and finally, after endocytosis into tumor cells, the polymers once again open the pores in the silica core, releasing the drug directly into endo-/lysosomes (see Figure 11).

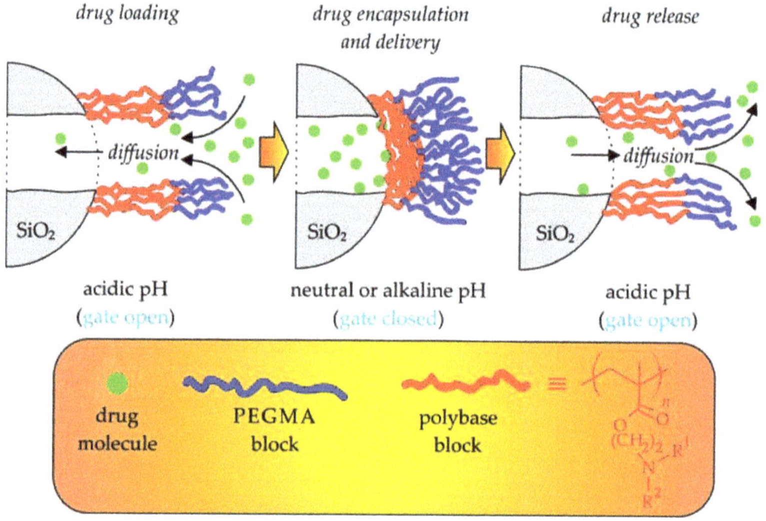

Figure 11. Schematic illustration of drug loading, delivery, and release from the surface-modified HNPs containing mesoporous silica cores.

It is worth noting that although the properties of the above-mentioned systems (e.g., their pH-responsive behavior) depend on the type of monomer and processing condition chosen for ATRP, they can be also tuned during post-polymerization chemical modification. For that purpose, one can utilize a well-known quaternization reaction with alkyl iodide to introduce new organic substituents and/or cationic sites to the amine-containing pendant groups of polymer chains grafted from a silica surface via ATRP. An example of such modification was reported by Alswieleh and coworkers in the case of the PTBAEMA, which was reacted with 2-bioethanol [224]. Depending on the reaction time and the number of quaternized amine units, HNPs behave differently at acidic conditions.

Li and coworkers proposed a different version of mesoporous silica HNPs for the delivery of DOX·HCl [225]. Instead of one amphiphilic LBCP, they used two separate homopolymers (hydrophilic MPEG and hydrophobic poly(2-(1-piperidino)ethyl methacrylate) synthesized via ATRP) for the modification of the surface of silica nanoparticles. Nevertheless, the mechanism of drug loading/release (the "gatekeeper model") did not change. The authors obtained DOX·HCl-loaded HNPs with hydrodynamic diameters below 100 nm and a high drug encapsulation efficiency (ca. 65%). At physiological pH, these DDSs were stable, showed no signs of an unfavorable aggregation, and only minimal drug release

(15% after 40 h of incubation), whereas at acidic conditions, they liberated DOX·HCl in a much-accelerated manner (the cumulative release after 40 h was 68% and 84% at pHs of 6.5 and 5.0, respectively). They were also characterized by very good cytotoxicity against HeLa tumor cells.

Zhang and coworkers extended the "gatekeeper" strategy on HNPs containing pH-cleavable linkers between their inorganic, mesoporous cores and polymeric coronas [226]. They loaded DOX (loading content of ca. 14%) into the pores of APTES-modified mesoporous SiO_2 nanoparticles and subsequently closed the pores by covalently grafting their surface with the chains of an amphiphilic copolymer containing a poly[p-(2-methacryloxyethoxy)benzaldehyde] (PMAEBA) hydrophobic block and PPEGMA hydrophilic block. pH labile imine linkages were formed in the reaction of the copolymer's pendant aldehyde groups with amine groups located on the surface of silica. ARGET ATRP was used to copolymerize PEGMA and the benzaldehyde-bearing monomer, in the presence of ethyl 2-bromoisobutyrate initiator and tin(II) 2-ethylhexanoate. At physiological pH, the DOX-loaded HNPs showed a very limited drug release (the cumulative DOX release was less than 20% after 3 days of incubation), whereas at a pH of 5.0, this process accelerated more than three times due to the imine bond-breaking, detachment of the copolymer "gatekeeper", and opening of the DOX-loaded pores. Tests showed that this DDS exhibited a good cellular up-take by the HepG2 liver tumor cells, as well as increased cytotoxicity toward them (the viability of HepG2 cells cultured at the 20 mg/L concentration of the DDS decreased to less than 20% after 48 h of incubation) [226].

It should be noted that a physical entrapment of the drug within the pores of mesoporous silica is not the only pathway leading to drug-loaded silica HNPs. Wei and coworkers showed that the latter can be obtained by a simple complexation of amine-bearing drug molecules (e.g., cisplatin) with polyacid copolymers attached to the surface of mesoporous silica nanoparticles [227]. They covalently attached BIBB on the surface of the APTES-modified silica and then utilized a surface-initiated metal-free ATRP procedure to copolymerize itaconic acid (a dicarboxylic unsaturated acid) with PEGMA. Interestingly, this polymerization was induced by an organic catalyst (10-phenylphenothiazine) and visible light instead of the conventional metal/amine ligand catalyst system. The pendant carboxylic groups in the copolymer shell of the obtained HNPs strongly complexed cisplatin at physiological pH and easily released it upon acidification: a cumulative cisplatin release after 48 h increased from less than 10% at a pH of 7.4 to ca. 60% at a pH of 5.5.

Alswieleh and coworkers also studied magnetic mesoporous HNPs, although they utilized them as solely pH-responsive DOX-delivering nanocarriers [228]. Their preparation included the coating of Fe_3O_4 nanoparticles with mesoporous silica, surface functionalization with an ATRP initiator, the grafting of PDEAEMA chains via the ARGET ATRP technique, and the optional capping of the PDEAEMA chain-ends with folic acid. The authors showed that due to the protonation of tertiary amine in the PDEAEMA units, the nanoparticle's dimensions increased to ca. 750 nm in acidic media (from the initial size of ca 450 nm in a neutral or slightly alkaline environment) and the entrapped DOX was released at an accelerated rate (16% at pH = 5.0 vs. <6% at pH > 7) [228]. The double responsivity of the Fe-containing HNPs to the pH and magnetic field was experimentally proven by He and coworkers, who investigated water-soluble Fe_2O_3 nanoparticles with a dendritic–linear-brush-like triblock copolymer located on their surface [229]. ATRP was utilized for a sequential synthesis of the linear part of the copolymer—first the hydrophobic block of PDMAEMA, and then the hydrophilic block of PPEGMA. Field-dependent magnetization tests showed that at room temperature, the obtained HNPs exhibited superparamagnetic properties (e.g., no hysteresis on the magnetization–magnetic field curves) and their saturation magnetizations were within the limits usually accepted for magnetic particles destined for biomedical applications. As for the drug release properties, these Fe_2O_3-based HNPs could be loaded with DOX up to a loading capacity of ca. 7% and showed a prolonged profile of DOX release in HeLa-line tumor cells, while maintaining good biocompatibility and a very low cytotoxicity against healthy cells.

5.2. Metal-Based Nanocarriers

Over the last decade of scientific research, two main types of metal-containing polymeric DDSs have been investigated. Depending on the location of the zero-valent metal component within these nanoassemblies, one can distinguish HNPs with a central metal core or micelles having metal aggregates in their hydrophilic coronas. Their structures and synthetic strategies are shown in Figure 12.

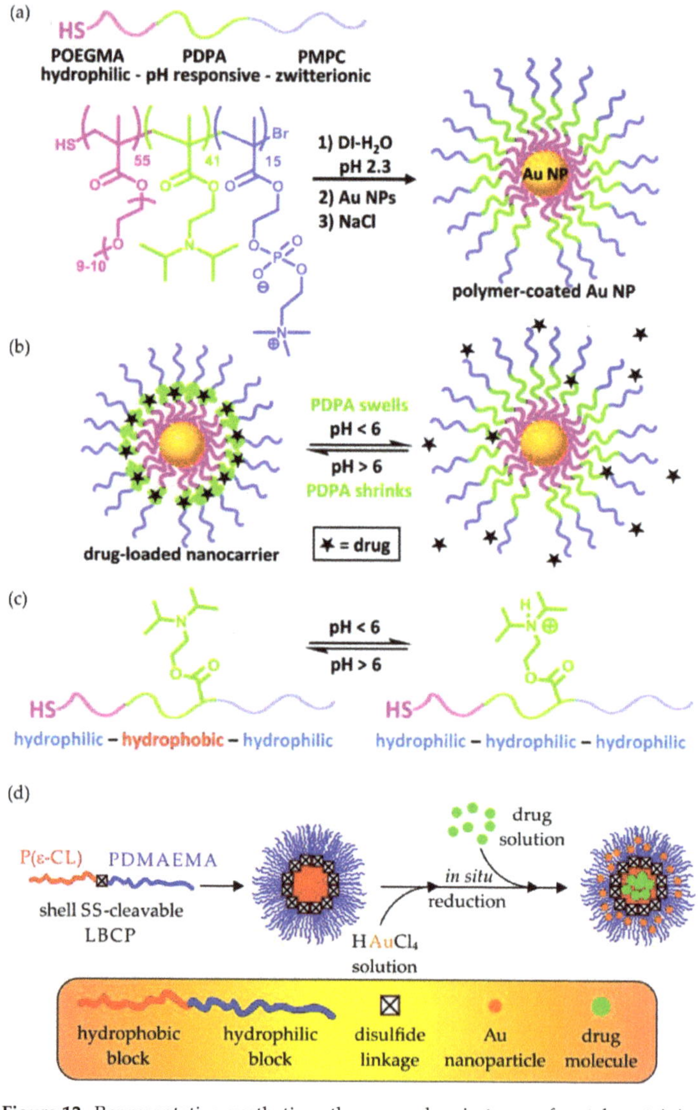

Figure 12. Representative synthetic pathways and main types of metal-containing drug delivery nanocarriers: (**a**–**c**) polymer-coated HNPs (reproduced from Ellis, E.; Zhang, K.; Lin, Q.; Ye, E.; Poma, A.; Battaglia, G.; Loh, X. J.; Lee, T.-C., Biocompatible pH-responsive nanoparticles with a core-anchored multilayer shell of triblock copolymers for enhanced cancer therapy. Journal of Materials Chemistry B 2017, 5 (23), 4421–4425, DOI: 10.1039/c7tb00654c, https://pubs.rsc.org/en/content/articlehtml/2017/tb/c7tb00654c (accessed on 26 February 2023) with permission from the Royal Society of Chemistry); (**d**) micelles with metal nanoparticle-containing coronas.

The utilization of metal nanoparticles as the inorganic cores of HNPs used for drug delivery was demonstrated by Lee and coworkers in their example of polymer-coated gold nanoparticles (hydrodynamic diameters of ca. 60 nm) subjected to loading with DOX (47% of LC, encapsulation efficiency ca. 37%) [230]. The organic coating was made of a double hydrophilic triblock copolymer, namely poly(oligo(ethylene glycol) methyl ether methacrylate)-*b*-poly(2-(*N*,*N*′-diisopropylamino)ethyl methacrylate)-*b*-poly(2-(methacryloyloxy)ethyl phosphorylcholine) (POEGMA-*b*-PDPAEMA-*b*-PMPC), anchored to the surface of the gold nanoparticles through the sulfide bond (the "grafting onto" approach). POEGMA-*b*-PDPAEMA-*b*-PMPC chains were synthesized via a reductive cleavage of the RS–SR moiety present in a symmetrical, hexablock copolymer produced through a sequential ATRP, which was started on a bifunctional initiator containing a disulfide bridge. The synthesized Au HNPs exhibited much better protein antifouling properties than their analogs coated with PEG chains, whereas their cytotoxicity against the MCF-7 breast cancer cells was greatly enhanced compared to the free DOX at the same concentration (the normalized cell viability was ca. 25% vs. ca. 80%, respectively). The pH-triggered DOX release from such HNPs (in an acidic environment) was explained by the protonation of the pendant tertiary amine groups, present within the central block of the POEGMA-*b*-PDPAEMA-*b*-PMPC copolymer, leading to its hydrophilization and electrostatic swelling [230].

Micelles with zero-valent gold nanoaggregates in their hydrophilic coronas have been reported by Zhang and coworkers [231,232]. The authors proposed that the starting point for the synthesis of the drug nanocarrier should be the preparation of either dynamic or static copolymeric micelles containing PDMAEMA blocks [231,232]. The former were obtained via the conventional self-assembling of amphiphilic LBCP chains containing redox-responsive disulfide linkages at the PCL/PDMAEMA block junction. This shell-sheddable copolymer was synthesized from a bifunctional initiator, 2-hydroxyethyl-2′-(bromoisobutyryl) ethyl disulfide via the ROP of ε-CL followed by the ARGET ATRP of DMAEMA [231]. On the other hand, single-molecule micelles resistant to dilution (unimolecular micelles) were obtained from amphiphilic 21-arm star-like copolymers composed of poly(lactide)-*b*-poly(2-(*N*,*N*-dimethylamino)ethyl methacrylate)-*b*-poly[oligo(2-ethyl-2-oxazoline) methacrylate] chains connected to β-cyclodextrin. First, β-cyclodextrin was used as a macroinitiator in the ROP of lactide and then, after esterification with BIBB of the hydroxyl chain-ends in the newly formed PLA, as a macroinitiator during a sequential ARGET ATRP of DMAEMA and oligo(2-ethyl-2-oxazoline) methacrylate [232]. The presence of the pendant tertiary amine moieties in both types of micelles was crucial for the next step of the synthesis (i.e., the formation of gold nanostructures) since they were able to actively reduce $[AuCl_4]^-$ ions to Au^0. Thus, after the infusion of the PDMAEMA-containing micelles with an aqueous solution of $HAuCl_4$, an in situ $[AuCl_4]^- \rightarrow Au^0$ reduction proceeded within the internal PDMAEMA layer of the micelle's hydrophilic corona. In both systems, the obtained gold nanoparticles had a uniform distribution and diameters of less than 10 nm [231,232]; however, their sizes strictly depended on the $HAuCl_4$ and copolymer concentration (a higher concentration of reagents promoted the generation of larger Au structures), PDMAEMA block length (the longer it was, the smaller the gold nanoparticles were), and the tertiary amine/$HAuCl_4$ molar ratio (a higher molar excess of the reducing groups resulted in smaller gold nanoparticles) [232]. Zhang's group reported that, regardless of their type, the Au-bearing micelles showed no significant cytotoxicity since the cell viability was over 80% even at the highest concentration of the micelles [231,232]. Moreover, the dynamic micelles made of the sulfide-containing LBCP could be loaded with DOX and exhibited an accelerated DOX release at the lysosomal pH and reductive microenvironment [231].

6. Bioconjugates

A large number of the drugs currently used are small-molecule compounds, which means that therapy using this form of the drug may have disadvantages, such as limited solubility, drug aggregation, low bioavailability, poor biodistribution, lack of selectivity, difficulties with targeting the therapeutic effect, and the troublesome side effects of therapeutic drugs [233]. These difficulties are being solved by the development of new preparations with better effects, acting in accordance with the principles of the DDS. An ideal DDS should allow the conjugate to find the target cell and freely penetrate the cell membrane, resulting in entry into the cell nucleus. In addition, the active substance should not be released until it has found its target cell. Through appropriate conjugation, the therapeutic efficacy of a drug can be improved, and toxic effects can be significantly reduced by increasing the amount and persistence of drugs in the vicinity of target cells while reducing drug exposure to non-target cells [234]. Drugs can be taken in different ways, such as by mouth, by inhalation, by skin absorption, or intravenously. Each method of drug delivery to the body has pros and cons. In addition, the method of delivery is strictly related to the type of therapeutic agent. Administering drugs topically, rather than systemically (affecting the whole body), is a common way to reduce the side effects and toxicity of drugs while maximizing the impact of treatment [235,236].

One of the most common methods of drug delivery is the conjugation of an active substance on a polymer carrier. This technique offers several benefits, including improved drug solubility, prolonged circulation, reduced immunogenicity, controlled release, and increased safety. In addition, it is possible to create an advanced complex DDS that, in addition to the polymer and the active substance, may contain other active ingredients that enhance the activity of the main drug [237–239]. Furthermore, polymeric materials are widely used in biomedical applications, such as implants, surgical sutures, tissue engineering, and many others [240–242]. The development of polymers as carriers of bioactive pharmaceuticals started relatively recently. In the past, it was believed that polymers were too heterogeneous in terms of molar mass, composition, and structure to be useful in the production of therapeutics. This approach changed in 1975 with the development of the first polymer anti-cancer drug by Ringsdorf, which ushered in a new era of polymer conjugate research. Ringsdorf proposed a macromolecular conjugate model that consists of a polymer backbone with three distinct regions. The first region contains moieties that modify the solubility of the conjugate, the second contains the drug (attached via a biodegradable linker), and the third contains tropic molecules (responsible for target cell recognition) [243–245]. One of the flagship, highly versatile, efficient, and sustainable controlled radical polymerization techniques is ATRP, which allows for obtaining functional polymers with well-defined structural parameters, such as molar mass and its distribution, as well as a specific architecture [246]. It is widely used as a technique for designing and obtaining multifunctional, nanostructured materials for various applications in the pharmaceutical, medical, and biotechnology industries, including drug delivery systems [35]. In the further course of this work, examples of polymer conjugates with active substances, such as proteins or therapeutic drugs, obtained using the ATRP technique will be presented.

6.1. Protein–Polymer Conjugates

Due to several advantages mentioned in the previous subchapters, ATRP techniques have found wide application in the pharmaceutical, medical, and biotechnological industries. In this section, examples from the literature of the synthesis of polymer–protein conjugates using the ATRP technique will be presented. Systems with this structure are produced to improve the efficiency of drug delivery and operation, as well as to improve their pharmacological properties. Proteins have found use as therapeutics due to some of their specific features, such as a relatively large size, high degree of structural definition, biocompatibility, and a range of diverse biological functions. However, some specific characteristics of proteins (short half-life, poor stability, low solubility, and immunogenicity) limit their wide application, making their attachment to a polymer matrix essential in the preparation of effective therapeutic drugs [247–249].

In laboratory practice, two main strategies for obtaining polymer–protein conjugates can be distinguished [250], which are presented in Figure 13. The first widely used strategy is the "grafting to" method. It consists of the initial synthesis of the polymer, which in the next stage is directly attached to the protein structure. This strategy clearly has its advantages because the polymer can be synthesized under any conditions before the final step of protein conjugation. However, disadvantages such as the low efficiency of the reaction between two large molecules and difficulties in purifying the products limit its further use [248,249,251]. The second widely used strategy is the "grafting from" method. It consists in transforming the structure of the protein, creating a macroinitiator capable of initiating the processes of controlled living polymerization, including the ATRP technique widely used in this strategy. In the next stage, the actual process of obtaining the conjugate takes place through the process of the polymerization of individual classes of monomers. The main advantage of this method is the ease of separation of small monomer molecules from protein–polymer conjugates after polymerization [252,253].

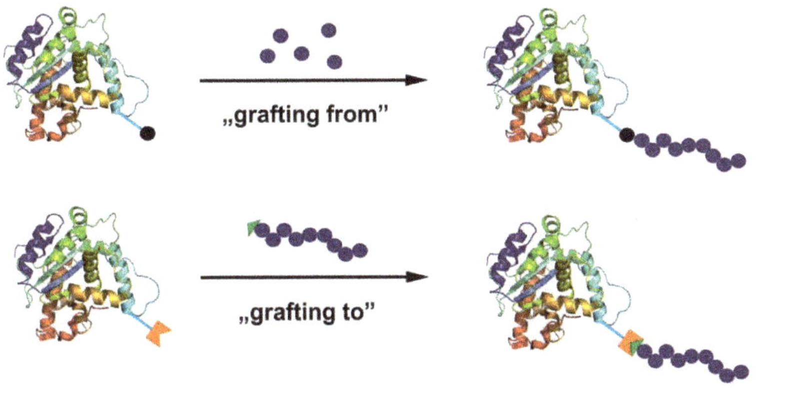

Figure 13. Two main strategies for obtaining polymer–protein conjugates.

Bontempo and co-workers have proposed a method for the synthesis of polymer–protein conjugates using the "grafting to" strategy, utilizing the ATRP technique in one of the stages and more specifically its classic variant. The first step is the ATRP of the HEMA monomer on a pyridyl initiator, containing disulfide groups, at room temperature. The PHEMA polymer is widely used in biomedical applications due to its easy and controlled polymerization process and biocompatibility, having a hydrophilic group, and forming a gel form when in contact with water, i.e., in the human body. The polymer synthesized in this way was reacted with protein and more specifically with bovine serum albumin (BSA), which is a standard reference protein used for various studies due to its availability and relatively low price. In its structure, it has cysteine, capable of forming sulfide bridges. This ability was used in the final conjugation process by introducing the previously obtained PHEMA polymer into the structure of the BSA. According to the authors, this strategy can be applied to the production of a wide range of polymer–protein conjugates without the need for post-polymerization modification of the polymers [254]. Another example of the application of the ATRP technique in the synthesis of protein–polymer conjugates by the "grafting to" strategy is the work of Sayers et al. In the first step, by polymerizing the PEGMA monomer on an aldehyde initiator with ATRP, several well-defined PPEGMA polymers were obtained. PPEGMAs were then conjugated to salmon calcitonin, a calcitropic hormone currently administered to treat a range of hypercalcemia-related diseases, by forming an N-terminal Schiff base followed by a reduction with sodium cyanoborohydride [255]. In vitro biological tests have shown that polymer conjugation does not affect the biological activity of the protein. According to the authors, the approach developed in this study appears to be of general application and could potentially open the door to the use of α-aldehyde coupling materials with different architectures for the N-terminal conjugation of a wider range of biologically relevant therapeutic proteins [256].

Notwithstanding, due to its undeniable advantages, in recent years, the strategy path has been mainly developed using the "grafting from" method, in which the processes of controlled radical polymerization play a key role, of which the lion's share are syntheses using the ATRP and RAFT techniques [67,257,258]. An interesting way of synthesizing polymer–protein conjugates using the "grafting from" strategy, as well as using ATRP techniques, has been proposed by Cummings et al. [259]. They presented a three-step synthesis of a block copolymer using an initiator with a protein in its structure. The present paper describes a novel method of protein permeation enhancement through a polymeric additive. The first step was the synthesis of an initiator based on the BSA protein, which would gain a halogen atom capable of initiating the ATRP process of the PEGMA monomer in the next act. Then, in the third step, also via ATRP, a block of N-(3-(4-phenylpiperanysyl)propyl)acrylamide was added, which, according to previous reports, may be useful as an intestinal permeation enhancer. The authors proved, that, by incorporating a block of permeation-enhancing polymer, absorption through the intestinal monolayers was increased up to 35 times compared to that of unmodified protein. The team led by Professor Dworak also researched the synthesis of polymer–protein conjugates [260]. They proposed the synthesis of an enzymatically cleavable hybrid biomaterial—poly(N-isopropylacrylamide)-pentapeptide (Gly-Arg-Lys-Phe-Gly-dansyl) conjugate, using the ATRP technique. The researchers showed that due to the hydrophilic nature of the pentapeptide bound to the polymer chain, the bioconjugate exhibited a higher phase transition temperature than that of the corresponding homopolymer. Moreover, the bioconjugate chains were able to form small-sized mesoglobules by rapidly heating the bioconjugate solution. It was also shown that the peptides formed the outer layer of the mesoglobula, which made them fully accessible to the enzyme, and the introduction of arginine or lysine into the bioconjugate structure provided the possibility of cleaving the peptide segment from the polymer anchor, which could be useful for peptide release.

Protein–polymer conjugates using the "grafting from" strategy can also be prepared by various variants of the ATRP technique. Cohen-Karni et al. proposed the synthesis of conjugates based on the well-defined acrylamide, N, N-dimethylacrylamide, and N-vinyl imidazole homo and block copolymers from a model protein BSA initiator under bio-relevant conditions, using the ICAR ATRP technique [261]. This technique allows for a significant reduction in the amount of copper catalyst needed, even to a level below 100 ppm [262]. In addition, using N-vinyl imidazole as a catalytic ligand, the authors prepared, by loading palladium, a biohybrid catalyst that successfully catalyzes the Suzuki-Miyaura coupling in an aqueous environment under aerobic conditions [261]. Moncalvo et al. synthesized and characterized various lysozyme-PPEGMA and lysozyme-poly(glycerol monomethacrylate) (PGMMA) conjugates in terms of topology (linear or bi-armed) and molar mass [263]. The process was carried out using the ARGET ATRP variant. These results highlighted the potential of PGMMA as an alternative to polyethylene glycol in extending the half-life of biotherapeutics. PGMMA is a hydrophilic synthetic polymer with a low toxicity and very limited interactions with proteins. In addition, the two hydroxyl groups it has in its structure can be easily functionalized to obtain various variants of conjugates. Researchers have also shown that appropriate polymer architecture design can help reduce enzymatic degradation.

Table 4 below presents a review of the literature on the synthesis of modern polymer-protein conjugates using the "grafting from" strategy and various ATRP techniques, depending on the type of conjugated protein and monomers used. It shows the enormity of work recently put in by scientists in the development of this type of conjugates, as well as the commonness and versatility of various variants of the ATRP technique.

Table 4. A review of the literature on the synthesis of modern polymer–protein conjugates using the "grafting from" strategy and ATRP techniques.

Type of Protein	Type of Monomers [1]	The Variant of the ATRP Technique Used	Applications and Conclusions	Ref.
Chymotrypsin	CBMA, AMA	normal	Modifying the structure by adding a polymer significantly increased protein stability and reduced protein–protein interactions.	[253]
Chymotrypsin-α Trypsin	CBMA PEGMA	normal	Protein–polymer conjugates, which can exist as a prodrug until the activator is introduced, can be used in enzyme-based biosensors and drug delivery for cancer treatment.	[264]
Chymotrypsin-α	CBMA PEGMA 3-SPMA DMAEMA	normal	Covalently attached synthetic polymers are able to modulate protein folding, emulating molecular chaperones.	[265]
Human serum albumin	DPA	normal	Promising as a new class of tumor microenvironment responsive nanocarriers for improved tumor imaging and therapy.	[266]
Interferon-α	HPMAPEGMA	normal	Promising next-generation technology that will significantly improve the pharmacological performance of therapeutic proteins with a short circulating half-life.	[267]
Lysozyme	CBMA PEGMA	normal	The covalent attachment of polymers to a protein can significantly change the protein solubility, which can be adjusted by changing the polymer type, grafting density, and polymer length. Polymer attachment increases the resistance to unfavorable environments and the thermostability of the protein.	[268]
Horseradish peroxidase	ACR	AGET	The resulting conjugates essentially retained the catalytic properties of the protein and showed significantly improved thermal stability to high temperature and trypsin digestion.	[269]
Green fluorescent protein	PEGMA	ARGET	The protein retained its bio-fluorescent properties during the process, indicating the utility of ARGET ATRP for the preparation of protein–polymer conjugates.	[270]
Lipase	DMAPAA	ICAR	A ubiquitous class of amino acid residues can be modified by ATRP initiators without affecting enzyme activity. This new amino acid modification strategy can be applied to other enzymes, providing access to new biohybrid modification schemes.	[271]
Bovine serum albumin	OEOMA	Photo	The first example of photo-ATRP using blue LED irradiation in an aquatic environment. Compared to more energetic light sources, blue light is more friendly to biological systems and allows enzymes to survive and maintain their structure and functions.	[272]
Bovine serum albumin	MSEAM	PICAR	A new sulfoxide-functional acrylamide monomer was synthesized as an alternative to PEG in some biomedical applications. It was used in the PICAR ATRP process under biologically relevant conditions without degassing the reaction mixture.	[273]
β-barrel transmembrane	NIPAM	SARA	The first example of the use of a transmembrane protein in the production of conjugates by the "grafting from" strategy, using ATRP techniques. Thanks to the preserved pore geometry, transmembrane protein–polymer conjugates can be used as building blocks of functional polymer membranes, drug and gene carriers, and nanoreactors.	[274]

[1] Monomer abbreviations not introduced in the text: carboxybetaine methacrylate (CBMA), azide methacrylate (AMA), 3-Sulfopropyl methacrylate (3-SPMA), 2-(diisopropylamino)ethyl methacrylate (DPA), 2-hydroxypropyl methacrylate (HPMA), acrylamide (ACR), N-[3-(N,N-dimethylamino)propyl] acrylamide (DMAPAA), 2-(methylsulfinyl)ethyl acrylamide (MSEAM).

6.2. Drug–Polymer Conjugates

The development of polymer carrier-based drug delivery vehicles is a very fast-growing field that has many advantages, such as the selective targeting and prolonged circulation of the therapeutic. The delivery of new therapeutic agents, combination therapies, and novel polymer architectures are very exciting and promising areas. Undoubtedly, the ATRP technique helps to create more and more new solutions in this field. In this section, examples from the literature of the synthesis of polymer–drug conjugates using ATRP techniques will be presented [5,35,237].

In laboratory practice, there are three types of strategies for obtaining covalently bound polymer–drug conjugates using ATRP techniques, which are presented in Figure 14. The key aspect is the creation of unstable covalent bonds between the drug and the macromolecular backbone. Various ways can be used to chemically link bioactive molecules to polymer chains through hydrolyzable or biodegradable bonds (for example, ester or carbonate bonds) [275].

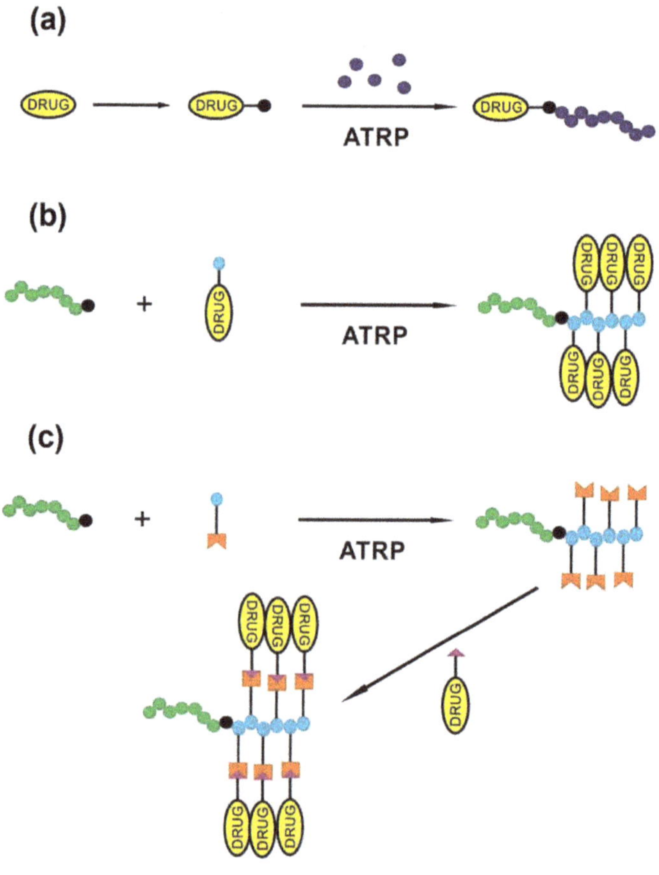

Figure 14. Three main strategies for obtaining covalently linked polymer–drug conjugates: (**a**) active substance as an ATRP initiator, (**b**) active substance as (meth)acrylate monomer, and (**c**) post-polymerization conjugation.

The first strategy is the possibility of introducing a terminal halogen group into the chemical structure of the drug, capable of initiating the ATRP reaction, similar to the "grafting from" strategy shown in the example of polymer–protein conjugates. Cohen-Karni et al. developed the synthesis of a polymer–drug conjugate based on a derivative of fentanyl acting as an AGET ATRP initiator, classified as a narcotic analgesic [216]. The paper is discussed herein in Section 4.2 due to the ability of the diblock co-polymer to form polyplexes with siRNA; however, formally, the copolymer initiated with the fentanyl derivative is a polymer–drug covalent conjugate. The mechanism of action of the drug consists in the fact that the active substance contained in the drug binds to opioid receptors in the body. As a result of this connection, these receptors are stimulated [276,277]. The main task of fentanyl is to efficiently target the MOR receptor for neuronal targeting. The introduced polymer matrix retains a high degree of binding to the receptors and allows the modification of its structure by reacting the functional groups of the monomer used, GMA in this case, in order to attach a near-infrared fluorescent dye (ADS790WS) or to build a targeted siRNA delivery system by modifying groups with secondary amines. The results obtained support the possible use of this system for delivery to MOR-expressing cells. Li et al. have developed dual-sensitive and time-controlled cationic liposomes based on a conjugate of CPT with polymeric carriers for the co-delivery of siRNA for anticancer therapy [278]. CPT is a monoterpene-indole alkaloid of natural origin, with a strong anticancer effect. This substance inhibits the activity of Topoisomerase I, an enzyme that is involved in the process of DNA replication and transcription, causing damage to the genetic material, which leads to cell death [279,280]. The pH-sensitive zwitterionic poly(carboxybetaine) polymer was conjugated to CPT via the ATRP process, using a CPT derivative having a halogen atom in its structure as the initiator. CPT-based cationic liposomes, consisting of the prepared conjugate and a cationic lipid, were then constructed for the co-delivery of siRNA for combination therapy. The double-sensitive lipoplexes simultaneously delivered two drugs to the tumor cells and enabled time-controlled drug release, such that siRNA was released rapidly after a 4 h incubation and CPT was released in a sustained manner.

The second strategy is the possibility of refunctionalization of the therapeutic by introducing into its structure (meth)acrylic moieties which are able to undergo ATRP processes to form polymer–drug conjugates. Plichta et al. proposed a method for the synthesis of polymer–drug conjugates using ATRP macroinitiators based on PLA and the produced methacrylic derivative of CPT, which was conjugated on a polymer matrix [281]. In addition, in some syntheses, an additional PEGMA block was added. This process and the structures of the obtained conjugates are shown in Figure 15. The great strength of this type of conjugation is the possibility of introducing more than one molecule of the active substance per chain of the polymer matrix, in contrast to the first presented strategy, in which only one molecule could be introduced per entire polymer chain. The CPT content of the conjugates was determined using three techniques and ranged from 8 to 16.9 wt%. The release profile of CPT was also examined, with which it was shown that the more D-LA units in the structure, the slower the release of the active substance, while the PEGMA groups acted antagonistically towards D-LA. Gao et al. developed an amphiphilic copolymer based on a hydrophilic beta-cyclodextrin derivative used as the initiator of the ATRP process to embed a methacrylic derivative of the hydrophilic anticancer drug irinotecan onto a polymer matrix [282]. The obtained star-shaped amphiphilic copolymer had the ability to form stable monomolecular micelles in an aqueous solution, the reducing properties of which contributed to the controlled release of the drug and reduced toxicity to healthy tissues. The nanoparticles can achieve targeted release due to the presence of disulfide bonds found in the irinotecan derivative. Furthermore, the cytotoxicity assay showed a higher antitumor efficacy of the conjugate, compared to the free drug, against the two types of tumor cells tested.

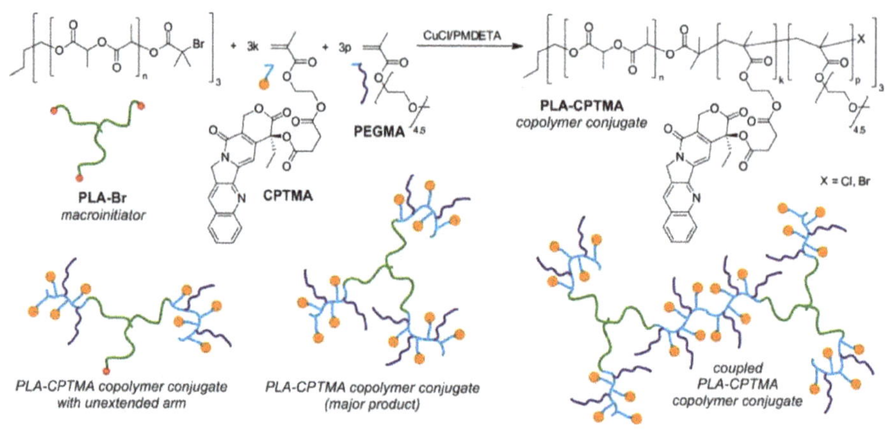

Figure 15. Synthesis of conjugates of PLA and CPT block copolymers proposed by Plichta et al. [281]. Reprinted with permission from Plichta, A.; Kowalczyk, S.; Kamiński, K.; Wasyłeczko, M.; Więckowski, S.; Olędzka, E.; Nałęcz-Jawecki, G.; Zgadzaj, A.; Sobczak, M., ATRP of Methacrylic Derivative of Camptothecin Initiated with PLA toward Three-Arm Star Block Copolymer Conjugates with Favorable Drug Release. Macromolecules 2017, 50 (17), 6439–6450, DOI: 10.1021/acs.macromol.7b01350 (accessed on 26 February 2023). Copyright 2017 American Chemical Society.

The third strategy is the possibility of obtaining a (co)polymer, whose repeat units will have additional functional groups capable of binding to the drug molecule in a post-polymerization act, similar to the "grafting to" strategy shown in the example of polymer–protein conjugates. Chen et al. synthesized a redox-responsive polymer–drug conjugate based on a hydrophilic diblock copolymer covalently linked to a sulfide-bridged derivative of the anti-cancer drug PTX [283]. The hydrophilic diblock copolymer PEG-*b*-PHEMA was synthesized via the ATRP process using 2-(trimethylsilyloxy)ethyl methacrylate and PEG-Br as a macroinitiator to then, in the next step, selectively hydrolyze the trimethylsilane group to hydroxyl groups. Utilizing the generated hydroxyl functionalities, PTX was covalently coupled to the polymer matrix resulting in an LC of 18.4 wt%. The authors demonstrated the possibility of the self-assembly of the conjugate into spherical micelles in an aqueous solution, with hydrophobic paclitaxel as the core and hydrophilic PEG chains as the shell. Most importantly, the results of cytotoxicity indicate that the obtained conjugates can effectively inhibit the proliferation of tumor cells. Dong et al. designed the synthesis of a polymer–drug conjugate based on a diblock copolymer via the ATRP process of GMA with an initiator based on PEG-Br and the post-polymerization aldehyde modification and conjugation of DOX via an acid labile imine bond [284]. The amphiphilic conjugate can self-assemble into nanoparticles with a core-shell structure, whereas the PEG block is a hydrophilic shell and the block containing DOX is a hydrophobic core. The authors showed that the conjugate produced can effectively deliver the active substance to the cell nuclei and shows a more effective anticancer effect compared to that of free DOX.

Table 5 presents an overview of the literature on the synthesis of covalently bound polymer–drug conjugates using three strategies utilizing various ATRP techniques, depending on the type of therapeutic agent used and the composition of the polymer matrix. The table shows the work that has recently been put into the development of this type of conjugate, as well as the universality, commonness, and versatility of various variants of the ATRP technique.

Table 5. A review of the literature on the synthesis of polymer–drug conjugates using ATRP techniques.

Strategy	(Co)Polymers	Active Substance	Synthesis Techniques	Ref.
Active substance as an ATRP initiator	Poly(carboxybetaine)	CPT	ATRP	[278]
	Poly(methacryloyloxyethyl phosphorylcholine)	CPT	ATRP	[68]
	Poly(oligo(ethylene oxide) methacrylate)-b-(glycidyl methacrylate)	Fentanyl	AGET ATRP	[216]
	Poly(di(ethylene glycol) methyl ether methacrylate) Poly(di(ethylene glycol) methyl ether methacrylate)-b-poly(methyl methacrylate)	Inositol (vitamin B_8)	ARGET ATRP, SARA ATRP, seATRP	[183]
	Poly(methyl methacrylate-co-2-hydroxyethyl methacrylate)	Retinol (vitamin A)	ATRP	[285]
	Poly(n-butyl acrylate) Poly(methyl methacrylate) Poly(N-isopropylacrylamide) Poly(N-isopropylacrylamide)-b-poly(oligo(ethylene glycol) acrylate) Poly(N-isopropylacrylamide)-b-poly(2-hydroxyethyl acrylate)	Riboflavin (vitamin B_2)	ARGET ATRP, Metal-free ATRP, Photo ATRP, seATRP	[286]
Active substance as (meth)acrylate monomer	Poly(lactic acid)-b-poly(camptothecin mono-2-(methacryloyloxy)ethyl succinate) Poly(lactic acid)-b-poly(camptothecin mono-2-(methacryloyloxy)ethyl succinate-co-poly(ethylene glycol) methyl ether methacrylate)	CPT	ATRP	[281,287]
	Poly(hydroxypropyl methacrylate-co-Methacryloyloxy-3-thiohexanoyl camptothecin-co-2-(2′-Bromoisobutyryloxy)ethyl-2′′-methacryloyl oxyethyl disulfide) Poly(hydroxypropyl methacrylate-co-Methacryloyloxy-3-thiohexanoyl camptothecin-co-2-(2′-Bromoisobutyryloxy)ethyl-2′′-methacryloyl oxyethyl disulfide)(poly(poly(ethylene glycol) methyl ether methacrylate))	CPT	ATRP	[288]
	Cellulose-g-poly(methacrylate derivative of camptothecin)-b- poly(ethylene glycol) methyl ether methacrylate)	CPT	ATRP	[289]
	Dextran-poly(methacrylate derivative of camptothecin)-b-poly(ethylene glycol) methyl ether methacrylate)	CPT	ATRP	[290]
	α-cyclodextrin- poly(ethylene glycol) polyrotaxanes-poly(methacrylate derivative of camptothecin)-b-poly(ethylene glycol) methyl ether methacrylate)	CPT	ATRP	[291]
	Poly(ethylene glycol)-b-poly(2-([2-4-(2-methylpropil)phenyl]propionyl]oxy) ethyl methacrylate	Ibuprofen	ATRP	[96]
	β-cyclodextrin-poly(methacrylate derivative of irinotecan-co-poly(ethylene glycol) methyl ether methacrylate)	Irinotecan	ATRP	[282]

Table 5. *Cont.*

Strategy	(Co)Polymers	Active Substance	Synthesis Techniques	Ref.
Post-polymerization conjugation	Poly(methacryloyloxyethyl phosphorylcholine)-graft-camptothecin	CPT	ATRP, Click Chemistry	[68]
	Poly(glycidyl methacrylate) Poly(poly(ethylene glycol) methyl ether methacrylate-co-glycidyl methacrylate)	Ciprofloxacin	AGET ATRP, ICAR ATRP, ROP, Click Chemistry	[218]
	Poly(methacryloyloxyethyl phosphorylcholine)-graft-doxorubicin Poly(methacryloyloxyethyl phosphorylcholine-co-2-tert-butoxy-2-oxoethyl methacrylate)	DOX	ATRP, Click Chemistry, Acylhydrazine formation	[292]
	Poly(poly(ethylene glycol) methacrylate)–b-poly(caprolactone)–b-poly(poly(ethylene glycol) methacrylate)	DOX	ATRP, Acylhydrazine formation	[293]
	Poly(methacryloyloxyethyl phosphorylcholine)-b-poly(2-methoxy-2-oxoethyl methacrylate)	DOX	ATRP, Acylhydrazine formation	[294]
	Poly(ethylene oxide)-b-poly(glycidyl methacrylate)	DOX	ATRP, Imine formation	[284]
	Poly(2-(2-bromoisobutyryloxy)ethyl methacrylate)-co-poly[poly(ethylene glycol) methacrylate-co-3-vinyl benzaldehyde]	DOX	ATRP, Imine formation	[295]
	Poly(methacrylic acid)	Estradiol Tamoxifen	ATRP, N-alkylation of amines with carboxylic acid	[296]
	Poly(ethylene oxide)-b-poly-(n-butyl methacrylate-co-4-methyl-[7-(methacryloyl)-oxyethyloxy]coumarin))	5-fluorouracil	ATRP, Photochemically induced [2 + 2] cycloaddition reaction	[297]
	Poly(ethylene oxide)-b-poly(glycerol monomethacrylate)	Indomethacin	ATRP, Steglich esterification	[298]
	Poly(ethylene glycol)-b-poly(2-(trimethylsilyloxyl) ethyl methacrylate)	Paclitaxel	ATRP, Esterification	[283]

7. Concluding Remarks

Although this review is limited to selected DDS-related papers published in the last decade, it demonstrates the enormous potential of ATRP in developing new classes of block copolymers, bioconjugates, and hybrid inorganic–organic particles that can serve as carriers for small drugs, proteins, and nucleic acids. Current research makes it possible to obtain smart carriers in the form of a wide range of stimuli-responsive micelles, polymersomes, polyplexes, hybrid inorganic–organic particles, or implantable wafers with a controlled delivery of selected active substances. One can expect that among these, the ones susceptible to the internal stimuli present in the human body subjected to a disease (e.g., pH changes or redox conditions) seem to have a particular potential for quick implementation in medical treatment due to their simplicity of use by patients (e.g., no or only limited requirements for specialized instrumentation, much less need to engage the patient's or physician's attention to control treatment conditions). Nevertheless, scaling up the synthesis of many of the described carriers may be a difficult problem and significant work should be directed toward more efficient, as well as less costly and time-consuming, procedures. As a result, it remains a very important task to develop new ATRP methods to effectively control the polymerization process using trace amounts of metal catalysts. Photochemical variants of metal-free ATRP also appear to be of interest, especially those in which the photo-initiator and the catalyst providing the balance between active and dormant species are fully biocompatible organic compounds (e.g., vitamin B2) [299]. It should also be noted that the RAFT polymerization technique has also been used extensively in drug delivery systems in recent years and offers a large library of alternative approaches, which can be applied to future technologies [300]. One of the next big challenges is to better understand the cytocompatibility of ATRP- and RAFT-based materials and to confirm their applicability in various animal models and ultimately in humans.

Author Contributions: Conceptualization, Z.F.; writing—original draft preparation, Z.F., A.P., A.I., M.D., M.S.-Ł. and S.K.; writing—review and editing, M.S.-Ł.; visualization S.K., M.D., A.P. and M.S.-Ł.; supervision, Z.F. All authors have read and agreed to the published version of the manuscript.

Funding: This research was funded by Warsaw University of Technology (Faculty of Chemistry), grant no. 504/04109/1020/44.000000.

Institutional Review Board Statement: Not applicable.

Data Availability Statement: Not applicable.

Conflicts of Interest: The authors declare no conflict of interest.

References

1. Widder, K.J. *Controlled Release Delivery Systems*; Wiley Online Library: New York, NY, USA; Marcel Dekker, Inc.: New York, NY, USA, 1983. [CrossRef]
2. Uhrich, K.E.; Cannizzaro, S.M.; Langer, R.S.; Shakesheff, K.M. Polymeric systems for controlled drug release. *Chem. Rev. Columb.* **1999**, *99*, 3181–3198. [CrossRef] [PubMed]
3. Duncan, R. The dawning era of polymer therapeutics. *Nat. Rev. Drug. Discov.* **2003**, *2*, 347–360. [CrossRef] [PubMed]
4. Garnett, M.C. Targeted drug conjugates: Principles and progress. *Adv. Drug Deliv. Rev.* **2001**, *53*, 171–216. [CrossRef] [PubMed]
5. Greco, F.; Vicent, M.J. Polymer-drug conjugates: Current status and future trends. *Front. Biosci.-Landmark* **2008**, *13*, 2744–2756. [CrossRef]
6. Pasut, G.; Veronese, F. Polymer–drug conjugation, recent achievements and general strategies. *Prog. Polym. Sci.* **2007**, *32*, 933–961. [CrossRef]
7. Goodman, L.S.; Gilman, A.G. *Goodman & Gilman's Pharmacological Basis of Therapeutics*, 13th ed.; McGraw-Hill Education LLC: New York, NY, USA, 2017.
8. Elsabahy, M.; Wooley, K.L. Design of polymeric nanoparticles for biomedical delivery applications. *Chem. Soc. Rev.* **2012**, *41*, 2545–2561. [CrossRef]
9. Hubbell, J.A.; Chilkoti, A. Nanomaterials for drug delivery. *Science* **2012**, *337*, 303–305. [CrossRef]
10. Kowalczuk, A.; Trzcinska, R.; Trzebicka, B.; Müller, A.H.; Dworak, A.; Tsvetanov, C.B. Loading of polymer nanocarriers: Factors, mechanisms and applications. *Prog. Polym. Sci.* **2014**, *39*, 43–86. [CrossRef]
11. Jahangirian, H.; Lemraski, E.G.; Webster, T.J.; Rafiee-Moghaddam, R.; Abdollahi, Y. A review of drug delivery systems based on nanotechnology and green chemistry: Green nanomedicine. *Int. J. Nanomed.* **2017**, *12*, 2957. [CrossRef]
12. Deng, C.; Jiang, Y.; Cheng, R.; Meng, F.; Zhong, Z. Biodegradable polymeric micelles for targeted and controlled anticancer drug delivery: Promises, progress and prospects. *Nano Today* **2012**, *7*, 467–480. [CrossRef]
13. Wei, H.; Zhuo, R.-X.; Zhang, X.-Z. Design and development of polymeric micelles with cleavable links for intracellular drug delivery. *Prog. Polym. Sci.* **2013**, *38*, 503–535. [CrossRef]
14. Lee, J.S.; Feijen, J. Polymersomes for drug delivery: Design, formation and characterization. *J. Control. Release* **2012**, *161*, 473–483. [CrossRef] [PubMed]
15. Lefley, J.; Waldron, C.; Becer, C.R. Macromolecular design and preparation of polymersomes. *Polym. Chem.* **2020**, *11*, 7124–7136. [CrossRef]
16. Chauhan, A.S. Dendrimers for drug delivery. *Molecules* **2018**, *23*, 938. [CrossRef]
17. Chacko, R.T.; Ventura, J.; Zhuang, J.; Thayumanavan, S. Polymer nanogels: A versatile nanoscopic drug delivery platform. *Adv. Drug Deliv. Rev.* **2012**, *64*, 836–851. [CrossRef]
18. Anglin, E.J.; Cheng, L.; Freeman, W.R.; Sailor, M.J. Porous silicon in drug delivery devices and materials. *Adv. Drug Deliv. Rev.* **2008**, *60*, 1266–1277. [CrossRef]
19. Vallet-Regí, M.; Balas, F.; Arcos, D. Mesoporous materials for drug delivery. *Angew. Chem. Int. Ed.* **2007**, *46*, 7548–7558. [CrossRef]
20. Li, Z.; Barnes, J.C.; Bosoy, A.; Stoddart, J.F.; Zink, J.I. Mesoporous silica nanoparticles in biomedical applications. *Chem. Soc. Rev.* **2012**, *41*, 2590–2605. [CrossRef]
21. Fouassier, J.; Allonas, X.; Lalevée, J. *Macromolecular Engineering: From Precise Macromolecular Synthesis to Macroscopic Materials Properties and Applications*, 2nd ed.; Wiley-VCH: Weinheim, Germany, 2022.
22. Corbin, D.A.; Miyake, G.M. Photoinduced organocatalyzed atom transfer radical polymerization (O-ATRP): Precision polymer synthesis using organic photoredox catalysis. *Chem. Rev.* **2021**, *122*, 1830–1874. [CrossRef]
23. Kreutzer, J. Atom-transfer radical polymerization: New method breathes life into ATRP. *Nat. Rev. Chem.* **2018**, *2*, 0111. [CrossRef]
24. Pan, X.; Fantin, M.; Yuan, F.; Matyjaszewski, K. Externally controlled atom transfer radical polymerization. *Chem. Soc. Rev.* **2018**, *47*, 5457–5490. [CrossRef] [PubMed]
25. Chmielarz, P.; Fantin, M.; Park, S.; Isse, A.A.; Gennaro, A.; Magenau, A.J.; Sobkowiak, A.; Matyjaszewski, K. Electrochemically mediated atom transfer radical polymerization (eATRP). *Prog. Polym. Sci.* **2017**, *69*, 47–78. [CrossRef]

26. Dadashi-Silab, S.; Atilla Tasdelen, M.; Yagci, Y. Photoinitiated atom transfer radical polymerization: Current status and future perspectives. *J. Polym. Sci. Part A Polym. Chem.* **2014**, *52*, 2878–2888. [CrossRef]
27. Tsarevsky, N.V.; Matyjaszewski, K. Atom transfer radical polymerization (ATRP). In *Fundamentals of Conrolled/Living Radical Polymerization*; Royal Society of Chemistry: Cambridge, UK, 2013; pp. 287–357. [CrossRef]
28. Ayres, N. Atom transfer radical polymerization: A robust and versatile route for polymer synthesis. *Polym. Rev.* **2011**, *51*, 138–162. [CrossRef]
29. Pintauer, T.; Matyjaszewski, K. Atom transfer radical addition and polymerization reactions catalyzed by ppm amounts of copper complexes. *Chem. Soc. Rev.* **2008**, *37*, 1087–1097. [CrossRef]
30. Tsarevsky, N.V.; Matyjaszewski, K. "Green" atom transfer radical polymerization: From process design to preparation of well-defined environmentally friendly polymeric materials. *Chem. Rev.* **2007**, *107*, 2270–2299. [CrossRef]
31. Faucher, S.; Zhu, S. Fundamentals and development of high-efficiency supported catalyst systems for atom transfer radical polymerization. *J. Polym. Sci. Part A Polym. Chem.* **2007**, *45*, 553–565. [CrossRef]
32. Tsarevsky, N.V.; Matyjaszewski, K. Environmentally benign atom transfer radical polymerization: Towards "green" processes and materials. *J. Polym. Sci. Part A Polym. Chem.* **2006**, *44*, 5098–5112. [CrossRef]
33. Matyjaszewski, K.; Xia, J. Atom transfer radical polymerization. *Chem. Rev.* **2001**, *101*, 2921–2990. [CrossRef]
34. Boyer, C.; Bulmus, V.; Davis, T.P.; Ladmiral, V.; Liu, J.; Perrier, S. Bioapplications of RAFT polymerization. *Chem. Rev.* **2009**, *109*, 5402–5436. [CrossRef]
35. Siegwart, D.J.; Oh, J.K.; Matyjaszewski, K. ATRP in the design of functional materials for biomedical applications. *Prog. Polym. Sci.* **2012**, *37*, 18–37. [CrossRef] [PubMed]
36. Wang, J.S.; Matyjaszewski, K. Controlled/"living" radical polymerization. atom transfer radical polymerization in the presence of transition-metal complexes. *J. Am. Chem. Soc.* **1995**, *117*, 5614–5615. [CrossRef]
37. Beers, K.L. The first dive into the mechanism and kinetics of ATRP. *Macromolecules* **2020**, *53*, 1115–1118. [CrossRef]
38. Curran, D.P. The Design and Application of Free Radical Chain Reactions in Organic Synthesis. Part 1. *Synthesis* **2002**, *1998*, 417–439. [CrossRef]
39. Hawker, C.J.; Barclay, G.G.; Orellana, A.; Dao, J.; Devonport, W. Initiating systems for nitroxide-mediated "living" free radical polymerizations: Synthesis and evaluation. *Macromolecules* **1996**, *29*, 5245–5254. [CrossRef]
40. Goto, A.; Sato, K.; Tsujii, Y.; Fukuda, T.; Moad, G.; Rizzardo, E.; Thang, S.H. Mechanism and kinetics of RAFT-based living radical polymerizations of styrene and methyl methacrylate. *Macromolecules* **2001**, *34*, 402–408. [CrossRef]
41. Zhou, Y.-N.; Li, J.-J.; Wang, T.-T.; Wu, Y.-Y.; Luo, Z.-H. Precision Polymer Synthesis by Controlled Radical Polymerization: Fusing the progress from Polymer Chemistry and Reaction Engineering. *Prog. Polym. Sci.* **2022**, 101555. [CrossRef]
42. Matyjaszewski, K.; Xia, J. Fundamentals of Atom Transfer Radical Polymerization. In *Handbook of Radical Polymerization*; Matyjaszewski, K.D., Thomas, P., Eds.; John Wiley & Sons, Inc.: Hoboken, NJ, USA, 2002; pp. 523–628. [CrossRef]
43. Lorandi, F.; Fantin, M.; Matyjaszewski, K. Atom Transfer Radical Polymerization: A Mechanistic Perspective. *J. Am. Chem. Soc.* **2022**, *144*, 15413–15430. [CrossRef]
44. Fung, A.K.; Coote, M.L. A mechanistic perspective on atom transfer radical polymerization. *Polym. Int.* **2021**, *70*, 918–926. [CrossRef]
45. Xia, J.; Zhang, X.; Matyjaszewski, K. The effect of ligands on copper-mediated atom transfer radical polymerization. In *Transition Metal Catalysis in Macromolecular Design*; ACS Publications: Washington, DC, USA, 2000; pp. 207–223. [CrossRef]
46. Matyjaszewski, K.; Göbelt, B.; Paik, H.-j.; Horwitz, C.P. Tridentate nitrogen-based ligands in Cu-based ATRP: A structure–activity study. *Macromolecules* **2001**, *34*, 430–440. [CrossRef]
47. Chen, X.P.; Qiu, K.Y. 'Living'radical polymerization of styrene with AIBN/FeCl3/PPh3 initiating system via a reverse atom transfer radical polymerization process. *Polym. Int.* **2000**, *49*, 1529–1533. [CrossRef]
48. Dadashi-Silab, S.; Matyjaszewski, K. Iron catalysts in atom transfer radical polymerization. *Molecules* **2020**, *25*, 1648. [CrossRef] [PubMed]
49. Zhu, S.; Xiao, G.; Yan, D. Synthesis of aromatic polyethersulfone-based graft copolyacrylates via ATRP catalyzed by FeCl2/isophthalic acid. *J. Polym. Sci. Part A Polym. Chem.* **2001**, *39*, 2943–2950. [CrossRef]
50. Xue, Z.; He, D.; Xie, X. Iron-catalyzed atom transfer radical polymerization. *Polym. Chem.* **2015**, *6*, 1660–1687. [CrossRef]
51. Plichta, A.; Li, W.; Matyjaszewski, K. ICAR ATRP of styrene and methyl methacrylate with Ru (Cp*) Cl (PPh3) 2. *Macromolecules* **2009**, *42*, 2330–2332. [CrossRef]
52. He, D.; Noh, S.K.; Lyoo, W.S. In situ-generated Ru (III)-mediated ATRP from the polymeric Ru (III) complex in the absence of activator generation agents. *J. Polym. Sci. Part A Polym. Chem.* **2011**, *49*, 4594–4602. [CrossRef]
53. Yang, X.; Yu, Y.; Lai, Q.; Yang, X.; Luo, P.; Zhang, B.; Zhang, X.; Wei, Y. Recent development and advances on fabrication and biomedical applications of Ga-based liquid metal micro/nanoparticles. *Compos. Part B Eng.* **2022**, 110384. [CrossRef]
54. Noto, N.; Saito, S. Arylamines as More Strongly Reducing Organic Photoredox Catalysts than fac-[Ir (ppy) 3]. *ACS Catal.* **2022**, *12*, 15400–15415. [CrossRef]
55. Bai, L.; Zhang, L.; Cheng, Z.; Zhu, X. Activators generated by electron transfer for atom transfer radical polymerization: Recent advances in catalyst and polymer chemistry. *Polym. Chem.* **2012**, *3*, 2685–2697. [CrossRef]
56. Mueller, L.; Jakubowski, W.; Tang, W.; Matyjaszewski, K. Successful chain extension of polyacrylate and polystyrene macroinitiators with methacrylates in an ARGET and ICAR ATRP. *Macromolecules* **2007**, *40*, 6464–6472. [CrossRef]

57. Zhang, L.; Miao, J.; Cheng, Z.; Zhu, X. Iron-Mediated ICAR ATRP of Styrene and Methyl Methacrylate in the Absence of Thermal Radical Initiator. *Macromol. Rapid Commun.* **2010**, *31*, 275–280. [CrossRef] [PubMed]
58. Konkolewicz, D.; Wang, Y.; Zhong, M.; Krys, P.; Isse, A.A.; Gennaro, A.; Matyjaszewski, K. Reversible-deactivation radical polymerization in the presence of metallic copper. A critical assessment of the SARA ATRP and SET-LRP mechanisms. *Macromolecules* **2013**, *46*, 8749–8772. [CrossRef]
59. Matyjaszewski, K.; Magenau, A.; Gennaro, A.; Strandwitz, N.C. Electrochemically mediated atom transfer radical polymerization. *Science* **2017**, *332*, 81–84. [CrossRef]
60. Mohapatra, H.; Kleiman, M.; Esser-Kahn, A.P. Mechanically controlled radical polymerization initiated by ultrasound. *Nat. Chem.* **2017**, *9*, 135–139. [CrossRef]
61. Wang, Z.; Wang, Z.; Pan, X.; Fu, L.; Lathwal, S.; Olszewski, M.; Yan, J.; Enciso, A.E.; Wang, Z.; Xia, H. Ultrasonication-induced aqueous atom transfer radical polymerization. *ACS Macro Lett.* **2018**, *7*, 275–280. [CrossRef] [PubMed]
62. Soly, S.; Mistry, B.; Murthy, C. Photo-mediated metal-free atom transfer radical polymerization: Recent advances in organocatalysts and perfection towards polymer synthesis. *Polym. Int.* **2022**, *71*, 159–168. [CrossRef]
63. Dworakowska, S.; Lorandi, F.; Gorczyński, A.; Matyjaszewski, K. Toward Green Atom Transfer Radical Polymerization: Current Status and Future Challenges. *Adv. Sci.* **2022**, 2106076. [CrossRef]
64. Berbigier, J.F.; Teixeira Alves Duarte, L.G.; Zawacki, M.F.; de Araujo, B.B.; Moura Santos, C.d.; Atvars, T.D.Z.; Gonçalves, P.F.B.; Petzhold, C.L.; Rodembusch, F.S. ATRP initiators based on proton transfer benzazole dyes: Solid-state photoactive polymer with very large Stokes shift. *ACS Appl. Polym. Mater.* **2020**, *2*, 1406–1416. [CrossRef]
65. Qi, X.; Yan, H.; Li, Y. ATRP-based synthesis of a pH-sensitive amphiphilic block polymer and its self-assembled micelles with hollow mesoporous silica as DOX carriers for controlled drug release. *RSC Adv.* **2021**, *11*, 29986–29996. [CrossRef]
66. Sumerlin, B.S. Proteins as initiators of controlled radical polymerization: Grafting-from via ATRP and RAFT. *ACS Macro Lett.* **2012**, *1*, 141–145. [CrossRef]
67. Messina, M.S.; Messina, K.M.M.; Bhattacharya, A.; Montgomery, H.R.; Maynard, H.D. Preparation of biomolecule-polymer conjugates by grafting-from using ATRP, RAFT, or ROMP. *Prog. Polym. Sci.* **2020**, *100*, 101186. [CrossRef] [PubMed]
68. Chen, X.; McRae, S.; Parelkar, S.; Emrick, T. Polymeric Phosphorylcholine−Camptothecin Conjugates Prepared by Controlled Free Radical Polymerization and Click Chemistry. *Bioconjugate Chem.* **2009**, *20*, 2331–2341. [CrossRef] [PubMed]
69. Cheng-Mei, L.; Rui, B.; Jin-Jun, Q.; Fen, H.; Yan, X.; Chen, Z.; Yun, Z. Coumarin end-capped polystyrene by ATRP and photodimerization reaction. *Polym. Bull.* **2006**, *57*, 139–149. [CrossRef]
70. Mansfeld, U.; Pietsch, C.; Hoogenboom, R.; Becer, C.R.; Schubert, U.S. Clickable initiators, monomers and polymers in controlled radical polymerizations—A prospective combination in polymer science. *Polym. Chem.* **2010**, *1*, 1560–1598. [CrossRef]
71. Coad, B.R.; Styan, K.E.; Meagher, L. One step ATRP initiator immobilization on surfaces leading to gradient-grafted polymer brushes. *ACS Appl. Mater. Interfaces* **2014**, *6*, 7782–7789. [CrossRef]
72. Ślusarczyk, K.; Flejszar, M.; Chmielarz, P. Less is more: A review of μL-scale of SI-ATRP in polymer brushes synthesis. *Polymer* **2021**, *233*, 124212. [CrossRef]
73. Matyjaszewski, K. Advanced materials by atom transfer radical polymerization. *Adv. Mater.* **2018**, *30*, 1706441. [CrossRef]
74. Boyce, J.R.; Shirvanyants, D.; Sheiko, S.S.; Ivanov, D.A.; Qin, S.; Börner, H.; Matyjaszewski, K. Multiarm molecular brushes: Effect of the number of arms on the molecular weight polydispersity and surface ordering. *Langmuir* **2004**, *20*, 6005–6011. [CrossRef]
75. Nasrullah, M.J.; Vora, A.; Webster, D.C. Block copolymer synthesis via a combination of ATRP and RAFT using click chemistry. *Macromol. Chem. Phys.* **2011**, *212*, 539–549. [CrossRef]
76. Dau, H.; Jones, G.R.; Tsogtgerel, E.; Nguyen, D.; Keyes, A.; Liu, Y.-S.; Rauf, H.; Ordonez, E.; Puchelle, V.; Basbug Alhan, H. Linear block copolymer synthesis. *Chem. Rev.* **2022**, *122*, 14471–14553. [CrossRef]
77. Plichta, A.; Jaskulski, T.; Lisowska, P.; Macios, K.; Kundys, A. Elastic polyesters improved by ATRP as reactive epoxy-modifiers of PLA. *Polymer* **2015**, *72*, 307–316. [CrossRef]
78. Karayianni, M.; Pispas, S. Block copolymer solution self-assembly: Recent advances, emerging trends, and applications. *J. Polym. Sci.* **2021**, *59*, 1874–1898. [CrossRef]
79. Plichta, A.; Zhong, M.; Li, W.; Elsen, A.M.; Matyjaszewski, K. Tuning dispersity in diblock copolymers using ARGET ATRP. *Macromol. Chem. Phys.* **2012**, *213*, 2659–2668. [CrossRef]
80. Listak, J.; Jia, X.; Plichta, A.; Zhong, M.; Matyjaszewski, K.; Bockstaller, M.R. Effect of block molecular weight distribution on the structure formation in block copolymer/homopolymer blends. *J. Polym. Sci. Part B Polym. Phys.* **2012**, *50*, 106–116. [CrossRef]
81. Listak, J.; Jakubowski, W.; Mueller, L.; Plichta, A.; Matyjaszewski, K.; Bockstaller, M.R. Effect of symmetry of molecular weight distribution in block copolymers on formation of "metastable" morphologies. *Macromolecules* **2008**, *41*, 5919–5927. [CrossRef]
82. Oliveira, A.S.; Mendonça, P.V.; Simões, S.; Serra, A.C.; Coelho, J.F. Amphiphilic well-defined degradable star block copolymers by combination of ring-opening polymerization and atom transfer radical polymerization: Synthesis and application as drug delivery carriers. *J. Polym. Sci.* **2021**, *59*, 211–229. [CrossRef]
83. Mühlebach, A.; Gaynor, S.G.; Matyjaszewski, K. Synthesis of amphiphilic block copolymers by atom transfer radical polymerization (ATRP). *Macromolecules* **1998**, *31*, 6046–6052. [CrossRef]
84. Hua, M.; Kaneko, T.; Liu, X.-Y.; Chen, M.-q.; Akashi, M. Successful ATRP syntheses of amphiphilic block copolymers poly (styrene-block-N, N-dimethylacrylamide) and their self-assembly. *Polym. J.* **2005**, *37*, 59–64. [CrossRef]

85. Wei, H.; Perrier, S.; Dehn, S.; Ravarian, R.; Dehghani, F. One-pot ATRP synthesis of a triple hydrophilic block copolymer with dual LCSTs and its thermo-induced association behavior. *Soft Matter* **2012**, *8*, 9526–9528. [CrossRef]
86. Kashapov, R.; Gaynanova, G.; Gabdrakhmanov, D.; Kuznetsov, D.; Pavlov, R.; Petrov, K.; Zakharova, L.; Sinyashin, O. Self-assembly of amphiphilic compounds as a versatile tool for construction of nanoscale drug carriers. *Int. J. Mol. Sci.* **2020**, *21*, 6961. [CrossRef]
87. He, W.; Jiang, H.; Zhang, L.; Cheng, Z.; Zhu, X. Atom transfer radical polymerization of hydrophilic monomers and its applications. *Polym. Chem.* **2013**, *4*, 2919–2938. [CrossRef]
88. Colombani, O.; Ruppel, M.; Schubert, F.; Zettl, H.; Pergushov, D.V.; Müller, A.H. Synthesis of poly (n-butyl acrylate)-block-poly (acrylic acid) diblock copolymers by ATRP and their micellization in water. *Macromolecules* **2007**, *40*, 4338–4350. [CrossRef]
89. Kuperkar, K.; Patel, D.; Atanase, L.I.; Bahadur, P. Amphiphilic Block Copolymers: Their Structures, and Self-Assembly to Polymeric Micelles and Polymersomes as Drug Delivery Vehicles. *Polymers* **2022**, *14*, 4702. [CrossRef] [PubMed]
90. Feng, H.; Lu, X.; Wang, W.; Kang, N.-G.; Mays, J.W. Block copolymers: Synthesis, self-assembly, and applications. *Polymers* **2017**, *9*, 494. [CrossRef]
91. Prasad, P.V.; Purkayastha, K.; Sharma, U.; Barik, M. Ph-sensitive nanomedicine for treating gynaecological cancers. *J. Womans Reprod. Health* **2020**, *2*, 35. [CrossRef]
92. Cabral, H.; Kataoka, K. Progress of drug-loaded polymeric micelles into clinical studies. *J. Control. Release* **2014**, *190*, 465–476. [CrossRef]
93. Wang, G.; Zhang, L. Synthesis, self-assembly and pH sensitivity of PDEAEMA–PEG–PDEAEMA triblock copolymer micelles for drug delivery. *React. Funct. Polym.* **2016**, *107*, 1–10. [CrossRef]
94. Zhang, L.; Zhang, C.; Gu, X.; Wang, G. Self-assembly, pH-responsibility and controlled release of doxorubicin of PDEAEMA-PEG-PDEAEMA triblock copolymers: Effects of PEG length. *J. Polym. Res.* **2021**, *28*, 1–12. [CrossRef]
95. Biswas, D.; An, S.Y.; Li, Y.; Wang, X.; Oh, J.K. Intracellular delivery of colloidally stable core-cross-linked triblock copolymer micelles with glutathione-responsive enhanced drug release for cancer therapy. *Mol. Pharm.* **2017**, *14*, 2518–2528. [CrossRef]
96. Zeng, Z.; Wei, Z.; Ma, L.; Xu, Y.; Xing, Z.; Niu, H.; Wang, H.; Huang, W. pH-Responsive nanoparticles based on ibuprofen prodrug as drug carriers for inhibition of primary tumor growth and metastasis. *J. Mater. Chem. B* **2017**, *5*, 6860–6868. [CrossRef]
97. Zhang, X.; Yuan, T.; Dong, H.; Xu, J.; Wang, D.; Tong, H.; Ji, X.; Sun, B.; Zhu, M.; Jiang, X. Novel block glycopolymers prepared as delivery nanocarriers for controlled release of bortezomib. *Colloid Polym. Sci.* **2018**, *296*, 1827–1839. [CrossRef] [PubMed]
98. Yang, C.; Liu, W.; Xiao, J.; Yuan, C.; Chen, Y.; Guo, J.; Yue, H.; Zhu, D.; Lin, W.; Tang, S. pH-sensitive mixed micelles assembled from PDEAEMA-PPEGMA and PCL-PPEGMA for doxorubicin delivery: Experimental and DPD simulations study. *Pharmaceutics* **2020**, *12*, 170. [CrossRef] [PubMed]
99. Hao, D.; Zhang, Z.; Ji, Y. Responsive polymeric drug delivery systems for combination anticancer therapy: Experimental design and computational insights. *Int. J. Polym. Mater. Polym. Biomater.* **2022**, *71*, 1221–1239. [CrossRef]
100. Chu, S.; Shi, X.; Tian, Y.; Gao, F. pH-Responsive Polymer Nanomaterials for Tumor Therapy. *Front. Oncol.* **2022**, *12*, 855019. [CrossRef] [PubMed]
101. Koltai, T. The Ph paradigm in cancer. *Eur. J. Clin. Nutr.* **2020**, *74*, 14–19. [CrossRef]
102. Kumar, R.; Santa Chalarca, C.F.; Bockman, M.R.; Bruggen, C.V.; Grimme, C.J.; Dalal, R.J.; Hanson, M.G.; Hexum, J.K.; Reineke, T.M. Polymeric delivery of therapeutic nucleic acids. *Chem. Rev.* **2021**, *121*, 11527–11652. [CrossRef]
103. Kennedy, L.; Sandhu, J.K.; Harper, M.-E.; Cuperlovic-Culf, M. Role of glutathione in cancer: From mechanisms to therapies. *Biomolecules* **2020**, *10*, 1429. [CrossRef]
104. Casado, N.; Hernandez, G.; Sardon, H.; Mecerreyes, D. Current trends in redox polymers for energy and medicine. *Prog. Polym. Sci.* **2016**, *52*, 107–135. [CrossRef]
105. Chen, Q.; Lin, W.; Wang, H.; Wang, J.; Zhang, L. PDEAEMA-based pH-sensitive amphiphilic pentablock copolymers for controlled anticancer drug delivery. *RSC Adv.* **2016**, *6*, 68018–68027. [CrossRef]
106. Chen, Q.; Zheng, J.; Yuan, X.; Wang, J.; Zhang, L. Folic acid grafted and tertiary amino based pH-responsive pentablock polymeric micelles for targeting anticancer drug delivery. *Mater. Sci. Eng. C* **2018**, *82*, 1–9. [CrossRef]
107. Yang, C.; Xiao, J.; Xiao, W.; Lin, W.; Chen, J.; Chen, Q.; Zhang, L.; Zhang, C.; Guo, J. Fabrication of PDEAEMA-based pH-responsive mixed micelles for application in controlled doxorubicin release. *RSC Adv.* **2017**, *7*, 27564–27573. [CrossRef]
108. Li, Y.; Yu, A.; Li, L.; Zhai, G. The development of stimuli-responsive polymeric micelles for effective delivery of chemotherapeutic agents. *J. Drug Target.* **2018**, *26*, 753–765. [CrossRef] [PubMed]
109. Jazani, A.M.; Oh, J.K. Development and disassembly of single and multiple acid-cleavable block copolymer nanoassemblies for drug delivery. *Polym. Chem.* **2020**, *11*, 2934–2954. [CrossRef]
110. Hu, X.; Jazani, A.M.; Oh, J.K. Recent advances in development of imine-based acid-degradable polymeric nanoassemblies for intracellular drug delivery. *Polymer* **2021**, *230*, 124024. [CrossRef]
111. Gannimani, R.; Walvekar, P.; Naidu, V.R.; Aminabhavi, T.M.; Govender, T. Acetal containing polymers as pH-responsive nano-drug delivery systems. *J. Control. Release* **2020**, *328*, 736–761. [CrossRef] [PubMed]
112. Zhang, L.-J.; Dong, B.-T.; Du, F.-S.; Li, Z.-C. Degradable thermoresponsive polyesters by atom transfer radical polyaddition and click chemistry. *Macromolecules* **2012**, *45*, 8580–8587. [CrossRef]
113. Qiao, Z.-Y.; Ji, R.; Huang, X.-N.; Du, F.-S.; Zhang, R.; Liang, D.-H.; Li, Z.-C. Polymersomes from dual responsive block copolymers: Drug encapsulation by heating and acid-triggered release. *Biomacromolecules* **2013**, *14*, 1555–1563. [CrossRef]

114. Zheng, L.; Zhang, X.; Wang, Y.; Liu, F.; Peng, J.; Zhao, X.; Yang, H.; Ma, L.; Wang, B.; Chang, C. Fabrication of acidic pH-cleavable polymer for anticancer drug delivery using a dual functional monomer. *Biomacromolecules* **2018**, *19*, 3874–3882. [CrossRef]
115. Khorsand, B.; Oh, J.K. pH-responsive destabilization and facile bioconjugation of new hydroxyl-terminated block copolymer micelles. *J. Polym. Sci. Part A Polym. Chem.* **2013**, *51*, 1620–1629. [CrossRef]
116. Cao, H.; Chen, C.; Xie, D.; Chen, X.; Wang, P.; Wang, Y.; Song, H.; Wang, W. A hyperbranched amphiphilic acetal polymer for pH-sensitive drug delivery. *Polym. Chem.* **2018**, *9*, 169–177. [CrossRef]
117. Song, C.-C.; Su, C.-C.; Cheng, J.; Du, F.-S.; Liang, D.-H.; Li, Z.-C. Toward tertiary amine-modulated acid-triggered hydrolysis of copolymers containing pendent ortho ester groups. *Macromolecules* **2013**, *46*, 1093–1100. [CrossRef]
118. Deng, H.; Zhao, X.; Liu, J.; Zhang, J.; Deng, L.; Liu, J.; Dong, A. Synergistic dual-pH responsive copolymer micelles for pH-dependent drug release. *Nanoscale* **2016**, *8*, 1437–1450. [CrossRef] [PubMed]
119. Fang, Y.; Xue, J.; Gao, S.; Lu, A.; Yang, D.; Jiang, H.; He, Y.; Shi, K. Cleavable PEGylation: A strategy for overcoming the "PEG dilemma" in efficient drug delivery. *Drug Deliv.* **2017**, *24*, 22–32. [CrossRef]
120. Zalba, S.; Ten Hagen, T.L.; Burgui, C.; Garrido, M.J. Stealth nanoparticles in oncology: Facing the PEG dilemma. *J. Control. Release* **2022**, *351*, 22–36. [CrossRef]
121. Patil, S.S.; Wadgaonkar, P.P. Temperature and pH dual stimuli responsive PCL-b-PNIPAA m block copolymer assemblies and the cargo release studies. *J. Polym. Sci. Part A Polym. Chem.* **2017**, *55*, 1383–1396. [CrossRef]
122. Babikova, D.; Kalinova, R.; Momekova, D.; Ugrinova, I.; Momekov, G.; Dimitrov, I. Multifunctional polymer nanocarrier for efficient targeted cellular and subcellular anticancer drug delivery. *ACS Biomater. Sci. Eng.* **2019**, *5*, 2271–2283. [CrossRef]
123. Jazani, A.M.; Shetty, C.; Movasat, H.; Bawa, K.K.; Oh, J.K. Imidazole-Mediated Dual Location Disassembly of Acid-Degradable Intracellular Drug Delivery Block Copolymer Nanoassemblies. *Macromol. Rapid Commun.* **2021**, *42*, 2100262. [CrossRef]
124. Hao, Y.; He, J.; Li, S.; Liu, J.; Zhang, M.; Ni, P. Synthesis of an acid-cleavable and fluorescent amphiphilic block copolymer as a combined delivery vector of DNA and doxorubicin. *J. Mater. Chem. B* **2014**, *2*, 4237–4249. [CrossRef]
125. Babikova, D.; Kalinova, R.; Zhelezova, I.; Momekova, D.; Konstantinov, S.; Momekov, G.; Dimitrov, I. Functional block copolymer nanocarriers for anticancer drug delivery. *RSC Adv.* **2016**, *6*, 84634–84644. [CrossRef]
126. Kamenova, K.; Grancharov, G.; Kortenova, V.; Petrov, P.D. Redox-Responsive Crosslinked Mixed Micelles for Controllable Release of Caffeic Acid Phenethyl Ester. *Pharmaceutics* **2022**, *14*, 679. [CrossRef]
127. Mirhadi, E.; Mashreghi, M.; Maleki, M.F.; Alavizadeh, S.H.; Arabi, L.; Badiee, A.; Jaafari, M.R. Redox-sensitive nanoscale drug delivery systems for cancer treatment. *Int. J. Pharm.* **2020**, *589*, 119882. [CrossRef] [PubMed]
128. Sun, W.; Yang, Y. Recent advances in redox-responsive nanoparticles for combined cancer therapy. *Nanoscale Adv.* **2022**, *4*, 3504–3516. [CrossRef]
129. Zhang, M.; Song, C.-C.; Ji, R.; Qiao, Z.-Y.; Yang, C.; Qiu, F.-Y.; Liang, D.-H.; Du, F.-S.; Li, Z.-C. Oxidation and temperature dual responsive polymers based on phenylboronic acid and N-isopropylacrylamide motifs. *Polym. Chem.* **2016**, *7*, 1494–1504. [CrossRef]
130. Zhang, M.; Song, C.-C.; Su, S.; Du, F.-S.; Li, Z.-C. ROS-activated ratiometric fluorescent polymeric nanoparticles for self-reporting drug delivery. *ACS Appl. Mater. Interfaces* **2018**, *10*, 7798–7810. [CrossRef] [PubMed]
131. Stubelius, A.; Lee, S.; Almutairi, A. The chemistry of boronic acids in nanomaterials for drug delivery. *Acc. Chem. Res.* **2019**, *52*, 3108–3119. [CrossRef] [PubMed]
132. Song, C.-C.; Ji, R.; Du, F.-S.; Liang, D.-H.; Li, Z.-C. Oxidation-accelerated hydrolysis of the ortho ester-containing acid-labile polymers. *ACS Macro Lett.* **2013**, *2*, 273–277. [CrossRef]
133. Hermanson, G. The Reactions of Bioconjugation. In *Bioconjugate Techniques*, 3rd ed.; Academic Press: Cambridge, MA, USA, 2013; pp. 229–258. [CrossRef]
134. Oh, J.K. Disassembly and tumor-targeting drug delivery of reduction-responsive degradable block copolymer nanoassemblies. *Polym. Chem.* **2019**, *10*, 1554–1568. [CrossRef]
135. Ko, N.R.; Oh, J.K. Glutathione-triggered disassembly of dual disulfide located degradable nanocarriers of polylactide-based block copolymers for rapid drug release. *Biomacromolecules* **2014**, *15*, 3180–3189. [CrossRef]
136. Chan, N.; An, S.Y.; Oh, J.K. Dual location disulfide degradable interlayer-crosslinked micelles with extended sheddable coronas exhibiting enhanced colloidal stability and rapid release. *Polym. Chem.* **2014**, *5*, 1637–1649. [CrossRef]
137. Liu, Y.-S.; Huang, S.-J.; Huang, X.-S.; Wu, Y.-T.; Chen, H.-Y.; Lo, Y.-L.; Wang, L.-F. The synthesis and comparison of poly (methacrylic acid)–poly (ε-caprolactone) block copolymers with and without symmetrical disulfide linkages in the center for enhanced cellular uptake. *RSC Adv.* **2016**, *6*, 75092–75103. [CrossRef]
138. Lo, Y.-L.; Huang, X.-S.; Chen, H.-Y.; Huang, Y.-C.; Liao, Z.-X.; Wang, L.-F. ROP and ATRP fabricated redox sensitive micelles based on PCL-SS-PMAA diblock copolymers to co-deliver PTX and CDDP for lung cancer therapy. *Colloids Surf. B Biointerfaces* **2021**, *198*, 111443. [CrossRef]
139. Li, S.-X.; Liu, L.; Zhang, L.-J.; Wu, B.; Wang, C.-X.; Zhou, W.; Zhuo, R.-X.; Huang, S.-W. Synergetic enhancement of antitumor efficacy with charge-reversal and reduction-sensitive polymer micelles. *Polym. Chem.* **2016**, *7*, 5113–5122. [CrossRef]
140. Ko, N.R.; Cheong, J.; Noronha, A.; Wilds, C.J.; Oh, J.K. Reductively-sheddable cationic nanocarriers for dual chemotherapy and gene therapy with enhanced release. *Colloids Surf. B Biointerfaces* **2015**, *126*, 178–187. [CrossRef]
141. Zhang, Q.; Ko, N.R.; Oh, J.K. Modulated morphologies and tunable thiol-responsive shedding of aqueous block copolymer aggregates. *RSC Adv.* **2012**, *2*, 8079–8086. [CrossRef]

142. An, S.Y.; Hong, S.H.; Tang, C.; Oh, J.K. Rosin-based block copolymer intracellular delivery nanocarriers with reduction-responsive sheddable coronas for cancer therapy. *Polym. Chem.* **2016**, *7*, 4751–4760. [CrossRef]
143. Zhang, Q.; Aleksanian, S.; Noh, S.M.; Oh, J.K. Thiol-responsive block copolymer nanocarriers exhibiting tunable release with morphology changes. *Polym. Chem.* **2013**, *4*, 351–359. [CrossRef]
144. Khorsand, B.; Lapointe, G.; Brett, C.; Oh, J.K. Intracellular drug delivery nanocarriers of glutathione-responsive degradable block copolymers having pendant disulfide linkages. *Biomacromolecules* **2013**, *14*, 2103–2111. [CrossRef]
145. Kumar, P.; Behl, G.; Kaur, S.; Yadav, N.; Liu, B.; Chhikara, A. Tumor microenvironment responsive nanogels as a smart triggered release platform for enhanced intracellular delivery of doxorubicin. *J. Biomater. Sci. Polym. Ed.* **2021**, *32*, 385–404. [CrossRef]
146. Tian, K.; Jia, X.; Zhao, X.; Liu, P. pH/Reductant Dual-Responsive Core-Cross-Linked Micelles via Facile in Situ ATRP for Tumor-Targeted Delivery of Anticancer Drug with Enhanced Anticancer Efficiency. *Mol. Pharm.* **2016**, *13*, 2683–2690. [CrossRef]
147. Chan, N.; Khorsand, B.; Aleksanian, S.; Oh, J.K. A dual location stimuli-responsive degradation strategy of block copolymer nanocarriers for accelerated release. *Chem. Commun.* **2013**, *49*, 7534–7536. [CrossRef]
148. Huang, Y.; Moini Jazani, A.; Howell, E.P.; Oh, J.K.; Moffitt, M.G. Controlled Microfluidic Synthesis of Biological Stimuli-Responsive Polymer Nanoparticles. *ACS Appl. Mater. Interfaces* **2019**, *12*, 177–190. [CrossRef]
149. Huang, Y.; Jazani, A.M.; Howell, E.P.; Reynolds, L.A.; Oh, J.K.; Moffitt, M.G. Microfluidic Shear Processing Control of Biological Reduction Stimuli-Responsive Polymer Nanoparticles for Drug Delivery. *ACS Biomater. Sci. Eng.* **2020**, *6*, 5069–5083. [CrossRef]
150. Jazani, A.M.; Oh, J.K. Dual location, dual acidic pH/reduction-responsive degradable block copolymer: Synthesis and investigation of ketal linkage instability under ATRP conditions. *Macromolecules* **2017**, *50*, 9427–9436. [CrossRef]
151. Shetty, C.; Noronha, A.; Pontarelli, A.; Wilds, C.J.; Oh, J.K. Dual-location dual-acid/glutathione-degradable cationic micelleplexes through hydrophobic modification for enhanced gene silencing. *Mol. Pharm.* **2020**, *17*, 3979–3989. [CrossRef]
152. Akimoto, J.; Ito, Y.; Okano, T.; Nakayama, M. Controlled aggregation behavior of thermoresponsive polymeric micelles by introducing hydrophilic segments as corona components. *J. Polym. Sci. Part A Polym. Chem.* **2018**, *56*, 1695–1704. [CrossRef]
153. Dong, Y.; Ma, X.; Huo, H.; Zhang, Q.; Qu, F.; Chen, F. Preparation of quadruple responsive polymeric micelles combining temperature-, pH-, redox-, and UV-responsive behaviors and its application in controlled release system. *J. Appl. Polym. Sci.* **2018**, *135*, 46675. [CrossRef]
154. Ma, X.; Liu, J.; Lei, L.; Yang, H.; Lei, Z. Synthesis of light and dual-redox triple-stimuli-responsive core-crosslinked micelles as nanocarriers for controlled release. *J. Appl. Polym. Sci.* **2019**, *136*, 47946. [CrossRef]
155. Yuan, W.; Guo, W. Ultraviolet light-breakable and tunable thermoresponsive amphiphilic block copolymer: From self-assembly, disassembly to re-self-assembly. *Polym. Chem.* **2014**, *5*, 4259–4267. [CrossRef]
156. Jazani, A.M.; Oh, J.K. Synthesis of multiple stimuli-responsive degradable block copolymers via facile carbonyl imidazole-induced postpolymerization modification. *Polym. Chem.* **2022**, *13*, 4557–4568. [CrossRef]
157. Sharma, A.K.; Prasher, P.; Aljabali, A.A.; Mishra, V.; Gandhi, H.; Kumar, S.; Mutalik, S.; Chellappan, D.K.; Tambuwala, M.M.; Dua, K. Emerging era of "somes": Polymersomes as versatile drug delivery carrier for cancer diagnostics and therapy. *Drug Deliv. Transl. Res.* **2020**, *10*, 1171–1190. [CrossRef]
158. Trombino, S.; Curcio, F.; Cassano, R. Polymersomes as a promising vehicle for controlled drug delivery. In *Stimuli-Responsive Nanocarriers*; Elsevier: Amsterdam, The Netherlands, 2022; pp. 351–366. [CrossRef]
159. Rodrigues, P.R.; Vieira, R.P. Advances in atom-transfer radical polymerization for drug delivery applications. *Eur. Polym. J.* **2019**, *115*, 45–58. [CrossRef]
160. Hasannia, M.; Aliabadi, A.; Abnous, K.; Taghdisi, S.M.; Ramezani, M.; Alibolandi, M. Synthesis of block copolymers used in polymersome fabrication: Application in drug delivery. *J. Control. Release* **2022**, *341*, 95–117. [CrossRef] [PubMed]
161. Mohammadi, M.; Ramezani, M.; Abnous, K.; Alibolandi, M. Biocompatible polymersomes-based cancer theranostics: Towards multifunctional nanomedicine. *Int. J. Pharm.* **2017**, *519*, 287–303. [CrossRef]
162. Kim, M.S.; Lee, D.S. Biodegradable and pH-sensitive polymersome with tuning permeable membrane for drug delivery carrier. *Chem. Comm.* **2010**, *46*, 4481–4483. [CrossRef] [PubMed]
163. Villani, S.; Adami, R.; Reverchon, E.; Ferretti, A.M.; Ponti, A.; Lepretti, M.; Caputo, I.; Izzo, L. pH-sensitive polymersomes: Controlling swelling via copolymer structure and chemical composition. *J. Drug Target.* **2017**, *25*, 899–909. [CrossRef] [PubMed]
164. Wei, P.; Sun, M.; Yang, B.; Xiao, J.; Du, J. Ultrasound-responsive polymersomes capable of endosomal escape for efficient cancer therapy. *J. Control. Release* **2020**, *322*, 81–94. [CrossRef]
165. Miele, Y.; Mingotaud, A.-F.; Caruso, E.; Malacarne, M.C.; Izzo, L.; Lonetti, B.; Rossi, F. Hybrid giant lipid vesicles incorporating a PMMA-based copolymer. *Biochim. Biophys. Acta BBA Gen. Subj.* **2021**, *1865*, 129611. [CrossRef]
166. Dadhwal, S.; Lee, A.; Goswami, S.K.; Hook, S.; Gamble, A.B. Synthesis and formulation of self-immolative PEG-aryl azide block copolymers and click-to-release reactivity with trans-cyclooctene. *J. Polym. Sci.* **2021**, *59*, 646–658. [CrossRef]
167. Vasile, C. Polymeric nanomaterials: Recent developments, properties and medical applications. In *Polymeric Nanomaterials in Nanotherapeutics*; Vasile, C., Ed.; Elsevier: Amsterdam, The Netherlands, 2019; pp. 1–66. [CrossRef]
168. Pergushov, D.V.; Müller, A.H.; Schacher, F.H. Micellar interpolyelectrolyte complexes. *Chem. Soc. Rev.* **2012**, *41*, 6888–6901. [CrossRef]
169. Ita, K. Polyplexes for gene and nucleic acid delivery: Progress and bottlenecks. *Eur. J. Pharm. Sci.* **2020**, *150*, 105358. [CrossRef]
170. Gombotz, W.R.; Pettit, D.K. Biodegradable polymers for protein and peptide drug delivery. *Bioconjugate Chem.* **1995**, *6*, 332–351. [CrossRef] [PubMed]

171. Bus, T.; Traeger, A.; Schubert, U.S. The great escape: How cationic polyplexes overcome the endosomal barrier. *J. Mater. Chem. B* **2018**, *6*, 6904–6918. [CrossRef] [PubMed]
172. Machtakova, M.; Thérien-Aubin, H.; Landfester, K. Polymer nano-systems for the encapsulation and delivery of active biomacromolecular therapeutic agents. *Chem. Soc. Rev.* **2022**, *51*, 128–152. [CrossRef]
173. Uchida, S.; Kataoka, K. Design concepts of polyplex micelles for in vivo therapeutic delivery of plasmid DNA and messenger RNA. *J. Biomed. Mater. Res. Part A* **2019**, *107*, 978–990. [CrossRef] [PubMed]
174. De Ávila Gonçalves, S.; Vieira, R.P. Current status of ATRP-based materials for gene therapy. *React. Funct. Polym.* **2020**, *147*, 104453. [CrossRef]
175. Xu, F.; Yang, W. Polymer vectors via controlled/living radical polymerization for gene delivery. *Prog. Polym. Sci.* **2011**, *36*, 1099–1131. [CrossRef]
176. Fliervoet, L.A.; van Nostrum, C.F.; Hennink, W.E.; Vermonden, T. Balancing hydrophobic and electrostatic interactions in thermosensitive polyplexes for nucleic acid delivery. *Multifunct. Mater.* **2019**, *2*, 024002. [CrossRef]
177. Fliervoet, L.A.; Zhang, H.; van Groesen, E.; Fortuin, K.; Duin, N.J.; Remaut, K.; Schiffelers, R.M.; Hennink, W.E.; Vermonden, T. Local release of siRNA using polyplex-loaded thermosensitive hydrogels. *Nanoscale* **2020**, *12*, 10347–10360. [CrossRef]
178. Zhang, Y.; He, J.; Cao, D.; Zhang, M.; Ni, P. Galactosylated reduction and pH dual-responsive triblock terpolymer Gal-PEEP-a-PCL-ss-PDMAEMA: A multifunctional carrier for the targeted and simultaneous delivery of doxorubicin and DNA. *Polym. Chem.* **2014**, *5*, 5124–5138. [CrossRef]
179. Wang, X.; Liow, S.S.; Wu, Q.; Li, C.; Owh, C.; Li, Z.; Loh, X.J.; Wu, Y.L. Codelivery for Paclitaxel and Bcl-2 Conversion Gene by PHB-PDMAEMA Amphiphilic Cationic Copolymer for Effective Drug Resistant Cancer Therapy. *Macromol. Biosci.* **2017**, *17*, 1700186. [CrossRef]
180. Sun, R.; Wang, Y.; Gou, P.; Zuo, M.; Li, X.; Zhu, W.; Shen, Z. Amphiphilic seven-arm star triblock copolymers with diverse morphologies in aqueous solution induced by crystallization and pH. *Chem. Res. Chin. Univ.* **2018**, *34*, 132–137. [CrossRef]
181. Huang, B.; Chen, M.; Zhou, S.; Wu, L. Synthesis and properties of clickable A(B-b-C)20 miktoarm star-shaped block copolymers with a terminal alkyne group. *Polym. Chem.* **2015**, *6*, 3913–3917. [CrossRef]
182. Zhang, Y.; Bradley, M.; Geng, J. Photo-controlled one-pot strategy for the synthesis of asymmetric three-arm star polymers. *Polym. Chem.* **2019**, *10*, 4769–4773. [CrossRef]
183. Chmielarz, P. Synthesis of inositol-based star polymers through low ppm ATRP methods. *Polym. Adv. Technol.* **2017**, *28*, 1804–1812. [CrossRef]
184. Zhou, P.; Liu, Y.-Y.; Niu, L.-Y.; Zhu, J. Self-assemblies of the six-armed star triblock ABC copolymer: pH-tunable morphologies and drug release. *Polym. Chem.* **2015**, *6*, 2934–2944. [CrossRef]
185. Huang, L.-m.; Li, L.-d.; Shang, L.; Zhou, Q.-h.; Lin, J. Preparation of pH-sensitive micelles from miktoarm star block copolymers by ATRP and their application as drug nanocarriers. *React. Funct. Polym.* **2016**, *107*, 28–34. [CrossRef]
186. Neugebauer, D.; Odrobińska, J.; Bielas, R.; Mielańczyk, A. Design of systems based on 4-armed star-shaped polyacids for indomethacin delivery. *New J. Chem.* **2016**, *40*, 10002–10011. [CrossRef]
187. Liu, X.; Tian, Z.; Chen, C.; Allcock, H.R. UV-cleavable unimolecular micelles: Synthesis and characterization toward photocontrolled drug release carriers. *Polym. Chem.* **2013**, *4*, 1115–1125. [CrossRef]
188. Mendrek, B.; Fus, A.; Klarzyńska, K.; Sieroń, A.L.; Smet, M.; Kowalczuk, A.; Dworak, A. Synthesis, Characterization and Cytotoxicity of Novel Thermoresponsive Star Copolymers of N,N′-Dimethylaminoethyl Methacrylate and Hydroxyl-Bearing Oligo(Ethylene Glycol) Methacrylate. *Polymers* **2018**, *10*, 1255. [CrossRef]
189. Zheng, A.; Xue, Y.; Wei, D.; Guan, Y.; Xiao, H. Amphiphilic star block copolymers as gene carrier Part I: Synthesis via ATRP using calix[4]resorcinarene-based initiators and characterization. *Mater. Sci. Eng. C* **2013**, *33*, 519–526. [CrossRef]
190. Mendrek, B.; Sieroń, Ł.; Żymełka-Miara, I.; Binkiewicz, P.; Libera, M.; Smet, M.; Trzebicka, B.; Sieroń, A.L.; Kowalczuk, A.; Dworak, A. Nonviral Plasmid DNA Carriers Based on N,N′-Dimethylaminoethyl Methacrylate and Di(ethylene glycol) Methyl Ether Methacrylate Star Copolymers. *Biomacromolecules* **2015**, *16*, 3275–3285. [CrossRef] [PubMed]
191. Cho, H.Y.; Averick, S.E.; Paredes, E.; Wegner, K.; Averick, A.; Jurga, S.; Das, S.R.; Matyjaszewski, K. Star Polymers with a Cationic Core Prepared by ATRP for Cellular Nucleic Acids Delivery. *Biomacromolecules* **2013**, *14*, 1262–1267. [CrossRef] [PubMed]
192. Kamaly, N.; Yameen, B.; Wu, J.; Farokhzad, O.C. Degradable Controlled-Release Polymers and Polymeric Nanoparticles: Mechanisms of Controlling Drug Release. *Chem. Rev.* **2016**, *116*, 2602–2663. [CrossRef]
193. Yilmaz, G. One-Pot Synthesis of Star Copolymers by the Combination of Metal-Free ATRP and ROP Processes. *Polymers* **2019**, *11*, 1577. [CrossRef] [PubMed]
194. Mielańczyk, A.; Kupczak, M.; Klymenko, O.; Mielańczyk, Ł.; Arabasz, S.; Madej, K.; Neugebauer, D. The structure–self-assembly relationship in PDMAEMA/polyester miktoarm stars. *Polym. Chem.* **2022**, *13*, 4763–4775. [CrossRef]
195. Xu, F.; Zheng, S.-Z.; Luo, Y.-L. Thermosensitive t-PLA-b-PNIPAAm tri-armed star block copolymer nanoscale micelles for camptothecin drug release. *J. Polym. Sci. Part A Polym. Chem.* **2013**, *51*, 4429–4439. [CrossRef]
196. Li, C.; Wang, B.; Liu, Y.; Cao, J.; Feng, T.; Chen, Y.; Luo, X. Synthesis and evaluation of star-shaped poly(ε-caprolactone)-poly(2-hydroxyethyl methacrylate) as potential anticancer drug delivery carriers. *J. Biomater. Sci. Polym. Ed.* **2013**, *24*, 741–757. [CrossRef]
197. Xiong, D.; Yao, N.; Gu, H.; Wang, J.; Zhang, L. Stimuli-responsive shell cross-linked micelles from amphiphilic four-arm star copolymers as potential nanocarriers for "pH/redox-triggered" anticancer drug release. *Polymer* **2017**, *114*, 161–172. [CrossRef]

198. Yuan, H.; Chi, H.; Yuan, W. A star-shaped amphiphilic block copolymer with dual responses: Synthesis, crystallization, self-assembly, redox and LCST–UCST thermoresponsive transition. *Polym. Chem.* **2016**, *7*, 4901–4911. [CrossRef]
199. Mielańczyk, A.; Kupczak, M.; Burek, M.; Mielańczyk, Ł.; Klymenko, O.; Wandzik, I.; Neugebauer, D. Functional (mikto)stars and star-comb copolymers from d-gluconolactone derivative: An efficient route for tuning the architecture and responsiveness to stimuli. *Polymer* **2018**, *146*, 331–343. [CrossRef]
200. Huang, X.; Xiao, Y.; Lang, M. Synthesis and self-assembly behavior of six-armed block copolymers with pH- and thermo-responsive properties. *Macromol. Res.* **2011**, *19*, 113–121. [CrossRef]
201. Teng, X.; Zhang, P.; Liu, T.; Xin, J.; Zhang, J. Biobased miktoarm star copolymer from soybean oil, isosorbide, and caprolactone. *J. Appl. Polym. Sci.* **2020**, *137*, 48281. [CrossRef]
202. Milner, S.T. Polymer brushes. *Science* **1991**, *251*, 905–914. [CrossRef]
203. Bousquet, A.; Boyer, C.; Davis, T.P.; Stenzel, M.H. Electrostatic assembly of functional polymer combs onto gold nanoparticle surfaces: Combining RAFT, click and LbL to generate new hybrid nanomaterials. *Polym. Chem.* **2010**, *1*, 1186–1195. [CrossRef]
204. Peng, S.; Bhushan, B. Smart polymer brushes and their emerging applications. *RSC Adv.* **2012**, *2*, 8557–8578. [CrossRef]
205. Sun, W.; Liu, W.; Wu, Z.; Chen, H. Chemical surface modification of polymeric biomaterials for biomedical applications. *Macromol. Rapid Commun.* **2020**, *41*, 1900430. [CrossRef]
206. Celentano, W.; Ordanini, S.; Bruni, R.; Marocco, L.; Medaglia, P.; Rossi, A.; Buzzaccaro, S.; Cellesi, F. Complex poly(ε-caprolactone)/poly(ethylene glycol) copolymer architectures and their effects on nanoparticle self-assembly and drug nanoencapsulation. *Eur. Polym. J.* **2021**, *144*, 110226. [CrossRef]
207. Tu, X.Y.; Meng, C.; Zhang, X.L.; Jin, M.G.; Zhang, X.S.; Zhao, X.Z.; Wang, Y.F.; Ma, L.W.; Wang, B.Y.; Liu, M.Z. Fabrication of Reduction-Sensitive Amphiphilic Cyclic Brush Copolymer for Controlled Drug Release. *Macromol. Biosci.* **2018**, *18*, 1800022. [CrossRef]
208. Cheng, R.; Meng, F.; Deng, C.; Klok, H.-A.; Zhong, Z. Dual and multi-stimuli responsive polymeric nanoparticles for programmed site-specific drug delivery. *Biomaterials* **2013**, *34*, 3647–3657. [CrossRef]
209. Kumar, A.; Lale, S.V.; Mahajan, S.; Choudhary, V.; Koul, V. ROP and ATRP fabricated dual targeted redox sensitive polymersomes based on pPEGMA-PCL-ss-PCL-pPEGMA triblock copolymers for breast cancer therapeutics. *ACS Appl. Mater. Interfaces* **2015**, *7*, 9211–9227. [CrossRef]
210. Nehate, C.; Moothedathu Raynold, A.A.; Haridas, V.; Koul, V. Comparative assessment of active targeted redox sensitive polymersomes based on pPEGMA-SS-PLA diblock copolymer with marketed nanoformulation. *Biomacromolecules* **2018**, *19*, 2549–2566. [CrossRef]
211. Zhao, P.; Deng, M.; Yang, Y.; Zhang, J.; Zhang, Y. Synthesis and Self-Assembly of Thermoresponsive Biohybrid Graft Copolymers Based on a Combination of Passerini Multicomponent Reaction and Molecular Recognition. *Macromol. Rapid Commun.* **2021**, *42*, 2100424. [CrossRef] [PubMed]
212. Rodrigues, P.R.; Wang, X.; Li, Z.; Lyu, J.; Wang, W.; Vieira, R.P. A new nano hyperbranched β-pinene polymer: Controlled synthesis and nonviral gene delivery. *Colloids Surf. B Biointerfaces* **2023**, *222*, 113032. [CrossRef] [PubMed]
213. Newland, B.; Tai, H.; Zheng, Y.; Velasco, D.; Di Luca, A.; Howdle, S.M.; Alexander, C.; Wang, W.; Pandit, A. A highly effective gene delivery vector–hyperbranched poly(2-(dimethylamino)ethyl methacrylate) from in situ deactivation enhanced ATRP. *Chem. Comm.* **2010**, *46*, 4698–4700. [CrossRef] [PubMed]
214. Flejszar, M.; Chmielarz, P.; Smenda, J.; Wolski, K. Following principles of green chemistry: Low ppm photo-ATRP of DMAEMA in water/ethanol mixture. *Polymer* **2021**, *228*, 123905. [CrossRef]
215. Bouzenna, H.; Hfaiedh, N.; Giroux-Metges, M.-A.; Elfeki, A.; Talarmin, H. Potential protective effects of alpha-pinene against cytotoxicity caused by aspirin in the IEC-6 cells. *Biomed. Pharmacother.* **2017**, *93*, 961–968. [CrossRef]
216. Cohen-Karni, D.; Kovaliov, M.; Li, S.; Jaffee, S.; Tomycz, N.D.; Averick, S. Fentanyl initiated polymers prepared by atrp for targeted delivery. *Bioconjugate Chem.* **2017**, *28*, 1251–1259. [CrossRef]
217. Lutz, J.F. Polymerization of oligo (ethylene glycol)(meth) acrylates: Toward new generations of smart biocompatible materials. *J. Polym. Sci. Part A Polym. Chem.* **2008**, *46*, 3459–3470. [CrossRef]
218. Li, S.; Cohen-Karni, D.; Kallick, E.; Edington, H.; Averick, S. Post-polymerization functionalization of epoxide-containing copolymers in trifluoroethanol for synthesis of polymer-drug conjugates. *Polymer* **2016**, *99*, 59–62. [CrossRef]
219. Zhang, Y.; Jiang, Q.; Wojnilowicz, M.; Pan, S.; Ju, Y.; Zhang, W.; Liu, J.; Zhuo, R.; Jiang, X. Acid-sensitive poly (β-cyclodextrin)-based multifunctional supramolecular gene vector. *Polym. Chem.* **2018**, *9*, 450–462. [CrossRef]
220. Nehate, C.; Moothedathu Raynold, A.A.; Koul, V. ATRP fabricated and short chain polyethylenimine grafted redox sensitive polymeric nanoparticles for codelivery of anticancer drug and siRNA in cancer therapy. *ACS Appl. Mater. Interfaces* **2017**, *9*, 39672–39687. [CrossRef]
221. Ensafi, A.A.; Khoddami, E.; Nabiyan, A.; Rezaei, B. Study the role of poly (diethyl aminoethyl methacrylate) as a modified and grafted shell for TiO2 and ZnO nanoparticles, application in flutamide delivery. *React. Funct. Polym.* **2017**, *116*, 1–8. [CrossRef]
222. Alotaibi, K.M.; Almethen, A.A.; Beagan, A.M.; Alfhaid, L.H.; Ahamed, M.; El-Toni, A.M.; Alswieleh, A.M. Poly (oligo (ethylene glycol) methyl ether methacrylate) Capped pH-Responsive Poly (2-(diethylamino) ethyl methacrylate) Brushes Grafted on Mesoporous Silica Nanoparticles as Nanocarrier. *Polymers* **2021**, *13*, 823. [CrossRef]

223. Alswieleh, A.M.; Beagan, A.M.; Alsheheri, B.M.; Alotaibi, K.M.; Alharthi, M.D.; Almeataq, M.S. Hybrid mesoporous silica nanoparticles grafted with 2-(tert-butylamino) ethyl methacrylate-b-poly (ethylene glycol) methyl ether methacrylate diblock brushes as drug nanocarrier. *Molecules* **2020**, *25*, 195. [CrossRef] [PubMed]
224. Alswieleh, A.M.; Alshahrani, M.M.; Alzahrani, K.E.; Alghamdi, H.S.; Niazy, A.A.; Alsilme, A.S.; Beagan, A.M.; Alsheheri, B.M.; Alghamdi, A.A.; Almeataq, M.S. Surface modification of pH-responsive poly (2-(tert-butylamino) ethyl methacrylate) brushes grafted on mesoporous silica nanoparticles. *Des. Monomers Polym.* **2019**, *22*, 226–235. [CrossRef] [PubMed]
225. Chen, T.; Wu, W.; Xiao, H.; Chen, Y.; Chen, M.; Li, J. Intelligent drug delivery system based on mesoporous silica nanoparticles coated with an ultra-pH-sensitive gatekeeper and poly (ethylene glycol). *ACS Macro Lett.* **2016**, *5*, 55–58. [CrossRef] [PubMed]
226. Peng, S.; Yuan, X.; Lin, W.; Cai, C.; Zhang, L. pH-responsive controlled release of mesoporous silica nanoparticles capped with Schiff base copolymer gatekeepers: Experiment and molecular dynamics simulation. *Colloids Surf. B Biointerfaces* **2019**, *176*, 394–403. [CrossRef]
227. Huang, L.; Liu, M.; Mao, L.; Xu, D.; Wan, Q.; Zeng, G.; Shi, Y.; Wen, Y.; Zhang, X.; Wei, Y. Preparation and controlled drug delivery applications of mesoporous silica polymer nanocomposites through the visible light induced surface-initiated ATRP. *Appl. Surf. Sci.* **2017**, *412*, 571–577. [CrossRef]
228. Beagan, A.M.; Alghamdi, A.A.; Lahmadi, S.S.; Halwani, M.A.; Almeataq, M.S.; Alhazaa, A.N.; Alotaibi, K.M.; Alswieleh, A.M. Folic acid-terminated poly (2-diethyl amino ethyl methacrylate) brush-gated magnetic mesoporous nanoparticles as a smart drug delivery system. *Polymers* **2020**, *13*, 59. [CrossRef]
229. He, X.; Wu, X.; Cai, X.; Lin, S.; Xie, M.; Zhu, X.; Yan, D. Functionalization of magnetic nanoparticles with dendritic–linear–brush-like triblock copolymers and their drug release properties. *Langmuir* **2012**, *28*, 11929–11938. [CrossRef]
230. Ellis, E.; Zhang, K.; Lin, Q.; Ye, E.; Poma, A.; Battaglia, G.; Loh, X.J.; Lee, T.-C. Biocompatible pH-responsive nanoparticles with a core-anchored multilayer shell of triblock copolymers for enhanced cancer therapy. *J. Mater. Chem. B* **2017**, *5*, 4421–4425. [CrossRef]
231. Xiong, D.; Zhang, X.; Peng, S.; Gu, H.; Zhang, L. Smart pH-sensitive micelles based on redox degradable polymers as DOX/GNPs carriers for controlled drug release and CT imaging. *Colloids Surf. B: Biointerfaces* **2018**, *163*, 29–40. [CrossRef] [PubMed]
232. Yao, N.; Lin, W.; Zhang, X.; Gu, H.; Zhang, L. Amphiphilic β-cyclodextrin-based star-like block copolymer unimolecular micelles for facile in situ preparation of gold nanoparticles. *J. Polym. Sci. Part A Polym. Chem.* **2016**, *54*, 186–196. [CrossRef]
233. Li, C.; Wang, J.; Wang, Y.; Gao, H.; Wei, G.; Huang, Y.; Yu, H.; Gan, Y.; Wang, Y.; Mei, L.; et al. Recent progress in drug delivery. *Acta Pharm. Sin. B* **2019**, *9*, 1145–1162. [CrossRef] [PubMed]
234. Paolino, D.; Sinha, P.; Fresta, M.; Ferrari, M. Drug Delivery Systems. In *Encyclopedia of Medical Devices and Instrumentation*; John Wiley & Sons, Ltd.: Hoboken, NJ, USA, 2006. [CrossRef]
235. Huang, Y.; Cole, S.P.C.; Cai, T.; Cai, Y. Applications of nanoparticle drug delivery systems for the reversal of multidrug resistance in cancer (Review). *Oncol. Lett.* **2016**, *12*, 11–15. [CrossRef]
236. Leppert, W.; Malec–Milewska, M.; Zajaczkowska, R.; Wordliczek, J. Transdermal and Topical Drug Administration in the Treatment of Pain. *Molecules* **2018**, *23*, 681. [CrossRef] [PubMed]
237. Khandare, J.; Minko, T. Polymer–drug conjugates: Progress in polymeric prodrugs. *Prog. Polym. Sci.* **2006**, *31*, 359–397. [CrossRef]
238. Neeraj Agrawal, R.; Alok Mukerji, A.J. Polymeric Prodrugs: Recent Achievements and General Strategies. *J. Antivir. Antiretrovir.* **2013**, *15*, 1–12. [CrossRef]
239. Ekladious, I.; Colson, Y.L.; Grinstaff, M.W. Polymer–drug conjugate therapeutics: Advances, insights and prospects. *Nat. Rev. Drug Discov.* **2019**, *18*, 273–294. [CrossRef]
240. Ulery, B.D.; Nair, L.S.; Laurencin, C.T. Biomedical applications of biodegradable polymers. *J. Polym. Sci. Part B Polym. Phys.* **2011**, *49*, 832–864. [CrossRef]
241. Dong, R.; Zhou, Y.; Huang, X.; Zhu, X.; Lu, Y.; Shen, J. Functional Supramolecular Polymers for Biomedical Applications. *Adv. Mater.* **2015**, *27*, 498–526. [CrossRef] [PubMed]
242. Singhvi, M.S.; Zinjarde, S.S.; Gokhale, D.V. Polylactic acid: Synthesis and biomedical applications. *J. Appl. Microbiol.* **2019**, *127*, 1612–1626. [CrossRef] [PubMed]
243. Ringsdorf, H. Structure and properties of pharmacologically active polymers. *J. Polym. Sci. Polym. Symp.* **1975**, *51*, 135–153. [CrossRef]
244. Liu, S.; Maheshwari, R.; Kiick, K.L. Polymer-Based Therapeutics. *Macromolecules* **2009**, *42*, 3–13. [CrossRef]
245. Parveen, S.; Arjmand, F.; Tabassum, S. Clinical developments of antitumor polymer therapeutics. *RSC Adv.* **2019**, *9*, 24699–24721. [CrossRef]
246. Ribelli, T.G.; Lorandi, F.; Fantin, M.; Matyjaszewski, K. Atom Transfer Radical Polymerization: Billion Times More Active Catalysts and New Initiation Systems. *Macromol. Rapid Commun.* **2019**, *40*, 1800616. [CrossRef]
247. Briand, V.A.; Kumar, C.V.; Kasi, R.M. Protein-Polymer Conjugates. In *Encyclopedia of Polymer Science and Technology*; John Wiley & Sons, Ltd.: Hoboken, NJ, USA, 2011. [CrossRef]
248. Zhao, W.; Liu, F.; Chen, Y.; Bai, J.; Gao, W. Synthesis of well-defined protein–polymer conjugates for biomedicine. *Polymer* **2015**, *66*, A1–A10. [CrossRef]
249. Wang, Y.; Wu, C. Site-Specific Conjugation of Polymers to Proteins. *Biomacromolecules* **2018**, *19*, 1804–1825. [CrossRef]
250. Kovaliov, M.; Cohen-Karni, D.; Burridge, K.A.; Mambelli, D.; Sloane, S.; Daman, N.; Xu, C.; Guth, J.; Kenneth Wickiser, J.; Tomycz, N.; et al. Grafting strategies for the synthesis of active DNase I polymer biohybrids. *Eur. Polym. J.* **2018**, *107*, 15–24. [CrossRef]

251. Silva, A.R.P.; Guimarães, M.S.; Rabelo, J.; Belén, L.H.; Perecin, C.J.; Farías, J.G.; Santos, J.H.P.M.; Rangel-Yagui, C.O. Recent advances in the design of antimicrobial peptide conjugates. *J. Mater. Chem. B* **2022**, *10*, 3587–3600. [CrossRef]
252. Pelegri-O'Day, E.M.; Maynard, H.D. Controlled Radical Polymerization as an Enabling Approach for the Next Generation of Protein–Polymer Conjugates. *Acc. Chem. Res.* **2016**, *49*, 1777–1785. [CrossRef] [PubMed]
253. Kaupbayeva, B.; Boye, S.; Munasinghe, A.; Murata, H.; Matyjaszewski, K.; Lederer, A.; Colina, C.M.; Russell, A.J. Molecular Dynamics-Guided Design of a Functional Protein–ATRP Conjugate That Eliminates Protein–Protein Interactions. *Bioconjug. Chem.* **2021**, *32*, 821–832. [CrossRef] [PubMed]
254. Bontempo, D.; Heredia, K.L.; Fish, B.A.; Maynard, H.D. Cysteine-Reactive Polymers Synthesized by Atom Transfer Radical Polymerization for Conjugation to Proteins. *J. Am. Chem. Soc.* **2004**, *126*, 15372–15373. [CrossRef] [PubMed]
255. Patel, A.; Graeff-Armas, L.; Ross, M.; Goldner, W. 35—Hypercalcemia. In *Abeloff's Clinical Oncology*, 6th ed.; Niederhuber, J.E., Armitage, J.O., Kastan, M.B., Doroshow, J.H., Tepper, J.E., Eds.; Elsevier: Philadelphia, PA, USA, 2020; pp. 565–571.e1. [CrossRef]
256. Sayers, C.T.; Mantovani, G.; Ryan, S.M.; Randev, R.K.; Keiper, O.; Leszczyszyn, O.I.; Blindauer, C.; Brayden, D.J.; Haddleton, D.M. Site-specific N-terminus conjugation of poly(mPEG1100) methacrylates to salmon calcitonin: Synthesis and preliminary biological evaluation. *Soft Matter* **2009**, *5*, 3038–3046. [CrossRef]
257. Baker, S.L.; Kaupbayeva, B.; Lathwal, S.; Das, S.R.; Russell, A.J.; Matyjaszewski, K. Atom Transfer Radical Polymerization for Biorelated Hybrid Materials. *Biomacromolecules* **2019**, *20*, 4272–4298. [CrossRef] [PubMed]
258. Yoshihara, E.; Sasaki, M.; Nabil, A.; Iijima, M.; Ebara, M. Temperature Responsive Polymer Conjugate Prepared by "Grafting from" Proteins toward the Adsorption and Removal of Uremic Toxin. *Molecules* **2022**, *27*, 1051. [CrossRef]
259. Cummings, C.S.; Fein, K.; Murata, H.; Ball, R.L.; Russell, A.J.; Whitehead, K.A. ATRP-grown protein-polymer conjugates containing phenylpiperazine selectively enhance transepithelial protein transport. *J. Control. Release* **2017**, *255*, 270–278. [CrossRef]
260. Trzebicka, B.; Robak, B.; Trzcinska, R.; Szweda, D.; Suder, P.; Silberring, J.; Dworak, A. Thermosensitive PNIPAM-peptide conjugate—Synthesis and aggregation. *Eur. Polym. J.* **2013**, *49*, 499–509. [CrossRef]
261. Cohen-Karni, D.; Kovaliov, M.; Ramelot, T.; Konkolewicz, D.; Graner, S.; Averick, S. Grafting challenging monomers from proteins using aqueous ICAR ATRP under bio-relevant conditions. *Polym. Chem.* **2017**, *8*, 3992–3998. [CrossRef]
262. Konkolewicz, D.; Magenau, A.J.D.; Averick, S.E.; Simakova, A.; He, H.; Matyjaszewski, K. ICAR ATRP with ppm Cu Catalyst in Water. *Macromolecules* **2012**, *45*, 4461–4468. [CrossRef]
263. Moncalvo, F.; Lacroce, E.; Franzoni, G.; Altomare, A.; Fasoli, E.; Aldini, G.; Sacchetti, A.; Cellesi, F. Selective Protein Conjugation of Poly(glycerol monomethacrylate) and Poly(polyethylene glycol methacrylate) with Tunable Topology via Reductive Amination with Multifunctional ATRP Initiators for Activity Preservation. *Macromolecules* **2022**, *55*, 7454–7468. [CrossRef]
264. Kaupbayeva, B.; Murata, H.; Rule, G.S.; Matyjaszewski, K.; Russell, A.J. Rational Control of Protein–Protein Interactions with Protein-ATRP-Generated Protease-Sensitive Polymer Cages. *Biomacromolecules* **2022**, *23*, 3831–3846. [CrossRef]
265. Baker, S.L.; Munasinghe, A.; Murata, H.; Lin, P.; Matyjaszewski, K.; Colina, C.M.; Russell, A.J. Intramolecular Interactions of Conjugated Polymers Mimic Molecular Chaperones to Stabilize Protein–Polymer Conjugates. *Biomacromolecules* **2018**, *19*, 3798–3813. [CrossRef]
266. Li, P.; Sun, M.; Xu, Z.; Liu, X.; Zhao, W.; Gao, W. Site-Selective in Situ Growth-Induced Self-Assembly of Protein–Polymer Conjugates into pH-Responsive Micelles for Tumor Microenvironment Triggered Fluorescence Imaging. *Biomacromolecules* **2018**, *19*, 4472–4479. [CrossRef] [PubMed]
267. Liu, X.; Sun, M.; Sun, J.; Hu, J.; Wang, Z.; Guo, J.; Gao, W. Polymerization Induced Self-Assembly of a Site-Specific Interferon α-Block Copolymer Conjugate into Micelles with Remarkably Enhanced Pharmacology. *J. Am. Chem. Soc.* **2018**, *140*, 10435–10438. [CrossRef] [PubMed]
268. Baker, S.L.; Munasinghe, A.; Kaupbayeva, B.; Rebecca Kang, N.; Certiat, M.; Murata, H.; Matyjaszewski, K.; Lin, P.; Colina, C.M.; Russell, A.J. Transforming protein-polymer conjugate purification by tuning protein solubility. *Nat. Commun.* **2019**, *10*, 1–12. [CrossRef] [PubMed]
269. Zhu, B.; Lu, D.; Ge, J.; Liu, Z. Uniform polymer–protein conjugate by aqueous AGET ATRP using protein as a macroinitiator. *Acta Biomater.* **2011**, *7*, 2131–2138. [CrossRef] [PubMed]
270. Averick, S.E.; Bazewicz, C.G.; Woodman, B.F.; Simakova, A.; Mehl, R.A.; Matyjaszewski, K. Protein–polymer hybrids: Conducting ARGET ATRP from a genetically encoded cleavable ATRP initiator. *Eur. Polym. J.* **2013**, *49*, 2919–2924. [CrossRef]
271. Kovaliov, M.; Cheng, C.; Cheng, B.; Averick, S. Grafting-from lipase: Utilization of a common amino acid residue as a new grafting site. *Polym. Chem.* **2018**, *9*, 4651–4659. [CrossRef]
272. Fu, L.; Wang, Z.; Lathwal, S.; Enciso, A.E.; Simakova, A.; Das, S.R.; Russell, A.J.; Matyjaszewski, K. Synthesis of Polymer Bioconjugates via Photoinduced Atom Transfer Radical Polymerization under Blue Light Irradiation. *ACS Macro Lett.* **2018**, *7*, 1248–1253. [CrossRef]
273. Olszewski, M.; Jeong, J.; Szczepaniak, G.; Li, S.; Enciso, A.; Murata, H.; Averick, S.; Kapil, K.; Das, S.R.; Matyjaszewski, K. Sulfoxide-Containing Polyacrylamides Prepared by PICAR ATRP for Biohybrid Materials. *ACS Macro Lett.* **2022**, *11*, 1091–1096. [CrossRef] [PubMed]
274. Charan, H.; Kinzel, J.; Glebe, U.; Anand, D.; Garakani, T.M.; Zhu, L.; Bocola, M.; Schwaneberg, U.; Böker, A. Grafting PNIPAAm from β-barrel shaped transmembrane nanopores. *Biomaterials* **2016**, *107*, 115–123. [CrossRef] [PubMed]
275. Elvira, C.; Gallardo, A.; Roman, J.; Cifuentes, A. Covalent Polymer-Drug Conjugates. *Molecules* **2005**, *10*, 114–125. [CrossRef] [PubMed]

276. Stanley, T.H. The Fentanyl Story. *J. Pain* **2014**, *15*, 1215–1226. [CrossRef]
277. Stanley, T.H. The History of Opioid Use in Anesthetic Delivery. In *The Wondrous Story of Anesthesia*; Eger Ii, E.I., Saidman, L.J., Westhorpe, R.N., Eds.; Springer: New York, NY, USA, 2014; pp. 641–659. [CrossRef]
278. Li, Y.; Liu, R.; Yang, J.; Ma, G.; Zhang, Z.; Zhang, X. Dual sensitive and temporally controlled camptothecin prodrug liposomes codelivery of siRNA for high efficiency tumor therapy. *Biomaterials* **2014**, *35*, 9731–9745. [CrossRef]
279. Mei, C.; Lei, L.; Tan, L.-M.; Xu, X.-J.; He, B.-M.; Luo, C.; Yin, J.-Y.; Li, X.; Zhang, W.; Zhou, H.-H.; et al. The role of single strand break repair pathways in cellular responses to camptothecin induced DNA damage. *Biomed. Pharmacother.* **2020**, *125*, 109875. [CrossRef]
280. Fan, X.; Lin, X.; Ruan, Q.; Wang, J.; Yang, Y.; Sheng, M.; Zhou, W.; Kai, G.; Hao, X. Research progress on the biosynthesis and metabolic engineering of the anti-cancer drug camptothecin in Camptotheca acuminate. *Ind. Crops Prod.* **2022**, *186*, 115270. [CrossRef]
281. Plichta, A.; Kowalczyk, S.; Kamiński, K.; Wasyłeczko, M.; Więckowski, S.; Olędzka, E.; Nałęcz-Jawecki, G.; Zgadzaj, A.; Sobczak, M. ATRP of Methacrylic Derivative of Camptothecin Initiated with PLA toward Three-Arm Star Block Copolymer Conjugates with Favorable Drug Release. *Macromolecules* **2017**, *50*, 6439–6450. [CrossRef]
282. Gao, Y.-E.; Bai, S.; Shi, X.; Hou, M.; Ma, X.; Zhang, T.; Xiao, B.; Xue, P.; Kang, Y.; Xu, Z. Irinotecan delivery by unimolecular micelles composed of reduction-responsive star-like polymeric prodrug with high drug loading for enhanced cancer therapy. *Colloids Surf. B: Biointerfaces* **2018**, *170*, 488–496. [CrossRef]
283. Chen, W.; Shah, L.A.; Yuan, L.; Siddiq, M.; Hu, J.; Yang, D. Polymer–paclitaxel conjugates based on disulfide linkers for controlled drug release. *RSC Adv.* **2014**, *5*, 7559–7566. [CrossRef]
284. Dong, Y.; Du, P.; Pei, M.; Liu, P. Design, postpolymerization conjugation and self-assembly of a di-block copolymer-based prodrug for tumor intracellular acid-triggered DOX release. *J. Mater. Chem. B* **2019**, *7*, 5640–5647. [CrossRef] [PubMed]
285. Odrobińska, J.; Neugebauer, D. Retinol derivative as bioinitiator in the synthesis of hydroxyl-functionalized polymethacrylates for micellar delivery systems. *Express Polym. Lett.* **2019**, *13*, 806–817. [CrossRef]
286. Zaborniak, I.; Chmielarz, P.; Matyjaszewski, K. Synthesis of Riboflavin-Based Macromolecules through Low ppm ATRP in Aqueous Media. *Macromol. Chem. Phys.* **2020**, *221*, 1900496. [CrossRef]
287. Plichta, A.; Kowalczyk, S.; Olędzka, E.; Sobczak, M.; Strawski, M. Effect of structural factors on release profiles of camptothecin from block copolymer conjugates with high load of drug. *Int. J. Pharm.* **2018**, *538*, 231–242. [CrossRef] [PubMed]
288. Qiu, L.; Liu, Q.; Hong, C.-Y.; Pan, C.-Y. Unimolecular micelles of camptothecin-bonded hyperbranched star copolymers via β-thiopropionate linkage: Synthesis and drug delivery. *J. Mater. Chem. B* **2015**, *4*, 141–151. [CrossRef]
289. Bai, S.; Jia, D.; Ma, X.; Liang, M.; Xue, P.; Kang, Y.; Xu, Z. Cylindrical polymer brushes-anisotropic unimolecular micelle drug delivery system for enhancing the effectiveness of chemotherapy. *Bioact. Mater.* **2021**, *6*, 2894–2904. [CrossRef]
290. Bai, S.; Gao, Y.-E.; Ma, X.; Shi, X.; Hou, M.; Xue, P.; Kang, Y.; Xu, Z. Reduction stimuli-responsive unimolecular polymeric prodrug based on amphiphilic dextran-framework for antitumor drug delivery. *Carbohydr. Polym.* **2018**, *182*, 235–244. [CrossRef]
291. Bai, S.; Hou, M.; Shi, X.; Chen, J.; Ma, X.; Gao, Y.-E.; Wang, Y.; Xue, P.; Kang, Y.; Xu, Z. Reduction-active polymeric prodrug micelles based on α-cyclodextrin polyrotaxanes for triggered drug release and enhanced cancer therapy. *Carbohydr. Polym.* **2018**, *193*, 153–162. [CrossRef]
292. Chen, X.; Parelkar, S.S.; Henchey, E.; Schneider, S.; Emrick, T. PolyMPC–Doxorubicin Prodrugs. *Bioconjugate Chem.* **2012**, *23*, 1753–1763. [CrossRef]
293. Lale, S.V.; Ravindran Girija, A.; Aravind, A.; Kumar, D.S.; Koul, V. AS1411 Aptamer and Folic Acid Functionalized pH-Responsive ATRP Fabricated pPEGMA–PCL–pPEGMA Polymeric Nanoparticles for Targeted Drug Delivery in Cancer Therapy. *Biomacromolecules* **2014**, *15*, 1737–1752. [CrossRef]
294. Wang, H.; Xu, F.; Li, D.; Liu, X.; Jin, Q.; Ji, J. Bioinspired phospholipid polymer prodrug as a pH-responsive drug delivery system for cancer therapy. *Polym. Chem.* **2013**, *4*, 2004–2010. [CrossRef]
295. Pelras, T.; Duong, H.T.T.; Kim, B.J.; Hawkett, B.S.; Müllner, M. A 'grafting from' approach to polymer nanorods for pH-triggered intracellular drug delivery. *Polymer* **2017**, *112*, 244–251. [CrossRef]
296. Rickert, E.L.; Trebley, J.P.; Peterson, A.C.; Morrell, M.M.; Weatherman, R.V. Synthesis and Characterization of Bioactive Tamoxifen-Conjugated Polymers. *Biomacromolecules* **2007**, *8*, 3608–3612. [CrossRef] [PubMed]
297. Jin, Q.; Mitschang, F.; Agarwal, S. Biocompatible Drug Delivery System for Photo-Triggered Controlled Release of 5-Fluorouracil. *Biomacromolecules* **2011**, *12*, 3684–3691. [CrossRef] [PubMed]
298. Giacomelli, C.; Schmidt, V.; Borsali, R. Nanocontainers Formed by Self-Assembly of Poly(ethylene oxide)-b-poly(glycerol monomethacrylate)–Drug Conjugates. *Macromolecules* **2007**, *40*, 2148–2157. [CrossRef]
299. Zaborniak, I.; Chmielarz, P.; Wolski, K. Riboflavin-induced metal-free ATRP of (meth) acrylates. *Eur. Polym. J.* **2020**, *140*, 110055. [CrossRef]
300. Fairbanks, B.D.; Gunatillake, P.A.; Meagher, L. Biomedical applications of polymers derived by reversible addition–fragmentation chain-transfer (RAFT). *Adv. Drug Deliv. Rev.* **2015**, *91*, 141–152. [CrossRef]

polymers

Review

Recent Advances in Biomedical Applications of Polymeric Nanoplatform Assisted with Two-Photon Absorption Process

Subramaniyan Ramasundaram [1,*,†], Sivasangu Sobha [2,†], Gurusamy Saravanakumar [2] and Tae Hwan Oh [1,*]

1 School of Chemical Engineering, Yeungnam University, Gyeongsan 38436, Republic of Korea
2 OmniaMed Co., Ltd., Pohang 37666, Republic of Korea
* Correspondence: ramasundaram79@hotmail.com (S.R.); taehwanoh@ynu.ac.kr (T.H.O.)
† These authors contributed equally to this work.

Abstract: Polymers are well-recognized carriers useful for delivering therapeutic drug and imaging probes to the target specified in the defined pathophysiological site. The functional drug molecules and imaging agents were chemically attached or physically loaded in the carrier polymer matrix via cleavable spacers. Using appropriate targeting moieties, these polymeric carriers (PCs) loaded with functional molecules were designed to realize target-specific delivery at the cellular level. The biodistribution of these carriers can be tracked using imaging agents with suitable imaging techniques. The drug molecules can be released by cleaving the spacers either by endogenous stimuli (e.g., pH, redox species, glucose level and enzymes) at the targeted physiological site or exogenous stimuli (e.g., light, electrical pulses, ultrasound and magnetism). Recently, two-photon absorption (2PA)-mediated drug delivery and imaging has gained significant attention because TPA from near-infrared light (700–950 nm, NIR) renders light energy similar to the one-photon absorption from ultraviolet (UV) light. NIR has been considered biologically safe unlike UV, which is harmful to soft tissues, cells and blood vessels. In addition to the heat and reactive oxygen species generating capability of 2PA molecules, 2PA-functionalized PCs were also found to be useful for treating diseases such as cancer by photothermal and photodynamic therapies. Herein, insights attained towards the design, synthesis and biomedical applications of 2PA-activated PCs are reviewed. In particular, specific focus is provided to the imaging and drug delivery applications with a special emphasis on multi-responsive platforms.

Keywords: polymeric carrier; two-photon absorption; near infrared; fluorophore; drug delivery; imaging; photodynamic therapy; theranostic

Citation: Ramasundaram, S.; Sobha, S.; Saravanakumar, G.; Oh, T.H. Recent Advances in Biomedical Applications of Polymeric Nanoplatform Assisted with Two-Photon Absorption Process. *Polymers* **2022**, *14*, 5134. https://doi.org/10.3390/polym14235134

Academic Editor: Marek Kowalczuk

Received: 31 October 2022
Accepted: 22 November 2022
Published: 25 November 2022

1. Introduction

The wide possibilities and realization witnessed in engineering a variety of structures and properties have made polymers a promising carrier matrix for functional biomedical applications such as drug delivery and imaging. The recent advances in modern polymerization techniques enables the synthesis of well-defined polymers with precise control over chemical composition, molecular weight, architecture and other functional properties such as stimuli-responsivity. Various strategies have advanced the biomedical applications of polymer-based drug delivery and imaging platforms, including chemical conjugation of active agents to the polymer scaffold, or physical encapsulation of active agents into self-assembled polymeric carriers (PCs) such as nanocapsules, micelles and polymersomes. Compared to free active agents, these PC-based systems endow several advantages, including improved solubility, long circulation and targeting [1–3]. Moreover, after exerting their functions, PCs based on biodegradable polymers can be easily degraded into non-toxic small molecules and exerted from the body [1]. A wide variety of natural and synthetic biodegradable polymers, such as polysaccharides, polyesters, polycarbonate, poly(amino acid), poly(aminoester), polyamides and polyurethane, have been used to design carriers for efficient and targeted delivery of active agents [4]. Over the past few decades, with appropriate design and fabrication, these PCs have found potential applications in the imaging and treatment of various diseases [5]. The specific advantages such as extravasation from the blood vessels and accumulation at tumor sites make nano-sized PCs a versatile and powerful platform for tumor-targeted drug delivery and imaging [6]. Recently, many PCs have also been explored for delivery of vaccines for coronaviruses (e.g., SARS-CoV, MERS-CoV, or hCoV). A small number of PC-based nanomedicines are under FDA approval, and several other are undergoing clinical trials and believed to be hopeful for future clinical use [7–9]. Remarkably, as they allow explicit spatial and temporal control of cargos at the desired sites, stimuli-responsive PCs have been emerging as on-demand drug delivery systems for effective therapy and precise imaging.

Light as external medical stimulus is an easy and convenient tool useful for noninvasive therapy, image guided surgery, control over spatial resolution, localized polymerization and degradation of tissue engineering scaffolds [10,11]. When exposed to light, the photo-responsive PCs undergo chemical bond rupture and induce the dissociation of the carrier structure, thereby facilitating the release of encapsulated drugs. In some other cases, light exposure activates the imaging probes and photosensitizing agents for non-invasive imaging and photodynamic therapy (PDT), respectively [12,13]. Typically, PCs are designed to be responsive conventional one-photon absorption system using high-energy ultraviolet (UV) and visible (Vis.) radiations. The short wavelength (350–650 nm) of light use in the one-proton absorption system has the following shortcomings: limited penetration depth in tissues, photobleaching of probes meant for imaging, photodamaging of tissues and interference of autofluorescence of biological species, which are greatly limiting applications in the biomedical field. To surmount these issues, recently, photo-responsive PCs based on a two-photon absorption (2PA) system using low-energy near-infrared (NIR) radiation (700–950 nm) have been developed. When compared to UV and Vis radiations, NIR can deeply penetrate into tissues without any destruction or scattering [14,15]. Several studies have indicated that the depth of penetration of NIR radiation is largely dependent on the wavelength and also varies with type of tissue [16]. Henderson and Morries reported that about 0.45–2.90% of 810 nm light penetrated through 3 cm depth of brain tissue [17]. The use of a cheap continuous-wave laser with a low power density source, instead of a femtosecond pulse laser with high power density, was also found to be appealing for 2PA-based biomedical application using NIR radiation. Therefore, PCs responsive to 2PA possess tremendous potential in clinical medicine [18–20]. After a brief introduction of basic principle and relevance of 2PA in biomedical applications, this review will comprehensively summarize the recent advances in the two-photon-activated PCs for drug delivery, imaging, PDT and theranostic applications.

2. Basic Principle and Relevance of Two-Photon Absorption in Biomedical Applications

In 1931, Maria Goppert-Mayer postulated the process of 2PA. Then, in 1961, 2PA was experimentally confirmed by Kaiser and Garret [21]. Figure 1 depicts the principle of 2PA and its biomedical applications. Constructive interference of light waves corresponding to two photons which are absorbed simultaneously by an ion, atom or molecule results in the promotion of electrons from ground state to excited state. The process by which two photons of the same frequency (degenerate 2PA) or different frequencies (non-degenerate) are absorbed by the molecular system proceeds through a stepwise process. In this process, the system first absorbs one photon and gets excited from the ground state (S_0) to a temporary virtual state of higher energy. Before relaxing back to its initial state, the system immediately absorbs the second photon and gets excited to the real final electronic state (S_1). When an excited electron returns to the ground state, the generation of a photon with energy greater than either two of the absorbed photons occurs. This phenomenon is known as 2PA [19,22,23]. The 2PA allows for the excitation of electrons in a molecule confined to volume of closer to the point of focus. Thus, pinpoint three-dimensional imaging can be realized by the polymeric nanoplatforms integrated with two-photon imaging probes [24,25]. A number of two-photon-responsive drug delivery systems that can accomplish the spatiotemporal control of drug accumulation at desired disease sites have also been developed to improve the therapeutic effect of the active agents. In these systems, a 2PA-generated photon induces the scission of incorporated photolabile groups within the carriers, resulting in the triggered release of drugs from the carriers to exert their therapeutic action [20]. Similarly, polymeric carriers encapsulated with photosensitizers (PS) that exhibit a large 2PA cross section have been utilized for two-photon excited photodynamic therapy. The irradiation of an NIR two-photon light source activates the PS to its excited state, which can subsequently interact with molecular oxygen to produce reactive oxygen species (ROS), leading to the destruction of the plasma membrane of tumor by localized oxidative stress. Notably, NIR radiation used in 2PA penetrates more deeply into tissues than the conventional one-photon process, where UV and visible light have been used. Though visible light is less harmful than UV, it can be scattered by tissues and absorbed strongly by blood. NIR is considered safer as it causes less or no damage to the normal tissues. Thus, 2PA-based systems are used to treat multidrug-resistant tumor cells by PDT alone or in combination with chemotherapy [3,26].

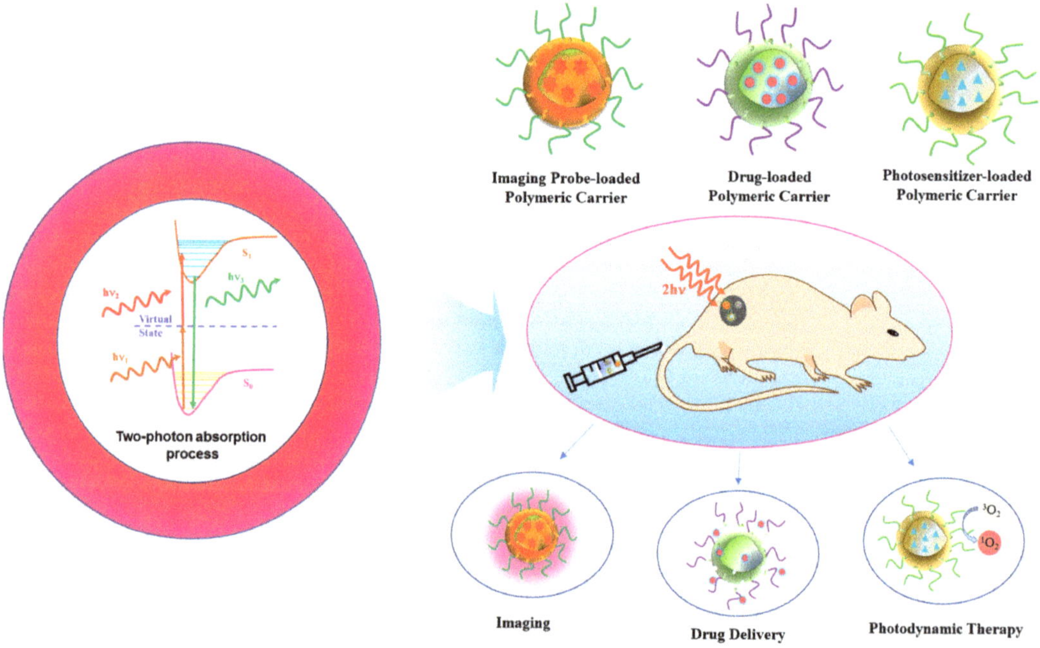

Figure 1. Basic principle of two-photon absorption process. Schematic representations of polymeric carriers loaded with imaging probe or therapeutic agents (drugs or photosensitizer) for two-photon-assisted imaging, drug delivery and photodynamic therapy.

In the literature, in the biomedical field, 2PA is used for either one of the three applications mentioned previously or in a multifunction platform combining at least two applications and other methods of therapy such as chemotherapy. The recent development witnessed in each of the above applications of 2PA and its multifunctional platforms have been discussed in subsequent sections.

3. Polymeric Nanoplatforms for Two-Photon-Assisted Imaging

PC-based two-photon imaging platforms have recently attracted significant interest due to their numerous advantages such as signal amplification, high photostability and low toxicity, compared with their small-molecule counterparts. There are two main strategies adopted in the development of PC-based systems for two-photon bioimaging. The first approach includes the physical encapsulation of small molecule two-photon imaging probes into the PCs, while the second one involves utilization of conjugated polymeric nanoassembly with large 2PA cross sections (Figure 2).

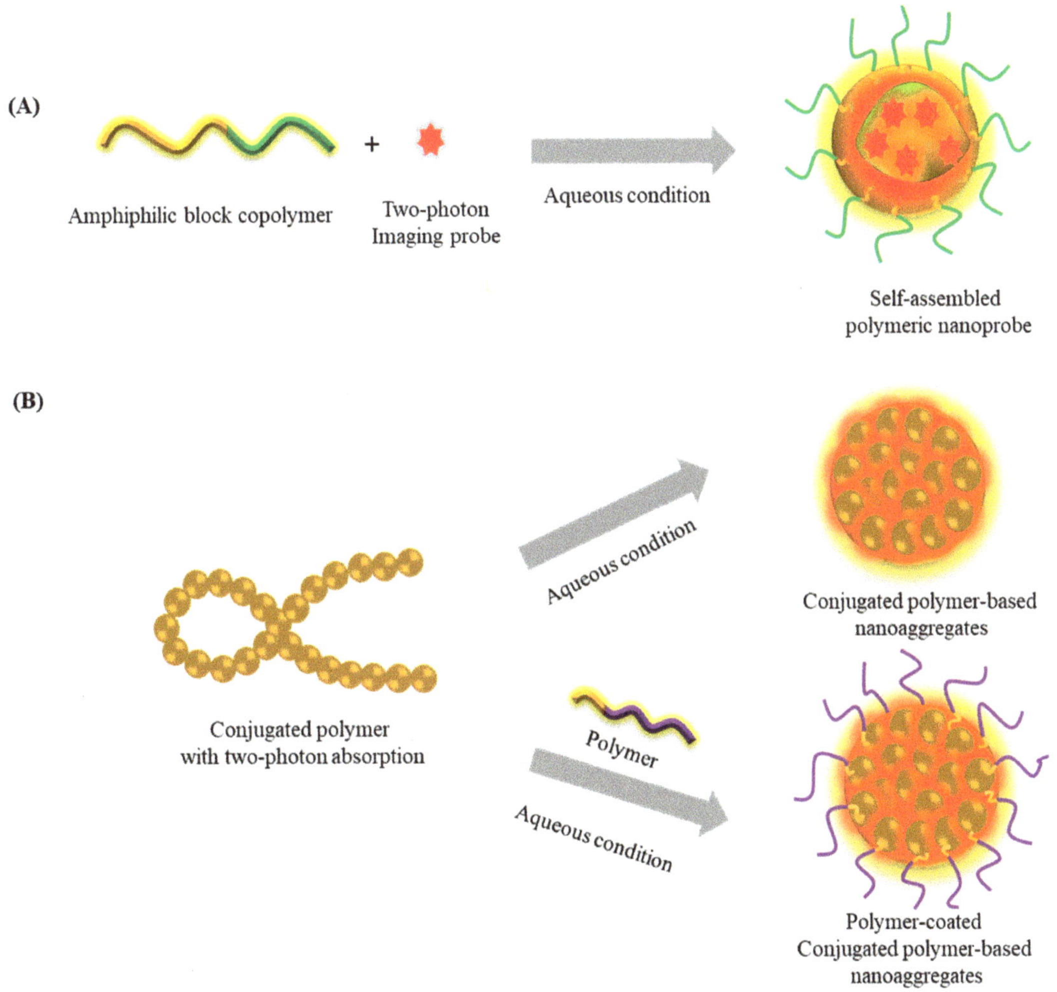

Figure 2. Schematic illustration of (**A**) two-photon imaging probe encapsulated polymeric nanoparticles and (**B**) conjugated polymer-based nanoaggregates with or without polymer coating for two-photon imaging.

3.1. Small Molecule Probe-Loaded Systems

Given the excellent physical stability, facile fabrication and the opportunity to encapsulate multiple functional agents into the core along with the two-photon imaging probes, these physically loaded PCs have received significant attention towards both bioimaging and theranostic applications [27]. Although various conventional small molecule organic probes-loaded PCs have been prepared, the emission of these probes is often weakened in aggregates due to the phenomenon known as the aggregation-induced quenching (ACQ) effect. Increasing the loading content of these probes leads to reduced fluorescence and poor sensitivity. The discovery of fluorogens with aggregation-induced emission (AIE) addressed this ACQ issues [28]. Unlike traditionally fluorogenic probes, these AIE probes emit strong fluorescence in the aggregated state compared to the molecularly dissolved state. Although a number of possible mechanisms were hypothesized and proposed, the restriction of intramolecular motion has been widely considered the AIE working mechanism. Thus, encapsulating AIE probes within the core of polymeric nanoparticles could induce aggregation and suppress the intramolecular motion of the loaded probes, and the resulting nanoparticles can emit strong fluorescence and exhibit a relatively high photoluminescence quantum yield [29]. By taking advantage of this desired property, several AIE probe-loaded polymeric nanoparticles have been prepared for two-photon bioimaging.

Samanta et al. synthesized four acrylonitrile-based AIE-active two-photon fluorescence (AIETP) probes and encapsulated each into FDA-approved amphiphilic block copolymer Pluronic F127-based nanoparticles [30]. Then, by utilizing these four AIE probe-loaded nanoparticles, they comprehensively studied how the structural variations of the AIE probes influence the two-photon brain vascular imaging. Among the four synthesized AIE probes, the one with the phenyl-thiazole unit in the structure (AIETP, Figure 3A) facilitated the formation of excellent water-dispersible polymeric nanoparticles and showed improved photostability and a better 2PA cross section. In addition, this AIETP polymeric nanoparticle demonstrated two-photon NIR-II (1040 nm) excitability, which enables high-contrast vascular brain imaging with increased penetration depth. The NIR-II biological window (1000–1700 nm) is more promising for two-photon bioimaging because tissue components scatter and absorb less at the longer wavelengths, resulting in deeper penetration depth and higher spatial resolution, compared to the visible or conventional NIR-I window [31,32]. However, the design and synthesis of efficient fluorophores are crucial in NIR-II bioimaging. A crab-shaped donor-acceptor AIE active probe with a high quantum yield was synthesized and loaded into Pluronic F127 [33]. This polymer-loaded AIE nanoprobe showed high stability and a large two-photon absorption cross section ($\delta = 1.22 \times 10^3$ GM). In vivo two-photon fluorescence imaging under NIR-II excitation (1300 nm) enables us to image 3D vascular information with high spatial resolution (sub 3.5 μm) and to visualize small blood vessels of ~5 μm as deep as 1065 μm in mouse brain. For two-photon fluorescence lifetime imaging (TP-FLIM), a thioxanthone-based AIE probe (TXO) was encapsulated within Pluronic F127 nanoparticles. As the TXO exhibits thermally activated delayed fluorescence properties and AIE features, the resulting polymeric nanoparticles exhibited ultralong fluorescence lifetime and efficient 2PA, making them useful for both in vitro and in vivo TP-FLIM [34]. Although the physical loading of AIE probes into a hydrophobic core of polymeric nanoparticles enhances its fluorescence intensity due to the spatial confinement that restricts intramolecular rotation of the probes, the chemical composition of the polymeric core also considerably influences the photophysical properties of AIE probes. In particular, AIE probes loaded into amphiphilic block copolymer nanoparticles with high hydrophobicity showed an improved fluorescence quantum yield compared to the one with less hydrophobicity. For example, owing to the increased hydrophobicity of poly(styrene) (PS) compared to that of poly(caprolactone) (PCL), the AIE probe loaded into poly(ethylene glycol)-*b*-PS (PEG-b-PS) copolymer nanoparticles exhibited higher fluorescence quantum yield compared to PEG-*b*-PCL [35]. Besides the polymeric core, engineering the surface of AIE probe-loaded polymeric nanoparticles also improved the fluorescence performance. When the shell of the AIE probe-loaded Pluronic F127 nanoparticles is coated with silica, their fluorescence quantum yield was greatly improved compared to the one without silica coating [36]. This enhanced characteristic of the silica shelled nanoparticles is ascribed to the non-polar microenvironment provided by the silica shell and reduced water and oxygen attack to the loaded AIE probes. Furthermore, encapsulating the AIE probe into folate conjugated poly([lactide-co-glycolide]-*b*-PEG (PLGA-*b*-PEG-Folate) nanoparticles showed potential for targeted cellular imaging [37]. In a similar way to small molecular fluorescent probe-loaded nanoparticles, several two-photon emissive polymers are also physically encapsulated into the PCs to improve optical stability and biocompatibility. Alifu et al. synthesized 2PA emissive polymer dots, encapsulated in poly(styrene-co-maleic anhydride) (PSMA) grafted with PEG matrix, and used them for imaging mouse ear and brain angiography [38]. When excited with 720–960 nm 2PA light, images with a penetration depth of 720 μm were obtained. For single-particle imaging in mouse brain, Khalin et al. prepared poly(methyl methacrylate)-sulfonate (PMMA-SO_3H)-based nanoparticles loaded with probe octadecyl rhodamine B along with a bulky hydrophobic perfluorinated tetraphenylboronate as a fluorophore insulator to avoid the ACQ effect [39]. To impart stealth characteristics, these nanoparticles were further coated with pluronic F-127 and F-68. The as-prepared nanoparticles with 74 nm size and 20% loading content exhibited 150-fold higher single-particle brightness than 39 nm Fluospheres loaded with Nile Red. Through two-photon intravital microscopy, these stealth nanoparticles were able to track for 1 h in the vessels of mouse brain, while the nanoparticles without the stealth layer were readily eliminated. In particular, the dynamics of stealth nanoparticles can be tracked in real time in the microvasculature of living animals with subcellular resolution. Similarly, noninvasive and real-time two-photon imaging of thoracoepigastric veins of living balb/c nude mice has been realized by preparing PMMA-poly(methacrylic acid)-based micelles loaded with europium(Eu)-luminescent complexes with large 2PA cross sections at NIR photosensitization wavelengths [40]. Without the interference of autofluorescence from tissues, the spatiotemporal evolution of these nanoparticles in veins was observed for 2 h with 1–2 mm imaging depth and 80 μm lateral resolution.

In addition to the block copolymer nanoparticles, several polymeric lipid nanoparticles and polymer supramolecular nanoassemblies have also been employed for encapsulating two-photon small molecule imaging probes [41–43]. Recently, Shen et al. fabricated supramolecular NIR emission nanoparticles using a tetraphenylethene derivative with a methoxyl and vinyl pyridine salt-based AIE fluoroprobe, cucurbit [8] uril, and cancer cell targeting component hyaluronic acid modified cyclodextrin (β-CD) for cell targeting imaging. These nanoparticles showed less cytotoxicity when tested in A549 cells by with CCK-8 assay. These supramolecular nanoparticles solved the short wavelength excitation and emission problems of conventional supramolecular imaging systems and demonstrated deep tissue penetration and mitochondrial-targeted two-photon imaging [42].

Besides physical encapsulation, AIE-active small molecules are chemically conjugated to the amphiphilic copolymers and self-assembled into nanoprobe. Xiao et al. synthesized amphiphilic block copolymer covalently conjugated with a two-photon AIE probe (TPBP) and magnetic resonance contrast agent Gd-DOTA (Figure 3B) [44]. In aqueous condition, this copolymer formed self-assembled nanoparticles of about 20 nm, which can be circulated in blood for a long time. This polymeric nanoprobe provides an improved dual fluorescence and magnetic resonance imaging.

3.2. *Conjugated Polymer-Based Systems*

Conjugated polymers (CPs) consist of alternating σ and π-bonds on their polymer backbones, and thus, electrons can delocalize throughout the entire polymer chain, which produces interesting and specific photophysical and electrical properties. Compared to the traditional fluorescence materials, CPs show several distinguished features such as large extinction coefficient, high fluorescence efficiency and excellent photostability. In general, CPs are highly hydrophobic, and thus, to realize these materials for in vivo biomedical applications requires significant improvement of their aqueous solubility. In this direction, the following two strategies have been considered [45]. The first one is formulating CPs into nanoparticles (CPNs) by using nanoengineering techniques or integrating it into amphiphilic block copolymer nanoassemblies using appropriate methods, such as the physical encapsulation method discussed above. The second one is incorporating charged (anionic or cationic) or hydrophilic side chains to the CP hydrophobic backbone. Owing to advantages such as large 2PA cross section, high single-particle brightness, low cytotoxicity and facile surface modification, CPNs have been emerging as promising sensing platforms for both in vitro and in vivo two-photon bioimaging.

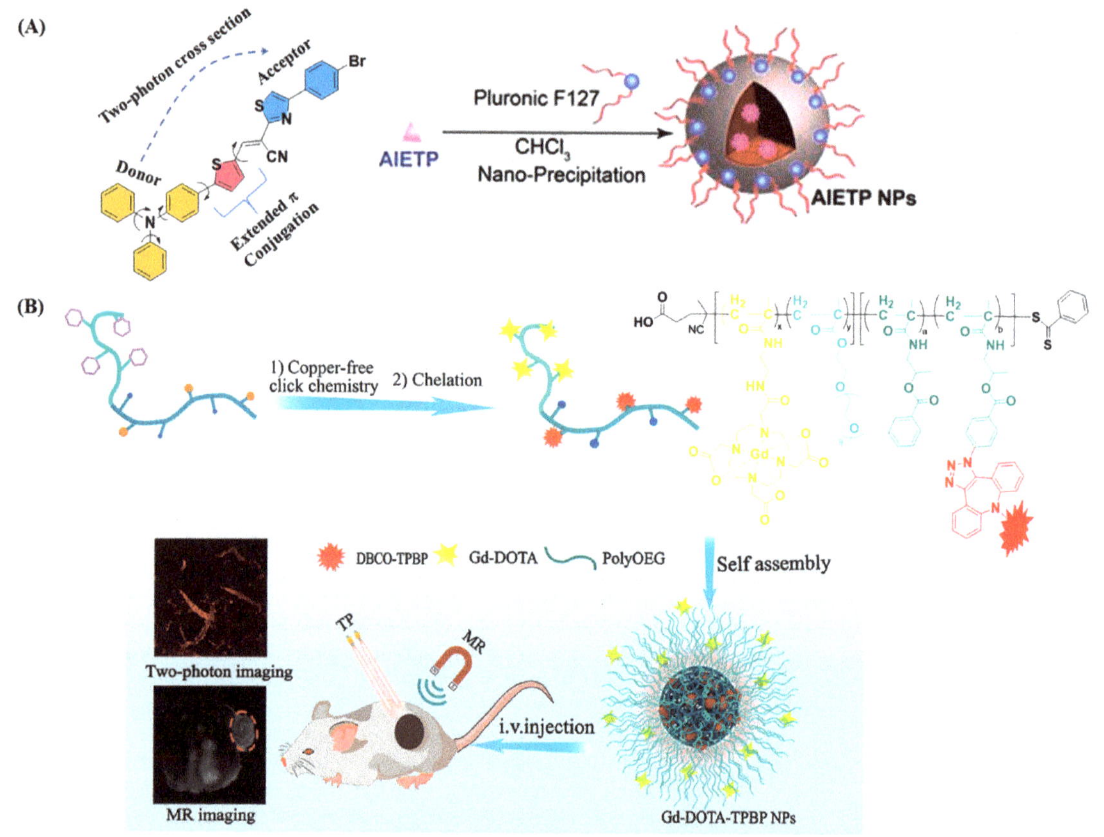

Figure 3. (**A**) Chemical structure of AIETP probe, and illustration for the preparation of AIETP-loaded Pluronic F127 nanoparticles. Adapted with permission from ref. [30] Copyright 2019 The Authors, Published by Theranostics. (**B**) Schematic representation of synthesis of amphiphilic block copolymer containing AIE probe TPBP and Gd-DOTA contrast agent for two-photon fluorescence and magnetic resonance imaging. Adopted with permission from ref. [44] Copyright 2022 The Authors. Published by KeAi Communications Co., Ltd.

McNeil and coworkers prepared CNPs from a series of CPs (polyfluorene derivative, polyfluorene copolymer and poly(phenylelenvinylene) derivative) by a reprecipitation method (Figure 4A) [46]. The two-photon action cross section of the as-prepared CNPs were found to be 3–4 orders of magnitude higher than the values of traditional fluorescence dyes, and an order of magnitude higher than those of inorganic quantum dots of comparable size. Single-particle fluorescence imaging was realized using relatively low laser power, demonstrating the potential of these CNPs for multiphoton fluorescence microscopy applications. In addition to the fabrications of CNPs using only CPs, the PC matrix encapsulated method has also been developed. Liu et al. prepared water dispersible CNPS by encapsulating red-emitting bis(diphenylaminostyryl)benzene (DPSB)-based conjugated polymers into the poly(styrene-co-maleic anhydride) (PSMA) matrix (Figure 4B) [47]. The size of CNPs was controlled by changing the feed ratio of DPSB to PSMA. By optimizing the feed ratio and initial solution concentration, a high quantum yield of 24% was obtained for the PSMA-CNPs and employed as a two-photon excitation nanoprobe for cell membrane imaging.

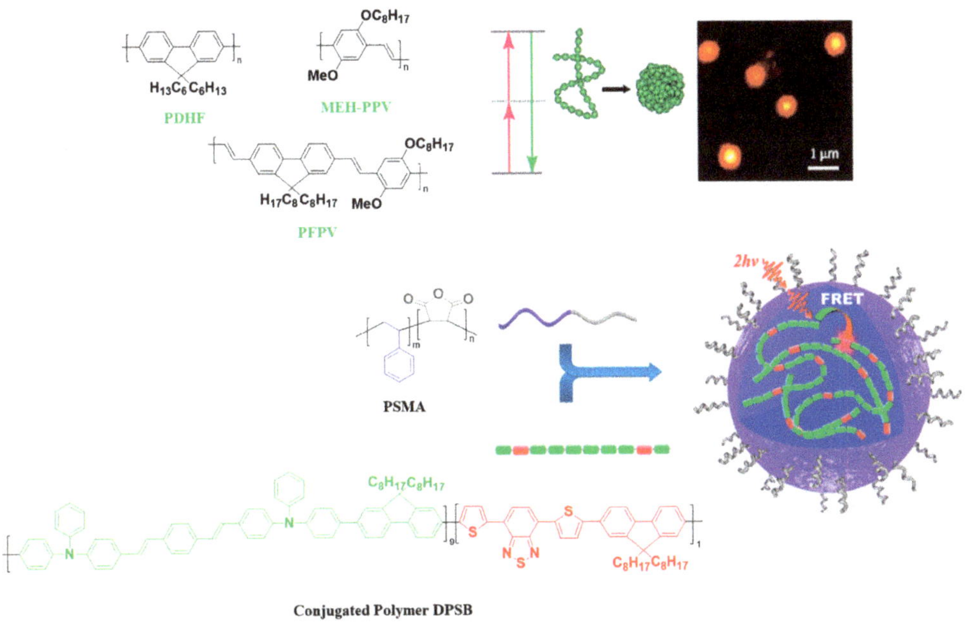

Figure 4. (**A**) Chemical structure of conjugated polymers (PDHF, MEH-PPV, and PFPV), and schematic illustration of formation of CNPs and fluorescence image of single PFPV particle immobilized on a glass coverslip obtained using two-photon excitation (800 nm). Adapted with permission from ref. [46]. Copyright 2007 American Chemical Society. (**B**) Chemical structures of conjugated polymer DPSB, amphiphilic block copolymer PSMA, and schematic illustration of DPSB-loaded CNPs. Adapted with permission from ref. [47] Copyright 2015 American Chemical Society.

4. Polymeric Nanoplatforms for Two-Photon-Assisted Drug Delivery

Owing to the tunable physicochemical characteristics and facile incorporation of photo-responsive functional groups, PCs have attracted much interest in the field of light-responsive targeted drug delivery systems. In general, most of the photo-responsive moieties incorporated into PCs respond to UV or visible light, which has limited penetration depth and thus cannot induce photoresponsive changes to trigger drug release. However, as an exogeneous stimulus, two-photon light irradiation can penetrate more deeply into biological tissues and facilitate the precise spatiotemporal control of drug release.

4.1. 2-Diazo-1,2-Napthoquinone (DNQ)-Based Systems

Among the various photoresponsive functional groups, DNQ is one of the most extensively studied units. This functional group induces the solubility switch and triggers drug release from the PCs. Goodwin et al. synthesized a functional PEG-lipid nanoparticle with 2-diazo-1,2-napthoquinones (DNQ) [48]. When this nanoparticle was irradiated with a 795 nm laser, the encapsulated Nile red dye was released from the nanoparticles through Wolff rearrangement of DNQ group via a two-photon process. These findings inspired many researchers to design two-photon responsive polymeric nanoparticles systems by judiciously incorporating DNQ moiety [49]. Yuan et al. developed a nanoplatform consisting of doxorubicin (DOX) and conjugated polymer (CP) meant for synchronous combination of photothermal therapy and chemotherapy (Figure 5A) [50]. The nanoparticle was formulated using an amphiphilic copolymer poly-L-lysine (PLL)-graft-PEG/DNQ (Figure 5B). A cyclic arginine-glycine-aspartic acid (cRGD) tripeptide was also attached to facilitate the accumulation of this nanosystem on cancer cells, where integrin $\alpha v\beta 3$ gene overexpressed. When incubated with MDA-MB-231 cells and exposed to 800 nm laser irradiation for 20 min, DOX was released through disassembly of the nanoparticles by Wolff rearrangement induced conversion of DNQ moieties from hydrophobic to hydrophilic. Further, the release profile could be controlled by tuning the power density of the laser. The loaded CPs also endow in photothermal effect. Moreover, (IC50) of the combined chemotherapy and photothermal therapy (PTT) was 13.7 $\mu g\ mL^{-1}$. The IC50 was 147.8 $\mu g\ mL^{-1}$ for chemotherapy and 36.2 $\mu g\ mL^{-1}$ for PTT. The combination index (C.I.) is 0.48 (<1), which indicates the synergistic effect for chemotherapy and PTT. By taking advantage of the DNQ group transformation, Sun et al. also constructed a two-photon-sensitive and sugar-targeted nanocarriers from degradable and dendritic amphiphiles [51].

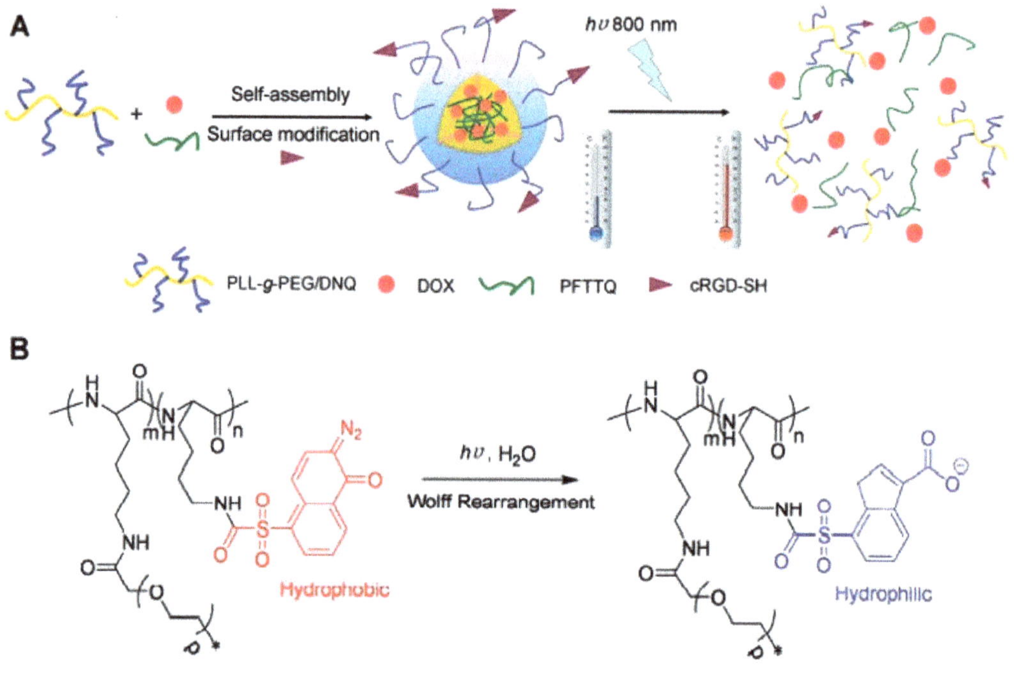

Figure 5. (**A**) Schematic illustration of preparation of PLL-g-PEG/DNQ nanoparticles and two-photon-mediated disassembly and concomitant drug release, and (**B**) chemical structure of PLL-g-PEG/DNQ and light-induced Wolff rearrangement. Adapted with permission from ref. [50] Copyright 2015 The Royal Society of Chemistry.

4.2. Coumarin-Based Systems

The coumarinyl ester can be more easily cleaved by two-photon NIR light compared to the o-nitrobenzyl group. Hartner et al. studied the use of 2PA-induced cycloreversion of photodimerized 7-(*tert*-butyldimethylsilyloxy)-coumarin (TBS-coumarin) derivatives in both solution and incorporated into PMMA matrix [52]. When exposed to two-photon excitation, TBS-protected 7-hydroxycourmarin monomer was released from both conditions, indicating the potential of this transformation for two-photon controlled drug delivery applications. Along this line of research, they synthesized PMMA-based copolymer with photocleavable dicoumarin moieties and subsequently conjugated with an alkylating neoplastic drug molecule, chlorambucil [53]. As expected, when exposed to two-photon excitation with a 532 nm laser, the drug chlorambucil was released from the polymer matrix. Kim et al. conjugated a precures of the cataract-healing drug 5-florouracil (5-FU) to the backbone of polymers used for making intraocular lenses (IOL) via a coumarin linker [54]. The excitation of two-photon light (532 nm) triggered the release of 5-FU from the lens polymer matrix, which is useful for preventing the occurrence of secondary cataracts after implantation of IOL.

A hybrid coumarin-containing photo-responsive nanocomposite was prepared for NIR light-controlled drug release via a two-photon process [55]. This photo-responsive nanocomposite was fabricated by coating an NIR-light-responsive block copolymer containing photoactive coumarin moiety and tumor-targeting folic acid group onto the surface of DOX-loaded octadecyltrimethoxysilane (C18)-modified hollow mesoporous silica nanoparticles (HMS@C18) through self-assembly (Figure 6). The folic acid moiety conferred selective folate receptor-mediated endocytosis to the nanocomposites, which showed good selectivity to the folate receptor (FR(+)) KB cells compared to FR(-) A549 cancer cells. Because of the high 2PA cross section of the coumarin moiety, the copolymers were disrupted when exposed to NIR femtosecond laser (800 nm), leading to the triggered release of DOX over the targeted tumor cells. Greater than 50% of the DOX loaded on the nanocomposites was released upon NIR light irradiation.

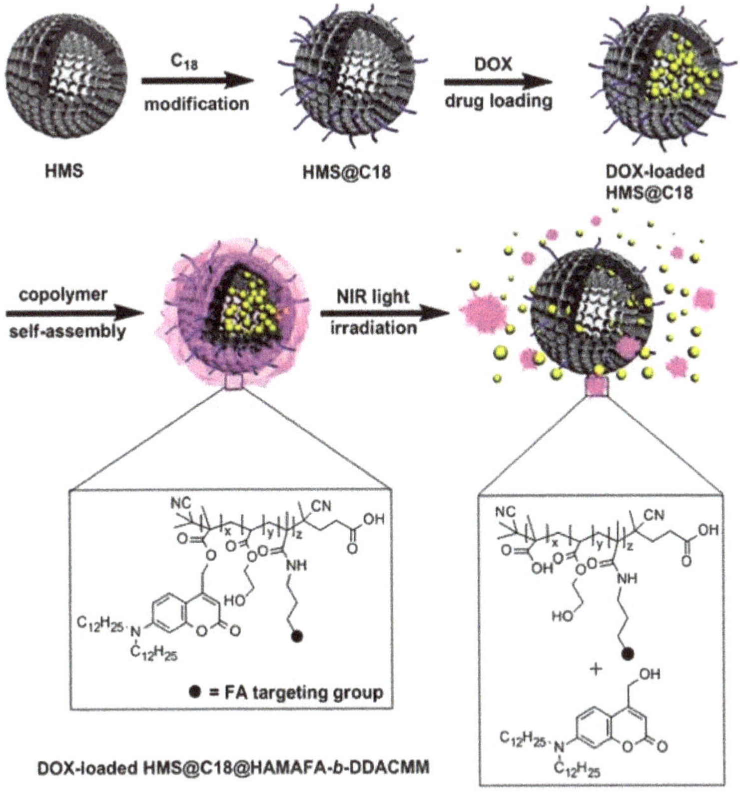

Figure 6. Schematic illustration of the preparation of HMS@C18@HAMAFA-*b*-DDACMM for drug delivery and controlled release by degradation upon NIR light exposure. The HAMAFA-*b*-DDACMM is a coumarin and folic acid conjugated copolymer, where the abbreviation HAMAFA denotes hydroxyethylacrylate and N-(3-aminopropyl) methacrylamide-conjugated folic acid segment, and DDACMM denotes 7-(didodecylamino) coumarin-4-yl] methyl methacrylate. Adapted with permission from ref. [55]. Copyright 2013 The Royal Society of Chemistry.

4.3. o-Nitrobenzyl-Based Systems

Although the o-nitrobenzyl group is less sensitive to NIR light compared to UV light, this group can respond to NIR light via a two-photon absorption process. In response to two-photon irradiation, this group induced a polarity change and triggered the release of active agents. Since UV absorbers are used in IOL, they may diminish or suppress the two-photon-triggered drug release from the polymer. Kehrloesser et al. addressed this problem by choosing a suitable UV absorber that is compatible with 2PA and does not interfere with two-photon-mediated drug release [56]. They selected 2(4-benzoyl-3-hydroxy phenoxy) ethyl acrylate (BHP-EA) as the UV absorber and *o*-nitrobenzyl moiety as the photo-cleavable linker. IOLs were prepared using acrylic polymer with *o*-nitrobenzyl group carrying BHP-EA and the drug 5-FU. When evaluated for 2PA-triggered drug delivery, no photochemical degradation of the UV absorber was observed, and the release rate of 5-FU was not significantly affected.

4.4. Other Systems

Using a popular light-sensitive azobenzene moiety, Huang et al. reported supramolecular conjugated unimicelles for two-photon-triggered drug release [57]. These unimicelles were constructed by taking the advantages of host–guest interactions between a β-CD-grafted hyperbranched conjugated polymer and azobenzene-functionalized PEG. DOX was effectively loaded into these micelles with high loading efficiency. Upon NIR-light irradiation (800 nm), these micelles triggered the release of DOX in cancer cells through photoisomerization of azobenzene via a two-photon excited fluorescence energy transfer process, resulting in effective antitumor therapy. Oleincizak et al. synthesized a two-photon degradable copolymer containing photocleavable moieties on the backbone by polymerizing adipoyl chloride, 1,6-hexanediol and a light-sensitive monomer with high two-photon absorbing efficiency [58]. The model dye molecule Nile Red was readily released from this polymer nanoparticle through backbone degradation when exposed to NIR light radiation, demonstrating its potential in two-photon drug delivery.

5. Polymeric Nanoplatforms for Two-Photon-Assisted Gene Delivery

Apart from the delivery of small molecule drugs, multifunctional polymeric gene delivery carriers assisted with two-photon irradiation have been developed for the precise and effective treatment of both acquired and genetic diseases [10,59]. In recent years, polymer-based non-viral gene delivery systems are emerging as a potential alterative to viral-based systems because of the improved biocompatibility and minimal immunogenicity [4,60]. Among these, cationic polymers, which can easily form electrostatic complexes with anionic genetic materials including antisense oligodeoxynucelotides, small interfering RNA (siRNA) and other forms of nucleic acid, have been widely investigated as non-viral systems. Thus, semiconducting CPs bearing positive charges are also studied as non-viral gene carriers [61,62]. Although these cationic CPs can be complex negatively charged therapeutic nucleic acids and allow us to monitor the cellular internalization, their low charge density and lack of intracellular release mechanism severely limited gene transfection efficiency. To address this issue, two-photon-induced charge-variable conjugated polyelectrolyte brushes with high-density cationic charges and photodegradable *o*-nitrobenzyl groups on the side chain were prepared using poly(phenylene ethynylene)-based semiconducting polymer backbone, which also served as an upconversion agent [62]. The dense cationic charges enabled remarkable complex formation with siRNA. The large 2PA cross section of conjugated polyelectrolyte prompted two-photon-induced photolysis of photoresponsive side chains and facilitated charge transformation from cations to zwitterions, which triggered the efficient intracellular release of siRNA. In vitro studies demonstrate that these polyelectrolyte brushes can efficiently knock out of targeted Plk1 mRNA to 24.7% under 720 nm illumination. The positively charged CP-based carriers are also used for the delivery of plasmid DNA. The gene expression efficiency of CP-based systems was greatly improved compared to that of lipofectamine, a commonly used gene delivery vector [61].

Instead of utilizing positively charged CPs, few studies have conjugated two-photon active agents to cationic polyethyleneimine (PEI), another widely used gene carrier, for efficient two-photon-assisted gene delivery. For example, Hayek et al. labeled PEI (25kDa) with a bis-stilbenyl-based two-photon active dye molecule and employed for gene delivery and imaging [63]. The dye-labeled PEI complexes showed comparable transfection efficiency and cytotoxicity to those of unlabeled complexes. Moreover, the size of the complexes did not affect much when both labeled and unlabeled PEI was combined at a ratio of 1:3. In vitro two-photon imaging of HeLa cells demonstrated efficient internalization of the complexes by the cells and their accumulation in perinuclear compartments, particularly at the endosomes. In general, the molecular weight of PEI strongly correlates with its toxicity and transfection efficiency. It has been reported that low molecular weight PEI (2kDa) modified with hydrophobic moieties could exert high gene transfection efficiency and good biocompatibility [64]. Thus, amphiphilic alkyl-PEG(2k) was synthesized and coated on the surface of photonic carbon dots (Cdots), and the resulting PEI-Cdots were evaluated as a carrier for the delivery of siRNA and DNA. The results demonstrated effective in vitro and in vivo gene transfection of PEI-Cdots. Moreover, the photonic Cdot endows two-photon imaging capabilities.

6. Polymeric Nanoplatforms for Two-Photon-Assisted Photodynamic Therapy

PDT is a minimally invasive treatment approved for certain cancers and non-cancerous disease states. It involves the administration of a photosensitizer, which is activated by light at specific wavelength, and generates cytotoxic species in the presence of oxygen.

Luo et al. prepared a thermosensitive nanocomposite hydrogel from methoxy poly(ethylene glycol)-polylactide copolymer (mPEG-PDLLA) and Pluronic F127, which is approved for clinical use [65]. A two-photon active imidazole derivative with large TPA cross section in the NIR region and a second-generation photosensitizer pyropheophorbide a (PPa) were loaded into these mPEG-PDLLA micelles. Upon injection, at body temperature, these micelles were transformed into hydrogel and displayed a long-term retention within the tumor. This composite hydrogel was tested using 4T1 murine breast cancer bearing BALB/c mice. When the thermosensitive hydrogel is activated using NIR, a significant amount of ROS was generated in deep tissue and inhibited the growth of tumor. Further, when the thermosensitive hydrogel-treated cells were observed through two-photon confocal microscope, obituary morphology and bubbling were noticed in the regions of cell membrane and nucleus, which is the indication for the occurrence of PDT. In another study, Luo et al. encapsulated PPa and a two-photon active imidazole derivative into dendritic prodrug-based nanoparticles obtained using poly[oligo(ethylene glycol)methyl ether methacrylate]-functionalized dendritic paclitaxel (PTX) (Figure 7) [66]. The branches of the dendritic polymers and PTX were conjugated to the polymer via a short peptide Gly-Phe-Leu- Gly that was sensitive to cathepsin B, a lysosomal cysteine protease overexpressed in tumor cells. This peptide linkage protected the drug from undesired premature leakage during circulation in the blood. After passive accumulation at the tumor site, the drug was released in response to the tumor microenvironment. The efficacy of this polymeric prodrug-based nanoparticle was evaluated using BALB/c mice bearing 4T1 murine breast cancer. The chemo/two-photon combined nanosystems showed a much higher tumor growth inhibition (67.23%) compared to that of the group treated with PTX and PPa (48.92%). Further, this nanosystem also allowed for the tracking of drug localization by two-photon bioimaging.

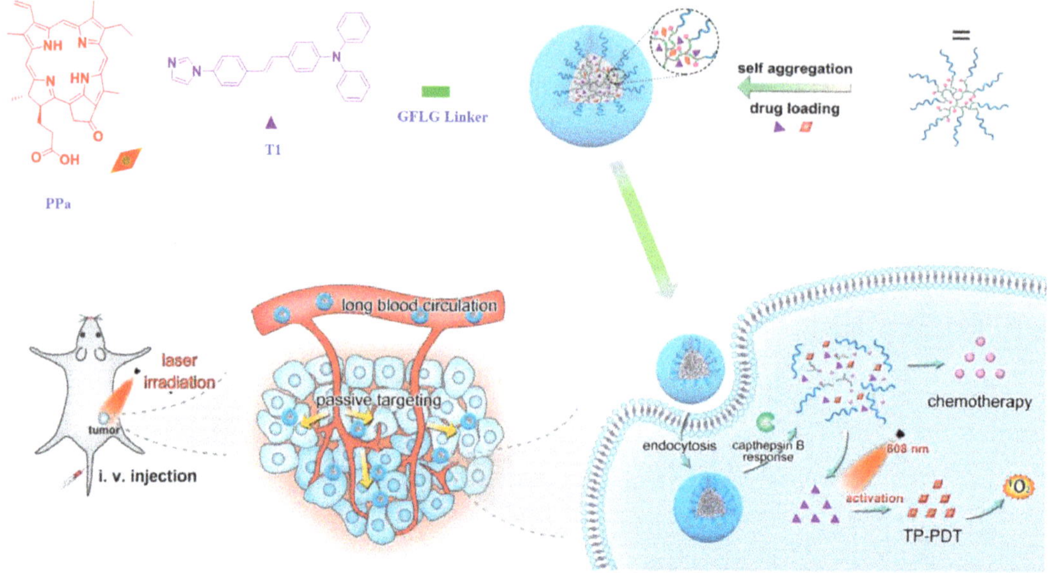

Figure 7. Preparation of cathepsin B-responsive dendritic prodrug and dendritic-[(GFLG-polyPTX)-block-polyOEGMA]. The self-assembled dendritic prodrug is loaded with TPA and PPa, leading to synergistic effect of chemotherapy and two-photon photodynamic therapy for breast tumor. Adapted with permission from ref. [66] Copyright 2021 Elsevier Ltd.

Kandoth et al. developed a three-component self-assembled polymeric nanosystem comprising an epichlorohydrin-β-CD copolymer as the carrier, a zinc phthalocyanine as a photosensitizer and a tailored nitroaniline derivative as a nitric oxide photodonor [67]. When encapsulated into the nanosystems, these two photoresponsive guest molecules did not interfere with each other, which was confirmed using steady-state and time-resolved spectroscopic and photochemical techniques. Thus, each of these molecules can be operated in parallel under the light stimuli. The tissue distribution of the system can be imaged using the two-photon fluorescence microscopy. When irradiated with visible light, this nanosystem simultaneously triggered generation of cytotoxic species $^1O_2$ and NO and resulted in death of carcinoma cells.

7. Polymeric Nanoplatforms for Two-Photon-Assisted Theranostic Applications

As discussed in the previous sections, PCs equipped with the 2PA probe enable imaging and drug delivery capabilities and PDT. Recently, significant efforts have been devoted to integrating the above modalities into one single theranostic platform. These integrated systems may allow for a safe, efficient and minimally invasive tool for personalized medicine. In this section, we will discuss a few representative theranostic polymeric systems for two-photon bioimaging, chemotherapy, PTT and PDT.

7.1. Cancer Theranostics

To date, polymeric nanoparticles are widely studied for cancer theranostics. Wang et al. prepared hybrid nanogels by immobilizing graphitic carbon dots (GCDs) within the PC matrix comprising a crosslinked PEG network and interpenetrated chitosan chains via an aqueous one-pot surfactant-free precipitation polymerization [68]. The biocompatible GCDs not only served as multifunctional confocal and two-photon imaging agent and pH-sensing probe, but also enhanced the loading capacity of hydrophobic anticancer drug, DOX, via π stacking mechanism. Due to their ability to absorb NIR and upconverting photoluminescent properties, GCDs provided the nanogels with high photothermal conversion ability and two-photon imaging ability. While the chitosan-induced swelling/deswelling regulated the drug release over the physiologically important pH range of 5.0–7.4, the thermoresponsive PEG network of the nanogels promoted the drug release through the local heat produced by GCDs under NIR irradiation. These results show the combined synergistic effect of the hybrid nanogels for treating cancer by chemo-photothermal modalities, as well for diagnostic imaging. Ardekani et al. synthesized photoresponsive nanoparticles based on the nitrogen-doped surface passivated (PEG, MW~200) carbon nanodots (CND-P) and anticancer drug DOX [69]. The nitrogen doping and surface passivation greatly improved their properties. For example, the quantum yield of CND-P was enhanced compared to the undoped CND and non-surface passivated CND by factors of 12.6 and 4.4, respectively. Under similar test conditions, the up-converted emission intensity of CND-P was increased by a factor of 4.5 compared to the non-surface passivated one. The CND-P was able to transport and release DOX into MCF-7 cancer cell through two-photon excitation (780 nm and 35 mW). The two-photon bioimaging indicated the internalization of CND-P in cytoplasm, nucleolus and nuclei of the MCF-7 cells. Due to the heat generation capability of CND-P, the DOX-mediated apoptosis of MCF-7 cells was increased through the combined chemo and photothermal effect.

Y. Wang and co-workers developed several polymeric systems that are capable of simultaneous two-photon imaging and stimuli-responsive drug delivery. By employing two-photon AIE-based imaging probe and judiciously placing stimuli-responsive units within the amphiphilic block copolymers either at the main backbone or side chain, they prepared number of theranostic polymeric platforms for enhanced two-photon imaging and effective therapy. Using the combination of RAFT polymerization and chemical conjugation methods, they synthesized a prodrug block copolymer featuring a two-photon AIE fluorophore and an anticancer drug capecitabine (Cap), which has been applied for the treatment of multiple malignant tumors, alone or in combination regimens [70]. In brief, first an amine-end functionalized hydrophilic poly(2-methacryloyloxyethyl phosphorylcholine) (PMPC-NH_2) was linked to an aldehyde-end functionalized poly(methacryloyl phenylboronic acid pinacol ester-*co*-TPMA)-*b*-poly(2-azepane ethylmethacrylate) (CHO-P(MPA-*co*-TPMA)-*b*-PAEMA) to introduce an acid-labile benzoyl imide bond on the main backbone. Second, the drug Cap was conjugated to MPA unit via ROS-responsive phenylboronate ester bond to obtain the final amphiphilic prodrug copolymer ($PMMTA_b$-Cap), which can form self-assembled micelles under aqueous conditions (Figure 8). The TPMA unit in the copolymer is a two-photon fluorophore. The PAEMA segment is a charge-convertible unit that can become electropositive and induce hydrophobic to hydrophilic solubility switch under mild acidic (pH 6.8) tumor microenvironment. Due to AIE effect of the TPMA, these prodrug micelles exhibited two-photon imaging capability and were able to obtain deep tissue imaging with a depth of 150 μm. After accumulation at the acidic tumor tissue, the outer shell PMPC can be detached and decrease the size of the micelles while the inner PAEMA become hydrophilic and electropositive, leading to effective internalization of the micelles into the tumor cells. After cellular uptake, the elevated intracellular ROS can disrupt the boronic ester bond and trigger the release of Cap for effective tumor therapy. In a subsequent study, they have also reported another prodrug micellar system with a neutral hydrophilic pH-detachable PEG shell and a dimethylmaleic anhydride-grafted-polyethyleneimine segment charge reversable segment [71]. While the AIE two-photon probe was chemically conjugated to the hydrophobic chain end, the therapeutic drug DOX was conjugated to the main backbone via an acid-labile imine bond. When incubated with 4T1 cells, these micelles produced fluorescence image with good quality, in response to excitations with one photon (405 nm) and two-photon (800 nm) light irradiations. The in vivo potential of these micelles was investigated using BALB/c mice bearing 4T1 breast cancer tumors. Under 800 two-photon excitation, bright fluorescence signals were observed in hepatic and nephric tissues, and the images were visualized even at the depth of 150 μm, whereas the florescence signals under one-photon excitation rapidly declined with increased in scanning depth. After accumulation at the tumor tissues by the enhanced permeation and retention effect, the detachment of the PEG shell enhanced the uptake by charge reversal and subsequently released the DOX, leading to improved antitumor efficacy. Along this line of research, redox-responsive theranostic prodrug micelles were also prepared that were capable of releasing anticancer drug gemcitabine under elevated glutathione (GSH) [72]. As the concentration of the GSH is at least 4 times higher in some tumors compared to the normal cells, these micelles demonstrated good in vivo tumor-suppression ability. In another study, oxidation-responsive PEG-*b*-poly(L-glutamic acid)-selenide micelles equipped with AIE-active two-photon probe were prepared and physically loaded with curcumin (Cur), a well-known herbal compound that exerts several anticancer effects in various types of cancers [73]. These micelles also showed good deep two-photon tissue imaging. Under the oxidative environments of the tumor, these micelles were readily dissociated and released the drug Cur. The prolonged circulation and efficient accumulation of these micelles at the tumor tissue greatly improved their tumor inhibition ability. Similarly, the anticancer drug DOX was physically loaded into redox-responsive theranostic micelles with two-photon imaging capability, which can induce charge conversion under acidic condition and detach PEG shell and concomitant DOX release in response to high concentration of GSH in the tumor cells [74]. In vivo, ex vivo imaging and in vivo pharmacokinetic study demonstrated the potential of these micelles for deep-tissue imaging and cancer therapy.

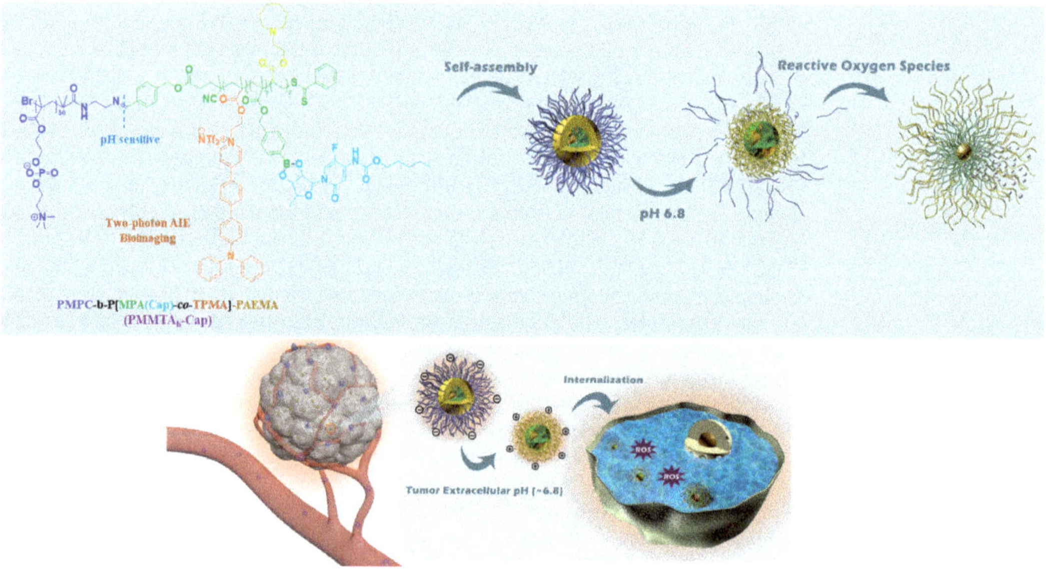

Figure 8. Schematic illustration for preparation of polymeric prodrug PMMTAb-Cap micelles with two-photon bioimaging, pH-triggered size shrinkage and charge-reversal, and ROS-triggered drug release. Adapted with permission from ref. [70]. Copyright 2019 American Chemical Society.

7.2. *Atherosclerosis Theranostics*

Atherosclerosis is the gradual narrowing of the arteries caused by the buildup of plaque in the inner lining of an artery. It is one of the leading causes of death worldwide. The rupture of vulnerable atherosclerosis plaque is the major cause of sudden deaths. Therefore, imaging the structural and anatomic features of atherosclerotic plaque is crucial to inhibit the vulnerable atherosclerotic plaque progress. Recently, Ma et al. prepared supramolelcular theranostic micelles for combined two-photon AIE imaging of atherosclerosis diagnosis and two-pronged therapy [75]. These micelles are prepared by following the three steps. First, a two-photon fluorophore (TP) was covalently linked to β-CD via a ROS-responsive peroxalate ester to obtain a conjugate TPCD. Second, prednisolone (PRDL), a poorly soluble anti-inflammatory glucocorticoid, was encapsulated into β-CD via supramolecular interaction (Figure 9). Finally, the PRDL-loaded TPCD (TPCDP) was packed inside an amphiphilic poly(2-methyltho ethanol methacrylate)-*b*-PMPC (PMM) to form the theranostic micelles (TPCDP@PMM). The outer polymeric PMM shell can improve in vivo blood circulation and provide a chance for the accumulation of these micelles at the atherosclerotic lesion through the damaged vascular endothelium. The in vivo two-photon bioimaging of TPCD@PMMA on atherosclerotic ApoE-/- mice exhibited stronger fluorescence and more plaques were recognized. This confirmed the continuous accumulation of micelles. After accumulation, the overexpressed ROS level at the inflammatory tissue triggers a hydrophobic-to-hydrophilic switch of the PMEMA segment of the copolymer, leading to the disassembly of the structure. This process exposes the TPCDP to a high level of ROS and disrupts the peroxalate ester bond and facilitates the release of PRDL as well as the β-CD, resulting in two-pronged therapy via anti-inflammatory activity of PRDL and lipid removal feature of β-CD for atherosclerosis inhibition. These results suggest that these micelles can be promising nanoplatform for atherosclerosis theranostics. By employing a polymeric β-CD conjugate and a nitric oxide (NO) photodonor, a multifunctional biocompatible nanoconstruct has been synthesized for photo-controlled NO release with two-photon fluorescence imaging [76]. After synthesizing the polymeric β-CD conjugate using β-CD with epichlorohydrin under alkaline conditions, the NO photodonor with anthracene and nitroaniline structural unit was loaded into the polymeric conjugate to obtain the nanoconstruct. The NO release from the nanoconstruct was triggered by two-photon excitation using NIR laser (700 nm). The uncaged fluorescent co-product acted as a two-photon emission fluorescence reporter for the concomitant NO release. The release of NO in human squamous carcinoma cells was monitored by two-photon NIR fluorescence microscopy, which indicated the localization of fluorescent reporter in the cytoplasm. This nanoconstruct has potential for combined photo-activated therapies and bioimaging.

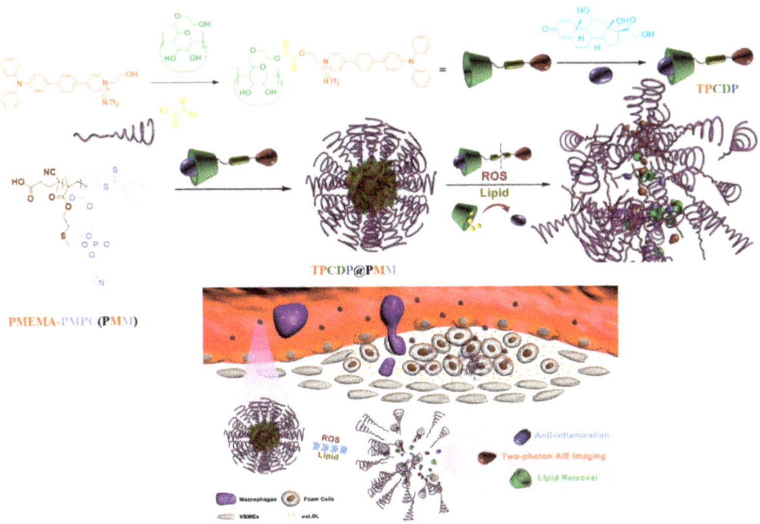

Figure 9. Schematic illustration of TPCDP@PMM theranostic micelles preparation and its responsive behaviors under overexpressed ROS and rich lipid. The TPCDP@PMM is designed to realize ROS-triggered micellar dissociation, rich lipid mediated drug Pred release and two-photon AIE imaging, which result in anti-tumor activity, lipid removal and clear dimensional two-photon imaging of atherosclerosis plaques. Adapted with permission from ref. [75]. Copyright 2020, Wiley-VCH GmbH.

8. Conclusions and Outlook

The capability of excitation light utilized in 2PA on penetrating deep tissue without damaging normal tissue is providing impetus for the development of two-photon-based systems for biomedical applications. By integrating diverse modularly designed copolymers and two-photon AIE probe or conjugated polymers with large 2PA cross section, several polymeric nanoprobes have been developed for imaging various diseases. In most of the cases the well-resolved images were obtained with a tissue penetration depth of 150 μm. With appropriately designed two-photon probe with a large two-photon absorption cross section, under NIR-II excitation, small blood vessels as deep as 1064 μm in mouse brain are visualized. In recent years, 2PA-based bioimaging has been combined with nanoplatforms that perform smart drug delivery and ROS-mediated PDT. In most cases, NIR radiation of 800 nm was used as the 2PA excitation light. The combination of 2PA-based PDT and stimuli-responsive chemotherapeutic drugs have rendered effective cancer treatment. The photo-responsive functional moieties such as DNQ, coumarin and *o*-nitrobenzyl were carefully placed within the polymeric nanoparticles to trigger on-demand photo-regulated drug release for enhanced therapy. Most of these systems are limited to cancer therapy. Thus, more two-photon-based systems should be designed and evaluated for other diseases. Despite the potential benefits and numerous advantages of these two-photon-activated polymeric systems, a number of studies conducted were mostly limited to in vivo animal models. The findings from animal studies are often difficult to transfer to humans. Therefore, studies related to preclinical efficacy and safety in humans are needed. Nonetheless, it was evident from the literature that 2PA responsive bioimaging can be effectively combined with other stimuli-responsive drug delivery and therapeutic applications. Additionally, it is certain that, in the future, the prime focus of 2PA-based research will move towards the development of a multifunctional theranostic system capable of performing imaging, drug delivery and ROS-induced therapy.

Author Contributions: Writing—original draft preparation, S.R. and S.S.; conceptualization and writing—revision, G.S.; supervision and review, T.H.O. All authors have read and agreed to the published version of the manuscript.

Funding: This research was supported by the Korean Ministry of Trade, Industry, and Energy (Project number: 20008490).

Institutional Review Board Statement: Not applicable.

Data Availability Statement: Upon reasonable request, the data supporting this investigation are available from the corresponding authors.

Acknowledgments: The authors thank the Core Research Support Center for Natural Products and Medical Materials (CRCNM) in Yeungnam University.

Conflicts of Interest: The authors declare no conflict of interest.

References

1. Mitchell, M.J.; Billingsley, M.M.; Haley, R.M.; Wechsler, M.E.; Peppas, N.A.; Langer, R. Engineering precision nanoparticles for drug delivery. *Nat. Rev. Drug Discov.* **2021**, *20*, 101–124. [CrossRef]
2. Li, J.; Yu, F.; Chen, Y.; Oupický, D. Polymeric drugs: Advances in the development of pharmacologically active polymers. *J. Control. Release* **2015**, *219*, 369–382. [CrossRef] [PubMed]
3. Das, S.S.; Bharadwaj, P.; Bilal, M.; Barani, M.; Rahdar, A.; Taboada, P.; Bungau, S.; Kyzas, G.Z. Stimuli-Responsive Polymeric Nanocarriers for Drug Delivery, Imaging, and Theragnosis. *Polymers* **2020**, *12*, 1397. [CrossRef]
4. Sung, Y.K.; Kim, S.W. Recent advances in the development of gene delivery systems. *Biomater. Res.* **2019**, *23*, 8. [CrossRef]
5. Ali, I.; Alsehli, M.; Scotti, L.; Tullius Scotti, M.; Tsai, S.-T.; Yu, R.-S.; Hsieh, M.F.; Chen, J.-C. Progress in Polymeric Nano-Medicines for Theranostic Cancer Treatment. *Polymers* **2020**, *12*, 598. [CrossRef] [PubMed]
6. Khan, M.I.; Hossain, M.I.; Hossain, M.K.; Rubel, M.H.K.; Hossain, K.M.; Mahfuz, A.M.U.B.; Anik, M.I. Recent Progress in Nanostructured Smart Drug Delivery Systems for Cancer Therapy: A Review. *ACS Appl. Bio Mater.* **2022**, *5*, 971–1012. [CrossRef]
7. Vu, M.N.; Kelly, H.G.; Kent, S.J.; Wheatley, A.K. Current and future nanoparticle vaccines for COVID-19. *eBioMedicine* **2021**, *74*, 103699. [CrossRef] [PubMed]

8. Park, H.; Otte, A.; Park, K. Evolution of drug delivery systems: From 1950 to 2020 and beyond. *J. Control. Release* **2022**, *342*, 53–65. [CrossRef]
9. Zhu, M.; Whittaker, A.K.; Han, F.Y.; Smith, M.T. Journey to the Market: The Evolution of Biodegradable Drug Delivery Systems. *Appl. Sci.* **2022**, *12*, 935. [CrossRef]
10. Zhou, Y.; Ye, H.; Chen, Y.; Zhu, R.; Yin, L. Photoresponsive Drug/Gene Delivery Systems. *Biomacromolecules* **2018**, *19*, 1840–1857. [CrossRef]
11. Azagarsamy, M.A.; Anseth, K.S. Wavelength-Controlled Photocleavage for the Orthogonal and Sequential Release of Multiple Proteins. *Angew. Chem. Int. Ed.* **2013**, *52*, 13803–13807. [CrossRef] [PubMed]
12. Sanchis, A.; Salvador, J.P.; Marco, M.P. Light-induced mechanisms for nanocarrier's cargo release. *Colloids Surf. B. Biointerfaces* **2019**, *173*, 825–832. [CrossRef]
13. Linsley, C.S.; Wu, B.M. Recent advances in light-responsive on-demand drug-delivery systems. *Ther. Deliv.* **2017**, *8*, 89–107. [CrossRef]
14. Hudson, D.E.; Hudson, D.O.; Wininger, J.M.; Richardson, B.D. Penetration of Laser Light at 808 and 980 nm in Bovine Tissue Samples. *Photomed. Laser Surg.* **2013**, *31*, 163–168. [CrossRef] [PubMed]
15. Chen, G.; Shen, J.; Ohulchanskyy, T.Y.; Patel, N.J.; Kutikov, A.; Li, Z.; Song, J.; Pandey, R.K.; Ågren, H.; Prasad, P.N.; et al. (α-NaYbF4:Tm3+)/CaF2 Core/Shell Nanoparticles with Efficient Near-Infrared to Near-Infrared Upconversion for High-Contrast Deep Tissue Bioimaging. *ACS Nano* **2012**, *6*, 8280–8287. [CrossRef] [PubMed]
16. Padalkar, M.V.; Pleshko, N. Wavelength-dependent penetration depth of near infrared radiation into cartilage. *Analyst* **2015**, *140*, 2093–2100. [CrossRef] [PubMed]
17. Henderson, T.A.; Morries, L.D. Near-infrared photonic energy penetration: Can infrared phototherapy effectively reach the human brain? *Neuropsychiatr. Dis. Treat.* **2015**, *11*, 2191–2208. [CrossRef]
18. Liu, G.; Liu, W.; Dong, C.-M. UV- and NIR-responsive polymeric nanomedicines for on-demand drug delivery. *Polym. Chem.* **2013**, *4*, 3431–3443. [CrossRef]
19. Zhu, X.; Su, Q.; Feng, W.; Li, F. Anti-Stokes shift luminescent materials for bio-applications. *Chem. Soc. Rev.* **2017**, *46*, 1025–1039. [CrossRef] [PubMed]
20. Yang, G.; Liu, J.; Wu, Y.; Feng, L.; Liu, Z. Near-infrared-light responsive nanoscale drug delivery systems for cancer treatment. *Coord. Chem. Rev.* **2016**, *320–321*, 100–117. [CrossRef]
21. Pawlicki, M.; Collins, H.A.; Denning, R.G.; Anderson, H.L. Two-Photon Absorption and the Design of Two-Photon Dyes. *Angew. Chem. Int. Ed.* **2009**, *48*, 3244–3266. [CrossRef] [PubMed]
22. Bort, G.; Gallavardin, T.; Ogden, D.; Dalko, P.I. From One-Photon to Two-Photon Probes: "Caged" Compounds, Actuators, and Photoswitches. *Angew. Chem. Int. Ed.* **2013**, *52*, 4526–4537. [CrossRef] [PubMed]
23. Alam, M.M.; Chattopadhyaya, M.; Chakrabarti, S.; Ruud, K. Chemical Control of Channel Interference in Two-Photon Absorption Processes. *Acc. Chem. Res.* **2014**, *47*, 1604–1612. [CrossRef] [PubMed]
24. Potter, S.M. Vital imaging: Two photons are better than one. *Curr. Biol.* **1996**, *6*, 1595–1598. [CrossRef] [PubMed]
25. Wu, L.; Liu, J.; Li, P.; Tang, B.; James, T.D. Two-photon small-molecule fluorescence-based agents for sensing, imaging, and therapy within biological systems. *Chem. Soc. Rev.* **2021**, *50*, 702–734. [CrossRef] [PubMed]
26. Xu, L.; Zhang, J.; Yin, L.; Long, X.; Zhang, W.; Zhang, Q. Recent progress in efficient organic two-photon dyes for fluorescence imaging and photodynamic therapy. *J. Mater. Chem. C* **2020**, *8*, 6342–6349. [CrossRef]
27. Yang, P.-P.; Yang, Y.; Gao, Y.-J.; Wang, Y.; Zhang, J.-C.; Lin, Y.-X.; Dai, L.; Li, J.; Wang, L.; Wang, H. Unprecedentedly High Tissue Penetration Capability of Co-Assembled Nanosystems for Two-Photon Fluorescence Imaging In Vivo. *Adv. Opt. Mater.* **2015**, *3*, 646–651. [CrossRef]
28. Hong, Y.; Lam, J.W.Y.; Tang, B.Z. Aggregation-induced emission: Phenomenon, mechanism and applications. *Chem. Commun.* **2009**, 4332–4353. [CrossRef] [PubMed]
29. Wang, Y.; Han, X.; Xi, W.; Li, J.; Roe, A.W.; Lu, P.; Qian, J. Bright AIE Nanoparticles with F127 Encapsulation for Deep-Tissue Three-Photon Intravital Brain Angiography. *Adv. Healthc. Mater.* **2017**, *6*, 1700685. [CrossRef] [PubMed]
30. Samanta, S.; Huang, M.; Li, S.; Yang, Z.; He, Y.; Gu, Z.; Zhang, J.; Zhang, D.; Liu, L.; Qu, J. AIE-active two-photon fluorescent nanoprobe with NIR-II light excitability for highly efficient deep brain vasculature imaging. *Theranostics* **2021**, *11*, 2137–2148. [CrossRef] [PubMed]
31. Shaw, P.A.; Forsyth, E.; Haseeb, F.; Yang, S.; Bradley, M.; Klausen, M. Two-Photon Absorption: An Open Door to the NIR-II Biological Window? *Front. Chem.* **2022**, *10*, 921354. [CrossRef] [PubMed]
32. Chowdhury, P.; Chan, Y.-H. Recent advances in D–A–D based Pdots with NIR-II fluorescence for deep-tissue imaging. *Mol. Syst. Des. Eng.* **2022**, *7*, 702–719. [CrossRef]
33. Qi, J.; Sun, C.; Li, D.; Zhang, H.; Yu, W.; Zebibula, A.; Lam, J.W.Y.; Xi, W.; Zhu, L.; Cai, F.; et al. Aggregation-Induced Emission Luminogen with Near-Infrared-II Excitation and Near-Infrared-I Emission for Ultradeep Intravital Two-Photon Microscopy. *ACS Nano* **2018**, *12*, 7936–7945. [CrossRef]
34. Hu, W.; Guo, L.; Bai, L.; Miao, X.; Ni, Y.; Wang, Q.; Zhao, H.; Xie, M.; Li, L.; Lu, X.; et al. Maximizing Aggregation of Organic Fluorophores to Prolong Fluorescence Lifetime for Two-Photon Fluorescence Lifetime Imaging. *Adv. Healthc. Mater.* **2018**, *7*, 1800299. [CrossRef] [PubMed]

35. Wu, W.-C.; Chen, C.-Y.; Tian, Y.; Jang, S.-H.; Hong, Y.; Liu, Y.; Hu, R.; Tang, B.Z.; Lee, Y.-T.; Chen, C.-T.; et al. Enhancement of Aggregation-Induced Emission in Dye-Encapsulating Polymeric Micelles for Bioimaging. *Adv. Funct. Mater.* **2010**, *20*, 1413–1423. [CrossRef]
36. Geng, J.; Goh, C.C.; Qin, W.; Liu, R.; Tomczak, N.; Ng, L.G.; Tang, B.Z.; Liu, B. Silica shelled and block copolymer encapsulated red-emissive AIE nanoparticles with 50% quantum yield for two-photon excited vascular imaging. *Chem. Commun.* **2015**, *51*, 13416–13419. [CrossRef]
37. Geng, J.; Li, K.; Qin, W.; Ma, L.; Gurzadyan, G.G.; Tang, B.Z.; Liu, B. Eccentric Loading of Fluorogen with Aggregation-Induced Emission in PLGA Matrix Increases Nanoparticle Fluorescence Quantum Yield for Targeted Cellular Imaging. *Small* **2013**, *9*, 2012–2019. [CrossRef]
38. Alifu, N.; Sun, Z.; Zebibula, A.; Zhu, Z.; Zhao, X.; Wu, C.; Wang, Y.; Qian, J. Deep-red polymer dots with bright two-photon fluorescence and high biocompatibility for in vivo mouse brain imaging. *Opt. Commun.* **2017**, *399*, 120–126. [CrossRef]
39. Khalin, I.; Heimburger, D.; Melnychuk, N.; Collot, M.; Groschup, B.; Hellal, F.; Reisch, A.; Plesnila, N.; Klymchenko, A.S. Ultrabright Fluorescent Polymeric Nanoparticles with a Stealth Pluronic Shell for Live Tracking in the Mouse Brain. *ACS Nano* **2020**, *14*, 9755–9770. [CrossRef]
40. Chuan-Xi, W.; Zhi-Yue, G.; Xin, W.; Can, K.; Zhuo, Z.; Chao-Jie, Z.; Li-Min, F.; Yuan, W.; Jian-Ping, Z. Noninvasive and real-time pharmacokinetics imaging of polymeric nanoagents in the thoracoepigastric vein networks of living mice. *J. Biomed. Opt.* **2019**, *24*, 066009. [CrossRef]
41. Liu, J.; Evrard, M.; Cai, X.; Feng, G.; Tomczak, N.; Ng, L.G.; Liu, B. Organic nanoparticles with ultrahigh quantum yield and aggregation-induced emission characteristics for cellular imaging and real-time two-photon lung vasculature imaging. *J. Mater. Chem. B* **2018**, *6*, 2630–2636. [CrossRef] [PubMed]
42. Shen, F.-F.; Chen, Y.; Xu, X.; Yu, H.-J.; Wang, H.; Liu, Y. Supramolecular Assembly with Near-Infrared Emission for Two-Photon Mitochondrial Targeted Imaging. *Small* **2021**, *17*, 2101185. [CrossRef] [PubMed]
43. Geng, J.; Li, K.; Ding, D.; Zhang, X.; Qin, W.; Liu, J.; Tang, B.Z.; Liu, B. Lipid-PEG-Folate Encapsulated Nanoparticles with Aggregation Induced Emission Characteristics: Cellular Uptake Mechanism and Two-Photon Fluorescence Imaging. *Small* **2012**, *8*, 3655–3663. [CrossRef] [PubMed]
44. Xiao, X.; Cai, H.; Huang, Q.; Wang, B.; Wang, X.; Luo, Q.; Li, Y.; Zhang, H.; Gong, Q.; Ma, X.; et al. Polymeric dual-modal imaging nanoprobe with two-photon aggregation-induced emission for fluorescence imaging and gadolinium-chelation for magnetic resonance imaging. *Bioact. Mater.* **2023**, *19*, 538–549. [CrossRef]
45. Li, S.; Jiang, X.-F.; Xu, Q.-H. Conjugated Polymers for Two-Photon Live Cell Imaging. In *Conjugated Polymers for Biological and Biomedical Applications*; Wiley-VCH: Weinheim, Germany, 2018; pp. 135–170.
46. Wu, C.; Szymanski, C.; Cain, Z.; McNeill, J. Conjugated Polymer Dots for Multiphoton Fluorescence Imaging. *J. Am. Chem. Soc.* **2007**, *129*, 12904–12905. [CrossRef]
47. Liu, P.; Li, S.; Jin, Y.; Qian, L.; Gao, N.; Yao, S.Q.; Huang, F.; Xu, Q.-H.; Cao, Y. Red-Emitting DPSB-Based Conjugated Polymer Nanoparticles with High Two-Photon Brightness for Cell Membrane Imaging. *ACS Appl. Mater. Interfaces* **2015**, *7*, 6754–6763. [CrossRef]
48. Goodwin, A.P.; Mynar, J.L.; Ma, Y.; Fleming, G.R.; Fréchet, J.M.J. Synthetic Micelle Sensitive to IR Light via a Two-Photon Process. *J. Am. Chem. Soc.* **2005**, *127*, 9952–9953. [CrossRef]
49. Babin, J.; Pelletier, M.; Lepage, M.; Allard, J.-F.; Morris, D.; Zhao, Y. A New Two-Photon-Sensitive Block Copolymer Nanocarrier. *Angew. Chem. Int. Ed.* **2009**, *48*, 3329–3332. [CrossRef]
50. Yuan, Y.; Wang, Z.; Cai, P.; Liu, J.; Liao, L.-D.; Hong, M.; Chen, X.; Thakor, N.; Liu, B. Conjugated polymer and drug co-encapsulated nanoparticles for Chemo- and Photo-thermal Combination Therapy with two-photon regulated fast drug release. *Nanoscale* **2015**, *7*, 3067–3076. [CrossRef]
51. Sun, L.; Yang, Y.; Dong, C.-M.; Wei, Y. Two-Photon-Sensitive and Sugar-Targeted Nanocarriers from Degradable and Dendritic Amphiphiles. *Small* **2011**, *7*, 401–406. [CrossRef]
52. Härtner, S.; Kim, H.-C.; Hampp, N. Photodimerized 7-hydroxycoumarin with improved solubility in PMMA: Single-photon and two-photon-induced photocleavage in solution and PMMA films. *J. Photochem. Photobiol. A Chem.* **2007**, *187*, 242–246. [CrossRef]
53. Härtner, S.; Kim, H.-C.; Hampp, N. Phototriggered release of photolabile drugs via two-photon absorption-induced cleavage of polymer-bound dicoumarin. *J. Polym. Sci. Part A Polym. Chem.* **2007**, *45*, 2443–2452. [CrossRef]
54. Kim, H.-C.; Härtner, S.; Behe, M.; Behr, T.; Hampp, N. Two-photon absorption-controlled multidose drug release: A novel approach for secondary cataract treatment. *J. Biomed. Opt.* **2006**, *11*, 034024. [CrossRef] [PubMed]
55. Ji, W.; Li, N.; Chen, D.; Qi, X.; Sha, W.; Jiao, Y.; Xu, Q.; Lu, J. Coumarin-containing photo-responsive nanocomposites for NIR light-triggered controlled drug release via a two-photon process. *J. Mater. Chem. B* **2013**, *1*, 5942–5949. [CrossRef] [PubMed]
56. Kehrloesser, D.; Behrendt, P.J.; Hampp, N. Two-photon absorption triggered drug delivery from a polymer for intraocular lenses in presence of an UV-absorber. *J. Photochem. Photobiol. A Chem.* **2012**, *248*, 8–14. [CrossRef]
57. Huang, Y.; Shen, L.; Guo, D.; Yasen, W.; Wu, Y.; Su, Y.; Chen, D.; Qiu, F.; Yan, D.; Zhu, X. A NIR-triggered gatekeeper of supramolecular conjugated unimicelles with two-photon absorption for controlled drug release. *Chem. Commun.* **2019**, *55*, 6735–6738. [CrossRef] [PubMed]
58. Olejniczak, J.; Sankaranarayanan, J.; Viger, M.L.; Almutairi, A. Highest Efficiency Two-Photon Degradable Copolymer for Remote Controlled Release. *ACS Macro Lett.* **2013**, *2*, 683–687. [CrossRef] [PubMed]

59. Ma, L.-L.; Liu, M.-X.; Liu, X.-Y.; Sun, W.; Lu, Z.-L.; Gao, Y.-G.; He, L. Macrocyclic polyamine [12]aneN3 modified triphenylamine-pyrazine derivatives as efficient non-viral gene vectors with AIE and two-photon imaging properties. *J. Mater. Chem. B* **2020**, *8*, 3869–3879. [CrossRef] [PubMed]
60. Pack, D.W.; Hoffman, A.S.; Pun, S.; Stayton, P.S. Design and development of polymers for gene delivery. *Nat. Rev. Drug Discov.* **2005**, *4*, 581–593. [CrossRef] [PubMed]
61. Wei, L.; Zhang, D.; Zheng, X.; Zeng, X.; Zeng, Y.; Shi, X.; Su, X.; Xiao, L. Fabrication of Positively Charged Fluorescent Polymer Nanoparticles for Cell Imaging and Gene Delivery. *Nanotheranostics* **2018**, *2*, 157–167. [CrossRef] [PubMed]
62. Zhao, H.; Tao, H.; Hu, W.; Miao, X.; Tang, Y.; He, T.; Li, J.; Wang, Q.; Guo, L.; Lu, X.; et al. Two-Photon-Induced Charge-Variable Conjugated Polyelectrolyte Brushes for Effective Gene Silencing. *ACS Appl. Bio Mater.* **2019**, *2*, 1676–1685. [CrossRef] [PubMed]
63. Hayek, A.; Ercelen, S.; Zhang, X.; Bolze, F.; Nicoud, J.-F.; Schaub, E.; Baldeck, P.L.; Mély, Y. Conjugation of a New Two-Photon Fluorophore to Poly(ethylenimine) for Gene Delivery Imaging. *Bioconjugate Chem.* **2007**, *18*, 844–851. [CrossRef] [PubMed]
64. Liu, G.; Xie, J.; Zhang, F.; Wang, Z.; Luo, K.; Zhu, L.; Quan, Q.; Niu, G.; Lee, S.; Ai, H.; et al. N-Alkyl-PEI-Functionalized Iron Oxide Nanoclusters for Efficient siRNA Delivery. *Small* **2011**, *7*, 2742–2749. [CrossRef] [PubMed]
65. Luo, L.; Zhang, Q.; Luo, Y.; He, Z.; Tian, X.; Battaglia, G. Thermosensitive nanocomposite gel for intra-tumoral two-photon photodynamic therapy. *J. Control. Release* **2019**, *298*, 99–109. [CrossRef] [PubMed]
66. Luo, L.; Yin, Z.; Qi, Y.; Liu, S.; Yi, Y.; Tian, X.; Wu, Y.; Zhong, D.; Gu, Z.; Zhang, H.; et al. An intracellular enzyme-responsive polymeric prodrug with synergistic effect of chemotherapy and two-photon photodynamic therapy. *Appl. Mater. Today* **2021**, *23*, 100996. [CrossRef]
67. Kandoth, N.; Kirejev, V.; Monti, S.; Gref, R.; Ericson, M.B.; Sortino, S. Two-Photon Fluorescence Imaging and Bimodal Phototherapy of Epidermal Cancer Cells with Biocompatible Self-Assembled Polymer Nanoparticles. *Biomacromolecules* **2014**, *15*, 1768–1776. [CrossRef] [PubMed]
68. Wang, H.; Di, J.; Sun, Y.; Fu, J.; Wei, Z.; Matsui, H.; del, C. Alonso, A.; Zhou, S. Biocompatible PEG-Chitosan@Carbon Dots Hybrid Nanogels for Two-Photon Fluorescence Imaging, Near-Infrared Light/pH Dual-Responsive Drug Carrier, and Synergistic Therapy. *Adv. Funct. Mater.* **2015**, *25*, 5537–5547. [CrossRef]
69. Ardekani, S.M.; Dehghani, A.; Hassan, M.; Kianinia, M.; Aharonovich, I.; Gomes, V.G. Two-photon excitation triggers combined chemo-photothermal therapy via doped carbon nanohybrid dots for effective breast cancer treatment. *Chem. Eng. J.* **2017**, *330*, 651–662. [CrossRef]
70. Ma, B.; Zhuang, W.; Xu, H.; Li, G.; Wang, Y. Hierarchical Responsive Nanoplatform with Two-Photon Aggregation-Induced Emission Imaging for Efficient Cancer Theranostics. *ACS Appl. Mater. Interfaces* **2019**, *11*, 47259–47269. [CrossRef] [PubMed]
71. Xu, H.; Ma, B.; Jiang, J.; Xiao, S.; Peng, R.; Zhuang, W.; Li, G.; Wang, Y. Integrated prodrug micelles with two-photon bioimaging and pH-triggered drug delivery for cancer theranostics. *Regen. Biomater.* **2019**, *7*, 171–180. [CrossRef] [PubMed]
72. Yu, T.; Zhuang, W.; Su, X.; Ma, B.; Hu, J.; He, H.; Li, G.; Wang, Y. Dual-Responsive Micelles with Aggregation-Induced Emission Feature and Two-Photon Aborsption for Accurate Drug Delivery and Bioimaging. *Bioconjugate Chem.* **2019**, *30*, 2075–2087. [CrossRef] [PubMed]
73. He, H.; Zhuang, W.; Ma, B.; Su, X.; Yu, T.; Hu, J.; Chen, L.; Peng, R.; Li, G.; Wang, Y. Oxidation-Responsive and Aggregation-Induced Emission Polymeric Micelles with Two-Photon Excitation for Cancer Therapy and Bioimaging. *ACS Biomater. Sci. Eng.* **2019**, *5*, 2577–2586. [CrossRef] [PubMed]
74. Zhuang, W.; Ma, B.; Hu, J.; Jiang, J.; Li, G.; Yang, L.; Wang, Y. Two-photon AIE luminogen labeled multifunctional polymeric micelles for theranostics. *Theranostics* **2019**, *9*, 6618–6630. [CrossRef] [PubMed]
75. Ma, B.; Xu, H.; Zhuang, W.; Wang, Y.; Li, G.; Wang, Y. ROS Responsive Nanoplatform with Two-Photon AIE Imaging for Atherosclerosis Diagnosis and "Two-Pronged" Therapy. *Small* **2020**, *16*, 2003253. [CrossRef] [PubMed]
76. Kirejev, V.; Kandoth, N.; Gref, R.; Ericson, M.B.; Sortino, S. A polymer-based nanodevice for the photoregulated release of NO with two-photon fluorescence reporting in skin carcinoma cells. *J. Mater. Chem. B* **2014**, *2*, 1190–1195. [CrossRef] [PubMed]

polymers

Article

Amphiphilic Copolymer-Lipid Chimeric Nanosystems as DNA Vectors

Varvara Chrysostomou [1,2], Aleksander Foryś [3], Barbara Trzebicka [3], Costas Demetzos [1] and Stergios Pispas [2,*]

1 Section of Pharmaceutical Technology, Department of Pharmacy, School of Health Sciences, National and Kapodistrian University of Athens, Panepistimioupolis Zografou, 15771 Athens, Greece
2 Theoretical and Physical Chemistry Institute, National Hellenic Research Foundation, 48 Vassileos Constantinou Avenue, 11635 Athens, Greece
3 Centre of Polymer and Carbon Materials, Polish Academy of Sciences, 34 ul. M. Curie-Skłodowskiej, 41-819 Zabrze, Poland
* Correspondence: pispas@eie.gr; Tel.: +30-2107273824

Abstract: Lipid-polymer chimeric (hybrid) nanosystems are promising platforms for the design of effective gene delivery vectors. In this regard, we developed DNA nanocarriers comprised of a novel poly[(stearyl methacrylate-co-oligo(ethylene glycol) methyl ether methacrylate] [P(SMA-co-OEGMA)] amphiphilic random copolymer, the cationic 1,2-dioleoyl-3-(trimethylammonium) propane (DOTAP), and the zwitterionic L-α-phosphatidylcholine, hydrogenated soybean (soy) (HSPC) lipids. Chimeric HSPC:DOTAP:P[(SMA-co-OEGMA)] nanosystems, and pure lipid nanosystems as reference, were prepared in several molar ratios of the components. The colloidal dispersions obtained presented well-defined physicochemical characteristics and were further utilized for the formation of lipoplexes with a model DNA of linear topology containing 113 base pairs. Nanosized complexes were formed through the electrostatic interaction of the cationic lipid and phosphate groups of DNA, as observed by dynamic, static, and electrophoretic light scattering techniques. Ultraviolet–visible (UV–Vis) and fluorescence spectroscopy disclosed the strong binding affinity of the chimeric and also the pure lipid nanosystems to DNA. Colloidally stable chimeric/lipid complexes were formed, whose physicochemical characteristics depend on the N/P ratio and on the molar ratio of the building components. Cryogenic transmission electron microscopy (Cryo-TEM) revealed the formation of nanosystems with vesicular morphology. The results suggest the successful fabrication of these novel chimeric nanosystems with well-defined physicochemical characteristics, which can form stable lipoplexes.

Keywords: non-viral vectors; gene delivery; cationic lipids; amphiphilic random copolymers; chimeric/hybrid lipoplexes

Citation: Chrysostomou, V.; Foryś, A.; Trzebicka, B.; Demetzos, C.; Pispas, S. Amphiphilic Copolymer-Lipid Chimeric Nanosystems as DNA Vectors. *Polymers* **2022**, *14*, 4901. https://doi.org/10.3390/polym14224901

Academic Editor: Marek Kowalczuk

Received: 14 October 2022
Accepted: 11 November 2022
Published: 13 November 2022

1. Introduction

Recently, gene therapy has gained prominence as one of the most promising therapeutic approaches for the treatment of genetic-based diseases [1]. The implementation of gene therapy involves the delivery of therapeutic nucleic acid-based medicines into cells to correct a cellular dysfunction or to provide a new cellular function in order to treat or prevent disorders [2,3]. A determinant factor for the successful application of this promising therapeutic strategy is the use of efficient carriers for the effective nucleic acid transfer and cellular uptake [4]. These gene delivery carriers, also known as vectors, are broadly classified into viral and non-viral vectors [5]. Viral vectors including viruses such as adenovirus, adeno-associated virus, lentivirus, retrovirus, herpes simplex virus, poxvirus, etc. have been employed for the delivery of therapeutic agents. Moreover, viral vectors are considered an effective means for the delivery of nucleic acids and have shown success in vivo and in clinical trials [6,7]. However, viral vectors are associated with serious safety issues, including acute immune response and insertional mutagenesis [4–6].

Although non-viral vectors are less efficient, their advantages regarding low biosafety risk, potential for large-scale production, controllable chemical structure, wide material source, multifunctionality, and no capacity limits in gene encapsulation have brought them at the forefront of gene delivery research [4–8].

Non-viral gene carriers based on cationic lipids or polymers or even more interesting a combination of these are now among the most promising technologies for gene therapy and therapeutics [9]. In particular, the cationic lipid-based liposomes represent one of the most intensively studied and clinically advanced platforms in the field of the non-viral vectors [10–12]. Cationic liposomal formulations utilized for gene delivery are frequently composed of a neutrally charged lipid and a cationic lipid. Neutral lipids, often called helper lipids, are components for a cationic liposome formulation in which they play an assistant role in stabilizing bilayer membranes [13,14]. Furthermore, the combination of a cationic lipid with a neutral lipid has conferred improved efficiency to the cationic lipids [14]. Cationic lipids are amphiphilic molecules consisting of a positive charged headgroup, covalently bound through a linker to a hydrophobic tail [15]. They can be easily synthesized and are considered as one of the most versatile tools for the delivery of nucleic acids and other therapeutic molecules [15]. The positively charged head groups are amines, quaternary ammoniums, guanidinium, or amino acids [10,16] which can electrostatically interact with the anionic phosphate groups of nucleic acids, leading to the formation of complexes containing condensed nucleic acids, namely lipoplexes [15–17]. The most frequent group is quaternary ammonium, due to its permanent positive charge that provides strong interaction with the nucleic acids and enhanced solubility in aqueous environments [10,16].

Lipoplexes protect the genetic material from enzymatic degradation, increase the stability of the vector, and interact with the cell membrane through electrostatic interactions [18]. Thus, the lipoplexes are typically formed with a slight excess of positive charge to permit their interaction with the negatively charged cell surface [19,20]. The transfection efficiency and stability of lipoplexes are strongly affected by a variety of formulation factors, including the lipid to nucleic acid charge ratio, the lipoplex size, the surface charge, and environmental conditions such as the ionic strength, temperature, and pH of the medium [16,21,22]. Compared to viral vectors, lipoplexes are formed spontaneously, their preparation is simple and cost-effective, and they do not present the risk of the insertion of genetic material into the host's genome [23]. However, lipoplexes are associated with cytotoxicity effects due to the permanent positive charge of the cationic lipids, which has become one of the main bottlenecks for their application, with their clinical utilization remaining a challenge [23–25]. Therefore, scientific research has been focused on developing delivery systems based on innovative nanomaterials that are able to overcome these hurdles.

The advances on the science of nanotechnology have greatly benefited progress in the exploration of new non-viral nanocarriers. The combination of different in nature materials can change their individual properties and generate hybrid nanostructures with new features. Particularly, the combination of lipid-based and polymer-based nanosystems is considered an innovative approach for biomedical applications [26–28]. A new generation of nanosystems has been created by harnessing the advantages of lipids, such as biomimetic nature and biocompatibility, with the advantages of polymers, such as versatility in chemical structure, chemical functionalities, and response to external stimuli [28,29]. The design of lipid and polymer-based nanoparticles as non-viral vectors, which can protect nucleic acids and ensure their targeted delivery, and also to promote controlled release and enhanced cellular uptake, has gained significant attention [30]. Regarding the lipoplexes, the combination of cationic lipids with polymers is more favorable, thus it can prevent the binding affinity of non-specific proteins to the lipoplexes and can improve circulatory half-lives [31–33]. In general, the synergistic effect of these two pillar classes of materials can lead to the generation of multifunctional nanocarriers capable of simultaneously delivering, in a single platform, different therapeutic compounds including hydrophobic drugs, nucleic acids, proteins/peptides, and diagnostic agents, e.g., magnetic nanoparticles and dyes for bioimaging, as a combinational therapeutic approach [7,29,33,34].

In this study, novel hybrid/chimeric (i.e., nanosystems composed of lipids and copolymers) [35] nanocarriers are developed for nucleic acid delivery utilizing the cationic 1,2-dioleoyl-3-(trimethylammonium) propane (DOTAP) lipid, which is one of the most widely used and efficient lipid transfection agents for gene delivery applications [36]. DOTAP consists of a quaternary ammonium salt as the cationic head group [16], which facilitates the spontaneous electrostatic interaction with nucleic acids forming lipoplexes, as well as the binding of the resulting lipoplexes to the negatively charged components of the cell membrane [37]. Different DOTAP-based lipoplexes have entered preclinical and clinical trials [16]. For instance, lipoplexes composed of DOTAP and cholesterol (DOTAP:chol) have been clinically investigated for the treatment of various diseases including several types of cancer [23]. As in most cationic lipid formulations, DOTAP, due to the increased density of positive charges on liposome surface, exhibits inefficient gene transfer and cannot be used alone [37,38]. Hence, to improve its gene transfection capabilities and to provide colloidal stability, DOTAP is utilized in combination with other neutral/helper lipids to self-assemble into cationic liposomes [15,16,39,40]. In this direction, we utilized the neutral lipid of L-α-phosphatidylcholine, hydrogenated soybean (soy) (HSPC), which has high melting temperature and can be used to construct highly stable liposomes and therefore to provide stability [40].

In the next step, employing the reversible addition fragmentation chain transfer (RAFT) polymerization, we synthesized the novel amphiphilic random copolymer of poly[(stearyl methacrylate-co-oligo(ethylene glycol) methyl ether methacrylate] [P(SMA-co-OEGMA)]. RAFT polymerization gives well-defined polymers with predetermined molecular characteristics for their implementation in gene delivery applications [41,42]. Moreover, the facile preparation of block and random copolymers through RAFT, permits scale-up production, which is important for the design of a gene delivery system [43]. The amphiphilic copolymer consists of two segments, the hydrophobic SMA and the hydrophilic OEGMA. The (P)SMA is considered a super hydrophobic polymer, possessing a long alkyl side chain of 18 CH_2 which can form crystalline domains [44–46]. The long hydrophobic chain in the SMA segment serves as an anchor for the incorporation of the copolymer inside the lipid bilayers, providing stability to the liposomal formulation [47,48]. On the other hand, (P)OEGMA is composed of a hydrophobic main chain and grafted hydrophilic side oligo(ethylene glycol) chains [43,49]. The toxicity of the cationic lipids remains an important issue, because of their high positive charge density. Accordingly, DOTAP also exhibits a cytotoxicity effect due to its quaternary amine headgroup [25,50]. The PEG-shielding strategy has been adopted in the cationic nanocarriers to mask the excessive positive charges [39]. Similar to a PEG-strategy, we utilized the non-ionic (P)OEGMA with average $M_n = 475\ g{\cdot}mol^{-1}$ and nine ethylene glycol repeated units as an alternative shielding agent, which displays similar properties to PEG [51,52], in an effort to mask the positive charges of DOTAP and therefore to reduce cytotoxicity effects and provide stealth properties, biocompatibility, and colloidal stability to the lipoplexes.

In this regard, we developed the novel HSPC:DOTAP:P[(SMA-co-OEGMA)] hybrid/chimeric liposomal nanosystem and we explored its potential to bind DNA and form lipoplexes. Chimeric nanosystems of HSPC:DOTAP:P[(SMA-co-OEGMA)] were prepared by the thin film hydration method, in several molar ratios of the copolymer and lipid components, including the ratios of 9:1:0.05, 9:1:0.1, 7:3:0.05, and 7:3:0.1. Furthermore, pure liposomal formulations of HSPC:DOTAP lipids were also prepared in molar ratios of 9:1 and 7:3 and were used as reference systems. The colloidal dispersions were examined by light scattering techniques for the determination of their physicochemical characteristics, such as their size and size distribution, morphology, and surface charge. Afterward, we investigated the ability of the chimeric nanosystems and also of the references pure liposomes, to electrostatically interact with nucleic acids, utilizing a double stranded model DNA from salmon testes of linear topology and length of 113 base pairs (bp). Nanosized complexes formed through the electrostatic interaction of the cationic quaternary amino group of the DOTAP component with the phosphate group of DNA, in a wide range of N/P ratios (nitrogen (N) of amine group of cationic lipid over phosphate (P) groups of DNA). The interaction process and the formed complexes were investigated using light scattering (dynamic, static and electrophoretic), fluorescence and UV–Vis spectroscopy and Cryo-TEM microscopy, in order to comprehensively study their physicochemical and morphological characteristics. In addition, the stability of the chimeric/lipid lipoplexes in increasing ionic strength, as a simulation of the physiological conditions of the biological fluids, was also evaluated by dynamic light scattering.

2. Materials and Methods

2.1. Materials

For the synthesis of the random copolymer, the monomer stearyl methacrylate (SMA, ≥89.5%) and the oligomer oligo(ethylene glycol) methyl ether methacrylate (OEGMA), with average M_n = 475 g/mol and 9 ethylene glycol units, were purchased from Sigma-Aldrich (Athens, Greece). The SMA and OEGMA monomers were purified by passing through columns packed with hydroquinone monomethyl ether (MEHQ) and butylated hydroxytoluene (BHT) inhibitor removers before the polymerization process. The MEHQ and BHT inhibitor removers, the chain transfer agent 4-cyano-4-(phenylcarbonothioylthio)pentanoic acid (CPAD), the radical initiator 2,2-azobis(isobutyronitrile) (AIBN), 1,4-dioxane (99.8%), tetrahydrofuran (THF, ≥99.9%), chloroform (99.9%), and deuterated chloroform ($CDCl_3$) (99.9%) were also obtained from Sigma-Aldrich (Athens, Greece) and used as received, except 1,4-dioxane, which was dried over molecular sieves before use. Moreover, AIBN was purified by recrystallization from methanol and subsequently used as a solution in 1,4-dioxane. Dialysis tubing membranes (MEMBRA-CEL®) from regenerated cellulose of MWCO 3500 and a diameter of 22 mm were obtained by SERVA (Heidelberg, Germany).

The lipids L-α-phosphatidylcholine, hydrogenated soybean (soy) (HSPC) (Scheme 1a) and 1,2-dioleoyl-3-(trimethylammonium) propane (chloride salt) (DOTAP) (Scheme 1b) were purchased from Avanti Polar Lipids Inc. (Alabaster, AL, USA) and used without further purification.

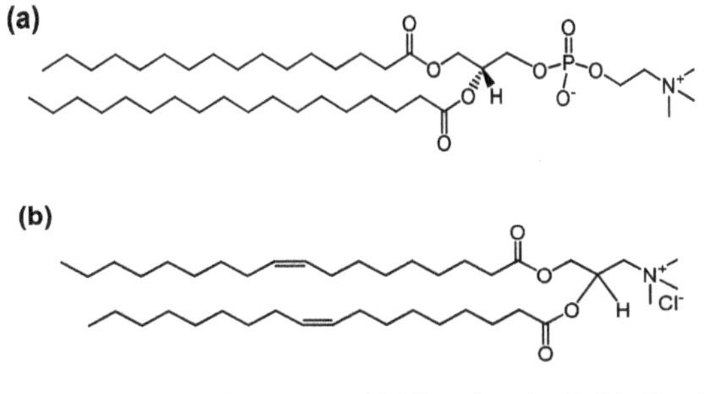

Scheme 1. Chemical structures of (**a**) L-α-phosphatidylcholine, hydrogenated soybean (soy) (HSPC) and (**b**) 1,2-dioleoyl-3-(trimethylammonium) propane (chloride salt) (DOTAP) lipids.

For the interaction of the lipid and chimeric nanosystems with nucleic acids, linear double stranded deoxyribonucleic acid sodium salt from salmon sperm of 113 base pairs length was purchased from Acros. Ethidium bromide (EtBr) dye for DNA quenching fluorescent assay and sodium chloride (≥99.0%), which was utilized for the preparation of NaCl solutions of 1 M, were received from Sigma-Aldrich (Athens, Greece). All the solutions were prepared using sterile water for injection (DEMO SA., Athens, Greece).

2.2. Synthesis of the Amphiphilic Random Copolymer

The synthesis of the novel amphiphilic random P(SMA-co-OEGMA) copolymer was achieved by employing reversible addition-fragmentation chain transfer (RAFT) polymerization. The copolymer was synthesized following a one-step synthetic procedure as described in detail below. The purified monomers SMA (0.6 g, 1.77 mmol) and OEGMA (1.4 g, 2.95 mmol), the CPAD (0.05 g, 0.20 mmol) chain transfer agent, and the AIBN (0.0065 g, 0.04 mmol) radical initiator were placed in a round bottom flask equipped with a magnetic stirrer and were dissolved in 10 mL of 1,4-dioxane (20 wt.% monomer solution) under stirring. The used CTA (CPAD) to initiator (AIBN) ratio ($[CTA]_0/[I]_0$) was 5:1. The flask was sealed with a rubber septum. The reaction solution was degassed by high purity nitrogen gas bubbling for 20 min and then immersed in a preheated oil bath at 70 °C for 24 h. The reaction was quenched by freezing the solution at −20 °C for 30 min and exposing it to air. Unreacted monomers or other impurities and 1,4-dioxane were removed by dialysis against deionized H_2O for 3 days with three exchanges per day. The purified copolymer was dried under vacuum oven for 48 h at room temperature and collected at >98% yield. The followed synthetic polymerization route and the chemical structure of the copolymer are depicted in Scheme 2.

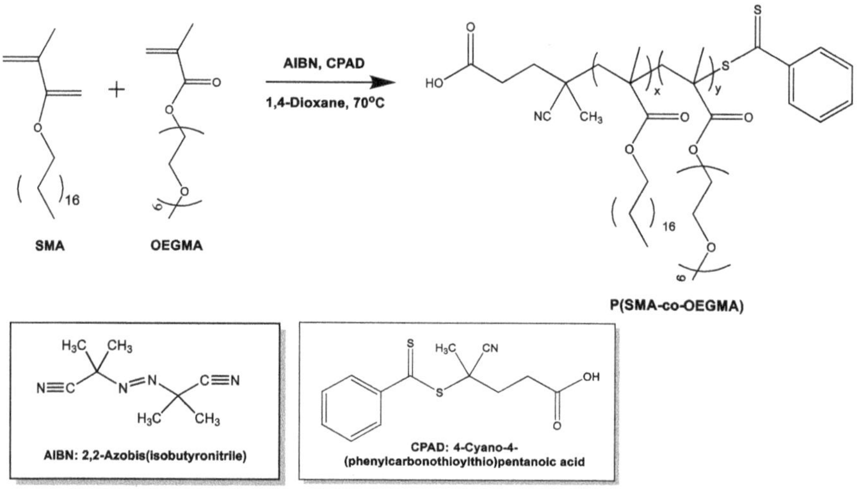

Scheme 2. Synthetic route for the synthesis of P(SMA-co-OEGMA) copolymer via RAFT polymerization.

2.3. *Size Exclusion Chromatography (SEC)*

A Waters size exclusion chromatography (SEC) instrument (Waters Corporation, Milford, MA, USA) was utilized for the determination of the molar mass (M_w), molar mass distributions, and dispersity index (M_w/M_n) of the synthesized P(SMA-co-OEGMA) copolymer. The chromatography system is equipped with a Waters 1515 isocratic pump (Waters Corporation, Milford, MA, USA), a set of three μ-Styragel mixed pore separation columns (pore size 10^2–10^6 Å), and a Waters 2414 differential refractive index detector (equilibrated at 40 °C) (Waters Corporation, Milford, MA, USA). The measurements and data analysis were conducted using the Breeze software (Waters Corporation, Milford, MA, USA). Tetrahydrofuran (THF) containing 5% *v/v* trimethylamine was the mobile phase, at a flow rate of 1 mL/min and temperature set at 30 °C. The calibration curve was set by utilizing linear polystyrene standards with average molecular mass in the range of 1200–152,000 g·mol^{-1} and narrow molecular mass distributions. The copolymer was dissolved in the mobile phase and measured at concentration of 1 mg mL^{-1}.

2.4. *Proton Nuclear Magnetic Resonance (^1H NMR) Spectroscopy*

^{1}H-NMR spectroscopy was implemented to confirm the chemical structure and to determine the mass composition (%wt.) of the synthesized copolymer. The spectrum was recorded using a Bruker AC 300 MHz FT-NMR spectrometer (Bruker, Billerica, MA, USA) and deuterated chloroform ($CDCl_3$) as the solvent. The chemical shifts are reported in parts per million (ppm) relative to tetramethylsilane (TMS) as the internal standard in $CDCl_3$.

^{1}H-NMR spectral peaks of P(SMA-co-OEGMA) copolymer (300 MHz, $CDCl_3$: 7.26 ppm, δ): 4.08 (peak c_2: 2H, -COOCH_2(CH_2)$_{16}$$CH_3$-), 3.89 (peak c_1: 2H, -(C=O)OC$H_2$$CH_2$O), 3.64 (peak f: 36H, -(CH_2CH_2O)$_9$$CH_3$-), 3.37 (peak g: 3H, -($CH_2$$CH_2$O)$_9CH_3$-), 1.60 (peak b: 4H, -C$H_2$C-), 1.25 (peak f: 32H, -CH_2(CH_2)$_{16}$$CH_3$-), 0.88 (peak a: 6H, -$CH_2CCH_3$, peak e: 3H, -$CH_2$ (CH_2)$_{16}$CH_3-).

2.5. Fourier-Transform Infrared (FT-IR) Spectroscopy

Fourier-transform infrared (FT-IR) spectroscopy in the form of ATR-FTIR was also employed to verify the chemical structure of the synthesized copolymer. The mid-infrared measurement was conducted at room temperature, in the spectral range of 5000–550 cm^{-1}, using a Fourier transform instrument (Bruker Equinox 55, Bruker Optics GmbH, Ettlingen, Germany) equipped with a single bounce attenuated total reflectance (ATR) diamond accessory (Dura-Samp1IR II by SensIR Technologies, Chapel Hill, NC, USA). The polymer sample was measured in the solid state and the spectrum was recorded after 64 scans with a resolution of 4 cm^{-1}.

ATR-FTIR spectral peaks of P(SMA-co-OEGMA), v (cm^{-1}), (s: stretching, b: bending): (CH_2): 2922 (s), 2854 (s) and 1454 (b), (C=O): 1728 (s), (C-O-C): 1105 (s).

2.6. Preparation of Pure Lipid and Chimeric Nanocarriers

Conventional liposomal nanocarriers of HSPC and DOTAP lipids and mixed/chimeric liposomes composed of HSPC, DOTAP lipids, and the amphiphilic random P(SMA-co-OEGMA) copolymer were prepared in different molar ratios by mixing the appropriate amounts of each component and following the thin film hydration method. Specifically, suitable amounts of HSPC, DOTAP, and P(SMA-co-OEGMA) were added in a round bottom flask and dissolved in chloroform. Subsequently, the mixture was placed under vacuum and heat using a rotary evaporator (Rotavapor R-114, Buchi, Switzerland), at 45 °C aqueous bath for 60 min, until the evaporation of the solvent and the formation of a thin film layer on the wall of the flask. The film was maintained under vacuum for 2 h and then in a desiccator for at least 24 h to remove possible traces of solvent. Afterward, the film was hydrated with water for injection by slowly spinning the round bottom flask in a water bath for 60 min, at a temperature 10 °C above the main phase transition of the main lipid component ($T_m \approx 52$ °C for HSPC [35], $T_m \approx -1$ °C for DOTAP [53]) to ensure complete hydration. The resultant structures were subjected to two 1-min sonication cycles (amplitude 70%, cycle 0.5 s) using a probe sonicator (UP 200 S, DrHielsher GmbH, Berlin, Germany) interrupted by a 3 min resting period in order to prevent sample overheating. Then, the obtained nanostructures were allowed to anneal for 30 min. The same procedure was followed for the preparation of both pure and chimeric liposomal formulations. Namely, two HSPC:DOTAP pure liposomal formulations with molar ratios 9:1 and 7:3 and four HSPC:DOTAP:P(SMA-co-OEGMA) chimeric nanosystems with molar ratios 9:1:0.05, 9:1:0.1, 7:3:0.05 and 7:3:0.1 were produced. The colloidal concentration was 10 mg·mL^{-1} in all the prepared dispersions. However, the physicochemical studies were performed on diluted samples with concentration of 0.5 mg·mL^{-1}.

2.7. Preparation of Pure and Chimeric Lipoplexes

Lipoplexes were formed through the electrostatic interaction between the cationic DOTAP and the DNA. The procedure followed for generating the lipoplexes includes as a first step the preparation of pure/chimeric liposomes (0.5 mg·mL^{-1}, same concentration for all liposomal formulations) and DNA 113 bp (concentration ranging from 0.018 to 0.065 mg·mL^{-1}) stock aqueous solutions. The utilized concentration of DNA stock solutions was calculated taking into consideration the moles of the cationic DOTAP in each liposomal formulation. Hence, DNA stock solutions with concentration of 0.018 mg·mL^{-1} were prepared in the case of liposomes with lower molar ratio in DOTAP, while stock solutions with concentration of 0.065 mg·mL^{-1} were prepared for liposomes with higher molar ratio in DOTAP. The preparation of the lipoplexes was achieved by mixing stock solutions of each liposomal formulation with different amounts of the relevant DNA stock solutions, under gentle stirring, at room temperature and neutral pH. The mixing amounts of the solutions are based upon calculations related to the molar ratio of nitrogen (N) from the positively charged quaternary amine group of DOTAP to the phosphate (P) from the negatively charged phosphate groups of DNA backbone, referred to as the Nitrogen-to-Phosphate ratio (N/P). The amounts of liposomes and DNA solutions were selected to give lipoplexes with N/P ratios ranging from 0.5 to 8. The formed lipoplexes were studied after allowing them to stand overnight at ambient temperature for equilibration.

2.8. Ultraviolet–Visible (UV–Vis) Spectroscopy

The interaction of the pure and chimeric liposomes with the DNA at different N/P ratios was explored by ultraviolet-visible (UV–Vis) spectroscopy. The absorption spectra of the lipoplexes were recorded on a Perkin Elmer (Lambda 19) UV–Vis–NIR spectrophotometer (Waltham, MA, USA) in the wavelength range of 200–600 nm. The recorded UV–Vis spectra of the lipoplexes as well as those of the pure and chimeric liposomes are presented in Figures S1 and S2 (Supplementary Materials), respectively.

2.9. Fluorescence Spectroscopy-Ethidium Bromide Quenching Assay

A standard fluorescence quenching assay based upon ethidium bromide (EtBr) exclusion was utilized to determine whether the pure/chimeric liposomes can bind DNA. Ethidium bromide was added in aqueous DNA stock solutions (0.01 mg·mL^{-1}) at a molar ratio, EtBr = P/4, where P corresponds to the molar concentration of DNA phosphate groups. The DNA solutions containing EtBr were left overnight to equilibrate and to ensure the complete intercalation of EtBr into the free DNA. Subsequently, the labeled DNA-EtBr solutions were titrated using concentrated aqueous solutions of liposomes (0.5 mg·mL^{-1}) in the range of N/P ratio 0 (neat DNA-EtBr) solution to 8. After the titration, the solutions at the studied N/P ratios were equilibrated for 15 min at 25 °C, before the operation of fluorescence spectroscopy measurements. The measurements were conducted on a Fluorolog-3 Jobin Yvon-Spex spectrofluorometer (model GL3–21). The excitation wavelength used for the recorded spectra was at 535 nm, while the emission was monitored at 600 nm. [54,55].

2.10. Light Scattering

The physicochemical characteristics regarding the size (hydrodynamic radius, R_h), the size distribution (Polydispersity index, PDI) the morphology, and the surface charge (zeta-potential, Z_p) of all the prepared pure/chimeric formulations and the formed pure/chimeric lipoplexes were determined by light scattering techniques including dynamic (DLS), static (SLS) and electrophoretic (ELS) light scattering, respectively. Prior to light scattering measurements, the solutions were filtered through 0.45 μm hydrophilic PVDF syringe filters (Membrane Solutions, Auburn, WA, USA) to remove large aggregates and dust particles.

DLS measurements were implemented on an ALV/CGS-3 compact goniometer system (obtained from ALV GmbH, Langen, Hessen, Germany), equipped with a cylindrical JDS Uniphase 22 mW He–Ne laser (ALV GmbH, Langen, Hessen, Germany), operating at 632.8 nm. The system was interfaced with an ALV-5000/EPP multi-τ digital correlator (ALV GmbH, Langen, Hessen, Germany) with 288 channels and an ALV/LSE-5003 light scattering electronics (ALV GmbH, Langen, Hessen, Germany) unit for stepper motor drive and limit switch control. Moreover, a Polyscience 9102A12E bath circulator (Polyscience, Illinois, USA) was utilized to regulate the temperature inside the measuring cell. Toluene was used as the calibration standard. The measurements were implemented on the angular range of 45° to 135°, at 25 °C. The scattered light intensity was simultaneously monitored. The autocorrelation functions were recorded five times for each angle and averaged. The obtained correlation functions were fitted and analyzed by the cumulants method and the

CONTIN algorithm. The apparent hydrodynamic radius, R_h, was calculated using the Stokes–Einstein equation. The presented data of R_h, PDI, and scattered light intensity (I) correspond to measurements at 90°.

Static light scattering (SLS) measurements were performed on the same instrument at 25 °C, in the angular range of 30–150°, at 10° intervals and using toluene as the calibration standard. SLS measurements were treated by the Zimm second order plot to estimate the radius of gyration R_g and therefore the R_g/R_{ho} ratio, after extrapolation to zero angle (R_{ho}). The R_g/R_h ratio provides useful information on the morphology and the shape of the nanoparticles.

Electrophoretic light scattering (ELS) measurements were also performed at 25 °C using a Nano Zeta Sizer (Malvern Instruments Ltd., Worcestershire, UK) composed of a 4 mW solid-state He–Ne laser, operating at 633 nm and at a fixed backscattering angle of 173°. Zeta-potential values were determined using the Henry approximation of the Smoluchowski equation. The recorded zeta-potential values were averages of 50 scans, with an error smaller than ±2 mV.

2.11. Cryogenic Transmission Electron Microscopy (Cryo-TEM)

Cryogenic Transmission Electron Microscopy (cryo-TEM) images were obtained using a Tecnai F20 X TWIN microscope (FEI Company, Hillsboro, OR, USA) equipped with a field emission gun, operating at an acceleration voltage of 200 kV. Images were recorded on the Gatan Rio 16 CMOS 4 k camera (Gatan Inc., Pleasanton, CA, USA) and processed with Gatan Microscopy Suite (GMS) software (Gatan Inc., Pleasanton, CA, USA). Specimen preparation was done by vitrification of the aqueous solutions on grids with holey carbon film (Quantifoil R 2/2; Quantifoil Micro Tools GmbH, Großlöbichau, Germany). Prior to use, the grids were activated for 15 s in oxygen plasma using a Femto plasma cleaner (Diener Electronic, Ebhausen, Germany). Cryo-samples were prepared by applying a droplet (3 μL) of the suspension to the grid, blotting with filter paper and immediate freezing in liquid ethane using a fully automated blotting device Vitrobot Mark IV (Thermo Fisher Scientific, Waltham, MA, USA). After preparation, the vitrified specimens were kept under liquid nitrogen until they were inserted into a cryo-TEM-holder Gatan 626 (Gatan Inc., Pleasanton, CA, USA) and analyzed in the TEM at −178 °C.

3. Results and Discussion

3.1. Synthesis and Molecular Characterization of Random Copolymer

The synthesis of the novel amphiphilic random poly[(stearyl methacrylate)-co-oligo(ethylene glycol)methyl ether methacrylate] P[(SMA-co-OEGMA)] copolymer was accomplished following a one-step RAFT polymerization procedure. In this regard, the 4-cyano-4-(phenylcarbonothioylthio)pentanoic acid (CPAD) was selected as the CTA, which is reactive and compatible with methacrylate monomers [49,56–58]. The applied polymerization conditions were also chosen based on literature data [49,58,59].

The copolymer was molecularly characterized by means of SEC chromatography and ^{1}H-NMR and ATR-FTIR spectroscopies (Figure 1). The chromatogram in Figure 1a, as obtained by SEC, depicts narrow, monomodal, and symmetric molar mass distribution. The molecular mass (M_w) of the copolymer and the polydispersity index (M_w/M_n) were also determined by SEC, with values of 12,240 g·mol^{-1} and 1.11 respectively. The resulting molecular mass is close to the stoichiometry and the polydispersity index is narrow.

^{1}H-NMR spectrum (Figure 1b) verified the expected chemical structure of the synthesized P(SMA-co-OEGMA) copolymer. The signals at δ 1.26 ppm corresponding to the methylene -CH_2 protons of the SMA alkyl side chain (peak d, 32H, $(CH_2)_{16}$) [46] and the one at δ 3.64 ppm to the -CH_2 protons of the OEGMA ethylene glycol side chain (peak f, 36H, -$CH_2CH_2O)_9$) [60] were utilized to quantify copolymer composition, which was found to be 38%wt. for the SMA segment and 62%wt. for the OEGMA one.

Along with ^{1}H-NMR, ATR-FTIR spectroscopy was also utilized. The ATR-FTIR spectrum of P(SMA-co-OEGMA) copolymer (Figure 1c) contains all the major chemical groups of which the copolymer is comprised, indicating the expected chemical structure. Specifically, the absorption bands observed at 2922 cm^{-1} and 2854 cm^{-1} correspond to C-H stretching vibrations of the $-CH_2$ groups of SMA and OEGMA side groups, while the band at 1454 cm^{-1} corresponds to bending vibrations also of the $-CH_2$ groups [61]. The strong stretching vibration band at 1722 cm^{-1} is ascribed to the C=O carbonyl of the ester group of both segments [61]. The broad band at 1105 cm^{-1} corresponds to C–O–C stretching vibrations of the ether group of OEGMA. [61,62]. Furthermore, the absence of vibrations of the C=C bond in the ATR-FTIR spectrum proves the successful polymerization at high monomer conversion.

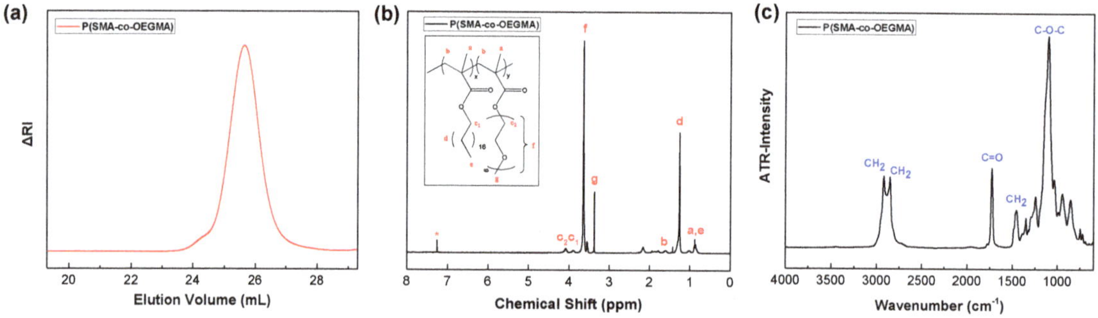

Figure 1. (**a**) SEC chromatogram of the synthesized P(SMA-co-OEGMA) copolymer in THF/5% *v/v* Et_3N. (**b**) ^{1}H-NMR spectrum of P(SMA-co-OEGMA) copolymer in $CDCl_3$. The peak at 7.26 ppm (*) is assigned to the solvent protons. (**c**) ATR-FTIR spectrum of P(SMA-co-OEGMA) copolymer in solid state.

The obtained results from SEC, ^{1}H-NMR, and ATR-FTIR indicate that the followed synthetic route and polymerization conditions worked satisfactorily. The followed RAFT methodology provided control of the molecular characteristics, which is an important issue in the design of copolymers for gene delivery.

3.2. Physicochemical Characterization of Pure and Chimeric Liposomes

The physicochemical characteristics of the developed nanosystems in aqueous solutions were examined by light scattering techniques. Dynamic light scattering (DLS) was employed to determine the size of the formed nanostructures by estimating the R_h and the size polydispersity index (PDI). The determined values are presented in Table 1. The hydrodynamic size distribution plots from CONTIN analysis in Figure 2 for pure HSPC:DOTAP liposomes and chimeric HSPC:DOTAP:P(SMA-co-OEGMA) liposomes revealed that both types of liposomes self-assemble into a single, uniform, relatively narrow, and well-defined nanoparticle population.

Table 1. Physicochemical characteristics of the developed nanosystems, as determined by light scattering techniques.

Sample	Molar Ratio	R_h (nm) [a]	PDI [a]	R_g/R_h [b]	Z_p (mV) [c]
HSPC:DOTAP	9:1	71	0.21	0.80	+44
HSPC:DOTAP:P(SMA-co-OEGMA)	9:1:0.05	46	0.45	1.10	+25
HSPC:DOTAP:P(SMA-co-OEGMA)	9:1:0.1	33	0.43	1.11	+22
HSPC:DOTAP	7:3	70	0.22	0.96	+56
HSPC:DOTAP:P(SMA-co-OEGMA)	7:3:0.05	55	0.29	1.08	+51
HSPC:DOTAP:P(SMA-co-OEGMA)	7:3:0.1	57	0.36	1.02	+45

[a] Determined by DLS and CONTIN analysis at 90°, [b] Determined by SLS, [c] Determined by ELS.

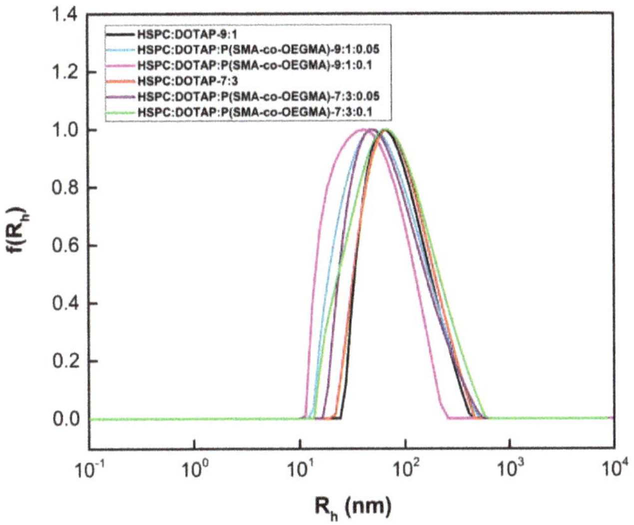

Figure 2. Size distributions from CONTIN analysis of HSPC:DOTAP 9:1 and 7:3 liposomes, HSPC:DOTAP:P(SMA-co-OEGMA)-9:1:0.05, 9:1:0.1, 7:3:0.5, and 7:3:0.1 chimeric liposomes in H_2O, C=0.5 mg·mL^{-1} at a scattering angle $\theta = 90°$.

All liposomal formulations presented sizes with hydrodynamic radius less than 100 nm and polydispersity index, PDI $\geq$ 0.45. However, according to the R_h values of Table 1 and the size distribution plots of Figure 2, it is evident that the incorporation of the amphiphilic random copolymer significantly reduced the size of the chimeric liposomes compared to that of the pure liposomes. This reduction can be attributed to the presence of hydrophobic and steric effects due to the anchoring of the long alkyl chains of the SMA segment into the lipid bilayers.

The zeta-potential values as determined by electrophoretic light scattering (ELS) are also listed in Table 1. Both types of liposomes exhibit positive zeta-potential values, with the liposomes comprised of higher molar ratio of DOTAP showing more positive absolute values. The increase of DOTAP ratio caused a great increase of the surface charge particularly in the HSPC:DOTAP-7:3 liposomes. It is evident that the higher ratio of the quaternary ammonium head group in the DOTAP lipid contributed to the strong positive charge of the liposomes. On the other hand, the incorporation of the amphiphilic copolymer led to a significant decrease of the zeta-potential values. This decrease was found to depend on copolymer molar ratio. The chimeric liposomes with molar ratio 0.1 presented lower zeta-potential values compared to those with 0.05. This reduction is assigned to the non-ionic OEGMA moieties, which are located on the outer liposomal surface and promote a shielding effect of the liposomes. Hence, these findings revealed that the incorporation of the amphiphilic copolymer and particularly the hydrophilic OEGMA chains efficiently reduced the strong cationic surface charge, which is important to prevent cytotoxicity effects and to produce effective gene delivery nanocarriers.

Static light scattering was utilized in order to extract information regarding the shape and morphology of the liposomal nanostructures, by determining the R_g/R_h ratio. The calculated values are summarized in Table 1. The liposomal nanoassemblies presented R_g/R_h values ranging from 0.80 to 1.11, indicating their assembly into vesicular morphologies [63,64]. Furthermore, in the case of the chimeric liposomes, the incorporation of the copolymer caused a slight increase of the R_g/R_h ratio, with the values remaining in the range of vesicles morphology.

Conclusively, light scattering techniques demonstrate the formation of liposomes with well-defined physicochemical characteristics, which are highly dependent on the molar ratio of each component which the liposomes consist of. Moreover, the obtained size values of the liposomes are in a satisfactory nanoscale range for gene nanocarriers. The efficient reduction of the high positive charges due to the shielding effect of OEGMA segment signifies that such segments are a useful tool for the control of the nanocarrier surface charge.

3.3. Ethidium Bromide Quenching Assay by Fluoresence Spectroscopy

To elucidate whether the pure and chimeric liposomes can bind DNA, fluorescence spectroscopy was employed by studying the quenching of ethidium bromide (EtBr) probe. Ethidium bromide is a cationic fluorescent probe, which intercalates between the adjacent base pairs of DNA double helix and exhibits a strong fluorescent intensity [65,66]. The electrostatic interaction between the positively charged liposomes and the DNA results to the exclusion of EtBr from DNA double helix to solution aqueous environment, accompanied by an evident decay of its fluorescence intensity. Therefore, fluorescence quenching of EtBr provides an indirect way to determine the binding affinity between the cationic liposomes and DNA and thus ascertain the formation of lipoplexes.

In this regard, the quenching of EtBr was examined in a range of N/P = 0 to N/P = 8, by titration of the liposomal solutions to the DNA-EtBr solution. The recorded spectra demonstrating the reduction in the fluorescence intensity of the intercalated EtBr upon the progressive addition of the cationic pure/chimeric liposomes are presented in Figure S3. Figure 3 depicts the typical curves of the relative fluorescence intensity of EtBr as a function of the N/P ratio, for the HSPC:DOTAP/DNA and HSPC:DOTAP:P(SMA-co-OEGMA)/DNA complexes. The curves exhibit a well-pronounced and gradual decrease of the fluorescence intensity for all the studied liposomal nanosystems.

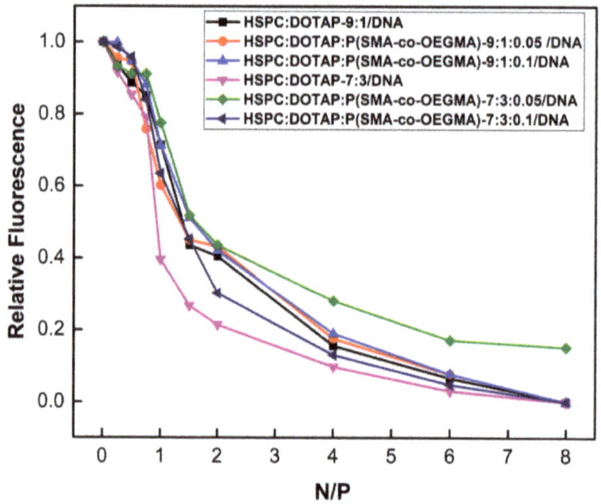

Figure 3. Ethidium bromide quenching in lipoplexes formed from the interaction of HSPC:DOTAP and HSPC:DOTAP:P(SMA-co-OEGMA) liposomes with the DNA of 113 bp, utilizing liposomes of different molar ratios.

Concerning the interaction of the pure HSPC:DOTAP liposomes with the DNA, the formed HSPC:DOTAP-7:3/DNA and the HSPC:DOTAP-9:1/DNA complexes present a similar trend in the decrease of the fluorescence intensity of the EtBr until the ratio N/P = 0.75. However, the decrease of the relative fluorescence intensity for the HSPC:DOTAP-7:3/DNA complexes at the ratio N/P = 1 and until the N/P = 8 is sharper, indicating the faster rate of the EtBr displacement and thus the better binding affinity of the HSPC:DOTAP-7:3 liposomes to DNA. The stronger binding affinity of the HSPC:DOTAP-7:3 nanocarriers was expected due to the higher molar content of these liposomes in DOTAP. Nevertheless, similar to HSPC:DOTAP 7:3 liposomes, the HSPC:DOTAP-9:1 liposomes also present sufficient complexation ability with DNA, due to the fact that at the ratio N/P = 8 the relative intensity was found equal to zero, denoting that the liposomes displaced efficiently the whole amount of the intercalated EtBr from the DNA double helix.

The HSPC:DOTAP:P(SMA-co-OEGMA)/DNA complexes formed by the chimeric liposomes and the DNA displayed similar behavior in EtBr exclusion. All the chimeric nanosystems exhibited the same displacement rate of EtBr with the relative fluorescent intensity reaching zero at N/P = 8, except for HSPC:DOTAP:P(SMA-co-OEGMA)-7:3:0.05 liposomes in which the relative fluorescence reached 0.15 at the same N/P ratio. The shielding effect of OEGMA is reflected in the rate of decrease of the relative fluorescence intensity, which in the case of the chimeric liposomes is somehow lower compared to the reduction rate for the HSPC:DOTAP-7:3 and HSPC:DOTAP-9:1 pure liposomes. Particularly, the HSPC:DOTAP:P(SMA-co-OEGMA) chimeric liposomes with molar ratio 9:1:0.05 and 9:1:0.1 present similar trend in the decrease of the relative fluorescence intensity, with the HSPC:DOTAP:P(SMA-co-OEGMA-9:1:0.05 liposomes displaying a slightly faster decrease from N/P = 0.75 to N/P = 8, in comparison with the 9:1:0.1 system. However, similar behavior is not observed in the case of the chimeric liposomes with molar ratios 7:3:0.05 and 7:3:0.1, with the latter presenting the more abrupt decrease of EtBr relative fluorescence intensity in comparison with all the chimeric nanosystems studied. This fact was expected due to the high molar ratio of DOTAP compared to the chimeric nanosystems with molar ratio 9:1:0.05 and 9:1:0.1. Nevertheless, all the chimeric nanosystems presented a strong binding affinity to the DNA of 113 bp, which was confirmed by the complete exclusion of EtBr from the DNA double helix.

Summarizing, the EtBr quenching assay provided better proof for lipoplex formation. The presence of the OEGMA chains probably affected the exclusion rate of EtBr; however, it is significant that at the ratio N/P = 8, total exclusion of the EtBr was observed for most of the investigated liposomal nanosystems, indicating their strong binding affinity to DNA. Hence, the liposomes can efficiently interact with the DNA, exhibiting strong binding affinity, which is a required parameter of an efficient gene nanocarrier.

3.4. *Lipoplexes Characterization by Light Scattering*

The physicochemical properties of the lipoplexes formed by the complexation of HSPC:DOTAP and HPSC:DOTAP:P(SMA-co-OEGMA) liposomes with the DNA of 113 bp, were further evaluated by light scattering techniques. The determination of their size and surface charge is essential for their efficient implementation as non-viral gene delivery nanocarriers since these parameters can affect the biological performance of the lipoplexes.

The lipoplexes were prepared and studied in a wide range of N/P ratios (N/P = 0.25 to N/P = 8), including ratios with an excess of DOTAP positive charges and an excess of DNA negatively charges, in an attempt to gain a better view on the complexation process and to find the ratios with the preferable complexation efficiency and colloidal stability. All the dispersions appeared colloidally stable by the naked eye. However, lipoplexes formed either by pure or chimeric liposomes presented partial precipitation at the ratio N/P = 1, in which ratio the charges of DOTAP are stoichiometrically equal to those of DNA. The precipitation of the lipoplexes probably occurred due to charge neutralization, which resulted to the decrease of their solubility and therefore their precipitation. Furthermore, it is noteworthy to mention that the partially precipitated chimeric lipoplexes appeared more stable by naked eye compared to the pure lipoplexes. This observation indicates that the presence of the non-ionic OEGMA segment as well as the long alkyl chain of the SMA segment probably contributed to the colloidal stability of the lipoplexes and prevented their total collapse. Light scattering measurements of the precipitated lipoplexes were performed on the supernatant of the solutions. The precipitation region is denoted in the presented data of the light scattering plots; however, the results for the ratio N/P = 1 are not taken into consideration.

Light scattering findings regarding the size, intensity, and zeta potential of the lipoplexes as a function of the N/P ratio are presented in Figure 4 for the HSPC:DOTAP-9:1/DNA, HSPC:DOTAP:P(SMA-co-OEGMA)-9:1:0.05/DNA and HSPC:DOTAP:P(SMA-co-OEGMA)-9:1:0.1/DNA lipoplexes. The lipoplexes formed by pure liposomes and DNA were utilized as reference in order to investigate whether the presence of the copolymer in different molar ratios affect the formation and properties of the chimeric lipoplexes. Both types of lipoplexes display monomodal and relatively narrow size distributions in the whole N/P range, indicating the homogeneity of the formed lipoplex nanostructures. The derived size distributions of the above-mentioned types of lipoplexes at the examined N/P ratios are included in Figures S4–S6, respectively. In general, the lipoplexes in Figure 4 formed either by pure or chimeric liposomes and DNA exhibit a similar behavior as far as the variations of the R_h (Figure 4a–c), the scattered intensity (Figure 4d–f) and the zeta-potential (Figure 4g–i) are concerned. In most cases, the size of the lipoplexes (R_h), as well as the scattered light intensity are gradually decreased as the N/P ratio increases from 0.25 to 8. Furthermore, the sizes of pure and chimeric lipoplexes at the whole range of the studied N/P ratios are larger compared to their parent liposomes, indicating the successful formation of lipoplexes in a wide range of N/P ratios. The decrease of the R_h, accompanied by the parallel decrease of the scattered light intensity upon increasing N/P ratio, denotes the formation of lipoplexes of smaller size and lower molar mass. Particularly, in the case of the chimeric lipoplexes, at low N/P ratios (N/P < 1), the excess of DNA phosphate groups compared to the positive charges of DOTAP, favors the formation of larger nanostructures of high molar mass. In contrast, going to higher N/P ratios (N/P > 1), the number of available positive charges increase, thus the electrostatic interactions between DOTAP and DNA are more intense, resulting in the formation of more compact lipoplex nanostructures. Furthermore, the chimeric HSPC:DOTAP:P(SMA-co-OEGMA)-9:1:0.1/DNA lipoplexes presented smaller R_h and lower values of the scattered intensity in the whole range of N/P ratios, in comparison with the HSPC:DOTAP:P(SMA-co-OEGMA)-9:1:0.05/DNA lipoplexes. In this case, it seems that the higher content of the lipoplexes in the amphiphilic copolymer prompted the formation of smaller and compact nanostructures.

The surface charge of the lipoplexes was estimated by electrophoretic light scattering, to elucidate the successful formation of the lipoplexes. As it can be seen in Figure 4g–i, the lipoplexes display similar behavior as the N/P ratio increases from 0.25 to 8. At the N/P ratios of 0.25 and 0.5 both types of lipoplexes present negative zeta-potential values, due to the prevailing negative charges of the phosphate groups. In contrast, upon increasing the N/P ratio above the neutralization point (N/P > 1), the transition of the surface charge from negative to positive values signifies that the majority of the available positive charges of liposomes have efficiently interacted with DNA. Furthermore, it is evident that the chimeric lipoplexes exhibit less positive zeta-potential values in comparison to those of the pure lipoplexes and even to those of their parent liposomes, which is evidence of the effective shielding effect provided by the OEGMA chains of the copolymer.

Figure 5 depicts the obtained DLS and ELS results, concerning the size, intensity, and zeta-potential of the HSPC:DOTAP-7:3/DNA, HSPC:DOTAP:P(SMA-co-OEGMA)-7:3:0.05/DNA, and HSPC:DOTAP:P(SMA-co-OEGMA)-7:3:0.1/DNA lipoplexes, as a function of the N/P ratio. The complexation between the liposomes with the molar ratio of DOTAP prevailing resulted in the formation of lipoplex nanostructures with homogeneity in the whole N/P range, as their monomodal and narrow size distributions revealed (Figures S7–S9, respectively).

According to Figure 5, the variations in the size (R_h) (Figure 5a–c), the scattered light intensity (Figure 5d–f) and the zeta-potential values (Figure 5g–i) of the pure and chimeric lipoplexes are generally following a pattern behavior similar to the one described in the case of the pure and chimeric lipoplexes composed of the lower molar ratio in DOTAP. Regarding the chimeric HSPC:DOTAP:P(SMA-co-OEGMA)-7:3:0.05/DNA and HSPC:DOTAP:P(SMA-co-OEGMA)-7:3:0.1/DNA lipoplexes, the decrease of the R_h along with the parallel decrease of the scattered light intensity, as the N/P ratio rises especially from 0.5 to 8, signals the assembly of lipoplexes with smaller size and lower molar mass. Furthermore, both HSPC:DOTAP:P(SMA-co-OEGMA)-7:3:0.05/DNA and HSPC:DOTAP:P(SMA-co-OEGMA)-7:3:0.1/DNA chimeric lipoplexes display similar values of the R_h and scattered light intensity at the majority of the N/P ratios. Moreover, their size, depending on the N/P ratio, is larger compared to the neat chimeric liposomes, indicating the efficacious formation of lipoplexes. The HSPC:DOTAP-7:3/DNA lipoplexes present higher values of scattered light intensity in comparison to the HSPC:DOTAP:P(SMA-co-OEGMA)-7:3:0.05/DNA, HSPC:DOTAP:P(SMA-co-OEGMA)-7:3:0.1/DNA chimeric lipoplexes, and also to the pure HSPC:DOTAP-9:1/DNA lipoplexes, implying that the increased molar ratio of DOTAP leads to the formation of nanostructures with higher molar mass.

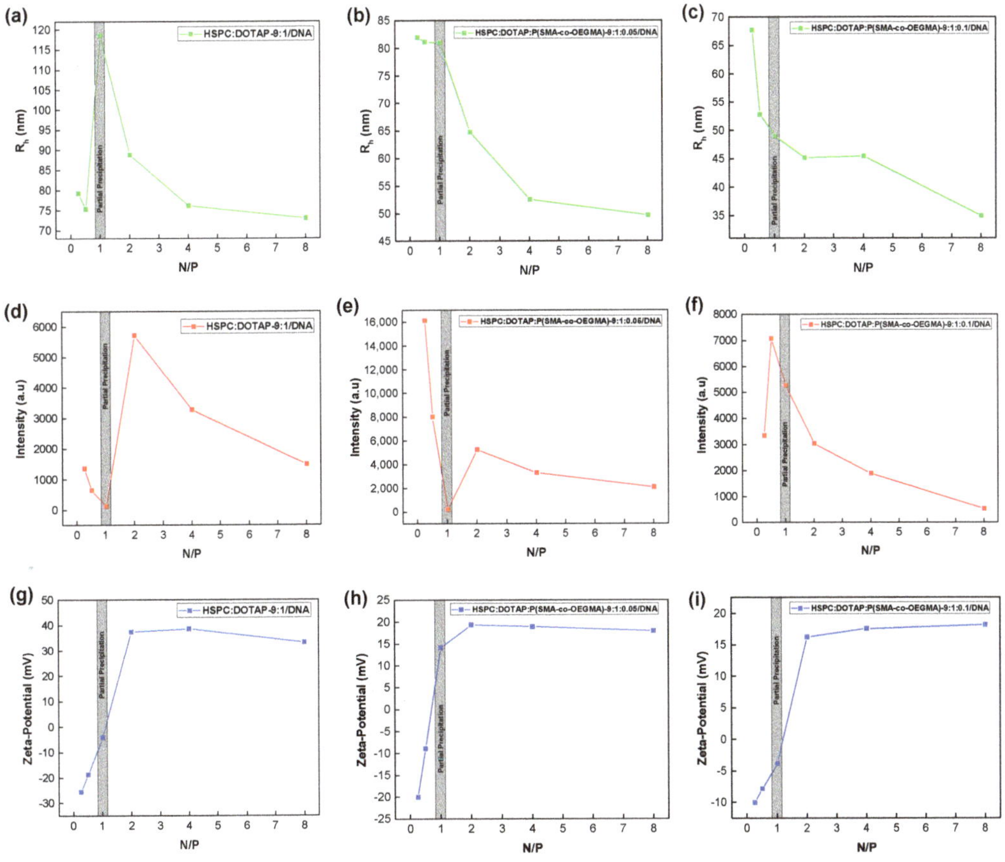

Figure 4. Variations of hydrodynamic radius (R_h) with N/P ratio for (**a**) HSPC:DOTAP-9:1/DNA, (**b**) HSPC:DOTAP-9:1:0.05/DNA and (**c**) HSPC:DOTAP-9:1:0.1/DNA lipoplexes. Variations of scattered light intensity with N/P ratio for (**d**) HSPC:DOTAP-9:1/DNA, (**e**) HSPC:DOTAP-9:1:0.05/DNA, and (**f**) HSPC:DOTAP-9:1:0.1/DNA lipoplexes. Variations of zeta-potential with N/P ratio for (**g**) HSPC:DOTAP-9:1/DNA, (**h**) HSPC:DOTAP-9:1:0.05/DNA and (**i**) HSPC:DOTAP-9:1:0.1/DNA lipoplexes.

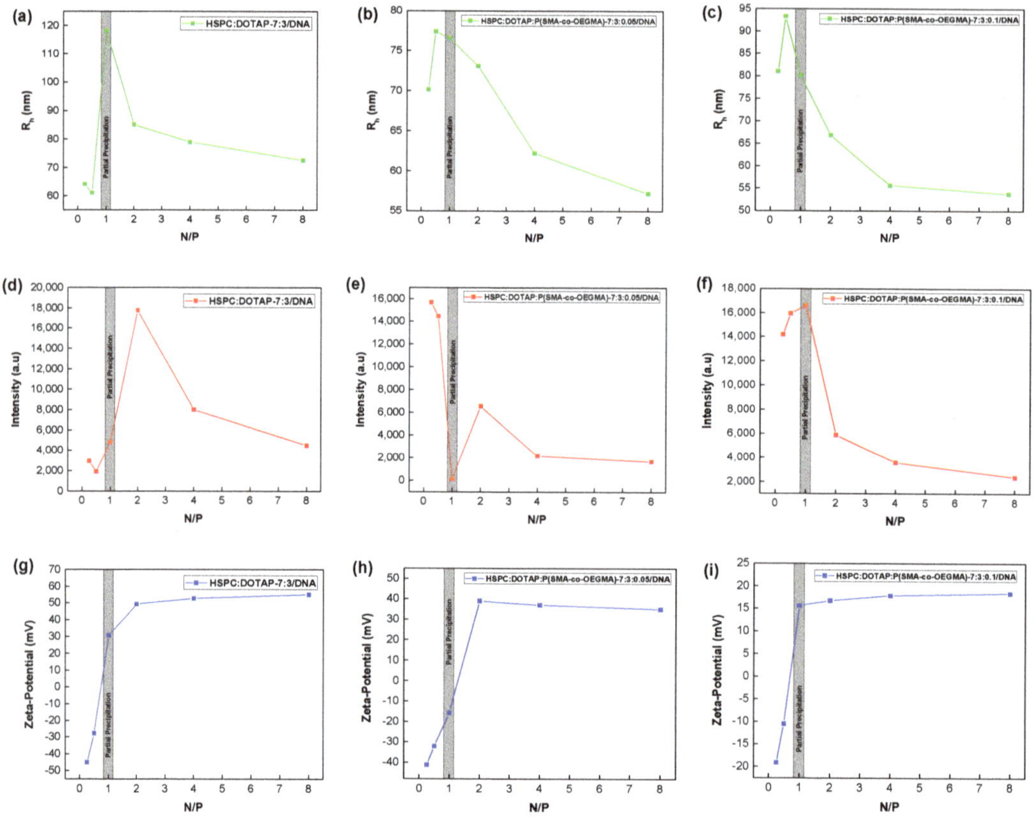

Figure 5. Variations of hydrodynamic radius (R_h) with N/P ratio for (**a**) HSPC:DOTAP-7:3/DNA, (**b**) HSPC:DOTAP-7:3:0.05/DNA and (**c**) HSPC:DOTAP-7:3:0.1/DNA lipoplexes. Variations of scattered light intensity with N/P ratio for (**d**) HSPC:DOTAP-7:3/DNA, (**e**) HSPC:DOTAP-7:3:0.05/DNA, and (**f**) HSPC:DOTAP-7:3:0.1/DNA lipoplexes. Variations of zeta-potential with N/P ratio for (**g**) HSPC:DOTAP-7:3/DNA, (**h**) HSPC:DOTAP-7:3:0.05/DNA, and (**i**) HSPC:DOTAP-7:3:0.1/DNA lipoplexes.

The surface charge of the pure HSPC:DOTAP-7:3/DNA and chimeric HSPC:DOTAP:P(SMA-co-OEGMA)-7:3:0.05/DNA, HSPC:DOTAP:P(SMA-co-OEGMA)-7:3:0.1/DNA lipoplexes, similarly to the lipoplexes comprised of the same molar ratio of DOTAP, display negative values of zeta-potential at N/P ratios with an excess of DNA phosphate groups (N/P < 1) and acquires positive values upon shifting to N/P ratios (N/P < 1) with a higher number of available DOTAP positive charges. Hence, bearing in mind the zeta-potential values of HSPC:DOTAP-7:3/DNA lipoplexes and observing the zeta-potential values of the chimeric lipoplexes in Figure 5h,i, it is discerned that even the small addition of the copolymer (molar ratio 0.05) drastically reduced the highly positive surface charge compared to the pure lipoplexes. Furthermore, this reduction is more distinct in the case of the chimeric HSPC:DOTAP:P(SMA-co-OEGMA)-7:3:0.1/DNA system, which is comprised of higher molar ratio of the copolymer and thus greater content of the OEGMA segments. Undoubtedly, the presence of the OEGMA moieties efficiently decreased the surface charge of the lipoplexes by the shielding of the positive charges. Furthermore, it is worth noticing that although the HSPC:DOTAP:P(SMA-co-OEGMA)-7:3:0.1/DNA lipoplexes have larger positive charge due to the higher molar ratio of DOTAP (molar ratio 3), the presence of the copolymer in the HSPC:DOTAP:P(SMA-co-OEGMA)-7:3:0.1/DNA lipoplexes managed to tremendously reduce the surface charge by reaching values similar to those of HSPC:DOTAP:P(SMA-co-OEGMA)-9:1:0.1/DNA lipoplexes (Figure 5i), which are comprised of a lower molar ratio of DOTAP (molar ratio 1) and thus present fewer positive charges. This finding reveals the efficacious contribution of the copolymer in the fabrication of liposomes with strong binding affinity to DNA and then the formation of lipoplexes with controllable surface charge. Hence, the decrease of the positive charges without affecting drastically the binding affinity and transfection efficiency is an issue of great importance in the design of effective and safe non-viral gene delivery vectors. Therefore, the choice of the OEGMA segment as an alternate to PEG was successful, due to the efficient decrease of the highly positive surface charge without affecting the binding affinity of the chimeric liposomes to the DNA, as was revealed by the fluorescence spectroscopy and the total displacement of the EtBr from the DNA double helix. This finding is also of great importance due to the fact that usually the shielding of the positive charges leads to weakening of the binding affinity of the nanocarrier to the nucleic acid.

Additionally, static light scattering measurements were performed aiming to gain a first view on the morphology of the formed lipoplexes at the different N/P ratios, by determining the R_g/R_h ratio. All the examined lipoplexes displayed similar behavior in the variations of the R_g/R_h upon increasing N/P ratio. Representative plots presenting the variations of the R_g/R_h ratio as a function of the N/P ratio are given in Figure 6a–c for the HSPC:DOTAP-7:3/DNA, HSPC:DOTAP-7:3:0.05/DNA, and HSPC:DOTAP-7:3:0.1/DNA lipoplexes, respectively. In the case of the pure HSPC:DOTAP-7:3/DNA (Figure 6a), as the N/P ratio rises from 0.25 to 8, accordingly the values of the R_g/R_h ratio increase from approximately 0.84 to 0.95. These values indicate the formation of lipoplexes with mostly vesicular morphology at these N/P ratios. Similarly to HSPC:DOTAP-7:3/DNA lipoplexes, the R_g/R_h ratio of the chimeric HSPC:DOTAP-7:3:0.05/DNA lipoplexes (Figure 6b) also increases by going to higher N/P ratios The obtained values suggest the formation of more compact spherical nanostructures at the N/P ratios with excess of DNA phosphate groups and the transition to more loose vesicular nanostructures upon increasing the N/P ratio. On the other hand, the HSPC:DOTAP:P(SMA-co-OEGMA)-7:3:0.1/DNA lipoplexes (Figure 6c) exhibit a decrease of their R_g/R_h values as the N/P ratio increases, particularly from 0.5 to 8. In this N/P ratio range, the R_g/R_h ratio obtained values ranging from ca. 1.23 at N/P = 0.5 to 0.78 at N/P = 8. These values show the formation of lipoplexes with loose conformation [63] at low N/P ratios, while at higher N/P ratios the lipoplexes assemble into more compact nanostructures with overall globular morphology.

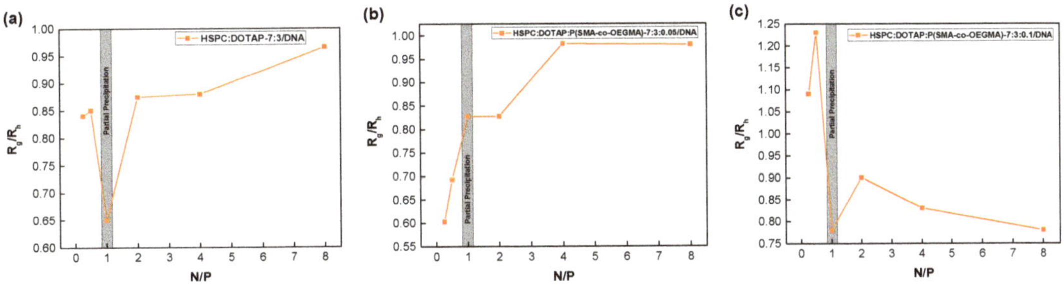

Figure 6. Variations of R_g/R_h ratio as a function of N/P ratio for (**a**) HSPC:DOTAP-7:3/DNA, (**b**) HSPC:DOTAP-7:3:0.05/DNA, and (**c**) HSPC:DOTAP-7:3:0.1/DNA lipoplexes.

Summarizing, light scattering findings revealed the efficient interaction of all the examined pure and chimeric liposomes with the DNA and therefore the successful formation of lipoplexes, whose physicochemical characteristics, colloidal stability, and morphology strongly depend on and are affected by the N/P ratio and especially on the molar ratio of the cationic lipid and the amphiphilic copolymer. Hence, the appropriate design focusing on the delicate balance of these parameters which affect the formation and the performance of the lipoplexes can provide the perspectives to the HSPC:DOTAP: P(SMA-co-OEGMA) liposomal formulations to be utilized as non-viral vectors for DNA delivery as well as other types of nucleic acids.

3.5. Influence of Ionic Strength on the Stability of the Lipoplexes

The ionic strength of the biological fluids affects the physicochemical characteristics of the nanocarriers by causing changes in their size, surface charge, and therefore to their colloidal stability [16]. Hence, the influence of the physiological conditions of the biological fluids on the nanocarriers is significant and it should be considered during the fabrication of effective gene nanocarriers.

The behavior of the lipoplexes in the presence of salt was examined by adding NaCl of 1 M and gradually increasing the salt concentration from 0 M to 0.5 M. DLS was employed to monitor changes on the size and scattered light intensity of the lipoplexes.

The tolerance of the pure and chimeric lipoplexes on increasing ionic strength was selected to be examined at N/P ratios above the precipitation region (N/P > 1) with optimal colloidal stability for at least one week. Furthermore, nanocarriers with a small excess of positive charges are more suitable for nucleic acid delivery due to the fact that the free positive charges facilitate the intracellular uptake by interacting with the negative charges of the cellular membrane [19,20]. In this regard, the influence of the ionic strength on the lipoplexes was investigated at the ratio N/P = 4.

The variations of the hydrodynamic radius (R_h) and the scattered light intensity of the pure and chimeric lipoplexes as a function of the ionic strength, at N/P = 4, are depicted on Figure 7. In the case of the pure HSPC:DOTAP-9:1/DNA lipoplexes (Figure 7a), R_h gradually increases reaching almost 1000 nm at the final salt concentration of 0.5 M. On the other hand, a parallel increase of the scattered light intensity is observed up to the salt concentration of 0.05 M and thereafter declines until the concentration of 0.5 M. The parallel increase of the R_h and the scattered intensity denote the growth of the lipoplexes in size and mass, while the decrease of the intensity from the concentration of 0.05 M till 0.5 M indicates the reduction in their mass. Particularly, the simultaneous increase in size and decrease in mass as the salt concentration rises, signifies the disintegration and destabilization of the lipoplexes in the presence of higher salt concentrations.

Similar tendency in the variations of the R_h and the scattered intensity was exhibited by the HSPC:DOTAP-7:3/DNA lipoplexes (Figure 7b). Specifically, R_h along with the scattered intensity simultaneously increases until 0.05 M, then the intensity decreases and the R_h rises, reaching ca. 750 nm at the concentration of 0.5 M NaCl. The addition of salt induces charge screening effects and leads to the weaking of the affinity between the cationic liposomes and the DNA. In this way, the solubility of the lipoplexes is enhanced, resulting in their swelling due to the insertion of water molecules within their structures. Hence, the addition of salt caused the rapid increase of the R_h of both pure lipoplexes and resulted in large sizes and high intensities which is evidence of lack of their colloidal stability, even at low NaCl concentrations.

As far as the chimeric lipoplexes are concerned, it can be observed in Figure 7c that the HSPC:DOTAP:P(SMA-co-OEGMA)-9:1:0.05/DNA lipoplexes display a slight decrease of the R_h, accompanied by the decrease of the intensity until 0.15 M. Nevertheless, these variations of R_h and intensity are small, thus the lipoplexes remain essentially stable until 0.15 M NaCl. However, the parallel increase of R_h and I after 0.15 M until 0.5 M evidences the increase of lipoplexes in size and mass, but not their decomposition. In contrast, HSPC:DOTAP:P(SMA-co-OEGMA)-7:3:0.05/DNA lipoplexes (Figure 7d) present simultaneous increase of the R_h and scattered light intensity even from the first addition of salt, demonstrating that the structure of the lipoplexes may became looser and a tendency to aggregation. Similar behavior pattern is also observed in the case of HSPC:DOTAP:P(SMA-co-OEGMA)-9:1:0.1/DNA (Figure 7e) and HSPC:DOTAP:P(SMA-co-OEGMA)-7:3:0.1/DNA (Figure 7f) chimeric lipoplexes. The stability of the lipoplexes in the presence of salt is more evident in the case of the HSPC:DOTAP:P(SMA-co-OEGMA)-7:3:0.1/DNA (Figure 7f). The

R_h and intensity remain rather constant until 0.2 M and thereafter increase sharply from 0.3 M to 0.5 M, indicating farther aggregation and destabilization.

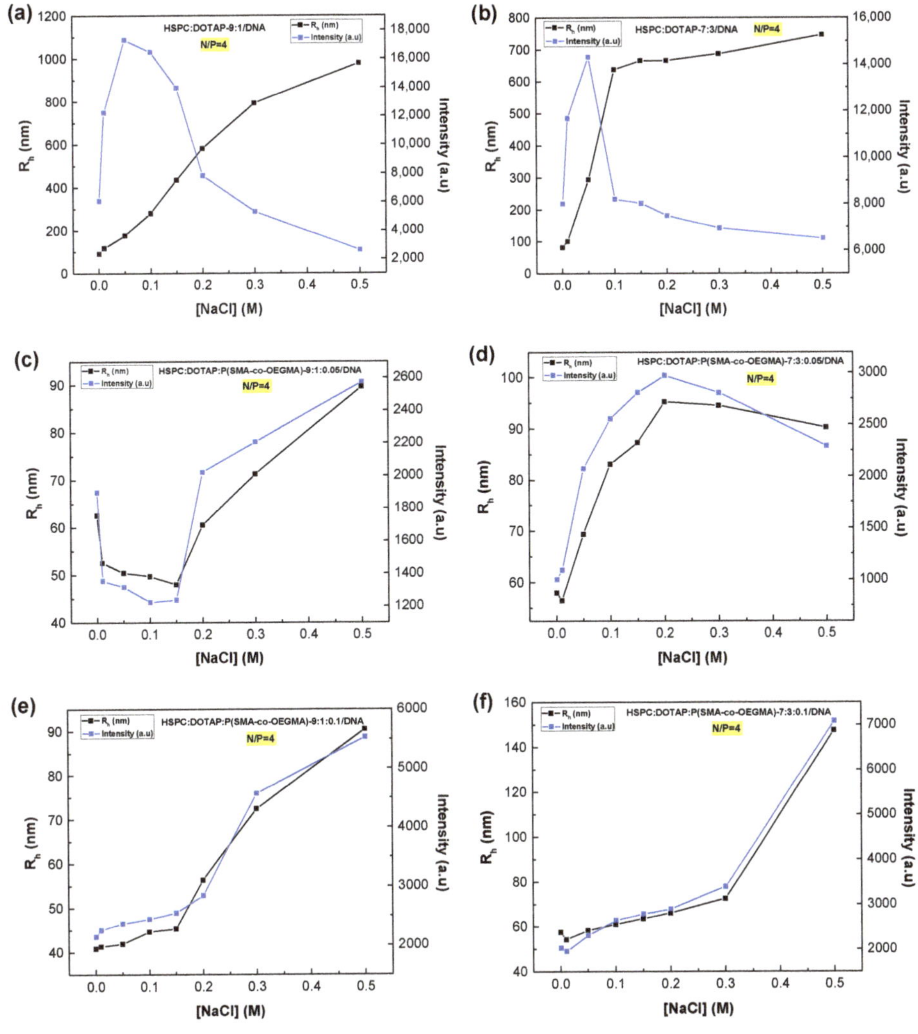

Figure 7. Hydrodynamic radius (R_h) and scattered light intensity as a function of ionic strength for (**a**) HSPC:DOTAP-9:1/DNA, (**b**) HSPC:DOTAP-7:3/DNA, (**c**) HSPC:DOTAP-9:1:0.05/DNA, (**d**) HSPC:DOTAP-7:3:0.05/DNA, (**e**) HSPC:DOTAP-9:1:0.1/DNA, and (**f**) HSPC:DOTAP-9:1:0.05/DNA lipoplexes at the ratio N/P = 4.

Overall, the chimeric lipoplexes were found to partially tolerate an increase in ionic strength, and to retain their complexation ability and colloidal stability under physiological salinity (equivalent to ca. 0.15 M NaCl). In contrast, even at low salt concentration, the pure lipoplexes exhibited large R_h and intensity values. Hence, the increase of ionic strength strongly affected the stability of the pure lipoplexes, resulting in their swelling and thereafter their decomposition.

In conclusion, the tolerance of the chimeric lipoplexes under the influence of the ionic strength is attributed to the presence of the copolymer, which noticeably contributed to lipoplexes stability. Particularly, the hydrophobic interactions provided by the long alkyl chains of the SMA segments, as well as the stealth/stabilization effect of the OEGMA segments, enhanced the colloidal stability of the chimeric lipoplexes and prevented their total collapse at higher salt concentrations. Moreover, the chimeric lipoplexes with the higher copolymer molar ratio (0.1) displayed smaller variations on their size and mass and thus exhibited a better stability profile under physiological salinity, compared to the chimeric lipoplexes with the lower copolymer molar ratio (0.05). Consequently, these findings evidence the efficient contribution of the amphiphilic copolymer in the lipoplex formulations.

3.6. *Morphological Characterization of Chimeric Liposomes and Lipoplexes via Cryo-TEM*

The morphology and the internal structure of the prepared chimeric liposomes as well as the chimeric lipoplexes were evaluated by Cryo-TEM. Representative Cryo-TEM images of the obtained nanostructures are provided in Figure 8. In the case of HSPC:DOTAP:P(SMA-co-OEGMA)-9:1:0.1 liposomes, the cryo-TEM micrograph (Figure 8a) evidenced the formation of vesicular structures of sizes varying from 20–290 nm with average size 58 nm (measured for 100 objects). Furthermore, the liposomes present unilamellar structure and circular shape (long white arrow in Figure 8a). The lipid membrane can be easily distinguished due to the different contrast between the periphery and the cavity of the liposomes. Moreover, the thickness of the membrane is 5 to 8 nm. It should be noted that the HSPC:DOTAP:P(SMA-co-OEGMA)-9:1:0.05 liposomes (data not shown) displayed similar morphology and structure, which indicates that the copolymer did not dramatically affect the morphological characteristics of the chimeric liposomes. Figure 8b depicts the lipoplexes formed by the electrostatic interaction of the HSPC:DOTAP:P(SMA-co-OEGMA)-9:1:0.1 liposomes with the DNA of 113 bp, at the N/P = 4 ratio. The formation of structures with morphology different than the precursor chimeric liposomes is observed. The sizes of lipoplexes range from 20 to 200 nm, with an average size of 73 nm (measured for 100 objects) and the thickness of their membrane is 5–7 nm. Furthermore, the outer surface of the lipoplexes presents a dot-like halo shape (long white arrow in Figure 8b) with several dark spots on the periphery of the membrane (short white arrow in Figure 8b). These dark spots of size 5–10 nm can be probably assigned to domains where DNA molecules are complexed, which may be visible due to the high contrast of the phosphorus atoms of DNA. Moreover, these spots are not detected in the precursor liposomes. This observation indicates the presence of DNA and thus the successful formation of lipoplexes. HSPC:DOTAP:P(SMA-co-OEGMA)-9:1:0.05/DNA lipoplexes presented the same dot-like membrane shape with the dark spots, as the one observed for the HSPC:DOTAP:P(SMA-co-OEGMA)-9:1:0.1/DNA lipoplexes (Figure 8b).

The chimeric liposomes with higher lipid molar ratio in DOTAP, namely the HSPC:DOTAP:P(SMA-co-OEGMA)-7:3:0.1 (Figure 8c), exhibited vesicular and unilamellar morphology with different shapes. Particularly, the co-existence of nanostructures with spherical (short white arrows in Figure 8c) and polygon-like (long white arrows in Figure 8c) shape is detected, in comparison to the HSPC:DOTAP:P(SMA-co-OEGMA)-9:1:0.1 liposomes which formed only spherical structures. The sizes of the liposomes were between 20 and 145 nm and average size of 66 nm (measured for 100 objects), while the thickness of the membrane was 5–8 nm. Furthermore, the HSPC:DOTAP:P(SMA-co-OEGMA)-7:3:0.05 liposomes presented the same spherical and polygon-like shape with the HSPC:DOTAP:P(SMA-co-OEGMA)-7:3:0.1 liposomes. It is also evident that the molar ratio of the copolymer did not affect the obtained liposomes morphology, in contrast to the molar ratio of DOTAP, which its increase caused differences in the liposomes shape. The HSPC:DOTAP:P(SMA-co-OEGMA)-7:3:0.1/DNA chimeric lipoplexes formed at the ratio N/P = 4 are presented in Figure 8d. The formation of mainly vesicular structures with thick lipid membranes is observed. Moreover, the lipoplexes exhibit mainly spherical shape and larger sizes of 30–265 nm (average size: 122 nm, measured for 100 objects), compared to their precursor liposomes. The thickness of the membrane for the lipoplexes ranges from 5–8 nm. Similar to the HSPC:DOTAP:P(SMA-co-OEGMA)-9:1:0.1/DNA lipoplexes, dark spots on the periphery of the lipid membrane of HSPC:DOTAP:P(SMA-co-OEGMA)-7:3:0.1/DNA lipoplexes are also observed. Moreover, these dark spots with sizes between 5–10 nm are more intense in some sites of the membrane surface (short white arrows). The high density of these spots maybe indicates that larger amount of DNA molecules is located/complexed on these sites.

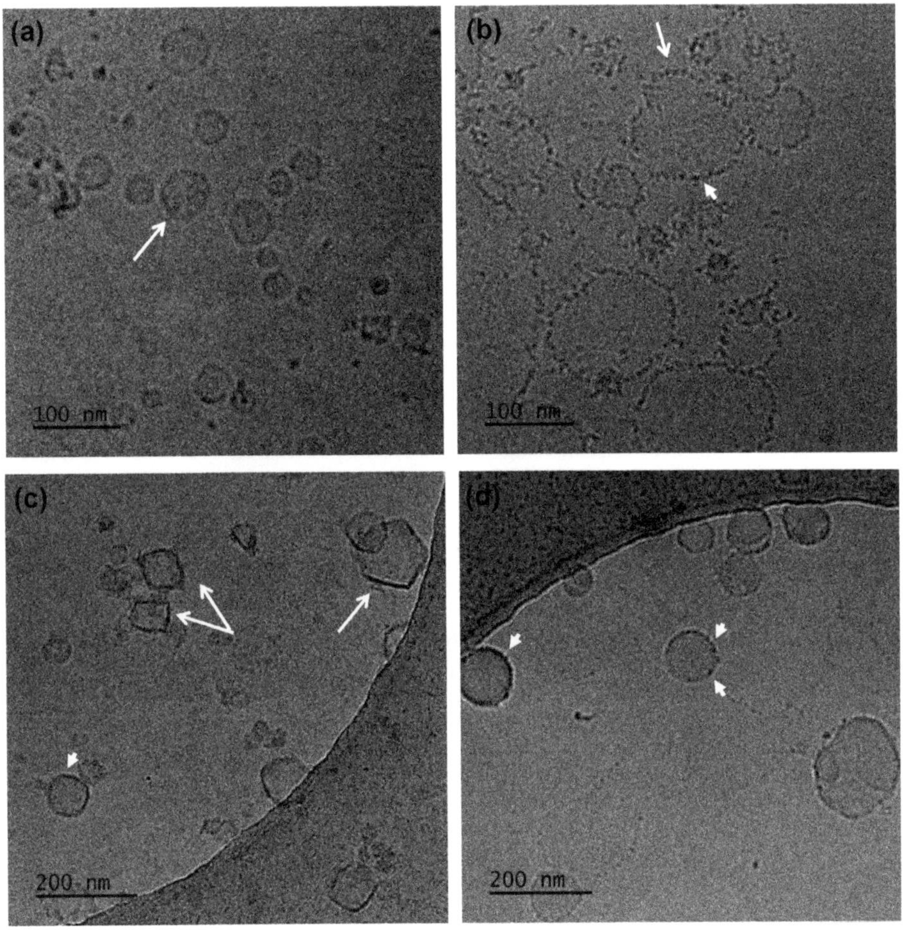

Figure 8. Representative Cryo-TEM images of (**a**) HSPC:DOTAP:P(SMA-co-OEGMA)-9:1:0.1 chimeric liposomes, (**b**) HSPC:DOTAP:P(SMA-co-OEGMA)-9:1:0.1/DNA chimeric lipoplexes at ratio N/P = 4, (**c**) HSPC:DOTAP:P(SMA-co-OEGMA)-7:3:0.1 chimeric liposomes, and (**d**) HSPC:DOTAP:P(SMA-co-OEGMA)-7:3:0.1/DNA chimeric lipoplexes at ratio N/P = 4.

Consequently, the morphological and structural characteristics of chimeric liposomes and lipoplexes are strongly dependent on the lipid ratio. Additionally, it should be noted that the obtained vesicular morphologies for most of the studied nanosystems by Cryo-TEM, are in accordance with the determined R_g/R_h ratios by light scattering. Hence, the lipoplexes, due to their small size and vesicular morphology, are expected to facilitate the delivery of DNA.

4. Conclusions

In this study, a novel hybrid-chimeric liposomal nanoplatform composed of the HSPC, DOTAP lipids, and the P(SMA-co-OEGMA) copolymer was developed and was further investigated as a potential nanocarrier for nucleic acid delivery. The implementation of RAFT polymerization technique resulted in the fabrication of a well-defined amphiphilic random copolymer P(SMA-co-OEGMA) with desirable and controllable molecular characteristics. The hybrid/chimeric liposomal nanosystems were successfully prepared by thin film method in several molar ratios of the copolymer and lipid components and compared to pure liposomes composed of HSPC and DOTAP lipids. Light scattering techniques revealed that all the prepared liposomes presented well-defined physicochemical characteristics. The chimeric liposomes as well as the pure liposomes were successfully explored for their ability to electrostatically interact with a model DNA of 113 bp, by forming lipoplexes. UV–Vis and especially fluorescence spectroscopy proved the strong binding affinity of the cationic liposomes to DNA and the formation of lipoplexes, by the total displacement of the intercalated EtBr from the double helix of DNA, in the latter studies. DLS, ELS, and SLS measurements proved that nanosized pure/chimeric lipoplexes were formed in a wide range of N/P ratios. The findings proved that the utilization of OEGMA is a valuable alternative to PEG and can be utilized as an approach to reduce possible cytotoxicity effects due to the incorporated cationic lipids. The stability of the lipoplexes under the influence of increasing solution ionic strength, as a preliminary simulation of the physiological conditions of the biological fluids, was investigated by dynamic light scattering. The chimeric lipoplexes were found to tolerate the increase of the ionic strength and to retain their colloidal stability and complexation state under physiological salinity. In contrast, the pure lipoplexes presented large sizes, indicating their disintegration and destabilization in the presence of salt, even at low salt concentrations. Consequently, the amphiphilic copolymer efficiently contributed and enhanced the stabilization of the chimeric lipoplexes in the presence of salt. Cryo-TEM images revealed the formation of vesicular nanostructures, with their shape depending on the lipid ratio. The presence of DNA molecules on the periphery of the lipid membrane evidenced the successful formation of lipoplexes, with their shape also depending on the lipid ratio.

In conclusion, the fabrication of a novel chimeric liposomal nanosystem in different molar ratios of its components and the interaction with DNA resulted in colloidally stable chimeric lipid-copolymer lipoplexes with well-defined physicochemical characteristics. Their size, surface charge, and morphology are strongly depending on the N/P ratio and particularly on the molar ratio of each component. The contribution of the copolymer on the well-defined characteristics was instrumental. The significant progress in polymer chemistry has opened the access to the design and fabrication of advanced polymeric materials with well-defined characteristics and desirable features. With the present study, we highlight the importance of mixing different materials. i.e., polymers and lipids. The synergistic effect of different in nature materials paves the road on the exploration and development of multifunctional nanocarriers as a new promising therapeutic approach in gene therapy.

Supplementary Materials: The following supporting information can be downloaded at: https://www.mdpi.com/article/10.3390/polym14224901/s1, Figure S1: UV–vis absorption spectra of (a) HSPC:DOTAP-9:1/DNA, (b) HSPC:DOTAP-7:3/DNA, (c) HSPC:DOTAP:P(SMA-co-OEGMA)-9:1:0.05/DNA, (d) HSPC:DOTAP:P(SMA-co-OEGMA)-7:3:0.05/DNA, (e) HSPC:DOTAP:P(SMA-co-OEGMA)-9:1:0.1/DNA, (f) HSPC:DOTAP:P(SMA-co-OEGMA)-7:3:0.1/DNA lipoplexes, for N/P ratios ranging from 0.25 to 8, Figure S2: UV–Vis spectra of pure HSPC:DOTAP and chimeric HSPC:DOTAP:P(SMA-co-OEGMA) liposomes in aqueous media, Figure S3: Fluorescence spectra of the intercalated EtBr for (a) HSPC:DOTAP-9:1/DNA, (b) HSPC:DOTAP-7:3/DNA, (c) HSPC:DOTAP:P(SMA-co-OEGMA)-9:1:0.05/DNA, (d) HSPC:DOTAP:P(SMA-co-OEGMA)-7:3:0.05/DNA, (e) HSPC:DOTAP:P(SMA-co-OEGMA)-9:1:0.1/DNA, and (f) HSPC:DOTAP:P(SMA-co-OEGMA)-7:3:0.1/DNA lipoplexes at different N/P ratios, Figure S4. Size distributions from CONTIN analysis of HSPC:DOTAP-9:1/DNA lipoplexes, at ratios (a) N/P = 0.25, (b) N/P = 0.5, (c) N/P = 1 (partial precipitation takes place at

N/P = 1, data shown are from supernatant solution), (d) N/P = 2, (e) N/P = 4, and (f) N/P = 8, Figure S5. Size distributions from CONTIN analysis of HSPC:DOTAP-9:1:0.05/DNA chimeric lipoplexes, at ratios (a) N/P = 0.25, (b) N/P = 0.5, (c) N/P = 1 (partial precipitation takes place at N/P = 1, data shown are from supernatant solution), (d) N/P = 2, (e) N/P = 4, and (f) N/P = 8, Figure S6. Size distributions from CONTIN analysis of HSPC:DOTAP-9:1:0.1/DNA chimeric lipoplexes, at ratios (a) N/P = 0.25, (b) N/P = 0.5, (c) N/P = 1 (partial precipitation takes place at N/P = 1, data shown are from supernatant solution), (d) N/P = 2, (e) N/P = 4 and (f) N/P = 8, Figure S7. Size distributions from CONTIN analysis of HSPC:DOTAP-7:3/DNA lipoplexes, at ratios (a) N/P = 0.25, (b) N/P = 0.5, (c) N/P = 1 (partial precipitation takes place at N/P = 1, data shown are from supernatant solution), (d) N/P = 2, (e) N/P = 4, and (f) N/P = 8, Figure S8. Size distributions from CONTIN analysis of HSPC:DOTAP-7:3:0.05/DNA chimeric lipoplexes, at ratios (a) N/P = 0.25, (b) N/P = 0.5, (c) N/P = 1 (partial precipitation takes place at N/P = 1, data shown are from supernatant solution), (d) N/P = 2, (e) N/P = 4, and (f) N/P = 8, Figure S9. Size distributions from CONTIN analysis of HSPC:DOTAP-7:3:0.1/DNA chimeric lipoplexes, at ratios (a) N/P = 0.25, (b) N/P = 0.5, (c) N/P = 1 (partial precipitation takes place at N/P = 1, data shown are from supernatant solution), (d) N/P = 2, (e) N/P = 4, and (f) N/P = 8.

Author Contributions: Conceptualization, S.P.; methodology, S.P., C.D. and B.T.; validation, V.C.; formal analysis, V.C. and A.F.; investigation, V.C. and A.F.; resources, S.P., C.D. and B.T.; data curation, V.C. and A.F.; writing—original draft preparation, V.C.; writing—review and editing, V.C., A.F., B.T., C.D. and S.P.; supervision, S.P. and C.D.; funding acquisition, V.C., C.D. and S.P. All authors have read and agreed to the published version of the manuscript.

Funding: This research work was supported by the Hellenic Foundation for Research and Innovation (HFRI) under the HFRI PhD Fellowship grant (Fellowship Number: 907).

Institutional Review Board Statement: Not applicable.

Informed Consent Statement: Not applicable.

Data Availability Statement: The data presented in this study are available upon request.

Conflicts of Interest: The authors declare no conflict of interest.

References

1. Sayed, N.; Allawadhi, P.; Khurana, A.; Singh, V.; Navik, U.; Pasumarthi, S.K.; Khurana, I.; Banothu, A.K.; Weiskirchen, R.; Bharani, K.K. Gene therapy: Comprehensive overview and therapeutic applications. *Life Sci.* **2022**, *294*, 120375. [CrossRef] [PubMed]
2. Kumar, S.R.; Markusic, D.M.; Biswas, M.; High, K.A.; Herzog, R.W. Clinical development of gene therapy: Results and lessons from recent successes. *Mol. Ther. Methods Clin. Dev.* **2016**, *3*, 16034. [CrossRef] [PubMed]
3. Hardee, C.L.; Arévalo-Soliz, L.M.; Hornstein, B.D.; Zechiedrich, L. Advances in Non-Viral DNA Vectors for Gene Therapy. *Genes* **2017**, *8*, 65. [CrossRef] [PubMed]
4. Liu, C.; Zhang, N. Chapter 13-Nanoparticles in gene therapy: Principles, prospects, and challenges. In *Nanoparticles in Translational Science and Medicine*; Villaverde, A., Ed.; Academic Press: Cambridge, MA, USA, 2011; Volume 104, pp. 509–562. [CrossRef]
5. Bono, N.; Ponti, F.; Mantovani, D.; Candiani, G. Non-viral in vitro gene delivery: It is now time to set the bar! *Pharmaceutics* **2020**, *12*, 183. [CrossRef]
6. Ghidini, M.; Silva, S.G.; Evangelista, J.; do Vale, M.L.C.; Farooqi, A.A.; Pinheiro, M. Nanomedicine for the Delivery of RNA in Cancer. *Cancers* **2022**, *14*, 2677. [CrossRef]
7. Lin, G.; Li, L.; Panwar, N.; Wang, J.; Tjin, S.C.; Wang, X.; Yong, K.T. Non-viral gene therapy using multifunctional nanoparticles: Status, challenges, and opportunities. *Coord. Chem. Rev.* **2018**, *374*, 133–152. [CrossRef]
8. Mintzer, M.A.; Simanek, E.E. Nonviral Vectors for Gene Delivery. *Chem. Rev.* **2009**, *109*, 259–302. [CrossRef]
9. Safinya, C.R.; Ewert, K.K.; Majzoub, R.N.; Leal, C. Cationic liposome-nucleic acid complexes for gene delivery and gene silencing. *New J. Chem.* **2014**, *38*, 5164–5172. [CrossRef]
10. Luiz, M.T.; Dutra, J.A.P.; Tofani, L.B.; de Araújo, J.T.C.; Di Filippo, L.D.; Marchetti, J.M.; Chorilli, M. Targeted Liposomes: A Nonviral Gene Delivery System for Cancer Therapy. *Pharmaceutics* **2022**, *14*, 821. [CrossRef]
11. Wahane, A.; Waghmode, A.; Kapphahn, A.; Dhuri, K.; Gupta, A.; Bahal, R. Role of Lipid-Based and Polymer-Based Non-Viral Vectors in Nucleic Acid Delivery for Next-Generation Gene Therapy. *Molecules* **2020**, *25*, 2866. [CrossRef]
12. Barba, A.A.; Bochicchio, S.; Dalmoro, A.; Lamberti, G. Lipid Delivery Systems for Nucleic-Acid-Based-Drugs: From Production to Clinical Applications. *Pharmaceutics* **2019**, *11*, 360. [CrossRef] [PubMed]
13. Ma, K.; Mi, C.L.; Cao, X.X.; Wang, T.Y. Progress of cationic gene delivery reagents for non-viral vector. *Appl. Microbiol. Biotechnol.* **2021**, *105*, 525–538. [CrossRef] [PubMed]

14. Zhang, S.; Xu, Y.; Wang, B.; Qiao, W.; Liu, D.; Li, Z. Cationic compounds used in lipoplexes and polyplexes for gene delivery. *J. Control. Release* **2004**, *100*, 165–180. [CrossRef] [PubMed]
15. Zhi, D.; Bai, Y.; Yang, J.; Cui, S.; Zhao, Y.; Chen, H.; Zhang, S. A review on cationic lipids with different linkers for gene delivery. *Adv. Colloid Interface Sci.* **2018**, *253*, 117–140. [CrossRef] [PubMed]
16. Ponti, F.; Campolungo, M.; Melchiori, C.; Bono, N.; Candiani, G. Cationic lipids for gene delivery: Many players, one goal. *Chem. Phys. Lipids* **2021**, *235*, 105032. [CrossRef]
17. Zhou, Z.; Liu, X.; Zhu, D.; Wang, Y.; Zhang, Z.; Zhou, X.; Qiu, N.; Chen, X.; Shen, Y. Nonviral cancer gene therapy: Delivery cascade and vector nanoproperty integration. *Adv. Drug Deliv. Rev.* **2017**, *115*, 115–154. [CrossRef]
18. Keles, E.; Song, Y.; Du, D.; Dong, W.J.; Lin, Y. Recent progress in nanomaterials for gene delivery applications. *Biomater. Sci.* **2016**, *4*, 1291–1309. [CrossRef]
19. Zhang, X.X.; McIntosh, T.J.; Grinstaff, M.W. Functional lipids and lipoplexes for improved gene delivery. *Biochimie* **2012**, *94*, 42–58. [CrossRef]
20. De Ilarduya, C.T.; Sun, Y.; Düzgünes, N. Gene delivery by lipoplexes and polyplexes. *Eur. J. Pharm. Sci.* **2010**, *40*, 159–170. [CrossRef]
21. Masotti, A.; Mossa, G.; Cametti, C.; Ortaggi, G.; Bianco, A.; Del Grosso, N.; Malizia, D.; Esposito, C. Comparison of different commercially available cationic liposome-DNA lipoplexes: Parameters influencing toxicity and transfection efficiency. *Colloids Surf. B Biointerfaces* **2009**, *68*, 136–144. [CrossRef]
22. Kim, B.K.; Hwang, G.B.; Seu, Y.B.; Choi, J.S.; Jin, K.S.; Doh, K.O. DOTAP/DOPE ratio and cell type determine transfection efficiency with DOTAP-liposomes. *Biochim. Biophys. Acta-Biomembr.* **2015**, *1848*, 1996–2001. [CrossRef] [PubMed]
23. Buck, J.; Mueller, D.; Mettal, U.; Ackermann, M.; Grisch-Chan, H.M.; Thöny, B.; Zumbuehl, A.; Huwyler, J.; Witzigmann, D. Improvement of DNA Vector Delivery of DOTAP Lipoplexes by Short-Chain Aminolipids. *ACS Omega* **2020**, *5*, 24724–24732. [CrossRef] [PubMed]
24. Cui, S.; Wang, Y.; Gong, Y.; Lin, X.; Zhao, Y.; Zhi, D.; Zhou, Q.; Zhang, S. Correlation of the cytotoxic effects of cationic lipids with their headgroups. *Toxicol. Res.* **2018**, *7*, 473–479. [CrossRef] [PubMed]
25. Lv, H.; Zhang, S.; Wang, B.; Cui, S.; Yan, J. Toxicity of cationic lipids and cationic polymers in gene delivery. *J. Control. Release* **2006**, *114*, 100–109. [CrossRef] [PubMed]
26. Mehta, S.; Suresh, A.; Nayak, Y.; Narayan, R.; Nayak, U.Y. Hybrid nanostructures: Versatile systems for biomedical applications. *Coord. Chem. Rev.* **2022**, *460*, 214482. [CrossRef]
27. Go, Y.K.; Leal, C. Polymer–lipid hybrid materials. *Chem. Rev.* **2021**, *121*, 13996–14030. [CrossRef]
28. Bochicchio, S.; Lamberti, G.; Barba, A.A. Polymer–Lipid Pharmaceutical Nanocarriers: Innovations by New Formulations and Production Technologies. *Pharmaceutics* **2021**, *13*, 198. [CrossRef]
29. Persano, F.; Gigli, G.; Leporatti, S. Lipid-polymer hybrid nanoparticles in cancer therapy: Current overview and future directions. *Nano Express* **2021**, *2*, 012006. [CrossRef]
30. Niculescu, A.-G.; Bîrcă, A.C.; Grumezescu, A.M. New Applications of Lipid and Polymer-Based Nanoparticles for Nucleic Acids Delivery. *Pharmaceutics* **2021**, *13*, 2053. [CrossRef]
31. Dave, V.; Tak, K.; Sohgaura, A.; Gupta, A.; Sadhu, V.; Reddy, K.R. Lipid-Polymer Hybrid Nanoparticles: Synthesis Strategies and Biomedical Applications. *J. Microbiol. Methods* **2019**, *160*, 130–142. [CrossRef]
32. Hadinoto, K.; Sundaresan, A.; Cheow, W.S. Lipid–polymer hybrid nanoparticles as a new generation therapeutic delivery platform: A review. *Eur. J. Pharm. Biopharm.* **2013**, *85*, 427–443. [CrossRef] [PubMed]
33. Byun, M.J.; Lim, J.; Kim, S.N.; Park, D.H.; Kim, T.H.; Park, W.; Park, C.G. Advances in Nanoparticles for Effective Delivery of RNA Therapeutics. *BioChip J.* **2022**, *16*, 128–145. [CrossRef] [PubMed]
34. Wang, C.; Shi, X.; Song, H.; Zhang, C.; Wang, X.; Huang, P.; Dong, A.; Zhang, Y.; Kong, D.; Wang, W. Polymer-lipid hybrid nanovesicle-enabled combination of immunogenic chemotherapy and RNAi-mediated PD-L1 knockdown elicits antitumor immunity against melanoma. *Biomaterials* **2021**, *268*, 120579. [CrossRef] [PubMed]
35. Pippa, N.; Stellas, D.; Skandalis, A.; Pispas, S.; Demetzos, C.; Libera, M.; Marcinkowski, A.; Trzebicka, B. Chimeric lipid/block copolymer nanovesicles: Physico-chemical and bio-compatibility evaluation. *Eur. J. Pharm. Biopharm.* **2016**, *107*, 295–309. [CrossRef]
36. Simberg, D.; Weisman, S.; Talmon, Y.; Barenholz, Y. DOTAP (and other cationic lipids): Chemistry, biophysics and transfection. *Crit. Rev. Ther. Drug Carrier Syst* **2004**, *21*, 257–317. [CrossRef]
37. Balazs, D.A.; Godbey, W.T. Liposomes for use in gene delivery. *J. Drug Deliv.* **2011**, *2011*, 326497. [CrossRef]
38. Choi, Y.S.; Lee, M.Y.; David, A.E.; Park, Y.S. Nanoparticles for gene delivery: Therapeutic and toxic effects. *Mol. Cell. Toxicol.* **2014**, *10*, 1–8. [CrossRef]
39. Yan, Y.; Liu, X.Y.; Lu, A.; Wang, X.Y.; Jiang, L.X.; Wang, J.C. Non-viral vectors for RNA delivery. *J. Control. Release* **2022**, *342*, 241–279. [CrossRef]
40. Cheng, X.; Lee, R.J. The role of helper lipids in lipid nanoparticles (LNPs) designed for oligonucleotide delivery. *Adv. Drug Deliv. Rev.* **2016**, *99*, 129–137. [CrossRef]
41. Rose, V.L.; Winkler, G.S.; Allen, S.; Puri, S.; Mantovani, G. Polymer siRNA conjugates synthesised by controlled radical polymerisation. *Eur. Polym. J.* **2013**, *49*, 2861–2883. [CrossRef]
42. Ahmed, M.; Narain, R. Progress of RAFT based polymers in gene delivery. *Prog. Polym. Sci.* **2013**, *38*, 767–790. [CrossRef]

43. Chrysostomou, V.; Katifelis, H.; Gazouli, M.; Dimas, K.; Demetzos, C.; Pispas, S. Hydrophilic Random Cationic Copolymers as Polyplex-Formation Vectors for DNA. *Materials* **2022**, *15*, 2650. [CrossRef] [PubMed]
44. Qin, S.; Matyjaszewski, K.; Xu, H.; Sheiko, S.S. Synthesis and visualization of densely grafted molecular brushes with crystallizable poly (octadecyl methacrylate) block segments. *Macromolecules* **2003**, *36*, 605–612. [CrossRef]
45. Semsarilar, M.; Penfold, N.J.; Jones, E.R.; Armes, S.P. Semi-crystalline diblock copolymer nano-objects prepared via RAFT alcoholic dispersion polymerization of stearyl methacrylate. *Polym. Chem.* **2015**, *6*, 1751–1757. [CrossRef]
46. Sentoukas, T.; Demetzos, C.; Pispas, S. Chimeric liposomes incorporating functional copolymers: Preparation and pH/thermo-responsive behaviour in aqueous solutions. *J. Liposome Res.* **2021**, *31*, 279–290. [CrossRef]
47. Amado, E.; Blume, A.; Kressler, J. Novel non-ionic block copolymers tailored for interactions with phospholipids. *React. Funct. Polym.* **2009**, *69*, 450–456. [CrossRef]
48. Naziris, N.; Pippa, N.; Chrysostomou, V.; Pispas, S.; Demetzos, C.; Libera, M.; Trzebicka, B. Morphological diversity of block copolymer/lipid chimeric nanostructures. *J. Nanopart. Res.* **2017**, *19*, 347–357. [CrossRef]
49. Vassiliadou, O.; Chrysostomou, V.; Pispas, S.; Klonos, P.A.; Kyritsis, A. Molecular dynamics and crystallization in polymers based on ethylene glycol methacrylates (EGMAs) with melt memory characteristics: From linear oligomers to comb-like polymers. *Soft Matter* **2021**, *17*, 1284–1298. [CrossRef]
50. McNeil, S.E.; Perrie, Y. Gene delivery using cationic liposomes. *Expert Opin. Ther. Pat.* **2006**, *16*, 1371–1382. [CrossRef]
51. Hoang Thi, T.T.; Pilkington, E.H.; Nguyen, D.H.; Lee, J.S.; Park, K.D.; Truong, N.P. The Importance of Poly(ethylene glycol) Alternatives for Overcoming PEG Immunogenicity in Drug Delivery and Bioconjugation. *Polymers* **2020**, *12*, 298. [CrossRef]
52. Kalinova, R.; Valchanova, M.; Dimitrov, I.; Turmanova, S.; Ugrinova, I.; Petrova, M.; Vlahova, Z.; Rangelov, S. Functional Polyglycidol-Based Block Copolymers for DNA Complexation. *Int. J. Mol. Sci.* **2021**, *22*, 9606. [CrossRef] [PubMed]
53. Athmakuri, K.; Padala, C.; Litt, J.; Cole, R.; Kumar, S.; Kane, R.S. Controlling DNA adsorption and diffusion on lipid bilayers by the formation of lipid domains. *Langmuir* **2010**, *26*, 397–401. [CrossRef] [PubMed]
54. Chrysostomou, V.; Forys, A.; Trzebicka, B.; Demetzos, C.; Pispas, S. Structure of micelleplexes formed between QPDMAEMA-b-PLMA amphiphilic cationic copolymer micelles and DNA of different lengths. *Eur. Polym. J.* **2022**, *166*, 111048. [CrossRef]
55. Haladjova, E.; Mountrichas, G.; Pispas, S.; Rangelov, S. Poly (vinyl benzyl trimethylammonium chloride) Homo and Block Copolymers Complexation with DNA. *J. Phys. Chem. B* **2016**, *120*, 2586–2595. [CrossRef]
56. Moad, G. RAFT polymerization to form stimuli-responsive polymers. *Polym. Chem.* **2017**, *8*, 177–219. [CrossRef]
57. Kafetzi, M.; Pispas, S. Multifaceted pH and Temperature Induced Self-Assembly of P(DMAEMA-co-LMA)-b-POEGMA Terpolymers and Their Cationic Analogues in Aqueous Media. *Macromol. Chem. Phys.* **2021**, *222*, 2000358. [CrossRef]
58. Skandalis, A.; Pispas, S. PLMA-b-POEGMA amphiphilic block copolymers: Synthesis and self-assembly in aqueous media. *J. Polym. Sci. A* **2017**, *55*, 155–163. [CrossRef]
59. Pei, Y.; Thurairajah, L.; Sugita, O.R.; Lowe, A.B. RAFT dispersion polymerization in nonpolar media: Polymerization of 3-phenylpropyl methacrylate in n-tetradecane with poly (stearyl methacrylate) homopolymers as macro chain transfer agents. *Macromolecules* **2015**, *48*, 236–244. [CrossRef]
60. Haladjova, E.; Chrysostomou, V.; Petrova, M.; Ugrinova, I.; Pispas, S.; Rangelov, S. Physicochemical properties and biological performance of polymethacrylate based gene delivery vector systems: Influence of amino functionalities. *Macromol. Biosci.* **2020**, *21*, 2000352. [CrossRef]
61. Coates, J. Interpretation of Infrared Spectra, A practical Approach. In *Encyclopedia of Analytical Chemistry*; Meyers, R.A., Ed.; John Wiley & Sons Ltd.: Chichester, UK, 2000; pp. 10815–10837.
62. Cao, X.T.; Nguyen, V.C.; Nguyen, T.D.; Doan, V.D.; Tu, T.K.T.; Lim, K.T. Ketal core cross-linked micelles for pH-triggered release of doxorubicin. *Mol. Cryst. Liq. Cryst.* **2020**, *707*, 29–37. [CrossRef]
63. Burchard, W. Static and dynamic light scattering from branched polymers and biopolymers. In *Advances in Polymer Science*; Springer: Berlin/Heidelberg, Germany, 1983; Volume 48, pp. 1–124. [CrossRef]
64. Pippa, N.; Kaditi, E.; Pispas, S.; Demetzos, C. PEO-b-PCL-DPPC chimeric nanocarriers: Self-assembly aspects in aqueous and biological media and drug incorporation. *Soft Matter* **2013**, *9*, 4073–4082. [CrossRef]
65. Izumrudov, V.A.; Zhiryakova, M.V.; Goulko, A.A. Ethidium bromide as a promising probe for studying DNA interaction with cationic amphiphiles and stability of the resulting complexes. *Langmuir* **2002**, *18*, 10348–10356. [CrossRef]
66. Geall, A.J.; Blagbrough, I.S. Rapid and sensitive ethidium bromide fluorescence quenching assay of polyamine conjugate–DNA interactions for the analysis of lipoplex formation in gene therapy. *J. Pharm. Biomed. Anal.* **2000**, *22*, 849–859. [CrossRef]

polymers

Article

Functional Nanogel from Natural Substances for Delivery of Doxorubicin †

Katya Kamenova [1], Lyubomira Radeva [2], Krassimira Yoncheva [2], Filip Ublekov [1], Martin A. Ravutsov [3], Maya K. Marinova [3], Svilen P. Simeonov [3,4], Aleksander Forys [5], Barbara Trzebicka [5] and Petar D. Petrov [1,*]

1 Institute of Polymers, Bulgarian Academy of Sciences, 1113 Sofia, Bulgaria
2 Department of Pharmaceutical Technology and Biopharmaceutics, Faculty of Pharmacy, Medical University of Sofia, 1000 Sofia, Bulgaria
3 Institute of Organic Chemistry with Centre of Phytochemistry, Bulgarian Academy of Sciences, 1113 Sofia, Bulgaria
4 Research Institute for Medicines (iMed.ULisboa), Faculty of Pharmacy, Universidade de Lisboa, 1649-003 Lisbon, Portugal
5 Centre of Polymer and Carbon Materials, Polish Academy of Sciences, 41-819 Zabrze, Poland
* Correspondence: ppetrov@polymer.bas.bg; Tel.: +359-2-9796335
† In Memory of Professor Andrzej Dworak.

Abstract: Nanogels (NGs) have attracted great attention because of their outstanding biocompatibility, biodegradability, very low toxicity, flexibility, and softness. NGs are characterized with a low and nonspecific interaction with blood proteins, meaning that they do not induce any immunological responses in the body. Due to these properties, NGs are considered promising candidates for pharmaceutical and biomedical application. In this work, we introduce the development of novel functional nanogel obtained from two naturally based products—citric acid (CA) and pentane-1,2,5-triol (PT). The nanogel was synthesized by precipitation esterification reaction of CA and PT in tetrahydrofuran using N-ethyl-N′-(3-dimethylaminopropyl) carbodiimide (EDC) and 4-(dimethylamino)pyridine (DMAP) catalyst system. Dynamic light scattering (DLS), cryogenic transmission electron microscopy (cryo-TEM) and atomic force microscopy (AFM) analyses revealed formation of spherical nanogel particles with a negative surface charge. Next, the nanogel was loaded with doxorubicin hydrochloride (DOX) by electrostatic interactions between carboxylic groups present in the nanogel and amino groups of DOX. The drug-loaded nanogel exhibited high encapsulation efficiency (EE~95%), and a bi-phasic release behavior. Embedding DOX into nanogel also stabilized the drug against photodegradation. The degradability of nanogel under acidic and neutral conditions with time was investigated as well.

Keywords: nanogels; nanocarriers; citric acid; pentane-1,2,5-triol; doxorubicin; drug delivery

Citation: Kamenova, K.; Radeva, L.; Yoncheva, K.; Ublekov, F.; Ravutsov, M.A.; Marinova, M.K.; Simeonov, S.P.; Forys, A.; Trzebicka, B.; Petrov, P.D. Functional Nanogel from Natural Substances for Delivery of Doxorubicin. *Polymers* **2022**, *14*, 3694. https://doi.org/10.3390/polym14173694

Academic Editor: Dimitrios Bikiaris

Received: 11 August 2022
Accepted: 2 September 2022
Published: 5 September 2022

1. Introduction

Nanogels are hydrogel particles formed by physically or chemically crosslinked three-dimensional polymer networks with dimensions typically between 10 and 200 nm [1]. They possess the typical highly hydrated nature and shrinking/swelling properties of hydrogels under different environmental conditions [2]. In recent years, nanogels have attracted great attention because of their outstanding biocompatibility, biodegradability, large specific surface area, very low toxicity, flexibility, and softness. They are characterized with a low and nonspecific interaction with blood proteins, meaning that they do not induce any immunological responses in the body. This fact makes nanogels promising candidates for pharmaceutical and biomedical applications. NGs have high stability, small size, high loading capacity and responsiveness to environmental factors, such as pH, temperature, and ionic strength, which is essential for modern nanomedicine. Their porous 3D structure allows the encapsulation of hydrophobic and/or hydrophilic drugs, potentially protecting the bioactive substances (BAS) from degradation during storage or blood circulation (e.g., hydrolysis or enzymolysis), and reduces toxic side effects. The loading of BAS in NGs can be achieved by physical interactions between the drug molecules and functional groups existing in NGs and usually happens spontaneously through hydrogen bonding, electrostatic, van der Waals and/or hydrophobic interactions. The physicochemical properties of NGs (surface charge, size, crosslinking density, and softness) can be adjusted by playing with the synthetic conditions, including monomer selection, monomer/crosslinker ratio, concentration, etc. [3–6].

Nanogels have been fabricated by polymerization of monomers (in a homogeneous phase, precipitation polymerization, micro-template polymerization, emulsion polymerization, emulsifier-free emulsion polymerization, dispersion polymerization, etc.) or by crosslinking of preformed polymers [6]. The most common method for the synthesis of NGs is the microemulsion polymerization in which the gel is obtained by combining an appropriate amount of water, oil, and surfactant(s). However, the nanogels synthesized in this way may contain large amounts of organic solvents and surfactant residues that are difficult to separate from the system and can lead to toxic reactions [2]. Soap-free emulsion polymerization is one of the eco-friendly methods by which narrowly dispersed nanogels can be obtained without the addition of surfactant [7]. Unfortunately, this technology is difficult to implement in the industry because of low polymerization rates and poor stability of the obtained emulsion. Precipitation polymerization is an alternative to the emulsion polymerization, having the advantages of a homogeneous reaction mixture (the monomer(s), crosslinking agent and initiator are homogeneously dissolved in the reaction medium before the reaction). When the polymerization/crosslinking reaction begins, at a certain chain length, the generated phase becomes insoluble, and separates form the solvent to form polymer colloidal particles that are precursors of nanogels [8].

The use of natural precursors for preparing polymeric carriers, including nanogels, is a beneficial strategy for developing drug delivery systems which are non-toxic, biodegradable, and biocompatible. Indeed, NGs based on chitosan [9], alginate/keratin [10], and alginate/gelatin [11] have been synthesized and assessed as drug carriers. pH-responsive alginate nanogel carriers of the anti-cancer drug DOX, obtained through in situ crosslinking with ionic calcium under ultrasound, showed a significantly higher accumulative release at pH 5.0 than at pH 7.4 [12]. Gyawali et al. developed photo-cross-linkable nanogels from a biodegradable polymer template with intrinsic photoluminescence and high photostability [13]. These fluorescent nanogels displayed excellent biodegradability and cytocompatibility owing to the biocompatible monomers citric acid, maleic acid, L-cysteine, and poly(ethylene glycol). The nanogels were further surface-functionalized with biologically active RGD peptides and loaded with DOX, resulting in a pH-responsive system capable of releasing the drug in acidic pH, resembling tumor environments.

Citric acid is a safe, nontoxic, low cost, water-soluble, UV-resistant, and biocompatible multifunctional reagent and it is generally regarded as safe (GRAS) by the US Food and Drug Administration (FDA) [14]. Owing to its three carboxylic groups, CA has been used as a covalent crosslinker of cellulose derivatives [15]. Recently, the synthesis of highly elastic super-macroporous cryogels fabricated by thermally induced crosslinking of 2-hydroxyethylcellulose (HEC) with citric acid was reported [16]. The polymer network was formed at elevated temperature by successive reactions of CA-based anhydride intermediates with HEC hydroxyl groups in bulk. An alternative mechanism of forming ester bonds under mild conditions is the Steglich esterification reaction using 1-ethyl-3-(3-dimethylaminopropyl) carbodiimide hydrochloride (EDC). For instance, Steglich esterification has been exploited for preparing hydrogels from alginate, collagen, and chitosan [17–19].

Doxorubicin is an anthracycline that is an important drug for treating various types of tumors [20,21]. Some of the main limitations regarding the use of doxorubicin include cardiotoxicity, tumor cell resistance, hydrolytic and photolytic degradation. [22,23]. A prospective strategy to overcome such problems is the development of drug delivery systems that can protect DOX from environmental factors and to provide its controlled delivery to the tumor tissue [24–26]. In this paper, we describe the synthesis of novel nanogel from two natural products—citric acid and pentane-1,2,5-triol. Pentane-1,2,5-triol is a renewable chemical derived from the hydrogenation of furfural, a by-product of lignocellulose processing [27]. CA and PT were crosslinked by the Steglich esterification precipitation reaction in THF at room temperature with N-ethyl-N′-(3-dimethylaminopropyl) carbodiimide and 4-dimethyl aminopyridine as a catalyst system. To the best of our knowledge, this is the first report describing fabrication of nanogels from CA and PT. Particle size and shape, size distribution, and surface charge of the nanogel were determined by using DLS, AFM and cryo-TEM. Next, the nanogel was loaded with the model drug DOX via electrostatic interactions between carboxylic groups present in the carrier and amino groups of DOX. The drug-loading capacity, encapsulation efficiency, release profile, biodegradability of nanogel and protecting effect towards photodegradation of DOX were investigated as well.

2. Materials and Methods

2.1. Materials

Citric acid (99.5%, Sigma-Aldrich, FOT, Sofia, Bulgaria), 4-(dimethylamino)pyridine (99%, Sigma-Aldrich, FOT, Sofia, Bulgaria), N-(3-dimethylaminopropyl)-N′-ethylcarbodiimide hydrochloride (for synthesis, Sigma-Aldrich, FOT, Sofia, Bulgaria), Furfuryl alcohol (98%, Sigma Aldrich, FOT, Sofia, Bulgaria), and doxorubicin hydrochloride (98.0–102.0%, Sigma-Aldrich, FOT, Sofia, Bulgaria) were used as received. Tetrahydrofuran (HPLC grade, Fisher Chemical, Labimex, Sofia, Bulgaria) were stirred overnight over calcium hydride and distilled prior to use.

2.2. Synthesis of Pentane-1,2,5-triol

The synthesis of PT was described elsewhere [28]. Briefly, furfuryl alcohol (20 g), dissolved in acetonitrile (200 mL), catalyst TS-1 (2 g) and H_2O_2 (30 mL; 37%) were reacted at 40 °C for 5 h. After work up, 6-hydroxy-2H-pyran-3(6H)-one was isolated as pale-yellow oil that crystallized in the freezer (−20 °C). Next, 6-hydroxy-(2H)-pyran-3(6H)-one (12.59 g, 0.11 mol), EtOH (275 mL) and Pd/C (1.26 g, 10 wt.%) catalyst were allowed to react in H_2 atmosphere (balloon) at RT for 5 h. The reaction mixture was then filtered through a pad of Celite® and the solvent was removed under reduced pressure to give 5-hydroxy-4-oxopentanal as a pale-yellow oil. Finally, 5-hydroxy-4-oxopentanal (12 g, 0.103 mol) was dissolved in MeOH (200 mL), and $NaBH_4$ (11.72 g, 0.310 mol) was added slowly into portions at 0 °C. The reaction mixture was then warmed up to room temperature and stirred for 24 h. After that time a solution of iPrOH/HCl was slowly added to the suspension up to pH 2 to decompose the boronic complexes, followed by addition of K_2CO_3 (sat. sol. in MeOH) to neutralize the unreacted acid. Successive filtration through filter paper and nylon membrane filter (pore size 0.45 μm, diam. 47 mm), followed by solvent removal under vacuum yielded the desired product.

2.3. Synthesis of Nanogel from Penthane-1,2,5-triol and Citric Acid

Penthane-1,2,5-triol (0.1 g, 0.8 mmol, 1 eq.) and citric acid (0.46 g, 2.4 mmol, 3 eq.) were dried by azeotropic distillations with toluene and dissolved in anhydrous tetrahydrofuran (15 mL) under inert atmosphere in a 50 mL three-neck flask equipped with a magnetic stirrer. Then, EDC (0.23 g, 1.2 mmol, 1.5 eq.) and DMAP (0.12 g, 0.6 mmol, 0.75 eq.) were added into the solution. The esterification reaction was carried out at room temperature by stirring the solution for 72 h under inert atmosphere. The resulting solution of nanogel was filtered and additionally purified by dialysis (3500 MWCO, Spectrum Labs, New Brunswick, NJ, USA) against deionized water for removing THF for 5 days. Finally, the product was freeze-dried to obtain light-yellow powder. The synthesis procedure was repeated three times (to assess the reproducibility), resulting in samples denoted as NG1, NG2 and NG3.

2.4. Hydrolytic Degradation

The hydrolytic degradation of nanogel was assessed by DLS measurements carried out in neutral (pH = 6.5, deionized water) and acidic medium (pH = 3, hydrochloric acid).

2.5. Drug Loading of Nanogel

The nanogel was loaded with doxorubicin by the incubation method. Doxorubicin was added to an aqueous nanogel dispersion (1 mg/mL) at drug/nanocarrier mass ratio = 1:8.5 and 1:10. The dispersion was gently stirred (700 rpm) for 2 h and filtered (0.45 μm). The concentration of the loaded drug was determined spectrophotometrically at λ = 480 nm (Thermo Fisher Scientific, Waltham, MA, USA) and calculated according to a standard curve (10–80 μg/mL, r = 0.9991).

The drug-loading degree (*LD*) and encapsulation efficiency (*EE*) were calculated using Equations (1) and (2), respectively:

$$LD(\%) = \frac{Total\ mass\ of\ drug - free\ drug}{Mass\ of\ drug - loaded\ nanogel} \times 100 \tag{1}$$

$$EE(\%) = \frac{Total\ mass\ of\ drug - free\ drug}{Total\ mass\ of\ drug} \times 100 \tag{2}$$

2.6. In Vitro Release Test

In vitro release study was performed by the dialysis method. The nanogel dispersion (3 mL) was introduced into a dialysis membrane (10,000 MWCO, Spectrum Labs, New Brunswick, NJ, USA), which was placed in a buffer phase (20 mL) at 37 °C. Two buffer media were used as acceptor phase—citrate (pH = 5.0) and phosphate buffer (pH = 7.4). Samples of 3 mL were withdrawn from the buffer phase at predetermined times (1, 2, 4, 6, 8 and 24 h), and the concentration of the released doxorubicin was determined spectrophotometrically as described above.

2.7. DOX Stability Studies

Aqueous doxorubicin solution and dispersion of DOX-loaded nanogels in equimolar concentrations (0.3 mg/mL) were placed in glass vials and exposed to UV-irradiation (Dymax 5000-EC" UV equipment with a 400 W metal halide flood lamp) at a dose rate of 5.7 J/cm^2.min. Aliquots from both samples were withdrawn at certain intervals and the concentration of doxorubicin was determined as described.

2.8. Methods

FTIR spectra of freeze-dried nanogels were recorded from 600 to 4000 cm^{-1} using an attenuated total reflection (ATR) spectrometer (IRAffinity-1, Shimadzu, Kyoto, Japan). The size of nanoparticles (hydrodynamic diameter, D_h) was determined with Zetasizer NanoBrook 90Plus PALS instrument, equipped with a 35-mW red diode laser (λ = 640 nm) at a scattering angle of 90°. The zeta potential was determined by the phase analysis light scattering (PALS) method at a scattering angle of 15°. Sample concentration was = 1.0 g L^{-1}, and each measurement was performed in triplicate. Differential scanning calorimetry (DSC) analysis was carried out under a nitrogen atmosphere using a Perkin Elmer Differential Scanning Calorimeter, DSC-7, within the temperature range of 40–250 °C, at a heating rate of 10 °C/min. The ultraviolet–visible absorption spectra were recorded on a UV-vis spectrophotometer (Thermo Fisher Scientific, Waltham, MA, USA) using quartz cells with a path length of 1 cm. Cryogenic transmission electron microscopy (cryo-TEM) images were obtained using a Tecnai F20 X TWIN microscope (FEI Company, Hillsboro, OR, USA) equipped with field emission gun, operating at an acceleration voltage of 200 kV. Images were recorded on the Gatan Rio 16 CMOS 4k camera (Gatan Inc., Pleasanton, CA, USA) and processed with Gatan Microscopy Suite (GMS) software (Gatan Inc., Pleasanton, CA, USA). Specimen preparation was done by vitrification of the aqueous solutions on grids with holey carbon film (Quantifoil R 2/2; Quantifoil Micro Tools GmbH, Großlöbichau, Germany). Prior to use, the grids were activated for 15 s in oxygen plasma using a Femto plasma cleaner (Diener Electronic, Ebhausen, Germany). Cryo-samples were prepared by applying a droplet (3 μL) of the suspension to the grid, blotting with filter paper and immediate freezing in liquid ethane using a fully automated blotting device Vitrobot Mark IV (Thermo Fisher Scientific, Waltham, MA, USA). After preparation, the vitrified specimens were kept under liquid nitrogen until they were inserted into a cryo-TEM-holder Gatan 626 (Gatan Inc., Pleasanton, CA, USA) and analyzed in the TEM at −178 °C. Atomic force microscopy (AFM) analyses were conducted using a Bruker NanoScope V9 Instrument operating at 1.00 Hz scan rate under ambient conditions. Then, 2 μL of filtered colloid solution (1 g L^{-1}) was placed onto a freshly cleaned glass substrate and spin-casted at 2000 rpm for a minute. The measurements were performed in ScanAsyst (Peak Force Tapping) mode. Wide-angle X-ray diffraction (WAXD) patterns were obtained on a Bruker D8 Advance ECO diffractometer, operating at 40 kV and 25 mA in Bragg–Brentano geometry with Ni-filtered Cu K_{α} radiation and a LynxEye-XE detector over the 2θ range of 5–50°, with a scanning rate of 0.02° s^{-1}.

3. Results and Discussion

3.1. Synthesis of Nanogel

Both penthane-1,2,5-triol and citric acid are naturally based polyfunctional reagents, which makes them attractive for fabricating new biomaterials. Moreover, the ester bonds formed after reacting CA with PT can be further hydrolyzed under given environmental conditions and, therefore, such materials are considered biodegradable. In this work, novel nanogel based on penthane-1,2,5-triol and citric acid was obtained at mild conditions via Steglich esterification precipitation reaction in THF. The formation of polymer nanonetwork proceeded at room temperature with the aid of N-ethyl-N′-(3-dimethylaminopropyl) carbodiimide as a coupling reagent and 4-(dimethylamino)-pyridine as a catalyst. Initially, all reagents (CA, PT, EDC and DMAP) were homogeneously dissolved in THF. Hence, the esterification crosslinking reaction started in a solution, but at a certain time, the formed polymeric phase became insoluble in THF and tended to precipitate. The generated insoluble particles were separated, purified, and dispersed in deionized water to form the nanogel. The presence of free a carboxylic group in the nanogel was further exploited for drug loading. The nanogel was loaded with the antitumor agent doxorubicin hydrochloride by electrostatic interactions between the carboxylic groups and amino groups of DOX. The scheme of preparing nanogel carriers of doxorubicin is shown in Figure 1.

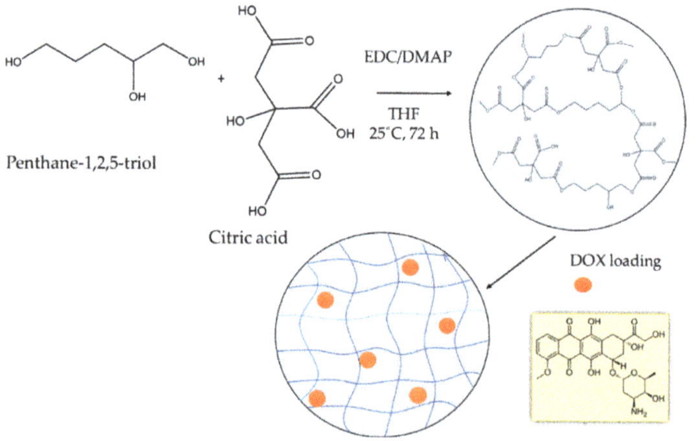

Figure 1. Preparation of DOX-loaded nanogel based on penthane-1,2,5-triol and citric acid.

Fourier-transform infrared spectroscopy was used in our study to gain an idea about the formation of ester bonds within the polymer network. Figure 2 shows the FTIR spectra of PA, CA and the nanogel formed on their basis. PT is characterized with the stretching vibration of the three -OH groups in the 3200–3500 cm^{-1} range; C-H stretching at 2820 and 2970 cm^{-1}; and C-O stretching of the primary alcohol at 1030 cm^{-1} [29]. The most characteristic bands in the infrared spectra of CA are the O-H and C-H stretching vibrations in the 3500–3000 cm^{-1} range; C=O stretching (C(O)-OH) at 1694 cm^{-1}; C-OH stretching at 1138 cm^{-1}; and CH_2 rocking at 781 cm^{-1} [30]. Two new intensive bands appeared in the FTIR spectrum of the nanogel—stretching vibrations of C=O groups at 1726 cm^{-1} and twisting vibration of CH_2 groups at 1192 cm^{-1}. These two bands are typical for polyesters [31]. Although the absorption intensity of the bands associated with -OH and -COOH groups was lower, the fact that they did not disappear completely suggests existence of free functional groups in the nanogel.

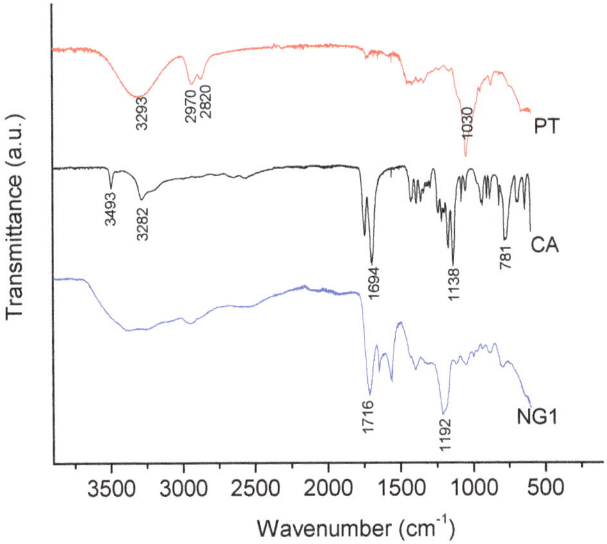

Figure 2. FTIR spectra of penthane-1,2,5-triol, citric acid and the nanogel (NG1) obtained on their basis.

3.2. Properties of Nanogel

The surfactant-free synthesis of nanogel was repeated three times at the same conditions to assess the reproducibility of the method. The main characteristics of purified nanogel systems were determined by dynamic light scattering and zeta potential measurements (Table 1). Nanogels exhibited monomodal particle size distribution, nanoscopic hydrodynamic diameter and relatively narrow dispersity index (DI). Zeta potential values were negative, most probably due to the presence of -COOH groups within the polymer network.

Table 1. Data from dynamic light scattering and PALS measurements of nanogels.

Sample Code	D_h (nm)	ζ-Potential (mV)	DI
NG1	153 ± 4	−13.0 ± 1.2	0.22 ± 0.017
NG2	172 ± 5	−12.8 ± 1.1	0.23 ± 0.015
NG3	173 ± 5	−13.2 ± 1.2	0.32 ± 0.019
NG1/DOX	146 ± 4	−8.9 ± 1.0	0.40 ± 0.020

The structural stability of blank nanogels dispersed in deionized water (pH~6.5) over a prolonged period (3 months) was monitored by DLS (Figures 3 and S1). Samples were measured every 10 days. In the first 30 days the nanogel dispersity index increased slightly. After that, the plot of the autocorrelation function tended to change its typical sigmoid shape and reached a nearly linear shape (after 70 days), while some smaller particles appeared in the intensity-weighted size distribution plot (Figure 3). The predominance of smaller fragments/monomers in the sample is evident from the number-weighted plot (Figure S1). These results indicated that the nanogel particles comprising ester bonds degrade in water with time. We hypothesize that the degradation mechanism follows the well-known reaction pathway of neutral hydrolysis of aliphatic polyesters, starting with nucleophilic attack of the labile ester bonds by water molecules [32]. In the beginning, the cleavage of ester links decreased the density of the polymer network, and the gel swelled more. Next, the network was gradually fragmented into smaller pieces, until its complete decomposition within 90 days (Figure S2). The process of gel degradation was much faster in an acidic environment. DLS results for nanogel dispersed in acidic water (pH = 3, HCl) revealed that the nanogel particles disappeared for a period of 20 days. The degradation under acidic conductions starts with protonation of the carbonyl oxygen atom of the ester group, which makes the carbonyl carbon atom more electrophilic. This facilitates the cleavage of C-O bonds of the main chain and accelerates the hydrolysis.

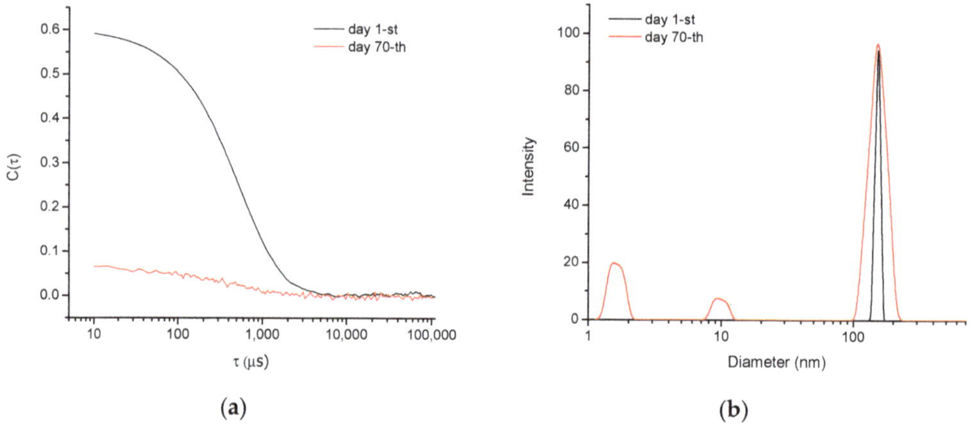

Figure 3. Autocorrelation function (**a**) and hydrodynamic diameter distribution (**b**) plots of NG1 in water (pH~6.5). Measurements were made on the first and seventieth day of sample preparation.

The morphology of nanogel particles (NG1) was visualized by cryogenic transmission electron microscopy and atomic force microscopy. Cryo-TEM micrographs of vitrified aqueous colloid (Figure 4a,b) showed that the main population of particles has a rather spherical shape and nanoscale size. However, some of the objects consist of two or three particles, most likely fused during the synthesis procedure. AFM images were obtained by spin-coating nanogel solution on a glass substrate and, therefore, represent the morphology of dry nanogel. In this case, the particles are smaller due to the dehydration of the gel, but the spherical shape is also well visible.

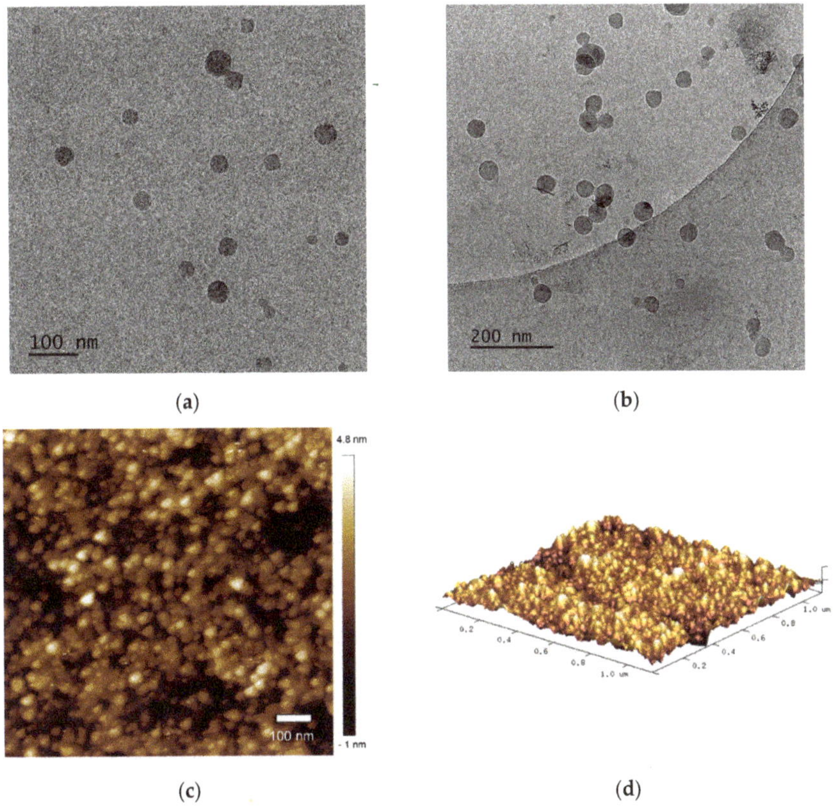

Figure 4. Representative cryo-TEM (**a**,**b**) macrographs and AFM 2D (**c**) and 3D (**d**) height images of NG1.

3.3. DOX-Loaded Nanogel

Loading of nanogel particles with the model drug doxorubicin was performed applying the incubation method. The expected driving force for embedding DOX molecules in the confined space of nanogel particles is the electrostatic interaction between carboxyl and amino groups from polymer network and drug, respectively. DOX-loaded nanogels maintained the initial size of blank carrier, but an increase of the dispersity and a decrease of zeta potential were recorded (Table 1). Such decrease of the surface charge of the nanogel is associated with carrier–DOX electrostatic interactions. This fact together with the increased solubility of DOX (a much smaller fraction of crystals was formed in the nanogel formulation compared to the free drug) confirmed that DOX was successfully embedded in the nanogel. The calculated encapsulation efficiency for the system obtained with 10-fold excess of polymer was higher (95%) than EE of the formulation prepared at a mass ratio 1:8.5 (Figure 5). In addition, the drug-loading degree of the two formulations was identical (~10%). Based on these results, DOX-loaded nanogel obtained at a mass ratio 1:10 was used in our further experiments.

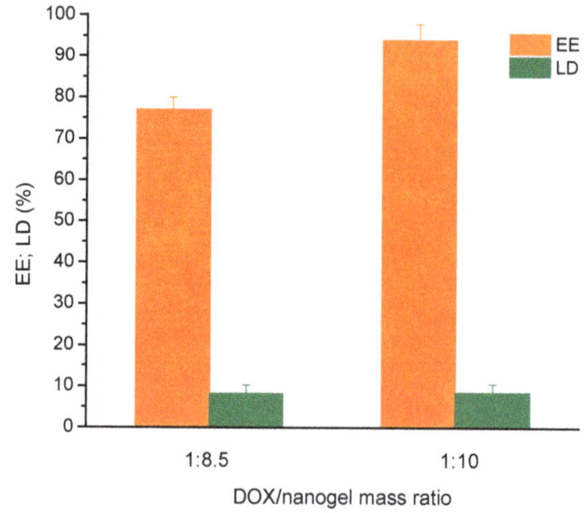

Figure 5. Drug-loading degree and encapsulation efficiency of doxorubicin in nanogel, prepared at mass ratio 1:8.5 and 1:10, respectively.

DSC and WAXD analyses revealed that DOX embedded in the nanogel carrier is in amorphous state (Figure 6). The pure DOX is a crystalline substance which melted at about 178 °C (Figure 6a). As can be seen in the same figure, the melting peak was missing in the thermogram of DOX-loaded nanogel, confirming that the drug, loaded in the nanogel, did not form crystals. The XRD patterns of pure DOX exhibited many sharp diffraction peaks, ranging from 13–45°, indicating its crystalline structure (Figure 6b). In contrast, the pattern of lyophilized DOX-loaded nanogel is only a halo, confirming the amorphous state of DOX. The fact that DOX embedded in the nanocarrier is amorphous might be advantageous regarding its release behavior.

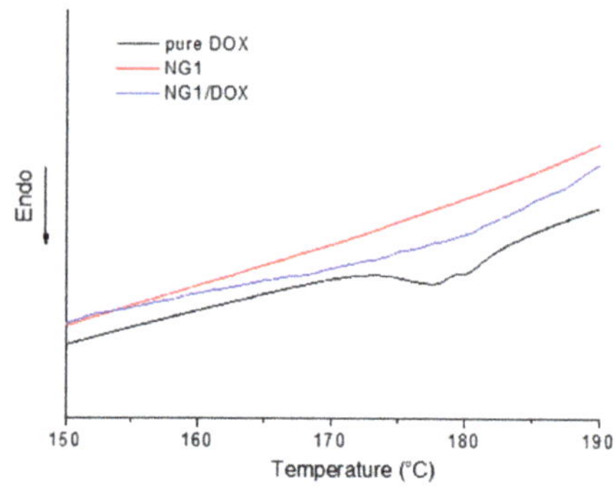

(**a**)

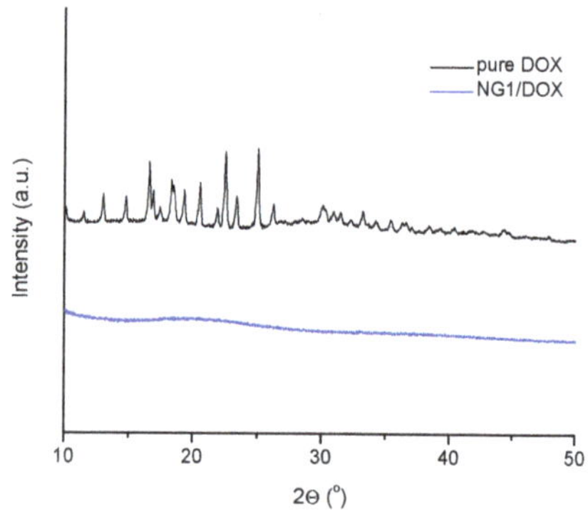

(**b**)

Figure 6. (**a**) DSC thermograms of pure doxorubicin, nanogel, and DOX-loaded nanogel, and (**b**) Wide-angle X-ray scattering spectra of pure doxorubicin and DOX-loaded nanogel.

In vitro release studies were performed in buffer media with pH values of 7.4 and 5.0, which resemble the slightly alkaline pH of most body fluids and the acidic pH of intracellular endosomal and lysosomal compartments, respectively. The release studies showed bi-phasic profiles (an initial burst release and then a gradual sustained release) of DOX in the two buffers (Figure 7). Kinetics analysis of the process revealed that the release followed the first-order model (Figure S3). This pattern describes the release of water-soluble drugs from porous matrices, suggesting formation of pores in the nanogel particles.

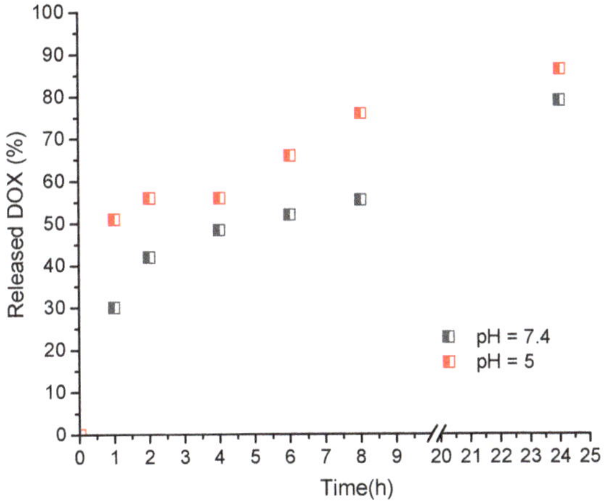

Figure 7. In vitro release of doxorubicin from the nanogel (NG1/DOX) in buffer media with pH-values 5.0 and 7.4.

The burst effect was less pronounced in the medium with pH 7.4. In particular, 30% of doxorubicin was released for 1 h in the slightly alkaline medium vs. 51% in the acidic medium. Furthermore, the release rate in the acidic medium was higher than in the neutral one. Since DOX has a pH-dependent solubility we have performed a dissolution test of free DOX under the same conditions. The results revealed that at the applied concentration (equal to that of the encapsulated DOX) the drug was rapidly dissolved in both media for less than 30 min. The latter indicated that the reason for the observed different release profiles was the influence of the developed nanogel. A similar pattern of release profile of doxorubicin from pH-sensitive nanogels was reported by Jayakumar et al. [33]. It is attributed to the fact that in acidic conditions the amino groups of doxorubicin and the carboxyl groups of nanogel are protonated, which weakens the interaction between the drug and nanocarrier and leads to a fast drug release [34]. Drug delivery systems possessing such behavior might be beneficial in cancer therapy; however, more in vitro and in vivo studies are needed to assess the full capacity of the NG/DOX system for solid tumor curing.

The low stability of doxorubicin in aqueous media (mainly hydrolytic and photolytic degradation) can be a limitation for its use in practice. For instance, Prokopowicz et al., observed a gradual irreversible photodegradation of doxorubicin in solution after only 1 h of exposure to UV–vis light, while the encapsulation of doxorubicin in a solid gel matrix prolonged the time of its use [35]. Recently, some of us reported that the formulation of DOX in appropriate drug delivery systems allows us to avoid the formation of degradation derivatives [25]. In the present work, we assumed that the inclusion of doxorubicin in nanogels can have a protective effect against light-induced degradation. Thus, the DOX-loaded nanogel and a referent DOX solution were exposed to UV-irradiation and the concentration of non-degraded drugs in both formulations was evaluated with time by spectrophotometry. The results showed a well pronounced decrease of doxorubicin concentration in the referent solution with the time, whereas DOX loaded into nanogel remained stable for 40 min (Figure 8). In fact, at the end of the experiment (60 min) 10% of the encapsulated DOX and 66% of the free DOX were degraded. This result is direct proof that the loading of doxorubicin into nanogel stabilized the drug against photodegradation.

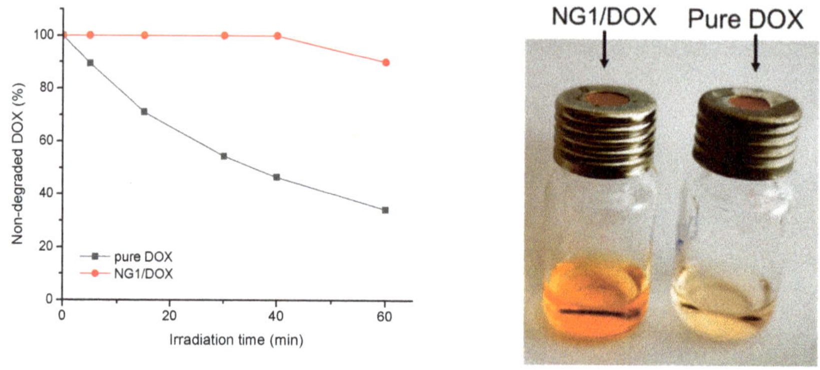

(a) (b)

Figure 8. (**a**) UV-induced degradation of doxorubicin as a function of the irradiation time in DOX-loaded nanogel dispersion and pure DOX aqueous solution, and (**b**) digital image of the two samples after 60 min irradiation with UV light.

4. Conclusions

Novel functional nanogel carriers of doxorubicin, based on the natural reagent citric acid and pentane-1,2,5-triol, were synthesized at mild conditions by nanoprecipitation esterification reaction without any surfactant. The synthesis procedure afforded fabrication of nano-sized gels of monomodal particle size distribution, negative zeta potential and relatively low dispersity index, in a reproducible way. The resulted nanogel particles were considered as an advantageous drug delivery system due to their biodegradability, high encapsulation capacity for doxorubicin and efficient drug protection against photolysis. Furthermore, the nanogel exhibited faster drug release in an acidic medium than in a neutral medium.

Supplementary Materials: The following supporting information can be downloaded at: https://www.mdpi.com/article/10.3390/polym14173694/s1, Figure S1: Hydrodynamic diameter distribution plot of NG1 in water (pH~6.5); Figure S2: Schematic representation of the hydrolysis of nanogel in aqueous media; Figure S3: Kinetics analysis of DOX release from NG1 in buffer media with pH-values 5.0 and 7.4.

Author Contributions: Conceptualization, K.K., S.P.S. and P.D.P.; methodology, K.K., P.D.P., B.T. and K.Y.; analysis, K.K., A.F., F.U. and L.R.; investigation, K.K., M.A.R. and M.K.M.; resources, S.P.S. and P.D.P.; writing—original draft preparation, P.D.P., K.K. and K.Y.; writing—review and editing, B.T., S.P.S., A.F. and F.U.; visualization, K.K., A.F. and F.U.; supervision, P.D.P., K.Y. and S.P.S.; project administration, S.P.S. and P.D.P.; funding acquisition, S.P.S. All authors have read and agreed to the published version of the manuscript.

Funding: This research was funded by the Bulgarian National Science Fund, grant number КП-06-ОПР 01/2-14.12.2018.

Institutional Review Board Statement: Not applicable.

Informed Consent Statement: Not applicable.

Data Availability Statement: Not applicable.

Acknowledgments: P.D.P., K.K. and F.U. are grateful to INFRAMAT project (part of the Bulgarian National Roadmap for research infrastructure, supported by the Bulgarian Ministry of Education and Science) for the research equipment that was used in this investigation. S.P.S. thanks European Union's Horizon 2020 research and innovation programme under grant agreement No 951996.

Conflicts of Interest: The authors declare no conflict of interest. The funders had no role in the design of the study; in the collection, analyses, or interpretation of data; in the writing of the manuscript; or in the decision to publish the results.

References

1. Kabanov, A.; Vinogradov, S. Nanogels as pharmaceutical carriers: Finite networks of infinite capabilities. *Angew. Chem. Int. Ed.* **2009**, *48*, 5418–5429. [CrossRef] [PubMed]
2. Kwon, J.; Drumright, O.R.; Siegwart, D.J.; Matyjaszewski, K. The development of microgels/nanogels for drug delivery applications. *Prog. Polym. Sci.* **2008**, *33*, 448–477.
3. Takeuchi, T.; Kitayama, Y.; Sasao, R.; Yamada, T.; Toh, K.; Matsumoto, Y.; Kataoka, K. Molecularly imprinted nanogels acquire stealth in situ by cloaking themselves with native dysopsonic proteins. *Angew. Chem. Int. Ed.* **2017**, *56*, 7088–7092. [CrossRef]
4. Schötz, S.; Reisbeck, F.; Schmitt, A.-C.; Dimde, M.; Quaas, E.; Achazi, K.; Tunable, R.H. Polyglycerol-based redox-responsive nanogels for efficient cytochrome c delivery. *Pharmaceutics* **2021**, *13*, 1276. [CrossRef] [PubMed]
5. Basak, S. The age of multistimuli-responsive nanogels: The finest evolved nano delivery system in biomedical sciences. *Biotechnol. Bioprocess Eng.* **2020**, *25*, 655–669. [CrossRef]
6. Sharma, A.; Garg, T.; Aman, A.; Panchal, K.; Sharma, R.; Kumar, S.; Markandeywar, T. Nanogel—An advanced drug delivery tool: Current and future. *Artif. Cells Nanomed. Biotechnol.* **2016**, *44*, 165–177. [CrossRef] [PubMed]
7. Lee, K.; Choo, H. Preparation of poly(BMA-co-MMA) particles by soap-free emulsion polymerization and its optical properties as photonic crystals. *J. Nanosci. Nanotechnol.* **2014**, *14*, 8279–8287. [CrossRef]
8. Zhang, R.; Gao, R.; Gou, Q.; Lai, J.; Li, X. Precipitation polymerization: A powerful tool for preparation of uniform polymer particles. *Polymers* **2022**, *14*, 1851. [CrossRef]
9. Schmitt, F.; Lagopoulos, L.; Käuper, P.; Rossi, N.; Busso, N.; Barge, J.; Wagnières, G.; Laue, C.; Wandrey, C.; Juillerat-Jeanneret, L. Chitosan-based nanogels for selective delivery of photosensitizers to macrophages and improved retention in and therapy of articular joints. *J. Controlled Release* **2010**, *144*, 242–250. [CrossRef]
10. Sun, Z.; Yi, Z.; Zhang, H.; Ma, H.; Su, W.; Sun, X.; Li, X. Bio-responsive alginate-keratin composite nanogels with enhanced drug loading efficiency for cancer therapy. *Carbohydr. Polym.* **2017**, *175*, 159–169. [CrossRef]
11. Sarika, P.R.; James, N.R.; Kumar, P.R.A.; Raj, D.K. Preparation, characterization and biological evaluation of curcumin loaded alginate aldehyde-gelatin nanogels. *Mater. Sci. Eng. C* **2016**, *68*, 251–257.
12. Xue, Y.; Xia, X.; Yu, B.; Luo, X.; Cai, N.; Long, S. Yu, F. A green and facile method for the preparation of a pH-responsive alginate nanogel for subcellulardelivery of doxorubicin. *RSC Adv.* **2015**, *5*, 73416–73423. [CrossRef]
13. Gyawali, D.; Kim, J.P.; Yang, J. Highly photostable nanogels for fluorescence-based theranostics. *Bioact. Mater.* **2018**, *3*, 39–47. [CrossRef] [PubMed]
14. Salihu, R.; Razak, S.I.A.; Zawawi, N.A.; Kadir, M.R.A.; Ismail, N.I.; Jusoh, N.; Mohamad, M.R.; Nayan, N.H.M. Citric acid: A green cross-linker of biomaterials for biomedical applications. *Eur. Polym. J.* **2021**, *146*, 110271. [CrossRef]
15. Gorgieva, S.; Kokol, V. Synthesis and application of new temperature-responsive hydrogels based on carboxymethyl and hydroxyethyl cellulose derivatives for the functional finishing of cotton knitwear. *Carbohydr. Polym.* **2011**, *85*, 664–673. [CrossRef]
16. Bozova, N.; Petrov, P.D. Highly elastic super-macroporous cryogels fabricated by thermally induced crosslinking of 2-hydroxyethylcellulose with citric acid in solid state. *Molecules* **2021**, *26*, 6370. [CrossRef]
17. Chhatbar, M.U.; Prasad, K.; Chejara, D.R.; Siddhanta, A.K. Synthesis of sodium alginate based sprayable new soft gel system. *Soft Matter* **2012**, *8*, 1837. [CrossRef]
18. Rafat, M.; Li, F.F.; Fagerholm, P.; Lagali, N.S.; Watsky, M.A.; Munger, R.; Matsuura, T.; Griffith, M. PEG-stabilized carbodiimide crosslinked collagen–chitosan hydrogels for corneal tissue engineering. *Biomaterials* **2008**, *29*, 3960. [CrossRef]
19. Nam, K.; Kimura, T.; Kishida, A. Preparation and characterization of cross-linked collagen–phospholipid polymer hybrid gels. *Biomaterials* **2007**, *28*, 1. [CrossRef] [PubMed]
20. Abdullah, C.S.; Ray, P.; Alam, S.; Kale, N.; Aishwarya, R.; Morshed, M.; Dutta, D.; Hudziak, C.; Banerjee, S.K.; Mallik, S. Chemical architecture of block copolymers differentially abrogate cardiotoxicity and maintain the anticancer efficacy of doxorubicin. *Mol. Pharmaceutics* **2020**, *17*, 4676–4690. [CrossRef]
21. Fleige, E.; Achazi, K.; Schaletzki, K.; Triemer, T.; Haag, R. pH-responsive dendritic core-multishell nanocarriers. *J. Control. Release.* **2014**, *185*, 99–108. [CrossRef] [PubMed]
22. Nawara, K.; Krysinski, P.; Blanchard, G.J. Photoinduced reactivity of doxorubicin: Catalysis and degradation. *J. Phys. Chem. A* **2012**, *116*, 4330–4337. [CrossRef] [PubMed]
23. Kaushik, D.; Bansal, G. Four new degradation products of doxorubicin: An application of forced degradation study and hyphenated chromatographic techniques. *J. Pharm. Analysis* **2015**, *5*, 285–295. [CrossRef]
24. Bandak, S.; Ramu, A.; Barenholz, Y.; Gabizon, A. Reduced UV-induced degradation of doxorubicin encapsulated in polyethyleneglycol-coated liposomes. *Pharm. Res.* **1999**, *16*, 841–846. [CrossRef]
25. Yoncheva, K.; Tzankov, B.; Yordanov, Y.; Spassova, I.; Kovacheva, D.; Frosini, M.; Valoti, M.; Tzankova, V. Encapsulation of doxorubicin in chitosan-alginate nanoparticles improves its stability and cytotoxicity in resistant lymphoma L5178 MDR cell. *J. Drug Deliv. Sci. Technol.* **2020**, *59*, 101870. [CrossRef]

26. Sumitha, N.S.; Prakash, P.; Nair, B.N.; Sailaja, G.S. Degradation-dependent controlled delivery of doxorubicin by glyoxal cross-linked magnetic and porous chitosan microspheres. *ACS Omega* **2021**, *6*, 21472–21484. [CrossRef] [PubMed]
27. Zhang, Z.; O'Hara, I.M.; Orlando, W.; Doherty, S.; Rackemann, D. Methods for converting lignocellulosic materials to useful products. US 2014/0093918 A1, 2014.
28. Simeonov, S.P.; Ravutsov, M.A.; Mihovilovic, M.D. Biorefinery via achmatowicz rearrangement: Synthesis of pentane-1,2,5-triol from furfuryl alcohol. *Chem.Sus.Chem.* **2019**, *12*, 2748–2754. [CrossRef] [PubMed]
29. Danish, M.; Mumtaz, M.W.; Fakhar, M.; Rashid, U. Response surface methodology: An imperative tool for the optimized purification of the residual glycerol from biodiesel production process. *Chiang Mai J. Sci.* **2016**, *44*, 1–13.
30. Pimpan, P.; Sumang, T.; Ch, S. Effect of concentration of citric acid on size and optical properties of fluorescence graphene quantum dots prepared by tuning carbonization degree. *Chiang Mai J. Sci.* **2018**, *45*, 2005–2014.
31. Koto, N.; Soegijono, B. Effect of rice husk ash filler of resistance against of high-speed projectile impact on polyester-fiberglass double panel composites. *J. Phys. Conf. Ser.* **2019**, *1191*, 012058. [CrossRef]
32. Rydz, J.; Sikorska, W.; Kyulavska, M.; Christova, D. Polyester-based (bio)degradable polymers as environmentally friendly materials for sustainable development. *Int. J. Mol. Sci.* **2015**, *16*, 564–596. [CrossRef] [PubMed]
33. Jayakumar, R.; Nair, A.; Rejinold, N.S.; Maya, S.; Nair, S.V. Doxorubicin-loaded pH-responsive chitin nanogels for drug delivery to cancer cells. *Carbohydr. Polym.* **2021**, *87*, 2352–2356. [CrossRef]
34. Abedi, F.; Davaran, S.; Hekmati, M.; Akbarzadeh, A.; Baradaran, B.; Moghaddam, S.V. An improved method in fabrication of smart dual-responsive nanogels for controlled release of doxorubicin and curcumin in HT-29 colon cancer cells. *J. Nanobiotechnol.* **2021**, *19*, 18. [CrossRef] [PubMed]
35. Prokopowicz, M.; Lukasiak, J.; Przyjazny, A. Synthesis and application of doxorubicin-loaded silica gels as solid materials for spectral analysis. *Talanta* **2005**, *65*, 663–671. [CrossRef] [PubMed]

polymers

Article

Microbial Poly-γ-Glutamic Acid (γ-PGA) as an Effective Tooth Enamel Protectant

Mattia Parati [1,*], Louisa Clarke [2], Paul Anderson [2,*], Robert Hill [2], Ibrahim Khalil [1], Fideline Tchuenbou-Magaia [1], Michele S. Stanley [3], Donal McGee [4], Barbara Mendrek [5], Marek Kowalczuk [5] and Iza Radecka [1,*]

1 Faculty of Science and Engineering, University of Wolverhampton, Wolverhampton WV1 1 LY, UK; ibrahim.khalil2@wlv.ac.uk (I.K.); f.tchuenbou-magaia@wlv.ac.uk (F.T.-M.)
2 Institute of Dentistry, Barts and The London School of Medicine and Dentistry, Queen Mary University of London, London E1 2AD, UK; louisaclarke35@hotmail.co.uk (L.C.); r.hill@qmul.ac.uk (R.H.)
3 Scottish Association for Marine Science, Scottish Marine Institute, Oban PA37 1QA, UK; michele.stanley@sams.ac.uk
4 AlgaeCytes Limited, Discovery Park House, Ramsgate Road, Sandwich, Kent CT13 9ND, UK; donalmcgee@algaecytes.com
5 Centre of Polymer and Carbon Materials, Polish Academy of Sciences, M. Curie-Sklodowskiej 34, 41-819 Zabrze, Poland; bmendrek@cmpw-pan.edu.pl (B.M.); marek.kowalczuk@cmpw-pan.edu.pl (M.K.)
* Correspondence: m.parati2@wlv.ac.uk (M.P.); p.anderson@qmul.ac.uk (P.A.); i.radecka@wlv.ac.uk (I.R.)

Abstract: Poly-γ-glutamic acid (γ-PGA) is a bio-derived water-soluble, edible, non-immunogenic nylon-like polymer with the biochemical characteristics of a polypeptide. This *Bacillus*-derived material has great potential for a wide range of applications, from bioremediation to tunable drug delivery systems. In the context of oral care, γ-PGA holds great promise in enamel demineralisation prevention. The salivary protein statherin has previously been shown to protect tooth enamel from acid dissolution and act as a reservoir for free calcium ions within oral cavities. Its superb enamel-binding capacity is attributed to the L-glutamic acid residues of this 5380 Da protein. In this study, γ-PGA was successfully synthesised from *Bacillus subtilis* natto cultivated on supplemented algae media and standard commercial media. The polymers obtained were tested for their potential to inhibit demineralisation of hydroxyapatite (HAp) when exposed to caries simulating acidic conditions. Formulations presenting 0.1, 0.25, 0.5, 0.75, 1, 2, 3 and 4% (w/v) γ-PGA concentration were assessed to determine the optimal conditions. Our data suggests that both the concentration and the molar mass of the γ-PGA were significant in enamel protection ($p = 0.028$ and $p < 0.01$ respectively). Ion Selective Electrode, combined with Fourier Transform Infra-Red studies, were employed to quantify enamel protection capacity of γ-PGA. All concentrations tested showed an inhibitory effect on the dissolution rate of calcium ions from hydroxyapatite, with 1% (wt) and 2% (wt) concentrations being the most effective. The impact of the average molar mass (M) on enamel dissolution was also investigated by employing commercial 66 kDa, 166 kDa, 440 kDa and 520 kDa γ-PGA fractions. All γ-PGA solutions adhered to the surface of HAp with evidence that this remained after 60 min of continuous acidic challenge. Inductively Coupled Plasma analysis showed a significant abundance of calcium ions associated with γ-PGA, which suggests that this material could also act as a responsive calcium delivery system. We have concluded that all γ-PGA samples tested (commercial and algae derived) display enamel protection capacity regardless of their concentration or average molar mass. However, we believe that γ-PGA D/L ratios might affect the binding more than its molar mass.

Keywords: γ-PGA; oral biology; biotechnology; *Bacillus*; statherin; delivery systems; polymer engineering

check for updates

Citation: Parati, M.; Clarke, L.; Anderson, P.; Hill, R.; Khalil, I.; Tchuenbou-Magaia, F.; Stanley, M.S.; McGee, D.; Mendrek, B.; Kowalczuk, M.; et al. Microbial Poly-γ-Glutamic Acid (γ-PGA) as an Effective Tooth Enamel Protectant. *Polymers* **2022**, *14*, 2937. https://doi.org/10.3390/polym14142937

Academic Editor: Giulio Malucelli

Received: 3 June 2022
Accepted: 16 July 2022
Published: 20 July 2022

1. Introduction

Over the years, the oral health care industry has attempted to develop therapeutic oral hygiene products to alleviate the symptoms of xerostomia and maintain the integrity of the dental hard tissues within the oral cavity [1,2]. Xerostomia is prolonged dry mouth, a condition which occurs as a result of decreased salivary flow (below 40–50% of its average value) [1,3]. Although there are many causes of xerostomia; the use of multiple medicines in the aging population is one of the most common causes of this condition [4]. Xerostomia can also arise following radiotherapy treatment, especially for head and neck cancers; following the development of atrophic glandular tissue [5]. It is well established that following radiotherapy, caries progress rapidly as a result of the loss of protection from saliva and alterations in enamel composition with loss of calcium and phosphorous ions [5]. In addition, autoimmune conditions (e.g., Sjogren's syndrome) can also cause xerostomia [4]. The incidence of xerostomia is estimated to be 22% [6]. Although there are commercial products available to ease the symptoms derived from xerostomia, none of them are capable of dealing with dental enamel demineralisation [7].

Saliva plays a fundamental role in oral health. Saliva is the transparent biofluid secreted into the mouth via the ductal systems associated with predominately three pairs of major salivary glands; with approximately 99% of it being water. The remaining 1% is proteins, mineral ions, electrolytes, immunoglobulins, and large mucopolysaccharides [8]. Saliva exhibits specific and non-specific physico-chemical properties and it is essential for lubrication, cleansing of the oral cavity, antimicrobial and antiviral activity, wound healing, buffering of acids, and supersaturated concentrations of calcium and phosphate ions, which are essentially the mineral calcium hydroxyapatite (HAp) required for the protection of dental hard tissues [9]. The rheological properties displayed by saliva are provided by mucopolysaccharide fraction of saliva and not by statherin or other proteins [10].

A fundamental role of saliva is its association in the formation of the acquired enamel pellicle (AEP) [11]. The AEP is formed by the adsorption of four salivary phosphoproteins, the first being statherin, in sequence onto the enamel surface, protecting the enamel from acid dissolution and acting as a reservoir for free calcium ions adjacent to the surface [12]. The mode of action of statherin is related to the high hydroxyapatite binding propensity of glutamic acid residues near its N-terminus [12,13].

The mineral in the enamel is a calcium-deficient carbonated hydroxyapatite ($Ca_{10}(PO_4)_6OH_2$) with some extrinsically sourced fluoride. From a chemical point of view, the solubility isotherm for HAp, calculated from the solubility of calcium hydroxyapatite, shows that when the pH of the surrounding solution is less than the "critical pH" (i.e., the pH at which a solution is just saturated with the enamel mineral) the solution is undersaturated and therefore demineralisation will occur [14]. These conditions can be simulated in in vitro models using cariogenic or erosiogeneic simulating artificial demineralisation solutions. Clinically, erosion (and caries) of enamel occurs when the pH of the surrounding solution in the oral cavity is below the critical pH at normal salivary calcium concentrations [15]. Statherin prevents such mechanisms by effectively shielding the enamel layer and thus, modifying, the effective solubility of calcium hydroxyapatite, reducing the interaction between the low pH, calcium and phosphate deficient solution and the hydroxyapatite layer [12,13].

An unfortunate and costly consequence of xerostomia is severe dental hard tissue destruction, leading to a clinical condition termed "rampant dental caries" [16]. This condition can render the patient edentulous, requiring significant dental intervention [16]. Therefore, investigation into new biopolymeric materials, such as poly(D/L- γ-glutamic acid), could provide novel solutions to tackle these issues.

Poly(D/L-γ-glutamic acid) is an extracellularly secreted protein-like polymeric material synthesised by an array of Prokaryotic and Eukaryotic organisms [17,18]. In contrast to other polypeptides, the biosynthesis of γ-PGA occurs through an enzymatic complex which is responsible for the polymerisation of both D- and L-glutamic acid monomers (see Figure 1). By employing an enzymatic complex, producers can greatly change the properties of the polymer in response to changing environmental conditions [19].

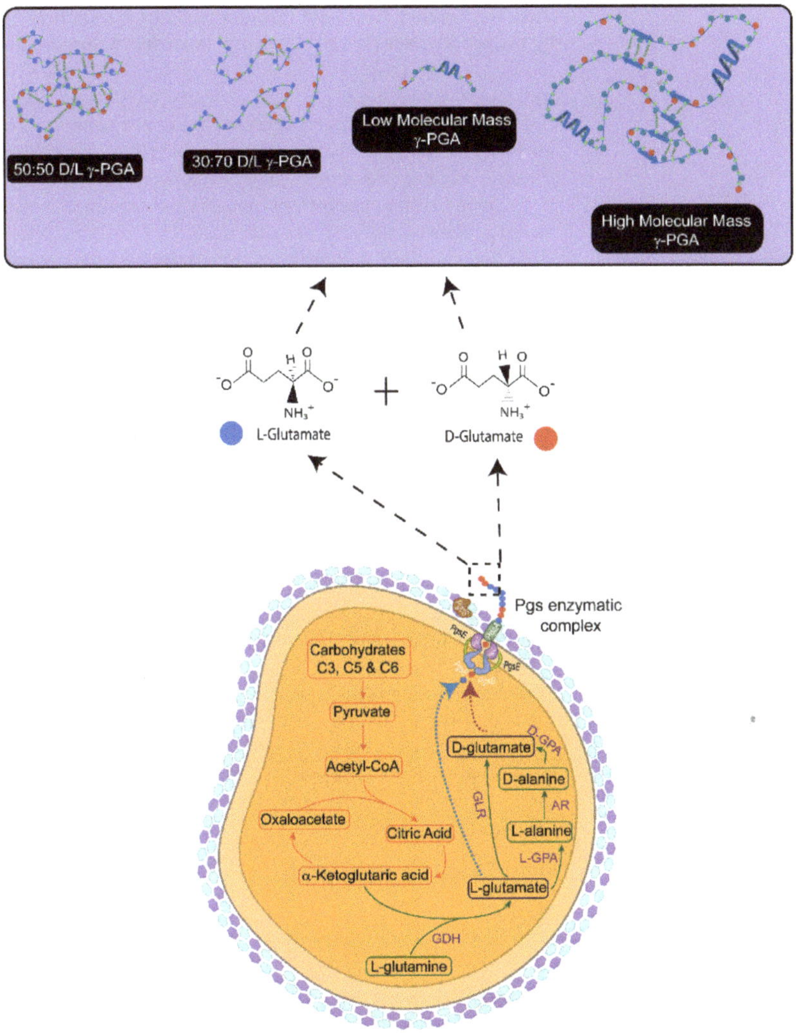

Figure 1. Factors and pathways involved in the biosynthesis of poly (γ-glutamic acid) (γ-PGA). Herein, the metabolic pathway (in red) of γ-PGA producers has been summarised. In red, the energy generation pathways have been shown. Similarly, γ-PGA generation intermediates/components have been presented in green. The figure also illustrates polymerisation of D- and L-glutamic acid monomers (red and blue dots respectively) into a polymeric chain by the enzymatic complex. The physical arrangement of the polymer, with varying D/L ratios, in solution has also been summarised. Light blue and purple hexagons represent cross-linked NAM and NAG moieties of Gram-positive peptidoglycan. GDH—Glutamate dehydrogenase; L-GPA—L-glutamic acid:pyruvate aminotransferase; AR—alanine racemase; D-GPA—D-glutamic acid:pyruvate aminotransferase; GLR—glutamate racemase.

Both the biochemical properties and physical properties of γ-PGA suggest its suitability as an anti-cariogenic agent, as well as a therapeutic treatment to prevent rampant caries associated with xerostomia [12]. It has been shown that γ-PGA has saliva-like viscosity and calcium ion binding capability and shows great similarity to statherin, which hints at its suitability for xerostomia treatment [20]. Further, it has been shown that γ-PGA can increase salivary flow, and gamma-peptidic bonds help γ-PGA to prevent degradation by microbial proteases [17,18,20,21]. In addition, the presence of carboxylic acid-rich pendant groups (Figure 1) helps γ-PGA to ionically interact with protons and other charged molecules/compounds [20]. In this study, we propose the supplementation of such material within acidic drinks, to decrease the overall enamel demineralisation which occurs a great deal from the ever-increasing consumption of such beverages, or its supplementation through other therapeutic delivery systems.

Given that the biosynthesis of γ-PGA is regulated by enzymatic activity (Figure 1), produced polymers can vary in molar mass, D/L ratio and chain association with salts [22,23]. Each one can have a significant impact on chain physical arrangement, and ultimately on polymeric behaviour [24]. In nature, variation of polymeric chemical structure helps microorganisms survive in challenging conditions, including nutrient shortage [25,26], high salt and/or heavy metal concentrations [27–29].

The biosynthesis of γ-PGA is greatly dependent on the strain, cultivation time, cultivation conditions and substrate. The fermentation system, mode and conditions also play an important role in determining polymeric yields and characteristics [18]. The understanding of these, and how they are further altered by the substrate and strain, is vital in obtaining the maximum amounts of specific polymeric composites in a consistent manner [30]. Another great barrier towards the large-scale commercialisation of γ-PGA is the cost of production. Currently, expensive standard media components are employed for the synthesis of this biopolymer [31,32]. To our knowledge, no complex waste media is currently being used for commercial synthesis of this valuable biopolymer. However, with the rapid utilisation of available land for human activities, relatively unexplored oceans are gaining attention as a source for food or novel biomaterials [33,34]. As a result of this activity, non-edible, nutrient-rich seaweed and algae industrial by-products are created, which could potentially be utilised as substrates for cost-effective biosynthesis of γ-PGA [32,34,35]. Herein, we attempted to synthesise γ-PGA using both micro- and macro-algae complex media. Further, as production methods are developed to deliver cost-effective γ-PGA, here we assess whether γ-PGA synthesised from complex waste media, could protect the enamel from acid dissolution and, therefore, reduce the overall enamel demineralisation in patients presenting xerostomia.

2. Materials and Methods

2.1. Biosynthesis of γ-PGA

2.1.1. Microorganism

Bacillus subtilis natto (ATCC15245) obtained from the National Collection of Industrial and Marine Bacteria (NCIMB) were freeze-dried and kept at −20 °C. Before use, cultures were resuscitated and grown on Tryptone Soya Agar TSA (Lab M, Heywood, UK) overnight at 37 °C. Highly mucoid colonies were selected and grown aerobically in shake flasks containing 100 mL of TSB medium (Lab M, Heywood, UK) at 37 °C for 24 h.

2.1.2. Fermentation Media

Standard GS Medium (50 g/L NaCl) was composed of 20 g/L L-glutamate purchased from Fischer Chemicals Ltd. (Loughborough, UK), 50 g/L sucrose, 2.7 g/L KH_2PO_4, 4.2 g/L Na_2HPO_4, 50 g/L NaCl, 5 g/L $MgSO_4 \cdot 7H_2O$, and 1 mL/L of Murashige-Skoog vitamin solution, all purchased from Sigma-Aldrich (Irvine, UK). The pH of this medium was adjusted to 6.8 using NaOH purchased from Fischer Chemicals Ltd. (Loughborough, UK).

Modified GS Medium (0 g/L NaCl) was composed of 20 g/L L-glutamate purchased from Fischer Chemicals Ltd. (Loughborough, UK), 50 g/L sucrose, 2.7 g/L KH_2PO_4, 4.2 g/L Na_2HPO_4, 5 g/L $MgSO_4 \cdot 7H_2O$, and 1 mL/L of Murashige-Skoog vitamin solution, purchased from Sigma-Aldrich (Irvine, UK). The pH of this medium was adjusted to 6.8 using NaOH purchased from Fischer Chemicals Ltd. (Loughborough, UK).

Macroalgal medium was composed of 40 g/L *Laminaria digitata* flakes (Cornish Seaweed Ltd., Helston, UK) or *Laminaria digitata* powder (Scottish Association of Marine

Science, Oban, Scotland, UK), and 10 g/L L-glutamate (Fischer Chemicals Ltd., Loughborough, UK), 25 g/L Sucrose (Sigma-Aldrich, Irvine, UK). The media was pre-treated with the aid of high shear mixing (Silverson, East Longmeadow, MA, USA) at 3000 rpm for 3 min. *Laminaria digitata*, was cultivated at the Port-a-Bhuiltin seaweed farm, operated by the Scottish Association for Marine Science (SAMS, Oban, Scotland, UK). Appropriate thalli of *Laminaria digitata* were hand-harvested using knives to cut the stipe above the blade, and placed into clean sampling containers (60 L plastic boxes). Samples were collected between 7 April 2021 and 1 July 2021. Seaweed thalli were macerated and homogenised using an industrial-sized mincer (Hobart, Model E4522) and frozen immediately at −20 °C, freeze dried and the dried material milled to <1 mm in size using a coffee grinder.

Microalgal medium was composed of 40 g/L *Nanochloropsis oceanica* (Algaecytes Ltd., Discovery Park, Kent, UK) previously ethanol extracted (Tennants Distribution, Manchester, UK), 10 g/L L-glutamate (Fischer Chemicals Ltd., Loughborough, UK), and 25 g/L Sucrose (Sigma-Aldrich, Irvine, UK). The post-oil extracted spent microalgal biomass was provided courtesy of AlgaeCytes Limited (Algaecytes Ltd., Discovery Park, Kent, UK). To generate the spent biomass, the spray dried *Nanochloropsis oceanica* biomass underwent AlgaeCytes' in-house omega-3 oil extraction process to generate their AVEPA^{TM} range of eicosapentaenoic acid (C20:5$_{(n-3)}$, EPA) vegan friendly enriched algal oil. The post-oil extracted biomass was transferred to a vacuum oven for 18 h to vent off residual ethanol solvent. The proximate biochemical composition of the spent biomass was determined using FT-IR analysis [36]; comprising of approximately 3.7% DW lipids, 13.7% DW protein and 18.0% carbohydrates.

All media were prepared using deionised water and sterilised by autoclaving at 121 °C for 15 min. The sucrose solution was sterilised separately and vitamin solution was filter sterilised (0.22 μm, Fischer Chemicals Ltd., Loughborough, UK) and added separately to the medium.

2.1.3. Cultivation Parameters

Batch cultures were carried out in 250 mL shake flasks. The cultures were inoculated by thawing the frozen cells in a 37 °C bath (Grant Instruments, Fischer Chemicals Ltd., Loughborough, UK) and 5% (*v*/*v*) of inoculum was added to the cultivation media. Growth temperature was kept at 37 °C, agitation was set 150 rpm for the 96 h of cultivation period.

Bacterial growth monitoring was carried out periodically by aseptically removing 0.5 mL aliquots and diluting them sequentially in one-quarter strength Ringer solutions (Lab M, Heywood, UK). For each sample, 0.5 mL was serially diluted in 10-fold steps to 10^{-7}. Ringer solution was prepared by dissolving 1 tablet in 500 mL of deionised water in a 1-L flask. Cell viability was determined by serial dilutions onto Petri dishes containing nutrient agar. Each dilution was plated following the Miles and Misra method [37] and employing 20 μL of each dilution. Colonies were counted after overnight incubation at 37 °C and organismal concentration was expressed as Log Colony Forming Units/mL. All results were statistically analysed using Microsoft Excel and SPSS 25.

2.1.4. Statistical Analysis

All cultivation parameters and yield presented were undertaken in triplicates. Results were statistically analysed by means of standard deviation, standard error and two-sided *t*-test using SPSS 25.

2.1.5. Isolation and Purification of γ-PGA

For material isolation, the culture broth was withdrawn from the fermentation vessel and centrifuged at 17,000× *g* for 30 min to remove cells, by employing a ERMLE Z 300K centrifuge (Wehingen, Germany). The supernatant was poured into 3 volumes of cold ethanol and left overnight at 5 °C to precipitate. The resultant precipitate was collected by filtration over a 0.22 μm paper filter (Fischer Chemicals Ltd., Loughborough, UK). The precipitate was subsequently lyophilised (Alpha 1–4 LSC plus Christ Freeze Dryer, SciQuip

Ltd. Bomere Heath, UK) with a 36 h run. The white/green/brown, dry powder was stored in a desiccator at room temperature until further use.

For further purification, the obtained dry powder was re-dissolved into deionised water and dialysed against deionised water within a cross flow system purchased from Repligen, US with a 20 cm MidiKros column of 30 KDa cut off from Repligen. The purified media was re-precipitated with 3 volumes of cold ethanol and left at 5 °C overnight. After precipitation the polymer was collected and lyophilised.

2.1.6. Characterisation of γ-PGA

The purified polymer isolated form was identified by FT-IR with Nicolet 380 FT-IR (Thermo Fisher Scientific Inc., Wilmington, DE, USA) with 32 scans and 4 cm^{-1} resolution. The measurements were over 100 scans and wave number range of 400–4000 cm^{-1}. FT-IR for γ-PGA-HAp interaction was performed using the Perkin Elmer FT-IR System Spectrum GX equipment and software. The dry HAp powder (Plasma Biotal, Tideswell, UK) was placed on a clean stage, where scans were performed in the wavelength range of 500–2000 cm^{-1}. Each spectrum was obtained based on an average of 10 scans. HAp disc (Plasma Biotal, Tideswell, UK) and loose HAp powder (Plasma Biotal, Tideswell, UK) references were recorded. Spectral scans were recorded of the exposed surface of the HAp discs after being treated. Treatment with HAp powder, was undertaken by placing 0.7 g (the equivalent weight of a HAp 20% porous disc) in 10 mL of 2% γ-PGA-Na (Table 1) and gently agitated in the formulation for 2 min before re-pelleting in a centrifuge for 2 min at 4000 rpm. The remaining γ-PGA was drawn out and the powder dried in a heat dryer for 1 h before a spectral scan was run.

Table 1. Summary of γ-PGA sample codes and relevant description.

Code	Description	Substrate
Xi'an γ-PGA	Commercial γ-PGA, hydrolysed to 66 kDa	Not specified
Natto γ-PGA	166 kDa commercial γ-PGA	Not specified
CS γ-PGA	520 kDa commercial γ-PGA	Not specified
YR Spec γ-PGA	440 kDa commercial γ-PGA	Not specified
Hydrolysed YR spec γ-PGA	102 kDa commercial hydrolysed γ-PGA	Not specified
GS, 0 g/L NaCl PTF	Modified commercial substrate	20 g/L L-glut, 50 g/L sucrose, 2.7 g/L KH_2PO_4, 4.2 g/L Na_2HPO_4, 0 g/L NaCl, 5 g/L $MgSO_4 \cdot 7H_2O$, 1 mL/L of Murashige-Skoog vitamin solution
GS, 50 g/L NaCl PTF	Commercial substrate	20 g/L L-glut, 50 g/L sucrose, 2.7 g/L KH_2PO_4, 4.2 g/L Na_2HPO_4, 50 g/L NaCl, 5 g/L $MgSO_4 \cdot 7H_2O$, 1 mL/L of Murashige-Skoog vitamin solution
L.d. Commercial flakes	Flakes from Cornish Seaweed	High shear mixing 3000 rpm for 3 min, 10 g/L L-glut, 25 g/L sucrose
L.d. Commercial flakes PTF		
L.d. SAMS PTF	Powder from Scottish Association for Marine Science ltd.	
N.o. Commercial Powder PTF	Powder from Algaecytes Ltd.	Ethanol extraction for 24 h, 10 g/L L-glut, 25 g/L sucrose

The average molar mass and molar mass distributions of the polymers were determined by gel permeation chromatography (GPC) with a differential refractive index detector (Δn-2010 RI WGE Dr. Bures, Berlin, Germany) and a multiangle laser light scattering detector (DAWN EOS, Wyatt Technologies, Santa Barbara, CA, USA) in buffer (0.15M $NaNO_3$, 0.01M EDTA, 0.02% NaN_3 and pH = 6 adjusted with NaOH). The following columns were used: guard PSS SUPREMA 10 μm and PSS SUPREMA analytical Linear XL 10 μm (Polymer Standards Service, Mainz, Germany). Measurements with nominal flow rate of 0.5 mL/min, at 40 °C were performed. ASTRA 4 software (Wyatt Technologies, Santa Barbara, CA, USA) was used to evaluate the results. All samples were filtered through the 0.45 μm PES syringe filters (Graphic Controls, DIA-Nielsen, Düren, Germany) before measurements. The refractive index increment of commercial γ-PGA (dn/dc = 0.142 mL/g) was estimated in independent measurement in buffer using a SEC-3010 dn/dc WGE Dr. Bures (Berlin, Germany) differential refractive index detector.

To attain high quality diffraction data for γ-PGA samples, a PANanalytical Empyrean X-ray diffractometer was employed. For sample analysis, approximately 200 mg of pure γ-PGA powder was homogeneously flattened within the holder with the aid of a glass slide. Analysis conditions employed were: Scan Axis: Gonio, Start Position [°2Th.]: 5.0090, End Position [°2Th.]: 99.9870, Step Size [°2Th.]: 0.0130, Scan Step Time [s]: 8.6700, Irradiated Length [mm]: 15.00, Specimen Length [mm]: 10.00, Measurement Temperature [°C] 25.00, Generator Settings: 40 mA, 40 kV, Goniometer Radius [mm]: 240.00, Dist. Focus-Diverge. Slit [mm]: 100.00.

2.1.7. Formulation of γ-PGA

The γ-PGA solutions were prepared using commercial and algal-derived γ-PGA powder from *Bacillus subtilis* natto (see Table 1). The molar mass of these γ-PGAs ranged from 4 to 3000 kDa. Solutions were made in different concentrations in the ranges of 0.25 to 4% (wt), with deionised water. Equal amounts of NaOH solutions in deionised water were prepared and added to the γ-PGA solutions to produce a soluble γ-PGA with pH ~ 7.00. Solutions using the sodium salt γ-PGA samples, with M of 66 kDa (hydrolysed from material obtained by Xi'an Zhongyum Biotechnology Co., Ltd., Xi'an, China), 166 kDa (Nippon Poly-Glu Co., Ltd., Osaka, Japan), 520 kDa γ-PGA (CS Innovation LLC., Schenectady, NY, USA) and 440 kDa (YRSpec, Tianjin, China) were prepared using deionised water.

2.1.8. Artificial Demineralisation Monitored Using ISE

Synthetic HAp discs (20% porosity—obtained from Plasma Biotal (Tideswell, UK)) γ-PGA protected or not, were immersed in a 0.1M acetic acid solution. The demineralisation solution of 0.1M acetic acid was prepared and buffered to pH 4.0 with 1M Sodium Hydroxide (NaOH) using a calibrated pH meter (Mettler Toledo: inLab Expert Go-ISM).

HAp discs were immersed in 5 mL of γ-PGA solutions for 2 or 5 min. A sealed test tube was used and manually shaken to replicate the swishing motion of a mouthwash. The discs were then immersed in 50 mL of 0.1M acetic acid, pH 4, for 1 h at 37.0 (±1.0 °C)

2.1.9. ISE or Real-Time Monitoring of Artificial Caries

ISEs have been used to measure the rate of loss of calcium from enamel or HAP discs as a proxy for the rate of demineralisation as described by [38]. A Ca^{2+} ion selective electrode (ISE) was paired with a double-junction lithium acetate reference electrode. The electrodes were calibrated using a serial dilution technique with a solution of pH4 and at 37 (±1.0 °C) while using a magnetic stirrer. A calibration curve was obtained by plotting the logarithm of calcium activity in millimols against the ISE readings in millivolts. For ISE readings, Ion-Selective Electrode for calcium ion (ELIT 8041 PVC membrane), a reference electrode: single junction silver chloride (ELIT 001n) was mounted on a dual electrode head (ELIT 201) obtained from Nico2000 (London, UK). An ELIT software was used to record the data Nico2000 (London, UK).

3. Results

3.1. Biosynthesis of γ-PGA from Macro-Algae and Micro-Algae Waste Fraction

The strong need for novel sustainable materials and products has significantly boosted algal agriculture. Herein, we tested the potential of algal substrate as a source of nutrients for the biosynthesis of value-added materials. FT-IR data [22] suggested that the brown macro-algae *Laminaria digitata* and the post-oil extracted micro-algae *Nanochloropsis oceanica* are suitable substrates for *Bacillus subtilis* natto biosynthesis of γ-PGA (Figure 2) [24].

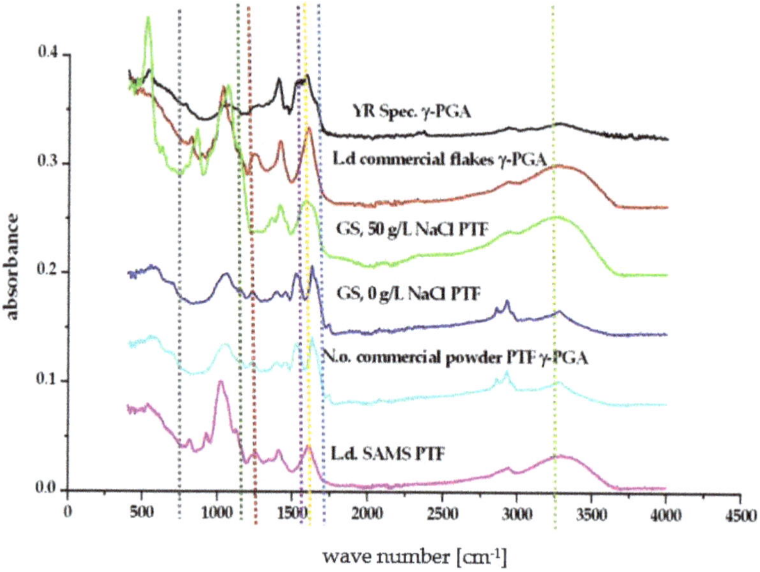

Figure 2. Fourier Transform Infra-Red spectra of γ-PGA obtained from different standard and complex waste substrates. The blue vertical line represents the C=O stretch, the orange line represents the Amide I N-H bend, the purple line represents the Amide II stretch, the red line represents the C=O symmetric stretch of γ-PGA in its sodium (1402 cm^{-1}) or calcium (1412 cm^{-1}) isoform, the dark green line represents the C-N stretch, the grey line represents the N-H bending, the light green line represents the O-H stretch.

Herein, *B. subtilis* natto cultivated on supplemented *L. digitata* flakes/powder and pre-extracted *N. oceanica* displayed significant biosynthesis of γ-PGA (Table 1, Figure 3). The data presented within Figure 3 suggested that variation in *Laminaria digitata* processing (commercial versus experimental treatments) significantly ($p < 0.001$) impacted polymeric yields after tangential flow purification (*L. digitata* commercial vs. *L. digitata* SAMS within Figure 3). Post-tangential flow yields of γ-PGA obtained from commercial *L. digitata* flakes were significantly higher compared to those obtained with standard GS media ($p = 0.003$). Differently, yields of γ-PGA obtained from commercial *L. digitata* flakes were approximately 22.5% lower compared to those obtained with standard GS media at 0 g/L NaCl. This difference was found to be significant ($p < 0.001$). The yields of γ-PGA from pre-extracted, supplemented *N. oceanica* were significantly lower ($p = 0.003$) than those obtained with either supplemented macro-algae media or standard GS media.

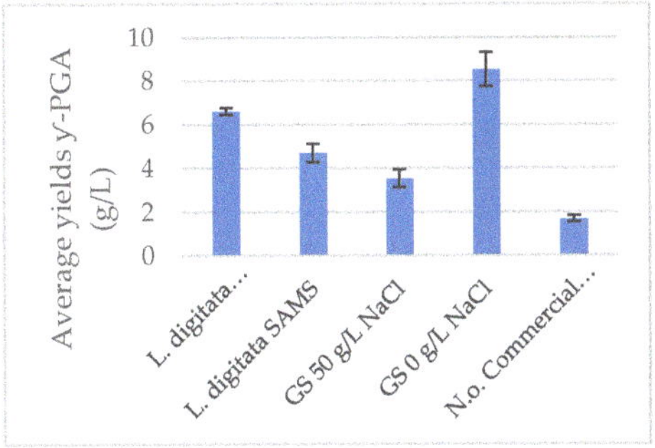

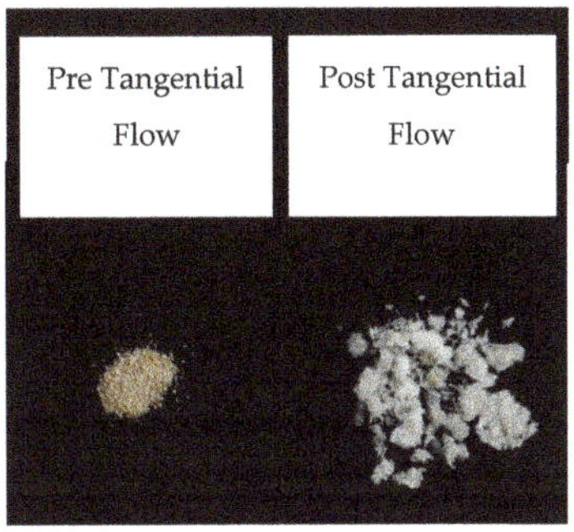

(a) (b)

Figure 3. (**a**) Average yields of γ-PGA obtained post-tangential flow with supplemented Laminaria digitata substrate (in flakes from commercial supplier and in powder from Scottish Association for Marine Science), pre-extracted and supplemented Nanochloropsis oceanica media, standard GS media (50 g/L NaCl) and modified GS media (0 g/L NaCl). (**b**) Pre- and post-tangential flow of purified γ-PGA obtained from L.d. commercial substrate. n = 3. Error bars indicate standard error values.

3.2. Physico-Chemical Characteristics of γ-PGA Synthesised from Macro-Algae and Micro-Algae Waste Fraction

As presented in Table 2, the molar mass of γ-PGA produced from *L. digitata* commercial flakes was lower compared to that obtained from SAMS monthly collected *L. digitata* samples. The molar mass of γ-PGA produced from *L. digitata* samples remained relatively constant across the months assessed. Surprisingly, the molar mass of SAMS *L. digitata* samples was comparable to that obtained with standard GS medium at 0 g/L NaCl. The molar mass of both standard GS media and GS media 0 g/L NaCl were significantly higher than those reported in literature for members of *Bacillus* sp. The molar mass for PTF *N. oceanica* was extremely low compared to other γ-PGAs isolated.

Table 2. Physico-chemical properties of commercial γ-PGA and γ-PGA synthesised from pre-extracted supplemented micro-algae media and pre-treated supplemented macro-algae media. A detailed summary of the cultivation media components (where applicable) can be found within Table 1. L.d.—*Laminaria digitata*, PTF: post-tangential flow purification.

Production Method	M_n [g/mol]	M_w [g/mol]	M_w/M_n	XRD
Xi'an γ-PGA	N.S.	66,000	Not Specified	Amorphous
Natto γ-PGA	47,800	166,000	3.5	Crystalline
CS γ-PGA	N.S.	520,000	Not Specified	Amorphous
Commercial γ-PGA	250,000	440,000	1.8	Amorphous
Hydrolysed commercial γ-PGA	33,200	102,000	3.1	Amorphous
GS, 0 g/L NaCl PTF	3,320,000	3,700,000	1.1	Amorphous
GS, 50 g/L NaCl PTF	1,810,000	2,700,000	1.5	Crystalline
L.d. Commercial flakes	10,900	145,000	13.3	Amorphous
L.d. Commercial flakes PTF	59,000	183,000	3.1	Amorphous
L.d. SAMS PTF	1,760,000	2,700,000	1.5	Amorphous
N.o. Commercial Powder PTF	1500	4600	3.1	Amorphous

3.3. The Impact of γ-PGA Concentrations on HAp Demineralisation

Ion selective electrode investigations are a generally accepted model employed to assess the demineralisation of human enamel in response to varying treatments [38]. The rate of HAp dissolution with increasing concentrations of γ-PGA was assessed with the aid of porous HAp discs. To this end, 0.5 to 4% (*w*/*v*) formulations of γ-PGA, with molar mass 66 kDa, were investigated and the results of this investigation are summarised in Figure 4.

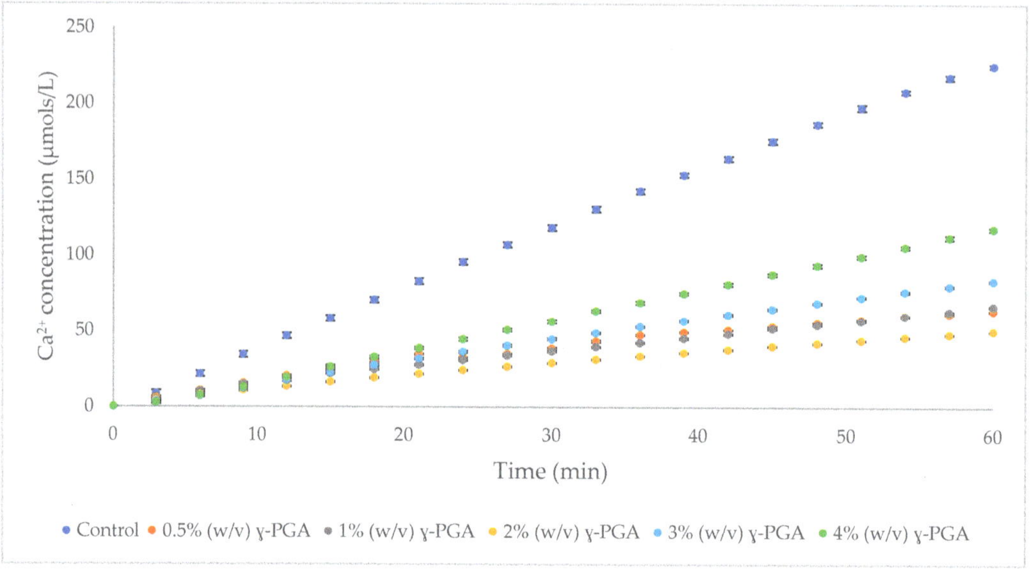

Figure 4. ISE study on the Ca^{2+} dissolution from HAp discs (20% porous) after being treated for 2 min with analogous γ-PGA solutions at different concentrations. n = 3. Error bars indicate standard error values.

All γ-PGA formulations tested imparted significant demineralisation inhibition on the HAp discs (Figure 4), with the greatest protection between 1 and 2% (*w*/*v*). The γ-PGA protection, from high to low, was observed as follows: 1% (*w*/*v*) > 2% (*w*/*v*) > 0.5% (*w*/*v*) > 3% (*w*/*v*) > 4% (*w*/*v*) γ-PGA formulations.

3.4. Comparison of γ-PGA Average Molar Mass on HAp Protection

Having established that the best HAp disc protection occurred with formulations presenting γ-PGA at 1% (*w*/*v*), commercial γ-PGAs with varying molar masses were assessed. The results of this investigation are schematically summarised in Figure 5.

There was little difference between the γ-PGA molecular mass tested and demineralisation inhibition (Figure 5). To assess whether this behaviour is concentration specific, or observed across all γ-PGA containing samples, 0.1 and 0.25% (*w*/*v*) formulations of γ-PGA were tested on HAp discs.

The overall optimum of HAp protection was achieved with 1% (*w*/*v*) γ-PGA (Figure 6). Significant variation was observed when both γ-PGA molar masses and its concentrations were compared. In this respect, variation in molar mass provided significant variation ($p < 0.05$). Further, different concentrations also presented significant variation in HAp protection ($p = 0.028$) (Two-way ANOVA).

From Figure 6 it appeared that better protection was achieved with 1% (*w*/*v*) formulations compared to 0.25% (*w*/*v*) formulations. Hence, we came to the conclusion that, regardless of chemical properties (Table 2), 1% (*w*/*v*) γ-PGA formulations were the most suitable in protecting HAp from dissolution across the range tested (0.1, 0.25, 0.5, 1, 2, 3 and 4% (*w*/*v*)).

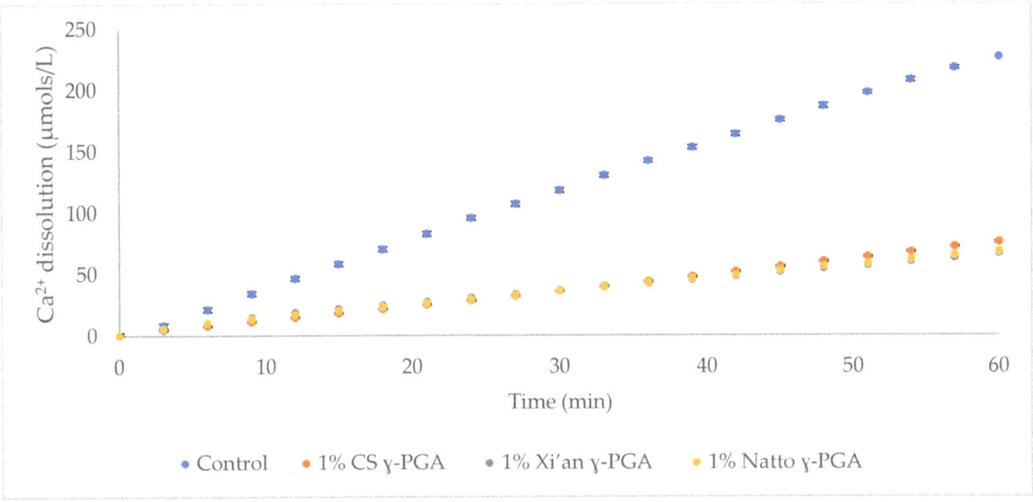

Figure 5. Linear regression analysis of the ISE pilot study result showing the Ca^{2+} activity from HAp disc (20% porous) protected by different commercial γ-PGA at 1% (*w*/*v*) with disc immersion for 2 min. n = 3. Error bars indicate standard error values.

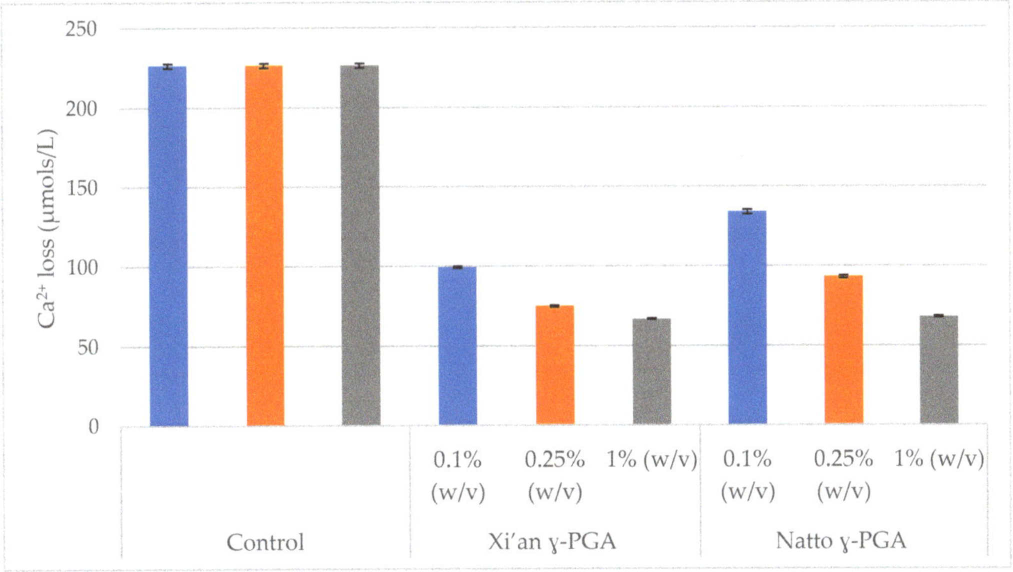

Figure 6. Comparison of the ISE pilot study result showing the Ca^{2+} activity from HAp disc (20% porous) protected by different commercial γ-PGAs at 0.1, 0.25 and 1% (*w*/*v*). n = 3. Error bars indicate standard error values.

To further investigate the association between γ-PGA and teeth enamel, FT-IR assessment was carried out for 0.25% (*w*/*v*) and 1% (*w*/*v*) formulations (see Figure 7). A small peak at the 1500–1600 cm^{-1} region of the spectra was visible on all HAp discs treated with both 0.25% (*w*/*v*) and 1% (*w*/*v*) γ-PGA after being immersed in 0.1M acetic acid for 1 h.

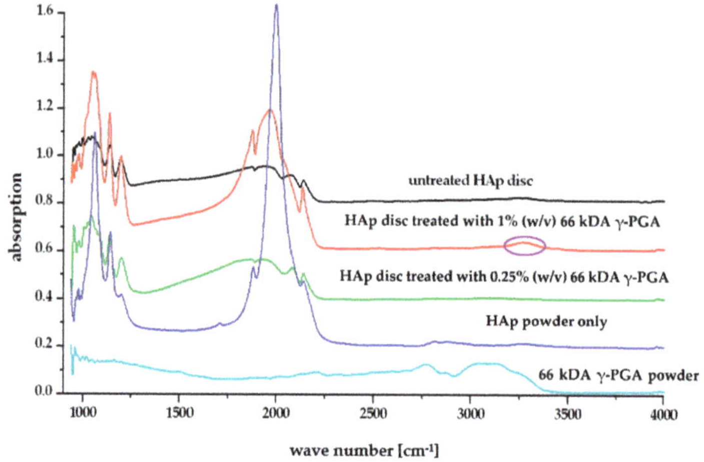

Figure 7. FT-IR profiles of HAp disc before and after 1 h acidic challenge in the presence of γ-PGA. Interaction peak, between γ-PGA and HAp disc, circled in purple.

3.5. Enamel Protection from Micro- and Macro-Algal Produced γ-PGA

Cost effective biosynthesis of raw materials is fundamental in product development. For this reason, sustainable and efficient synthesis of γ-PGA is necessary. To this end micro- and macro-algae substrates were employed for the synthesis of γ-PGA. Figure 8 summarises the HAp protection capability of γ-PGA synthesised from micro- and macro-algal substrates.

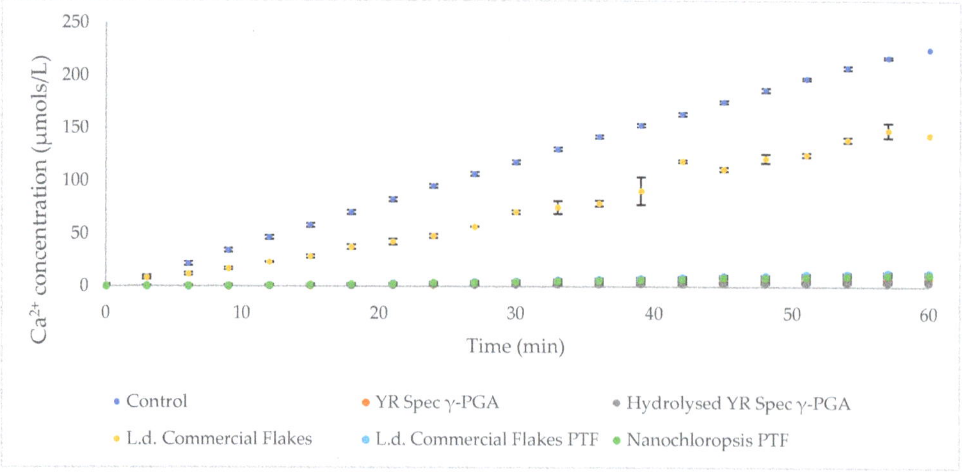

Figure 8. Protective effect of 1% (w/v) γ-PGA formulations obtained from micro- and macro-algal substrate towards HAp discs. The γ-PGA formulations tested were comprised of water only (control), commercial -440 KDa- γ-PGA (YR Spec γ-PGA), Commercial γ-PGA hydrolytically degraded (Hydrolysed YR Spec γ-PGA), γ-PGA produced from algal biomass raw precipitate, before tangential flow (L.d. Commercial Flakes), γ-PGA produced from algal biomass after tangential flow (L.d. Commercial Flakes PTF), γ-PGA produced from pre-solvent extracted, waste, micro-algal biomass fraction after tangential flow (N.o Commercial Powder PTF). A detailed summary of the cultivation media components (where applicable) can be found within Table 1. All HAp discs were immersed in the formulation for 5 min. n = 3. Error bars indicate standard error values.

The data again showed that all of the 1% (w/v) formulations tested demonstrated HAp protection, compared to control. The formulation which displayed the least protection was L.d. Commercial Flakes. All the other formulations displayed similar demineralisation inhibition.

4. Discussion

4.1. Variation in γ-PGA Yields and Physico-Chemical Properties with Micro- and Macro-Algal Substrate

Cost effective biosynthesis of ultra-pure γ-PGA is fundamental for its use in the biomedical field. To this end, we employed wild-cultivated macro-algae and solvent extracted micro-algae and tested their potential as substrates for γ-PGA biosynthesis. Our data suggested that both supplemented macro-algae and pre-extracted, supplemented micro-algae allowed for γ-PGA biosynthesis. Conventionally, γ-PGA is synthesised by members of the *Bacillus* species with average yields ranging from 4 g/L to 35 g/L [39–45]. Depending upon the nature of the substrate, yields reported can be of raw precipitates. In our study we reported yields of γ-PGA raw precipitate obtained from standard GS media of 9.8 g/L for 50 g/L NaCl and 14.5 g/L for 0 g/L NaCl, with 3.5 and 8.5 g/L from tangential flow purified GS media and GS media at 0 g/L NaCl. For complex waste media we reported raw yields at 10.3 g/L for *L. digitata* collected by SAMS and at 8.8 g/L for commercially available *L. digitata* flakes (data not shown). As summarised in Figure 3, purified *L. digitata* yielded 4.8 g/L and 6.6 g/L γ-PGA for samples collected by SAMS and commercial samples, respectively. These results were in line with the only other report of γ-PGA production from macro-algae (*Ulva* spp.) by Kim et al., (2019) [32]. Therein Kim et al., (2019) [32] reported a yield of 6.29 g/L of purified γ-PGA, in line with 6.6 and 4.78 g/L reported herein. The significant variation in yields observed can be explained by the variation in micronutrient content present within the algal substrate. It has been widely reported that the concentrations of Ca^{2+}, Zn^{2+}, Co^{2+}, Zn^{2+}, Mg^{2+}, Na^{2+}, Fe^{2+} and Mn^{2+} impact the yields, molar mass and D/L ratios of the resulting material [32,39,40,43,44,46–49]. The variation in concentration of these minerals could be greatly affected by the location of cultivation or by the pre-treatment method (sterilisation) commercial *L. digitata* could have incurred before packaging and distribution [50]. In this case, complex carbohydrates and proteinaceous material would have been mostly degraded into simple sugars and amino acids, respectively. These building blocks could have stimulated growth of *B. subtilis* natto more, compared to the SAMS *L. digitata* substrate. To the best of our knowledge, this is the first report of successful γ-PGA production from micro-algae. Herein, we reported the synthesis of 1.68 g/L of γ-PGA from supplemented, solvent extracted *Nanochloropsis oceanica*.

Although the yields obtained from pre-extracted micro-algae were relatively low, this cheap waste source could be particularly interesting for its valorisation, especially because the polymer produced presents relatively low molar mass of 4.6 kDa, often difficult to obtain with conventional substrates. Furthermore, another advantage of micro-algae-based media is the greater biomass consistency obtained, when cultivated in photobioreactors, compared to macro-algae, conventionally cultivated in ocean plots. In that respect, molar mass obtained with standard GS media and GS media without NaCl were 2700 kDa and 3700 kDa, respectively, which are significantly higher than those reported in literature for members of the *Bacillus* sp., which commonly range from 100 kDa to 2100 kDa [34,39–42,44,45], but analogous to the 2575 kDa reported by Bhat et al., (2013) [51] for *Bacillus subtilis* natto cultivated in standard GS media (50 g/L NaCl). Surprisingly, the variation in molar mass between commercial and SAMS samples of *L. digitata* observed was 16-fold. In fact, all samples of *L. digitata* collected by SAMS between 07.04.2021 and 01.07.2021 led to γ-PGA masses above 3000 kDa, compared to 184 kDa obtained from commercial samples. This result further supported the hypothesis previously formulated, for which it is suggested that variation in concentrations of Ca^{2+}, Zn^{2+}, Co^{2+}, Zn^{2+}, Mg^{2+}, Na^{2+}, Fe^{2+} and Mn^{2+} impact the yields, molar mass and D/L ratios of the resulting material [32,40,41,45,47–49].

4.2. Affinity and Protection of γ-PGA towards HAp

Alternative substrate engineering for the biosynthesis of γ-PGA, and increased understanding of its ability to maintain the integrity of dental enamel at low pH conditions, provides novel insights into the behaviour and physico-chemical interaction of this biomaterial. To assess the proposed protective effect of γ-PGA towards enamel, we used a synthetic enamel-like model system (20% porous calcium HAp discs), that has been used in many artificial caries studies [39,52,53]. Although these analogous materials do not have the carbonated HAp of enamel, chemical differences are small, However, structurally they are very different. Human tooth enamel consists of non-homogenous carbonated hydroxyapatite and contains ionic substitutions within the crystal lattice [8]. A pure calcium HAp has a higher solubility product than a non-homogenous carbonated HAp. Human enamel has a typical porosity of 1%. During acidic challenge, disassociated protons penetrate these pores causing dissolution of the tightly packed, parallel HAp enamel crystallites in the subsurface layers [20]. The synthetic HAp discs used here have a greater porosity, 20%, allowing for a much greater amount of demineralisation solution to penetrate the structure of the disc. The porous structure is not uniform throughout the disc and may interconnect throughout the entire structure [20]. This leads to greater surface contact of the protons with the HAp crystals, increasing the rate of Ca^{2+} dissolution released from the discs [20]. Untreated HAp discs released Ca^{2+} at a rate of 3 μmol/L/min in acetic acid of pH4, as extrapolated from Figure 4.

However, when 20% HAp discs were immersed in 0.1, 0.25, 0.5, 1, 2, 3 and 4% (*w*/*v*) formulations of γ-PGA for 2 min, Ca^{2+} dissolution could be reduced below 2 μmol/L/min, regardless of the concentration of γ-PGA. Figure 4 suggested that the lowest Ca^{2+} dissolution rate was observed with 1% (*w*/*v*) γ-PGA formulations. This result was not surprising, as lower γ-PGA concentrations struggle to effectively coat the surface of the disc, whereas higher concentrations are hypothesised to display greater chain to chain interaction, bridged by Ca^{2+} ions, rather than chain to HAp interaction. Although, chain length and enantiomeric profile are known to play a crucial role in polymeric interactions [54], this did not appear to be the case for commercial γ-PGA formulations at 1% (*w*/*v*) of different molar mass (Figure 5). Nonetheless, when the concentration of γ-PGA formulations was reduced, a distinct separation in protective effect was noticed. In fact, Figure 6 suggested that a significant variation in protective effect was observed with variation in polymeric molar mass. Our data suggested that at 0.25% (*w*/*v*) γ-PGA concentrations, higher molar mass provided the lowest Ca^{2+} dissolution, whereas at 0.1% (*w*/*v*), the lowest molar mass material displayed the greatest protection. This apparent dichotomy could be explained as follows: at 0.25% (*w*/*v*), the greater the molar mass of the material, more surface area is covered and lower demineralisation observed. In contrast, with lower molar mass polymers, enantiomeric properties significantly modulate the attachment of the material and ultimately lower Ca^{2+} dissolution. As discussed earlier, previous research suggests that the L-glutamic acid component of statherin is responsible for its attachment to the enamel [12,13], and, thus, it is possible that samples presenting a higher content of L-glutamic acid units had better affinity compared to samples that presented higher D-glutamic acid units. When the concentration of γ-PGA was reduced to 0.25% (*w*/*v*), the variation in protection was not limited to molar mass but could also possibly be affected by D/L enantiomeric distribution. Both the molar mass and the enantiomeric ratio are assumed to affect the protection as a consequence of γ-PGA binding to HAp disc. This multivariate behaviour can also be complicated by the presence of ions associated with the γ-PGA. A similar molar mass-protection pattern was also observed with 0.1 *w*/*v*% γ-PGA, but, in this case, 520 kDa and 200–400 kDa γ-PGA displayed analogous protection with 66 kDa γ-PGA displaying higher protection. These chemico-physical aspects have to be further investigated.

At both 0.25% (*w*/*v*) and 1% (*w*/*v*) concentration, the FT-IR data suggested that γ-PGA was chemically bound to HAp as novel peaks were observed at 1580 and 1620 cm^{-1} (Figure 7). These peaks were characteristic of γ-PGA and represented the α- and γ- amide linkages in the γ-PGA and were contributed to by -C=O stretching and -N-H bending modes [24]. These peaks were not visible in the spectra for untreated HAp powder; therefore, their presence on the treated HAp spectra indicated that γ-PGA had bound to the surface of the HAp [20]. The binding of γ-PGA to HAp has been suggested to occur due to the high affinity of the α-carboxyl groups of the residual side chains to calcium, forming stable ionic complexes via chelating ligands [55]. As the proposed HAp-γ-PGA and HAp-statherin binding mechanisms are suggested to be analogous, the observed reduction in Ca^{2+} dissolution could be a direct consequence of the shielding behaviour carried out by both γ-PGA and statherin. The shielding proposed effectively reduced the interaction between the acidic environment (H^+) and the HAp layer thereby reducing the concentration of soluble products of calcium hydroxyapatite, and ultimately preventing bioerosion. Although very weak, small peaks in this region were also visible on all spectra of the HAp discs after they had been treated with γ-PGA at different concentration and molar mass (Figure 7) following constant acidic challenge. This indicated that the concentration and the average molar mass did not affect the binding ability of γ-PGA to the HAp surface and that binding remained after a constant acidic challenge at pH, further supporting the shielding behaviour of γ-PGA.

4.3. HAp Protection through Micro- and Macro-Algal Produced γ-PGA

The γ-PGA synthesised from micro- and macro-algal substrate was further assessed by protection capability for HAp protection. In this case, five 1% (*w*/*v*) γ-PGA formulations were prepared and tested. The data summarised within Figure 8 suggested that, when HAp discs are exposed to γ-PGA formulations for 5 min, all tangential flow-purified polymers displayed >92% reduction in Ca^{2+} dissolution. Differently from the other formulations tested, sample 'L.d commercial' only displayed 31% reduction in Ca^{2+} dissolution, compared to the control. This variation in protection was attributed to the fact that the material produced had not undergone tangential flow purification. Being produced from a carbohydrate and protein rich substrate, non-γ-PGA-selective ethanol precipitation led to an inhomogeneous material. This hypothesis was greatly supported by the GPC investigations which suggested a dispersity index of 13, compared to the 3.2 dispersity index obtained with the tangential flow-purified sample (Table 2). Within this pre-purification sample, it was very likely that γ-PGA chains' available pendant carboxylic acids excessively interacted with other molecules and proteins, compared to HAp disc.

5. Conclusions

To conclude, we successfully demonstrated the biosynthesis of γ-PGA from cheap, complex sources (namely micro-algae and macro-algae) and its ability to protect enamel dissolution. The yields of pure polymer obtained from these sources was found to be higher compared to the costly GS media. However, that was not the case when the NaCl concentration of the GS media was reduced from 50 g/L to 0 g/L, in which case higher yields were obtained with the standard media. Interestingly, γ-PGA obtained by SAMS *Laminaria digitata* presented very high molar mass of 3000 kDa, analogous to those obtained with both GS media and GS media 0 g/L NaCl.

Investigations into γ-PGA-HAp interaction suggested that 1% (*w*/*v*) γ-PGA formulations presented the greatest HAp protecting properties. Although the molar mass did not appear to affect HAp protection capabilities at 1% (*w*/*v*), we showed that there was significant variation in HAp protection at lower γ-PGA concentration (0.25 and 0.1% (*w*/*v*)) ($p = 0.028$). The variation in such protective effect was attributed to both molar mass and, potentially, to the enantiomeric properties of the polymers. To further assess the effect of enantiomeric properties towards enamel attachment, further investigation will focus on the D- and L-glutamic acid residue and their effects towards HAp binding.

FT-IR analysis suggested that there was physical interaction between γ-PGA and HAp for all formulations tested. This interaction was evidenced by a small peak in the

1500–1600 cm^{-1} region. We aim to further confirm and visualise such interaction through micro-CT and other imaging techniques to establish the bonding and its efficiency over time.

Further, we have shown that by treating HAp discs with purified, micro- and macro-algae-derived γ-PGA significant reduction in enamel dissolution was achieved. The combination of purified polymers, alongside a five minute exposure of HAp discs to the γ-PGA formulation displayed a reduction in Ca^{2+} dissolution from 227μmol/L (control) to 11.4 and 14.1 μmol/L for N.o commercial PTF and L.d. commercial PTF, respectively.

Author Contributions: M.P., P.A., R.H., L.C. and I.R. were the main people involved in the planning of the experiments, interpretation of the data and γ-PGA characterization. M.P. and I.R. were responsible for the bacterial production of a γ-PGA and initial characterization. B.M. and M.K. were responsible for GPC analysis. M.P., B.M. and L.C. were responsible for the FT-IR analysis. P.A., L.C., R.H. and M.P. were responsible for γ-PGA's affinity experiments. M.P., I.R., P.A., L.C., I.K., R.H., and F.T.-M. helped with manuscript preparation. D.M. and M.S.S. were responsible for algal biomass preparation and manuscript preparation. All authors have read and agreed to the published version of the manuscript.

Funding: This work was partially supported the University of Wolverhampton Research Investment Fund (RIF4); ERDF Science in Industry Research Centre (SIRC 01R19P03464) project and BBSRC Algae-UK for Proof of Concept project BB/S009825/1; UCL Ref: 5749484.

Institutional Review Board Statement: Not applicable.

Informed Consent Statement: Not applicable.

Acknowledgments: We would like to thank Mandeep Kaur for the Institute of Dentistry, Barts and The London School of Medicine and Dentistry, Queen Mary University of London for her help with the use of the ion selective electrode, David Townrow and Diane Spencer from Faculty of Science and Engineering, University of Wolverhampton for the technical support.

Conflicts of Interest: The authors declare no conflict of interest.

References

1. Villa, A.; Connell, C.; Abati, S. Diagnosis and Management of Xerostomia and Hyposalivation. *Ther. Clin. Risk Manag.* **2014**, *11*, 45. [CrossRef] [PubMed]
2. Pedersen, A.M.L.; Sørensen, C.E.; Proctor, G.B.; Carpenter, G.H.; Ekström, J. Salivary Secretion in Health and Disease. *J. Oral Rehabil.* **2018**, *45*, 730–746. [CrossRef] [PubMed]
3. Villa, A.; Wolff, A.; Narayana, N.; Dawes, C.; Aframian, D.; Lynge Pedersen, A.; Vissink, A.; Aliko, A.; Sia, Y.; Joshi, R.; et al. World Workshop on Oral Medicine VI: A Systematic Review of Medication-Induced Salivary Gland Dysfunction. *Oral Dis.* **2016**, *22*, 365–382. [CrossRef] [PubMed]
4. Neville, B.W.; Damm, D.D.; Allen, C.M.; Chi, A.C. Physical and Chemical Injuries. In *Color Atlas of Oral and Maxillofacial Diseases*; Elsevier: New York, NY, USA, 2019.
5. Reed, R.; Xu, C.; Liu, Y.; Gorski, J.P.; Wang, Y.; Walker, M.P. Radiotherapy Effect on Nano-Mechanical Properties and Chemical Composition of Enamel and Dentine. *Arch. Oral Biol.* **2015**, *60*, 690–697. [CrossRef]
6. Anderson, P.; Hector, M.P.; Rampersad, M.A. Critical PH in Resting and Stimulated Whole Saliva in Groups of Children and Adults. *Int. J. Paediatr. Dent.* **2008**, *11*, 266–273. [CrossRef]
7. Söderling, E.; Le Bell, A.; Kirstilä, V.; Tenovuo, J. Betaine-Containing Toothpaste Relieves Subjective Symptoms of Dry Mouth. *Acta Odontol. Scand.* **1998**, *56*, 65–69. [CrossRef]
8. Fehrenbach, M.J.; Popowics, T. *Illustrated Dental Embryology, Histology, and Anatomy*; Missouri Elsevier: St. Louis, MO, USA, 2020. ISBN 9780323611084.
9. Dawes, C.; Pedersen, A.M.L.; Villa, A.; Ekström, J.; Proctor, G.B.; Vissink, A.; Aframian, D.; McGowan, R.; Aliko, A.; Narayana, N.; et al. The Functions of Human Saliva: A Review Sponsored by the World Workshop on Oral Medicine VI. *Arch. Oral Biol.* **2015**, *60*, 863–874. [CrossRef]
10. van der Reijden, W.A.; Veerman, E.C.I.; Nieuw Amerongen, A.V. Rheological Properties of Commercially Available Polysaccharides with Potential Use in Saliva Substitutes. *Biorheology* **1994**, *31*, 631–642. [CrossRef]
11. Chawhuaveang, D.D.; Yu, O.Y.; Yin, I.X.; Lam, W.Y.-H.; Mei, M.L.; Chu, C.-H. Acquired Salivary Pellicle and Oral Diseases: A Literature Review. *J. Dent. Sci.* **2021**, *16*, 523–529. [CrossRef]
12. Qamar, Z.; Haji, Z.B.; Rahim, A.; Chew, H.P.; Fatima, T. Poly-γ-Glutamic Acid a Substitute of Salivary Protein Statherin? *J. Chem. Soc. Pak.* **2016**, *38*, 730–736.

13. Goobes, G.; Goobes, R.; Shaw, W.J.; Gibson, J.M.; Long, J.R.; Raghunathan, V.; Schueler-Furman, O.; Popham, J.M.; Baker, D.; Campbell, C.T.; et al. The Structure, Dynamics, and Energetics of Protein Adsorption—Lessons Learned from Adsorption of Statherin to Hydroxyapatite. *Magn. Reson. Chem.* **2007**, *45*, S32–S47. [CrossRef] [PubMed]
14. Lussi, A.; Schlueter, N.; Rakhmatullina, E.; Ganss, C. Dental Erosion—An Overview with Emphasis on Chemical and Histopathological Aspects. *Caries Res.* **2011**, *45*, 2–12. [CrossRef] [PubMed]
15. Agostini, B.A.; Cericato, G.O.; da Silveira, E.R.; Nascimento, G.G.; Costa, F.d.S.; Thomson, W.M.; Demarco, F.F. How Common Is Dry Mouth? Systematic Review and Meta-Regression Analysis of Prevalence Estimates. *Braz. Dent. J.* **2018**, *29*, 606–618. [CrossRef]
16. Namita; Rai, R. Adolescent Rampant Caries. *Contemp. Clin. Dent.* **2012**, *3*, 122. [CrossRef] [PubMed]
17. Ogunleye, A. *Bacterial Poly-Gamma-Glutamic Acid (γ-pga)—A Promising Biosorbent of Heavy Metals*; University of Wolverhampton: Wolverhampton, UK, 2015.
18. Luo, Z.; Guo, Y.; Liu, J.; Qiu, H.; Zhao, M.; Zou, W.; Li, S. Microbial Synthesis of Poly-γ-Glutamic Acid: Current Progress, Challenges, and Future Perspectives. *Biotechnol. Biofuels* **2016**, *9*, 1–12. [CrossRef] [PubMed]
19. Ashiuchi, M.; Misono, H. Biochemistry and Molecular Genetics of Poly-γ-Glutamate Synthesis. *Appl. Microbiol. Biotechnol.* **2002**, *59*, 9–14. [CrossRef]
20. Qamar, Z.; Haji Abdul Rahim, Z.B.; Neon, G.S.; Chew, H.P.; Zeeshan, T. Effectiveness of Poly-γ-Glutamic Acid in Maintaining Enamel Integrity. *Arch. Oral Biol.* **2019**, *106*, 104482. [CrossRef]
21. Buescher, J.M.; Margaritis, A. Microbial Biosynthesis of Polyglutamic Acid Biopolymer and Applications in the Biopharmaceutical, Biomedical and Food Industries. *Crit. Rev. Biotechnol.* **2007**, *27*, 1–19. [CrossRef]
22. Ho, G.-H.; Ho, T.-I.; Hsieh, K.-H.; Su, Y.-C.; Lin, P.-Y.; Yang, J.; Yang, K.-H.; Yang, S.-C. γ-Polyglutamic Acid Produced ByBacillus Subtilis(Natto): Structural Characteristics, Chemical Properties and Biological Functionalities. *J. Chin. Chem. Soc.* **2006**, *53*, 1363–1384. [CrossRef]
23. Tanimoto, H.; Fox, T.; Eagles, J.; Satoh, H.; Nozawa, H.; Okiyama, A.; Morinaga, Y.; Fairweather-Tait, S.J. Acute Effect of Poly-γ-Glutamic Acid on Calcium Absorption in Post-Menopausal Women. *J. Am. Coll. Nutr.* **2007**, *26*, 645–649. [CrossRef]
24. Tanimoto, H.; Mori, M.; Motoki, M.; Torii, K.; Kadowaki, M.; Noguchi, T. Natto Mucilage Containing Poly-γ-Glutamic Acid Increases Soluble Calcium in the Rat Small Intestine. *Biosci. Biotechnol. Biochem.* **2001**, *65*, 516–521. [CrossRef] [PubMed]
25. Kimura, K. Characterization of Bacillus Subtilis Gamma-Glutamyltransferase and Its Involvement in the Degradation of Capsule Poly-Gamma-Glutamate. *Microbiology* **2004**, *150*, 4115–4123. [CrossRef] [PubMed]
26. Moradali, M.F.; Rehm, B.H.A. Bacterial Biopolymers: From Pathogenesis to Advanced Materials. *Nat. Rev. Microbiol.* **2020**, *18*, 195–210. [CrossRef] [PubMed]
27. Hezayen, F.F.; Rehm, B.H.; Tindall, B.J.; Steinbüchel, A. Transfer of Natrialba Asiatica B1T to Natrialba Taiwanensis Sp. Nov. And Description of Natrialba Aegyptiaca Sp. Nov., a Novel Extremely Halophilic, Aerobic, Non-Pigmented Member of the Archaea from Egypt That Produces Extracellular Poly(Glutamic Acid). *Int. J. Syst. Evol. Microbiol.* **2001**, *51*, 1133–1142. [CrossRef] [PubMed]
28. McLean, R.J.C.; Beauchemin, D.; Clapham, L.; Beveridge, T.J. Metal-Binding Characteristics of the Gamma-Glutamyl Capsular Polymer of Bacillus Licheniformis ATCC 9945. *Appl. Environ. Microbiol.* **1990**, *56*, 3671–3677. [CrossRef]
29. Kaplan, D. *Biopolymers from Renewable Resources*; Springer: Berlin/Heidelberg, Germany; New York, NY, USA, 1998. ISBN 9783540635673.
30. Bajaj, I.; Singhal, R. Poly (Glutamic Acid)—An Emerging Biopolymer of Commercial Interest. *Bioresour. Technol.* **2011**, *102*, 5551–5561. [CrossRef]
31. Zhang, D.; Feng, X.; Zhou, Z.; Zhang, Y.; Xu, H. Economical Production of Poly(γ-Glutamic Acid) Using Untreated Cane Molasses and Monosodium Glutamate Waste Liquor by Bacillus Subtilis NX-2. *Bioresour. Technol.* **2012**, *114*, 583–588. [CrossRef]
32. Kim, J.; Lee, J.M.; Jang, W.J.; Park, H.D.; Kim, Y.; Kim, C.; Kong, I. Efficient Production of Poly γ-D-Glutamic Acid from the Bloom-Forming Green Macroalgae, Ulva Sp., by Bacillus Sp. SJ-10. *Biotechnol. Bioeng.* **2019**, *116*, 1594–1603. [CrossRef]
33. Duarte, C.M.; Bruhn, A.; Krause-Jensen, D. A Seaweed Aquaculture Imperative to Meet Global Sustainability Targets. *Nat. Sustain.* **2021**, 1–9. [CrossRef]
34. Jumaidin, R.; Sapuan, S.M.; Jawaid, M.; Ishak, M.R.; Sahari, J. Seaweeds as Renewable Sources for Biopolymers and Its Composites: A Review. *Curr. Anal. Chem.* **2018**, *14*, 249–267. [CrossRef]
35. Bajaj, I.B.; Lele, S.S.; Singhal, R.S. Enhanced Production of Poly (γ-Glutamic Acid) from Bacillus Licheniformis NCIM 2324 in Solid State Fermentation. *J. Ind. Microbiol. Biotechnol.* **2008**, *35*, 1581–1586. [CrossRef] [PubMed]
36. Mayers, J.J.; Flynn, K.J.; Shields, R.J. Rapid Determination of Bulk Microalgal Biochemical Composition by Fourier-Transform Infrared Spectroscopy. *Bioresour. Technol.* **2013**, *148*, 215–220. [CrossRef] [PubMed]
37. Miles, A.A.; Misra, S.S.; Irwin, J.O. The Estimation of the Bactericidal Power of the Blood. *Epidemiol. Infect.* **1938**, *38*, 732–749. [CrossRef] [PubMed]
38. Huang, W.-T.; Shahid, S.; Anderson, P. Validation of a Real-Time ISE Methodology to Quantify the Influence of Inhibitors of Demineralization Kinetics in Vitro Using a Hydroxyapatite Model System. *Caries Res.* **2018**, *52*, 598–603. [CrossRef]
39. Fang, J.; Huan, C.; Liu, Y.; Xu, L.; Yan, Z. Bioconversion of Agricultural Waste into Poly-γ-Glutamic Acid in Solid-State Bioreactors at Different Scales. *Waste Manag.* **2020**, *102*, 939–948. [CrossRef]

40. Kedia, G.; Hill, D.; Hill, R.; Radecka, I. Production of Poly-γ-Glutamic Acid by Bacillus Subtilis and Bacillus Licheniformis with Different Growth Media. *J. Nanosci. Nanotechnol.* **2010**, *10*, 5926–5934. [CrossRef]
41. Birrer, G.A.; Cromwick, A.-M.; Gross, R.A. γ-Poly(Glutamic Acid) Formation by Bacillus Licheniformis 9945a: Physiological and Biochemical Studies. *Int. J. Biol. Macromol.* **1994**, *16*, 265–275. [CrossRef]
42. Bajaj, I.B.; Lele, S.S.; Singhal, R.S. A Statistical Approach to Optimization of Fermentative Production of Poly(γ-Glutamic Acid) from Bacillus Licheniformis NCIM 2324. *Bioresour. Technol.* **2009**, *100*, 826–832. [CrossRef]
43. Ashiuchi, M. *Effective Synthesis of Bio-nylon Materials Using Bacillus Megaterium*; Kochi University Repository: Kochi, Japan, 2007.
44. Shimizu, K.; Nakamura, H.; Ashiuchi, M. Salt-Inducible Bionylon Polymer from Bacillus Megaterium. *Appl. Environ. Microbiol.* **2007**, *73*, 2378–2379. [CrossRef]
45. Ashiuchi, M.; Shimizu, K. Process for Producing Poly-y-Glutamic Acid Having High Optical Purity. U.S. 20100233764A1, 16 September 2010.
46. Ji, Y.; Song, W.; Xu, L.; Yu, D.-G.; Annie Bligh, S.W. A Review on Electrospun Poly(Amino Acid) Nanofibers and Their Applications of Hemostasis and Wound Healing. *Biomolecules* **2022**, *12*, 794. [CrossRef]
47. Ashiuchi, M.; Kamei, T.; Baek, D.-H.; Shin, S.-Y.; Sung, M.-H.; Soda, K.; Yagi, T.; Misono, H. Isolation of Bacillus Subtilis (Chungkookjang), a Poly-Gamma-Glutamate Producer with High Genetic Competence. *Appl. Microbiol. Biotechnol.* **2001**, *57*, 764–769. [CrossRef] [PubMed]
48. Shih, I.-L.; Van, Y.-T. The Production of Poly-(γ-Glutamic Acid) from Microorganisms and Its Various Applications. *Bioresour. Technol.* **2001**, *79*, 207–225. [CrossRef]
49. Yamashiro, D.; Yoshioka, M.; Ashiuchi, M. Bacillus Subtilis PgsE (Formerly YwtC) Stimulates Poly-γ-Glutamate Production in the Presence of Zinc. *Biotechnol. Bioeng.* **2010**, *108*, 226–230. [CrossRef] [PubMed]
50. Bhat, A.R.; Irorere, V.U.; Bartlett, T.; Hill, D.; Kedia, G.; Morris, M.R.; Charalampopoulos, D.; Radecka, I. Bacillus Subtilis Natto: A Non-Toxic Source of Poly-γ-Glutamic Acid That Could Be Used as a Cryoprotectant for Probiotic Bacteria. *AMB Express* **2013**, *3*, 36. [CrossRef]
51. Jens, B.H.-N.; Ehiaze, A.E. *Biomass Supply Chains for Bioenergy and Biorefining*; Woodhead Publishing: Oxford, UK, 2016; pp. 319–332, ISBN 9781782423669.
52. Kamarudin, A.; Anderson, P.; Hill, R. The Effect of Different Fluoride Varnishes on the Release of Calcium Ions from Hydroxyapatite Discs: An Ion-Selective Electrodes Study. *Padjadjaran J. Dent.* **2020**, *32*, 82–90. [CrossRef]
53. Abou Neel, E.; Aljabo, A.; Strange, A.; Ibrahim, S.; Coathup, M.; Young, A.; Bozec, L.; Mudera, V. Demineralization–Remineralization Dynamics in Teeth and Bone. *Int. J. Nanomed.* **2016**, *11*, 4743–4763. [CrossRef]
54. Gross, R.A. Bacterial Y-Poly(Glutamic Acid). In *Biopolymers from Renewable Resources*; Springer: Berlin/Heidelberg, Germany, 1998; pp. 195–219.
55. Park, S.-B.; Hasegawa, U.; van der Vlies, A.J.; Sung, M.-H.; Uyama, H. Preparation of Poly(γ-Glutamic Acid)/Hydroxyapatite Monolith via Biomineralization for Bone Tissue Engineering. *J. Biomater. Sci. Polym. Ed.* **2014**, *25*, 1875–1890. [CrossRef]

polymers

MDPI

Article

Multifunctional PEG Carrier by Chemoenzymatic Synthesis for Drug Delivery Systems: In Memory of Professor Andrzej Dworak

Judit E. Puskas [1,*], Gayatri Shrikhande [2], Eniko Krisch [1] and Kristof Molnar [1]

[1] Department of Food, Agricultural and Biological Engineering, College of Food, Agricultural, and Environmental Sciences, The Ohio State University, 222 FABE, 1680 Madison Avenue, Wooster, OH 44691, USA; molnarnekrisch.1@osu.edu (E.K.); molnar.182@osu.edu (K.M.)
[2] Dantari, Inc., 1290 Rancho Conejo Blvd, Suite 103, Thousand Oaks, CA 91320, USA; gayatri.shrikhande20@gmail.com
* Correspondence: puskas.19@osu.edu

Abstract: This paper describes the synthesis and characterization of new bivalent folate-targeted PEGylated doxorubicin (FA_2-dPEG-DOX_2) made by modular chemo-enzymatic processes using *Candida antarctica* lipase B (CALB) as a biocatalyst. Unique features are the use of monodisperse PEG (dPEG) and the synthesis of thiol-functionalized folic acid yielding exclusive γ-conjugation of folic acid (FA) to dPEG. The polymer-based drug conjugate is built up by a series of transesterification and Michael addition reactions all catalyzed be CALB. In comparison with other methods in the literature, the modular approach with enzyme catalysis leads to selectivity, full conversion and high yield, and no transition metal catalyst residues. The intermediate product with four acrylate groups is an excellent platform for Michael-addition-type reactions for a wide variety of biologically active molecules. The chemical structures were confirmed by nuclear magnetic resonance spectroscopy (NMR). Flow cytometry analysis showed that, at 10 μM concentration, both free DOX and FA_2-dPEG-DOX_2 were taken up by 99.9% of triple-negative breast cancer cells in 2 h. Fluorescence was detected for 5 days after injecting compound IV into mice. Preliminary results showed that intra-tumoral injection seemed to delay tumor growth more than intravenous delivery.

Keywords: enzyme catalysis; discrete poly(ethylene glycol) dPEG; polymer drug conjugate; modular assembly; doxorubicin; folic acid; Michael addition

Citation: Puskas, J.E.; Shrikhande, G.; Krisch, E.; Molnar, K. Multifunctional PEG Carrier by Chemoenzymatic Synthesis for Drug Delivery Systems: In Memory of Professor Andrzej Dworak. *Polymers* **2022**, *14*, 2900. https://doi.org/10.3390/polym14142900

Academic Editors: Marek Kowalczuk, Iza Radecka and Barbara Trzebicka

Received: 6 June 2022
Accepted: 11 July 2022
Published: 16 July 2022

1. Introduction

Targeted drug delivery systems promise to send cancer drugs to diseased cells without affecting healthy cells, thereby reducing cytotoxicity and minimizing devastating side-effects [1–6]. Such delivery systems consist of a drug or diagnostic agent (or both), a linker, a cleavable bond for the release of the drug and a targeting agent, all built into one molecule [7]. Diagnostic and therapeutic agents are being developed that target vitamin receptors (e.g., folate or biotin receptors) that are highly concentrated on the surface of cancer cells [8,9]. Most reports discuss compounds containing folic acid (FA) targeting folate receptors (FR) [10,11]. The two major groups of compounds studied are small molecule drug conjugates and polymeric drug conjugates (PDCs) [12,13]. This latter group showed promise due to increased water solubility and circulation time in the body and multivalent attachment to FRs [14–17]. However, as a recent review pointed out, the greatest challenge is the inherent heterogeneity of PDCs, coupled with uncontrolled conjugation of diagnostic and therapeutic agents, resulting in polydisperse polymer mixtures [6]. There are a wide variety of polymers used in the synthesis of PDCs with poly(amido amine) dendrimers and poly(ethylene glycol) (PEG) being the most well-known [18–20]. We investigated monodisperse PDCs based on PEG, specifically, dPEG (discrete PEG with Đ = 1). First, we synthesized fluorescein (FL)-labeled PEGs containing two FA (FA-FL-PEG-FL-FA, in comparison with compounds with one or two FA (FA-FL, and FA-FL-FA), all made by chemo-enzymatic methods with excellent yield (95%+) and selectivity (100%) [21].

FA-FL-FA with two FA showed better endocytosis in both MDA-MB-231 (Caucasian) and MDA-MB-468 (African American, less FR) triple-negative breast cancer (TNBC) cell lines than FA-FL with a single targeting group. dPEG20, with precisely 20 repeat units with no polydispersity and two FA in each molecule, demonstrated the best uptake, in comparison with polydisperse PEGs with M_n = 1050 and 2000 g/mol. This was the first instance of using dPEG in FR-targeted PDCs. The uptake of FA-FL-PEG-FL-FA was monitored in vivo using a rat liver cancer model [22]. For intravenous delivery, tissue autofluorescence interfered with monitoring. In contrast, intra-arterial delivery led to accumulation in the tumor. FL is used extensively in cell culture studies but it is less than optimal for in vivo monitoring [14]. Therefore, we designed a new PDC platform based on a four-functional dPEG core to which drug and diagnostic molecules could be attached via enzyme-catalyzed Michael addition. The first compound tested was a bivalent folate-targeted PEGylated doxorubicin (DOX) serving as both a drug and an imaging agent, made by modular chemo-enzymatic processes (FA_2-dPEG-DOX_2) [23]. DOX is a widely used chemotherapeutic drug, which prohibits cell division by blocking the topoisomerase 2 enzyme [24]. It is also one of the most often chosen drugs in the synthesis of PDCs [25,26]. It fluoresces in red, enabling in vitro and in vivo tracking of drug release and distribution by fluorescent imaging techniques [27]. Our synthetic strategy is shown in Scheme 1. Exclusive γ-conjugation of FA was achieved using FA-SH made with a chemo-enzymatic method [28]. Flow cytometry analysis showed that, at 10 μM concentration, both free DOX and FA_2-dPEG-DOX_2 would be taken up by 99.9% of TNBC cells in 2 h. However, no cytotoxicity was found in the first 24 h. Slow cytotoxicity development led us to the conclusion that DOX was released slowly from the compound. Preliminary testing revealed that intra-tumoral injection of mice seemed to delay tumor growth more than intravenous delivery. Thus, this PDC showed great promise.

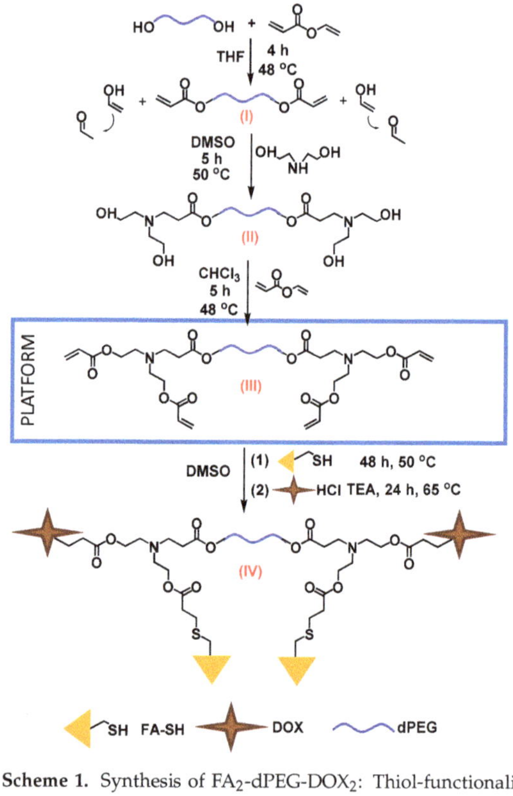

Scheme 1. Synthesis of FA_2-dPEG-DOX_2: Thiol-functionalized folic acid (FA-SH), Doxorubicin (DOX), discrete poly(ethylene glycol) (dPEG).

This paper discusses the synthesis of FA_2-dPEG-DOX_2 and the challenges associated with characterization of the compound.

2. Materials and Methods

Discrete poly(ethylene glycol) ($dPEG_{20}$, FW = 882 g/mol, Đ = 1.00) was purchased from Quanta Biodesign Limited (Plain City, OH, USA). Doxorubicin hydrochloride (DOX.HCl, CAS 25316-40-9) was purchased from AvaChem Scientific (San Antonio, TX, USA). DL-α-tocopherol (Vitamin E, purity 97+%) was obtained from Alfa Aesar (Tewksbury, MA, USA). Thiol-functionalized folic acid (FA-SH) was synthesized as reported in [28]. *Candida antarctica* lipase B immobilized on acrylic resin (CALB, Novozyme 235), Vinyl acrylate (VA, <600 ppm MEHQ as inhibitor), diethanolamine (DEA, 99%), tetrahydrofuran (THF, ACS reagent grade), *n*-hexane (Hexane, ACS reagent grade), and anhydrous dimethyl sulfoxide (DMSO, ≥99.9%) were purchased from Sigma-Aldrich (Darmstadt, Germany) and used without further purification. Other solvents, such as anhydrous diethyl ether (95.8%, BHT free ACS Certified), methanol (ACS Certified), and acetone (ACS Certified), were obtained from Fisher Chemicals (Pittsburgh, PA, USA). Deuterated solvents, such as dimethyl sulfoxide (DMSO *d6*, purity 99.9%), chloroform ($CDCl_3$, purity 99.8%), methylene chloride (CD_2Cl_2, purity 99.9%), and methanol (CD_3OD, purity 99.8%) were purchased from Cambridge Isotope Laboratories (Tewksbury, MA, USA).

2.1. Synthesis

2.1.1. Synthesis of Compound I: Acrylate-$dPEG_{20}$-Acrylate

dPEG (1.4398 g, 0.0016 mol, 1 equivalent) was placed into a 25 mL round-bottom flask and dried under vacuum on a Schlenk line at 65 °C until bubble formation ceased. It was then cooled to room temperature and VA (0.3928 g, 0.0040 mol, 2.5 equivalents), CALB (0.1332 g @ 20 wt.% enzyme, 3×10^{-4} mol/L) and vitamin E (antioxidant) were added to the reaction mixture which was heated to 48 °C in an oil bath. After 4 h, the reaction mixture was diluted with 10 mL of dried THF. CALB was filtered over a Q5 filter paper and THF and VA were removed by a rotary evaporator under reduced pressure. The product was then dried in a vacuum oven. An amount of 1.3191 g (1.31 mmol) diacrylated product was obtained (82% yield).

2.1.2. Synthesis of Compound II: $(HO)_2$-$dPEG_{20}$-$(OH)_2$

Acrylate-$dPEG_{20}$-acrylate (1.3191 g, 0.0013 mol, 1 equivalent), DEA (0.2782 g, 0.0026 mol, 2.02 equivalents) and 0.4 mL of DMSO were added to a 25 mL round-bottom flask and stirred at room temperature for 10 min. CALB (0.1089 g, 20 wt.% enzyme, 3×10^{-4} mol/L). One drop of vitamin E (antioxidant) was added to the reaction mixture which was then heated in an oil bath for 5 h at 50 °C. The reaction mixture was then taken out of the oil bath and diluted with 10 mL of THF. CALB was filtered over a Q5 filter paper and THF was removed using a rotary evaporator under reduced pressure. The product was then precipitated twice in 150 mL of hexane to remove excess DEA and DMSO, followed by drying of the product in a vacuum oven for 2 days. An amount of 1.0365 g (0.851 mmol) product was obtained (65% yield).

2.1.3. Synthesis of Compound III: $(Acr)_2$-dPEG-$(Acr)_2$

$(HO)_2$-$dPEG_{20}$-$(OH)_2$ (1.0365 g, 0.0009 mol, 1 equivalent) was mixed with VA (0.3427 g, 0.0035 mol, 4.10 equivalent) and 1.5 mL of $CHCl_3$ and stirred at room temperature for 10 min. CALB (0.1133 g, 20 wt.% enzyme, 4×10^{-4} mol/L) and a drop of vitamin E (antioxidant) were added to the mixture which was kept in an oil bath for 5 h at 48 °C. Then the reaction mixture was diluted with 15 mL of THF. CALB was filtered over a Q5 filter paper and THF and VA were then removed by a rotary evaporator under reduced pressure. The product was then dried in a vacuum oven. An amount of 0.3290 g (0.229 mmol) $(Acrylate)_2$-$dPEG_{20}$-$(Acrylate)_2$ was obtained (26% yield).

2.1.4. Synthesis of Compound IV: FA_2-dPEG-DOX_2

Considering that the FA-SH used contained ~28 mol% of unreacted FA, FA-SH (0.3830 g, 0.0007 mol, 2.84 equivalents) was reacted with $(Acrylate)_2$-$dPEG_{20}$-$(Acrylate)_2$ (0.3290 g, 0.0002 mol, 1 equivalent) using CALB (20 wt.% enzyme, 3×10^{-4} mol/L) in 1.4 mL of DMSO and a drop of vitamin E. The progress of the reaction was monitored by ^{1}H-NMR. After 3 days, DOX (0.2659 g, 0.0005 mol, 2.13 equivalents) was first desalted using TEA and then added to the reaction mixture of the previous reaction. After running the reaction for 24 h at 65 °C, CALB was filtered out and the product was obtained by precipitation in 300 mL diethyl ether.

2.2. Characterization

Nuclear Magnetic Resonance (NMR) Spectroscopy

A Varian Mercury 300 MHz spectrometer was used to record the ^{1}H-NMR spectra at 40 mg/mL concentration with the following parameters: 2 s relaxation time, 64 scans, and a 45° half-angle.

3. Results and Discussion

3.1. Synthesis of Acrylate-dPEG$_{20}$-Acrylate (I)

Figure 1 shows the ^{1}H-NMR of dPEG. The integral ratio of protons (c + d) with respect to proton (b) was 4.01:76.34, in excellent agreement with the theoretical 4:76 ratio.

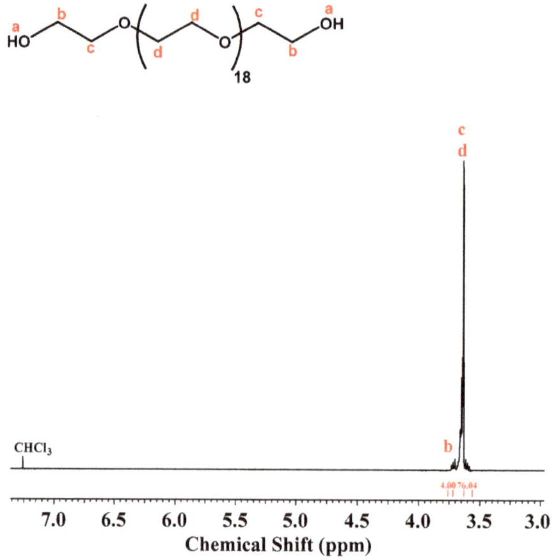

Figure 1. ^{1}H-NMR of dPEG$_{20}$ ([δ 3.73 ppm ((b), s, 4H), 3.64 ((c, d), 76H)]).

dPEG was reacted with VA in bulk; no solvent was necessary because the liquified dPEG was miscible with VA. This reaction is irreversible because the vinyl alcohol product immediately tautomerizes into acetaldehyde that evaporates from the system. The ^{1}H-NMR spectrum of the Acr-dPEG-Acr product after purification is shown in Figure 2. Resonance b shifted from 3.73 ppm to 4.30 ppm (b′). The integral ratio of the methylene (g, g′) and methine (f) protons of the acrylate group and proton (b′) relative to the reference protons of dPEG (c + d) was 2.20: 1.96: 2.16: 3.97: 80.01. This demonstrated successful transesterification between VA and dPEG and confirmed the structure of the diacrylate product.

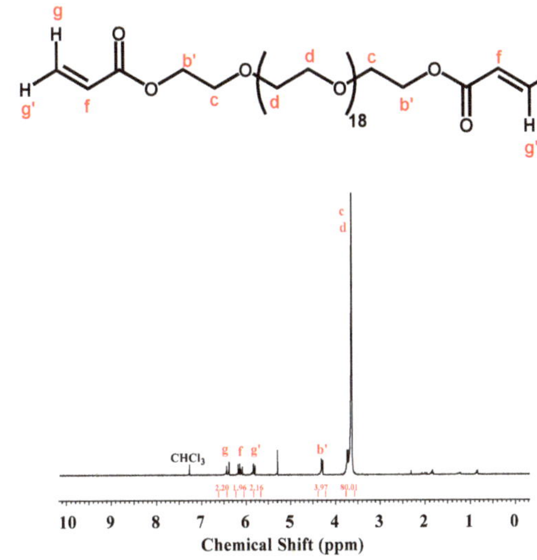

Figure 2. ^{1}H-NMR of the Acrylate-dPEG$_{20}$-Acrylate in $CDCl_3$ ([δ 6.50–6.38 ppm ((g), 2H), 6.21–6.07 ((f), 76H), 5.94–5.80 ((g′), 2H), 4.30 ((b′), 4H), 77–3.59 ((c,d), 80H)]).

3.2. Synthesis of (HO)$_2$-dPEG$_{20}$-(OH)$_2$

Acrylate-dPEG20-Acrylate was reacted with DEA in DMSO using CALB catalysis. Figure 3 shows the ^{1}H-NMR of the product after purification and drying. Signal (i) of the DEA shifted from 3.42 to 3.56 ppm (i′), and signal (h) shifted from 2.54 ppm to 2.59 ppm (h′) after the reaction. No methylene and methine protons due to the acrylate were present. The integral ratio of proton (h′) and newly generated signals (g) and (f) with respect to the reference proton (b) was 8.25: 4.02: 4.03: 4.06 which demonstrated successful Michael addition between DEA and dPEG-diacrylate and confirmed the structure of the product.

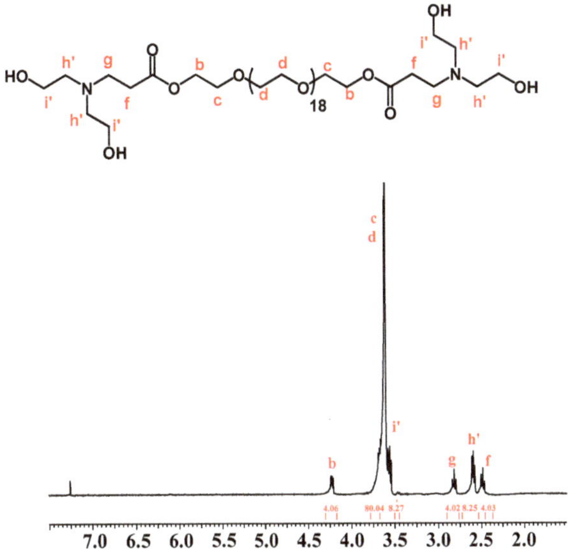

Figure 3. ^{1}H NMR spectrum of *(HO)$_2$-dPEG$_{20}$-(OH)$_2$* ([δ 4.24 ((b), 4H), δ 3.77–3.59 ((c,d), 80H), δ 3.56 ((i′), 8H), δ 2.83 ((g), 4H), δ 2.59 ((h′), 8H), δ 2.48 ((f), 4H)]).

3.3. *Synthesis of (Acrylate)$_2$-dPEG$_{20}$-(Acrylate)$_2$*

(HO)$_2$-dPEG$_{20}$-(OH)$_2$ was reacted with four equivalents of VA using CALB catalysis. As mentioned before, this transesterification reaction is irreversible. Figure 4 shows the ^{1}H-NMR spectrum of the product after purification and drying. Signal (i′) moved from 3.56 ppm to 4.21 ppm (i″) and signal (h′) moved from 2.59 ppm to 2.69 ppm (h″) after the reaction.

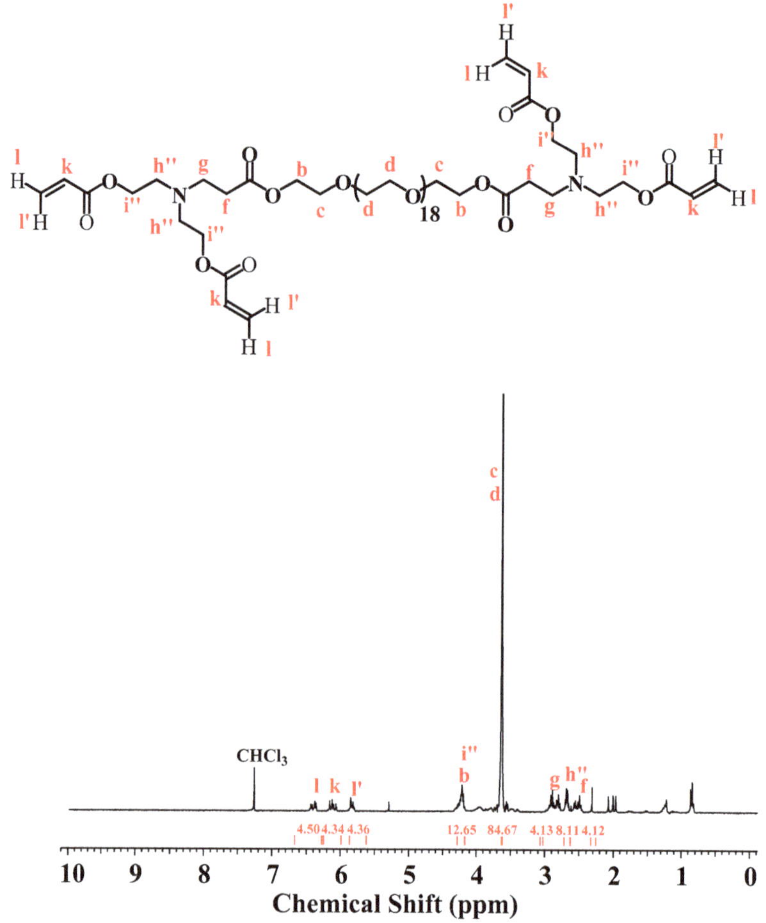

Figure 4. ^{1}H-NMR of the (Acrylate)$_2$-dPEG$_{20}$-(Acrylate)$_2$ in CDCl$_3$ ([δ 6.39 ppm ((l), 4H), δ 6.13 ppm ((k), 4H), δ 5.38 ppm ((l′), 4H), δ 4.21 ppm ((i″, b), 12H), δ 3.64 ppm ((c, d), 80H), δ 2.81 ppm ((g), 4H), δ 2.69 ppm ((h″), 8H), δ 2.54 ppm ((f), 4H)]).

New methylene and methine protons were generated at 6.39 ppm (l), 6.13 ppm (k) and δ 5.38 ppm (l′). The integral ratio of newly generated methylene and methine protons to the signal (i + b) was 4.50: 4.34: 4.36: 12.65. This demonstrated the successful transesterification reaction and confirmed the structure of the tetra-acrylated product. **This product (III) is the platform to which drugs and diagnostic agents can be attached to form PDCs. The first PDC made and tested was FA$_2$-dPEG$_{20}$-DOX$_2$.**

3.4. *Synthesis of FA$_2$-dPEG$_{20}$-DOX$_2$*

FA-SH was attached to the (Acrylate)$_2$-dPEG$_{20}$-(Acrylate)$_2$ by CALB-catalyzed Michael addition. The ^{1}H NMR spectrum with assignments of FA-SH prepared as reported in [28] is shown in Figure 5. The ^{1}H NMR spectrum of the product of the reaction can be seen in Figure 6.

Capital letters are used to mark the protons of the FA in Figure 6 as the spectra are quite complicated. The integral ratio of methylene protons (l) (6.39 ppm), (l') (5.38 ppm) and methine protons (k) (6.13 ppm) with respect to the reference proton E of FA-SH was 2:2.20:2.28:2.18. Since the integral of the methylene and methine proton signals (l, l', k) were reduced from 4 to 2, it was concluded that two of the acrylate groups reacted with FA-SH. The spectrum is very complicated with many overlaps, so only resonances assigned to protons l, k, l', b, r, n, i, v, E, Z, A and C are identified in the spectrum.

The structure of DOX.HCl is shown in Figure 7 (See the ^{1}H NMR spectrum in Figure S1) Whereas the ^{1}H NMR spectrum of FA_2-$dPEG_{20}$-DOX_2 can be seen in Figure 8. Proper assignment of the NMR signals of DOX.HCl and some conjugates was published in 2017, correcting some errors in earlier publications [29]. When DOX is attached via amide bond formation from the *primary* amine after removal of the HCl, the signals associated with the protons 1′ through 6′ shift (see Table 1). Especially important is the proton in the position marked 3′ at 3.37 ppm which was shown to shift to 3.94 ppm upon formation of an amide bond. However, this overlapped with the methyl protons of the methoxy group of DOX marked OCH_3, also at 3.94 ppm, that remained in its original position.

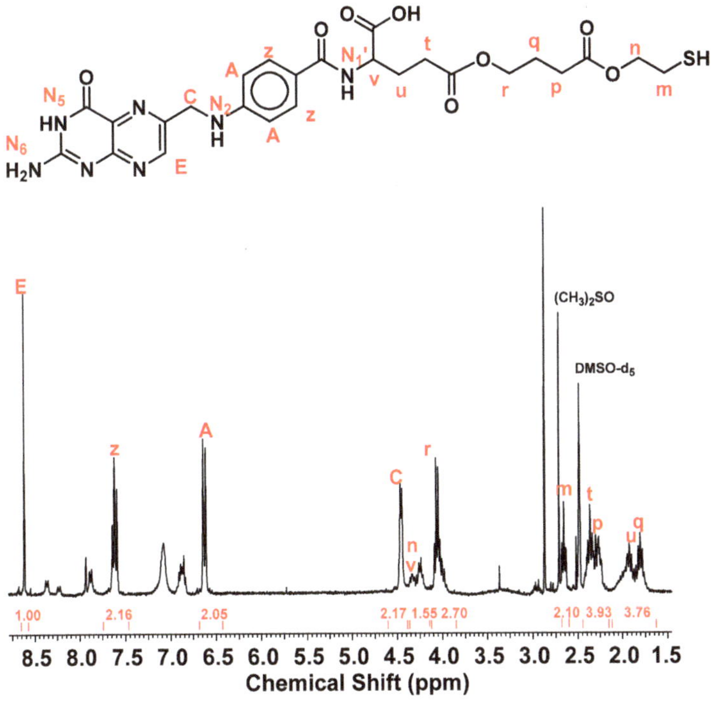

Figure 5. ^{1}H-NMR of the FA-SH in DMSO *d6* ([δ 8.60 ppm ((E), 1H), δ 7.60 ((z), 2H), δ 6.60 ((A), 2H), δ 4.40 ((C), 2H), δ 4.30-4.20 ((n, v), 3H), δ 3.95 ((r), 2H), δ 2.65 ((m), 2H), δ 2.25 ((t, p), 4H), 1.90 ((u), 2H), 1.50 ((q), 2H)]).

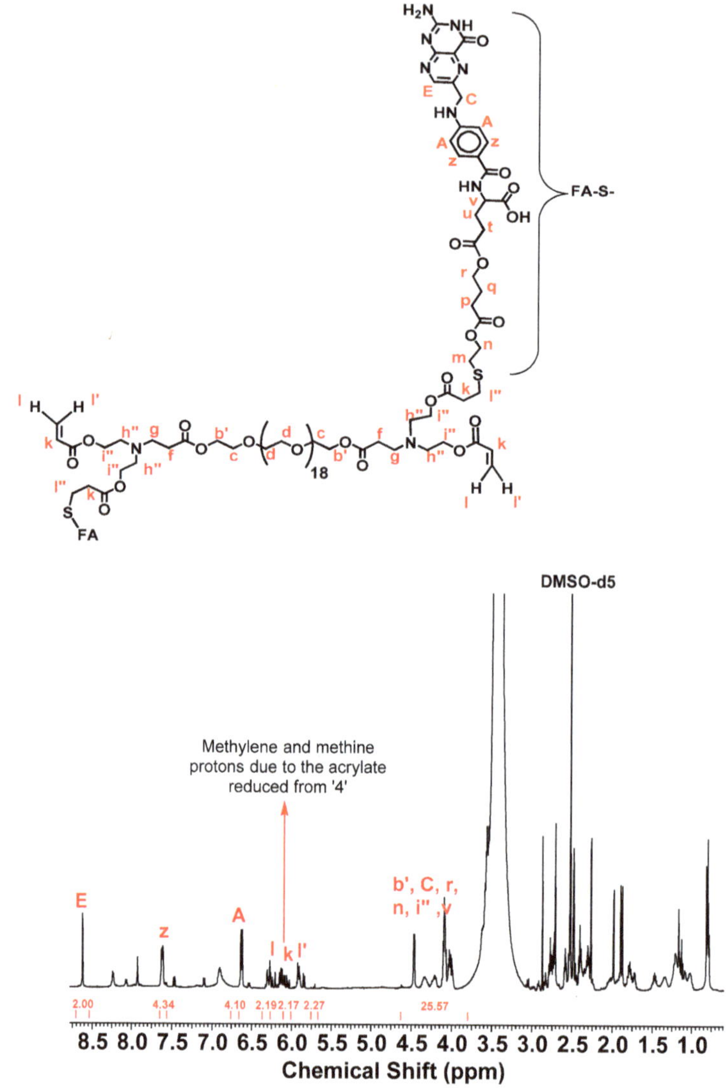

Figure 6. ^{1}H NMR of the raw sample of Michael addition between FA-SH and (Acrylate)$_2$-dPEG$_{20}$-(Acrylate)$_2$ in DMSO *d6*.

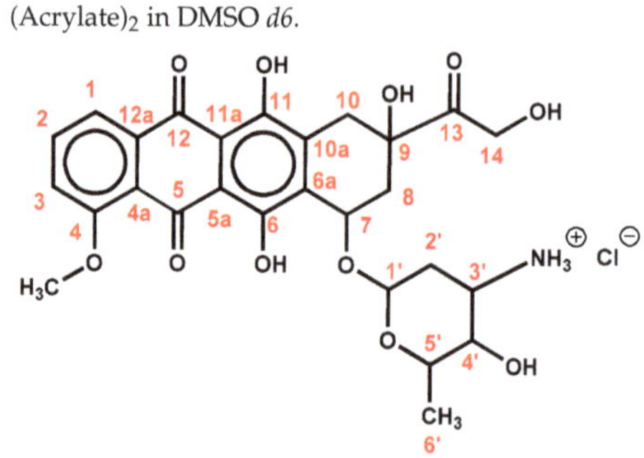

Figure 7. Structure of DOX.HCl: the NMR assignments listed in Table 1.

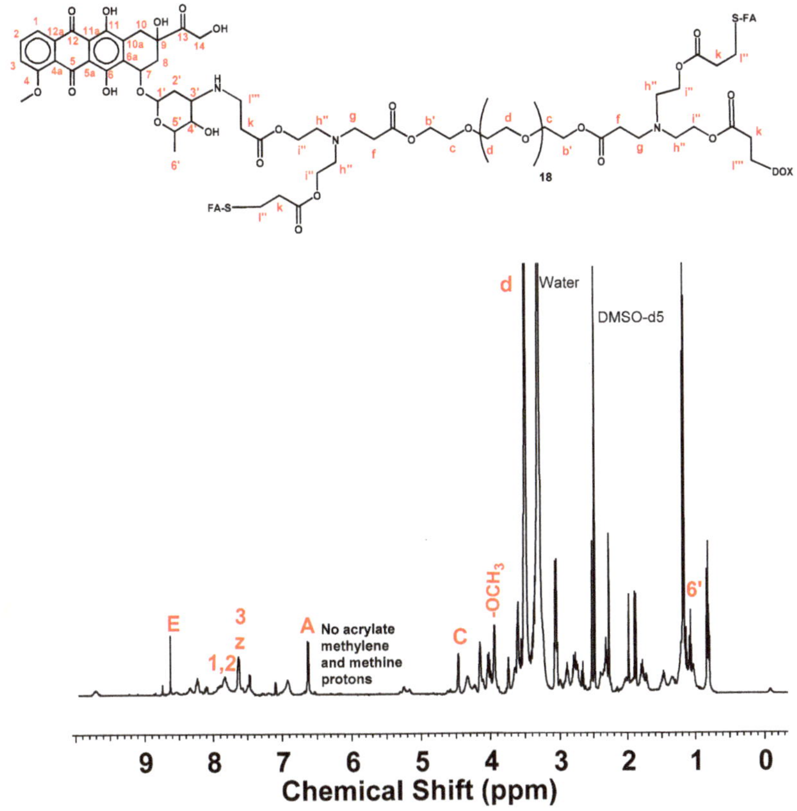

Figure 8. ^{1}H-NMR FA_2-dPEG$_{20}$-DOX_2 in DMSO *d6*.

Table 1. NMR assignments from reference [29] in comparison with FA_2-dPEG$_{20}$-DOX_2.

	DOX.HCl		DOC-NH-(C = O)-R		FA_2-dPEG$_{20}$-DOX_2	
Group	**H**	**C**	**H**	**C**	**H**	**C**
OCH_3 (4)	3.94	57.0	3.94	57.0	3.94	56.6
1′	5.25	99.7	5.18	100.9	5.18	-
2′	1.67; 1.87	28.6	(1.38); 1.79	30.2	1.79	30.2
3′	**3.37**	**46.9**	**3.94**	**45.3**	**3.94**	**45.9**
4′	3.61	66.6	3.37	68.6	3.35	69.7
4′-OH	5.46	-	4.70	-	4.46	-
5′	4.19	66.5	4.12	67.2	4.16	68.1
6′	1.14	17.2	1.19	17.6	1.17	19.6
			FA-SH		**FA_2-dPEG$_{20}$-DOX_2**	
E			8.6		8.63	
z			7.6		7.63	
A			6.6		6.63	
C			**4.4**		**4.47**	
m			2.7			

The integral ratio of E from FA (see Figure 6), 6′ from DOX, and the dPEG main chain protons observed of 2:6:80, indicated that the desired FA_2-dPEG$_{20}$-DOX_2 was obtained.

Some characteristic signals were able to be identified: E at 8.63 ppm from FA (see Figure 6), and 5′ at 4.16 ppm and 6′ (methyl protons) at 1.14 ppm from DOX. The ^{13}C NMR was also crowded but shifts in some characteristic signals supported DOX conjugation: the 3′ signals shifted from 46.50 to 45.62 ppm, while the 2′ and 4′ signals shifted from 28.76 and 66.80 to 30.23 and 68.10 ppm, respectively (see Table 1 and Figure S2). The characteristic signal of γ-substituted FA was seen at 172.3 ppm, while the α carbonyl signal appeared at 173.7 ppm [28].

3.5. *In Vitro and In Vivo Testing*

FA_2-$dPEG_{20}$-DOX_2 was tested in vitro and in vivo in comparison with free DOX at the same concentration [22,23]. Free DOX already showed a cytotoxic effect after 24 h at 0.1 μM concentration, while no toxicity was observed with FA_2-dPEG-DOX_2. After 48 h treatment, the viability of the cells was reduced to 75% of the untreated control, even at the lowest (0.01 μM) concentration, and remained below the control level at all other concentrations applied. In comparison, the cytotoxicity of free DOX increased with increasing concentration, killing all cells at 100 μM. Preliminary testing in a live nude mouse model showed localization in an induced prostate cancer (PC3-PSCA-PSMA) tumor when delivered via intra-tumoral injection (Figure 9). The increase in the tumor volumes slowed down until Day 21 (see Figure 10). After this time point, the intravenously injected mouse tumor grew in a faster manner than the tumor of the intra-tumorally injected animal. Throughout the study, the intra-tumoral injection seemed to delay tumor growth more than the intravenous route of delivery. The results appear to indicate that DOX was released relatively slowly from the FA_2-$dPEG_{20}$-DOX_2.

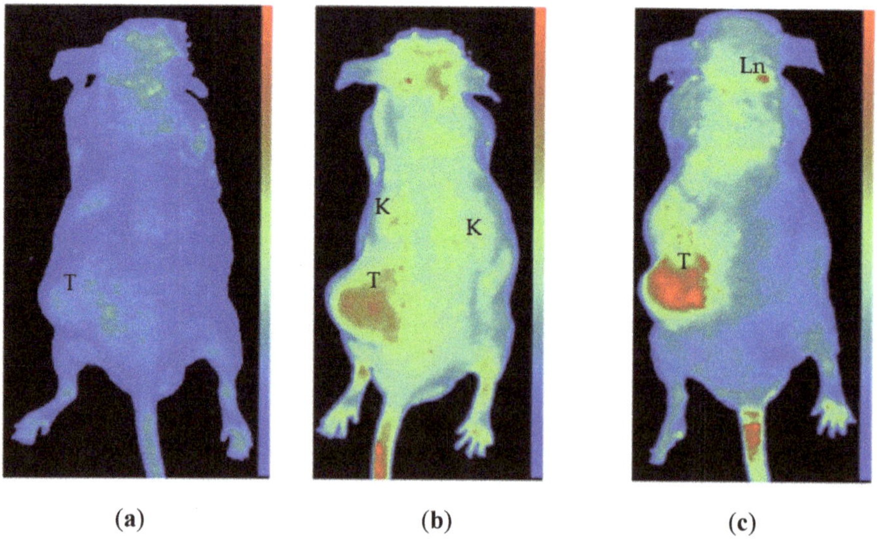

Figure 9. In vivo fluorescent images from the dorsal view of $Foxn^{Nu/Nu}$ nude mice bearing LNCaP prostate xenograft tumors. Images before (**a**) and 24 h after IV (**b**) in one mouse, and after 24 h after intra-tumoral injection in another mouse (**c**). All images are standardized to identical light radiance minima and maxima in relative light intensity per pixel arbitrary units to allow for direct comparison. Fluorescence of the PDC is identified in the tumor (T) tail vein, kidneys (K), and capillary-rich head nuchal skin in (**b**), while a distinct fluorescent signal is observed in the tumor (T) and some lymph nodes of the neck (Ln) in the case of the intra-tumoral PDC injected animal (**c**). The slight autofluorescence, as seen in (**a**) before injection, is clearly different from the fluorescent signals of injected animals. Reproduced with permission from [23]. 2022, MDPI.

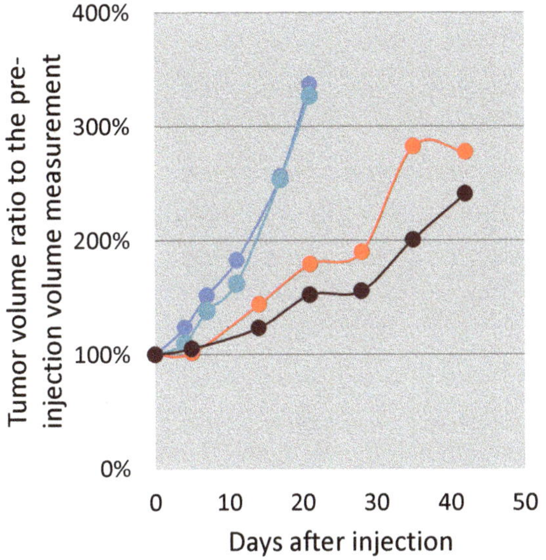

Figure 10. Follow-up of tumor volume increase in percentage of the initial tumor volume in a PC3-PSMA overexpressing cell line xenograft in two saline control (blue and turquoise) Nu/Nu nude mice, one injected intra-tumorally (brown) and one injected intravenously (red) with the PDC. Reproduced with permission from [23]. 2022, MDPI.

In summary, FA_2-$dPEG_{20}$-DOX_2 seems to be a promising candidate as a folate-targeted cancer diagnostic and therapeutic agent, but more investigations are necessary in vitro and in vivo of this, and similar dPEG-based compounds, made by chemo-enzymatic synthesis.

We will continue research in the spirit of Andrzej's legacy—another great scientist lost [2,30,31].

4. Conclusions

The chemo-enzymatic esterification and Michael addition reactions catalyzed by CALB are excellent selective reactions for the modification of PEGs. As was shown, a platform based on dPEG with four reactive acrylate groups was built with high selectivity. This is an excellent platform for the synthesis of a variety of polymer-based drug carriers. In the current investigation, we have shown that two folic acid groups and two doxorubicin groups could be attached to the platform using a CALB-catalyzed Michael addition reaction. We will expand this concept in the future for the synthesis of compounds with a variety of drugs and targeting agents.

Supplementary Materials: The following supporting information can be downloaded at: https://www.mdpi.com/article/10.3390/polym14142900/s1, Figure S1: 1H NMR spectrum of DOX.HCl; Figure S2. ^{13}C NMR spectrum of FA2-dPEG-DOX2

Author Contributions: Conceptualization, J.E.P.; methodology, G.S.; validation, G.S.; formal analysis, G.S.; investigation, G.S., K.M. and E.K.; writing—original draft preparation, J.E.P., K.M. and E.K.; writing—review and editing, J.E.P., K.M. and E.K.; visualization, G.S., K.M. and E.K.; supervision, J.E.P.; funding acquisition, J.E.P. All authors have read and agreed to the published version of the manuscript.

Funding: The authors acknowledge the Ohio State University USDA-NIFA under Hatch project number OHO01417 and the Breast Cancer Innovation Foundation for financial support.

Data Availability Statement: Raw data can be requested from the corresponding author.

Conflicts of Interest: The authors declare no conflict of interest.

References

1. Zhong, L.; Li, Y.; Xiong, L.; Wang, W.; Wu, M.; Yuan, T.; Yang, W.; Tian, C.; Miao, Z.; Wang, T.; et al. Small molecules in targeted cancer therapy: Advances, challenges, and future perspectives. *Signal. Transduct. Target. Ther.* **2021**, *6*, 201. [CrossRef] [PubMed]
2. Lipowska-Kur, D.; Szweda, R.; Trzebicka, B.; Dworak, A. Preparation and characterization of doxorubicin nanocarriers based on thermoresponsive oligo(ethylene glycol) methyl ether methacrylate polymer-drug conjugates. *Eur. Polym. J.* **2018**, *109*, 391–401. [CrossRef]
3. Bae, Y.H.; Park, K. Targeted drug delivery to tumors: Myths, reality and possibility. *J. Control. Release* **2011**, *153*, 198–205. [CrossRef] [PubMed]
4. Tewabe, A.; Abate, A.; Tamrie, M.; Seyfu, A.; Abdela Siraj, E. Targeted Drug Delivery—From Magic Bullet to Nanomedicine: Principles, Challenges, and Future Perspectives. *J. Multidiscip. Healthc.* **2021**, *14*, 1711–1724. [CrossRef] [PubMed]
5. Manzari, M.T.; Shamay, Y.; Kiguchi, H.; Rosen, N.; Scaltriti, M.; Heller, D.A. Targeted drug delivery strategies for precision medicines. *Nat. Rev. Mater.* **2021**, *6*, 351–370. [CrossRef] [PubMed]
6. Ekladious, I.; Colson, Y.L.; Grinstaff, M.W. Polymer–drug conjugate therapeutics: Advances, insights and prospects. *Nat. Rev. Drug Discov.* **2019**, *18*, 273–294. [CrossRef]
7. Srinivasarao, M.; Low, P.S. Ligand-Targeted Drug Delivery. *Chem. Rev.* **2017**, *117*, 12133–12164. [CrossRef]
8. Jurczyk, M.; Jelonek, K.; Musiał-Kulik, M.; Beberok, A.; Wrześniok, D.; Kasperczyk, J. Single- versus Dual-Targeted Nanoparticles with Folic Acid and Biotin for Anticancer Drug Delivery. *Pharmaceutics* **2021**, *13*, 326. [CrossRef]
9. Frigerio, B.; Bizzoni, C.; Jansen, G.; Leamon, C.P.; Peters, G.J.; Low, P.S.; Matherly, L.H.; Figini, M. Folate receptors and transporters: Biological role and diagnostic/therapeutic targets in cancer and other diseases. *J. Exp. Clin. Cancer Res.* **2019**, *38*, 125. [CrossRef]
10. Puskas, J.E.; Molnar, K.; Krisch, E. Toward the effective synthesis of bivalent Folate-targeted PEGylated cancer diagnostic and therapeutic agents using chemo-enzymatic processes. *J. Mol. Liq.* **2020**, *310*, 113218–113227. [CrossRef]
11. Ullah, S.; Azad, A.K.; Nawaz, A.; Shah, K.U.; Iqbal, M.; Albadrani, G.M.; Al-Joufi, F.A.; Sayed, A.A.; Abdel-Daim, M.M. 5-Fluorouracil-Loaded Folic-Acid-Fabricated Chitosan Nanoparticles for Site-Targeted Drug Delivery Cargo. *Polymers* **2022**, *14*, 2010. [CrossRef] [PubMed]
12. Alshamrani, M. Broad-Spectrum Theranostics and Biomedical Application of Functionalized Nanomaterials. *Polymers* **2022**, *14*, 1221. [CrossRef] [PubMed]
13. Casi, G.; Neri, D. Antibody–Drug Conjugates and Small Molecule–Drug Conjugates: Opportunities and Challenges for the Development of Selective Anticancer Cytotoxic Agents. *J. Med. Chem.* **2015**, *58*, 8751–8761. [CrossRef] [PubMed]
14. Baker, J.R. Dendrimer-based nanoparticles for cancer therapy. *Hematology* **2009**, *2009*, 708–719. [CrossRef] [PubMed]
15. Singh, P.; Gupta, U.; Asthana, A.; Jain, N.K. Folate and Folate−PEG−PAMAM Dendrimers: Synthesis, Characterization, and Targeted Anticancer Drug Delivery Potential in Tumor Bearing Mice. *Bioconjug. Chem.* **2008**, *19*, 2239–2252. [CrossRef]
16. Majoros, I.J.; Myc, A.; Thomas, T.; Mehta, C.B.; Baker, J.R. PAMAM Dendrimer-Based Multifunctional Conjugate for Cancer Therapy: Synthesis, Characterization, and Functionality. *Biomacromolecules* **2006**, *7*, 572–579. [CrossRef]
17. van Dongen, M.A.; Dougherty, C.A.; Banaszak Holl, M.M. Multivalent Polymers for Drug Delivery and Imaging: The Challenges of Conjugation. *Biomacromolecules* **2014**, *15*, 3215–3234. [CrossRef]
18. Marcinkowska, M.; Sobierajska, E.; Stanczyk, M.; Janaszewska, A.; Chworos, A.; Klajnert-Maculewicz, B. Conjugate of PAMAM Dendrimer, Doxorubicin and Monoclonal Antibody—Trastuzumab: The New Approach of a Well-Known Strategy. *Polymers* **2018**, *10*, 187. [CrossRef]
19. Shiraishi, K.; Yokoyama, M. Toxicity and immunogenicity concerns related to PEGylated-micelle carrier systems: A review. *Sci. Technol. Adv. Mater.* **2019**, *20*, 324–336. [CrossRef]
20. Rahme, K.; Dagher, N. Chemistry Routes for Copolymer Synthesis Containing PEG for Targeting, Imaging, and Drug Delivery Purposes. *Pharmaceutics* **2019**, *11*, 327. [CrossRef]
21. Das, D.; Koirala, N.; Li, X.; Khan, N.; Dong, F.; Zhang, W.; Mulay, P.; Shrikhande, G.; Puskas, J.; Drazba, J.; et al. Screening of Polymer-Based Drug Delivery Vehicles Targeting Folate Receptors in Triple-Negative Breast Cancer. *J. Vasc. Interv. Radiol.* **2020**, *31*, 1866–1873.e2. [CrossRef] [PubMed]
22. Koirala, N.; Das, D.; Fayazzadeh, E.; Sen, S.; McClain, A.; Puskas, J.E.; Drazba, J.A.; McLennan, G. Folic acid conjugated polymeric drug delivery vehicle for targeted cancer detection in hepatocellular carcinoma. *J. Biomed. Mater. Res. Part A* **2019**, *107*, 2522–2535. [CrossRef] [PubMed]
23. Nagy, K.S.; Toth, K.; Pallinger, E.; Takacs, A.; Kohidai, L.; Jedlovszky-Hajdu, A.; Mathe, D.; Kovacs, N.; Veres, D.S.; Szigeti, K.; et al. Folate-Targeted Monodisperse PEG-Based Conjugates Made by Chemo-Enzymatic Methods for Cancer Diagnosis and Treatment. *Int. J. Mol. Sci.* **2021**, *22*, 10347. [CrossRef]
24. Christidi, E.; Brunham, L.R. Regulated cell death pathways in doxorubicin-induced cardiotoxicity. *Cell Death Dis.* **2021**, *12*, 339. [CrossRef] [PubMed]
25. Zhao, N.C.; Woodle, M.; Mixson, A.J. Advances in Delivery Systems for Doxorubicin. *J. Nanomed. Nanotechnol.* **2018**, *9*, 519. [CrossRef]
26. Psarrou, M.; Kothri, M.G.; Vamvakaki, M. Photo- and Acid-Degradable Polyacylhydrazone–Doxorubicin Conjugates. *Polymers* **2021**, *13*, 2461. [CrossRef]
27. Xue, X.; Wu, Y.; Xu, X.; Xu, B.; Chen, Z.; Li, T. pH and Reduction Dual-Responsive Bi-Drugs Conjugated Dextran Assemblies for Combination Chemotherapy and In Vitro Evaluation. *Polymers* **2021**, *13*, 1515. [CrossRef]

28. Puskas, J.E.; Castano, M.; Mulay, P.; Dudipala, V.; Wesdemiotis, C. Method for the Synthesis of γ-PEGylated Folic Acid and Its Fluorescein-Labeled Derivative. *Macromolecules* **2018**, *51*, 9069–9077. [CrossRef]
29. Piorecka, K.; Stanczyk, W.; Florczak, M. NMR analysis of antitumor drugs: Doxorubicin, daunorubicin and their functionalized derivatives. *Tetrahedron Lett.* **2017**, *58*, 152–155. [CrossRef]
30. Kowalczuk, A.; Trzcinska, R.; Trzebicka, B.; Müller, A.H.E.; Dworak, A.; Tsvetanov, C.B. Loading of polymer nanocarriers: Factors, mechanisms and applications. *Prog. Polym. Sci.* **2014**, *39*, 43–86. [CrossRef]
31. Haladjova, E.; Toncheva-Moncheva, N.; Apostolova, M.D.; Trzebicka, B.; Dworak, A.; Petrov, P.; Dimitrov, I.; Rangelov, S.; Tsvetanov, C.B. Polymeric Nanoparticle Engineering: From Temperature-Responsive Polymer Mesoglobules to Gene Delivery Systems. *Biomacromolecules* **2014**, *15*, 4377–4395. [CrossRef] [PubMed]

polymers

Review

Polyglycerols as Multi-Functional Platforms: Synthesis and Biomedical Applications †

Paria Pouyan, Mariam Cherri and Rainer Haag *

Institute of Chemistry and Biochemistry, Freie Universität Berlin, Takustr. 3, 14195 Berlin, Germany; paria.pouyan@fu-berlin.de (P.P.); mcherri6@zedat.fu-berlin.de (M.C.)
* Correspondence: haag@chemie.fu-berlin.de
† In Honorable Memory of Professor Andrzej Dworak.

Abstract: The remarkable and unique characteristics of polyglycerols (PG) have made them an attractive candidate for many applications in the biomedical and pharmaceutical fields. The presence of multiple hydroxy groups on the flexible polyether backbone not only enables the further modification of the PG structure but also makes the polymer highly water-soluble and results in excellent biocompatibility. In this review, the polymerization routes leading to PG with different architectures are discussed. Moreover, we discuss the role of these polymers in different biomedical applications such as drug delivery systems, protein conjugation, and surface modification.

Keywords: linear polyglycerol; hyperbranched polyglycerol; ring-opening polymerization of glycidol

Citation: Pouyan, P.; Cherri, M.; Haag, R. Polyglycerols as Multi-Functional Platforms: Synthesis and Biomedical Applications. *Polymers* **2022**, *14*, 2684. https://doi.org/10.3390/polym14132684

Academic Editor: Marek Kowalczuk

Received: 8 June 2022
Accepted: 26 June 2022
Published: 30 June 2022

1. Introduction

Since the introduction of "macromolecules" in the 1920′s by Hermann Staudinger, the field of chemistry has been tremendously transformed, and the resulting polymer science has revolutionized modern life in many different aspects, especially in life sciences and medicine. The chemical foundation of polymers has led to the design of an endless variety of polymeric structures with different physical and chemical properties [1]. The immense impact of polymers on the field of medicine and pharmaceuticals has resulted in the development of many advanced biomaterials for prevalent applications such as carriers in smart drug delivery [2], multivalent antiviral agents [3,4], and anti-inflammatory [5] agents. Biocompatibility, water-solubility, and non-toxicity are the requirements for polymers in biomedical and pharmaceutical applications. Furthermore, in order to adjust their properties, introduction of necessary functional groups such as targeting moieties, and degradable units, these polymers need to be suitable for modification. In this sense, polyglycerol (PG, also known as polyglycidol), which is highly biocompatible and possesses all the mentioned requirements, has been thoroughly studied as an important polymer candidate in biomedical fields [6,7]. PG can be synthesized with different architectures via cationic, anionic, or coordination ring-opening polymerization of glycidol. Polyglycerol is highly water-soluble and biocompatible, and the hydroxy groups on each repeating unit enable numerous subsequent modifications [8].

In this review, we highlight the most common synthetic routes for obtaining polyglycerol with different architectures and highlight their application as carriers for active ingredients in different biomedical and pharmaceutical applications. A comprehensive and detailed overview of different polymerization methods of epoxide monomers has already been published elsewhere [9].

2. Synthetic Strategies for Polyglycidols

2.1. Cationic Ring-Opening Polymerization

2.1.1. Hyperbranched Polyglycidol

Cationic ring-opening polymerization (CROP) of heterocycles, which exhibits oxygen, is of high importance in the industrial production of engineering plastics such as poly(oxymethylene). However, due to a process called "back-biting", which is intramolecular chain transfer, high amounts of cyclic byproducts can be formed. The first polymerization of glycidol was reported in 1966 when Sandler and Berg polymerized the AB_2 type monomer (the theoretical representation suggested by Flory [10]) at room temperature using various catalysts. After that, Dworak et al., suggested, for the first time, the ring-opening polymerization of glycidol through a cationic route [11]. They proposed two possible mechanisms, namely active chain (AC) and activated monomer (AM) (Figure 1) [11]. The typical initiators for cationic ring-opening polymerization of glycidol were Lewis acids (BF_3OEt_2 or $SnCl_4$) or Brønsted acids (CF_3COOH or CF_3SO_3H) [11,12]. However, the main challenge was controlling the reaction kinetics since multiple side reactions can occur, which hinders propagation. Most HPGs synthesized following this route resulted in low molecular weights (less than 10,000 Da) and broad dispersities [11]. Nevertheless, the ease and simplicity of this method still engages researchers to explore possible ways to employ cationic ring-opening polymerization in investigating new HPGs architectures. Mohammadifar et al., reported a green synthesis route for synthesizing HPG using citric acid as an initiator [13]. They proved that not only the citric acid can initiate the propagating chain, but it is also incorporated into the polymeric backbone as it acts as a proton donor [13]. They suggested that the mechanism is based on AM cationic ring-opening polymerization [13]. With the citric acid molecule incorporated in the backbone, the HPG was degradable under neutral and acidic pH [13]. Although the polymerization did not lead to molecular weights higher than 1.5 kDa, the complete green synthesis and impuritiy-free polymers meant that it qualified as a good candidate for biomedical applications [13]. Recently, Kim et al., showed a recyclable metal-free catalytic system for the cationic ring-opening polymerization of glycidol also under ambient conditions using tris(pentafluorophenyl)borane as a catalyst [14]. The further propagation of polymerization, in this case, induced the precipitation of the higher molecular weight (1–4 kDa), highly hydrophilic HPG in nonpolar solvents, allowing the recycling of the catalyst and the solvent by the simple sequence of decantation of HPG [14].

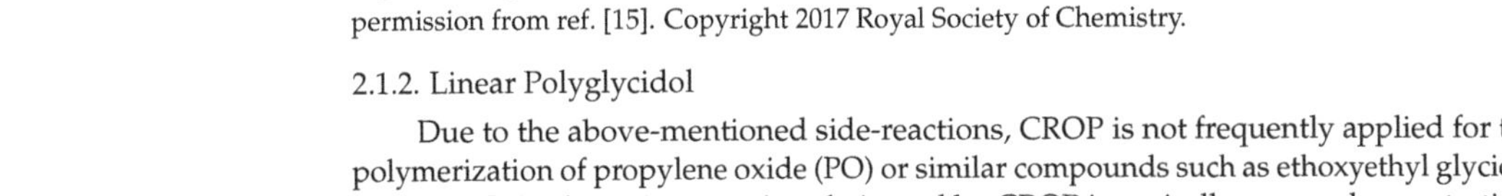

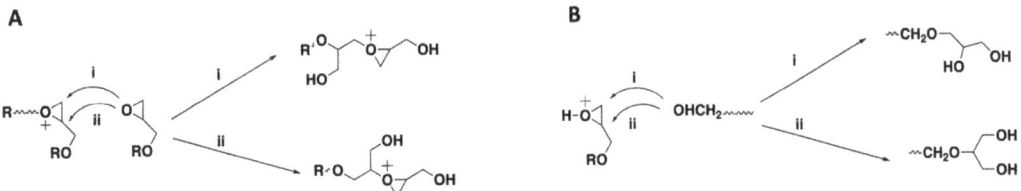

Figure 1. Proposed mechanism for CROP of glycidol based on (**A**) ACE and (**B**) AM. Adapted with permission from ref. [15]. Copyright 2017 Royal Society of Chemistry.

2.1.2. Linear Polyglycidol

Due to the above-mentioned side-reactions, CROP is not frequently applied for the polymerization of propylene oxide (PO) or similar compounds such as ethoxyethyl glycidyl ether (EEGE). The active growing chain-end by CROP is typically a secondary or tertiary oxonium cation, which can follow two discussed mechanisms: (i) ACE and nucleophilic attack of the oxygen of a cyclic monomer to a carbon atom of the oxygen which was formally bearing the positive charge which leads to propagation and chain growth. (ii) The AM mechanism, which is conducted in the presence of alcohols and allows better control over molecular weight while suppressing the cyclization issue [9]. In this case, the positive charge or the active end is on the monomer, and the polymer chain is inactive or dormant, which effectively hinders the back-biting side-reaction. However, the suitable reaction condition is to keep the instant concentration of monomer low; this can be achieved by slowly adding the epoxide monomer to the reaction mixture, which leads to prolonged reaction times and exact process control. Taken together, although AM CROP is a suitable method for the synthesis of telechelic polymers, substituted epoxide monomers such as PO are mostly polymerized with other techniques which can have more control over molecular weight and end-group fidelity [9].

2.2. Anionic Ring-Opening Polymerization

2.2.1. Dendritic and Hyperbranched Polyglycerol (dPG and HPG)

A powerful single-step alternative to multistep polyglycerol dendrimers synthesis was reported by Haag and co-workers [16] (Figure 2a). Another method proposed by Sunder et al., combined anionic multi-branch ring-opening polymerization and slow addition of the monomer to obtain hyperbranched polyglycerol (Figure 2b) [17]. These two approaches overcome the limitations of the cationic ring-opening polymerization of glycidol. The initiator employed was a partially deprotonated 1,1,1-tris(hydroxymethyl)propane (TMP), and the reaction was carried out at 90–100 °C. The slow addition of glycidol ensured the controlled propagation of the polymerization and avoided low molecular weight chains due to the limitation of the intra-cyclization of oligomers. The resulting polymer was a medium-sized HPG with a relatively narrow dispersity (Đ < 1.5) [17]. However, even if the polymerization method reported by Sunder et al., demonstrated a controlled propagation and an end-polymer, the challenge still remains with reaching molecular weights higher than 6500 Da since the polymerization is solvent-free and the increase in viscosity will eventually result in poor mixing in the system [15,17]. Hence, synthesizing a high molecular weight polyglycerol is essential, especially for drug delivery system applications. High molecular weight will ensure that HPG possesses a large hydrodynamic size, enhanced vascular retention, and the presence of an extensive number of functional groups [15]. Schmitt et al., showed the dependency of the blood circulation of nanocarriers on HPGs molecular weight [18]. They tested various molecular weight HPGs nanocarriers (25–500 kDa) covalently linked to chelator desferrioxamine (DFO) and radiolabeled with the gamma emitter ^{67}Ga [18]. Using qSPECT/CT imagining inside the heart of Rag2m mice, they proved that the blood circulation half-lives of the ^{67}Ga labeled HPGs increase from 9.9 to 47.8 h with increasing molecular weight [18]. Different approaches have been suggested to synthesize high molecular weight HPGs. These approaches were well described in a review written by Kizhakkedathu's group [15] that we would like to summarize in this review and describe other most recent methods. The first approach was reported by Frey and co-workers, which included the modification of stirring intensity, stirrer geometry, and monomer addition rate, as well as the addition of an inert emulsifying agent. This adjustment to the reaction parameters resulted in an HPG with a molecular weight of up to 20 kDa [19]. Recently, Haag's group reported on the automated solvent-free polymerization of HPGs that ensures high reproducibility and traceability in the system due to automation [20]. The group reported a linear correlation between the torque and the degree of polymerization that can be applied to monitor the molecular weight during the polymerization [20].

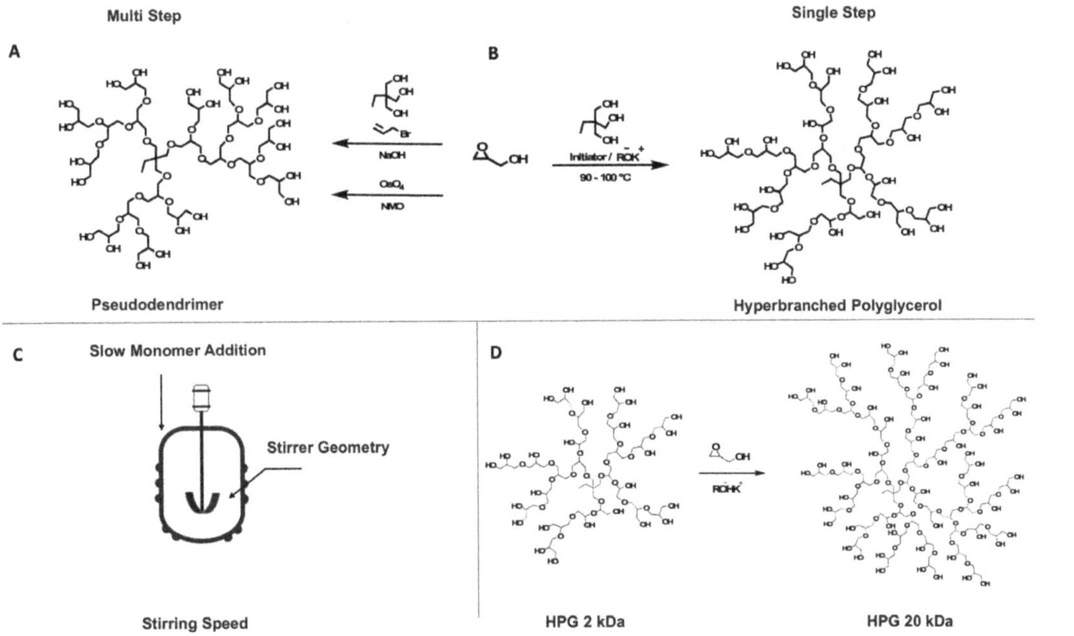

Figure 2. Top: (**A**) Pseudodendrimer, (**B**) HPG. **Bottom:** (**C**) reaction parameters, (**D**) (macromonomer, multistep).

The second approach is the macroinitiator approach, in which a reaction starts with the slow addition of glycidol to a partially deprotonated low molecular weight HPG acting as a macroinitiator. Wilms et al., employed a 500 and 1000 Da HPG-macroinitiator and reached a 24 kDa HPG [21], whereas Moore et al., started with a larger HPG-macroinitiator (2 kDa), which resulted in high molecular weight HPGs up to 100 kDa [22]. The third approach introduces a solvent as an emulsifying agent as another reaction parameter. Different solvents were screened, including dioxane [23], 1,4- dioxane, tetrahydropyran, ethylene glycol diethyl ether, and decane [24]. It was demonstrated that the use of the solvent does not hinder the properties nor the degree of branching of the resulting HPG [23]. In addition, the type of solvent can affect the exchange of counter cations with the propagating species, which can further affect the end-molecular weight [24]. Most recently, Kizhakkedathu's group reported a gram-scale synthesis of mega hyperbranched polyglycerols (mega HPGs) with molecular weights in million Daltons up to 9.3 MDa and a Đ as narrow as 1.2 [25]. Kizhakkedathu's group combined both the macroinitiator and the solvent-based approaches in the ROMP reaction. The macroinitiator was a partially deprotonated (10%) 840 kDa using KH and dissolved in DMF, where the slow monomer addition of glycidol was also followed to generate mega HPGs. The macroinitiator itself was synthesized using partially deprotonated TMP in dioxane at 90 °C with the slow addition of glycidol [25]. The mega HPGs conserved their properties of high water solubility, low intrinsic viscosity, and globular structure [25].

For HPGs to serve as a functional platform in biomedical applications and as drug delivery systems, the biodegradability aspect needs to be integrated into the HPG structure. Since it was proven that HPGs tend to accumulate in organs such as the liver in relatively high doses (10% of 500 kg/mol of the injected HPG) [26], For this purpose, the incorporation of physiologically degradable moieties into the HPG backbone will ensure the breakage of the structure when encountering the corresponding stimulus. Kizhakkedathu's group incorporated ketal degradable linkage in the dendritic polyether backbone of HPG [27]. The polymer was synthesized following the anionic ROMP approach, using glycidol and the comonomer 2-(1-(2-(oxiran-2-ylmethoxy)ethoxy)cyclohexyloxy)ethanol) [27]. The ketal moieties were proven to degrade under mildly acidic conditions both in solution and within the cells [27]. Since the tumor site is characterized by a more acidic pH when compared to healthy cells, systems that are pH-responsive can serve as tumor-targeting drug delivery systems [28]. Kim's group presented the anionic ROMP of a glycidol derivative molecule bearing a disulfide bond, i.e., 2-((2-(oxiran-2-ylmethoxy)ethyl)disulfanyl)ethan-1-ol (SSG) [29]. The polymerization resulted in a redox-degradable HPG. For an alternative synthetic approach, see chapter 2.3.1. The polymer was proved to be redox-responsive when treated with reducing agents, conserving the properties of HPG [29]. The cytosol of cancerous cells is characterized by an abundance of glutathione (GSH), 10 mM in comparison with <10 μM in healthy blood vessels and extracellular fluids; hence, a reductive sensitive drug delivery system is one of the approaches used to deliver active pharmaceutical ingredients (API) to the tumor site [30,31].

2.2.2. Linear Polyglycidol

By chemical protection of the hydroxy group in glycidol, the proton exchange during polymerization can be efficiently hampered to obtain linear structures. In the next step after polymerization, deprotection yields linear polyglycerol (LPG) with free hydroxy groups. This strategy enables a toolbox for the design of various compositions and architectures using a myriad of protected glycidol monomers [7,32,33]. The multi-functional backbone of LPG allows a wide range of substitution reactions, resulting in high degrees of functionalization [32].

LPG can be synthesized from glycidyl ethers, protected glycidol derivatives, by anionic ring-opening polymerization (AROP) [9], the classical way to polymerize epoxides to obtain polyethers and the main polymerization strategy since the 1930s [7,9]. Figure 3 represent the most common protected monomers applied for obtaining and preparing LPG.

ethoxyethyl glycidyl ether (EEGE) **trimethylsilyl glycidyl ether (TMSGE)** ***tert*-butyl glycidyl ether (*t*BGE)** **allyl glycidyl ether (AGE)**

Figure 3. Common protected glycidols for synthesis of linear polyglycerol LPG. Adapted with permission from ref. [7]. Copyright 2014 American Chemical Society.

tert-butyl glycidyl ether (tBGE) and allyl glycidyl ether (AGE) are commercially available. However, ethoxyethyl glycidyl ether (EEGE) is the most frequently used protected glycidol for LPG synthesis owing to its simple synthesis procedure and acidic deprotection of the protecting acetal groups [7,33]. EEGE was first synthesized by Fitton et al., in 1987 [34] and is prepared by the reaction of ethyl vinyl ether with glycidol in the presence of *p*-toluenesulfonic acid as a catalyst (Figure 4) [35,36].

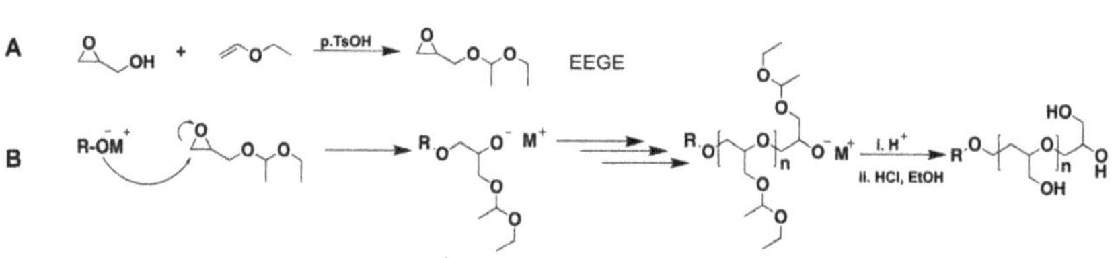

Figure 4. (**A**) Protection of glycidol, (**B**) polymerization of EEGE, and acidic deprotection of acetal groups. Adapted with permission from refs. [36,37]. Copyright 2010 American Chemical Society; 2021 Wiley Online Library.

In 1994 Taton et al., reported for the first time the successful polymerization of EEGE by applying cesium hydroxide as an initiator in bulk polymerization. The reaction resulted in poly(ethoxyethyl glycidyl ether) (PEEGE) with a molecular weight in the range of 30 kDa and a relatively broad dispersity (PDI: 1.5) [35].

Changing the initiator to potassium or cesium alkoxide resulted in the synthesis of polymers with narrow molecular weight distribution [38]. Until now, several initiators including potassium *tert*-Butoxide (t-BuOK) [39,40], potassium 3-phenyl propanolate (PPOK) [41,42], alkoxy ethanolates [43], potassium methoxide (MeOK) [44], and BuLi/phospahezene base (Li^+/t-BuP_4) [40,45] have been successfully applied for the controlled polymerization of EEGE. Nevertheless, for the polymerization of EEGE, alkali metal-based initiators lead to molecular weights limited to a maximum of about 30 kDa or a degree of polymerization (DP_n) of about 300 [36,46]. Möller et al., explained this to be due to the chain transfer reaction to a monomer either from the active chain-end or the oxyanion initiator [40]. Other monosubstituted epoxides such as PO or phenyl glycidyl ether are reported to undergo this chain transfer in the same way during the polymerization [47]. The proton substitution from the adjacent group to the epoxide ring results in the formation of an unsaturated allyl alkoxid. (Figure 5) [40].

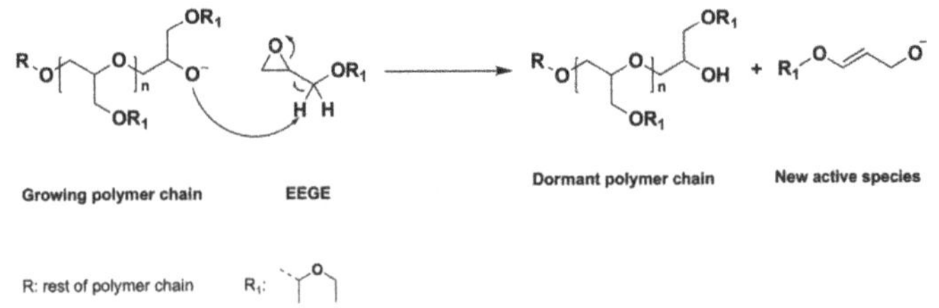

Figure 5. Possible mechanism of chain transfer reactions during the anionic polymerization of EEGE using alkali metal alkoxide initiators. Adapted with permission from ref. [40]. Copyright 2009 Elsevier.

With the increase in temperature and in higher monomer to initiator ratios, this transfer reaction was more noticeable. In order to obtain polyethers with high molecular weights, it is generally assumed that the nucleophilicity of the active propagating chain-end has to be high enough to utilize the ring-opening of the epoxide ring but have relatively low basicity to prevent proton transfer and side-reactions [7].

2.3. Coordination Ring-Opening Polymerization

2.3.1. Hyperbranched Polyglycidol

Harth et al. developed an alternative method to synthesize semi-branched polyglycidols to the traditional ionic polymerizations. They investigated the homopolymerization of glycidol by employing Tin(II) trifluoromothenesulfate ($Sn(OTf)_2$) as a catalyst and varying the temperature. The results showed control over the branching of the polyglycerol backbone by varying the temperature and creating different protein-glycidol bioconjugates as an alternative to pegylated biostructures [48].

The bulk polymerization of cyclic esters using a catalyst such as Tin(II) 2-ethylhexanoate ($Sn(Oct)_2$) to form polyesters has been proven to be controlled and pseudoliving [49]. Additionally, given the high epoxide ring strain of the glycidol monomer, a new strategy has emerged to synthesize biodegradable high molecular weight HPGs by copolymerizing cyclic esters and glycidol following the catalytic route and a coordination-insertion ring-opening polymerization mechanism. Haag's group was dedicated to synthesizing and investigating such systems in solvent-free one-pot synthesis. Figure 6 shows some of the strategies to obtain degradable HPG. Cherri et al., proved the scalability and reproducibility of the sulfated HPGs bearing bio-degradable caprolactone segments system with controlled molecular weights (20–60 kDa) in two-step synthesis [50]. The system showed the ability to encapsulate the hydrophobic tyrosine kinase inhibitor chemotherapeutical drug sunitinib [50]. The drug delivery system was able to release its guest molecule under acidic and enzymatic conditions, accumulate in the tumor site, and perform better than the free sunitinib in vivo [50].

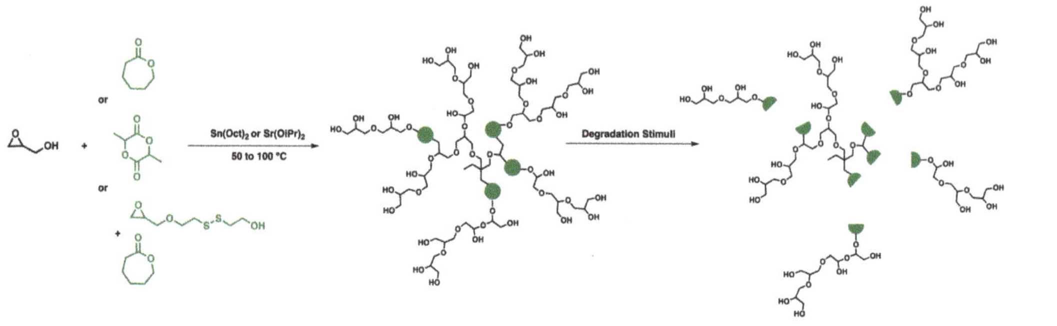

Figure 6. Synthesis of degradable HPG via coordination–insertion mechanism. The degradation follows different stimuli such as changes in pH, reductive environment, etc.

Another system developed by Zabihi et al., followed a similar strategy to synthesize poly(glycidol-lactide) up to 43 kDa [51]. Further, the system was then loaded with tacrolimus (TAC), a macrolide immunosuppressant which is used for the treatment of atopic dermatitis [51]. TAC has low bioavailability, and the system was proven to load TAC effectively (14.5% *w*/*w* loading capacity) and ensure its educated delivery into the stratum corneum, viable epidermis, and upper epidermis when compared with Protopic®® (containing 0.03% *w*/*w* of TAC) [51]. On the other hand, Reisbeck et al., developed an HPG sulfates drug delivery system that was dual degradable based on the copolymerization of glycidol, ε-caprolactone, and SSG, the glycidol derivative monomer bearing a disulfide bond in bulk screening different catalysts [52]. The catalyst that was most prominent for polymerization was strontium isopropoxide, leading to the highest molecular weight and degree of branching [52]. The system was degradable under enzymatic and reductive stimuli and was capable of encapsulating and releasing doxorubicin under the same conditions [52].

2.3.2. High Molecular Weight Linear Polyglycidol

To obtain high molecular weight LPG, coordination-type polymerizations using organo metallic catalysts were also conducted using EEGE as a monomer [53–55].

Haout et al. prepared LPG with high molecular weights of about 1 kDa by applying diethylzinc and water as an initiating system. However, poor control over the molecular weight leads to higher dispersities (PDI = 1.46–1.80) [54].

In 2007, using a dual initiating system composed of tetraoctylammonium bromide $(Oct)_4NBr$ and triisobutylaluminium $(i\text{-}Bu)_3Al$, Carlotti and Deffieux et al. reported a polymeri zation strategy, which was a breakthrough for the polymerization of many functional epoxides including EEEGE [56]. This strategy is based on parallel activation of the monomer towards nucleophiles while reducing the basicity of the growing chain-end through coordination with the Lewis acid (catalyst). The coordination of the catalyst with the epoxide ring results in the reduction of electron density in the ring and hence promoting ring-opening (monomer activation). At the same time, the catalyst and the initiating species (a weak nucleophile) form an "ate-complex", through which the initiation begins (Figure 7) [9]. This polymerization offers the advantage that the polymerization can be performed at lower temperatures (from −30 °C to R.T), and the chain transfer reactions of adjacent proton to epoxide ring are effectively suppressed, especially in systems where ammonium salts and $(i\text{-}Bu)_3$ Al are combined [9]. Gervais et al., further applied this system to prepare PEEGE to molecular weights up to 85 kDa with narrow molecular weight distributions (PDI $\geq$ 1.03) [36]. Although side reactions such as ring-opening via hydride or *iso*-butyl groups can cause lower molecular weights and ill-defined chain-ends, for the preparation of high molecular weight polyethers with defined structures, this "activated monomer" polymerization is a widely studied and applied method in comparison to the conventional AROP [9,57,58]. In a study aiming toward the synthesis of biodegradable LPG, Köhler et al., synthesized polyglycidol with graftings of polycaprolactone [59].

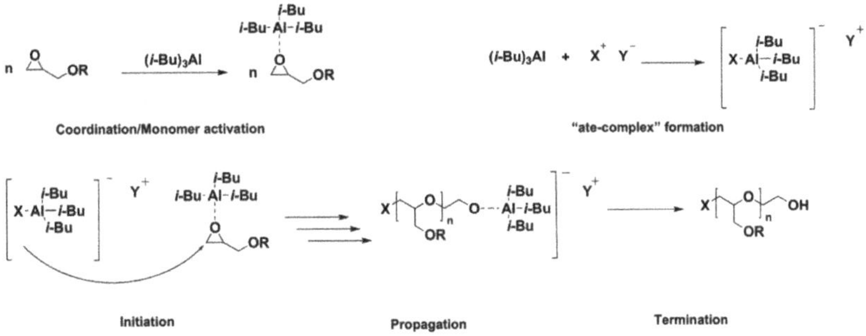

Figure 7. Reaction mechanism of activated monomer polymerization technique. Adapted with permission from ref. [9]. Copyright 2016 American Chemical Society.

3. Polyglycidols as a Multiplatform for Biomedical Applications

3.1. Drug Delivery Systems, Protein and Surface Conjugation

HPG is characterized by its high water-solubility and end group tune-functionality. Hence, its derivatives can be designed and functionalized to be used as a drug delivery platform when modified to include biodegradability to avoid in vivo accumulation and release its cargo and hydrophobicity to encapsulate hydrophobic cargos as additional properties. Nevertheless, HPG can also be integrated into other supramolecular structures such as polymeric micelles or nanogels to form drug delivery systems. As HPG-based drug delivery systems have been intensively discussed in previous reviews [15], we would like to summarize the most recent systems in the scope of this review.

Zhong et al. developed amphiphilic block copolymers that included sulfated HPG as its hydrophilic part, attached with a disulfide-bearing linker to polycaprolactone as the hydrophobic part [60]. The amphiphilic block copolymer underwent self-assembly to form micellar structures. The micellar system was then loaded with doxorubicin (DOX) and proved to have a long plasma half-life and significant tumor accumulation [60]. When tested on MCF-7 human mammary carcinoma mice models, the micelles induced complete tumor suppression and improved the survival rate [60]. Later, Braatz et al., assembled a toolbox of the same micellar system that contained various polyesters as the hydrophobic segment as well as two different molecular weights of the sulfated HPG hydrophilic segment [61]. The study also included determining the critical micelle concentration (CMC), testing the stability, drug release, and tumor targeting of the system when encapsulated with sunitinib [61]. When the loaded micelles were injected into an HeLa human cervical tumor-bearing mice model, a ten-fold lower dose of the micelle in comparison to the free drug was able to improve the antitumor efficacy of the API [61].

Baabur-Cohen et al. developed two polymeric carriers that have different supramolecular assembly architectures for the combination drug therapy of paclitaxel (PTX) and doxorubicin (DOX) [62]. The drugs were covalently bonded to the linear section of the polymer, polyglutamic acid (PGA), or to the hyperbranched scaffold of PG that was attached to polyethylene glycol (PEG), respectively, in both polymeric architectures [62]. The aim was to study the activity of both conjugates in a murine model of mammary adenocarcinoma in immunocompetent mice. In this regard, both conjugates showed superior antitumor activity when compared to the combination of the two free drugs [62]. Furthermore, researchers have investigated thermoresponsive nanogel (tNGs) drug delivery systems with HPGs platforms. One of the studies presented by Rancan et al., aimed to improve dermal and transdermal drug delivery using tNGs based on acrylated dendritic polyglycerol as a water-soluble crosslinker, combined with ethylene glycol methacrylate with thermoresponsive properties [63]. The characteristics of the tNGs were dependent on their cloud point temperature (Tcp), as well as their skin penetration and cellular uptake [63]. Gerecke et al., proved the intracellular localization of HPG-based tNGs, a combination of dPG with poly(glycidyl methyl ether-co-ethyl glycidyl ether), in keratinocytes [64]. The tNGs were able to encapsulate dexamethasone and tacrolimus, drugs used for skin disease treatment. By laser scanning confocal microscopy, it was proven that the fluorescently labeled system was able to localize predominantly within the lysosomal compartment [64]. In addition, the tNGs showed no cytotoxic or genotoxic effect, any induction, or reactive oxygen species when the MTT assay, comet assay, and carboxy-H2DCFDA assay were performed. Later, the same TNGs were tested for the dependency of the uptake mechanism on the cloud point temperature of the TNGs [65]. Intriguingly, it was shown that above the Tcp, the uptake mechanism was caveolae-mediated endocytosis; however, at the Tcp, micropinocytosis was also included as an uptake mechanism [65]. Other thermoresponsive nanogel carrier systems were developed for controlled delivery through the hair follicle by Sahle et al., The nanogels were synthesized by the precipitation polymerization technique using N-isopropylacrylamide as a monomer, acrylated dendritic polyglycerol as a crosslinker, VA-044 as an initiator, and sodium dodecyl sulphate as a stabilizer [66]. The follicular penetration of the labeled nanogels was assessed ex vivo using porcine ear skin [66]. Another TNGs was developed by Molina et al. and based on Poly-*N*-isopropylacrylamide–dendritic polyglycerol NG that was semi-interpenetrated with 2-acrylamido-2-methylpropane sulfonic acid or (2-dimethylamino) ethyl methacrylate [67]. The tNGs were loaded with doxorubicin and showed more efficiency in multidrug-resistant cancer cell proliferation inhibition studies. When admitted in vivo, the tNGs reduced the tumor volume to about 25% [67]. Miceli et al., combined the thermoresponsiveness with the pH responsiveness of the NGs based on dendritic polyglycerol (dPG) and pNIPAM that were semi-interpenetrated with poly(4-acryloylamine-4-(carboxyethyl)heptanodioic acid) (pABC) [68]. They showed the stability of the system under physiological conditions and tunable electrophoretic mobilities around the human body temperature [68]. The NGs were able to release their hosted molecule (cytochrome c) upon cooling and were able to deliver it to cancer cells and induce apoptosis at 30 °C [68].

Hofmann et al. developed polyethylene glycol (PEG)-substituted liposomes by developing multi-functional lipids based on the AROP of protected glycidols (EEGE and isopropylidene glyceryl glycidyl ether (IGG)) initiated by cholesterol and 1,2-bis-n-alkyl glyceryl ether. Due to the multi-functionality of these liposomes, they can be further functionalized. The authors used rhodamine B as an example for further modification [69]. For selective delivery and release of species for biomedical applications, Jamróz-Piegza et al., designed and synthesized deblock copolymers of PEG and LPG via AROP, where the LPG block was modified with cinnamic acid in a subsequent step [70]. After micelle formation, owing to the properties of glycidyl cinnamate groups, the core could be crosslinked by UV-irradiation, forming stable nanoparticles.

LPG was recently thoroughly studied as an alternative to PEG for the conjugation of biopharmaceuticals due to its structural similarity to PEG. In the last three decades, PEG has been the "gold standard" for conjugation to biopharmaceuticals (a process known as PEGylation) to address their intrinsic shortcomings such as instability, immunogenicity [71], and a short half-life [72,73]. To date, there are more than 18 PEGylated drugs approved by the Food and Drug Administration (FDA) on the market [74]. PEGylation can also be applied to lipids and nanoparticle (NP) drug delivery systems to enhance the stability of NPs [75]. This technology has been employed in the new mRNA vaccines against SARS-CoV-2 [76]. However, despite these major breakthroughs, there are some shortcomings associated with PEGylated systems which can limit their broad application. These include: induced anti-PEG antibodies upon repeated administration of PEGylated drugs, negatively altering their therapeutic efficiency and, in some cases, leading to life-threatening allergic reactions such as anaphylaxis [77]. Similar to PEG, LPG is based on a polyether backbone but exhibits hydroxy groups leading to much higher hydrophilicity of the polymer backbone; in contrast to conventional PEG, LPG enables the introduction of immobilization, targeting, or labeling moieties [78,79]. Additionally, esters of oligoglycerols with up to 10 repeating units have been approved by the FDA as pharmaceutical and food additives and have been commercially available for a few decades [7,32]. In a study comparing, LPG, hPG, and PEG, Imran Ul-haq et al., tested these polymers with molecular weights of about 100 kDa both in vitro and in vivo. They showed that the hydrodynamic size measurements confirmed the absence of intermolecular aggregation. It was observed that the intrinsic viscosity of LPG was about 25 times smaller than that of PEG of the same size. These characteristics play an important role in the application of these polymers in formulations where higher doses are needed. LPG was further tested in different hemocompatibility assays, such as red blood cell (RBC) aggregation and homolysis. They showed that LPG (and HPG), even in high concentrations (10 mg/mL), did not induce any unwanted RBC aggregation, while PEG induced a massive RBC aggregation at the same concentration [80]. This was in line with the result of this assay performed with lower molecular weight LPG (6400 g/mol) by Kainthan et al., [81]. Their hypothesis was that although LPG and PEG have similar structures, the more compact and hydrophilic structure of LPG limits its interaction with RBC. Furthermore, LPG did not induce any platelet or complement activation in the concentration range which was studied (up to 10 mg/mL). In contrast, a mild to moderate complement activation already at 5 mg/mL was induced by PEG. Furthermore, a longer half-life for LPG was observed in comparison to other polymers [80].

In a study conducted by Abu Lila et al. liposomes were modified with PEG and LPG. It was observed that replacing LPGylated liposomes enhances the in vivo performance in comparison to PEGylated counterparts, as LPG-modified liposomes did not induce accelerated blood clearance (ABC), a limitation PEGylated liposomes have upon repeated administration, which negatively affects their pharmaceutical activity [82]. The end-group of LPG can be easily modified to introduce the desired functional group for protein modification or other applications. This can be carried out either by applying different initiators or post-modification of the end group subsequent to the polymerization process [5,33,36,83–85]. LPG has been successfully conjugated to proteins via different chemistries, including random conjugation to bovine serum albumin (BSA) [33], site-specific CuAAC to Exenatide, *N*-terminally to Interleukin-4 [86], and Anakinra [87] and to IFN-α-2a via SPAAC [88]. These studies clearly demonstrated a comparable size of PEGylated and LPGylated proteins with a similar half-life extension in vivo. Owing to its excellent biocompatibility, ease of synthesis, multi-functionality, and low intrinsic viscosity, LPG is a promising candidate for various other biomedical applications.

3.2. Viral Infection Inhibition

Viral diseases are one of the major threats to our health, which are associated with morbidity, mortality, and serious socioeconomic consequences [89]. Vaccination, as the most effective solution, is only available against a limited number of viral infections [90]. Antiviral small molecule drugs are likewise effective against a limited number of viruses and target essential viral functions and proteins, meaning virus mutations can escape and make the virus resistant to drugs [91]. Therefore, the development of alternative ways to intervene with a broad spectrum of virus families is of high interest.

A primary mechanism that several viruses have developed for binding and delivering the viral genome is initial multivalent binding to different heparan sulfate proteoglycans (HSPGs) at the host cell surface [92–94]. HSPGs are composed of protein cores covalently linked to unbranched negatively charged polysaccharides named heparan sulfate (HS) [95]. The counter ion release is the main driving force for these electrostatic interactions. The positively charged patches on the virus surface play the role of many counterions for the negatively charged HS. This interaction is favored entropically due to the release of the counterions from the polyelectrolyte. It has recently been shown that interaction with HS is a necessary co-factor for SARS-CoV-2 cell entry and infection. Inspired by HS, negatively charged natural polysaccharides such as heparin [96,97] are used as viral infection inhibitors. Some biological activities of heparin include angiogenesis and tumor growth inhibition [98,99], complement system regulation [100,101], and antiviral [97,102] and anti-inflammatory [103] activity. It is approved by FDA for clinical use in the treatment of deep vein thrombosis. Furthermore, all these properties of heparin make it a unique and promising treatment candidate for COVID-19 patients due to thromboembolic events [104] and pathological inflammation [96,105]. Recently it was shown that heparin can effectively inhibit SARS-CoV-2 infection. The application of heparin for the treatment of COVID-19 patients is under intense investigation, and nebulized heparin in the treatment of the SARS-CoV-2 infection has reached clinical trials [4,100,106,107]. However, currently, the only source of heparin is animal tissues which raises the risk of virus contaminations, adverse effects, and batch-to-batch variability [92,108]. Additionally, heparin can be altered or degraded by heparinases which could result in loss of its activity [100]. Another issue is the anticoagulant activity of heparin which can lead to undesired side effects such as hemorrhagic complexities [109]. Therefore, designing heparin-mimicking compounds to overcome the limitations is intensively studied. Our group recently introduced linear polyglycerol sulfates (LPGS) as heparin mimetics and powerful SARS-CoV-2 inhibitors (Figure 8). Surprisingly the LPGS with a molecular weight of 40 kDa was almost 60 times more active than heparin but showed a much lower anticoagulant activity [110].

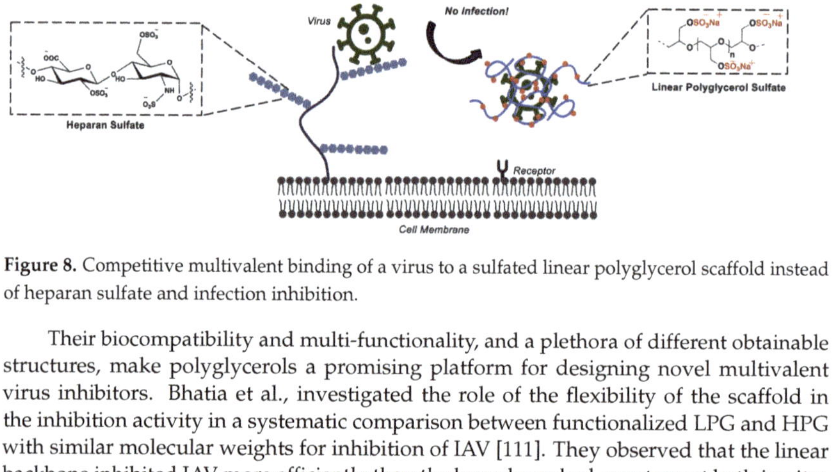

Figure 8. Competitive multivalent binding of a virus to a sulfated linear polyglycerol scaffold instead of heparan sulfate and infection inhibition.

Their biocompatibility and multi-functionality, and a plethora of different obtainable structures, make polyglycerols a promising platform for designing novel multivalent virus inhibitors. Bhatia et al., investigated the role of the flexibility of the scaffold in the inhibition activity in a systematic comparison between functionalized LPG and HPG with similar molecular weights for inhibition of IAV [111]. They observed that the linear backbone inhibited IAV more efficiently than the hyperbranched counterpart both in vitro and in vivo. They further investigated this observation in polyglycerol-based nanogel (NG) scaffolds [106]. NGs are cross-linked 3D constructs from water-soluble polymers and are swellable. For the introduction of different flexibilities into the gels, they applied LPG and HPG as flexible and rigid cross-linkers, respectively, to obtain flexible (F-NG) and rigid (R-NG) cross-linkers, and observed that the most flexible functionalized NG based on LPG can inhibit IAV infection 400 times better than the more rigid counterparts [106].

Dey et al. made this observation for the inhibition of HSV-1 by NGs. They prepared two classes of NGs based on sulfated HPG (HPGS) and used LPG or HPG as cross-linkers to develop scaffolds with distinctive rigidities. They further proved the flexibility differences between F-NGs and R-NGs by atomic force microscopy (AFlinearM), in which the negatively charged NGs were coated on a positively charged mica substrate, and it was observed that the F- NG, which was made with LPG linker, exhibited a smaller height and higher width in comparison to R-NG cross-linked with HPG. It was additionally observed that flexible NG had increased antiviral activity due to better shielding of the virus interaction with the cell surface. Their mathematical modeling supported this data by showing that to sterically shield a virus particle, six rigid NG are needed while only three of the soft NG of the same size are needed [107]. Pouyan et al. investigated the effect of backbone flexibility on the inhibition of HSV-1 by synthesizing a series of sulfated polyglycerols with different architectures, namely linear, dendronized, and hyperbranched (LPGS, DenPGS, HPGS) and compared it to Heparin as the natural sulfated polymer [112]. From the plaque reduction assay, an increase in IC_{50} values from 0.03 to 374 nM from the flexible backbone (LPGS) to less flexible ones (dendronized and HPGS) was observed. Knowing that all the polymers had the same density of negative charges, it was concluded that the more flexible the backbone, the more it can change conformation and shield the virus surface. To further evaluate the role of the scaffold's flexibility, Mohammadifar and Ahmadi et al. used HPGS to form 2D constructs by reversibly fixing HPG on a graphene sheet, crosslinking them together to form a 2D construct, separation from graphene sheet, followed by sulfation. This system was then compared with sulfated 3D NGs in viral infection inhibition. Due to higher flexibility, the 2D system outperformed the 3D system by showing four times better inhibition of HSV-1 and SARS-CoV-2 [113].

To investigate other heparin-like characteristics of the HPGS, Ferraro and Silberreis et al. investigated the anti-inflammatory properties of this scaffold. They observed that HPGS has anti-inflammatory activity and can regulate the complement response as good as heparin [101], and much less anticoagulation time for this material was detected, which is of high interest for applications of anti-inflammatory or antiviral compounds to reduce the risk of uncontrolled bleedings.

3.3. *Antifouling Coatings for Biomedical Application*

Unspecific biofouling or nonspecific protein adsorption on surfaces present a serious problem in biomedical applications such as medical implants, biosensors, and surgical equipment [114]. Due to its antifouling characteristics, PEG has been the focus of many studies for the development of non-fouling coatings for biomedical applications [115,116]. As PEG-modified surfaces have disadvantages such as the instability of the polymer upon heating and immunological challenges after repeated exposure [117], LPG has been investigated as an alternative to PEG for surface modification. Kulka et al., developed antifouling surface coatings based on mussel-inspired dendritic PG (MI-dPG) modified with LPG containing a block of oligo-amine (LPG-b-OA_{11}) and compared the surface characteristics such as cell fouling, protein fouling, and chemical stability with the MI-dPG surface modified with commercially available amine terminated PEG (HO-PEG-NH_2). The quartz crystal microbalance measurements with dissipation monitoring (QCM-D) revealed that the post functionalized surfaces with LPG outperformed the PEG modified surfaces in protein antifouling properties. In a follow-up study, the authors investigated the applicability of these coatings in regard to reducing shear and biomaterial-induced thrombosis on continuous-flow ventricular assist devices [118]. They observed that the post-modified surfaces with LPG outperformed PEG-modified surfaces in rejection of cellular adhesion and proliferation. With regard to cell toxicity, the LPG-b-OA_{11} showed no cytotoxicity up until 5 $mg \cdot mL^{-1}$ on A549 cell lines. In another study attempting to develop antifouling surfaces based on PG, the authors reported a simple and solvent-free surface-initiated polymerization from MI-dPG-coated TiO_2 (hydrophilic) and polydimethylsiloxane (PDMS, hydrophobic) [119]. They performed cell viability studies with two different cell lines (A549 and DF-1) on various coatings to observe the influence of the coating on the cell numbers. The results showed that the introduction of MI-dPG on both TiO_2 and PDMS led to a slight decrease in the cell number on the respective surface. However, after grafting dPG, a drastic decrease (>95%) in the cell number was observed for both cell lines and investigated surfaces. Their approach provided a successful strategy for developing a highly biocompatible but cell-repelling surface coating [119]. Figure 9 represents an overview of the properties of PG discussed in this and previous sections.

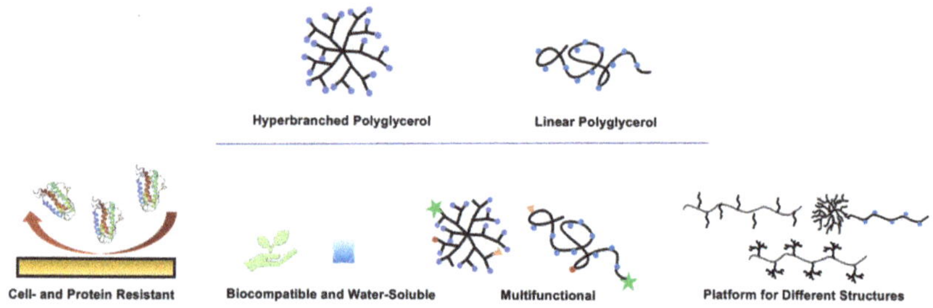

Figure 9. Properties of polyglycerols make them suitable candidates for biomedical applications.

4. Summary

Functional polymers are an indispensable tool in medicine and life sciences. PG, with different structures and properties such as water-solubility, biocompatibility, and multifunctionality, has become one of the most studied and applied polymers in these fields (Figure 9). In the past few decades, great efforts have been invested to optimize the synthesis and modification of these polymers for biomedical and pharmaceutical applications. HPG can be synthesized via cationic, anionic, or coordination ring-opening polymerizations from the monomer glycidol. By protecting the hydroxy group of the monomer prior to polymerization, linear structures (LPG) are obtained. The most common monomer for this purpose is EEGE, in which the hydroxy group of glycidol is protected by an acid-labile acetal group.

PGs have been successfully applied in targeted drug delivery systems as carriers and micelles. Furthermore, moieties with redox or pH-sensitive groups have been successfully introduced to the backbone, making these systems degradable under biological conditions. Due to similar characteristics of LPG to the gold standard PEG, it has been conjugated to biomacromolecules such as proteins to increase their stability, solubility, and half-life.

We believe that the characteristics, ease of synthesis, scalability, multi-functionality, and structural versatility of PGs and PG-based systems can make a significant impact on the development and application of these materials both in vitro and in vivo.

Author Contributions: Conceptualization, P.P., M.C. and R.H.; writing—original draft preparation, P.P. and M.C.; Scheme making: P.P. and M.C., writing—review and editing, R.H.; funding acquisition, R.H. All authors have read and agreed to the published version of the manuscript.

Funding: This research was funded by Deutsche Forschungsgemeinschaft: SFB 1449, TP B03.

Conflicts of Interest: The authors declare no conflict of interest.

References

1. Mülhaupt, R. Hermann Staudinger and the Origin of Macromolecular Chemistry. *Angew. Chem. Int. Ed.* **2004**, *43*, 1054–1063. [CrossRef] [PubMed]
2. Jiayi, P.; Rostamizadeh, K.; Filipczak, N.; Torchilin, V.P. Polymeric Co-Delivery Systems in Cancer Treatment: An Overview on Component Drugs' Dosage Ratio Effect. *Molecules* **2019**, *24*, 1035.
3. Bianculli, R.H.; Mase, J.D.; Schulz, M.D. Antiviral Polymers: Past Approaches and Future Possibilities. *Macromolecules* **2020**, *53*, 9158–9186. [CrossRef]
4. Zelikin, A.N.; Stellacci, F. Broad-Spectrum Antiviral Agents Based on Multivalent Inhibitors of Viral Infectivity. *Adv. Healthc. Mater.* **2021**, *10*, e2001433. [CrossRef]
5. Dernedde, J.; Rausch, A.; Weinhart, M.; Enders, S.; Tauber, R.; Licha, K.; Schirner, M.; Zügel, U.; von Bonin, A.; Haag, R. Dendritic polyglycerol sulfates as multivalent inhibitors of inflammation. *Proc. Natl. Acad. Sci. USA* **2010**, *107*, 19679. [CrossRef]
6. Bochenek, M.; Oleszko-Torbus, N.; Wałach, W.; Lipowska-Kur, D.; Dworak, A.; Utrata-Wesołek, A. Polyglycidol of Linear or Branched Architecture Immobilized on a Solid Support for Biomedical Applications. *Polym. Rev.* **2020**, *60*, 717–767. [CrossRef]
7. Thomas, A.; Müller, S.S.; Frey, H. Beyond Poly(ethylene glycol): Linear Polyglycerol as a Multifunctional Polyether for Biomedical and Pharmaceutical Applications. *Biomacromolecules* **2014**, *15*, 1935–1954. [CrossRef]
8. Wilms, D.; Stiriba, S.-E.; Frey, H. Hyperbranched Polyglycerols: From the Controlled Synthesis of Biocompatible Polyether Polyols to Multipurpose Applications. *Acc. Chem. Res.* **2010**, *43*, 129–141. [CrossRef]

9. Herzberger, J.; Niederer, K.; Pohlit, H.; Seiwert, J.; Worm, M.; Wurm, F.R.; Frey, H. Polymerization of Ethylene Oxide, Propylene Oxide, and Other Alkylene Oxides: Synthesis, Novel Polymer Architectures, and Bioconjugation. *Chem. Rev.* **2016**, *116*, 2170–2243. [CrossRef]
10. Flory, P.J. Molecular Size Distribution in Three Dimensional Polymers. VI. Branched Polymers Containing A—R—Bf-1 Type Units. *J. Am. Chem. Soc.* **1952**, *74*, 2718–2723. [CrossRef]
11. Dworak, A.; Walach, W.; Trzebicka, B. Cationic polymerization of glycidol. Polymer structure and polymerization mechanism. *Macromol. Chem. Phys.* **1995**, *196*, 1963–1970. [CrossRef]
12. Tokar, R.; Kubisa, P.; Penczek, S.; Dworak, A. Cationic polymerization of glycidol: Coexistence of the activated monomer and active chain end mechanism. *Macromolecules* **1994**, *27*, 320–322. [CrossRef]
13. Mohammadifar, E.; Bodaghi, A.; Dadkhahtehrani, A.; Nemati Kharat, A.; Adeli, M.; Haag, R. Green Synthesis of Hyperbranched Polyglycerol at Room Temperature. *ACS Macro Lett.* **2017**, *6*, 35–40. [CrossRef]
14. Kim, S.E.; Yang, H.J.; Choi, S.; Hwang, E.; Kim, M.; Paik, H.-J.; Jeong, J.-E.; Park, Y.I.; Kim, J.C.; Kim, B.-S.; et al. A recyclable metal-free catalytic system for the cationic ring-opening polymerization of glycidol under ambient conditions. *Green Chem.* **2022**, *24*, 251–258. [CrossRef]
15. Abbina, S.; Vappala, S.; Kumar, P.; Siren, E.M.J.; La, C.C.; Abbasi, U.; Brooks, D.E.; Kizhakkedathu, J.N. Hyperbranched polyglycerols: Recent advances in synthesis, biocompatibility and biomedical applications. *J. Mater. Chem. B* **2017**, *5*, 9249–9277. [CrossRef] [PubMed]
16. Haag, R.; Sunder, A.; Stumbé, J.-F. An Approach to Glycerol Dendrimers and Pseudo-Dendritic Polyglycerols. *J. Am. Chem. Soc.* **2000**, *122*, 2954–2955. [CrossRef]
17. Sunder, A.; Hanselmann, R.; Frey, H.; Mülhaupt, R. Controlled Synthesis of Hyperbranched Polyglycerols by Ring-Opening Multibranching Polymerization. *Macromolecules* **1999**, *32*, 4240–4246. [CrossRef]
18. Schmitt, V.; Rodríguez-Rodríguez, C.; Hamilton, J.L.; Shenoi, R.A.; Schaffer, P.; Sossi, V.; Kizhakkedathu, J.N.; Saatchi, K.; Häfeli, U.O. Quantitative SPECT imaging and biodistribution point to molecular weight independent tumor uptake for some long-circulating polymer nanocarriers. *RSC Adv.* **2018**, *8*, 5586–5595. [CrossRef]
19. Kautz, H.; Sunder, A.; Frey, H. Control of the molecular weight of hyperbranched polyglycerols. *Macromol. Symp.* **2001**, *163*, 67–74. [CrossRef]
20. Wallert, M.; Plaschke, J.; Dimde, M.; Ahmadi, V.; Block, S.; Haag, R. Automated Solvent-Free Polymerization of Hyperbranched Polyglycerol with Tailored Molecular Weight by Online Torque Detection. *Macromol. Mater. Eng.* **2021**, *306*, 2000688. [CrossRef]
21. Wilms, D.; Wurm, F.; Nieberle, J.; Böhm, P.; Kemmer-Jonas, U.; Frey, H. Hyperbranched Polyglycerols with Elevated Molecular Weights: A Facile Two-Step Synthesis Protocol Based on Polyglycerol Macroinitiators. *Macromolecules* **2009**, *42*, 3230–3236. [CrossRef]
22. Moore, E.; Zill, A.T.; Anderson, C.A.; Jochem, A.R.; Zimmerman, S.C.; Bonder, C.S.; Kraus, T.; Thissen, H.; Voelcker, N.H. Synthesis and Conjugation of Alkyne-Functional Hyperbranched Polyglycerols. *Macromol. Chem. Phys.* **2016**, *217*, 2252–2261. [CrossRef]
23. Kainthan, R.K.; Muliawan, E.B.; Hatzikiriakos, S.G.; Brooks, D.E. Synthesis, Characterization, and Viscoelastic Properties of High Molecular Weight Hyperbranched Polyglycerols. *Macromolecules* **2006**, *39*, 7708–7717. [CrossRef]
24. Ul-haq, M.I.; Shenoi, R.A.; Brooks, D.E.; Kizhakkedathu, J.N. Solvent-assisted anionic ring opening polymerization of glycidol: Toward medium and high molecular weight hyperbranched polyglycerols. *J. Polym. Sci. Part A Polym. Chem.* **2013**, *51*, 2614–2621. [CrossRef]
25. Anilkumar, P.; Lawson, T.B.; Abbina, S.; Mäkelä, J.T.A.; Sabatelle, R.C.; Takeuchi, L.E.; Snyder, B.D.; Grinstaff, M.W.; Kizhakkedathu, J.N. Mega macromolecules as single molecule lubricants for hard and soft surfaces. *Nat. Commun.* **2020**, *11*, 2139. [CrossRef]
26. Kainthan, R.K.; Brooks, D.E. In Vivo Biological Evaluation of High Molecular Weight Hyperbranched Polyglycerols. no. 0142-9612 (Print). *Biomaterials* **2007**, *28*, 4779–4787. [CrossRef]
27. Shenoi, R.A.; Narayanannair, J.K.; Hamilton, J.L.; Lai, B.F.L.; Horte, S.; Kainthan, R.K.; Varghese, J.P.; Rajeev, K.G.; Manoharan, M.; Kizhakkedathu, J.N. Branched Multifunctional Polyether Polyketals: Variation of Ketal Group Structure Enables Unprecedented Control over Polymer Degradation in Solution and within Cells. *J. Am. Chem. Soc.* **2012**, *134*, 14945–14957. [CrossRef]
28. Hubbell, J.A.; Thomas, S.N.; Swartz, M.A. Materials engineering for immunomodulation. *Nature* **2009**, *462*, 449–460. [CrossRef]
29. Son, S.; Shin, E.; Kim, B.-S. Redox-Degradable Biocompatible Hyperbranched Polyglycerols: Synthesis, Copolymerization Kinetics, Degradation, and Biocompatibility. *Macromolecules* **2015**, *48*, 600–609. [CrossRef]
30. Mura, S.; Nicolas, J.; Couvreur, P. Stimuli-responsive nanocarriers for drug delivery. *Nat. Mater.* **2013**, *12*, 991–1003. [CrossRef]
31. Sun, H.; Meng, F.; Cheng, R.; Deng, C.; Zhong, Z. Reduction-sensitive degradable micellar nanoparticles as smart and intuitive delivery systems for cancer chemotherapy. *Expert Opin. Drug Deliv.* **2013**, *10*, 1109–1122. [CrossRef] [PubMed]
32. Knop, K.; Hoogenboom, R.; Fischer, D.; Schubert, U.S. Poly(ethylene glycol) in Drug Delivery: Pros and Cons as Well as Potential Alternatives. *Angew. Chem. Int. Ed.* **2010**, *49*, 6288–6308. [CrossRef] [PubMed]
33. Wurm, F.; Dingels, C.; Frey, H.; Klok, H.-A. Squaric Acid Mediated Synthesis and Biological Activity of a Library of Linear and Hyperbranched Poly(Glycerol)–Protein Conjugates. *Biomacromolecules* **2012**, *13*, 1161–1171. [CrossRef] [PubMed]
34. Fitton, A.O.; Hill, J.; Jane, D.E.; Millar, R. Synthesis of Simple Oxetanes Carrying Reactive 2-Substituents. *Synthesis* **1987**, *1987*, 1140–1142. [CrossRef]

35. Taton, D.; Le Borgne, A.; Sepulchre, M.; Spassky, N. Synthesis of chiral and racemic functional polymers from glycidol and thioglycidol. *Macromol. Chem. Phys.* **1994**, *195*, 139–148. [CrossRef]
36. Gervais, M.; Brocas, A.-L.; Cendejas, G.; Deffieux, A.; Carlotti, S. Synthesis of Linear High Molar Mass Glycidol-Based Polymers by Monomer-Activated Anionic Polymerization. *Macromolecules* **2010**, *43*, 1778–1784. [CrossRef]
37. Baek, J.; Kim, M.; Park, Y.; Kim, B.-S. Acetal-Based Functional Epoxide Monomers: Polymerizations and Applications. *Macromol. Biosci.* **2021**, *21*, 2100251. [CrossRef]
38. Walach, W.; Kowalczuk, A.; Trzebicka, B.; Dworak, A. Synthesis of High-Molar Mass Arborescent-Branched Polyglycidol via Sequential Grafting. *Macromol. Rapid Commun.* **2001**, *22*, 1272–1277. [CrossRef]
39. Dworak, A.; Panchev, I.; Trzebicka, B.; Walach, W. Hydrophilic and amphiphilic copolymers of 2,3-epoxypropanol-1. *Macromol. Symp.* **2000**, *153*, 233–242. [CrossRef]
40. Hans, M.; Keul, H.; Moeller, M. Chain transfer reactions limit the molecular weight of polyglycidol prepared via alkali metal based initiating systems. *Polymer* **2009**, *50*, 1103–1108. [CrossRef]
41. Hans, M.; Gasteier, P.; Keul, H.; Moeller, M. Ring-Opening Polymerization of ε-Caprolactone by Means of Mono- and Multifunctional Initiators: Comparison of Chemical and Enzymatic Catalysis. *Macromolecules* **2006**, *39*, 3184–3193. [CrossRef]
42. Erberich, M.; Keul, H.; Möller, M. Polyglycidols with Two Orthogonal Protective Groups: Preparation, Selective Deprotection, and Functionalization. *Macromolecules* **2007**, *40*, 3070–3079. [CrossRef]
43. Anja, T.; Niederer, K.; Wurm, F.; Frey, H. Combining Oxyanionic Polymerization and Click-Chemistry: A General Strategy for the Synthesis of Polyether Polyol Macromonomers. *Polym. Chem.* **2014**, *5*, 899–909.
44. Gesine, G.; Weinhart, M.; Becherer, T.; Haag, R.; Huck, W.T.S. Effect of Polymer Brush Architecture on Antibiofouling Properties. *Biomacromolecules* **2011**, *12*, 4169–4172.
45. Toy, A.A.; Reinicke, S.; Müller, A.H.E.; Schmalz, H. One-Pot Synthesis of Polyglycidol-Containing Block Copolymers with Alkyllithium Initiators Using the Phosphazene Base t-BuP4. *Macromolecules* **2007**, *40*, 5241–5244. [CrossRef]
46. Dimitrov, I.; Tsvetanov, C.B. 4.21—High-Molecular-Weight Poly(ethylene oxide). In *Polymer Science: A Comprehensive Reference*; Matyjaszewski, K., Möller, M., Eds.; Elsevier: Amsterdam, The Netherlands, 2012; pp. 551–569.
47. Brocas, A.-L.; Mantzaridis, C.; Tunc, D.; Carlotti, S. Polyether synthesis: From activated or metal-free anionic ring-opening polymerization of epoxides to functionalization. *Prog. Polym. Sci.* **2013**, *38*, 845–873. [CrossRef]
48. Spears, B.R.; Waksal, J.; McQuade, C.; Lanier, L.; Harth, E. Controlled branching of polyglycidol and formation of protein–glycidol bioconjugates via a graft-from approach with "PEG-like" arms. *Chem. Commun.* **2013**, *49*, 2394–2396. [CrossRef]
49. Nijenhuis, A.J.; Grijpma, D.W.; Pennings, A.J. Lewis acid catalyzed polymerization of L-lactide. Kinetics and mechanism of the bulk polymerization. *Macromolecules* **1992**, *25*, 6419–6424. [CrossRef]
50. Cherri, M.; Ferraro, M.; Mohammadifar, E.; Quaas, E.; Achazi, K.; Ludwig, K.; Grötzinger, C.; Schirner, M.; Haag, R. Biodegradable Dendritic Polyglycerol Sulfate for the Delivery and Tumor Accumulation of Cytostatic Anticancer Drugs. *ACS Biomater. Sci. Eng.* **2021**, *7*, 2569–2579. [CrossRef]
51. Zabihi, F.; Graff, P.; Schumacher, F.; Kleuser, B.; Hedtrich, S.; Haag, R. Synthesis of poly(lactide-co-glycerol) as a biodegradable and biocompatible polymer with high loading capacity for dermal drug delivery. *Nanoscale* **2018**, *10*, 16848–16856. [CrossRef]
52. Reisbeck, F.; Ozimkovski, A.; Cherri, M.; Dimde, M.; Quaas, E.; Mohammadifar, E.; Achazi, K.; Haag, R. Gram Scale Synthesis of Dual-Responsive Dendritic Polyglycerol Sulfate as Drug Delivery System. *Polymers* **2021**, *13*, 982. [CrossRef] [PubMed]
53. Jamróz-Piegza, M.; Utrata-Wesołek, A.; Trzebicka, B.; Dworak, A. Hydrophobic modification of high molar mass polyglycidol to thermosensitive polymers. *Eur. Polym. J.* **2006**, *42*, 2497–2506. [CrossRef]
54. Haouet, A.; Sepulchre, M.; Spassky, N. Preparation et proprietes des poly(R)-glycidols. *Eur. Polym. J.* **1983**, *19*, 1089–1098. [CrossRef]
55. Utrata-Wesołek, A.; Żymełka-Miara, I.; Kowalczuk, A.; Trzebicka, B.; Dworak, A. Photocrosslinking of Polyglycidol and Its Derivative: Route to Thermoresponsive Hydrogels. *Photochem. Photobiol.* **2018**, *94*, 52–60. [CrossRef]
56. Labbé, A.; Carlotti, S.; Billouard, C.; Desbois, P.; Deffieux, A. Controlled High-Speed Anionic Polymerization of Propylene Oxide Initiated by Onium Salts in the Presence of Triisobutylaluminum. *Macromolecules* **2007**, *40*, 7842–7847. [CrossRef]
57. Kowalczuk, A.; Trzcinska, R.; Trzebicka, B.; Müller, A.H.E.; Dworak, A.; Tsvetanov, C.B. Loading of polymer nanocarriers: Factors, mechanisms and applications. *Prog. Polym. Sci.* **2014**, *39*, 43–86. [CrossRef]
58. Mohammadifar, E.; Nemati Kharat, A.; Adeli, M. Polyamidoamine and polyglycerol; their linear, dendritic and linear–dendritic architectures as anticancer drug delivery systems. *J. Mater. Chem. B* **2015**, *3*, 3896–3921. [CrossRef] [PubMed]
59. Koehler, J.; Marquardt, F.; Keul, H.; Moeller, M. Phosphonoethylated Polyglycidols: A Platform for Tunable Enzymatic Grafting Density. *Macromolecules* **2013**, *46*, 3708–3718. [CrossRef]
60. Zhong, Y.; Dimde, M.; Stöbener, D.; Meng, F.; Deng, C.; Zhong, Z.; Haag, R. Micelles with Sheddable Dendritic Polyglycerol Sulfate Shells Show Extraordinary Tumor Targetability and Chemotherapy in Vivo. *ACS Appl. Mater. Interfaces* **2016**, *8*, 27530–27538. [CrossRef]
61. Braatz, D.; Dimde, M.; Ma, G.; Zhong, Y.; Tully, M.; Grötzinger, C.; Zhang, Y.; Mavroskoufis, A.; Schirner, M.; Zhong, Z.; et al. Toolbox of Biodegradable Dendritic (Poly glycerol sulfate)–SS-poly(ester) Micelles for Cancer Treatment: Stability, Drug Release, and Tumor Targeting. *Biomacromolecules* **2021**, *22*, 2625–2640. [CrossRef]

62. Baabur-Cohen, H.; Vossen, L.I.; Krüger, H.R.; Eldar-boock, A.; Yeini, E.; Landa-Rouben, N.; Tiram, G.; Wedepohl, S.; Markovsky, E.; Leor, J.; et al. In vivo comparative study of distinct polymeric architectures bearing a combination of paclitaxel and doxorubicin at a synergistic ratio. *J. Control. Release* **2017**, *257*, 118–131. [CrossRef] [PubMed]
63. Rancan, F.; Asadian-Birjand, M.; Dogan, S.; Graf, C.; Cuellar, L.; Lommatzsch, S.; Blume-Peytavi, U.; Calderón, M.; Vogt, A. Effects of thermoresponsivity and softness on skin penetration and cellular uptake of polyglycerol-based nanogels. *J. Control. Release* **2016**, *228*, 159–169. [CrossRef] [PubMed]
64. Gerecke, C.; Edlich, A.; Giulbudagian, M.; Schumacher, F.; Zhang, N.; Said, A.; Yealland, G.; Lohan, S.B.; Neumann, F.; Meinke, M.C.; et al. Biocompatibility and characterization of polyglycerol-based thermoresponsive nanogels designed as novel drug-delivery systems and their intracellular localization in keratinocytes. *Nanotoxicology* **2017**, *11*, 267–277. [CrossRef] [PubMed]
65. Edlich, A.; Gerecke, C.; Giulbudagian, M.; Neumann, F.; Hedtrich, S.; Schäfer-Korting, M.; Ma, N.; Calderon, M.; Kleuser, B. Specific uptake mechanisms of well-tolerated thermoresponsive polyglycerol-based nanogels in antigen-presenting cells of the skin. *Eur. J. Pharm. Biopharm.* **2017**, *116*, 155–163. [CrossRef]
66. Sahle, F.F.; Giulbudagian, M.; Bergueiro, J.; Lademann, J.; Calderón, M. Dendritic polyglycerol and N-isopropylacrylamide based thermoresponsive nanogels as smart carriers for controlled delivery of drugs through the hair follicle. *Nanoscale* **2017**, *9*, 172–182. [CrossRef]
67. Molina, M.; Wedepohl, S.; Miceli, E.; Calderón, M. Overcoming drug resistance with on-demand charged thermoresponsive dendritic nanogels. *Nanomedicine* **2017**, *12*, 117–129. [CrossRef]
68. Miceli, E.; Wedepohl, S.; Osorio Blanco, E.R.; Rimondino, G.N.; Martinelli, M.; Strumia, M.; Molina, M.; Kar, M.; Calderón, M. Semi-interpenetrated, dendritic, dual-responsive nanogels with cytochrome c corona induce controlled apoptosis in HeLa cells. *Eur. J. Pharm. Biopharm.* **2018**, *130*, 115–122. [CrossRef]
69. Hofmann, A.M.; Wurm, F.; Frey, H. Rapid Access to Polyfunctional Lipids with Complex Architecture via Oxyanionic Ring-Opening Polymerization. *Macromolecules* **2011**, *44*, 4648–4657. [CrossRef]
70. Jamróz-Piegza, M.; Wałach, W.; Dworak, A.; Trzebicka, B. Polyether nanoparticles from covalently crosslinked copolymer micelles. *J. Colloid Interface Sci.* **2008**, *325*, 141–148. [CrossRef]
71. Haji Abdolvahab, M.; Venselaar, H.; Fazeli, A.; Arab, S.S.; Behmanesh, M. Point Mutation Approach to Reduce Antigenicity of Interferon Beta. *Int. J. Pept. Res. Ther.* **2020**, *26*, 1353–1361. [CrossRef]
72. Bailon, P.; Palleroni, A.; Schaffer, C.A.; Spence, C.L.; Fung, W.J.; Porter, J.E.; Ehrlich, G.K.; Pan, W.; Xu, Z.X.; Modi, M.W.; et al. Rational design of a potent, long-lasting form of interferon: A 40 kDa branched polyethylene glycol-conjugated interferon alpha-2a for the treatment of hepatitis C. no. 1043-1802 (Print). *Bioconjugate Chem.* **2001**, *12*, 195–202. [CrossRef] [PubMed]
73. Reddy, K.R.; Modi, M.W.; Pedder, S. Use of peginterferon alfa-2a (40 KD) (Pegasys) for the treatment of hepatitis C. no. 0169-409X (Print). *Adv. Drug Deliv. Rev.* **2002**, *54*, 571–586.
74. Kozma, G.T.; Shimizu, T.; Ishida, T.; Szebeni, J. Anti-PEG antibodies: Properties, formation, testing and role in adverse immune reactions to PEGylated nano-biopharmaceuticals. *Adv. Drug Deliv. Rev.* **2020**, *154–155*, 163–175. [CrossRef]
75. Aldosari, B.N.; Alfagih, I.M.; Almurshedi, A.S. Lipid Nanoparticles as Delivery Systems for RNA-Based Vaccines. *Pharmaceutics* **2021**, *13*, 206. [CrossRef] [PubMed]
76. Cao, Y.; Gao, G.F. mRNA vaccines: A matter of delivery. *EClinicalMedicine* **2021**, *32*, 100746. [CrossRef] [PubMed]
77. Lubich, C.; Allacher, P.; de la Rosa, M.; Bauer, A.; Prenninger, T.; Horling, F.M.; Siekmann, J.; Oldenburg, J.; Scheiflinger, F.; Reipert, B.M. The Mystery of Antibodies Against Polyethylene Glycol (PEG)—What do we Know? *Pharm. Res.* **2016**, *33*, 2239–2249. [CrossRef]
78. Licha, K.; Welker, P.; Weinhart, M.; Wegner, N.; Kern, S.; Reichert, S.; Gemeinhardt, I.; Weissbach, C.; Ebert, B.; Haag, R.; et al. Fluorescence Imaging with Multifunctional Polyglycerol Sulfates: Novel Polymeric near-IR Probes Targeting Inflammation. *Bioconjugate Chem.* **2011**, *22*, 2453–2460. [CrossRef]
79. Krüger, H.R.; Schütz, I.; Justies, A.; Licha, K.; Welker, P.; Haucke, V.; Calderón, M. Imaging of doxorubicin release from theranostic macromolecular prodrugs via fluorescence resonance energy transfer. *J. Control. Release* **2014**, *194*, 189–196. [CrossRef]
80. Imran ul-haq, M.; Lai, B.F.L.; Chapanian, R.; Kizhakkedathu, J.N. Influence of architecture of high molecular weight linear and branched polyglycerols on their biocompatibility and biodistribution. *Biomaterials* **2012**, *33*, 9135–9147. [CrossRef]
81. Kainthan, R.K.; Janzen, J.; Levin, E.; Devine, D.V.; Brooks, D.E. Biocompatibility Testing of Branched and Linear Polyglycidol. *Biomacromolecules* **2006**, *7*, 703–709. [CrossRef]
82. Abu Lila, A.S.; Nawata, K.; Shimizu, T.; Ishida, T.; Kiwada, H. Use of polyglycerol (PG), instead of polyethylene glycol (PEG), prevents induction of the accelerated blood clearance phenomenon against long-circulating liposomes upon repeated administration. *Int. J. Pharm.* **2013**, *456*, 235–242. [CrossRef] [PubMed]
83. Gervais, M.; Labbé, A.; Carlotti, S.; Deffieux, A. Direct Synthesis of α-Azido,ω-hydroxypolyethers by Monomer-Activated Anionic Polymerization. *Macromolecules* **2009**, *42*, 2395–2400. [CrossRef]
84. Bej, R.; Achazi, K.; Haag, R.; Ghosh, S. Polymersome Formation by Amphiphilic Polyglycerol-b-polydisulfide-b-polyglycerol and Glutathione-Triggered Intracellular Drug Delivery. *Biomacromolecules* **2020**, *21*, 3353–3363. [CrossRef] [PubMed]
85. Mendrek, S.; Mendrek, A.; Adler, H.-J.; Walach, W.; Dworak, A.; Kuckling, D. Synthesis of poly(glycidol)-block-poly(N-isopropylacrylamide) copolymers using new hydrophilic poly(glycidol) macroinitiator. *J. Polym. Sci. Part A Polym. Chem.* **2008**, *46*, 2488–2499. [CrossRef]

86. Tully, M.; Hauptstein, N.; Licha, K.; Meinel, L.; Lühmann, T.; Haag, R. Linear Polyglycerol for N-terminal-selective Modification of Interleukin-4. *J. Pharm. Sci.* **2021**, *111*, 1642–1651. [CrossRef]
87. Tully, M.; Dimde, M.; Weise, C.; Pouyan, P.; Licha, K.; Schirner, M.; Haag, R. Polyglycerol for Half-Life Extension of Proteins—Alternative to PEGylation? *Biomacromolecules* **2021**, *22*, 1406–1416. [CrossRef]
88. Hauptstein, N.; Pouyan, P.; Kehrein, J.; Dirauf, M.; Driessen, M.D.; Raschig, M.; Licha, K.; Gottschaldt, M.; Schubert, U.S.; Haag, R.; et al. Molecular Insights into Site-Specific Interferon-α2a Bioconjugates Originated from PEG, LPG, and PEtOx. *Biomacromolecules* **2021**, *22*, 4521–4534. [CrossRef]
89. Fonkwo, P.N. Pricing infectious disease. *EMBO Rep.* **2008**, *9*, S13–S17. [CrossRef]
90. Soria-Martinez, L.; Bauer, S.; Giesler, M.; Schelhaas, S.; Materlik, J.; Janus, K.; Pierzyna, P.; Becker, M.; Snyder, N.L.; Hartmann, L.; et al. Prophylactic Antiviral Activity of Sulfated Glycomimetic Oligomers and Polymers. *J. Am. Chem. Soc.* **2020**, *142*, 5252–5265. [CrossRef]
91. Strasfeld, L.; Chou, S. Antiviral drug resistance: Mechanisms and clinical implications. *Infect. Dis. Clin. N. Am.* **2010**, *24*, 413–437. [CrossRef]
92. Achazi, K.; Haag, R.; Ballauff, M.; Dernedde, J.; Kizhakkedathu, J.N.; Maysinger, D.; Multhaup, G. Understanding the Interaction of Polyelectrolyte Architectures with Proteins and Biosystems. *Angew. Chem. Int. Ed.* **2021**, *60*, 3882–3904. [CrossRef] [PubMed]
93. Clausen, T.M.; Sandoval, D.R.; Spliid, C.B.; Pihl, J.; Perrett, H.R.; Painter, C.D.; Narayanan, A.; Majowicz, S.A.; Kwong, E.M.; McVicar, R.N.; et al. SARS-CoV-2 Infection Depends on Cellular Heparan Sulfate and ACE2. *Cell* **2020**, *183*, 1043–1057.e1015. [CrossRef] [PubMed]
94. Zhang, Q.; Chen, C.Z.; Swaroop, M.; Xu, M.; Wang, L.; Lee, J.; Wang, A.Q.; Pradhan, M.; Hagen, N.; Chen, L.; et al. Heparan sulfate assists SARS-CoV-2 in cell entry and can be targeted by approved drugs in vitro. *Cell Discov.* **2020**, *6*, 80. [CrossRef] [PubMed]
95. Cagno, V.; Tseligka, E.D.; Jones, S.T.; Tapparel, C. Heparan Sulfate Proteoglycans and Viral Attachment: True Receptors or Adaptation Bias? *Viruses* **2019**, *11*, 596. [CrossRef]
96. Conzelmann, C.; Müller, J.A.; Perkhofer, L.; Sparrer, K.M.J.; Zelikin, A.N.; Münch, J.; Kleger, A. Inhaled and systemic heparin as a repurposed direct antiviral drug for prevention and treatment of COVID-19. *Clin. Med.* **2020**, *20*, e218. [CrossRef]
97. Nahmias, A.J.; Kibrick, S. Inhibitory effect of heparin on herpes simplex virus. *J. Bacteriol.* **1964**, *87*, 1060–1066. [CrossRef]
98. Folkman, J.; Langer, R.; Linhardt, R.J.; Haudenschild, C.; Taylor, S. Angiogenesis Inhibition and Tumor Regression Caused by Heparin or a Heparin Fragment in the Presence of Cortisone. no. 0036-8075 (Print). *Science* **1983**, *221*, 719–725. [CrossRef]
99. Lubor, B.; Wong, R.; Feramisco, J.; Nadeau, D.R.; Varki, N.M.; Varki, A. Heparin and Cancer Revisited: Mechanistic Connections Involving Platelets, P-Selectin, Carcinoma Mucins, and Tumor Metastasis. *Proc. Natl. Acad. Sci. USA* **2001**, *98*, 3352.
100. Paluck Samantha, J.; Nguyen, T.H.; Maynard, H.D. Heparin-Mimicking Polymers: Synthesis and Biological Applications. *Biomacromolecules* **2016**, *17*, 3417–3440. [CrossRef]
101. Kim, S.; Niesler, N.; Rades, N.; Haag, R.; Dernedde, J. Sulfated Dendritic Polyglycerol Is a Potent Complement Inhibitor. *Biomacromolecules* **2019**, *20*, 3809–3818.
102. Pujol, C.A.; Ray, S.; Ray, B.; Damonte, E.B. Antiviral Activity against Dengue Virus of Diverse Classes of Algal Sulfated Polysaccharides. *Int. J. Biol. Macromol.* **2012**, *51*, 412–416. [CrossRef] [PubMed]
103. Edward, Y. The Anti-Inflammatory Effects of Heparin and Related Compounds. *Thromb. Res.* **2008**, *122*, 743–752.
104. Francisco, P.; Arqué, G. Influence of Thromboembolic Events in the Prognosis of Covid-19 Hospitalized Patients. Results from a Cross Sectional Study. *PLoS ONE* **2021**, *16*, e0252351.
105. Miriam, M.; Martin, J.C. Pathological Inflammation in Patients with Covid-19: A Key Role for Monocytes and Macrophages. *Nat. Rev. Immunol.* **2020**, *20*, 355–362.
106. Sumati, B.; Hilsch, M.; Cuellar-Camacho, J.L.; Ludwig, K.; Nie, C.; Parshad, B.; Wallert, M.; Block, S.; Lauster, D.; Böttcher, C.; et al. Adaptive Flexible Sialylated Nanogels as Highly Potent Influenza a Virus Inhibitors. *Angew. Chem. Int. Ed.* **2020**, *59*, 12417–12422.
107. Pradip, D.; Bergmann, T.; Cuellar-Camacho, J.L.; Ehrmann, S.; Chowdhury, M.S.; Zhang, M.; Dahmani, I.; Haag, R.; Azab, W. Multivalent Flexible Nanogels Exhibit Broad-Spectrum Antiviral Activity by Blocking Virus Entry. *ACS Nano* **2018**, *12*, 6429–6442.
108. Blossom, D.B.; Kallen, A.J.; Patel, P.R.; Elward, A.; Robinson, L.; Gao, G.; Langer, R.; Perkins, K.M.; Jaeger, J.L.; Kurkjian, K.M.; et al. Outbreak of Adverse Reactions Associated with Contaminated Heparin. *N. Engl. J. Med.* **2008**, *359*, 2674–2684. [CrossRef]
109. Oduah, E.I.; Linhardt, R.J.; Sharfstein, S.T. Heparin: Past, Present, and Future. *Pharmaceuticals* **2016**, *9*, 38. [CrossRef]
110. Chuanxiong, N.; Pouyan, P.; Lauster, D.; Trimpert, J.; Kerkhoff, Y.; Szekeres, G.P.; Wallert, M.; Block, S.; Sahoo, A.K.; Dernedde, J.; et al. Polysulfates Block Sars-Cov-2 Uptake through Electrostatic Interactions. *Angew. Chem. Int. Ed.* **2021**, *60*, 15870–15878.
111. Sumati, B.; Lauster, D.; Bardua, M.; Ludwig, K.; Angioletti-Uberti, S.; Popp, N.; Hoffmann, U.; Paulus, F.; Budt, M.; Stadtmüller, M.; et al. Linear Polysialoside Outperforms Dendritic Analogs for Inhibition of Influenza Virus Infection In vitro and In vivo. *Biomaterials* **2017**, *138*, 22–34.
112. Pouyan, P.; Nie, C.; Bhatia, S.; Wedepohl, S.; Achazi, K.; Osterrieder, N.; Haag, R. Inhibition of Herpes Simplex Virus Type 1 Attachment and Infection by Sulfated Polyglycerols with Different Architectures. *Biomacromolecules* **2021**, *22*, 1545–1554. [CrossRef]
113. Mohammadifar, E.; Ahmadi, V.; Gholami, M.F.; Oehrl, A.; Kolyvushko, O.; Nie, C.; Donskyi, I.S.; Herziger, S.; Radnik, J.; Ludwig, K.; et al. Graphene-Assisted Synthesis of 2d Polyglycerols as Innovative Platforms for Multivalent Virus Interactions. *Adv. Funct. Mater.* **2021**, *31*, 2009003. [CrossRef] [PubMed]

114. Pavithra, D.; Doble, M. Biofilm Formation, Bacterial Adhesion and Host Response on Polymeric Implants—Issues and Prevention. *Biomed. Mater.* **2008**, *3*, 034003. [CrossRef] [PubMed]
115. Prime, K.L.; Whitesides, G.M. Self-Assembled Organic Monolayers: Model Systems for Studying Adsorption of Proteins at Surfaces. *Science* **1991**, *252*, 1164–1167. [CrossRef] [PubMed]
116. Gombotz, W.R.; Guanghui, W.; Horbett, T.A.; Hoffman, A.S. Protein adsorption to poly(ethylene oxide) surfaces. *J. Biomed. Mater. Res.* **1991**, *25*, 1547–1562. [CrossRef]
117. Han, S.; Kim, C.; Kwon, D. Thermal/oxidative degradation and stabilization of polyethylene glycol. *Polym. Int. J. Sci. Technol. Polym.* **1997**, *38*, 317–323. [CrossRef]
118. Kulka, M.W.; Smatty, S.; Hehnen, F.; Bierewirtz, T.; Silberreis, K.; Nie, C.; Kerkhoff, Y.; Grötzinger, C.; Friedrich, S.; Dahms, L.I.; et al. The Application of Dual-Layer, Mussel-Inspired, Antifouling Polyglycerol-Based Coatings in Ventricular Assist Devices. *Adv. Mater. Interfaces* **2020**, *7*, 2000272. [CrossRef]
119. Kulka, M.W.; Nie, C.; Nickl, P.; Kerkhoff, Y.; Garg, A.; Salz, D.; Radnik, J.; Grunwald, I.; Haag, R. Surface-Initiated Grafting of Dendritic Polyglycerol from Mussel-Inspired Adhesion-Layers for the Creation of Cell-Repelling Coatings. *Adv. Mater. Interfaces* **2020**, *7*, 2000931. [CrossRef]

polymers

MDPI

Article

Chitosan-Based Therapeutic Systems for Superficial Candidiasis Treatment. Synergetic Activity of Nystatin and Propolis

Andra-Cristina Humelnicu [1], Petrişor Samoilă [1], Corneliu Cojocaru [1], Raluca Dumitriu [1], Andra-Cristina Bostănaru [2,*], Mihai Mareş [2], Valeria Harabagiu [1,*] and Bogdan C. Simionescu [1]

1 "Petru Poni" Institute of Macromolecular Chemistry, 41A Grigore Ghica Voda Alley, 700487 Iasi, Romania; humelnicu.andra@icmpp.ro (A.-C.H.); samoila.petrisor@icmpp.ro (P.S.); cojocaru.corneliu@icmpp.ro (C.C.); rdumi@icmpp.ro (R.D.); bcsimion@icmpp.ro (B.C.S.)

2 Laboratory of Antimicrobial Chemotherapy, Faculty of Veterinary Medicine, "Ion Ionescu de la Brad" Iasi University of Life Sciences (IULS), 8 Mihail Sadoveanu Alley, 700489 Iasi, Romania; mycomedica@gmail.com

* Correspondence: acbostanaru@gmail.com (A.-C.B.); hvaleria@icmpp.ro (V.H.)

Abstract: The paper deals with new approaches to chitosan (CS)-based antifungal therapeutic formulations designed to fulfill the requirements of specific applications. Gel-like formulations were prepared by mixing CS dissolved in aqueous lactic acid (LA) solution with nystatin (NYS) powder and/or propolis (PRO) aqueous solution dispersed in glycerin, followed by water evaporation to yield flexible mesoporous (pore widths of 2–4 nm) films of high specific surfaces between 1×10^3 and 1.7×10^3 m^2/g. Morphological evaluation of the antifungal films showed uniform dispersion and downsizing of NYS crystallites (with initial sizes up to 50 μm). Their mechanical properties were found to be close to those of soft tissues (Young's modulus values between 0.044–0.025 MPa). The films presented hydration capacities in physiological condition depending on their composition, i.e., higher for NYS-charged (628%), as compared with PRO loaded films (118–129%). All NYS charged films presented a quick release for the first 10 min followed by a progressive increase of the release efficiency at 48.6%, for the samples containing NYS alone and decreasing values with increasing amount of PRO to 45.9% and 42.8% after 5 h. By in vitro analysis, the hydrogels with acidic pH values around 3.8 were proven to be active against *Candida albicans* and *Candida glabrata* species. The time-killing assay performed during 24 h on *Candida albicans* in synthetic vagina-simulative medium showed that the hydrogel formulations containing both NYS and PRO presented the faster slowing down of the fungal growth, from colony-forming unit (CFU)/mL of 1.24×10^7 to CFU/mL < 10 (starting from the first 6 h).

Keywords: chitosan; drug delivery systems; nystatin; propolis; antifungal activity

Citation: Humelnicu, A.-C.; Samoilă, P.; Cojocaru, C.; Dumitriu, R.; Bostănaru, A.-C.; Mareş, M.; Harabagiu, V.; Simionescu, B.C. Chitosan-Based Therapeutic Systems for Superficial Candidiasis Treatment. Synergetic Activity of Nystatin and Propolis. *Polymers* **2022**, *14*, 689. https://doi.org/10.3390/polym14040689

Academic Editors: Iza Radecka and Barbara Trzebicka

Received: 10 January 2022
Accepted: 9 February 2022
Published: 11 February 2022

1. Introduction

More than one billion people are affected by superficial *Candida* infections as oral candidiasis and vaginitis (vulvovaginal candidiasis) [1,2]. An improper treatment of these infections can lead to the fungi spreading from the surface of the body to the internal organs (kidneys, heart, brain) and to the blood, causing deadly invasive infections or candidemia [3]. Worldwide, the incidence of invasive *Candida* infections is increasing (700,000 cases annually), and is associated with considerable mortality [2,4]. The antagonistic augmentation of case numbers and geographical spread by the end of 2020 [5] requires an increased effort from scientists to promote appropriate prevention, protection and therapeutic systems, and to combat candidiasis in early stages.

The treatment of superficial candidiasis involves the development of mucoadhesive pharmaceutical systems for the local administration of antifungal agents. The polysaccharide-based therapeutic systems are known for achieving specific functions in a complex biological environment, since they are considered one of the most propitious subjects lying on the frontier between chemistry, biology, medicine, and bioengineering [6]. Undoubtedly, natural polymers are promptly recognized and embraced by the human body considering their biochemical similarity with extracellular matrix components [7]. Chitosan is a cationic amino-polysaccharide which possesses useful biological properties, such as mucoadhesion, antibacterial and antifungal activity, antioxidant, hemostatic, and antiseptic characteristics [8]. The reactive hydroxyl and amino groups on the chitosan chain facilitate different reactions with other active compounds and are able to perform a multitude of intermolecular and intramolecular interactions [9]. Moreover, chitosan-based therapeutic systems are known for their capacity to incorporate and release multiple active principles, either simultaneously or sequentially, for achieving more efficient associated therapy [10]. Therefore, this study is based on the well-known mucoadhesive properties of chitosan, but also on its resistance in contact with the oral and vaginal mucosa [11,12].

Among the few classes of active principles with antifungal effect (azoles, polyenes, echinocandins, allylamines, and pyrimidine analogs) [13,14], emphasis was put on polyenes characterized by a large spectrum of activity induced by their unusual mechanism of action. By comparison with other classes of antifungals, polyenes do not target a specific enzyme, but rather interact selectively with sterols (especially ergosterol) in the plasma membranes of fungi, thus causing loss of membrane function, altered permeability and nutrient damage [15]. Nystatin is a broad-spectrum polyene antifungal agent derived from *Streptomyces noursei*, known for increased susceptibility of the fungi and for its high efficacy rate in the prophylaxis and the treatment of superficial candidiasis [16,17].

In terms of antifungal efficiency, among the use of nystatin, this work aims to evaluate certain bioactive compounds derived from natural sources, namely propolis—a resinous mixture produced by the honey bees, also known for its antifungal activity—and to scrutinize the combination therapy [18]. The novelty of the approach consists in providing chitosan film formulations with mechanical properties close to those of the soft tissues, by taking advantage of chitosan increased flexibility and bioadhesivity induced by lactic acid pH regulator and of the plasticizing properties of both glycerin additive and propolis bioactive agent. Moreover, to the best of our knowledge, this work is the first to demonstrate the combined effect of nystatin and propolis aqueous solution against *Candida* species, when administered together, under relevant simulative conditions, on vulvovaginal candidiasis. Thus, chitosan-based therapeutic formulations containing either propolis or nystatin, either their mixture were prepared as hydrogels and films, and proved to be active in the treatment of vulvovaginal and oral candidiasis, respectively.

2. Materials and Methods

2.1. Materials

Chitosan (CS), with an average molar mass of 290 kDa and 81.6% deacetylation degree (as previously determined by viscometry, respectively by NMR analysis [19]) was provided by Merck Chemical (Saint Louis, MO, USA). L-(+)-Lactic acid (LA) and anhydrous glycerin (Gly) were purchased from Chemical Company (Iași, Romania). Nystatin (NYS) drug, internationally qualified by USP Reference Standard, with particle sizes under 50 μm (Figure S1 in Supplementary Materials) was supplied by Antibiotice SA, Iași, Romania. Aqueous Propolis 30% (*w*/*v*) (PRO) with 7.72 mg/mL polyphenols and 0.26 mg/mL flavones/flavonoids contents (spectrophotometrically determined according to Singleton et al., 1999 [20] at APHIS-DIA Laboratory, Cluj-Napoca, Romania) was purchased from Dapis Transilvania (Cluj-Napoca, Romania). All the analytical grade chemicals were used as received. *Candida albicans* ATCC 90028 and *Candida glabrata* ATCC 90030 were provided by American Type Culture Collection (Manassas, VA, USA). Yeast Nitrogen Base Agar with dextrose and Yeast Extract Peptone Dextrose Agar (YPD) were purchased from Merck Chemical (Saint Louis, MO, USA), while Sabouraud dextrose agar (SDA) was acquired from Biokar Diagnostics (Allonne, France).

2.2. Preparation of Chitosan Antifungal Therapeutic Formulations

Chitosan based hydrogels or flexible films of different contents of antifungal compounds (Table 1) were prepared in several steps, as described below. First, appropriate solutions or dispersions of each individual component were obtained. Thus, a chitosan stock solution of 3% (*w*/*v*) concentration was prepared by dissolving the biopolymer in a 2% (*w*/*v*) lactic acid solution under continuous stirring at 600 rpm and 40 °C for 24 h. NYS (15 mg) was dispersed in glycerin (0.4 g) by magnetic stirring of their mixture for 30 min, at 40 °C and 400 rpm, and a subsequent homogenization by ultrasonication (Emmi 12 HC ultrasonic bath, 100% ultrasonic efficiency) for 10 min. PRO aqueous solution (0.3 mL) was also mixed with Gly (0.4 g) under stirring at 40 °C and 400 rpm for 30 min.

Table 1. Composition of chitosan formulations * and the thickness of the films.

Hydrogel/Film Code	NYS (mg)	PRO (mL Solution/mg Solid Compound)	Films Thickness (μm)
CS-LA	-	-	49
CS-NYS	15	-	53
CS-PRO	-	0.30/90	210
CS-NYS-PRO1	15	0.15/45	166
CS-NYS-PRO2	15	0.30/90	220

* all hydrogel formulations also contain 0.3 g CS (in 10 mL 2% LA solution) and 0.4 g Gly; the films were obtained by hydrogel drying and each formulation has a composition identical to the corresponding hydrogel, without water.

CS-based formulations charged either with NYS (CS-NYS) or with PRO (CS-PRO) were obtained by adding 10 mL of the stock solution of chitosan over each of their mixtures with Gly and stirring for 1 h at room temperature.

Two chitosan formulations, CS-NYS-PRO1 and CS-NYS-PRO2, containing both antifungal agents in different ratios (PRO/NYS = 3/1 or 6/1 *w*/*w*) were prepared by subsequently adding calculated amounts of PRO aqueous solution and of 10 mL of the stock solution of chitosan onto NYS-Gly dispersion (previously prepared as described above) and stirring at room temperature till homogenization was achieved (about 1 h). All the prepared hydrogels were characterized by pH values of 3.85 ± 0.02, as measured by using a HANNA instruments HI8417pH meter (Amorim, Portugal).

An identical set of hydrogels was separately prepared and each sample was poured into a 5 cm diameter Petri dish. The drying at room temperature yielded transparent CS-NYS, CS-PRO, CS-NYS-PRO1 and CS-NYS-PRO2 films of micrometric thickness (measured with a Dial Thickness Gauge 7301 handheld micrometer—Mitoyuto Corporation, Kangagawa, Japan; accuracy of 1 μm), depending on composition (Table 1). For analytical purposes, uncharged chitosan hydrogel and film (CS-LA) were also prepared.

2.3. Methods of Characterization

2.3.1. Structural and Morphological Characterization of Film Formulations

Fourier transform (FTIR) spectra of film formulations were registered by using a Bruker Vertex 70 spectrophotometer (Bruker Optics, Ettlingen, Germany), in ATR (Attenuated Total Reflectance) mode (wavelength range of 4000–600 cm^{-1}, resolution of 2 cm^{-1} and 64 scans at room temperature). Scanning electron microscopy (SEM) images of cross-sections of the prepared films were obtained on an FEI QUANTA 200 electron scanning microscope (Brno, Czech Republic) with a resolution of 4 nm at 30 kV. The drug dispersion in the films was visualized on a Leica Microsystems Polarized Optical Microscope (Wetzlar, Germania).

2.3.2. Mechanical Properties of the Films

Tensile strength, elongation at break, and Young's modulus were determined on an Instron 3365 equipment (Norwood, MA, USA) with two columns and a 500 N force cell. Dumbbell-shaped samples for each film were cut using a press (length/width/active length = 50/4.1/35 mm) and the uniaxial stress–strain curves of the samples occurs at

50 mm/min elongation speed. The stress (σ_b) and strain (ε_b) at break were calculated according to Equations (1) and (2), respectively:

$$\sigma_b \text{ (MPa)} = F_b/A \tag{1}$$

$$\varepsilon_b \text{ (\%)} = \Delta l/l_0 \times 100 \tag{2}$$

where F_b is the breaking force, A is the cross-sectional area of the sample at time t, l_0 represents the initial length, and Δl represents the elongation at traction. Based on the specific deformation curve, the modulus of elasticity for each film was also determined as the ratio between the stress and the tensile deformation (1%).

2.3.3. Swelling Behavior of Film Formulations in Simulated Conditions

1 cm^2 of film samples were dried in an oven at 40 °C till constant weight, were immersed in 50 mL phosphate buffered saline (PBS) solution of pH 7.4 and were placed in an Orbital Shaker-Incubator ES-20/60 (Biosan—Riga, Latvia) at 37 °C and 80 rpm. At predetermined time intervals, the samples were removed from the immersion medium; the excess PBS solution was removed from the surface by buffering with filter paper and the samples were weighed. The kinetics of the film swelling was evaluated by a gravimetric method, according to Equation (3):

$$S_t \text{ (g/g)} = (w_t - w_0)/w_0 \tag{3}$$

where S_t (w/w) represents the value of swelling capacity at time t (min), w_0 is the initial mass of the dried sample, and w_t is the mass of the sample at time t after immersion in PBS. The data were plotted by using the average values of the three determinations for each film formulation. To establish the equilibrium swelling capacity, the experimental data were also analyzed by applying a pseudo-second order kinetic model (PSO), expressed by Equation (4) [21,22]:

$$S_t \text{ (g/g)} = k_s \times S_e^2/(1 + k \times S_e \times t) \tag{4}$$

where S_t (g/g) represents the swelling capacity at contact time t (min), k_s is the constant of the swelling rate and S_e represents the theoretical swelling capacity at equilibrium time.

The swelling dynamics (diffusion-controlled or relaxation-controlled) of the films into PBS up to 5 h was evaluated using the Equation (5) [23,24], adapted after Korsmeyer and Peppas (K-P) model [25]:

$$F = S_t/S_e = k_p \times t^n \tag{5}$$

where F is the swelling fraction, S_t and S_e are the swelling capacities at time t, and at equilibrium, respectively; k_p is a constant dependent on the polymeric network; n is the diffusion parameter of aqueous PBS in the formulation film [23].

The diffusion parameter value is characterizing the diffusion mechanism of aqueous molecules in the film formulations: $n < 0.5$ implies a Fickian diffusion-controlled mechanism; $0.5 < n < 1$ suggests anomalous non-Fickian diffusion; $n = 1$ indicates a relaxation-controlled water transport and $n > 1$ represents a supercase II of diffusion [22]. The experimental data were processed by using SCILAB 6.1.0 software.

2.3.4. In Vitro Nystatin/Propolis Release from Film Formulations

Pre-weighed dried samples of antifungal chitosan hydrogels were immersed each in 50 mL of neutral PBS solution (pH 7.4), placed in an Orbital Shaker-Incubator ES-20/60 (Biosan—Riga, Latvia) and maintained at 37 °C and 80 rpm. At predetermined time intervals, 0.5 mL were extracted from each solution (subsequently being replaced with the same volume of initial buffer solution) and were spectrophotometrically analyzed using a double-beam UV-VIS spectrophotometer Hitachi U-3900 (Hitachi High-Tech Europe GmbH—Krefeld, Germania). The drugs concentrations in solution were established based on previously performed calibration curves and the release efficiency was evaluated as a function of time. Nystatin characteristic absorption bands were identified at 293, 305, and 320 nm. The calibration curves were plotted and the concentrations were determined by using the band at 320 nm for nystatin, and the specific absorption band at 284 nm for propolis.

In order to determine the drugs release mechanism (Fickian or non-Fickian diffusion), the experimental data were mathematically fitted (SCILAB 6.1.0 software) using the semi-empirical equation (Equation (6)) proposed by Korsmeyer and Peppas (1981), which describes both in vitro drug release from thin plane films and the stability behavior [25]:

$$M_t/M_\infty = k \times t^n \quad (6)$$

where M_t/M_∞ refers to the fractional drug release at time t; M_t and M_∞ are the total amount of drug released at time t and at infinite time, respectively; k represents the transport constant, n is the diffusion exponent that indicates the type of drug release transport (Quasi-Fickian diffusion for $n < 0.5$, Fickian diffusion for $n = 0.5$, non-Fickian transport for $n > 0.5$, Case II transport for $n = 1$, and supercase II transport for $n > 1$) [26].

2.3.5. Dynamic Vapor Sorption Measurements

The water vapor sorption capacity of the hydrogel formulations was determined in dynamic regime by using a fully automated gravimetric device IGAsorp provided by Hiden Analytical (Warrington, UK). The samples were placed in a special container and dried at 25 °C using a nitrogen flow of 250 mL/min until their weights reached equilibrium at a relative humidity, RH, less than 1%. Subsequently, the RH was gradually increased from 0 to 90%, with humidity steps of 10% (at pre-established equilibrium time between 40–60 min) and the sorption equilibrium was registered for each step in the sorption curve. The RH was further decreased and desorption curves were also registered. To evaluate the specific surface area, Brunauer–Emmett–Teller kinetic model (BET) was applied by modeling the sorption isotherms registered under dynamic conditions, according to Equation (7) [27,28]:

$$W = W_m \times C \times RH/(1 - RH)(1 - RH + C \times RH) \quad (7)$$

where W and W_m represent the weight of absorbed water and the weight of water forming a monolayer, respectively; C is the sorption constant and RH represents the relative humidity.

The average pore size, r_{pm}, was estimated by applying Barrett, Joyner, and Halenda model (BJH) [28], based on calculation methods for cylindrical pores, in accordance with Equations (8) and (9):

$$V_{liq} = n/100 \times \rho_a \quad (8)$$

$$r_{pm} = 2V_{liq}/A \quad (9)$$

where V_{liq} is the liquid volume; n is the absorption percentage; ρ_a represents the adsorbed phase density and A is the specific surface area determined by BET method.

2.3.6. Rheological Properties of the Hydrogels

Rheological measurements were performed on a Physica MCR 301 Stress Controlled Rotational Rheometer (Anton Paar Company, Graz, Austria) with cone-plate geometry (50 mm diameter, an angle of 1°/99 μm truncation). The viscoelastic properties were analyzed based on the dynamic oscillatory and temperature sweep measurements. Preliminary tests were performed at 10 rad/s in the 0.1–200% deformation range with a constant shear displacement γ of 10%. The dynamic oscillatory tests were conducted at constant temperature of 35 ± 0.1 °C (accurately controlled by Peltier heating system) and angular frequency (ω) in the range of 0.05/0.1–100 rad/s. The elastic modulus, G′, which represents the solid-like component of viscoelastic behavior of the material, the viscous modulus, G″, representative for liquid-like component were investigated as functions of the angular frequency ω (rad/s).

The viscoelastic behavior was evaluated also by determining the tangent of δ phase angle (Equation (10)) and the complex viscosity (Equation (11)), which is related to complex modulus as described by Equation (12) [29]:

$$\tan \delta = G''/G' \tag{10}$$

$$\eta^* = G^*/\omega \tag{11}$$

$$G^* = G' + iG'' \tag{12}$$

where δ is the phase angle (a relative measure for the visco-elastic properties); G' and G'' are the elastic and viscous moduli; η^* is the complex viscosity and G^* represents the complex modulus, which is a complex number with i representing an imaginary number [30].

Temperature sweep measurements were realized at a constant shear rate of 10 s^{-1}, over the temperature range 5–40 °C with a heating rate of 2 °C/min. After loading and before measurement, a 5 min rest time for each sample was allowed to ensure stress relaxation and temperature equilibration. The viscosity of the hydrogels (η) was evaluated as a function of temperature and the activation energy (Eη) of the viscous flow was calculated by fitting the experimental data with Arrhenius exponential equation (Equation (13)), using a logarithmic form according to Equation (14). The natural logarithm of η was plotted as function of 1/T (K^{-1}) [31]:

$$\eta = A \exp(E\eta/RT) \tag{13}$$

$$\ln \eta = \ln A + E\eta/RT \tag{14}$$

where η is viscosity of the hydrogels; A represents the Arrhenius factor (a constant associated with the nature of the liquid); Eη is the viscous flow activation energy; R is the molar gas constant (1.9872 cal·K^{-1}·mol^{-1}) and T is the temperature (K).

2.3.7. Antifungal Activity of the Hydrogel

The prepared chitosan formulations (CS-NYS, CS-PRO, CS-NIS-PRO1 and CS-NIS-PRO2) as well as CS-LA solution were evaluated for antifungal activity against two reference strains from the American Type Culture Collection (*Candida albicans* and *Candida glabrata*), which are the main pathogens responsible for invasive candidiasis. Both *Candida* strains were stored in 20% glycerin at −80 °C. Prior to testing, each strain was refreshed on Sabouraud dextrose agar (SDA) and incubated for 48 h at 30 °C. The yeast strains were used when a maximal number of conidia were formed.

The antifungal activity of CS-based hydrogels was evaluated by using the agar disk diffusion method. Microbial suspensions were prepared in sterile saline solution to obtain an optical turbidity comparable to that of the 0.5 McFarland standards (each suspension contains 1 × 108 colony-forming units/mL (CFU/mL)). Volumes of 0.2 mL from each inoculum were taken and spread onto Yeast Nitrogen Base Agar with dextrose previously poured in Petri dishes. After the drying of the medium surface, each sample of chitosan hydrogels (10 μL) was added. The antifungal properties of the tested hydrogels were determined by measuring fungal growth inhibition, under standard conditions (after incubation for 48 h at a temperature of 30 °C). All experiments were performed in triplicate, in order to verify the final results. The diameter of the inhibition zone around the hydrogel samples was finally measured using a caliper with digital display.

2.3.8. Time–Kill Assay

Time–kill experiments were performed on Candida albicans ATCC MYA-2876 (SC5314 wild type) using a slightly modified previously described method [32]. Thus, a synthetic vagina-simulative medium (SVSM) was prepared [33,34] to assure specific biomimetic conditions. Before performing the tests, the strain was sub cultured at least twice and grown for 24 h at 35 °C on SDA plates.

To examine the rate of killing, from each strain, a 5 McFarland suspension in SVSM was prepared and adjusted to 1.24 × 10^7 CFU/mL using the TC20 automated cell counter

(Bio-rad, Hercules, CA, USA). Subsequently, equal volumes of yeast cell suspension and chitosan hydrogels dispersions in SVSM were mixed and incubated at 36 ± 1 °C, to obtain final mixtures for each compound. A drug free control was also prepared by mixing equal volumes of yeast suspension and SVSM. At predetermined time intervals (0, 6, 12, and 24 h), 1000 μL aliquot from each test and control tube was serially diluted in sterile distilled water, plated onto YPD agar medium and incubated 48 h at 36 ± 1 °C, in order to evaluate the number of CFU/mL. The reproducible detection limit for colony counts is 10^1 CFU/mL. All time–kill curves were plotted for the binary mixture (log10 CFU/mL against time) and the assays were conducted in duplicate and on two separate occasions.

3. Results and Discussions

3.1. Therapeutic Formulations Design and Preparative Protocol

Hydrogels and films formulations are considered to be the most suitable therapeutic systems for the treatment of superficial candidiasis, as they can cover a larger area for the administration of antifungal agents, offering, at the same time, physical protection [35]. Although, chitosan-based hydrogels with propolis and nystatin (among others active principles) were evaluated by Perchyonok et al. (2012–2014) as restorative materials for increasing the dentin bond strength capacity and for oral mucositis treatment [35–38], this work is a first report on evaluating and demonstrating the antifungal effect of nystatin and/or propolis loaded chitosan systems under relevant simulative conditions on vulvovaginal candidiasis. In addition, the use of L-lactic acid for chitosan solubilization and of an aqueous propolis extract in the preparation of chitosan-based nystatin and/or propolis delivery systems are for the first time investigated.

Glycerin was selected for nystatin dispersion and as an additive in antifungal formulations, having in mind its ability to solubilize lipophilic drugs in w/o emulsions, to promote local delivery of the drugs in topical and mucosal applications [39], as well as to act as a plasticizer for chitosan [40,41].

The dissolution of chitosan was performed in lactic acid solution, as it has been shown to induce to the CS formulations mechanical properties suitable for their use on soft tissues, by increasing flexibility (for the films) and bioadhesion (for hydrogels), as opposed to acetic acid [42]. Although DL-lactic acid was used in most pharmaceutical applications, we choose to use L-(+)-lactic acid, primarily because it is a known compound of human metabolism and secondly because it was proven to lower the tensile strength and to increase elongation at break of chitosan films [43]. The physicochemical characterization of film formulations (mechanical properties, swelling and drug release in simulative medium) was done to demonstrate their usefulness as buccal drug delivery systems. As the chitosan hydrogels were characterized by pH values of 3.85 ± 0.02, close to the value of vaginal fluids (pH = 4.2 [33]), they were tested as therapeutic systems under simulative conditions in the therapy of vulvovaginal candidiasis.

3.2. Structural Characterization

The structure of nystatin (NYS), its ^{13}C NMR and ^{1}H NMR spectra are given in Supplementary Materials, Figures S2 and S3, proving the high purity of the polyene macrolide [44–46]. The structure of chitosan and a comparison between the FTIR spectra of pristine chitosan powder and of the film prepared by its dissolving in aqueous lactic acid solution and drying are presented in Supplementary Materials, Figures S4 and S5, respectively. Polyphenols and flavonoids (not structurally identified) are present in propolis (PRO) solubilized in aqueous media, in the proportion mentioned in the experimental part, as specified by the supplier.

Figure 1 compares the characteristics FTIR absorptions of the drugs (NYS and PRO), CS-LA matrix and of the resulted film formulations (CS-NYS and CS-PRO, as typical exemples). As no visible modifications were found in the region of asymmetric and symmetric C–H stretching vibrations (2900–2800 cm^{-1}), for the sake of clarity, this region is not represented in Figure 1. NYS (Figure 1a) is characterized by a large number of hydroxyl groups whose stretching vibrations are found in the range 3200–3700 cm^{-1}, overlapping with the vibrations of the NH_2 group. Other characteristic bands are found at 1706 cm^{-1} (C=O stretching vibrations from ester and carboxylic acid groups), at 1575 cm^{-1} and 846 cm^{-1} (typical vibrations corresponding to the polyenes C=C units), in the intervals 1440–1320 cm^{-1} (deformation vibrations of CH_3) and 1180–1000 cm^{-1} (vibrations of the C–C–O and C–O–C groups, respectively) [47,48].

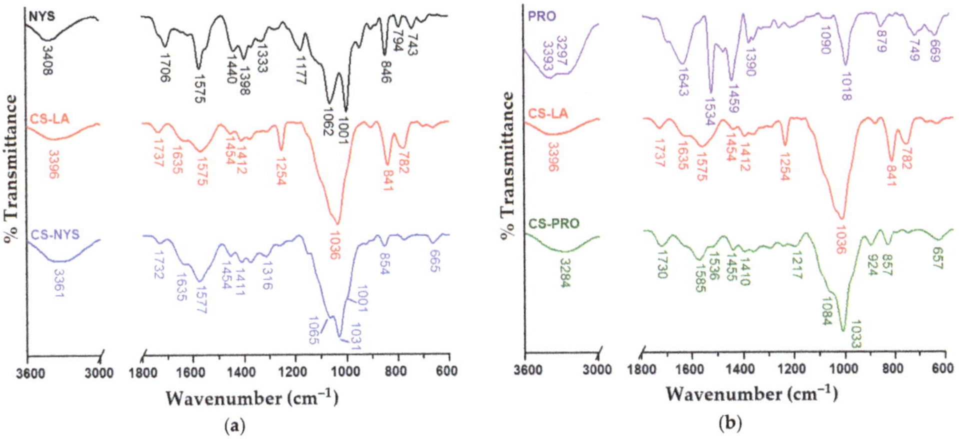

Figure 1. FTIR spectra of the individual components and: CS-NYS film (**a**); CS-PRO film (**b**).

Propolis (PRO) chemical composition is well-known for polyphenols and flavonoids content, with characteristic absorption bands clearly evidenced in the FTIR spectrum (Figure 1b). Wide absorption band centered at 3395 cm^{-1} is attributed to the OH stretching vibration of phenolic compounds and to the formation of inter(intra)molecular hydrogen bonds. Specific to PRO, the following absorption bands are identified: 1643 cm^{-1} (C=O flavonoid carbonyl stretching vibration), 1534 cm^{-1}, and 1459 cm^{-1} (aromatic rings of polyphenols and C=C groups of flavonoids) [49,50], 1390 cm^{-1} (C–O stretching vibrations and O-H deformation, specific to phenols), 1018 cm^{-1} (strong absorption band of C–O esters asymmetric stretching vibration), 879 cm^{-1}, 749 cm^{-1}, and 669 cm^{-1} (medium absorption bands corresponding to the out-of-plane deformation vibration of phenol CH, and CH_2 rocking of hydrocarbons) [50,51].

The chitosan film obtained from its aqueous solution (CS-LA) reveals the absorption band at 1737 cm^{-1} due to the carboxylate (COO^-) groups provided by lactic acid and a broad band centered at 1575 cm^{-1} that was attributed to the deformation vibrations of protonated amine groups (superimposed over amide vibrations), mainly confirming the electrostatic interactions between protonated chitosan amine groups and carboxylate ions ($NH_3{}^{+}{}^{-}OOCCH(OH)CH_3$) [52]. The presence of lactic acid determines the appearance of the band at 1454 cm^{-1}, associated with the CH_3 asymmetric deformation vibrations in its structure [53]. Moreover, the absorption band centered at 1036 cm^{-1} indicates the presence of stretching vibrations of the C–O groups of lactic acid, overlapped with the C–O–C vibrations of chitosan [54]. The asymmetric C–O stretching vibrations are found at 1254 cm^{-1}, while the C–C stretching and C=O deformation vibrations from the lactic acid structure appear at 841 cm^{-1} and 782 cm^{-1} [55]. The lack of a new amide absorption band at about 1540 cm^{-1} indicates the absence of covalently grafting of LA on CS [56].

FTIR spectra of the antifungal films of nystatin (CS-NYS) and propolis (CS-PRO) embedded into chitosan matrix can be observed in the bottom of the Figure 1a,b, respectively. Due to the structural similarities, the characteristic bands of nystatin and/or propolis are overlapping with polysaccharide bands. In addition, the low content of the drugs into the CS formulations makes rather difficult their identifications in the spectra of the corresponding formulations. However, an in-depth comparison between the spectra of the intermediates and of the corresponding film formulations reveals the presence of the drugs into CS matrix. Thus, as a consequence of the increased number of OH groups coming from NYS and PRO components, and of the larger diversity of hydrogen bonding, the typical broad O–H absorption bands in the region 3600–3200 cm^{-1} have higher intensities in the spectra of both CS-NYS and CS-PRO film formulations, as compared to that of CS-LA matrix, and are shifted at different wavenumbers, as compared with the spectra of the individual components. The embedment of the active principles in the polysaccharide matrix induced a slight displacement of the characteristic C=O lactate band from 1737 (CS-LA) to 1732 cm^{-1} (CS-NYS) and 1730 cm^{-1} (CS-PRO). Moreover, some of the drug bands are also visible in the spectra of the corresponding film formulations. Thusly, CS-NYS spectrum shows broader and a more intense band at 1577 cm^{-1} due to the superposition of the C=C band of NYS located at 1575 cm^{-1} over amide/amine bands of CS; the bands at 1062 and 1001 cm^{-1} of NYS are visible as shoulders (1065 and 1001 cm^{-1}) in the region of the C–O–C vibrations of CS-NYS spectrum. Similar traces of PRO are visible in the spectrum of CS-PRO film (e.g., the appearance of a shoulder at 1536 cm^{-1} induced by the contribution of the aromatic ring vibrations of polyphenols and the displacement of the amide/protonated amine band of CS from 1575 cm^{-1} to 1585 cm^{-1}).

3.3. Films Morphology

The morphology of chitosan antifungal films (CS-NYS, CS-PRO, CS-NYS-PRO1, and CS-NYS-PRO2) was examinated by reference to CS-LA film on SEM images of their cross-sections (Figure 2). The CS-LA sample presents a homogeneous, dense and cohesive structure, characteristic for unmodified chitosan films. The addition of nystatin to the polysaccharide matrix (CS-NYS) did not induce major morphological changes of the film. However, besides the dense and compact morphology, the dispersion of NYS as white nanoparticles in the chitosan matrix can be distinguished. The downsizing of NYS crystallites (with initial dimensions up to 50 µm as seen in Figure S1, Supplementary Materials), as an effect of ultrasonication in glycerin and incorporation into the chitosan matrix can be noticed. The introduction of propolis into the chitosan matrix induces film thickening, as compared to CS-LA and CS-NYS, and visible morphological changes. Most are evidenced for CS-PRO and CS-NYS-PRO2 samples, due to the higher amount of propolis. Thus, CS-PRO presents an homogenous wavy organized structure, while CS-NYS-PRO2 shows a discontinuous morphology with phase separation between chitosan and glycerin/propolis plasticizers. Similar morphology was observed when sorbitol was used as a plasticizer for chitosan films [57].

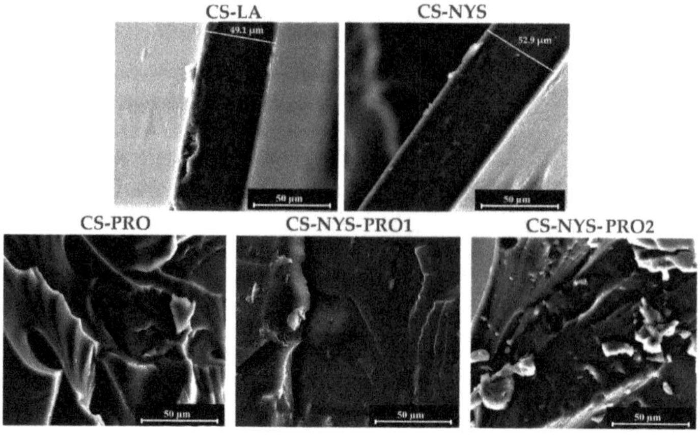

Figure 2. Scanning electron microscopy images of the cross-sections of CS-LA, CS-NYS, CS-PRO, CS-NYS-PRO1 and CS-NYS-PRO2 films.

The morphology of the prepared transparent films was also investigated by polarized light microscopy (Figure 3). The image obtained for CS-LA film is typical for a partially crystalline polymer with crystalline domains embedded in the amorphous ones. Even if nystatin is known for its insolubility in aqueous medium and its tendency to form aggregates, the uniform dispersion of the polyene in the CS-NYS, CS-NYS-PRO1 and CS-NYS-PRO2 chitosan films is easily visible.

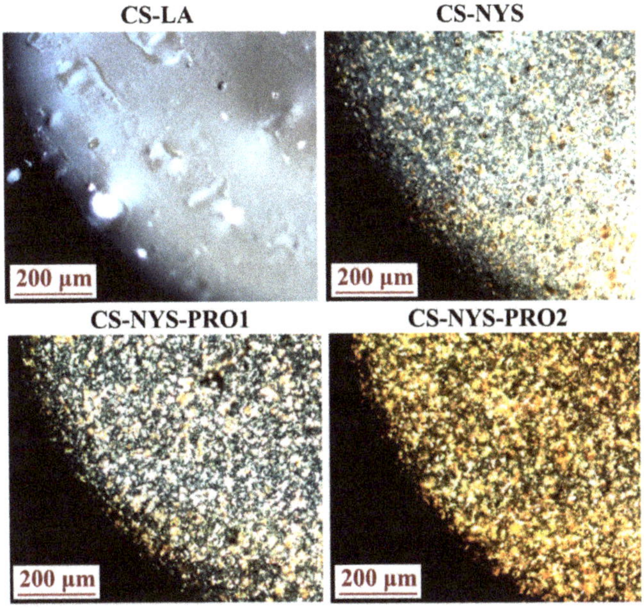

Figure 3. Polarized light microscopy images of CS-LA, CS-NYS, CS-NYS-PRO1 and CS-NYS-PRO2 films surface.

3.4. Mechanical Properties of Films

Chitosan with nystatin and/or propolis hydrogels were dried and the mechanical properties of the resulted micronic films (see the thickness in Table 1) were evaluated in order to prove potential similarities with the mechanical properties of the human soft tissues and the suitability of their application in the treatment of oral candidiasis. Mechanical properties of the antifungal films were investigated based on the characteristic stress–strain curves (Figure S6 in Supplementary Materials). The tensile strength at break (σ_b), elongation at break (ε_b) and modulus of elasticity (Y) were determined and evaluated by comparison with the those of CS-LA film (Table 2). All charged chitosan films are much ductile, contrary to CS-LA sample. They present more than 10 folds and about 2 order of magnitude decrease of the tensile strenght (from 73.3 MPa to 3.0–8.6 MPa) and of the Young's modulus (from 9.310 to 0.044–0.025 MPa), respectively, as well as a considerable increase of the elongation at break (from 7.7 to 68.9–113.9%). These characteristics of the charged CS films are mainly due to the plasticizing action synergetically induced by GLY and PRO to the CS matrix. Similar behavior was reported by others for Gly-CS mixtures and was explained by the formation of hydrogen bonds between the OH groups of the triol compound and the protonated amino groups (NH^{3+}) of the polysaccharide, leading to the reorganization of chitosan chains and formation of an orderly structure [40,41]. NYS alone provided to the film the highest elasticity (the higher elongation at break was registered for CS-NYS film), while PRO seems to provide most pronounced plasticizing effect (lower values of the three studied mechanical properties were registered for all PRO containing films). Propolis was also proposed as a bioplasticizer for starch [51]. All the prepared antifungal chitosan films are characterized by increased elasticity, malleability and mechanical properties close to those of the soft tissues (Young's modulus of soft biological tissues has values lower than <1 MPa [58]), being suitable for application on buccal mucosa.

Table 2. Mechanical properties * of CS films.

Film Code	σ_b (MPa)	ε_b (%)	Y (MPa)
CS-LA	73.3	7.7	9.310
CS-NYS	8.6	113.9	0.044
CS-PRO	7.5	88.1	0.031
CS-NYS-PRO1	5.4	74.8	0.036
CS-NYS-PRO2	3.0	68.9	0.025

* tensile strength at break—σ_b; elongation at break—ε_b; Young's modulus—Y.

3.5. Swelling Capacity of Films

Knowing that the excessive hydration of the films might lead to decreased bioadhesion at the interface with buccal mucosa [59], the swelling capacities of chitosan films charged with nystatin and/or propolis were evaluated in PBS, at physiological oral pH 7.4 and 37 °C. It should be noted that the pure chitosan film (CS-LA) was completely dissolved in the aqueous medium, while all chitosan charged films were recovered after 24 h contact with PBS solution as self-standing swollen hydrogels. Their increased hidrodynamic stability can be explained by the effect of the electrostatic interactions and/or intermolecular hydrogen bonding between polysacharide and the drug molecules. Figure 4 shows the 5 h swelling kinetics of the charged CS films and the plots for PSO and K-P kinetic models. For CS-NYS sample a rapid increase of the swelling capacity to more than 500% the first 2 min, followed by an importand decrease of the swelling rate was registered. The film reaches swelling degrees of 618 and 751% after 5 and 24 h, respectively of immersion in PBS (Table S1).

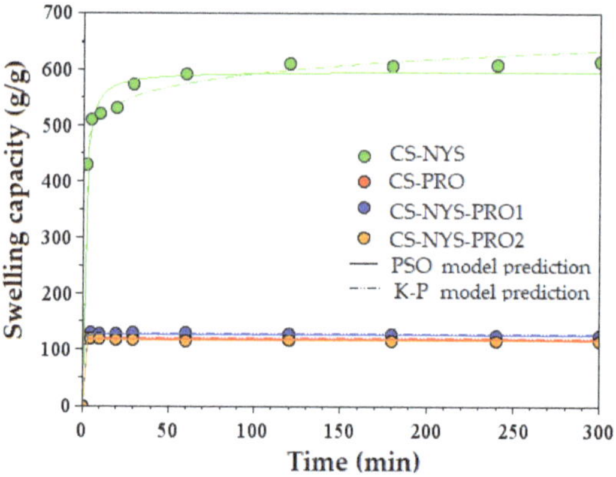

Figure 4. Swelling kinetics of chitosan films with nystatin (CS-NYS), propolis (CS-PRO), and both nystatin and propolis (CS-NYS-PRO1, CS-NYS-PRO2).

CS films charged with propolis (CS-PRO, CS-NYS-PRO1 and CS-NYS-PRO2) presented much lower capacities of welling, as compared to CS-NYS sample, probably due to the chemical crosslinking of the polyphenols found in the natural antifungal agent to the amino groups of the chitosan. All these three samples indicated almost identical swelling behavior, with an immediate (first 2 min) increase to practical maximum swelling capacity, between 117 and 122% (Table S1).

The kinetic parameters of the PSO and K-P models are given in Table S1 for all drug charged CS films. The theoretical equilibrium swelling capacity (S_e) values are close to those of the experimental swelling capacity after 5 h, except the CS-NYS sample. The diffusion exponents, n, have values lower than 0.5, suggesting a Fickian diffusion-controlled swelling kinetics, with the diffusion of the water molecules in the polymeric matrix faster than the chitosan chains relaxation [22,24].

3.6. In Vitro NYS and PRO Release from Antifungal Films

Nystatin and propolis release efficiencies (%) were calculated based on their concentrations in PBS imersion solutions over time (24 h), as determined from their UV-VIS spectra (Figure S7, Supplementary Materials). The release of the drugs from the antifungal films during 5 h is graphically represented in Figure 5. As one may see from Figure 5a, all NYS charged films presented a quick release the first 10 min and progressively increasing release efficiency up to 48.6% (CS-NYS), 45.9% (CS-NYS-PRO1), and 42.8% (CS-NYS-PRO2) after 5 h. Further, the NYS release continues to rise by 5–7% at 24 h (Table S2). The adding of propolis to CS-NYS formulation induced a lower NYS release efficiency, as the crosslinking between polyphenols and chitosan makes more difficult the diffusion of NYS through the resulted polymeric network (CS-NYS-PRO2 had higher propolis content and lowest release efficiency).

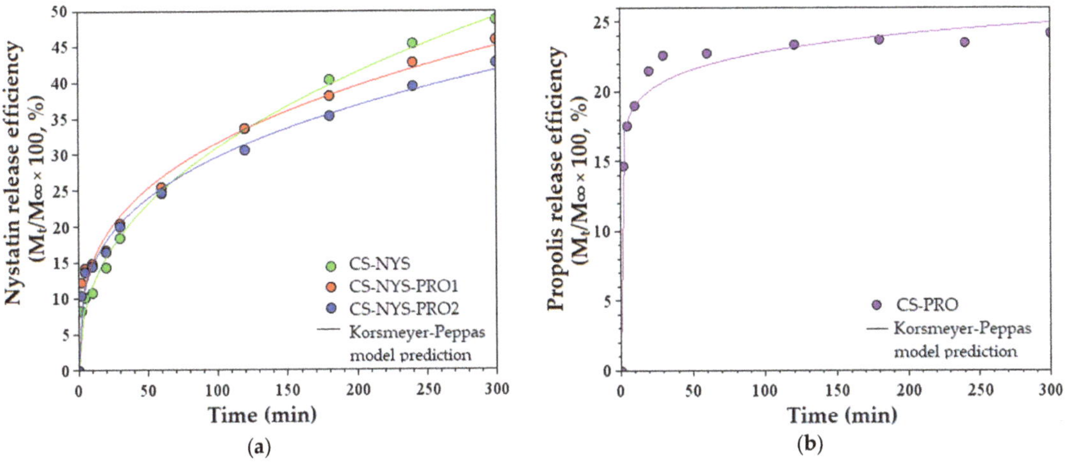

Figure 5. Release efficiency of: (**a**) Nystatin from CS-NYS, CS-PRO-NYS1 and CS-PRO-NYS2; (**b**) Propolis from CS-PRO; the experimental data were fitted based on Korsmeyer and Peppas kinetic model.

Due to the superposition of the absorptions of NYS on the characteristic UV-VIS band of PRO, the release efficiency of the later was represented for CS-PRO film only (Figure 5b). PRO release curve shows a burst effect the first 30 min (with a release efficiency of 22.58%), followed by a plateau located at about 24.2%, which lasts at the same value after 24 h.

Korsmeyer and Peppas kinetic release model was applied for fitting the in vitro NYS and PRO release experimental data (Figure 5). For NYS, the model provided a good fitting, indicating that the release mechanism is based on diffusion through the polysaccharide matrix. The fitting of PRO release experimental data showed a good fitting the first 20 min, followed by a satisfactory fitting up to 5 h. By calculating kinetics parameters of the semi-empirical matehmatical model (given in Table S2), it was established that the mechanims of both antifungal agents transport is based on Quasi-Fickian diffusion ($n < 0.5$) and the release mechanism occurs from non-swellable matrix diffusion [26].

3.7. Dynamic Vapor Sorption onto the Hydrogels

The behavior of the hydrogel formulations in the presence of moisture was investigated and the water vapor sorption/desorption isotherms were represented in Figure S8 (Supplementary Materials). Based on the IUPAC classification the isotherms are related to

Type V curves, specific to water sorption on hydrophobic micro- and mesoporous surfaces and indicate weak adsorbent-water interaction [27]. Similar isotherms were registered for other chitosan-based hydrogels [60,61].

The hydrogels containing both NYS and PRO presented enhanced water vapor barrier properties, with lower water sorption capacity values (89.8 for CS-NYS-PRO1 and 88.1% CS-NYS-PRO2) as compared to the uncharged CS-LA sample (94.1%) as given in Table S3. This may be explained by the simultaneous effects of hydrophobic NYS component and of chemical crosslinking of CS amino groups by polyphenols, which makes more difficult the diffusion of water vapors through the polymeric network. This behavior is supported by the results of swelling capacity of the films formulations (Figure 5).

The BET and BJH models were applied to evaluate the specific surface areas and the pore dimensions, respectively (Table S3, Supplementary Materials). The values obtained indicate the presence of larger mesopores for not charged CS-LA samples (width about 14 nm) as compared to NYS and PRO loaded samples (pore widths between about 2 and 4 nm). Moreover, the specific surface areas increased several times from 275 m^2/g for CS-LA sample to 1035, 1379, 1444 and 1727 m^2/g for CS-PRO, CS-NYS, CS-NYS-PRO1, and CS-NYS-PRO2 samples, respectively, as an effect of both the nature and content of functional groups and the pore size over the complex mode of the sorption capacity of the studied samples.

3.8. Rheological Properties of Chitosan Hydrogel Formulations

The rheological properties of chitosan hydrogels (with and without active principles) were evaluated as functions of angular frequency or temperature (Figure 6). Preliminary deformation tests at 0.1–200% deformation range (γ of 10%) confirmed that tests are in the linear viscoelastic regime (LVE). The dynamic oscillatory measurements were performed at 35 °C (temperature close to that of physiological body temperature of 37 °C) and the properties such as elasic modulus (G'), viscous modulus (G'') and complex viscosity (η^*) were investigated in the range of 0.01–100 rad/s.

To establish the type of the hydrogels, the evolution of G' and G'' at 35 °C and a γ = 10% vs. ω is represented in Figure 6a and their values at ω = 1 Hz are given in Table S4, Supplementary Materials. All the hydrogels are characterized by increasing G' and G'' moduli values with increasing angular frequency. As seen from this figure, the CS-LA, CS-NYS and CS-NYS-PRO1 hydrogels present a solid-like behavior (G' modulus higher than G'' modulus), which denotes a polymeric network structure that is self-dependent of the applied rheological stress parameters. On the contrary, the hydrogels with higher content of propolis (CS-PRO and CS-NYS-PRO2) exhibit a predominantly liquid-like behavior ($G'' > G'$), indicating properties of a more viscous than elastic polymeric network. The rheological behavior of the hydrogels was also verified by the calculation of the loss tangent (tan δ) as represented in Figure S9, Supplementary Materials. The subunitary values of the tan δ confirms the gel like structure of the CS-LA, CS-NYS and CS-NYS-PRO1, while values higher than 1 of the loss tangents are found for the liquid-like behavior of the CS-PRO and CS-NYS-PRO2 hydrogels (Table S4, Supplementary Materials).

All chitosan-based hydrogels presented a linear decrease of the complex viscosity (η^*) with the increase of the angular frequency (Figure 5b) induced by the breaking of the intermolecular forces and by the alignment of chitosan chains in the same direction with the applied stress. This behavior was observed by others in cationic hydrogel composites based on synthetic polymers and suggests a non-Newtonian behavior for all the range of angular frequencies, with a shear-thinning behavior [29].

The temperature sweep measurements were performed in the range of 5 to 40 °C, as it was proved that the storage temperature of chitosan solution is favorable at 5 °C as it reduces the polyssacharide decomposition [62] and the viscosity (η) of the hydrogels was evaluated up to the human body temperature (37 °C). Figure 6c shows an expected trend of decreasing viscosity of chitosan hydrogels when temperature increases, due to the augmentation of the polymeric thermal motion. However, all the samples present increased viscosity for temperatures exceeding 35 °C (more evident for CS-LA), mostly because of the moisture loss when subjected to higher temperatures [63].

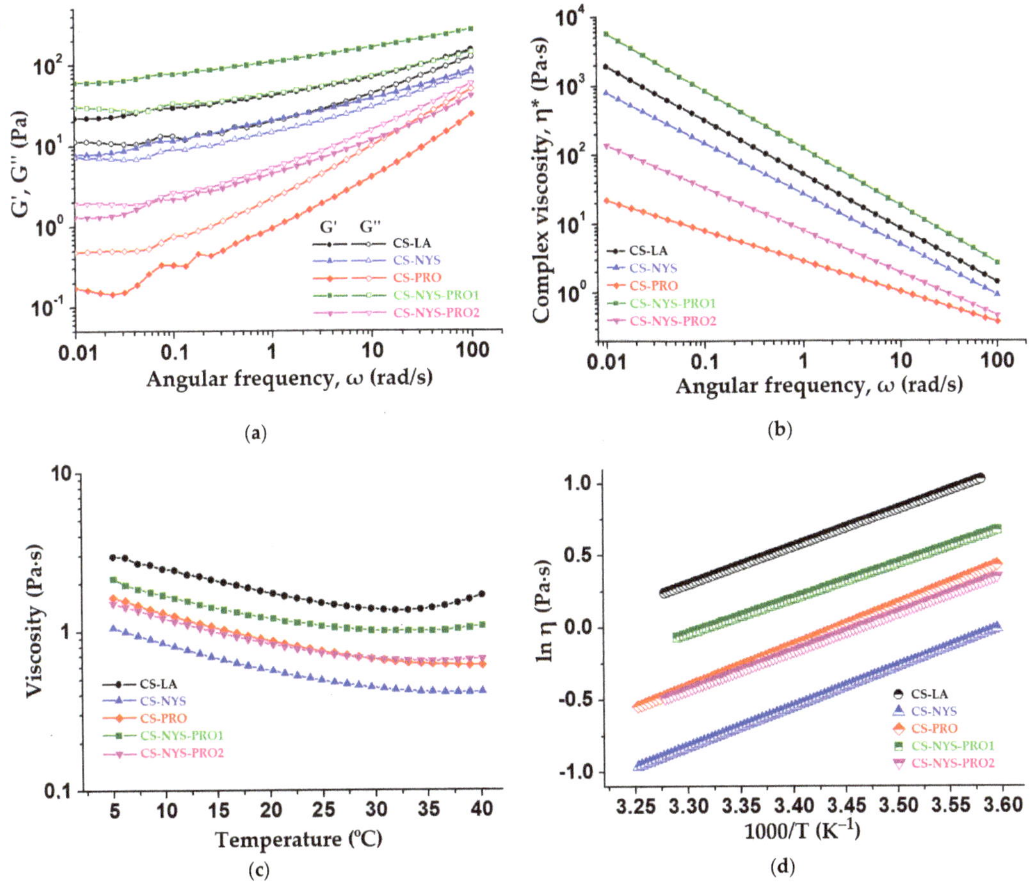

Figure 6. Rheological properties of chitosan hydrogel formulations: (**a**) dynamic viscoelastic moduli G′ (filled symbols) and G″ (open symbols) dependence on angular frequency at constant temperature of 35 °C and γ of 10%; (**b**) complex viscosity (η*) as a function of angular frequency (35 °C and γ = 10%); (**c**) temperature influence on the viscosity of hydrogel formulations; (**d**) Arrhenius fitting of the experimental data.

The activation energy (Eη) for the viscous flow was calculated from the regression results obtained by Arrhenius fitting of the temperature-dependence curve of the viscosity (Figure 6d). Close to each other and positive activation energy (Eη) values (Table S4, Supplementary Materials) were registered for all hydrogels, as expected for solutions with thermo-thinning behavior (viscosity decrease with temperature increase).

3.9. *Antifungal Activity of Hydrogel Formulations*

The in vitro antifungal activity was investigated against two Candida species (Candida albicans and Candida glabrata) on hydrogels samples, using the agar disk diffusion method which supposes the addition of the hydrogels on the culture medium pre-inoculated with the microbial suspension, and measuring the clear zone caused by fungal growth inhibition around the hydrogels, after 48 h of incubation. The images of the inhibition zones and the average inhibition diameters are given in Figure S10, Supplementary Materials, and

Figure 7, respectively. All the hydrogels killed the fungal cultures when placed in direct contact, but only chitosan hydrogels with NYS content (CS-NYS, CS-PRO-NYS1, CS-PRO-NYS2) presented inhibition of the fungal growth on a large area around the hydrogels, for both analyzed Candida strains. However, CS-NYS-PRO1 had the highest inhibition zone diameter (24 mm) compared with CS-NYS and CS-NYS-PRO2 (23, and 22 mm respectively) when evaluated against Candida albicans, which denotes the cumulative antifungal effect of NYS and PRO. The highest inhibition growth diameter against Candida glabrata was obtained for the CS-NYS hydrogel (19 mm), while CS-NYS-PRO1 and CS-NYS-PRO2 had diameters of 16, and 12 mm, respectively.

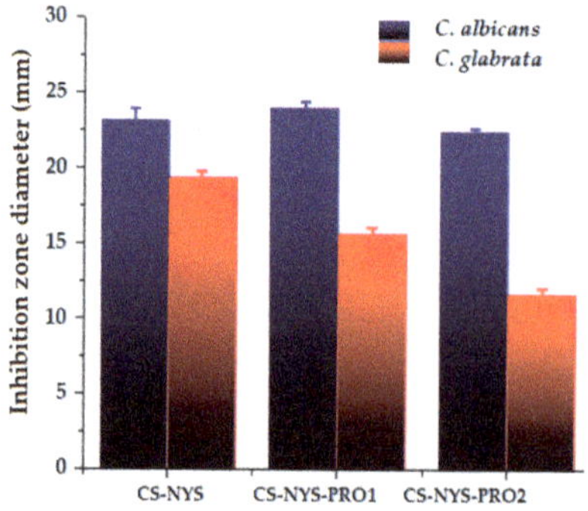

Figure 7. Diameters of the inhibition zones in the antifungal activity test of chitosan hydrogels against Candida albicans and Candida glabrata; error bars represent standard deviation (for n = 10).

3.10. Time–Kill Assay Test

The time–kill kinetics against the Candida albicans (Figure 8) was assessed for the hydrogels that presented higher antifungal activity (CS-NYS and CS-NYS-PRO1), as compared with the hydrogel with less antifungal inhibition (CS-PRO) and a control sample. The time–kill assay supports the results previously obtained in agar plate antifungal testing against C. albicans, showing that CS-PRO-NYS1 exhibit the strongest and faster slowing down of the fungal growth (from the first 6 h). A similar trend is identified also for the chitosan hydrogel charged with nystatin, which reveal fungal killing efficiency < 10^1 CFUs/mL in the first 12 h (Table S5). Corroborating data, it could be appreciated that the under-study chitosan hydrogels with nystatin and propolis have a potent fungicidal effect with a fungal burden reduction of 6-log killing (99.999% killing and 10 fungi left) after 24 h.

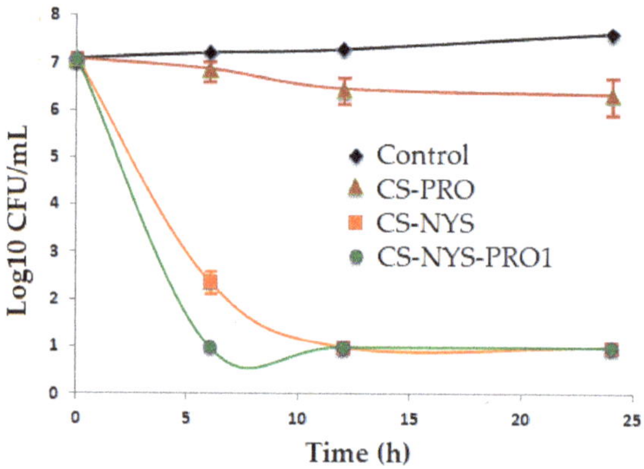

Figure 8. Time–kill curves for evaluation of the fungicidal activity of CS-NYS, CS-PRO and CS-NYS-PRO1 against yeast cells of *Candida albicans* as compared to the control assay; the error bars represent standard deviation for time killing kinetics (n = 4).

4. Conclusions

CS dissolved in aqueous lactic solution was demonstrated by structural (FTIR) and morphological (SEM, polarized light microscopy) investigations to successfully embed nystatin and/or propolis to yield gel-like or flexible film formulations. The antifungal films presented increased elasticity and mechanical properties close to those of soft tissues (Young's modulus lower than 1 MPa), adequate hydration capacities to maintain films bioadhesion and a progressive release of nystatin based on Quasi-Fickian diffusion mechanism, being suitable for application on buccal mucous membranes (cheek mucosa). As established by in vitro antifungal analysis, the samples containing higher amounts of propolis (CS-PRO and CS-NYS-PRO2) do not show noticeable activity, contrary to CS-NYS and CS-NYS-PRO1 (lower amount of PRO) samples, which proved to be efficient against *C. albicans* and *C. glabrata*. The results were confirmed by time–kill assay in synthetic vagina-simulative medium, where the hydrogel containing both NYS and smaller amounts of PRO (CS-PRO-NYS1) proved a combined effect on combating *C. albicans*, exhibiting the strongest and faster slowing down of fungal growth from the first 6 h of contact. Further studies are ongoing to demonstrate the additive or synergistic effect of these two active principles against candida species.

Supplementary Materials: The following supporting information can be downloaded at: https://www.mdpi.com/article/10.3390/polym14040689/s1, Figure S1: Nystatin powder: (**a**) Scanning electron microscopy images for nystatin; (**b**) particle size distribution histogram (particle sizes were determined using NIH Image J software); Figure S2: ^{13}C-RMN spectrum of nystatin; IUPAC Name: (1S,3R,4R,7R,9R,11R,15S,16R,17R,18S,19E,21E,25E,27E,29E,31E,33R,35S,36R,37S)-33-[(2R,3S,4S,5S,6R)-4-amino-3,5-dihydroxy-6-methyloxan-2-yl]oxy-1,3,4,7,9,11,17,37-octahydroxy-15,16,18-trimethyl-13-oxo-14,39-dioxabicyclo[3 3.3.1] nonatriaconta-19,21,25,27,29,31-hexaene-36-carboxylic acid); signal attributions made according to ref. [44–46]; Figure S3: ^{1}H-RMN spectrum of nystatin. IUPAC Name: 1S,3R,4R,7R,9R,11R,15S,16R,17R,18S,19E,21E,25E,27E,29E,31E,33R,35S,36R,37S)-33- [(2R,3S, 4S,5S,6R)-4-amino-3,5-dihydroxy-6-methyloxan-2-yl]oxy-1,3,4,7,9,11,17,37-octahydroxy- 15,16,18-trimethyl-13-oxo-14,39-dioxabicyclo[33.3.1]nonatriaconta-19,21,25,27,29,31-hexaene-36-carboxylic acid); signal attributions made according to ref. [44–46]; Figure S4: Chemical formula of chitosan with highlighting the functional groups: free primary amino groups (NH_2 to C2), primary hydroxyl group (OH to C6) and secondary hydroxyl groups (OH to C3); *m* and *n* represent the numbers of deacetylated (d-glucosamine) and N-acetyl-d-glucosamine repeating units, linked by (1–4)-β-glycosidic linkages, respectively; Figure S5: FTIR spectra of chitosan powder (CS) and chitosan gel (3%) after dispersion in 2% lactic acid solution (CS-LA); Figure S6: Characteristic stress-strain curves of the films: (**a**) without biological active compounds (CS-LA); (**b**) nystatin, propolis, and nystatin/propolis loaded chitosan films (CS-NYS, CS-PRO, CS-NYS-PRO1, CS-NYS-PRO2); Table S1: Parameters of pseudo-second order (PSO) and Korsmeyer–Peppas (K-P) kinetic models, where k_s is the constant of the swelling rate, S_e is the theoretical swelling capacity at equilibrium, k_p is a constant dependent on the polymeric network; n is the diffusion parameter of aqueous PBS in the formulation film and SD represents the standard deviation of n = 11); swelling capacity in PBS of chitosan charged films after 5 and 24 h; Figure S7: Evolution in time of UV-VIS spectra of: nystatin released from CS-NYS (**a**), CS-NYS-PRO1 (**b**) and CS-NYS-PRO2 (**c**); propolis released from CS-PRO (**d**); Table S2: NYS and PRO release from chitosan films: Korsmeyer-Peppas kinetics parameters (k represents the transport constant, n is the diffusion exponent and SD is standard deviation of n = 11), and release efficiency after 5 and 24 h.; Figure S8. Sorption/desorption isotherms for the hydrogel formulations; Table S3. Surface parameters of the hydrogels evaluated based on adsorption/desorption isotherms: water vapor sorption capacity, final weight (W); average pore size (r_{pm}) and BET data (surface area and monolayer weight); Figure S9: The loss tangent ($\tan \delta = G''/G'$) as a function of angular frequency; Table S4: Viscoelastic parameters (G', G'' and $\tan \delta$) at a frequency of 1 Hz and viscous flow activation energy ($E\eta$) calculated by fitting the temperature-dependence curve of the viscosity by Arrhenius equation; Figure S10. Antifungal activity of chitosan hydrogels against: (**a**) *C. albicans* and (**b**) *C. glabrata*; Table S5: Killing-time efficiency of CS-NYS, CS-PRO and CS-NYS-PRO1 hydrogels, against *Candida albicans* in 24 h as compared to the control test.

Author Contributions: Conceptualization, A.-C.H. and V.H.; methodology, A.-C.H., C.C., P.S. and M.M.; formal analysis, C.C.; investigation, A.-C.H., A.-C.B. and R.D.; resources, P.S, V.H. and M.M.; writing—original draft preparation, A.-C.H.; writing—review and editing, M.M. and V.H.; supervision, V.H. and B.C.S. All authors have read and agreed to the published version of the manuscript.

Funding: This research received no external funding.

Institutional Review Board Statement: Not applicable.

Informed Consent Statement: Not applicable.

Data Availability Statement: Not applicable.

Acknowledgments: One of the authors (A.-C.H.) acknowledges the partial support of the PhD—School of Advanced Studies of the Romanian Academy (SCOSAAR) for her PhD thesis entitled Chitosan based materials for biomedical, energy and environmental applications; PhD degree confirmation by order of the Romanian Minister of Education No 4640/18.08.2021.

Conflicts of Interest: The authors declare no conflict of interest.

References

1. Spampinato, C.; Leonardi, D. Candida infections, causes, targets, and resistance mechanisms: Traditional and alternative antifungal agents. *Biomed Res. Int.* **2013**, *2013*, 204237. [CrossRef] [PubMed]
2. Bongomin, F.; Gago, S.; Oladele, R.; Denning, D. Global and multi-national prevalence of fungal diseases—Estimate precision. *J. Fungi* **2017**, *3*, 57. [CrossRef] [PubMed]
3. Vallabhaneni, S.; Mody, R.K.; Walker, T.; Chiller, T. The global burden of fungal diseases. *Infect. Dis. Clin. North Am.* **2016**, *30*, 1–11. [CrossRef] [PubMed]
4. Mareković, I.; Pleško, S.; Rezo Vranješ, V.; Herljević, Z.; Kuliš, T.; Jandrlić, M. Epidemiology of candidemia: Three-year results from a croatian tertiary care hospital. *Fungi* **2021**, *7*, 267. [CrossRef] [PubMed]
5. Wiederhold, N. Antifungal resistance: Current trends and future strategies to combat. *Infect. Drug Resist.* **2017**, *10*, 249–259. [CrossRef]
6. Peptu, C.; Humelnicu, A.-C.; Rotaru, R.; Fortuna, M.E.; Patras, X.; Teodorescu, M.; Tamba, B.I.; Harabagiu, V. chitosan-based drug delivery systems. In *Chitin and Chitosan: Properties and Applications*, 1st ed.; Van den Broek, L.A.M., Boeriu, C.G., Eds.; John Wiley & Sons Ltd.: West Sussex, UK, 2020; pp. 259–289.
7. Shelke, N.B.; James, R.; Laurencin, C.T.; Kumbar, S.G. Polysaccharide bio-materials for drug delivery and regenerative engineering. *Polym. Adv. Technol.* **2014**, *25*, 448–460. [CrossRef]
8. Aranaz, I.; Alcántara, A.R.; Civera, M.C.; Arias, C.; Elorza, B.; Heras Caballero, A.; Acosta, N. Chitosan: An overview of its properties and applications. *Polymers* **2021**, *13*, 3256. [CrossRef]
9. Yu, J.; Wang, D.; Geetha, N.; Khawar, K.M.; Jogaiah, S.; Mujtaba, M. Current trends and challenges in the synthesis and applications of chitosan-based nanocomposites for plants: A review. *Carbohydr. Polym.* **2021**, *261*, 117904. [CrossRef]
10. Sood, A.; Gupta, A.; Agrawal, G. Recent advances in polysaccharides based bio-materials for drug delivery and tissue engineering applications. *Carbohydr. Polym. Technol. Appl.* **2021**, *2*, 100067. [CrossRef]
11. Martín-Villena, M.J.; Fernández-Campos, F.; Calpena-Campmany, A.C.; Bozal-de Febrer, N.; Ruiz-Martínez, M.A.; Clares-Naveros, B. Novel microparticulate systems for the vaginal delivery of nystatin: Development and characterization. *Carbohydr. Polym.* **2013**, *94*, 1–11. [CrossRef]
12. Martín, M.J.; Calpena, A.C.; Fernández, F.; Mallandrich, M.; Gálvez, P.; Clares, B. Development of alginate microspheres as nystatin carriers for oral mucosa drug delivery. *Carbohydr. Polym.* **2015**, *117*, 140–149. [CrossRef] [PubMed]
13. Fuentefria, A.M.; Pippi, B.; Dalla Lana, D.F.; Donato, K.K.; de Andrade, S.F. Antifungals discovery: An insight into new strategies to combat antifungal resistance. *Lett. Appl. Microbiol.* **2017**, *66*, 2–13. [CrossRef] [PubMed]
14. Revie, N.M.; Iyer, K.R.; Robbins, N.; Cowen, L.E. Antifungal drug resistance: Evolution, mechanisms and impact. *Curr. Opin. Microbiol.* **2018**, *45*, 70–76. [CrossRef]
15. Bell, A.S. Major Antifungal Drugs. In *Comprehensive Medicinal Chemistry II*, 8th ed.; Taylor, J.B., Triggle, D.J., Eds.; Elsevier: Oxford, UK, 2006; Volume 7, pp. 445–468.
16. Shaikh, M.S.; Alnazzawi, A.; Habib, S.R.; Lone, M.A.; Zafar, M.S. Therapeutic role of nystatin added to tissue conditioners for treating denture-induced stomatitis: A systematic review. *Prosthesis* **2021**, *3*, 61–74. [CrossRef]
17. Zager, R.A. Polyene antibiotics: Relative degrees of in vitro cytotoxicity and potential effects on tubule phospholipid and ceramide content. *Am. J. Kidney Dis.* **2000**, *36*, 238–249. [CrossRef] [PubMed]
18. Rathore, S.S.; Ramakishnan, J.; Raman, T. Recent advancement in combinational antifungal therapy and immunotherapy. In *Recent Trends in Antifungal Agents*, 1st ed.; Basak, A., Chakraborty, R., Mandal, S., Eds.; Springer: New Delhi, India, 2016; pp. 75–95.
19. Humelnicu, A.-C.; Samoila, P.; Asandulesa, M.; Cojocaru, C.; Bele, A.; Marinoiu, A.T.; Saccà, A.; Harabagiu, V. Chitosan-sulfated titania composite membranes with potential applications in fuel cell: Influence of cross-linker nature. *Polymers* **2020**, *12*, 1125. [CrossRef]

20. Singleton, V.L.; Orthofer, R.; Lamuela-Raventós, R.M. Analysis of total phenols and other oxidation substrates and antioxidants by means of Folin-Ciocalteu reagent. *Meth. Enzymol.* **1999**, *299*, 152–178. [CrossRef]
21. Li, X.; Wang, Y.; Li, A.; Ye, Y.; Peng, S.; Deng, M.; Jiang, B. A novel ph- and salt-responsive n-succinyl-chitosan hydrogel via a one-step hydrothermal process. *Molecules* **2019**, *24*, 4211. [CrossRef]
22. Budianto, E.; Amalia, A. Swelling behavior and mechanical properties of chitosan-poly(n-vinyl-pyrrolidone) hydrogels. *J. Polym. Eng.* **2020**, *40*, 551–560. [CrossRef]
23. Kipcak, A.S.; Ismail, O.; Doymaz, I.; Piskin, S. Modeling and investigation of the swelling kinetics of acrylamide-sodium acrylate hydrogel. *J. Chem.* **2014**, *2014*, 1–8. [CrossRef]
24. Peppas, N. Hydrogels in pharmaceutical formulations. *Eur. J. Pharm. Biopharm.* **2000**, *50*, 27–46. [CrossRef]
25. Korsmeyer, R.W.; Peppas, N.A. Effect of the morphology of hydrophilic polymeric matrices on the diffusion and release of water soluble drugs. *J. Membr. Sci.* **1981**, *9*, 211–227. [CrossRef]
26. Rehman, Q.; Hamid Akash, M.S.; Rasool, M.F.; Rehman, K. Role of kinetic models in drug stability. In *Drug Stability and Chemical Kinetics*; Akash, M.S.H., Rehman, K., Eds.; Springer: Singapore, 2020; pp. 155–166. [CrossRef]
27. Thommes, M.; Kaneko, K.; Neimark, A.V.; Olivier, J.P.; Rodriguez-Reinoso, F.; Rouquerol, J.; Sing, K.S.W. Physisorption of gases, with special reference to the evaluation of surface area and pore size distribution (IUPAC Technical Report). *Pure Appl. Chem.* **2015**, *87*, 1051–1069. [CrossRef]
28. Hernández-Castillo, D.J.; de la Cruz Hernández, E.N.; Frías Márquez, D.M.; Tilley, R.D.; Gloag, L.; Owen, P.Q.; López González, R.; Alvarez Lemus, M.A. Albendazole release from silica-chitosan nanospheres. In vitro study on cervix cancer cell lines. *Polymers* **2021**, *13*, 1945. [CrossRef]
29. Sun, M.; Sun, H.; Wang, Y.; Sánchez-Soto, M.; Schiraldi, D. The Relation between the rheological properties of gels and the mechanical properties of their corresponding aerogels. *Gels* **2018**, *4*, 33. [CrossRef]
30. Malkin, A.Y.; Isayev, A. Viscoelasticicty. In *Rheology: Concepts, Methods, and Applications*, 3rd ed.; Malkin, A.Y., Isayev, A., Eds.; ChemTec Publishing: Toronto, ON, Canada, 2017; pp. 45–49.
31. Rehman, T.U.; Shah, L.A. Rheological investigation of go doped p(aptmacl) composite hydrogel. *Z. Phys. Chem.* **2021**, *235*, 329–343. [CrossRef]
32. Canton, E.; Peman, J.; Gobernado, M.; Viudes, A.; Espinel-Ingroff, A. Patterns of amphotericin b killing kinetics against seven candida species. *Antimicrob. Agents Chemother.* **2004**, *48*, 2477–2482. [CrossRef]
33. Moosa, M.-Y.S.; Sobel, J.D.; Elhalis, H.; Du, W.; Akins, R.A. Fungicidal Activity of fluconazole against candida albicans in a synthetic vagina-simulative medium. *Antimicrob. Agents Chemother.* **2003**, *48*, 161–167. [CrossRef]
34. Marques, M.R.; Loebenberg, R.; Almukainzi, M. Simulated biological fluids with possible application in dissolution testing. *Dissolution Technol.* **2011**, *18*, 15–28. [CrossRef]
35. Perchyonok, V.T.; Zhang, S.; Oberholzer, T. Alternative chitosan based drug delivery system to fight oral mucositis: Synergy of conventional and bioactives towards the optimal solution. *Curr. Nanosci.* **2012**, *8*, 541–547. [CrossRef]
36. Perchyonok, V.T.; Zhang, S.; Grobler, S.; Oberholzer, T. Insights into and relative effect of chitosan-H, chitosan-H-propolis, chitosan-H-propolis-nystatin and chitosan-H-nystatin on dentine bond strength. *Eur. J. Dent.* **2013**, *7*, 412–418. [CrossRef] [PubMed]
37. Perchyonok, V.T.; Zhang, S.; Basson, N.; Grobler, S.; Oberholzer, T.; Massey, W. Insights into chitosan based gels as functional restorative biomaterials prototypes: In vitro approach. *Open J. Stomat.* **2013**, *3*, 22–30. [CrossRef]
38. Perchyonok, V.-T.; Zhang, S.; Oberholzer, T. Novel melatonin-chitosan hydrogels as suitable oral bio-drug delivery systems to fight oral mucositis: Synergy of antioxidants and bioactives in action. *Curr. Org. Chem.* **2012**, *16*, 2430–2436. [CrossRef]
39. Friedman, D.I. Stable Oil-in-Glycerin Emulsion. European Patent EP1077713A1, 28 February 2001.
40. Dallan, P.R.M.; da Luz Moreira, P.; Petinari, L.; Malmonge, S.M.; Beppu, M.M.; Genari, S.C.; Moraes, A.M. Effects of chitosan solution concentration and incorporation of chitin and glycerol on dense chitosan membrane properties. *J. Biomed. Mater. Res. B Appl. Biomater.* **2007**, *80B*, 394–405. [CrossRef] [PubMed]
41. Ma, Y.; Xin, L.; Tan, H.; Fan, M.; Li, J.; Jia, Y.; Ling, Z.; Chen, Y.; Hu, X. Chitosan membrane dressings toughened by glycerol to load antibacterial drugs for wound healing. *Mater. Sci. Eng. C* **2017**, *81*, 522–531. [CrossRef]
42. Peh, K.; Khan, T.; Ch'ng, H. Mechanical, bioadhesive strength and biological evaluations of chitosan films for wound dressing. *J. Pharm. Pharm. Sci.* **2000**, *3*, 303–311.
43. Niamsa, N.; Baimark, Y. Preparation and characterization of highly flexible chitosan films for use as food packaging. *Am. J. Food Technol.* **2009**, *4*, 162–169. [CrossRef]
44. Szwarc, K.; Płosiński, M.; Czerniejewska, K.; Laskowski, T.; Leniak, A.; Czub, J.; Kubica, P.; Sowiński, P.; Pawlaka, J.; Borowski, E. Intramolecular transformation of an antifungal antibiotic nystatin A1 into its isomer, iso -nystatin A1 structural and molecular modeling studies. *Magn. Reson. Chem.* **2016**, *54*, 953–961. [CrossRef]
45. Volpon, L.; Lancelin, J.-M. Solution NMR structure of five representative glycosylated polyene macrolide antibiotics with a sterol-dependent antifungal activity. *Eur. J. Biochem.* **2002**, *269*, 4533–4541. [CrossRef]
46. Sletta, H.; Borgos, S.E.F.; Bruheim, P.; Sekurova, O.N.; Grasdalen, H.; Aune, R.; Ellingsen, T.E.; Zotchev, S.B. Nystatin biosynthesis and transport: nysH and nysG genes encoding a Putative ABC transporter system in Streptomyces noursei ATCC 11455 are required for efficient conversion of 10-deoxynystatin to nystatin. *Antimicrob. Agents Chemother.* **2005**, *49*, 4576–4583. [CrossRef]

47. Benavent, C.; Torrado-Salmerón, C.; Torrado-Santiago, S. Development of a solid dispersion of nystatin with maltodextrin as a carrier agent: Improvements in antifungal efficacy against Candida spp. biofilm infections. *Pharmaceuticals* **2021**, *14*, 397. [CrossRef] [PubMed]
48. Mandru, M.; Ciobanu, C.; Ignat, M.E.; Popa, M.; Verestiuc, L.; Vlad, S. Sustained release of nystatin from polyurethane membranes for biomedical applications. *Dig. J. Nanomater. Biostruct.* **2011**, *6*, 1227–1238.
49. Rajczak, E.; Tylkowski, B.; Constantí, M.; Haponska, M.; Trusheva, B.; Malucelli, G.; Giamberini, M. Preparation and characterization of uv-curable acrylic membranes embedding natural antioxidants. *Polymers* **2020**, *12*, 358. [CrossRef] [PubMed]
50. Svečnjak, L.; Marijanović, Z.; Okińczyc, P.; Marek Kuś, P.; Jerković, I. Mediterranean propolis from the adriatic sea islands as a source of natural antioxidants: Comprehensive chemical biodiversity determined by GC-MS, FTIR-ATR, UHPLC-DAD-QqTOF-MS, DPPH and FRAP assay. *Antioxidants* **2020**, *9*, 337. [CrossRef]
51. Villalobos, K.; Rojas, H.; González-Paz, R.; Granados, D.B.; González-Masís, J.; Baudrit, J.V.; Corrales-Ureña, Y.R. Production of starch films using propolis nanoparticles as novel bioplasticizer. *J. Renew. Mater.* **2017**, *5*, 189–198. [CrossRef]
52. Velásquez-Cock, J.; Ramírez, E.; Betancourt, S.; Putaux, J.-L.; Osorio, M.; Castro, C.; Ganán, P.; Zuluaga, R. Influence of the acid type in the production of chitosan films reinforced with bacterial nanocellulose. *Int. J. Biol. Macromol.* **2014**, *69*, 208–213. [CrossRef]
53. Liu, Y.; Wang, S.; Zhang, R. Composite poly(lactic acid)/chitosan nanofibrous scaffolds for cardiac tissue engineering. *Int. J. Biol. Macromol.* **2017**, *103*, 1130–1137. [CrossRef]
54. He, Y.; Miao, J.; Chen, S.; Zhang, R.; Zhang, L.; Tang, H.; Yang, H. Preparation and characterization of a novel positively charged composite hollow fiber nanofiltration membrane based on chitosan lactate. *RSC Adv.* **2019**, *9*, 4361–4369. [CrossRef]
55. Bhattarai, N.; Ramay, H.R.; Chou, S.H.; Zhang, M. Chitosan and lactic acid-grafted chitosan nanoparticles as carriers for prolonged drug delivery. *Int. J. Nanomed.* **2006**, *1*, 181–187. [CrossRef]
56. Ambrosio-Martín, J.; Fabra, M.J.; Lopez-Rubio, A.; Lagaron, J.M. An effect of lactic acid oligomers on the barrier properties of polylactide. *J. Mater. Sci.* **2014**, *49*, 2975–2986. [CrossRef]
57. Campos, M.G.N.; Mei, L.H.I.; Santos, A.R., Jr. Sorbitol-plasticized and neutralized chitosan membranes as skin substitutes. *Mater. Res.* **2015**, *18*, 781–790. [CrossRef]
58. Gleadall, A.; Ruiz-Cantu, L. Transplantable scaffolds. In *3D Printing in Medicine and Surgery-Applications in Healthcare*, 1st ed.; Thomas, D.J., Singh, D., Eds.; Woodhead Publishing: Kidlington, UK, 2021; pp. 199–222.
59. Koland, M.; Vijayanarayana, K.; Charyulu, R.; Prabhu, P. In vitro and in vivo evaluation of chitosan buccal films of ondansetron hydrochloride. *Int. J. Pharm. Investig.* **2011**, *1*, 164–171. [CrossRef] [PubMed]
60. Ipate, A.-M.; Serbezeanu, D.; Bargan, A.; Hamciuc, C.; Ochiuz, L.; Gherman, S. Poly(vinylpyrrolidone)-chitosan hydrogels as matrices for controlled drug release. *Cellul. Chem. Technol.* **2021**, *55*, 63–73. [CrossRef]
61. Butnaru, E.; Stoleru, E.; Brebu, M.; Darie-Nita, R.; Bargan, A.; Vasile, C. Chitosan-based bionanocomposite films prepared by emulsion technique for food preservation. *Materials* **2019**, *12*, 373. [CrossRef]
62. Nguyen, T.T.B.; Hein, S.; Ng, C.-H.; Stevens, W.F. Molecular stability of chitosan in acid solutions stored at various conditions. *J. Appl. Polym. Sci.* **2007**, *107*, 2588–2593. [CrossRef]
63. Viljoen, J.M.; Steenekamp, J.H.; Marais, A.F.; Kotzé, A.F. Effect of moisture content, temperature and exposure time on the physical stability of chitosan powder and tablets. *Drug. Dev. Ind. Pharm.* **2013**, *40*, 730–742. [CrossRef]

polymers

MDPI

Article

Reversible Protein Capture and Release by Redox-Responsive Hydrogel in Microfluidics

Chen Jiao [1,2], Franziska Obst [3], Martin Geisler [1], Yunjiao Che [1], Andreas Richter [3], Dietmar Appelhans [1], Jens Gaitzsch [1,*] and Brigitte Voit [1,2,*]

1 Leibniz-Institut für Polymerforschung Dresden e.V., Hohe Straße 6, 01069 Dresden, Germany; jiao@ipfdd.de (C.J.); geisler@ipfdd.de (M.G.); yunjiao.che@gmail.com (Y.C.); applhans@ipfdd.de (D.A.)
2 Organische Chemie der Polymere, Technische Universität Dresden, Mommsenstraße 4, 01062 Dresden, Germany
3 Institut für Halbleiter- und Mikrosystemtechnik, Technische Universität Dresden, Nöthnitzer Straße 64, 01187 Dresden, Germany; franziska.obst@tu-dresden.de (F.O.); andreas.richter7@tu-dresden.de (A.R.)
* Correspondence: gaitzsch@ipfdd.de (J.G.); voit@ipfdd.de (B.V.)

Abstract: Stimuli-responsive hydrogels have a wide range of potential applications in microfluidics, which has drawn great attention. Double cross-linked hydrogels are very well suited for this application as they offer both stability and the required responsive behavior. Here, we report the integration of poly(*N*-isopropylacrylamide) (PNiPAAm) hydrogel with a permanent cross-linker (*N,N′*-methylenebisacrylamide, BIS) and a redox responsive reversible cross-linker (*N,N′*-bis(acryloyl)cystamine, BAC) into a microfluidic device through photopolymerization. Cleavage and re-formation of disulfide bonds introduced by BAC changed the cross-linking densities of the hydrogel dots, making them swell or shrink. Rheological measurements allowed for selecting hydrogels that withstand long-term shear forces present in microfluidic devices under continuous flow. Once implemented, the thiol-disulfide exchange allowed the hydrogel dots to successfully capture and release the protein bovine serum albumin (BSA). BSA was labeled with rhodamine B and functionalized with 2-(2-pyridyldithio)-ethylamine (PDA) to introduce disulfide bonds. The reversible capture and release of the protein reached an efficiency of 83.6% in release rate and could be repeated over 3 cycles within the microfluidic device. These results demonstrate that our redox-responsive hydrogel dots enable the dynamic capture and release of various different functionalized (macro)molecules (e.g., proteins and drugs) and have a great potential to be integrated into a lab-on-a-chip device for detection and/or delivery.

Keywords: microfluidics; disulfide bonds; redox-responsive; hydrogels; protein capture and release; swelling behaviors; mechanical properties

check for updates

Citation: Jiao, C.; Obst, F.; Geisler, M.; Che, Y.; Richter, A.; Appelhans, D.; Gaitzsch, J.; Voit, B. Reversible Protein Capture and Release by Redox-Responsive Hydrogel in Microfluidics. *Polymers* **2022**, *14*, 267. https://doi.org/10.3390/polym14020267

Academic Editors: Marek Kowalczuk, Iza Radecka and Barbara Trzebicka

Received: 13 December 2021
Accepted: 5 January 2022
Published: 10 January 2022

1. Introduction

Microfluidics is a technique for accurately processing and manipulating small amounts of fluid in a microscale device with channels ranging from tens to hundreds of micrometers [1]. Some of the main advantages of microfluidics are the low consumption of reagents, high repeatability, fast reaction rates, and accurate control of physical/chemical properties [1–6]. One application of microfluidics is the capture and release of biomolecules and drug delivery on the micrometer scale [7]. Key features of microfluidic capture and release devices are their small volumes, large specific surface areas and strong mass and heat transfer within the microchannels [8]. The specific surface can be increased even further when hydrogel dots are integrated into the microfluidic system. A high number of hydrogel dots in a microfluidic chamber reactor enlarge the specific surface area within the device. A hydrogel scaffold and their chemical units present an enhanced surface area for the fluidic components to interact with microfluidic compartments [9,10]. These compartments allow for the physical entrapment of proteins (e.g., enzymes) or a chemical reaction to induce the swelling of stimuli responsive hydrogels [11]. Despite the various advances from the last two decades, the challenge of constructing microfluidic systems with stimuli responsive hydrogels which are able to capture and release biomolecules based on chemical bonds rather than physical interactions still remains [12–17]. Microfluidic systems with integrated double cross-linked stimuli-responsive hydrogels which respond to chemical switches, can overcome this challenge and are the topic of the work presented here.

Stimuli-responsive hydrogels can shrink and swell through expelling and absorbing additional amount of water, triggered by different external stimuli such as temperature [18], light [19], redox potential [20,21], pH [22], salinity [23–25], or electric fields [26]. They have been of widespread interest in recent years due to their great potential in applications as biosensors [27,28], microfluidic systems [9,10,29], tissue scaffolds [30], cell culture [31], or drug release [32–34]. Within microfluidics, especially the thermally responsive poly(*N*-isopropylacrylamide) (PNiPAAm) hydrogel matrix is used for example as valves or micropumps [22,35–38]. There are, however, only few microfluidic devices containing hydrogels which exploit a redox reaction [7,39]. The disulfide bond is an extremely valuable redox-responsive functional group with high reactivity. It is well known that disulfide bonds can dissociate or re-form depending on the redox potential [40,41]. Additionally, disulfide bonds can react with thiol groups and make an exchange reaction [42]. Briefly, the thiolate, as a nucleophile, attacks the disulfide bond to form a new disulfide bond and produces a new thiol leaving group [43,44]. As thiol groups frequently occur in proteins and other biological samples, biological applications, such as chemosensors and nanomaterial carriers based on the functionalization of disulfides have been developed [40,45,46].

The introduction of disulfide bonds into hydrogels enables the formation of redox responsive systems [47–49] but also allows for the application of double cross-linked hydrogels containing a mixture of permanent and reversible cross-linking points. Combined with the commonly used *N,N′*-methylenebisacrylamide (BIS, Figure 1) as permanent cross-linker, the *N,N′*-bis(acryloyl)cystamine (BAC, Figure 1) is a very convenient disulfide-bearing redox responsive reversible cross-linker [50]. The double-crosslinked hydrogel system then shows an additional reversible swelling in response to changes in the redox potential while exposing thiols in the swollen state. When applied with P(NiP)AAm as a backbone, this BAC/BIS system has been reported and characterized for its viscoelastic properties as well as their ability to capture and release organophosphates [51–53]. Recently, our group also investigated the correlation between the mechanical properties, swelling behaviors and swelling kinetics of PNiPAAm-BAC-BIS hydrogels. Long-term cycle stability was achieved in bulk hydrogel and micro-structured hydrogel was realized by photopolymerization, laying a solid foundation for the application in microfluidics [54]. In order to integrate such a hydrogel into the microfluidic device, PDMS-on-glass device formed by a polydimethylsiloxane (PDMS) sheet and a glass holder has been developed [55–57]. PDMS can be shaped quickly by lithography, and fluid flow can easily be observed due to the high transparency of both PDMS sheet and glass slide [9,58]. In order to integrate a large number of hydrogels into a microfluidic device, the reported synthesis of an array of hydrogel dots proved to be a feasible platform [10,59].

Here, we report a simple and fast method to reversibly capture and release proteins using redox responsive hydrogels within a microfluidic device. Integration of PNiPAAm-BAC-BIS hydrogel dots into a microfluidic device was achieved by in-situ photopolymerization (Figure 1). Cleavage or re-formation of redox-responsive disulfide bonds changes the cross-linking densities of the hydrogel dots, making them swell or shrink in the microfluidic devices. The hydrogel dots were to capture and release the protein bovine serum albumin (BSA) modified with 2-(2-pyridyldithio)-ethylamine (PDA, source of disulfide bonds) and rhodamine B (RhB) via the thiol-disulfide exchange. Ideally, the protein capture and release could even be repeated over various cycles. Such a reproducible capture and release of a functionalized protein in microfluidics has great potential to be integrated in a lab-on-a-chip device for enzyme immobilization, rapid detection devices and/or delivery of captured macro (molecules).

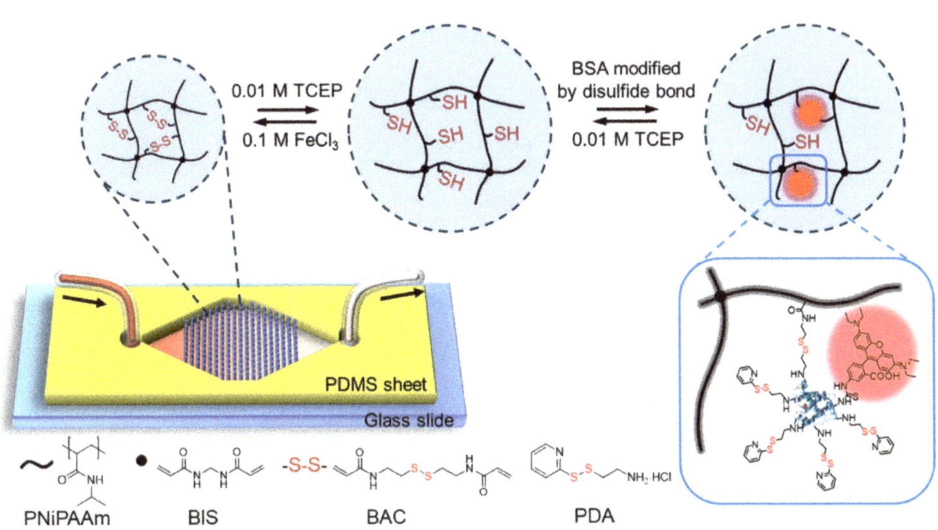

Figure 1. Schematic drawing of the microfluidic chip design and the reversible redox-responsive property of double cross-linked PNiPAAm hydrogel dots permanently cross-linked by BIS and reversibly cross-linked by the disulfide bonds of BAC. By the cleavage and re-formation of disulfide bonds in hydrogel dots, the capture and release of modified BSA can be achieved. BSA was modified with rhodamine B and PDA (bottom right), which introduced the disulfide bonds.

2. Materials and Methods

2.1. Materials

N-isopropylacrylamide (NiPAAm, ≥99%), *N,N′*-methylenebisacrylamide (BIS, 99%), lithium phenyl-2,4,6-trimethylbenzoylphosphinate (LAP, ≥95%), tris(2-carboxyethyl)phosphine hydrochloride (TCEP), dimethyl sulfoxide (DMSO), cystamine hydrochloride (≥98%), 2,2′-dipyridyl disulfide (98%), iron(III) chloride hexahydrate (≥99%), bovine serum albumin (BSA, ≥96%), rhodamine B isothiocyanate (RhB), phosphate buffered saline tablet (PBS), *N*-ethyl-*N′*-(3-dimethylaminopropyl)carbodiimide hydrochloride (EDAC, ≥99%), 2-(*N*-morpholino)ethanesulfonic acid (MES, ≥99%), sodium peroxodisulfate (99%), *N,N,N′,N′*-tetramethylethylenediamine (TMEDA, 99%) were purchased from Sigma-Aldrich (Damstadt, Germany). *N,N′*-Bis(acryloyl)cystamine (BAC, 98%) was purchased from Alfa Aesar (Kandel, Germany). Sodium bicarbonate (99.5%) was purchased from ACROS Organics (Geel, Belgium). Sodium carbonate anhydrous (≥99.5%) was purchased from Honeywell Fluka (Seelze, Germany). Ethanol absolute (≥99.5%) was purchased from VWR Chemicals BDH (London, UK). Silicone elastomer kit (PDMS) was purchased from DOW Corning (Midland, MI, USA). Deionized water was used for all experiments. All chemicals were used as received without further purification.

2.2. Synthesis of 2-(2-pyridyldithio)-ethylamine (PDA)

PDA was produced following a published procedure [60]. Aldrithiol (1.1 g, 5.0 mmol, 2 eq.) was dissolved in 10 mL of methanol in a round bottom flask and 400 µL acetic acid was added. Then the cysteamine hydrochloride (0.28 g, 2.5 mmol, 1 eq.) dissolved in 5 mL of methanol was added dropwise in 30 min. The reaction was performed under argon atmosphere. After stirring for 48 h, the solvent was evaporated from the reaction mixture and the residue was purified by precipitation in cold diethyl ether overnight followed by filtration (2 times). The product was dried overnight in an oven at 40 °C. Yield: 0.42 g, 75%. The ^{1}H NMR spectral data are consistent with the structure (Figure S1 in Supplementary Material). ^{1}H NMR (500 MHz, D_2O, δ): 8.47 (d, $^{3}J_{HH}$ = 4.2 Hz, 1H, Ar H), 7.86 (td, $^{3}J_{HH}$ = 7.7 Hz, $^{4}J_{HH}$ = 1.8 Hz, 1H, Ar H), 7.76 (d, $^{3}J_{HH}$ = 8.1 Hz, 1H, Ar H), 7.35 (td, $^{3}J_{HH}$ = 6.2 Hz, $^{4}J_{HH}$

= 1.0 Hz, 1H, Ar H), 3.37 (t, $^3J_{HH}$ = 6.3 Hz, 2H, $CH_2CH_2NH_2$), 3.13 (t, $^3J_{HH}$ = 6.3 Hz, 2H, SCH_2CH_2).

2.3. Synthesis of BSA-RhB

BSA (500 mg, 7.6 μmol, 1 eq.) was dissolved in 1 mM carbonate buffer (25 mL, pH 10) and stirred for 30 min before adding RhB (6.1 mg, 11.4 μmol, 1.5 eq.) dissolved in 1 mL DMSO. After 20 h of stirring at room temperature, the mixture was extensively dialyzed against 1 mM PBS buffer (pH 7.4) for 3 days to remove all unbound RhB. All processes were performed under light protection. Finally, the purified mixture was freeze-dried overnight to get the product BSA-RhB, which was stored at −20 °C. Molecular weight was confirmed by matrix assisted laser desorption ionization–time of flight mass spectrometry (MALDI-TOF-MS) (M_{BSA}: 66,400 Da and $M_{BSA\text{-}RhB}$: 66,600 Da).

2.4. Synthesis of BSA-RhB-PDA

BSA-RhB (400 mg, 6.0 μmol, 1 eq.) and EDAC (57.7 mg, 300.0 μmol, 50 eq.) were dissolved in 50 mM MES buffer (25 mL, pH 6.5) and stirred for 30 min before adding dropwise PDA (13.4 mg, 6 μmol, 10 eq.) dissolved in 50 mM MES buffer (5 mL, pH 6.5). After 20 h of stirring at room temperature, the mixture was extensively dialyzed against 1 mM PBS buffer (pH 7.4) for 3 days to remove all unbounded molecules. All processes were performed under light protection. Finally, the purified mixture was freeze-dried overnight to get the product BSA-RhB-PDA, which was stored at −20 °C. Molecular weight was confirmed by MALDI-TOF-MS (details in part 2.7) ($M_{BSA\text{-}RhB\text{-}PDA}$: 67,600 Da).

2.5. Preparation of Hydrogel Arrays

Photomask for patterning (array with diameter 350 μm, Figure S2a in Supplementary Material) was designed with the CAD 2021 software Autodesk Inventor (San Rafael, CA, USA) and produced on a black and white flat film by photo plotting (MIVA 26100 ReSolution, MIVA Technologies, Schönaich, Germany). Polyoxymethylene (POM) mold with single chamber (Figure S2c in Supplementary Material) used for the photopolymerization of gel precursor was produced in-house by milling on a four-axis CNC milling machine (DMU 50, DMG MORI, Bielefeld, Germany). The depth of the chamber was 151 ± 1 μm, confirmed by the confocal microscope (μsurf explorer, Nano Focus, Oberhausen, Germany). Before the photopolymerization of hydrogel arrays, the Menzel glass slide (76 mm × 26 mm × 1 mm) was cleaned with isopropanol, MilliQ and ethanol sequentially in an ultrasonic bath.

For the preparation of hydrogel precursor solution, the cross-linker BAC was first dissolved in ethanol and stirred for 20 min, then the monomer NiPAAm (930.8 mg, 12.5 wt %) and the other cross-linker BIS dissolved in deionized water were added. The entire amount of cross-linker is fixed at 2 mol% and with molar ratios (BAC:BIS) of 1:1, 1.5:1, 2:1, 3:1, 4:1, and 5:1, which were named as N 1:1, N 1.5:1, N 2:1, N 3:1, N 4:1, and N 5:1 hydrogel, respectively (Table S1). In order to dissolve all the cross-linkers, a higher amount of ethanol was used in the monomer solutions with a higher BAC ratio (0.71 mL ethanol used for N 1:1, N 1.5:1, N 2:1 hydrogels and 1.22 mL ethanol used for N 3:1, N 4:1, N 5:1 hydrogels). The amount of photoinitiator is 0.65 mol% to the monomer, which was added under light protection. The precursor solutions were purged with argon for 15 min to remove oxygen.

The preparation of the hydrogel dot arrays was divided into five steps (Figure S3 in Supplementary Material) and derived from a previously published procedure [10]. Firstly, the hydrogel precursor solution was transferred to the reaction chamber of the POM mold. Then, the washed glass slide was aligned on the chamber and patterned photomask was aligned on the glass slide. According to the design principle, an even layer without any trapped air was formed between the glass slide and the reaction chamber of POM mold. Afterwards, the reaction chamber was exposed to an UV lamp (DELOLUX 04, DELO, Windach, Germany) with a light power of 8 mW cm^{-2} on the sample surface. The emission spectrum for the photopolymerization ranges from 315 to 500 nm and the irradiation time is 7 s. Next, the glass slide with the covalently attached cylinder-shaped hydrogel dots was separated from the POM mould and immersed in the deionized water overnight to remove all the unbound precursor solution. After the as-prepared hydrogel dots reached equilibrium in deionized water, dried and then wiped the glass slide with isopropanol without touching the hydrogel dots. Finally, the microfluidic chip was sealed by aligning the glass slide on the PDMS sheet.

2.6. Microfluidic Testing

Reducing agent used was 0.01 M TCEP aqueous solution and oxidant used was 0.1 M $FeCl_3$ aqueous solution. By purging different solutions (0.01 M TECP aqueous solution, 0.1 M $FeCl_3$ aqueous solution) with different flow rate (1, 2, 5, 10, 50 μL min^{-1}) through the hydrogel arrays with different hydrogel contents (N 1:1 to N 5:1 hydrogels) in the microfluidic chip, the swelling behavior of the microstructured hydrogel dots was observed by an optical microscope (Leica S8APO, DFC295 camera, cold light source: KL 1500 LCD). The swelling ratio (S_R) of the hydrogel dots was calculated by Formula (1):

$$S_R = D_t / D_0 \tag{1}$$

where D_t and D_0 represent pending test diameter and original diameter of the hydrogel dot, respectively.

For protein capture and release test, 0.01 M TCEP aqueous solution was first purged at the flow rate of 1 μL min^{-1} for 60 min, followed by deionized water washing for 30 min. Afterwards, 50 μM BSA-RhB-PDA aqueous solution was purged at the flow rate of 1 μL min^{-1} for 90 min, followed by 2 mM PBS buffer (pH 7.4) washing for 40 min. For protein release, 0.01 M TCEP aqueous solution was purged at the flow rate of 2 μL min^{-1} overnight. The described cyclic capture and release process of protein was repeated three times. The whole processes were observed by confocal laser fluorescence microscope (CLSM, Leica SP5, Wetzlar, Germany) at the height of 80 μm under both laser field ($\lambda_{excitation}$: 561 nm, laser intensity: 15%) and bright field. The release ratios of protein were analyzed by ImageJ. One was purging 50 μM BSA-RhB-PDA aqueous solution at the flow rate of 1 μL min^{-1} for 90 min, followed by 2 mM PBS buffer (pH 7.4) washing for 40 min. The other was consistent with the purging experiment of the protein capture, except that BSA-RhB-PDA is replaced by BSA-RhB. The flow rate of all the chip washing was 5 μL min^{-1}.

2.7. MALDI-TOF

The MALDI-TOF mass spectra were measured in linear positive detection mode by an autoflex® speed MALDI-TOF system (Bruker Daltonics GmbH, Bremen, Germany) equipped with a smartbeam™ II (modified Nd:YAG) laser having a wavelength of 355nm. Sinapidic acid and α-cyano-4-hydroxycinnamic acid (HCCA) (both by Sigma Aldrich, Damstadt, Germany) were used as matrix (both 10 g L^{-1}) dissolved in Methanol and mixed at a ratio of 1:1 (v/v). 1 μL of a 0.5 g L^{-1} methanolic sodium trifluoroacetate solution was added per 100 μL matrix solution. The sample was prepared in deionized water at a concentration of 2 g L^{-1}, mixed with the matrix solution in a ratio of 1:1 (v/v) and spotted on the MALDI plate via the dried droplet method [61]. BSA of different batches was used for calibration and for verification. The samples were measured with an acceleration voltage of 19.5 kV, a laser attenuation of 30%, a laser repetition rate of 1 kHz and a detector gain of 70× (3.446 kV). Each mass spectrum was recorded by accumulation of 8000 shots.

2.8. NMR Measurement

^{1}H NMR spectra were recorded on the Bruker Advance III spectrometer (Bruker Biospin, Ettlingen, Germany) at 500 MHz using deuterium oxide as the solvent. Chemical shifts of ^{1}H NMR were referred to TMS (δ = 0).

3. Results and Discussion

3.1. Experimental Design and Microfluidic Chip Design

Precisely patterned cylinder hydrogel dot arrays with a diameter of 350 μm per dot were prepared through fast and efficient photopolymerization. Their height of 150 μm was determined by the POM mold (Figure S2) used for the polymerization. The preparation process and formation mechanism of PNiPAAm-based hydrogels are illustrated in Figure S3. Based on previous results, the re-oxidation of disulfide bonds in the double cross-linked hydrogel dots was based on a protocol for BAC/BIS double cross-linked macrogels [54]. 0.01 M TCEP aqueous solution has been reported to sufficiently reduce the disulfide bonds into thiol groups, and 0.1 M $FeCl_3$ aqueous solution was proven to oxidize thiol groups into disulfide bonds. The redox-responsive PNiPAAm hydrogels were hence ready to be transferred into a microfluidic chip. For this, a single-chamber microfluidic chip with a chamber width of 14 mm was selected and prepared for the capture of proteins. Based on previously published considerations, a conical widening inlet and conical narrowing outlet was designed with an opening angle of 58° (Figure 1) to lead to a chamber containing the hexagonally arranged hydrogel dots [10]. Removing any trapped air was achieved by a freeze-pump-thaw process and an integrated bubble trap in the microfluidic set up (Figure S4) before purging the water into the chip.

3.2. Optimizing the Cross-Linker Composition of the Hydrogel

The swelling ratio of hydrogel dots was expected to differ with different molar ratios of BAC to BIS. Thus, several molar ratios were tested (1:1, 1.5:1, 2:1, 3:1, 4:1, and 5:1) while keeping the overall amount of cross-linker constant at 2 mol% with respect to the monomer. All gels were named accordingly as N 1:1, N 1.5:1, N 2:1, N 3:1, N 4:1 and N 5:1. Kinetic measurements to note the swelling and de-swelling behaviors over time are shown in Figure 2b,c, respectively. TCEP and $FeCl_3$ aqueous solutions were purged at a flow rate of 1 and 2 μL min^{-1} in the microfluidic chip, respectively. Photos of the original, the reduced and the oxidized N 1:1 (lowest amount of responsive cross-linker) hydrogel dots are shown in Figure 2a, while the ones of N 5:1 (highest amount of responsive cross-linker) hydrogel dots are shown in Figure S5. Irrespective of the different molar ratios of BAC to BIS, the degree of swelling and deswelling remained constant after 60 min of reduction and 20 min of oxidation. All final swelling ratios of the hydrogel dots, determined by measuring the diameters of the dots, where thus validated after 60 min of reduction and 30 min of oxidation (summarized in Table 1). With the increase of BAC content, the swelling ratio of hydrogel dots in reducing agent increased from 1.09 of the N 1:1 hydrogel to 1.16 of the N 5:1 hydrogel. The oxidative shrinkage rose from 33% of the N 1:1 hydrogel to 44% of the N 5:1 hydrogel, except that N 3:1 and N 4:1 hydrogels were around 34%. This meant that with increasing of BAC content, more disulfide bonds were cleaved by TCEP reduction. Thus, a higher decrease in cross-linking density is given which leads to more water absorption and higher swelling ratios. Similarly, the presence of more thiol groups in reduced hydrogel dots also gave a higher ratio of reformed disulfide bonds and hence a higher shrinking during the oxidation. No hydrogel returned to the original size since the hydrogels deswelled to the thermodynamically favored state in water, at which no complete disulfide reformation is assumed under the given experimental conditions. Thus, the ideal BAC/BIS ratio could be identified with these experiments.

In addition to the BAC content, the influence of different flow rates on swelling ratios of hydrogel dots has been studied. Kinetics for the swelling ratios of N 1:1, N 1.5:1, and N 2:1 hydrogel dots under 0.01 M TCEP purged at flow rate of 1, 10, and 50 μL min^{-1}, respectively, are shown Figure S6. The hydrogel dots had a relatively low equilibrium swelling ratio (1.09 for N 2:1) at a high flow rate of 50 μL min^{-1}, while there was no significant difference between 1 and 10 μL min^{-1} (both are 1.12 for N 2:1). This behavior indicated that high flow rates could prevent a complete reduction of disulfide units and hence a complete swelling of the hydrogel dots. In the follow-up microfluidic tests, a flow rate below 10 μL min^{-1} (1–5 μL min^{-1}) was chosen to ensure a complete swelling.

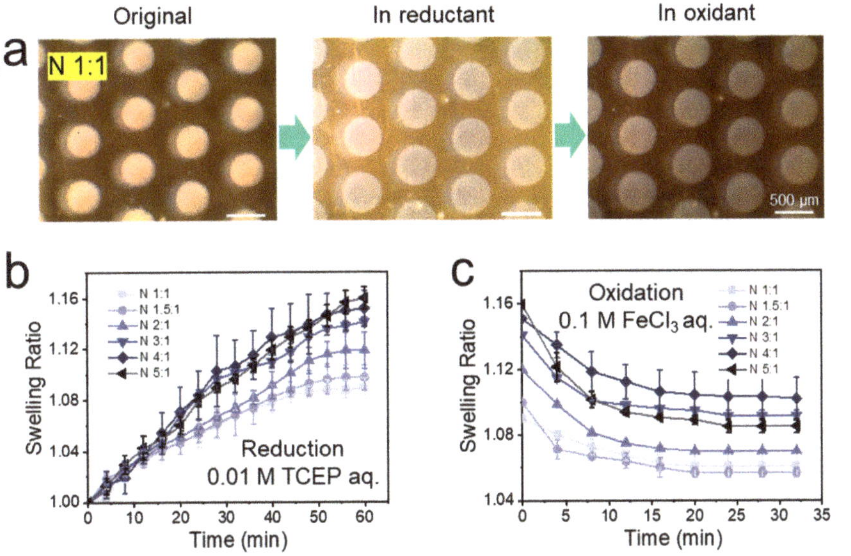

Figure 2. (**a**) Optical images of the original, the reduced (FR 1 μL min^{-1}) and the oxidized (FR 2 μL min^{-1}) N 1:1 hydrogel dots. (**b**) Swelling of PNiPAAm hydrogel dots with different molar ratios of cross-linkers BAC to BIS (from N 1:1 to N 5:1 hydrogel) in the microfluidic chip when 0.01 M TCEP aqueous solution perfused. (**c**) Shrinking of PNiPAAm hydrogel dots with different molar ratios of cross-linker BAC to BIS (from N 1:1 to N 5:1 hydrogel) in the microfluidic chip when 0.1 M $FeCl_3$ aqueous solution was purged. At least three hydrogel dot arrays were tested for each experimental point to obtain reliable data. Longer time for oxidation process did not result in larger degree on shrinking for hydrogel dots.

Table 1. Swelling behaviors and mechanical properties of PNiPAAm hydrogel dots in microfluidics.

Hydrogel	Swelling in 0.01 M TCEP aq. [a]	Residual Swelling in 0.1 M $FeCl_3$ aq. [b]	Percentage Reduction [c]	Compression Strength at 50% Strain (kPa) [d]	Elastic Modulus (Pa) [d]
N 1:1	1.09 ± 0.007	1.06 ± 0.007	33%	5.66 ± 0.17	43 ± 5
N 1.5:1	1.10 ± 0.010	1.06 ± 0.003	40%	13.03 ± 0.60	86 ± 3
N 2:1	1.12 ± 0.014	1.07 ± 0.001	42%	15.85 ± 0.55	143 ± 3
N 3:1	1.14 ± 0.004	1.09 ± 0.003	36%	7.42 ± 0.13	53 ± 2
N 4:1	1.15 ± 0.014	1.10 ± 0.013	33%	4.55 ± 0.08	23 ± 2
N 5:1	1.16 ± 0.009	1.09 ± 0.004	44%	1.47 ± 0.14	10 ± 1

[a] Compared to the original size after 60 min at 1 μL min^{-1} flow rate (flow rate reasoned in the main text); [b] compared to the original size after 30 min at 2 μL min^{-1} flow rate (flow rate reasoned in the main text); [c] percentage of oxidative shrinkage compared to reduction swelling; [d] determined at room temperature and at a 0.05 mm s^{-1} constant linear rate of compressive stress.

3.3. Mechanical Properties of Bulk Hydrogels

For long-term experiments, the resistance of the hydrogels to the fluid shear force and their ability to maintain their shape were particularly important. Therefore, the mechanical properties of the bulk hydrogels were studied.

The frequency dependence of storage modulus (G', Figure 3a) and loss modulus (G'', Figure 3b) was determined from all hydrogels alongside the loss factor (tan δ, Figure S7a). All hydrogels exhibited weak frequency-dependent viscoelastic behaviors. It was obvious that G' was always larger than G'' in all the hydrogels, which was also reflected in that the tan δ is always much lower than 1, suggesting an elastic and fully cross-linked nature of the hydrogels [62,63]. Since the tan δ of an ideal covalent gel is zero, meaning all network chains sustain the stress, the extremely low tan δ value of the hydrogel indicated that there were few defects in the hydrogels [64,65]. G' generally remained constant, while G'' increased with increasing frequency, which could be attributed to the cleavage of disulfide bonds at high frequencies. With increasing BAC content, the hydrogels possessed more reversible disulfide bonds that could be broken, resulting in higher G''. The latter became more pronounced at lower frequencies for N 4:1 and N 5:1, suggesting that these hydrogels were less feasible for an application in microfluidics.

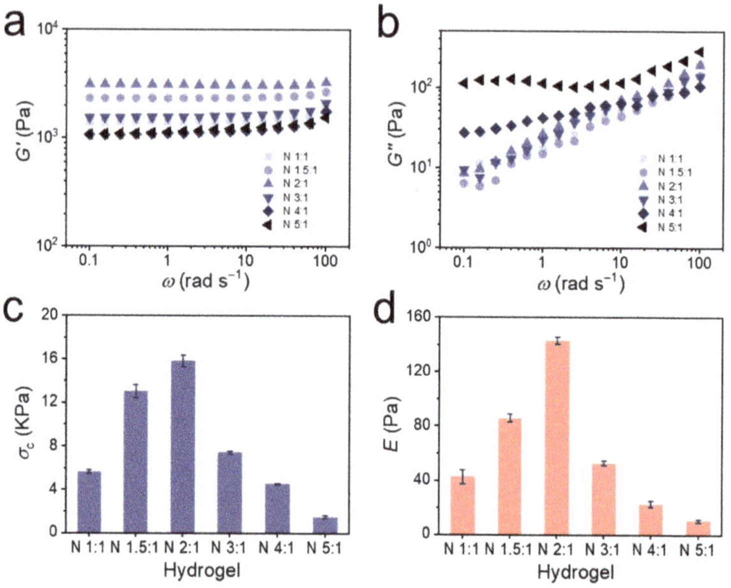

Figure 3. Frequency dependence of (**a**) storage modulus (G') and (**b**) loss modulus (G'') for the bulk PNiPAAm hydrogel (from N 1:1 to N 5:1 hydrogel) at a fixed temperature (T = 25 °C) and strain (γ = 1%). (**c**) Compression strengths at 50% strain and (**d**) elastic moduli (E) of bulk PNiPAAm hydrogels (from N 1:1 to N 5:1 hydrogel). At least three bulk hydrogels were tested for each experimental point to obtain reliable data.

In addition to rheological tests, compression tests also showed the same conclusion. The typical compression stress–strain (σ_c–ε_t, Figure S7b) curves of the bulk hydrogels were used to determine the compression strengths of 50% strain (Figure 3c) and elastic moduli (E, Figure 3d) were calculated from these measurements. All values are also summarized in Table 1. With the increasing BAC ratio, the compressive strength at 50% strain initially increased from 5.66 kPa (N 1:1) to 15.85 kPa (N 2:1), and then dropped over 7.43 kPa (N 3:1) to 1.47 kPa (N 5:1). Similarly, the elastic modulus of hydrogels increased from 43 Pa (N 1:1) to 143 Pa (N 2:1), and then dropped over 53 Pa (N 3:1) to 10 Pa (N 5:1). This behavior indicated that when subjected to external stress, the disulfide bonds broke and effectively dissipated the energy [66,67]. Thus, hydrogels with a higher BAC ratio first showed a higher compressive strength at 50% strain and elastic modulus until N 2:1. However, when the BAC:BIS ratio was higher than 2:1, the reduced amount of permanent cross-linker could not maintain the hydrogel network, and the large amount of reversible cross-linker then accelerated the fracture of the network. As a result, the elastic modulus of the N 3:1, N 4:1, and N 5:1 hydrogel dropped sharply and the hydrogel N 2:1 has proven to be the ideal candidate to be used in microfluidics due to the most suitable mechanical properties.

3.4. Modifications of BSA

In order to be captured by the hydrogels, BSA was modified with RhB for imaging purposes and with PDA to contain disulfide bonds (Figure 4a). PDA containing the disulfide bonds was synthesized following an established protocol (Figure S1) [60,68]. BSA is an economical and easy-to-obtain protein that contains a large amount of carboxyl and amino groups, which is why it was selected for the modification. BSA was first modified with RhB isothiocyanate, which reacted with the amino group of BSA. After that, BSA-RhB was modified by PDA through the condensation reaction of amino and carboxyl groups using EDAC activation [69]. Both modifications were analyzed by MALDI-TOF-MS (Figure 4b). Differences in molecular weights allowed to determine the number of RhB and PDA groups per BSA biomacromolecule. The difference between BSA and BSA-RhB was about 200 Da and since one RhB weighs about 500 DA, an average of 0.4 RhB molecule was bound per protein, i.e., 40% of the BSA biomacromolecules were modified with RhB. Partial modification was supported by the shape of the MALDI-TOF spectrum of BSA-RhB as it closely overlaps with the mass spectrum of pure BSA in the lower range until 66,400 m/z (maximum for BSA), but then departs from the BSA mass spectra in the region beyond the maximum intensity (Figure 4b). Similarly, the difference between BSA-RhB and BSA-RhB-PDA was about 1000 Da and since one PDA weighs about 200 DA, an average of 5 PDA molecules were bound per protein. Both degrees of modification were considered sufficient for good imaging (0.4 eq. RhB) and high binding affinity (5 eq. PDA) of the BSA-RhB-PDA.

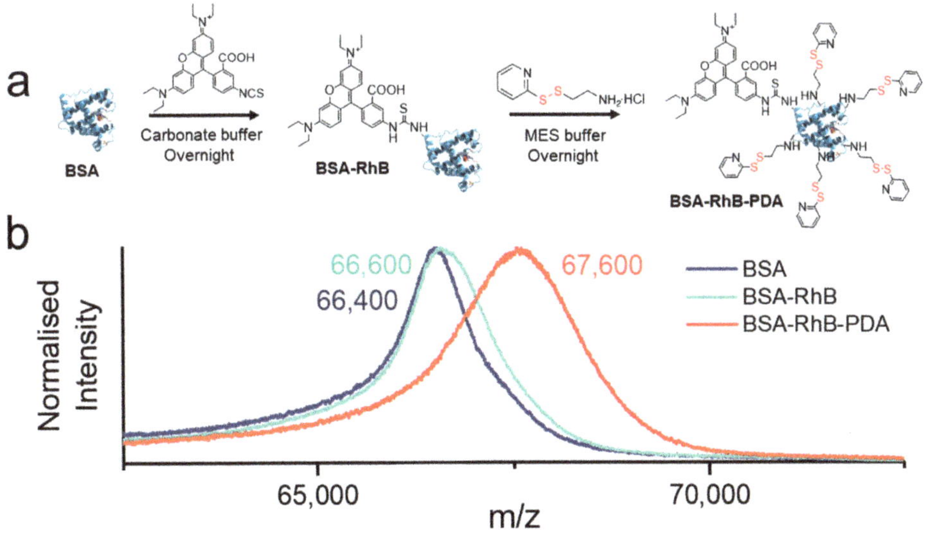

Figure 4. (**a**) Synthetic route of BSA modification with rhodamine B isothiocyanate (RhB) and disulfide bonds (PDA). (**b**) MALDI-TOF spectra in the $[M + X]^+$ (X: Na or H) range of BSA, BSA-RhB, and BSA-RhB-PDA, respectively.

3.5. Protein Capture and Release by Hydrogel Dots in Microfluidics

For microfluidic device, each hydrogel dot array contained 227 hydrogel dots of the N 2:1 gel (Figure 5b). The large number of hydrogel dots greatly increased the contact area between the hydrogel and the solution, ultimately leading to a more efficient reaction setup. The purging sequence of capturing and releasing protein is shown in Figure 5a. It should be noted that a compulsory cleaning step with a washing solution was inserted after all reaction steps to completely remove the reaction solution from the microfluidic chamber (Figure 5a,c, segments II, IV and VI) [9,59]. A low flow rate of 1 µL min^{-1} was chosen for all binding steps, a slightly higher flow rate of 2 µL min^{-1} for all unbinding steps to prevent the formation of bubbles, and the highest flow rate of 5 μL min^{-1} was used for all cleaning steps to cleave all material bound by non-specific interactions. As calculated in Table S2, the residence time of the fluid in the microfluidic chamber was 4.7 min under a flow rate of 5 μL min^{-1}, and 23.6 min when 1 μL min^{-1} was applied. In order to ensure a complete reaction and thorough removal of any cleaved biomacromolecules, a longer perfusion time than the theoretical value (20–40 min for 5 μL min^{-1} and 60–90 min for 1 μL min^{-1}) was used in each process step.

In the protein capture process, 0.01 M TCEP aqueous solution was first perfused at the flow rate of 1 μL min^{-1} for 60 min to break the disulfide bonds, followed by a washing step with deionized water for 20 min at the flow rate of 5 μL min^{-1} (Figure 5a, segments I and II). Afterwards, 50 μM BSA-RhB-PDA aqueous solution was perfused at the flow rate of 1 μL min^{-1} for 90 min to capture the protein, followed by a washing step with 2 mM PBS buffer (pH 7.4) for 40 min at the flow rate of 5 μL min^{-1} (Figure 5a, segments III and IV). This large excess ensured maximum binding of the modified BSA which was not quantified further. Since the protein has a higher solubility and stability in the buffer, the 2 mM PBS buffer (pH 7.4) was used to clean the chip after protein capture. As expected, the presence of BSA-RhB-PDA in the hydrogel dots could be verified with optical and fluorescence microscopy after this procedure (Figure 5b (optical), the entire process diagram is shown in Figure S8, and Figure 5d (fluorescence)). In addition to the protein capture, the swelling of the hydrogel dots was also monitored closely. Following the treatment with aqueous TCEP the hydrogel dots were swelling up to 1.12 times of the original diameter due to the breaking of disulfide bonds. (60 min in Figure 5c, segment I). It should be noted that this is the same value, the N 2:1 dots reached in the previous experiments, highlighting the reproducibility of the hydrogel dot formation process (Table 1). After removing the reducing agent with deionized water, BSA-RhB-PDA aqueous solution was administered to bind the protein to the hydrogel dots. The degree of swelling did not change over the washing and binding step, remaining at 1.13 after 220 min (Figure 5c, segment III). One BSA-RhB-PDA biomacromolecule had one or more binding sites to the hydrogel dots, depending on steric hindrance. These multiple binding sites and reformation of the disulfide bonds did hence not impact the degree of swelling of the hydrogel dots and the cross-linking density of the hydrogel dots remained unchanged.

Two controls were necessary to prove the successful capture of proteins. The first one was the same experiment, but without TCEP to break the disulfide bonds (procedure in Figure S9). As a result, no free thiol groups were available for the exchange with the disulfide bonds on the modified protein. Hence no protein was captured (Figure S10a) and the swelling ratios remained at 1.00 (Figure S11). The second control experiment did see TCEP reduction but then BSA-RhB aqueous solution was administered. Without modification by PDA, the protein did not possess the disulfide bonds required for covalent binding, resulting in a very low amount of captured protein (Figure S10b shows extremely low fluorescence intensity). Together with the controls, these experiments proved the successful thiol-disulfide exchange which enabled the protein BSA-RhB-PDA to be captured by the hydrogel dots.

In the protein release process, 0.01 M TCEP aqueous solution was purged at the flow rate of 2 μL min^{-1} overnight (Figure 5a, segment V). When TCEP was re-purged into the chip, the disulfide bonds between the hydrogel dots and the protein broke, ultimately releasing the protein. The longer time was necessary to dissociate the disulfide bonds and wash off as much BSA-RhB-PDA as possible. As a result, only few BSA-RhB-PDA remained after the release and the following washing step (fluorescence images in Figure 5e). The swelling ratio of hydrogel dots after protein release was 1.14 (after 1150 min in Figure 5c, segment V). Since there was no significant change in cross-linking density, the swelling ratio of the hydrogel dots remained almost constant (slight increase from 1.12 to 1.14) after the first purging with TCEP aqueous solution at the end of protein release (Figure 5c, segments II-VI). After cleaning with deionized water, the microfluidic chip was treated with 0.1M $FeCl_3$ aqueous solution to re-oxidize thiol groups and reform disulfide bonds in the hydrogel dots (Figure 5a,c, segments VI and VII). With the increasing cross-linking densities, the swelling ratio of the hydrogel dots decreased to 1.07 (after 1180 min in Figure 5c, segment VII). Conformational changes in the hydrogel dots after breaking the disulfide bonds presumably prevented a complete shrinking to the original size.

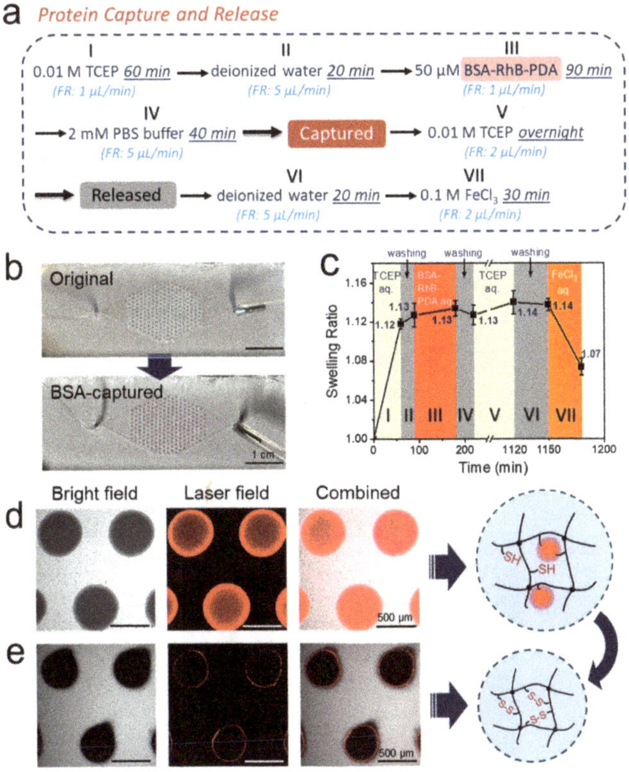

Figure 5. (**a**) Description of one measurement cycle to capture and release BSA-RhB-PDA, (**b**) Optical images of the original and BSA-captured hydrogel dot array in the microfluidic chip. (**c**) Swelling behavior of PNiPAAm hydrogel dots in one complete measurement cycle (**d**) Micrographs of the PNiPAAm hydrogel dots after BSA-RhB-PDA has been captured, (**e**) Micrographs of the PNiPAAm hydrogel dots after completing one whole cycle (i.e., after releasing the previously captured BSA-RhB-PDA of (**b**)). Both cartoons (**d**,**e**) refer to the composition of each hydrogel dots at the time of the measurements. All the hydrogel dots studies were performed in microfluidic chip at room temperature and observed by confocal laser fluorescence microscope under bright field, laser field and the fields combined at the height of 80 μm (channel height: 150 μm).

Release profile of BSA biomacromolecules was investigated next and was followed by confocal laser fluorescence microscopy (Figure 6a). The release ratio over time (Figure 6b) was analyzed from the remaining fluorescence intensity (Figure S12). A release of 10% hence corresponded to 90% remaining fluorescence. It could be noted that the fluorescence intensity increased initially within 10 min and then dropped dramatically, which was attributed to the short increase in background noise generated by the release of the protein and the enrichment in the solution. After 3 h of release (after 180 min in Figure 6b), 74.5% of the protein had been released from the hydrogel dots. The number increased to 83.6% after 15 h (after 900 min in Figure 6b), finally reaching 84.0% after 21 h (after 1260 min in Figure 6b). Thus, the release time of the protein was fixed at 15 h in the following cyclic tests due to the very low increase in release after 15 h.

Cyclic tests of protein capture and release have been performed following the complete characterization of the first cycle. Each cycle went through the seven steps of hydrogel dots swelling, protein capture, protein release, and hydrogel dots shrinking, including all washing steps. The release ratio of protein decreased from 70.1% in the second cycle to 59.0% in the third cycle (Figure 6c). Essentially, each step resulted in a release rate of about 85% of the previous value leading to a release rate of 61% after three cycles (as 0.85^3 = 61%), which is close to the observed 59.0%. This decay was likely to continue over possible next cycles, but we did not address this point. A repeated capture and release of the device was hence possible, but declined in quality with each cycle. Swelling ratios of the hydrogel dots in cyclic tests were measured as well (Figure 6d). Every cycle consists of two data points, beginning from the first purging of 0.01 M TCEP aqueous solution and ending after the purging of 0.1 M $FeCl_3$ aqueous solution. The swelling ratios of hydrogel dots under TCEP reduction decreased marginally from 1.13 to 1.10, while swelling ratios of hydrogel dots under $FeCl_3$ oxidation were 1.07, 1.03, and 1.05 in the three cycles, following no pattern. It could be concluded that the swelling and shrinking of the hydrogel dots in cyclic tests showed no clear trend, lacking complete reproducibility of the first cycle. Combined with the reduced protein release ratio, the hydrogel dots exhibited a decreasing response to redox stimuli under the long-term shearing stress. However, the cyclic tests still confirmed that it was possible to capture and release protein through the disulfide bonds on the hydrogel dots in microfluidics for at least three cycles. This thiol-based capture and release of proteins in microfluidics through hydrogel dots over at least three cycles showed that our previously reported redox responsive double cross-linked hydrogels are not only an interesting concept, but are also fit for applications.

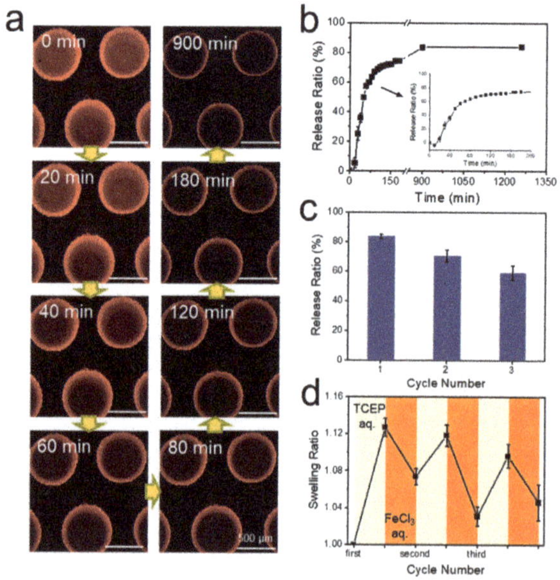

Figure 6. (**a**) Development of the fluorescence images of the PNiPAAm hydrogel dots over time in the microfluidic chip, during the release of BSA-RhB-PDA (image height set to 80 μm). (**b**) Release profile of BSA over time derived from the fluorescence images of Figure 6a, analyzed by ImageJ. (**c**) Release ratios of BSA captured by PNiPAAm hydrogel over three cycles. (**d**) Swelling ratios of the PNiPAAm hydrogel dots in each capture and release cycle. Every cycle consists of two data points, representing the highest swelling with broken disulfide bonds after treatment with 0.01 M TCEP and the reduced swelling after reforming these bonds with 0.1 M $FeCl_3$. For (**b**–**d**) at least three hydrogel dots were tested for each experimental point to obtain reliable data.

4. Conclusions

In summary, the PNiPAAm hydrogel dot arrays were successfully integrated into the PDMS-on-glass microfluidic device through photopolymerization. The hydrogel dot array inside the microfluidics was cross-linked by the permanent cross-linker BIS and the reversible cross-linker BAC. Cleavage or re-formation of redox-responsive disulfide bonds introduced by cross-linker BAC changed the cross-linking densities of the hydrogel dots, making them swell or shrink under redox conditions. Following rheological and compression measurements, hydrogel dots with a 2:1 ratio of BAC to BIS proved to be the mechanically most stable hydrogel dots with an appropriate degree of swelling. The thiol-disulfide exchange allowed the hydrogel dots to successfully capture and release the protein BSA modified by PDA containing disulfide bonds and dye RhB. The release ratio of protein reached 83.6% in the first cycle and proved to be reproducible on the same chip, reaching a release ratio of 59.0% after the third cycle. This selective capture and release of proteins on a microscopic scale through the redox-responsive hydrogel dots bares the advantages of minimal amount of sample and successful reusability. Thus, it has great potential for future applications as it opens up the possibility of capturing and releasing various, differently functionalized proteins, enzymes or drugs. The reported process has all prerequisites to become a lab-on-a-chip device for rapid detection and/or delivery of various (macro)molecules.

Supplementary Materials: The following are available online at https://www.mdpi.com/article/10.3390/polym14020267/s1. Table S1: Compositions for the synthesis of PNiPAAm hydrogels cross-linked by BAC and BIS. Table S2: Parameters and formulas applied for the calculation of the residence time of the substrates in the microfluidic device. Figure S1: Synthetic route and ^{1}H NMR of PDA incl. the peak assignment. Figure S2: (a) Photomask for structuring hydrogel dots. (b) Top view of master for production of single-chamber PDMS sheet. (c) Top view (photograph) and side view (schematic) of POM mold for production of hydrogel dots. Figure S3: Hydrogel arrays prepared by photopolymerization of monomer solutions in mould with single chamber. Figure S4: Photos of the entire microfluidic setup. Figure S5: Appearances of the original, the reduced and the oxidized N 5:1 hydrogel dots. Figure S6: Swelling behaviors of PNiPAAm hydrogel dots with different mole ratios of cross-linker BAC to BIS in the microfluidic chip at different flow rates. Figure S7: (a) Frequency dependence of the loss factor (tan δ) for the bulk PNiPAAm hydrogels. (b) Typical compression stress–strain curves of bulk PNiPAAm hydrogels. Figure S8: Photos of the process of capturing BSA in the microfluidic chip. Figure S9: The experimental pro-cedure of both controls. Figure S10: Micrographs of the hydrogel dots in two control tests. Figure S11: (a) Optical images of the hydrogel dots in the first control test and (b) the swelling ratios of them. Figure S12: Fluorescence intensity of the PNiPAAm hydrogel dots over time in the BSA re-leasing procedure.

Author Contributions: C.J., F.O., J.G. and D.A. designed the experiments; C.J., F.O., M.G. and Y.C. performed the experiments; C.J., M.G. and J.G. performed the data analysis; C.J., J.G., D.A., A.R. and B.V. wrote the manuscript. The paper was written through contributions of all the authors. All authors have read and agreed to the published version of the manuscript.

Funding: This research received no external funding.

Institutional Review Board Statement: Not applicable.

Informed Consent Statement: Not applicable.

Data Availability Statement: The data presented in this study are available on request from the corresponding author.

Acknowledgments: C. Jiao gratefully acknowledge the financial support by China Scholarship Council (CSC). We thank the workshop of Leibniz Institute of Polymer Research (IPF) for the production of POM molds. We thank Mohammed Hadi Shahadha (Technical University of Dresden, Germany) for manufacturing microfluidic chip molds and Gerald Hielscher (Technical University of Dresden, Dresden, Germany) for manufacturing photomasks.

Conflicts of Interest: The authors declare no conflict of interest.

References

1. Whitesides, G.M. The origins and the future of microfluidics. *Nature* **2006**, *442*, 368–373. [CrossRef] [PubMed]
2. Liu, Y.; Sun, L.; Zhang, H.; Shang, L.; Zhao, Y. Microfluidics for drug development: From synthesis to evaluation. *Chem. Rev.* **2021**, *121*, 7468–7529. [CrossRef]
3. Shang, L.; Cheng, Y.; Zhao, Y. Emerging droplet microfluidics. *Chem. Rev.* **2017**, *117*, 7964–8040. [CrossRef]
4. Li, W.; Zhang, L.; Ge, X.; Xu, B.; Zhang, W.; Qu, L.; Choi, C.H.; Xu, J.; Zhang, A.; Lee, H.; et al. Microfluidic fabrication of microparticles for biomedical applications. *Chem. Soc. Rev.* **2018**, *47*, 5646–5683. [CrossRef]
5. Choi, T.M.; Lee, G.H.; Kim, Y.S.; Park, J.G.; Hwang, H.; Kim, S.H. Photonic microcapsules containing single-crystal colloidal arrays with optical anisotropy. *Adv. Mater.* **2019**, *31*, e1900693. [CrossRef] [PubMed]
6. Wang, H.; Liu, Y.; Chen, Z.; Sun, L.; Zhao, Y. Anisotropic structural color particles from colloidal phase separation. *Sci. Adv.* **2020**, *6*, eaay1438. [CrossRef] [PubMed]
7. Kieviet, B.D.; Schon, P.M.; Vancso, G.J. Stimulus-responsive polymers and other functional polymer surfaces as components in glass microfluidic channels. *Lab Chip* **2014**, *14*, 4159–4170. [CrossRef]
8. Nge, P.N.; Rogers, C.I.; Woolley, A.T. Advances in microfluidic materials, functions, integration, and applications. *Chem. Rev.* **2013**, *113*, 2550–2583. [CrossRef] [PubMed]
9. Obst, F.; Beck, A.; Bishayee, C.; Mehner, P.J.; Richter, A.; Voit, B.; Appelhans, D. Hydrogel microvalves as control elements for parallelized enzymatic cascade reactions in microfluidics. *Micromachines* **2020**, *11*, 167. [CrossRef]
10. Obst, F.; Simon, D.; Mehner, P.J.; Neubauer, J.W.; Beck, A.; Stroyuk, O.; Richter, A.; Voit, B.; Appelhans, D. One-step photostructuring of multiple hydrogel arrays for compartmentalized enzyme reactions in microfluidic devices. *React. Chem. Eng.* **2019**, *4*, 2141–2155. [CrossRef]
11. Obst, F.; Mertz, M.; Mehner, P.J.; Beck, A.; Castiglione, K.; Richter, A.; Voit, B.; Appelhans, D. Enzymatic synthesis of sialic acids in microfluidics to overcome cross-inhibitions and substrate supply limitations. *ACS Appl. Mater. Interfaces* **2021**, *13*, 49433–49444. [CrossRef] [PubMed]
12. Shastri, A.; McGregor, L.M.; Liu, Y.; Harris, V.; Nan, H.; Mujica, M.; Vasquez, Y.; Bhattacharya, A.; Ma, Y.; Aizenberg, M.; et al. An aptamer-functionalized chemomechanically modulated biomolecule catch-and-release system. *Nat. Chem.* **2015**, *7*, 447–454. [CrossRef] [PubMed]
13. Sheng, W.; Chen, T.; Kamath, R.; Xiong, X.; Tan, W.; Fan, Z.H. Aptamer-enabled efficient isolation of cancer cells from whole blood using a microfluidic device. *Anal. Chem.* **2012**, *84*, 4199–4206. [CrossRef] [PubMed]
14. Adams, A.A.; Okagbare, P.I.; Feng, J.; Hupert, M.L.; Patterson, D.; Gottert, J.; McCarley, R.L.; Nikitopoulos, D.; Murphy, M.C.; Soper, S.A. Highly efficient circulating tumor cell isolation from whole blood and label-free enumeration using polymer-based microfluidics with an integrated conductivity sensor. *J. Am. Chem. Soc.* **2008**, *130*, 8633–8641. [CrossRef] [PubMed]
15. Sarioglu, A.F.; Aceto, N.; Kojic, N.; Donaldson, M.C.; Zeinali, M.; Hamza, B.; Engstrom, A.; Zhu, H.; Sundaresan, T.K.; Miyamoto, D.T.; et al. A microfluidic device for label-free, physical capture of circulating tumor cell clusters. *Nat. Methods* **2015**, *12*, 685–691. [CrossRef] [PubMed]
16. Shah, A.M.; Yu, M.; Nakamura, Z.; Ciciliano, J.; Ulman, M.; Kotz, K.; Stott, S.L.; Maheswaran, S.; Haber, D.A.; Toner, M. Biopolymer system for cell recovery from microfluidic cell capture devices. *Anal. Chem.* **2012**, *84*, 3682–3688. [CrossRef]
17. Hatch, A.; Hansmann, G.; Murthy, S.K. Engineered alginate hydrogels for effective microfluidic capture and release of endothelial progenitor cells from whole blood. *Langmuir* **2011**, *27*, 4257–4264. [CrossRef] [PubMed]
18. Hu, Z.; Zhang, X.; Li, Y. Synthesis and application of modulated polymer gels. *Science* **1995**, *269*, 525–527. [CrossRef] [PubMed]
19. Suzuki, A.; Tanaka, T. Phase transition in polymer gels induced by visible light. *Nature* **1990**, *346*, 345–347. [CrossRef]
20. Miyamae, K.; Nakahata, M.; Takashima, Y.; Harada, A. Self-healing, expansion-contraction, and shape-memory properties of a preorganized supramolecular hydrogel through host-guest interactions. *Angew. Chem. Int. Ed. Engl.* **2015**, *54*, 8984–8987. [CrossRef]
21. Lin, C.Y.; Battistoni, C.M.; Liu, J.C. Redox-responsive hydrogels with decoupled initial stiffness and degradation. *Biomacromolecules* **2021**, *22*, 5270–5280. [CrossRef] [PubMed]
22. Ni, H.; Kawaguchi, H.; Endo, T. Characteristics of pH-sensitive hydrogel microsphere of poly(acrylamide-*co*-methacrylic acid) with sharp pH–volume transition. *Colloid Polym. Sci.* **2007**, *285*, 873–879. [CrossRef]
23. Grafe, D.; Frank, P.; Erdmann, T.; Richter, A.; Appelhans, D.; Voit, B. Tetra-sensitive graft copolymer gels as active material of chemomechanical valves. *ACS Appl. Mater. Interfaces* **2017**, *9*, 7565–7576. [CrossRef] [PubMed]
24. Jiao, C.; Zhang, J.; Liu, T.; Peng, X.; Wang, H. Mechanically strong, tough, and shape deformable poly(acrylamide-*co*-vinylimidazole) hydrogels based on Cu^{2+} complexation. *ACS Appl. Mater. Interfaces* **2020**, *12*, 44205–44214. [CrossRef] [PubMed]
25. Chimisso, V.; Conti, S.; Kong, P.; Fodor, C.; Meier, W.P. Metal cation responsive anionic microgels: Behaviour towards biologically relevant divalent and trivalent ions. *Soft Matter* **2021**, *17*, 715–723. [CrossRef]
26. Tanaka, T.; Nishio, I.; Sun, S.T.; Ueno-Nishio, S. Collapse of gels in an electric field. *Science* **1982**, *218*, 467–469. [CrossRef] [PubMed]
27. Li, L.; Wang, Y.; Pan, L.; Shi, Y.; Cheng, W.; Shi, Y.; Yu, G. A nanostructured conductive hydrogels-based biosensor platform for human metabolite detection. *Nano Lett.* **2015**, *15*, 1146–1151. [CrossRef] [PubMed]

28. Li, J.; Ji, C.; Lu, B.; Rodin, M.; Paradies, J.; Yin, M.; Kuckling, D. Dually crosslinked supramolecular hydrogel for cancer biomarker sensing. *ACS Appl. Mater. Interfaces* **2020**, *12*, 36873–36881. [CrossRef] [PubMed]
29. Beebe, D.J.; Moore, J.S.; Bauer, J.M.; Yu, Q.; Liu, R.H.; Devadoss, C.; Jo, B.H. Functional hydrogel structures for autonomous flow control inside microfluidic channels. *Nature* **2000**, *404*, 588–590. [CrossRef] [PubMed]
30. Drury, J.L.; Mooney, D.J. Hydrogels for tissue engineering: Scaffold design variables and applications. *Biomaterials* **2003**, *24*, 4337–4351. [CrossRef]
31. Caliari, S.R.; Burdick, J.A. A practical guide to hydrogels for cell culture. *Nat. Methods* **2016**, *13*, 405–414. [CrossRef]
32. Vermonden, T.; Censi, R.; Hennink, W.E. Hydrogels for protein delivery. *Chem. Rev.* **2012**, *112*, 2853–2888. [CrossRef]
33. Hamidi, M.; Azadi, A.; Rafiei, P. Hydrogel nanoparticles in drug delivery. *Adv. Drug Deliv. Rev.* **2008**, *60*, 1638–1649. [CrossRef] [PubMed]
34. Elkassih, S.A.; Kos, P.; Xiong, H.; Siegwart, D.J. Degradable redox-responsive disulfide-based nanogel drug carriers via dithiol oxidation polymerization. *Biomater. Sci.* **2019**, *7*, 607–617. [CrossRef] [PubMed]
35. Richter, A.; Kuckling, D.; Howitz, S.; Gehring, T.; Arndt, K. Electronically controllable microvalves based on smart hydrogels: Magnitudes and potential applications. *J. Microelectromech. Syst.* **2003**, *12*, 748–753. [CrossRef]
36. Richter, A.; Howitz, S.; Kuckling, D.; Arndt, K.F. Influence of volume phase transition phenomena on the behavior of hydrogel-based valves. *Sens. Actuators B Chem.* **2004**, *99*, 451–458. [CrossRef]
37. Lee, E.; Lee, H.; Yoo, S.I.; Yoon, J. Photothermally triggered fast responding hydrogels incorporating a hydrophobic moiety for light-controlled microvalves. *ACS Appl. Mater. Interfaces* **2014**, *6*, 16949–16955. [CrossRef] [PubMed]
38. Richter, A.; Klatt, S.; Paschew, G.; Klenke, C. Micropumps operated by swelling and shrinking of temperature-sensitive hydrogels. *Lab Chip* **2009**, *9*, 613–618. [CrossRef] [PubMed]
39. Wang, M.; Xiao, Y.; Lin, L.; Zhu, X.; Du, L.; Shi, X. A microfluidic chip integrated with hyaluronic acid-functionalized electrospun chitosan nanofibers for specific capture and nondestructive release of CD44-overexpressing circulating tumor cells. *Bioconjug. Chem.* **2018**, *29*, 1081–1090. [CrossRef] [PubMed]
40. Lee, M.H.; Yang, Z.; Lim, C.W.; Lee, Y.H.; Dongbang, S.; Kang, C.; Kim, J.S. Disulfide-cleavage-triggered chemosensors and their biological applications. *Chem. Rev.* **2013**, *113*, 5071–5109. [CrossRef]
41. Roy, D.; Cambre, J.N.; Sumerlin, B.S. Future perspectives and recent advances in stimuli-responsive materials. *Prog. Polym. Sci.* **2010**, *35*, 278–301. [CrossRef]
42. Winther, J.R.; Thorpe, C. Quantification of thiols and disulfides. *Biochim. Biophys. Acta Gen. Subj.* **2014**, *1840*, 838–846. [CrossRef] [PubMed]
43. Bach, R.D.; Dmitrenko, O.; Thorpe, C. Mechanism of thiolate-disulfide interchange reactions in biochemistry. *J. Org. Chem.* **2008**, *73*, 12–21. [CrossRef] [PubMed]
44. Houk, J.; Whitesides, G.M. Structure-reactivity relations for thiol-disulfide interchange. *J. Am. Chem. Soc.* **2002**, *109*, 6825–6836. [CrossRef]
45. Pires, M.M.; Chmielewski, J. Fluorescence imaging of cellular glutathione using a latent rhodamine. *Org. Lett.* **2008**, *10*, 837–840. [CrossRef]
46. Xiao, Y.; Wang, M.; Lin, L.; Du, L.; Shen, M.; Shi, X. Integration of aligned polymer nanofibers within a microfluidic chip for efficient capture and rapid release of circulating tumor cells. *Mater. Chem. Front.* **2018**, *2*, 891–900. [CrossRef]
47. Chong, S.F.; Chandrawati, R.; Stadler, B.; Park, J.; Cho, J.; Wang, Y.; Jia, Z.; Bulmus, V.; Davis, T.P.; Zelikin, A.N.; et al. Stabilization of polymer-hydrogel capsules via thiol-disulfide exchange. *Small* **2009**, *5*, 2601–2610. [CrossRef]
48. An, S.Y.; Noh, S.M.; Oh, J.K. Multiblock copolymer-based dual dynamic disulfide and supramolecular crosslinked self-healing networks. *Macromol. Rapid Commun.* **2017**, *38*, 1600777. [CrossRef] [PubMed]
49. Yang, Y.; Gao, G. A stimuli-responsive hydrogel with reversible three-state transition controlled by redox stimulation. *Macromol. Chem. Phys.* **2017**, *218*, 1700002. [CrossRef]
50. Deng, G.; Li, F.; Yu, H.; Liu, F.; Liu, C.; Sun, W.; Jiang, H.; Chen, Y. Dynamic hydrogels with an environmental adaptive self-healing ability and dual responsive sol-gel transitions. *ACS Macro Lett.* **2012**, *1*, 275–279. [CrossRef]
51. Lee, W.F.; Lu, Y.Y. Influence of novel crosslinker on the properties of the degradable thermosensitive hydrogels. *Macromol. Symp.* **2015**, *358*, 41–51. [CrossRef]
52. Gaulding, J.C.; Smith, M.H.; Hyatt, J.S.; Fernandez-Nieves, A.; Lyon, L.A. Reversible inter- and intra-microgel cross-linking using disulfides. *Macromolecules* **2012**, *45*, 39–45. [CrossRef] [PubMed]
53. Gkikas, M.; Avery, R.K.; Mills, C.E.; Nagarajan, R.; Wilusz, E.; Olsen, B.D. Hydrogels that actuate selectively in response to organophosphates. *Adv. Funct. Mater.* **2016**, *27*, 1602784. [CrossRef]
54. Che, Y.; Zschoche, S.; Obst, F.; Appelhans, D.; Voit, B. Double-crosslinked reversible redox-responsive hydrogels based on disulfide–thiol interchange. *J. Polym. Sci. Part A Polym. Chem.* **2019**, *57*, 2590–2601. [CrossRef]
55. Eddington, D.T.; Beebe, D.J. Flow control with hydrogels. *Adv. Drug Deliv. Rev.* **2004**, *56*, 199–210. [CrossRef] [PubMed]
56. Dong, L.; Jiang, H. Autonomous microfluidics with stimuli-responsive hydrogels. *Soft Matter* **2007**, *3*, 1223–1230. [CrossRef] [PubMed]
57. Oh, K.W.; Ahn, C.H. A review of microvalves. *J. Micromech. Microeng.* **2006**, *16*, R13–R39. [CrossRef]
58. Mirhosseini Moghaddam, M.; Baghbanzadeh, M.; Sadeghpour, A.; Glatter, O.; Kappe, C.O. Continuous-flow synthesis of CdSe quantum dots: A size-tunable and scalable approach. *Chem. Eur. J.* **2013**, *19*, 11629–11636. [CrossRef]

59. Simon, D.; Obst, F.; Haefner, S.; Heroldt, T.; Peiter, M.; Simon, F.; Richter, A.; Voit, B.; Appelhans, D. Hydrogel/enzyme dots as adaptable tool for non-compartmentalized multi-enzymatic reactions in microfluidic devices. *Reat. Chem. Eng.* **2019**, *4*, 67–77. [CrossRef]
60. Battistella, C.; Klok, H.A. Reversion of P-gp-mediated drug resistance in ovarian carcinoma cells with PHPMA-zosuquidar conjugates. *Biomacromolecules* **2017**, *18*, 1855–1865. [CrossRef]
61. Hanton, S.D.; Owens, K.G. Polymer MALDI sample preparation. In *Mass Spectrometry in Polymer Chemistry*; Wily: Weinheim, Germany, 2012; pp. 119–147.
62. Jiao, C.; Chen, Y.; Liu, T.; Peng, X.; Zhao, Y.; Zhang, J.; Wu, Y.; Wang, H. Rigid and strong thermoresponsive shape memory hydrogels transformed from poly(vinylpyrrolidone-*co*-acryloxy acetophenone) organogels. *ACS Appl. Mater. Interfaces* **2018**, *10*, 32707–32716. [CrossRef]
63. Van Den Bulcke, A.I.; Bogdanov, B.; De Rooze, N.; Schacht, E.H.; Cornelissen, M.; Berghmans, H. Structural and rheological properties of methacrylamide modified gelatin hydrogels. *Biomacromolecules* **2000**, *1*, 31–38. [CrossRef] [PubMed]
64. Rubinstein, M.; Colby, R.H. *Polymer Physics*; Oxford University Press: New York, NY, USA, 2003; Volume 23.
65. Hao, J.; Weiss, R.A. Viscoelastic and mechanical behavior of hydrophobically modified hydrogels. *Macromolecules* **2011**, *44*, 9390–9398. [CrossRef]
66. Chen, T.; Chen, Y.; Rehman, H.U.; Chen, Z.; Yang, Z.; Wang, M.; Li, H.; Liu, H. Ultratough, self-healing, and tissue-adhesive hydrogel for wound dressing. *ACS Appl. Mater. Interfaces* **2018**, *10*, 33523–33531. [CrossRef] [PubMed]
67. Cao, J.; Zhao, X.; Ye, L. Facile method to fabricate superstrong and tough poly(vinyl alcohol) hydrogels with high energy dissipation. *Ind. Eng. Chem. Res.* **2020**, *59*, 10705–10715. [CrossRef]
68. Zugates, G.T.; Anderson, D.G.; Little, S.R.; Lawhorn, I.E.; Langer, R. Synthesis of poly(β-amino ester)s with thiol-reactive side chains for DNA delivery. *J. Am. Chem. Soc.* **2006**, *128*, 12726–12734. [CrossRef] [PubMed]
69. Huang, X.; Li, M.; Green, D.C.; Williams, D.S.; Patil, A.J.; Mann, S. Interfacial assembly of protein-polymer nano-conjugates into stimulus-responsive biomimetic protocells. *Nat. Commun.* **2013**, *4*, 2239. [CrossRef]

polymers
MDPI

Article

Copolymer Involving 2-Hydroxyethyl Methacrylate and 2-Chloroquinyl Methacrylate: Synthesis, Characterization and In Vitro 2-Hydroxychloroquine Delivery Application

Abeer Aljubailah [1], Wafa Nazzal Odis Alharbi [1], Ahmed S. Haidyrah [2], Tahani Saad Al-Garni [3], Waseem Sharaf Saeed [4,*], Abdelhabib Semlali [5], Saad M. S. Alqahtani [3], Ahmad Abdulaziz Al-Owais [3], Abdulnasser Mahmoud Karami [3] and Taieb Aouak [3,*]

1 Department of Chemistry, College of Science, Imam Mohammad Ibn Saud Islamic University (IMSIU), Riyadh 13623, Saudi Arabia; akaljubailah@imamu.edu.sa (A.A.); wnalharbi@imamu.edu.sa (W.N.O.A.)
2 Nuclear and Radiological Control Unit, King Abdulaziz City for Science and Technology (KACST), Riyadh 11442, Saudi Arabia; ahydrah@kacst.edu.sa
3 Chemistry Department, College of Science, King Saud University, Riyadh 11451, Saudi Arabia; tahanis@ksu.edu.sa (T.S.A.-G.); salqahtani2@ksu.edu.sa (S.M.S.A.); aowais@ksu.edu.sa (A.A.A.-O.); akarami@ksu.edu.sa (A.M.K.)
4 Engineer Abdullah Bugshan Research Chair for Dental and Oral Rehabilitation, College of Dentistry, King Saud University, Riyadh 11545, Saudi Arabia
5 Groupe de Recherche en Écologie Buccale, Faculté de Médecin Dentaire, Université Laval, Quebec City, QC G1V 0A6, Canada; abdelhabib.semlali@greb.ulaval.ca
* Correspondence: wsaeed@ksu.edu.sa (W.S.S.); taouak@ksu.edu.sa (T.A.)

Citation: Aljubailah, A.; Alharbi, W.N.O.; Haidyrah, A.S.; Al-Garni, T.S.; Saeed, W.S.; Semlali, A.; Alqahtani, S.M.S.; Al-Owais, A.A.; Karami, A.M.; Aouak, T. Copolymer Involving 2-Hydroxyethyl Methacrylate and 2-Chloroquinyl Methacrylate: Synthesis, Characterization and In Vitro 2-Hydroxychloroquine Delivery Application. *Polymers* **2021**, *13*, 4072. https://doi.org/10.3390/polym13234072

Academic Editors: Marek Kowalczuk, Iza Radecka and Barbara Trzebicka

Received: 14 October 2021
Accepted: 8 November 2021
Published: 23 November 2021

Abstract: The Poly(2-chloroquinyl methacrylate-*co*-2-hydroxyethyl methacrylate) (CQMA-*co*-HEMA) drug carrier system was prepared with different compositions through a free-radical copolymerization route involving 2-chloroquinyl methacrylate (CQMA) and 2-hydroxyethyl methacrylate) (HEMA) using azobisisobutyronitrile as the initiator. 2-Chloroquinyl methacrylate monomer (CQMA) was synthesized from 2-hydroxychloroquine (HCQ) and methacryloyl chloride by an esterification reaction using triethylenetetramine as the catalyst. The structure of the CQMA and CQMA-*co*-HEMA copolymers was confirmed by a CHN elementary analysis, Fourier transform infra-red (FTIR) and nuclear magnetic resonance (NMR) analysis. The absence of residual aggregates of HCQ or HCQMA particles in the copolymers prepared was confirmed by a differential scanning calorimeter (DSC) and XR-diffraction (XRD) analyses. The gingival epithelial cancer cell line (Ca9-22) toxicity examined by a lactate dehydrogenase (LDH) assay revealed that the grafting of HCQ onto PHEMA slightly affected (4.2–9.5%) the viability of the polymer carrier. The cell adhesion and growth on the CQMA-*co*-HEMA drug carrier specimens carried out by the (3-(4,5-dimethylthiazol-2-yl)-2,5-diphenyltetrazolium bromide) (MTT) assay revealed the best performance with the specimen containing 3.96 wt% HCQ. The diffusion of HCQ through the polymer matrix obeyed the Fickian model. The solubility of HCQ in different media was improved, in which more than 5.22 times of the solubility of HCQ powder in water was obtained. According to Belzer, the in vitro HCQ dynamic release revealed the best performance with the drug carrier system containing 4.70 wt% CQMA.

Keywords: preparation and characterization; poly(2-chloroquinyl methacrylate-*co*-2-hydroxyethyl methacrylate); 2-chloroquinyl methacrylate; drug carrier system; in vitro 2-hydroxychloroquine release

1. Introduction

2-Hydroxychloroquine (HCQ), also called "Plaquenil", is a weak basic drug belonging to the family of 4-aminoquinolines. This medication is an antimalarial widely used in the treatment of systemic lupus erythematosus, rheumatoid arthritis, malaria and other autoimmune diseases [1,2]. It is also recommended to take 2-hydroxychloroquine with food to reduce stomach irritation. According to Hedya et al. [3], the addition of HCQ to cytotoxic or antiangiogenic agents can dramatically improve antitumor activity. Many tests [4,5] combining different anticancer therapies, including chemo and radiation therapies with HCQ, have shown very satisfactory results. This drug is mainly found in a dicationic form in physiological pH media and is readily trapped in cellular tissues resulting in a tissue deposition effect. According to a report published in 2005 by Day et al. [6], the distribution of HCQ in the bloodstream is relatively slow, which delays the onset of an antirheumatic effect. Therefore, a high dose of HCQ is needed to reach a steady state more quickly. According to Ono et al. [7], a dose of 400 to 600 mg is usually prescribed for normal men, while 200 to 400 mg is the maintenance dose. A higher dose results in a greater magnitude of dose-dependent side effects such as retinopathy, which is mainly due to a build-up of threads in the cornea [8]. Antirheumatic drugs, of which HCQ is a part of, can cause serious gastrointestinal complications that become more acute at higher doses. As a result, symptomatic treatments with glucocorticoids and non-steroidal antirheumatic drugs (NSAIDs) are known to induce gastric or duodenal ulcers, especially in association with combination therapy [9]. The control of the HCQ amount released during gastrointestinal transit (GIT) in the different organs is necessary in order to minimize the release of HCQ in organs sensitive to unwanted side effects and, more particularly, in the stomach.

In the absence of an established treatment regimen, many drug reuse strategies have been emerging to treat corona virus disease (COVID-19) [10]. Indeed, among these drugs, HCQ in combination with azithromycin has been recommended to treat this virus, especially in older patients or patients with underlying conditions and severe symptoms [11,12]. Despite the promise of the reuse of this drug, there are concerns about its toxicity, because some in vitro studies suggest that the dose needed to be effective against COVID-19 may be higher than that used in malaria. The World Health Organization (WHO) is currently conducting clinical tests (SOLIDARITY) to assess the effectiveness of 2-hydroxychloroquine as a treatment for COVID-19, while some countries have already included treatment with 2-hydroxychloroquine in their clinical advice for patients with COVID-19 [13,14]. In addition, the HCQ base has a partition coefficient ranged between 2.89 and 3.87 and a water solubility of 26.1 $mg \cdot L^{-1}$ [15]. These two properties place this drug in the Biopharmaceutical Classification System-II (BCS-II); in other words; this medication is highly permeable but poorly soluble. In its sodium sulfate form, HCQ has excellent bioavailability with an average fraction of 0.74 absorbed doses [16]. On the other hand, this salt is highly soluble in gastric fluid but can recrystallize in the environment at a higher pH (small intestine), because the mother base is highly insoluble in this medium. In addition, an increase in the pH of the stomach in the postprandial state can also present a solubility challenge. In both cases, a decrease in the solubility of the drug is conceivable, which results in a lower in vivo exposure and, as such, an approach allowing the solubility can overcome this variation in solubility during gastrointestinal transit (GIT). In particular, and as recently described in the literature by Zhang et al., salt forms of drugs can experience a significant decrease in solubility upon transition to the higher pH of the small intestine [17].

Poly(2-hydroxyethyl methacrylate) (PHEMA) has exhibited very interesting properties in its application in the biomedical field due to its high water content, its non-toxicity and its favorable biocompatibility. This polymer is easily synthesized by free-radical polymerization involving 2-hydroethyl methacrylate as the monomer and azobisisobutyronitile (AIBN) as the initiator. The ease of handling by the formulation chemistry has allowed this polymer a wide application in the biomedical field, such as contact lenses [18,19], keratoprostheses and as orbital implants [20]. The presence of the hydroxyl and carboxyl groups on the substituent of this polymer give it compatibility with water, a hydrolytic stability, support for mechanical resistance [21] and a better adhesion of cells [22,23].

In this work, to increase the solubility of the HCQ base in pH-neutral media (intestines) and reduce the amount of HCQ released in acidic media (stomach), a new monomer (CQMA) involving HCQ and methacryloyl chloride was synthesized via a catalytic esterification reaction. The new monomer obtained was copolymerized at different ratios with 2-hydroxyethylmethacrylate (HEMA) using the radical polymerization route to obtain poly(2-chloroquinylmethacrylate-*co*-2-hydroxyethylmethacrylate) (CQMA-*co*-HEMA). The structures of monomers and copolymers obtained were characterized by the Fourier transform infrared (FTIR), nuclear magnetic resonance (NMR) and CHN elementary analyses. The absence of free drug particles aggregated incrusted in the copolymer matrix was confirmed by DSC and XRD methods. The cell toxicity was examined by the lactate dehydrogenase (LDH) assay on the gingival epithelial cancer cell line (Ca9-22) and the cell adhesion and growth on the CQMA-*co*-HEMA drug carrier was examined by the 3-(4,5-dimethylthiazol-2-yl)-2,5-diphenyltetrazolium bromide (MTT) test. The in vitro dynamic release of the HCQ base from these drug carrier systems occur in different pH media through a retroesterification reaction, in which the influence of the swelling capacity of the drug carrier system, the drug content grafting in the copolymer and pH media are investigated.

2. Materials and Methods

2.1. Chemicals

HEMA (purity 98%), triethylenetetramine (purity > 97%), methacryloyl chloride (purity $\geq$ 97%), AIBN (purity 99%) and chloroform (purity $\geq$ 99%) were supplied from Sigma-Aldrich (Taufkirchen, Germany). Plaquenil commercial trade tablets were manufactured and commercialized by Aventis Pharma Limited (Guildford, UK), Sanofi (Paris, France), Kyowa Hakko (Galashiels, UK). All the chemicals were used without further purification except the monomer which was purified by distillation under reduced pressure.

2.2. Extraction of 2-Hydroxychloroquine from the Plaquenil Commercial Tablets

A determined amount of 200 mg of Plaquenil tablets was finely ground using a quartz mortar, then added in small portions with continuous stirring in distilled water until complete dissolution (~72 h). Two drops of phenolphthalein indicator were added to the aqueous solution, then titrated by 2% by volume ammonia solution until pink color persisted. In order to extract HCQ from the aqueous solution, the obtained solution was then transferred in a separator funnel containing equivalent amount of diethyl ether. HCQ was extracted from the organic layer by vaporization of diethyl ether using a rotary evaporator. The purity of HCQ was confirmed by NMR analysis. The physicochemical characteristics of this molecule are gathered in Table 1.

Table 1. Some physicochemical characteristics of 2-hydroxychloroquine [24].

Formula	Molar Mass ($g\cdot mol^{-1}$)	Appearance	Density ($g\cdot L^{-1}$)	Water Solubility ($mg\cdot L^{-1}$)	Melting Temperature (°C)
$C_{18}H_{26}ClN_3O$	335.87	White crystals	1.2	26	90

2.3. Synthesis of 2-Chloroquinyl Methacrylate (CQMA)

CQMA was synthesized by an esterification reaction involving HCQ and methacryloyl chloride using triethylenetetramine as a catalyst according to the reaction in Scheme 1. In a 250 mL two-necked flask, 10 g of HCQ was dissolved by stirring in 100 mL of chloroform, and then triethylenetetramine was added to this mixture so that the 2-hydroxychloroquine/triethylenetetramine molar ratio was 1:3 in order to ensure complete consumption of HCQ during the esterification reaction. A reflux condenser was connected through the main opening of the flask containing the HCQ/triethylenetetramine mixture in chloroform, through which a stream of nitrogen gas at low flow (3 $mL\cdot min^{-1}$) passed. The whole was placed in an ice bath for 10 min. The excess of methacryloyl chloride diluted in chloroform was then added dropwise with moderate stirring to the preceding solution using an addition bulb. The reaction took place in a dark atmosphere under a stream of nitrogen and under reflux during the whole period of the reaction. CQMA was isolated from the reaction mixture by a complete evaporation of solvent and liquid residual reactants using rotary system. The residual HCQ was removed by washing the CQMA obtained three times by an excess of distilled water, then dried in a vacuum oven until constant mass.

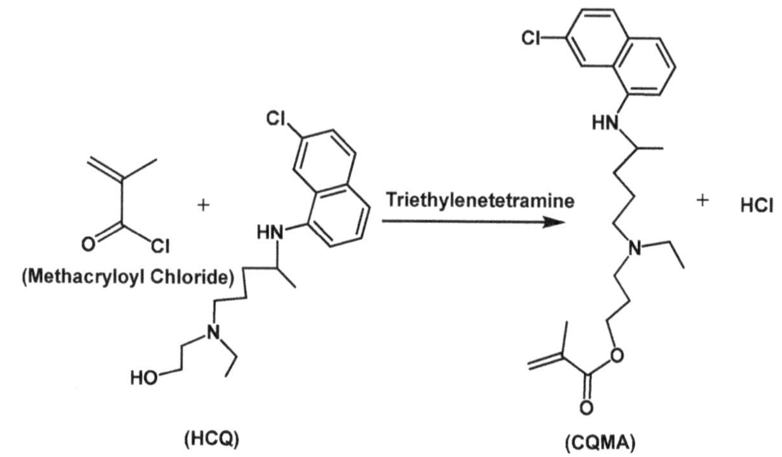

Scheme 1. Synthesis reaction of 2-chloroquinyl methacrylate.

2.4. Preparation of CQMA-co-HEMA Drug Carrier Systems

CQMA was copolymerized with HEMA at 60 °C through a free-radical polymerization route under nitrogen gas using AIBN as initiator (Scheme 2). Different copolymers containing different co-monomer ratios were synthesized and the preparation conditions are summarized in Table 2.

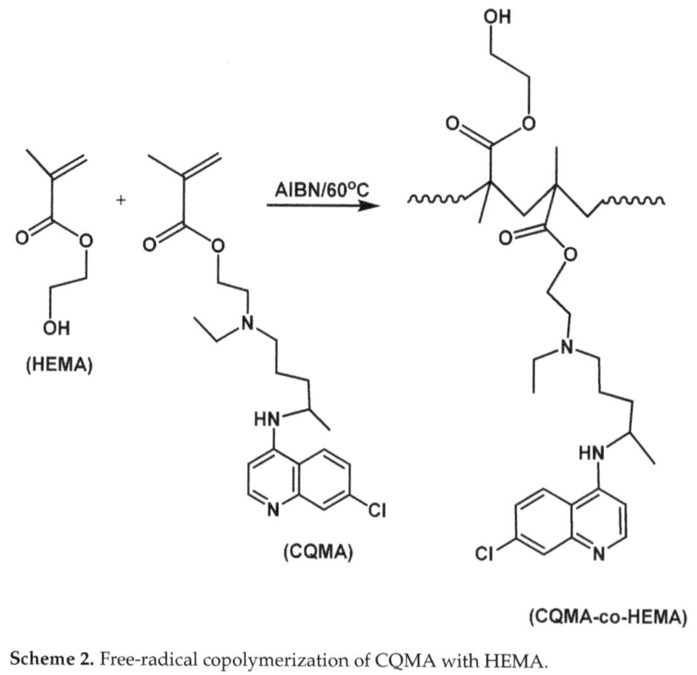

Scheme 2. Free-radical copolymerization of CQMA with HEMA.

Table 2. Experimental conditions of the synthesis of CQMA-*co*-HEMA copolymer with different CQMA content.

Copolymer	CQMA (g)	HEMA (g)	AIBN (wt%)	Monomers Composition (wt%)	Drug Content (wt%)
CQMA-*co*-HEMA5	0.20	9.80	0.1	10	2.0
CQMA-*co*-HEMA7	0.85	9.15	0.1	25	8.5
CQMA-*co*-HEMA10	1.00	9.00	0.1	50	10.0
CQMA-*co*-HEMA15	1.50	8.50	0.1	75	15.0

2.5. Characterization

The new synthesized monomer and copolymers were characterized by different techniques. The FTIR spectra of the samples were recorded on a Nicolet iS10 spectrometer (Thermo Scientific, Madison, WI, USA), equipped with an attenuated total reflection (ATR; diamond crystal) accessory. Spectra were obtained over a region of 4000–500 cm^{-1} at room temperature and acquired with a total of 32 scans per spectrum and resolution of 2 cm^{-1}. The ^{1}H NMR and ^{13}C NMR spectra of samples were taken at 400 and 100 MHz, respectively, on a spectrometer (JEOL Resonance, JEOL, Tokyo, Japan) using deuterated dimethyl sulfoxide (DMSO-d_6) as a solvent. The DSC thermograms were traced by a Shimadzu DSC 60A (Kyoto, Japan) system previously calibrated with indium. In total, 8–10 mg of monomer and copolymer powders was packed in aluminum DSC pans before being placed in a DSC cell, then heated under nitrogen gas from 30 to 200 °C at a heating rate of 20 °C·min^{-1}. Data were collected from the second scan run for all samples. No degradation phenomena of HCQ, CQMA and CQMA-*co*-HEMA samples were observed in all DSC thermograms in the investigated temperature range, noting that the T_g value was estimated as the midpoint of the heat capacity change with temperature and the T_m at the top of the melting peak. X-ray diffraction of pure HCQ and CQMA microparticles and copolymers was recorded by (Rigaku D_{max} 2000, The Woodlands, TX, USA) diffractometer using an anode tube of Cu working with voltage of 40 KV and a generator current of 100 mA.

The range of diffraction angle was 0–80 two theta. The samples were used as thin films, except that of pure HCQ and prepared CQMA which were analyzed as powder. SEM images of film samples before and after the release process were performed on a JEOL JSM-6610LV scanning electron microscope (SEM) (Tokyo, Japan) at an accelerating voltage of 10 kV. The surface and cross-sections of samples were first sputter-coated with a thin layer of gold and then observed at magnification range of 300–3000×.

2.6. Toxicity and Cell Adhesion

2.6.1. Cell Culture Conditions

The gingival epithelial cancer cell line (Ca9-22) was obtained from RIKEN BioResource Research Center, Tsukuba, Japan. Ca9-22 cells were cultured in RPMI-1640 medium (Thermo Fisher Scientific, Burlington, ON, Canada), supplemented with L-glutamine, 5% fetal bovine serum (FBS) (Gibco; Thermo Fisher Scientific, Burlington, ON, Canada) and antibiotics (Sigma-Aldrich, Oakville, ON, Canada). Cell cultures were performed at 37 °C in humidified incubator with 5%; CO_2 atmosphere conditions and cell culture medium were changed every two days.

2.6.2. Cell Adhesion by MTT Assay

Each biomaterial was placed in 24-well plates and cancer gingival cells at 100 × 10^3 cells/wells were seeded directly in biomaterial sample, then cultured for 24 h with RPMI medium. After 24 h of adhesion cells, an MTT assay was performed as described by Semlali et al. [25,26]. Briefly, each culture was supplemented with 10% of MTT (5 mg/mL) and incubated for 3 h at 37 °C. The stained cells were then lysed using 500 µL of isopropanol-HCl solution at 0.04 M with agitation for 15 min. An amount of 100 µL triplicate of lysis solution was transferred to 96-well plates to be read at 550 nm using an iMark microplate reader (Bio-Rad, Mississauga, ON, Canada). Cell adhesion levels were determined at 550 nm by means of the following formula:

$$Cell\ viability(\%) = \frac{OD(treated\ cells) - OD(Blank)}{OD(Control\ cell) - OD(Blank)} \times 100 \quad (1)$$

where *OD* is the optical density.

2.6.3. Cell Toxicity by LDH Assay

Ca9-22 cells were seeded at 10^5/well in 24-well cell culture plates for 48 h. The LDH activity was assessed in culture supernatants collected from cell adhered in different biomaterials for 48 h. LDH activity was measured using LDH (Sigma-Aldrich, Oakville, ON, Canada) [27]. Triton (1%) was used as positive control (100% of toxicity) and pure PHEMA as negative control (13% of toxicity).

2.7. Mass Transfer

The mass transfer of media in the polymers was carried out by gravimetric method [28]. CQMA-*co*-HEMA films of known masses and thicknesses were immersed in excess of water at 37 °C in media of pH 1, 3, 5 and 7. These samples were then removed after time intervals, wiped with tissue paper from water droplets deposited on both surfaces, then immediately weighed. Each operation lasted until the films were saturated (constant weight). Each process was triplicate and the masses of media absorbed were taken from the arithmetic average. The swelling of the film sample in weight percent was calculated according to Equation (2):

$$S\ (\text{wt}\%) = \frac{m_t + m_{\text{HCQ}} - m_o}{m_o} \times 100 \quad (2)$$

where m_t and m_o are the masses of the film sample at time *t* and t_o, respectively. m_{HCQ} is the mass of HCQ released during time *t*.

2.8. In Vitro Drug Release

CQMA-*co*-HEMA films with a square surface area of 4 cm^2 and an average thickness of 1.23 mm were suspended in 100 mL of water/hydrochloric acid solution at pH fixed at 1, 3, 5 and 7, and stirred at 100 rpm at 37 °C. Aliquots of 0.5 mL were withdrawn at time intervals and immediately replaced by water at the same pH media just after analysis. This operation kept a constant volume of media during the release process and reflected as much as possible what was actually happening in the intestines in which the HCQ released was absorbed gradually. The concentration of drug released during this period was calculated taking into account the volume of medium removed for quantitative analysis. The total mass of HCQ released during this period (m_t) was calculated from its concentration deduced from the UV–visible calibration curve indicating the change in the absorbance versus the concentration of medication. It is important to note that, during the drug release process, the pH of the medium was practically not affected by the small amount of HCQ released and, hence, the addition of a buffer solution to stabilize the pH of the medium was not necessary. The percentage of HCQ released, *R* (wt%), in the media during a certain time, *t*, was calculated from Equation (3):

$$R\ (\text{wt}\%) = \frac{m_t - m_o}{m_o} \times 100 \quad (3)$$

where m_o is the mass of the initial HCQ incorporated in the drug carrier sample.

3. Results and Discussions

3.1. Characterization

The chemical structures of the CQMA monomer and CQMA-*co*-HEMA copolymers were confirmed by FTIR, NMR and elemental analyses. The absence of HCQ and CQMA residual particles aggregated in the copolymer samples was highlighted by DSC, XRD and SEM analyses.

3.1.1. Elementary Analysis

The elemental composition of CQMA was confirmed by the CHN analysis through the consistent comparison of the experimental data with those calculated, and the results obtained are summarized in Table 3.

Table 3. Comparison between the experimental determination of the elementary composition of the CQMA and that calculated.

Element	Composition (wt%)					
	Sulfanilamid (Ref)		HCQ		CQMA	
	Exp.	Calc.	Exp.	Calc.	Exp.	Calc.
C	41.850	41.860	64.230	64.310	65.470	65.367
H	04.680	04.651	07.787	07.741	07.489	07.428
N	16.260	16.279	12.393	12.505	10.450	10.399

The composition of the copolymer in CQMA and HEMA monomeric units was also determined by this same method and the results obtained were gathered in Table 4. As these data show, the composition of these copolymers in the CQMA unit did not accurately reflect the composition in starting monomers. This seemed obvious, because the large steric hindrance of the CQMA monomer, which was much greater than that of HEMA, went in the sense of considerably reducing its reactivity with respect to HEMA.

Table 4. Composition of CQMA-*co*-HEMA determined by CHN elementary analysis.

System	Elementary Composition (wt%)			CQMA (wt%)	HCQ (mol%)
	C	H	N		
PHEMA	55.287	7.723	0	0	0
CQMA-*co*-HEMA5	56.747	7.602	01.412	04.70	1.61
CQMA-*co*-HEMA7	59.906	5.294	02.116	07.22	2.52
CQMA-*co*-HEMA10	59.496	7.720	02.673	09.68	3.44
CQMA-*co*-HEMA15	57.110	7.673	01.852	14.82	6.49
PCQMA	65.362	7.540	10.563	100	100

3.1.2. FTIR Analysis

The FTIR spectrum of CQMA in Figure 1 shows the presence of the combination of the different absorption bands belonging to the two reagents HCQ and methacryloyl chloride and the total disappearance of that at 3343 cm^{-1} assigned to the hydroxyl group of the pure HCQ indicating the formation of this monomer. The CQMA spectrum confirmed the structure of this monomer through the apparition of the same absorption bands attributed to both HCQ and mathacryloyl methacrylate, except that of –OH stretching frequency which appeared in the HCQ spectrum at 3385 cm^{-1}. Indeed, the main signals that appeared on the CQMA spectrum were the aromatic C–H stretching frequency at 2915 cm^{-1}, aromatic C=C stretching frequency in two regions at 1634 and 1457 cm^{-1}, the C–Cl stretching frequency at 1056 cm^{-1} and the C–N bending frequency at 1157 cm^{-1}. The two absorption bands observed at 2500 and 2600 cm^{-1} were probably assigned to residual quaternary ammonium salts, resulting from the reaction involving methacryloyl chloride and triethylenetetramine during the monomer preparation.

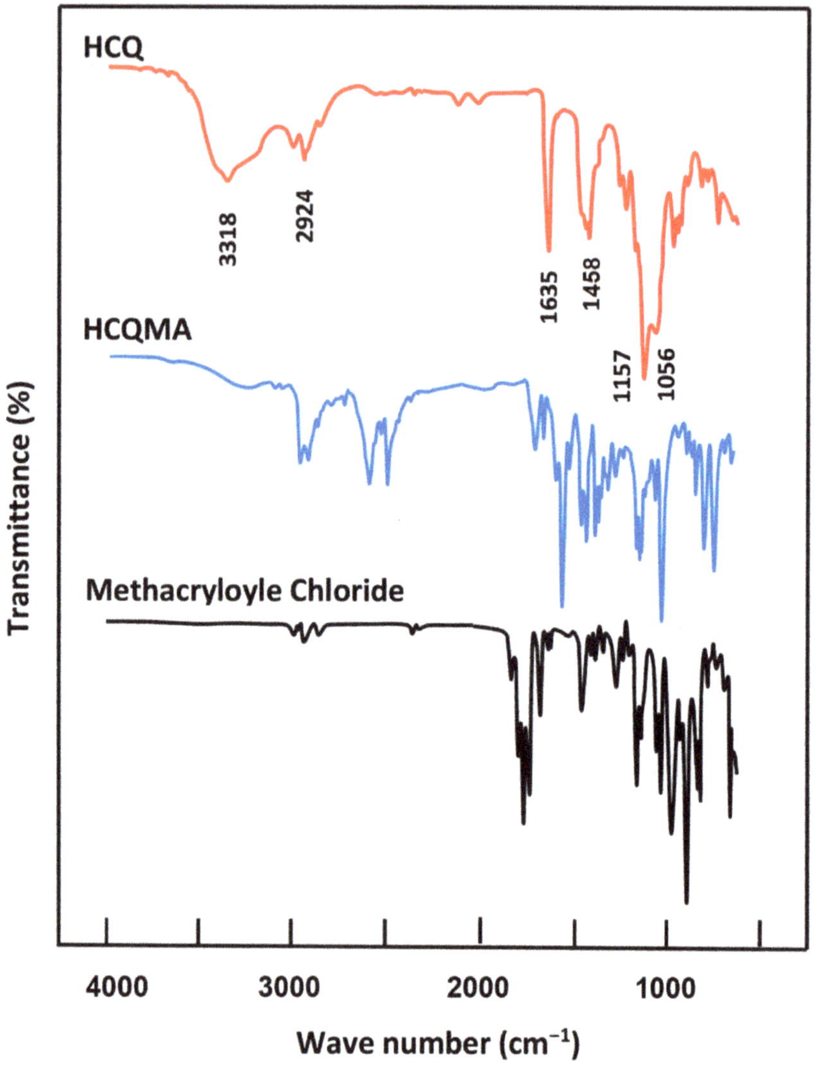

Figure 1. FTIR spectra of CQMA, HEMA monomers, pure HCQ and pure methacroyle chloride.

Figure 2 shows the FTIR spectra of the CQMA-*co*-HEMA copolymer of different compositions. These spectra showed the combination of the absorption bands attributed to the two monomers HEMA and CQMA, except that which characterized the vinyl C=C double bond of the two monomers of average intensity at 1635 cm^{-1} represented in Figure 1. The bands of absorption attributed to C–Cl stretches and CN bending shown at 1057 and 1157 cm^{-1} in the CQMA monomer (Figure 1) were covered in the copolymer spectra by the –C–O–C– bands of the HEMA units, which appeared between 1321 and 1032 cm^{-1} [29].

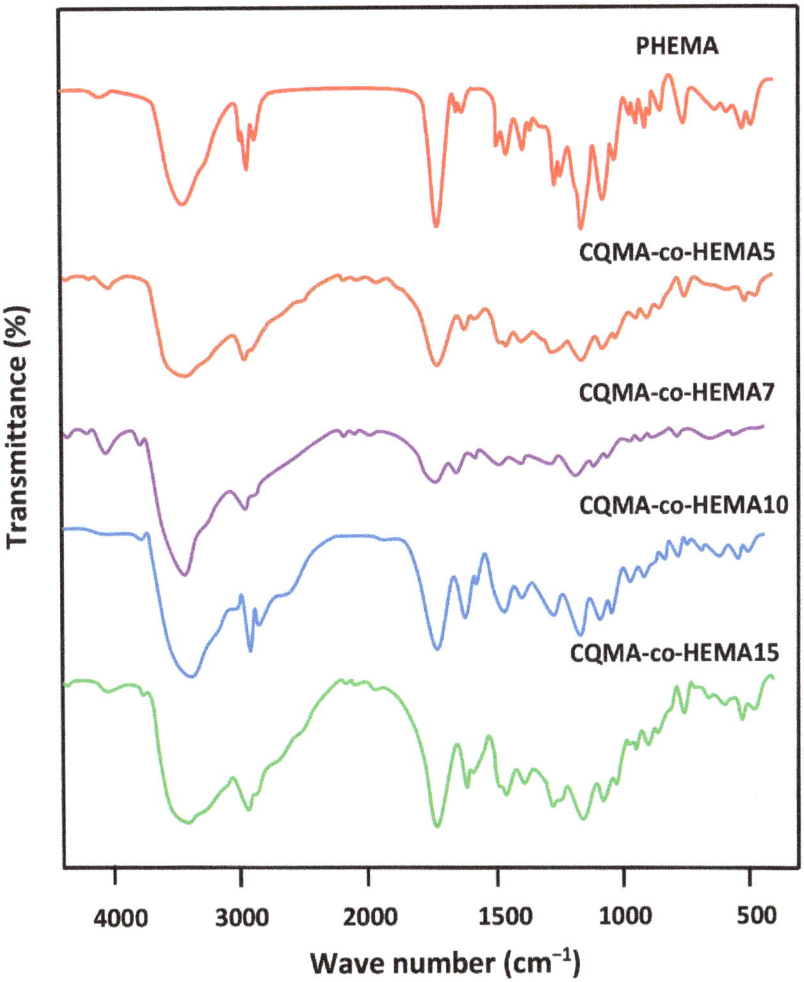

Figure 2. FTIR spectra of pure PHEMA, PCQMA and CQMA-*co*-HEMA copolymers.

3.1.3. NMR Analysis

The ^{1}H NMR spectrum of the CQMA monomer in Figure 3 presents the signals of protons involving HCQ and methacryloyl chloride with a significant shift toward the right from 3.45 to 4.32 ppm. This was due to the change in the environment of the two protons of the methylene group in position (1) belonging to HCQ to those of the ester group of the 2-chloroquinyl mathacrylate unit; thus, confirming the results of the FTIR.

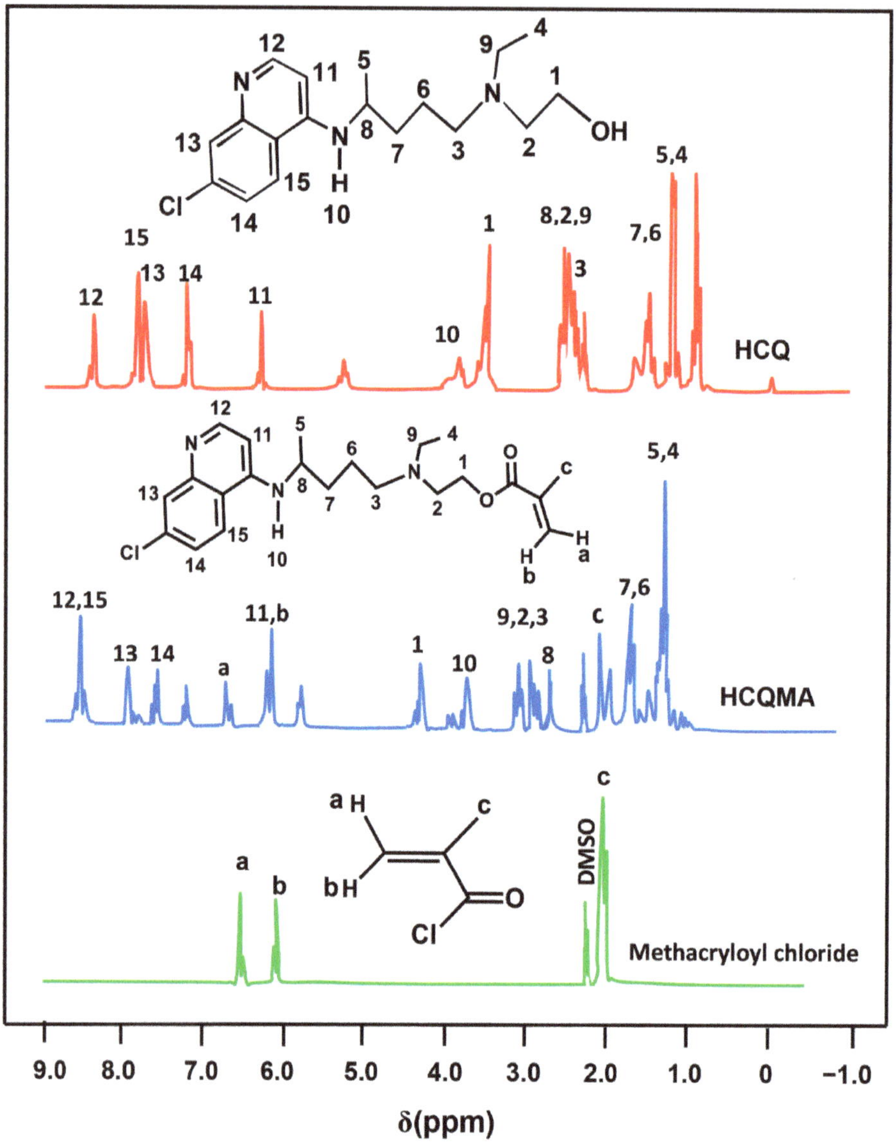

Figure 3. ^{1}H NMR spectra of pure HCQ, methacryloyl chloride and CQMA.

The ^{1}HNMR spectra of CQMA-*co*-HEMA15, selected among the other copolymers by their higher density in chloroquinyl groups, the PHEMA and PHCQMA homopolymers were grouped in Figure 4. As in the case of the FTIR analysis, the spectra of the copolymers revealed the presence of the combination of all the signals attributed to the two monomeric units, CQMA and HEMA, with the exception of those attributed to the protons (a) and (b) of the vinyl double bonds (sites of the polyaddition reaction), which appeared on the monomers spectra of Figure 3.

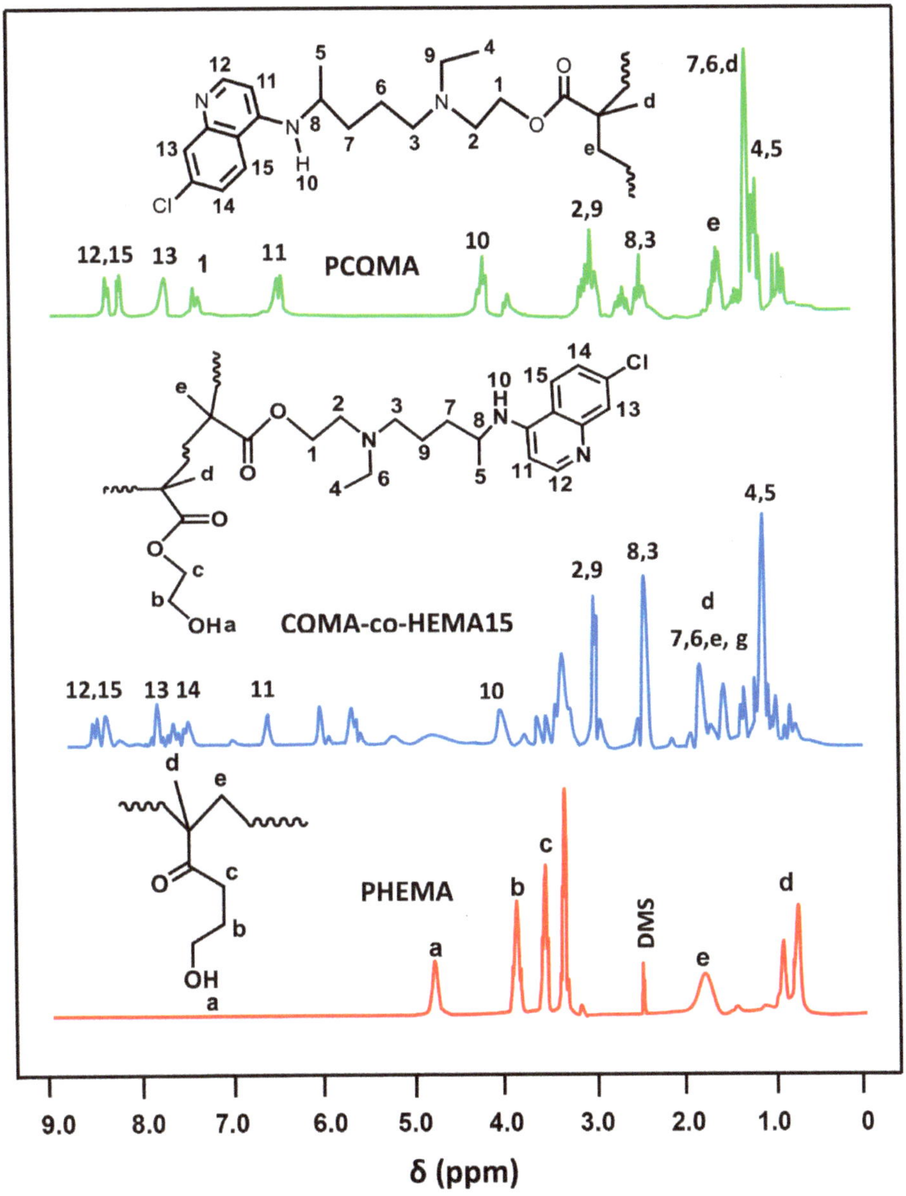

Figure 4. ^{1}H NMR spectra of PCQMA, PHEMA homopolymers and CQMA-*co*-HEMA15 copolymers.

Figure 5 shows the comparison of the ^{13}C NMR spectra of the CQMA monomer with those of its reagents HCQ and methacroyl chloride. As in the case of the ^{1}HNMR analysis, the structure of the CQMA monomer was demonstrated in its spectrum by the presence of all the signals attributed to the carbons of the two reagents, except that in position (1) observed at 52.4 ppm, which was directly linked to the hydroxyl group of HCQ, which shifted to the higher chemical shifts (63.5 ppm), indicating the formation of the ester group of the CQMA monomer.

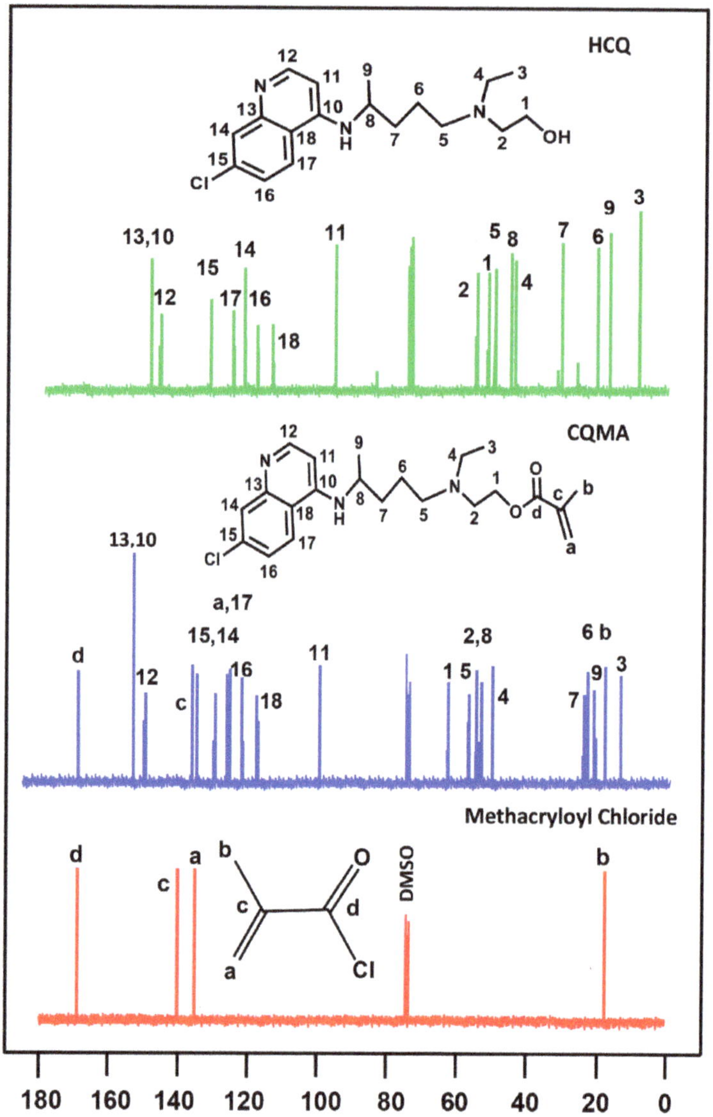

Figure 5. ^{13}C NMR spectra of pure HCQ, pure methacryloyl chloride and CQMA in DMSOd_6.

The structure of the CQMA-*co*-HEMA copolymer was confirmed by the ^{13}C NMR analysis from the comparison of their spectra with that of the PHEMA homopolymer. Indeed, as shown in Figure 6, the signals of carbons attributed to the CQMA and HEMA co-monomeric units were present in the spectra of the copolymer except those of the two ethylenic carbons (c) and (d) of their starting monomers (Figure 5), which were transformed into two ethenic carbons (i) and (j), respectively, during the polyaddition reaction. In addition, small shifts were observed on some signals of the carbons in the environment of sites subject to additional reactions.

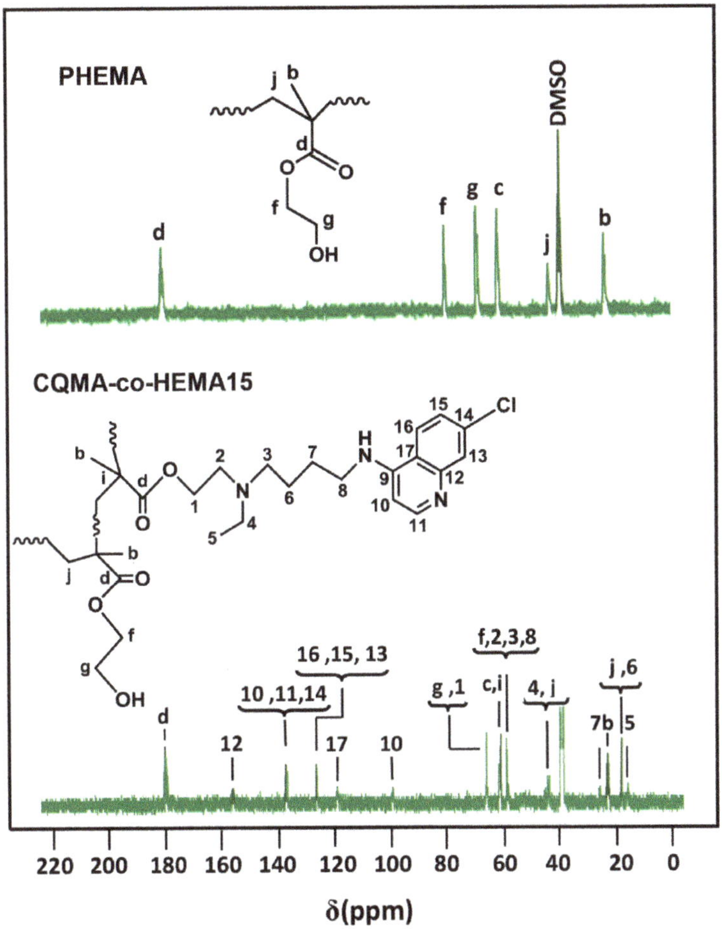

Figure 6. ^{13}C NMR spectra of PHEMA homopolymer and CQMA-*co*-HEMA15 copolymers.

3.1.4. DSC Analysis

The DSC thermograms of the pure HCQ and CQMA monomers were grouped for comparison in Figure 7. The profile of each thermal curve shows a sharp endothermic peak characterizing the absorption enthalpy during the melting process. The melting temperature of each sample was taken from the top of the peak, which was 90 °C for HCQ, which agreed with the literature [15], and 167 °C for CQMA. The slight change in the heat capacity of the CQMA sample observed at 117 °C seemed to indicate a glass transition temperature, revealing the presence of a polymer resulting from a thermal polymerization of a fraction of the monomer during the DSC heating process in a nitrogen gas atmosphere.

For CQMA-*co*-HEMA, Figure 8 shows a comparison between the thermograms of the PHEMA homopolymer and those of the copolymers involving CQMA and HEMA monomers with different compositions. As shown in these profiles, PHEMA presented a glass transition temperature, T_g, at 95 °C, which agreed with that of the literature [30], while the thermal curves of the copolymer showed a shift in the T_g towards low temperatures, which increased with the CQMA content. This appeared to be evident due to the effect of the steric hindrance of the chloroquinyl methacrylyl substituent of the CQMA units. Indeed, according to Reimschuessel [31], the more the substituent of the alkyl methacrylyl units in the homopolymer is hindered, the less the values of T_gs are important. The large spacing between the polymer chains caused by this substituent facilitated the sliding of the chains between them, and this reflected the reduction in T_g values. On the other hand, the more the number of these units increase in the copolymer, the less the T_g values.

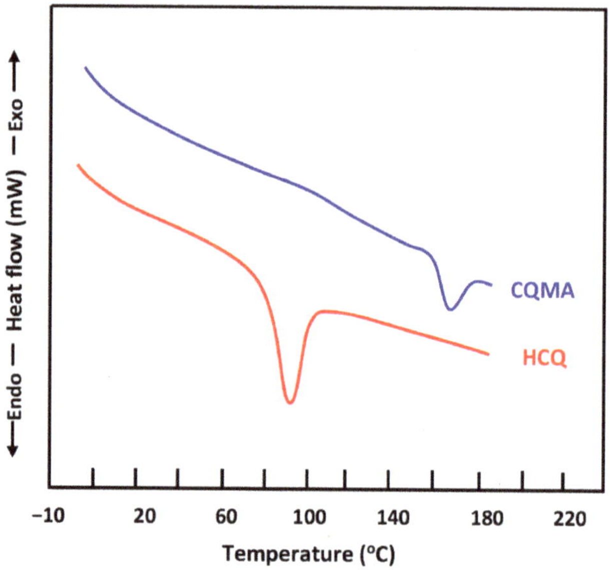

Figure 7. DSC thermograms of pure HCQ and CQMA monomers.

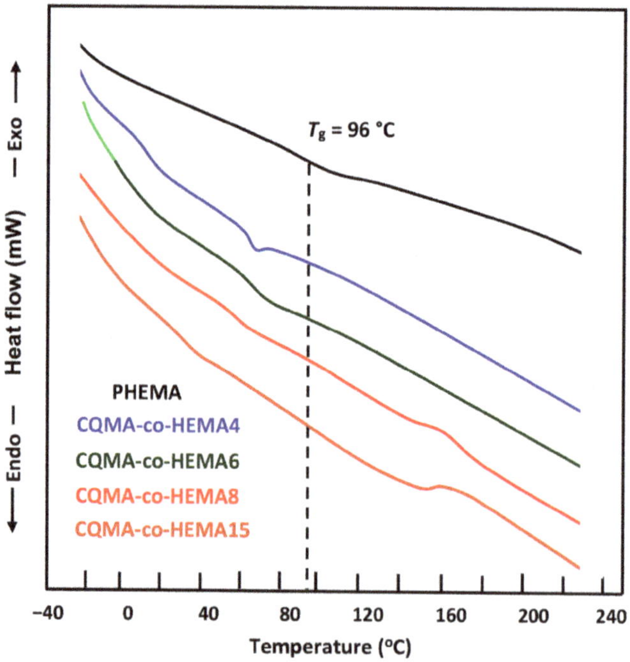

Figure 8. DSC thermograms of PHEMA and CQMA-*co*-HEMA with different compositions.

3.1.5. X-ray Diffraction Analysis

The X-ray diffractograms of the CQMA monomer, PHEMA homopolymer and CQMA-*co*-HEMA copolymers with different compositions are shown in Figure 9. The profiles of the curves attributed to the copolymers, as for the homopolymer, did not show any signs indicating a crystalline region. Indeed, the appearance of broad peaks in PHEMA centered at 19.5° 2 theta [32] and CQMA-*co*-HEMA, which slightly shifted from 20.0 to 21.2° 2 theta with the CQMA content, was due to the lack of crystallographic order of the polymeric chains; thus, leading to the formation of amorphous structures which was mainly caused by the steric hindrance of the substituent on both sides of the polymer chains. The absence of the main signals attributed to the CQMA monomer crystals observed in its spectrum at 12.3, 22.0 and 25.02 theta in the spectra of the copolymers indicated the absence of residual particles of chloroquinyl methacrylate non polymerized incrusted in the copolymer matrix; thus, confirming the results of DSC.

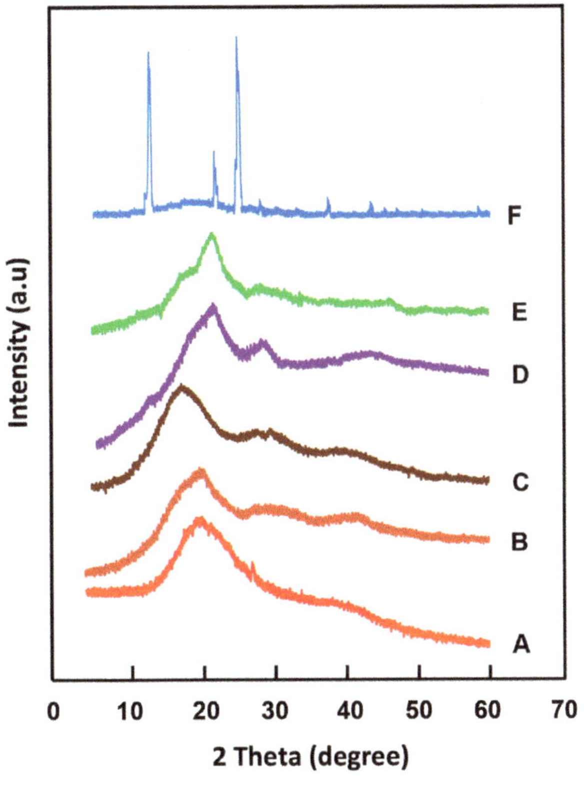

Figure 9. XRD patterns of: (A) PHEMA; CQMA-*co*-HEMA with different CQMA contents: (B) 5 wt%; (C) 7 wt%; (D) 10 wt%; (E) 15 wt% and (F) CQMA.

3.2. *Surface Morphology*

Figure 10 shows the SEM surface morphology of a virgin PHEMA film, HCQ particles and CQMA-*co*-HEMA copolymer films. As can be seen in the blank PHEMA image, there was a smooth, non-porous, wave-shaped surface devoid of any aggregated residual particles of HCQ or CQMA. The image (on the right) shows the HCQ particles clustered together as endangered snowflakes, indicating the presence of moderate attractive electrostatic forces between them. Micrographs of CQMA-*co*-HEMA films showed pores of different shapes and sizes on their surfaces, which varied depending on the amount of

HCQ grafted. Indeed, the CQMA-*co*-HEMA5 and CQMA-*co*-HEMA7 samples exhibited surface morphologies containing spherical pores of sizes between 3 and 70 μm, and were much denser in the case of the film sample containing 5% of CQMA by weight. The CQMA-*co*-HEMA10 image showed a smooth surface containing significantly fewer pores of similar sizes to the two previous copolymer films. The film surface of the highest HCQ-loaded copolymer sample (CQMA-*co*-HEMA15) exhibited fewer spherically shaped or compressed pores, probably due to the mechanical stress during film preparation.

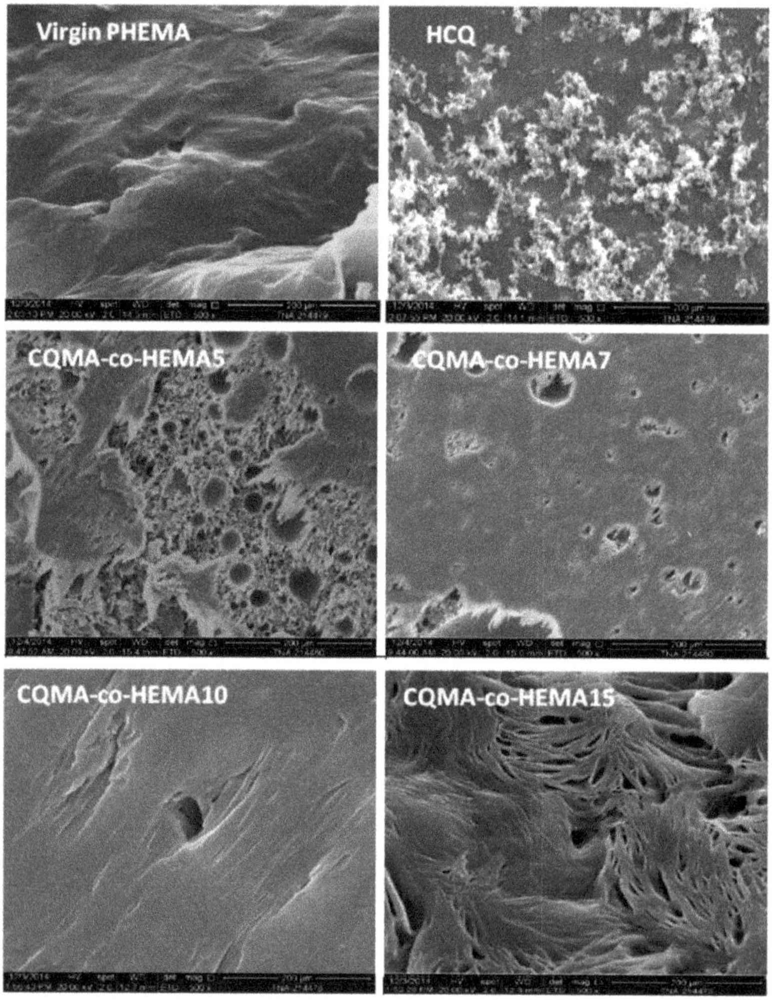

Figure 10. SEM micrographs of HCQ particles, surface morphologies of virgin PHEMA and CQMA-*co*-HEMA copolymers films.

3.3. *Swelling Behavior of CQMA-*co*-HEMA Systems*

The swelling behavior of any hydrogel is an important key which controls the amount and transfer of drug released from a drug carrier system to a target external medium. The mechanism of the swelling of a polymer gel in a liquid medium mainly involves two important factors, which are the affinity of the polymer–solvent type and the diffusion dynamics of the absorbate in the absorbent material. The affinity of the polymer with respect to the absorbate is governed, mainly, by the difference between their solubility parameters. On the other hand, the diffusion dynamics of the molecules of the absorbate in the absorbent polymer matrix are controlled by various parameters such as the degree of the crosslinking of the polymer [33,34], the temperature [35,36] and the pH of the medium in which the drug is to be released [37,38]. In this work, the swelling degree at the equilibrium (S) of CQMA-*co*-HEMA films was calculated at different pH media from Equation (3) and the results obtained are plotted in Figure 11.

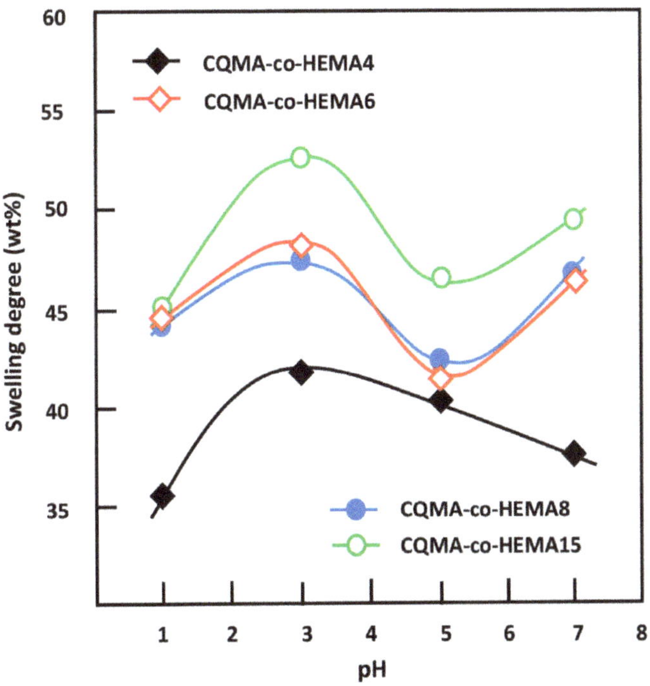

Figure 11. Variation of the swelling degree at equilibrium of CQMA-*co*-HEMA systems versus the media pH.

As can be observed from these curve profiles, the capacity of the CQMA-*co*-HEMA system to swell did not vary continuously with the pH medium, regardless of the copolymer composition. Indeed, for the drug carrier systems containing a CQMA content equal or superior to 6.22 wt%, the swelling degree at saturation reached a maximum in pH media 3 and 7, then passed by a minimum when the pH medium was 5, while for the CQMA-*co*-HEMA system containing less CQMA (4.70 wt%), the variation in the swelling rate at equilibrium reached its maximum at pH medium 3, then continuously decreased to reach a minimum at pH 7. Similar results were also obtained by different authors on the swelling capacity of PHEMA-containing amine groups [39,40]. These authors explain the increase in the swelling degree by the positive charge on the amine after protonation in the acidic medium. According to these authors, the electrostatic repulsion gives more volume to the hydrogel, allowing the diffusion of a higher water content. Similar results were also obtained by Kost et al. [40]. In this case, the greater the amount of CQMA in the copolymer, the greater the swelling capacity. This seemed to be true at a certain limit of CQMA content, but the three copolymers containing equal amounts of CQMA or more than 7.22 wt% practically had the same swelling capacities in pH medium 1. At pH 3, the swelling at equilibrium reached its maximum for all the samples. Similar results were also obtained by Sari et al. [39] in the case of copolymers involving HEMA and *N*,*N*-dimethylaminoethyl methacrylate (DMAEMA) monomers. At a high CQMA content (9.68 and 14.82 wt%), the increase in the release dynamic in the neutral pH medium was probably due to the elimination of HCQ from the copolymer through a retroesterification reaction, which was favorable in the medium at pH 7 as shown in Scheme 3. In this case, the hydrophilic contact density between hydroxyl (HEMA) and hydroxyl (water) increased at the expense of the hydrophobic contact density between carboxyl ester and hydroxyl water, resulting in a higher swelling of the system.

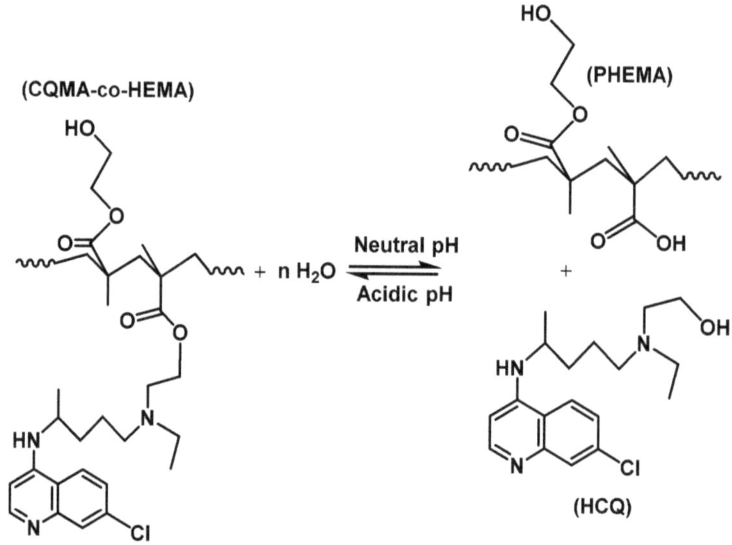

Scheme 3. Esterification/retroesterification reactions between PHEMA and HCQ.

Figure 12 shows the change in the swelling capacity of the CQMA-*co*-HEMA copolymer at equilibrium as a function of the CQMA content. These curve profiles revealed a linear dependence of the swelling capacity for the copolymer with the CQMA content in pH medium 5 and logarithmic in pH 1, 3 and 7. As can also be seen from these data, in any pH medium investigated, the swelling rate of copolymers increased quickly when the CQMA content in the copolymer ranged between 4.70 and 7.22 wt%, while an increased linearly in pH medium 5 in all the investigated composition ranges was seen. The increase in the swelling capacity of the copolymer with the CQMA content could be explained by two main factors: (i) the increase in the solvation of the chloroquinyl methacrylate substituent through the protonation of the amine group which occurred in acidic media and (ii) the increase in the density of the hydrogen bonds between the hydroxyl of water and that of the HEMA unit through an increase in the free volume between the copolymer chains created by the bulky chloroquinyl methacrylate substituent. At a certain limit of the CQMA content, the hydrophobic character of the ester substituent intervened by limiting the solvation of the copolymer, which reflected the slowing down of the swelling capacity of copolymers containing more than 7.22 wt% in the CQMA unit, notably in pH medium 1, in which a plateau was obtained resulting from a pseudo-equilibrium between a decrease in the hydrophilicity of the copolymer due to the formation of ester (chloroquinyl methacrylate) and, simultaneously, an increase in the solvation of HCQ due to the protonation of its amine groups.

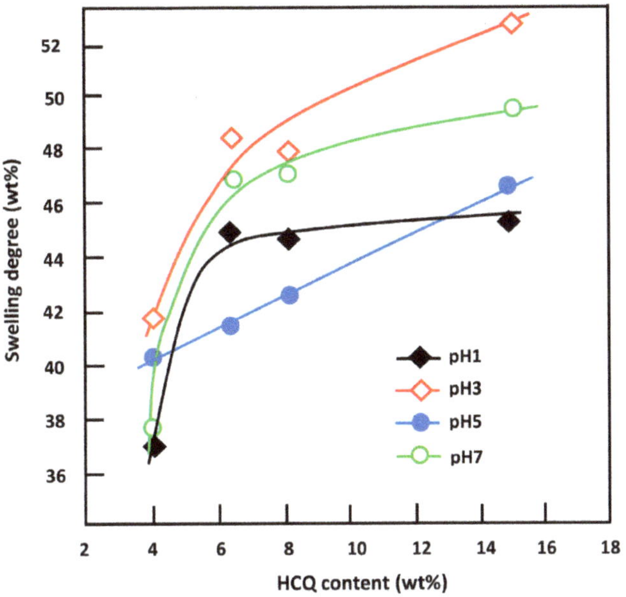

Figure 12. Variation of the swelling capacity of CQMA-*co*-HEMA systems in different pH media versus the HCQ content.

3.4. Cell Viability of HCQMA-co-HEMA on Ca9-22 Cells

As shown in Figure 13, Ca9-22 cells, when treated with the polymer, presented a low toxicity between 2% and 13%. However, when the cells were treated with the drug carrier systems, an increase in cell cytotoxicity was observed for CQMA-HEMA5 (A1) (8.2 ± 0.2%), CQMA-HEMA7 (A2) (12.28 ± 0.4%), CQMA-HEMA10 (A3) (7.36 ± 0.12%), CQMA-HEMA15 (A4) (12.68 ± 0.27%), CQMA-*co*-HEMA15 (A5) and 3.07 ± 0.10% for the virgin PHEMA used as the reference. The increase in cell cytotoxicity was due to the effect of HCQ released as an anti-oral cancer agent (Figure 14).

3.5. Cell Adhesion and Growth

Figure 14 shows the effect of the CQMA content in the CQMA-*co*-HEMA drug carrier system on Ca9-22 cell adhesion. As emerged from these histograms, the system containing the least CQMA (4.70 wt%) seemed to have the best performance (0.64 ± 0.19) compared to that of the reference film (virgin PHEMA) (0.23 ± 0.002). Beyond this amount grafted into the polymer, this performance almost returned again to that of the virgin polymer with a slight increase (0.27 ± 0.01) and a slight decrease (0.20 ± 0.008) when the CQMA content was 14.82 and 9.68 wt%, respectively. The adhesion of cells to a polymeric material could be managed by three essential factors which were toxicity, chemical affinity between the two target entities and surface porosity. In this case, according to the toxicity test (Figure 13), the grafting of HCQ onto PHEMA slightly affected (4.2–9.5%) the viability of the virgin material. Regarding the affinity between the polymer and the cell, this seemed to be favored by the presence of hydrogen bonds between the hydroxyl groups of the polymer and those of the cell. The size of the pores on the surface of the polymer carrier played an important role in cell adhesion and growth, because the larger the pore size, the greater the penetration of cells, the more it grows and the greater its adhesion. In this work, it was found that the best cell adhesion was observed in the system containing 4.70 wt% CQMA. This appeared to be quite consistent with the swelling results as a function of the amount of drug grafted onto the polymer (Figure 12). Because the more the swelling, the less the size of the pores as shown in Scheme 4.

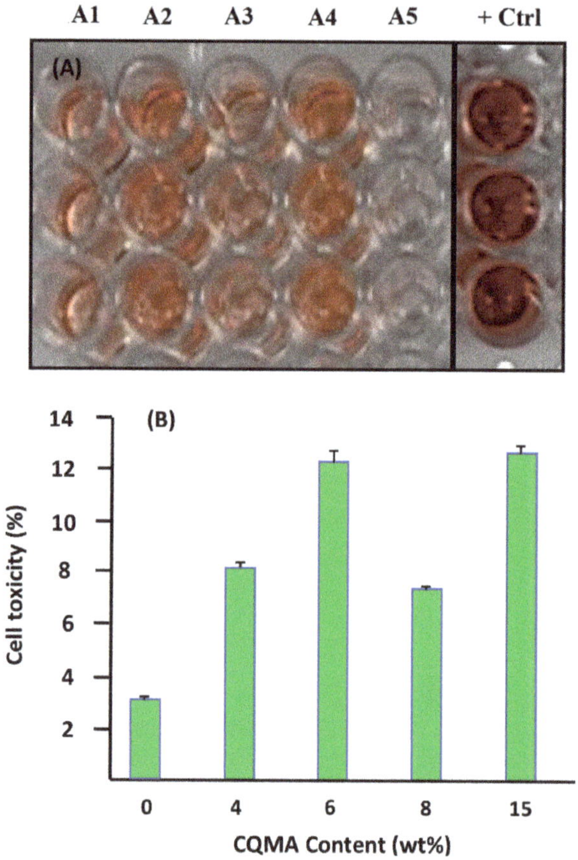

Figure 13. Effect of CQMA-*co*-HEMA on Ca9-22 cells cytotoxicity of different CQMA contents. (**A**) LDH assay. (**B**) Representative figure of percentage of cell toxicity with different CQMA contents.

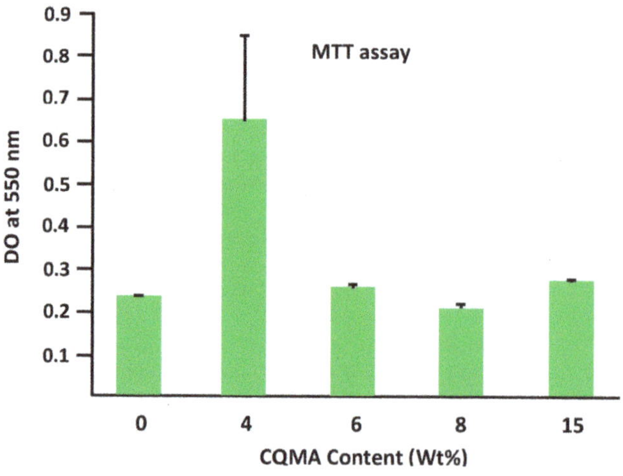

Figure 14. The effect of HCQ content in the CQMA-*co*-HEMA drug carrier system on Ca9-22 cell adhesion.

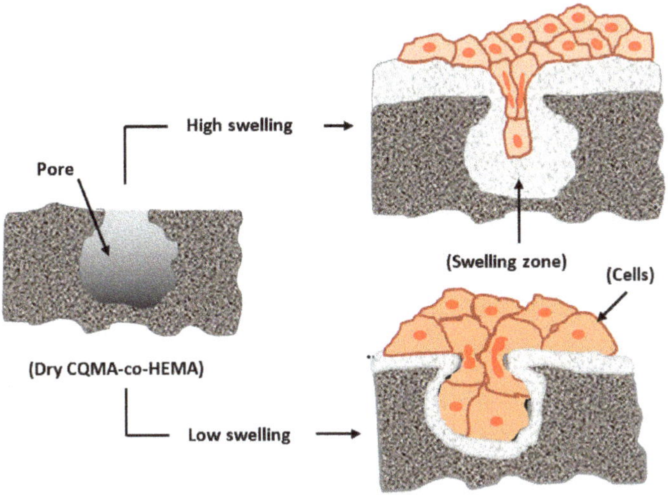

Scheme 4. Effect of the swelling on the cell adhesion and growth.

3.6. In Vitro HCQ Release

3.6.1. Release Kinetics of HCQ

Figure 15 shows the profiles of the curves indicating the change in HCQ released at 37 °C from CQMA-*co*-HEMA drug carrier systems in different pH media as a function of time over a week. As can be seen from these plots, the CQMA-*co*-HEMA10 system appeared to be the most efficient in releasing HCQ in all pH media. In fact, 62% of this drug by weight was released in pH medium 1, 68% by weight at pH medium 3, 69% by weight at pH medium 5 and 66% by weight at a neutral pH. Just after came the CQMA-*co*-HEMA7 and CQMA-*co*-HEMA15 systems, releasing during the same period 66% and 64% of 2-hydroxychloroquine by weight in pH medium 3, respectively, and 59% by weight for both in the neutral pH medium. On the other hand, CQMA-*co*-HEMA5, although it delivered much less of this drug, the maximum amount released was observed in the neutral pH medium with 27 wt% and between 24 and 25 wt% in acidic media.

3.6.2. HCQ Solubility Enhancement

The solubility of any drug in an aqueous medium is a key factor in any drug delivery process, because it governs its rate and kinetics of absorption by target organs. The solubility of pure HCQ in water reported in the literature was estimated at 26.1 mg·L^{-1} at 25 °C [15]. It is well known that the solubility of a poorly water-soluble drug increases with a decrease in the size of its dispersed particles [41–43]. Indeed, the decrease in the size of the drug particles leads to an increase in their total surface area, leading to an increase in the contact surface water molecule drug particles. This promotes the increase in the dissolution of a very large number of particles, particularly when they are slowly released into water in the molecular state through a retroesterification reaction. In this work, the solubility of HCQ in different pH media was approximately estimated through the cumulative amount of this drug released until saturation (equilibrium) and the results obtained are given in Table 5.

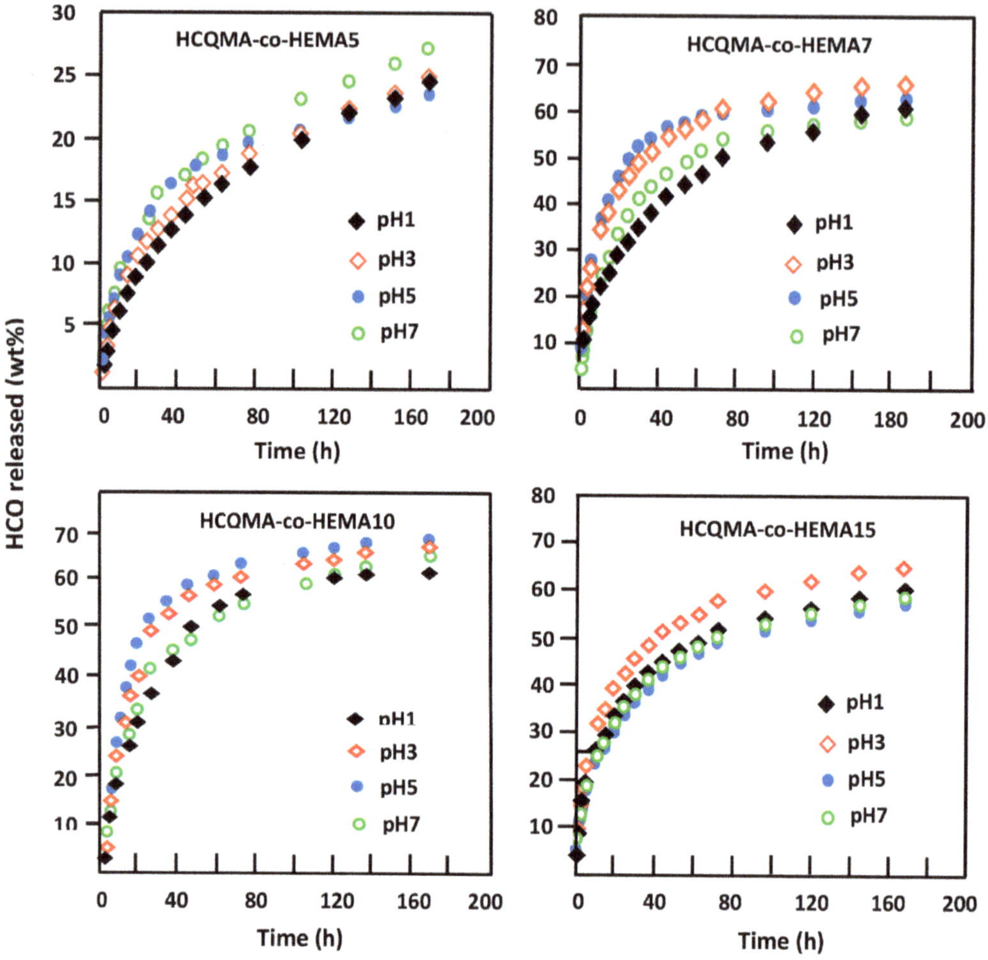

Figure 15. Release behavior of HCQ from CQMA-*co*-HEMA systems with different HCQ content versus time in different pH media.

Table 5. Solubility of HCQ in different pH media at 37 °C deducted from the maximum amount of HCQ released from CQMA-*co*-HEMA drug carrier systems.

Compound	Solubility ($mg \cdot L^{-1}$)			
pH	1	3	5	7
HCQ powder	35.62	31.17	28.08	27.82
CQMA-*co*-HEMA5	22.61	21.41	19.0	19.32
CQMA-*co*-HEMA7	168.0	149.0	137.0	167.0
CQMA-*co*-HEMA10	58.0	149.0	70.0	84.0
CQMA-*co*-HEMA15	186.0	249.0	252.0	261.0

In this work, the solubility of pure HCQ powder estimated at 37 °C in a neutral pH medium was 27.82 $mg \cdot L^{-1}$ and increased slowly when the pH of the medium decreased to reach 35.62 $mg \cdot L^{-1}$ in pH1. On the other hand, the concentrations of HCQ released in these solutions from the CQMA-*co*-HEMA drug carrier systems during 164 h were much higher, with the exception of the one containing the least HCQ (4.70% by weight), in which the concentration obtained during this same period did not exceed 22.61 mg·L^{-1}, regardless of the pH of the medium. This seemed to be obvious, since 4.70 wt% as the HCQ/HEMA starting composition was not sufficient to release a sufficient amount of HCQ to reach or approach the saturation of the medium. Compared to the solubility of dissolved HCQ in powder form, this value was slightly lower. For example, the maximum concentration of HCQ released in the water of the CQMA-*co*-HEMA15 system varied between 186 and 261 mg·L^{-1} when the pH of the medium went from 1 to 7 without observing any precipitation or cloudiness for one week. This represented an improvement of over 5.22 to 9.38 times the solubility of HCQ when dissolved in the powder form. In the case of the CQMA-*co*-HEMA5 system, in which the release dynamics continued to increase (Figure 16), the lower concentrations of HCQ obtained in the different pH media obtained at the end of the release process were due to the lower amount of HCQ initially incorporated into the copolymer.

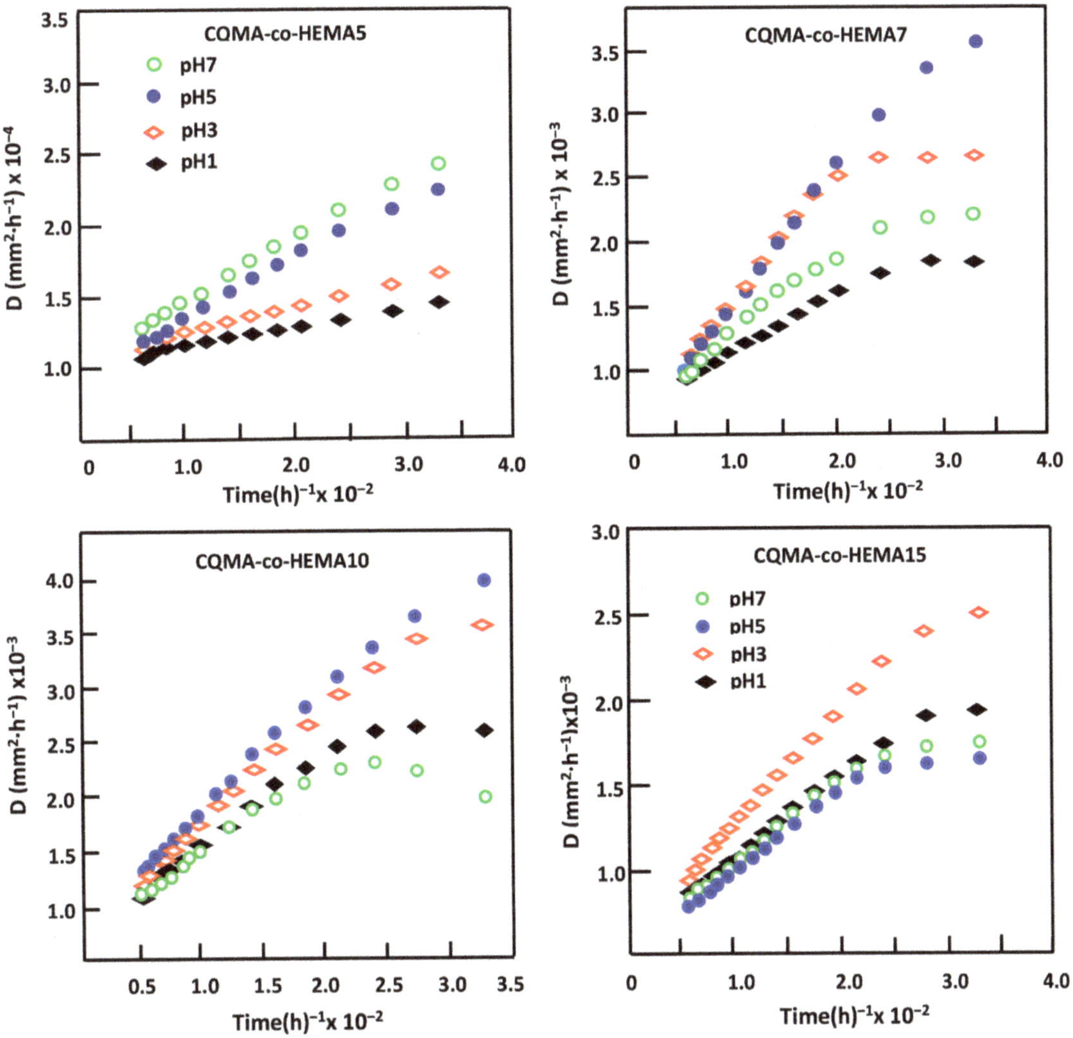

Figure 16. Variation of the diffusion factor of HCQ in HCQMA-*co*-HEMA systems versus reverse time.

3.6.3. Diffusion Behavior of HCQ

According to Lin et al. [44], for a release of less than 60 wt% from the total drug loaded, the diffusion of this substance through the carrier system follows a Fickian model. The value of the diffusion coefficient, D, can be calculated from Equation (4) [45–48]:

$$D = \frac{0.196 \times l^2}{t}\left(\frac{m_t}{m_o}\right)^2 \tag{4}$$

where l is the film thickness, m_o and m_t are as defined previously. The D value is determined when the permanent regime of the release process is reached and the HCQ particles deposited on the surface of the material are totally washed. Under such conditions, the profile of the D curve as a function of time would be significant and accurately reflect the dynamic of the drug released into the aqueous solution inside the carrier material. Figure 16 indicates, for all samples, the variation of D as a function of the reverse of time plotted from the data of Figure 15 and Equation (4). The profiles of the curves obtained clearly showed two types of diffusion which occurred during the HCQ release process. The first one was rapid, which took place during the first hours of the process, mainly due to a detachment and a leaching of HCQ particles from the surface of the sample; the second, which was long, described a straight line indicating an establishment of a permanent regime. This indicated that the HCQ diffusion process through CQMA-*co*-HEMA materials obeyed a Fickian model and also indicated that the release dynamics of this drug from these hybrid materials were mainly governed by a diffusion mechanism through the copolymer matrix. Under these conditions, the steady state of the liberation process was reached and it was possible to build our investigation on the second zone of the liberation process in which the steady state was reached.

3.6.4. Effect of the Swelling Degree of CQMA-*co*-HEMA Systems

The swelling capacity of a polymer in an aqueous medium is a fundamental property and considered as a key in the field of drug delivery. Indeed, this parameter regulates the amount of drug released, controls both the rate of diffusion of the penetrate into the polymeric carrier and its dissolution in the medium in which the drug is released [49–52]. In this work, the influence of the swelling capacity of the CQMA-*co*-HEMA system on the release behavior of hydroxychloroquine was carried out in different pH media during 72 h of the release process and the results obtained are plotted in Figure 17. As can be observed from these curve profiles, minimum releases of 53.52, 50.02 and 48.32 wt% were obtained for the drug carrier systems containing 9.68, 7.22 and 14.82 wt% of HCQ content, respectively, whereas for the system containing the lowest HCQ content (4.70% by weight), the dynamics behaved differently, which passed by a maximum of 23.70 wt% at a swelling saturation of 38.70% by weight.

3.6.5. Effect of the CQMA Content

Figure 18 shows the influence of the CQMA content in the CQMA-*co*-HEMA drug carrier system on the release dynamics during 24 and 72 h of the process. For both durations, these curve profiles revealed a significant increase in the HCQ released when the concentration of this medication in the drug carrier system did not exceed 7.22% by weight, then stabilized or decreased more or less quickly depending on the pH medium at a higher HCQ content. This could be explained mainly by two opposing factors which acted simultaneously on the swelling capacity of the drug vehicle system depending on their capacities to act and the content of CQMA in the copolymer. In fact, an increase in the hydroxychloroquinyl ester group as a substituent of the copolymer acted negatively on the hydrophilic power of the CQMA-*co*-HEMA system. This had an effect of reducing the quantity of water necessary for the reaction of retroesterification (drug regeneration reaction) and also reduced the dissolution of HCQ released inside the polymeric matrix.

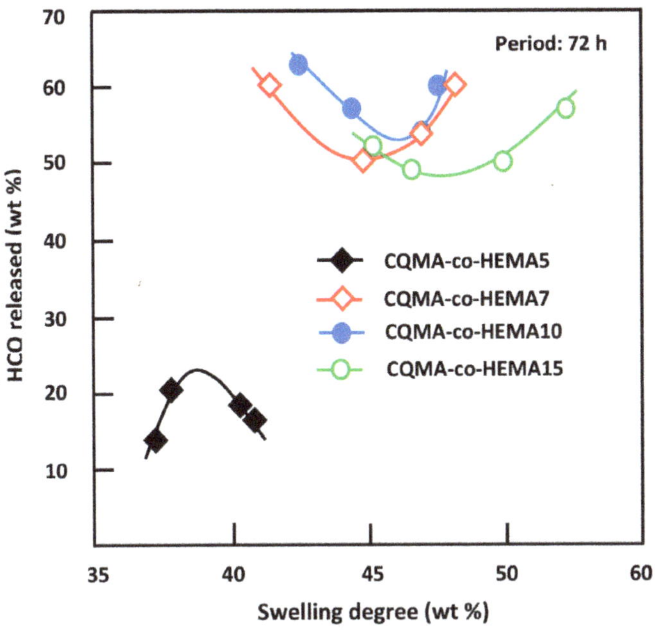

Figure 17. Release behavior of HCQ versus the swelling capacity of CQMA-*co*-HEMA systems in different pH media.

On the other hand, an increase in the steric hindrance of this substituent also promoted an increase in the free volume between the chains of the copolymer, facilitating the penetration of more water molecules. In this situation, a competition between these two opposite effects, which essentially depends on the amount of HCQ grafted into the copolymer, governs the dynamics of HCQ release in the medium. Finally, we could conclude from the results obtained that the ability to release HCQ in these different media was governed by the more or less hydrophilic character of the CQMA-*co*-HEMA system, which depended on its CCQMA/PHEMA composition.

3.6.6. Effect of pH Media

The nature of the environment in which drugs are released greatly influences the dynamics of drug release in a target organ. Indeed, various studies have been carried out on the influence of the medium such as the pH [53,54], the enzymes [54,55] and the bacteria [56,57] on the dynamics of the drug released by the drug delivery systems and the results obtained were very striking. In this work, the effect of pH media on the dynamics of HCQ released from CQMA-*co*-HEMA drug carrier systems was performed at pH media 1, 3, 5 and 7, and the results obtained for 24 and 72 h are plotted for comparison in Figure 19. As can be seen from the profiles of the curves obtained, for the drug carrier system containing the least CQMA (4.70% by weight), virtually no change in the release dynamics was observed, regardless of the pH of the medium. At 24 h of the release process, the two systems containing 7.22 and 9.68 wt% of CQMA content evolved according to the same trend, in which the release dynamics passed by a maximum of 54 wt% for the first system and 52 wt% for the second when the pH of the medium was 4.2. In contrast, for drug carrier systems containing 4.70 and 14.82% of CQMA by weight, the release dynamic continued to decrease slightly and linearly for the first drug carrier system to reach 10.35 wt% of HCQ in pH medium 7, while that of the second system passed through an inflection point at 37% of HCQ by weight released in pH medium 4.2. At 72 h of release process, the dynamics of HCQ released from the CQMA-*co*-HEMA7 and CQMA-*co*-HEMA10 drug carrier systems, which contained 7.22 and 9.68 wt% of CQMA, respectively, completely changed from those observed at 24 h. Indeed, two extremums were observed for the drug carrier system containing 7.22 wt% of CQMA content in which a maximum release of 29.7 wt% of HCQ was observed in Ph 3 and a minimum of 26.6 wt% in pH medium 5, while that with 9.68 wt% of CQMA content reached a maximum release of 63 wt% in this same pH medium. The explanation for such behavior appeared to be complicated by the fact that two main antagonistic factors may simultaneously intervene in the drug release mechanism: the first being chemical which resides in the reaction of retro-esterification which generates the HCQ and the second being physical which resides in the hydrophilic character of the copolymer. The intensity of each depends on the pH of the medium and the CQMA content in the drug carrier system. Indeed, the chemical factors were the reactions of retro-esterification which took place in a neutral pH medium which regenerated the HCQs in the presence of an excess of water, leading to the release of a large amount of drug, and the reaction of esterification which took place in an acidic medium, which, on the contrary, promoted the formation of insoluble ester, which disfavored the release of the drug as shown in Scheme 3. The physical factor that could affect the release dynamic of HCQ from the CQMA-*co*-HEMA system was already discussed in the previous section. Indeed, at a high CQMA content, the swelling capacity of the polymer was also affected by the pH medium, since the more acidic the medium the less swelling of the system. This was due to the formation of the ester which was insoluble in water. Contrarily, as can be revealed in the previous section, a minimal amount of CQMA in the copolymer, less than 7.22% by weight, promoted an increase in the swelling capacity.

3.6.7. Performance of CQMA-*co*-HEMA Drug Carrier System

The study of the performance of the CQMA-*co*-HEMA drug carrier system based on the drug amount released the instantaneous release rate and the duration of the release process, deduced from the slopes of the pseudo-linear portions of the kinetic curves of Figure 15, led to the results of Table 6.

Table 6. Performance of the CQMA-*co*-HEMA drug carrier systems.

System	pH	Stable Zone (h)	HCQ Released (wt%)	Release Rate (wt%/h)	System	Stable Zone (h)	HCQ Released (wt%)	Release Rate (wt%/h)
CQMA-*co*-HEMA5	1	0–27	12.0 ± 0.2	0.44 ± 0.04	CQMA-*co*-HEMA7	0–15	22.0 ± 0.8	1.46 ± 0.02
		30–164	12 ± 0.5	0.10 ± 0.03		15–62	22.0 ± 0.7	0.47 ± 0.02
						62–164	10.0 ± 0.3	010 ± 0.02
	3	0–30	13 ± 0.2	0.43 ± 0.04		0–10	26.03 ± 0.6	2.60 ± 0.04
		44–164	18.2 ± 0.5	0.15 ± 0.01		10–60	18.00 ± 0.6	0.36 ± 0.02
						60–164	10.0 ± 0.6	0.10 ± 0.06
	5	0–37	17.0 ± 0.6	0.46 ± 0.08		0–10	37.0 ± 0.6	3.70 ± 0.07
		50–164	6.0 ± 0.7	0.05 ± 0.04		10–60	17.0 ± 0.5	0.34 ± 0.03
						60–164	13.0 ± 0.6	0.13 ± 0.06
	7	0–30	16.0 ± 0.5	0.53 ± 0.04		0–22	38.0 ± 0.4	1.73 ± 0.10
		30–164	12.0 ± 0.8	0.09 ± 0.03		22–65	18.0 ± 0.5	0.42 ± 0.02
						65–164	7.0 ± 0.6	0.07 ± 0.06
CQMA-*co*-HEMA10	1	0–20	30.0 ± 0.3	1.5 ± 0.08	CQMA-*co*-HEMA15	0–10	26.0 ± 0.3	2.60 ± 0.04
		20–58	22.0 ± 0.4	0.58 ± 0.09		10–48	20.0 ± 0.3	0.53 ± 0.05
		58–164	7.0 ± 0.4	0.070 ± 0.04		48–164	13.0 ± 0.02	0.11 ± 0.04
	3	0–20	40.0 ± 0.3	2.3 ± 0.02		0–10	31 ± 0.4	3.10 ± 0.06
		20–60	9.0 ± 0.5	0.23 ± 0.02		10–42	20 ± 0.4	0.63 ± 0.04
		60–164	9.0 ± 0.4	0.09 ± 0.04		50–146	10.0 ± 0.4	0.06 ± 0.04
	5	0–20	46.0 ± 0.3	2.30 ± 0.02		0–12	24.0 ± 0.5	2.00 ± 0.06
		20–60	14.0 ± 0.4	0.35 ± 0.09		12–56	22.0 ± 0.5	0.50 ± 0.05
		60–164	9.0 ± 0.4	0.09 ± 0.04		56–164	10.0 ± 0.04	0.09 ± 0.04
	7	0–28	41.0 ± 0.4	1.46 ± 0.06		0–12	25.0 ± 0.5	1.93 ± 0.04
		28–73	13.0 ± 0.5	0.29 ± 0.02		12–54	21.0 ± 0.5	0.10 ± 0.05
		73–164	10.0 ± 0.4	0.11 ± 0.04		54–164	10.0 ± 0.04	0.09 ± 0.4

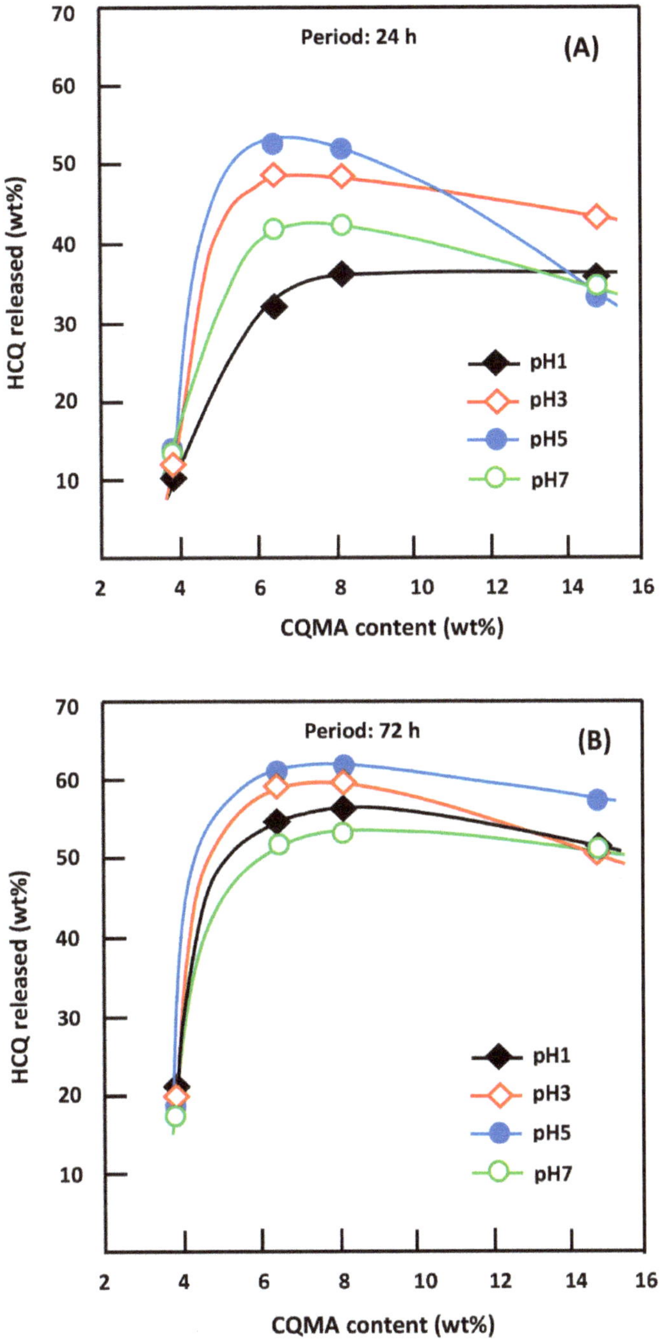

Figure 18. Variation of the HCQ released during 24 h (**A**) and 72 h (**B**) of the release process versus the CQMA content in the drug carrier system and in different media pHs.

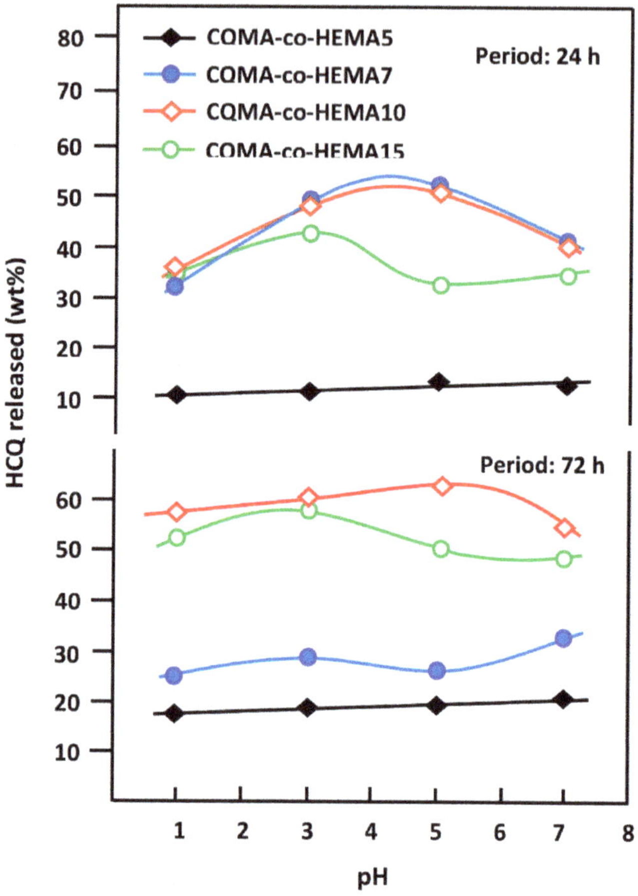

Figure 19. Variation of the cumulative HCQ released from the CQMA-*co*-HEMA drug carrier systems versus the media pH.

As can be seen from these curve profiles, the release dynamic of HCQ showed two principal pseudo stable zones for the specimen containing 4.70 wt% of CQMA and a supplementary transit zone for the CQMA-*co*-HEMA drug carrier systems with a higher CQMA content. The first zone observed during the first hours of the release process was between 10 and 30 h depending on the composition of the copolymer and the pH of the medium. During this period, a significant amount of HCQ was released, probably due to the large difference between the chemical potential for dissolving HCQ outside and inside the polymer matrix. For the CQMA-*co*-HEMA5 drug carrier system, a rapid release dynamic (0.43–0.53 wt%/h) was observed during the second zone which was located between 120 and 134 h in which 6.0 to 18.2% of HCQ by weight was released depending on the pH medium. For drug carrier systems initially containing a higher CQMA content, the second zone was short (32–43 h) and relatively fast (0.10–0.63 wt%/h). This step was considered as a transit zone in which 9.0–22.0% of drug by weight was released during this period depending on the CQMA content in the copolymer and the pH of the medium. The third zone observed for CQMA-*co*-HEMA drug carrier systems containing 4.70, 9.68 and 14.82 wt% CQMA was the longest (91–116 h) and also the slowest (0, 06–0.13 wt%/h), wherein 7.0–13.0 wt% HCQ was released depending on the initial content of CQMA in the drug carrier system and the pH medium. The gradual decrease in the release dynamics of these drug delivery systems can mainly be explained thermodynamically by a gradual decrease in the difference between the chemical potentials of the HCQ dissolution inside and outside the polymer matrix up until the equilibrium is reached, taking into account the various parameters of interaction between the various components of the copolymer and that of the medium. The performance of the drug carrier systems was determined from the criteria that stipulate: the maximum of drug released in neutral pH medium (intestinal transit), minimum drug release in acidic pH medium (stomach), stable release rate and longest release period. According to Belzer et al. [58], the mean total gastrointestinal transit time (GITT) was between 53 and 88 h distributed over three main stages: (i) stomach transit (pH $\approx$ 1.5–3.5), which lasts between one and 4 h, (ii) intestinal transit (pH $\approx$ 7–9), which varies between 4 and 12 h and, finally, (iii) transit in the colon (pH $\approx$ 5–7), which lasts between 48 and 72 h. Taking into account the pH of the medium and the transit times in different digestive organs, it was possible to approximately estimate from the data of Table 7 the distribution of the percentages of cumulative HCQ released in different organs and the average stomach/digestive organs rate (SDO) (Equation (5)), independently of the effects of enzymes and microorganisms and the results obtained are grouped in Table 7. These data revealed that the CQMA-*co*-HEMA5 system was the most efficient, because it was able to reduce the release of HCQ in the stomach to 2.40 wt% of the total quantity released for the fast GITTs and 6.69 wt% for the slow GITTs.

$$SDO(\text{wt\%}) = \frac{r_s}{r_s + r_{si} + r_c} \times 100, \tag{5}$$

where r_s, r_{si} and r_c are the percentages of HCQ released in the stomach, small intestine and colon, respectively, during a certain transit time.

Table 7. Estimated distribution of the cumulative HCQ released from CQMA-*co*-HEMA drug carrier systems on the principal digestive organs timed according to Belzer.

CQMA-*co*-HEMA System	**Stomach Transit (wt%)**		**Small Intestine Transit (wt%)**		**Colon Transit (wt%)**		**SDO (wt%)**	
Transit Time	**Min (1 h)**	**Max (4 h)**	**Min (4 h)**	**Max (12 h)**	**Min (48 h)**	**Max (72 h)**	**Min (48 h)**	**Max (72 h)**
CQMA-*co*-HEMA5	0.44	1.74	2.12	6.36	15.76	17.90	2.40	6.69
CQMA-*co*-HEMA7	2.03	8.12	6.92	20.76	43.68	49.86	3.86	10.31
CQMA-*co*-HEMA10	1.90	7.60	5.84	17.52	40.96	47.92	3.90	10.40
CQMA-*co*-HEMA15	2.85	11.4	7.72	23.16	20.98	23.38	9.03	19.68

4. Conclusions

Very interesting results were obtained from this investigation. Indeed, 2-chloroquinyl methacrylate as a new monomer could be easily synthesized through a catalytic reaction involving methacryloyl chloride in the presence of triethylenetetramine. Poly(2-chloroquinyl methacrylate-*co*-2-hydroxyethyl methacrylate drug carrier systems with different compositions were also successfully synthesized through a radical copolymerization route involving the synthesized 2-chloroquinyl methacrylate and 2-hydroxyethylmethacrylate. This method was able to obtain, in aqueous medium, a generator of a desired percentage of 2-hydroxychloroquine regularly and slowly released during a long period through a retro-esterification reaction. The cell toxicity examined by the LDH assay revealed that the grafting of HCQ onto PHEMA was slightly affected (4.2–9.5%) and the cell adhesion and growth on the CQMA-*co*-HEMA drug carrier specimens carried out by the MTT assay revealed the best performance with the specimen containing 4.70 wt% CQMA. A significant improvement of 5.22 to 9.38 times the solubility of HCQ over that in powder form was obtained when it was released from CQMA-*co*-HEMA systems. The results of the in vitro release dynamics of HCQ from these systems showed very encouraging results in which the system initially containing 4.70% of CQMA by weight showed the best performance.

Author Contributions: Data curation, A.A., W.N.O.A., W.S.S. and S.M.S.A.; Formal analysis, A.A., W.N.O.A., T.S.A.-G., W.S.S., A.S. and S.M.S.A.; Funding acquisition, A.S.H. and A.A.A.-O.; Investigation, A.S.H., W.S.S. and A.S.; Methodology, A.A., W.N.O.A., T.S.A.-G. and S.M.S.A.; Resources, A.S.H., A.A.A.-O. and A.M.K.; Software, project administration: T.A., T.S.A.-G., W.S.S. and S.M.S.A.; Visualization, A.M.K.; writing—original draft T.A.; writing—review and editing, T.A. All authors have read and agreed to the published version of the manuscript.

Funding: The authors are grateful to the Deanship of Scientific Research, King Saud University for funding through the Vice Deanship of Scientific Research Chairs, Engineer Abdullah Bugshan research chair for Dental and Oral Rehabilitation.

Institutional Review Board Statement: Not applicable.

Informed Consent Statement: Not applicable.

Data Availability Statement: The data presented in this study are available on request from the corresponding author.

Conflicts of Interest: The authors declare no conflict of interest.

References

1. Kravvariti, E.; Koutsogianni, A.; Samoli, E.; Sfikakis, P.P.; Tektonidou, M.G. The effect of hydroxychloroquine on thrombosis prevention and antiphospholipid antibody levels in primary antiphospholipid syndrome: A pilot open label randomized prospective study. *Autoimmun. Rev.* **2020**, *19*, 102491. [CrossRef] [PubMed]
2. Shipman, W.D.; Vernice, N.A.; Demetres, M.; Jorizzo, J.L. An update on the use of hydroxychloroquine in cutaneous lupus erythematosus: A systematic review. *J. Am. Acad. Dermatol.* **2020**, *82*, 709–722. [CrossRef] [PubMed]
3. Hedya, S.A.; Safar, M.M.; Bahgat, A.K. Hydroxychloroquine antiparkinsonian potential: Nurr1 modulation versus autophagy inhibition. *Behav. Brain Res.* **2019**, *365*, 82–88. [CrossRef]
4. Belizna, C.; Pregnolato, F.; Abad, S.; Alijotas-Reig, J.; Amital, H.; Amoura, Z.; Andreoli, L.; Andres, E.; Aouba, A.; Bilgen, S.A. HIBISCUS: Hydroxychloroquine for the secondary prevention of thrombotic and obstetrical events in primary antiphospholipid syndrome. *Autoimmun. Rev.* **2018**, *17*, 1153–1168. [CrossRef] [PubMed]
5. Li, J.; Yuan, X.; Tang, Y.; Wang, B.; Deng, Z.; Huang, Y.; Liu, F.; Zhao, Z.; Zhang, Y. Hydroxychloroquine is a novel therapeutic approach for rosacea. *Int. Immunopharmacol.* **2020**, *79*, 106178. [CrossRef]
6. Bothwell, B.; Furst, D. *Antirheumatic Therapy: Actions and Outcomes*; Springer: Berlin/Heidelberg, Germany, 2005.
7. Ono, C.; Yamada, M.; Tanaka, M. Absorption, distribution and excretion of 14C-chloroquine after single oral administration in albino and pigmented rats: Binding characteristics of chloroquine-related radioactivity to melanin in-vivo. *J. Pharm. Pharmacol.* **2003**, *55*, 1647–1654. [CrossRef]
8. Banks, C.N. Melanin: Blackguard or red herring? Another look at chloroquine retinopathy. *Aust. N. Z. J. Ophthalmol.* **1987**, *15*, 365–370. [CrossRef]
9. Schiemann, U.; Kellner, H. Gastrointestinale Nebenwirkungen der Therapie rheumatischer Erkrankungen. *Zeitschrift Für Gastroenterologie* **2002**, *40*, 937–943. [CrossRef]
10. Smith, T.; Bushek, J.; LeClaire, A.; Prosser, T. *COVID-19 Drug Therapy*; Elsevier: Amsterdam, The Netherlands, 2020.
11. Sung-sun, K. Physicians Work Out Treatment Guidelines for Coronavirus. Available online: https://www.koreabiomed.com/news/articleView.html?idxno=7428 (accessed on 17 March 2020).
12. Bukhari, M.H.; Mahmood, K.; Zahra, S.A. Over view for the truth of COVID-19 pandemic: A guide for the Pathologists, Health care workers and community. *Pak. J. Med. Sci.* **2020**, *36*, S111. [CrossRef]
13. Liu, J.; Cao, R.; Xu, M.; Wang, X.; Zhang, H.; Hu, H.; Li, Y.; Hu, Z.; Zhong, W.; Wang, M. Hydroxychloroquine, a less toxic derivative of chloroquine, is effective in inhibiting SARS-CoV-2 infection in vitro. *Cell Discov.* **2020**, *6*, 1. [CrossRef]
14. Cortegiani, A.; Ingoglia, G.; Ippolito, M.; Giarratano, A.; Einav, S. A systematic review on the efficacy and safety of chloroquine for the treatment of COVID-19. *Crit. Care* **2020**, *57*, 279–283. [CrossRef]
15. Da Silva, A.E.A.; de Abreu, P.M.B.; Geraldes, D.C.; de Oliveira Nascimento, L. Hydroxychloroquine: Pharmacological, physicochemical aspects and activity enhancement through experimental formulations. *J. Drug Deliv. Sci. Technol.* **2021**, *63*, 102512. [CrossRef]
16. Food and Drug Administration, FDA. *Plaquenil® Hydroxychloroquine Sulfate Tablets*; United States Pharmacopeia (USP): North Bethesda, MD, USA, 2018.
17. Zhang, D.; Lee, Y.-C.; Shabani, Z.; Frankenfeld Lamm, C.; Zhu, W.; Li, Y.; Templeton, A. Processing impact on performance of solid dispersions. *Pharmaceutics* **2018**, *10*, 142. [CrossRef]
18. Arıca, M.Y.; Bayramoğlu, G.; Arıca, B.; Yalçın, E.; Ito, K.; Yagci, Y. Novel Hydrogel Membrane Based on Copoly (hydroxyethyl methacrylate/p-vinylbenzyl-poly (ethylene oxide)) for Biomedical Applications: Properties and Drug Release Characteristics. *Macromol. Biosci.* **2005**, *5*, 983–992. [CrossRef]
19. Brigger, I.; Dubernet, C.; Couvreur, P. Nanoparticles in cancer therapy and diagnosis. *Adv. Drug Deliv. Rev.* **2012**, *64*, 24–36. [CrossRef]

20. Flynn, L.; Dalton, P.D.; Shoichet, M.S. Fiber templating of poly (2-hydroxyethyl methacrylate) for neural tissue engineering. *Biomaterials* **2003**, *24*, 4265–4272. [CrossRef]
21. Chou, K.; Lee, S.; Han, C. Water transport in crosslinked 2-hydroxyethyl methacrylate. *Polym. Eng. Sci.* **2000**, *40*, 1004–1014. [CrossRef]
22. Wang, L.; Abedalwafa, M.; Wang, F.; Li, C. Biodegradable poly-epsilon-caprolactone (PCL) for tissue engineering applications: A review. *Rev. Adv. Mater. Sci* **2013**, *34*, 123–140.
23. Alghamdi, A.A.; Alattas, H.; Saeed, W.S.; Al-Odayni, A.-B.; Alrahlah, A.; Aouak, T. Preparation and Characterization of Poly(ethylene-*co*-vinyl alcohol)/poly(ε-caprolactone) Blend for Bioscaffolding Applications. *Int. J. Mol. Sci.* **2020**, *21*, 5881. [CrossRef] [PubMed]
24. Drugbank Online. Hydroxychloroquine. Available online: https://go.drugbank.com/drugs/DB01611 (accessed on 17 March 2021).
25. Semlali, A.; Jacques, E.; Rouabhia, M.; Milot, J.; Laviolette, M.; Chakir, J. Regulation of epithelial cell proliferation by bronchial fibroblasts obtained from mild asthmatic subjects. *Allergy* **2010**, *65*, 1438–1445. [CrossRef]
26. Semlali, A.; Chakir, J.; Goulet, J.P.; Chmielewski, W.; Rouabhia, M. Whole cigarette smoke promotes human gingival epithelial cell apoptosis and inhibits cell repair processes. *J. Periodontal Res.* **2011**, *46*, 533–541. [CrossRef] [PubMed]
27. Semlali, A.; Chakir, J.; Rouabhia, M. Effects of whole cigarette smoke on human gingival fibroblast adhesion, growth, and migration. *J. Toxicol. Environ. Health Part A* **2011**, *74*, 848–862. [CrossRef] [PubMed]
28. Singh, B.; Sharma, N. Mechanistic implication for cross-linking in sterculia-based hydrogels and their use in GIT drug delivery. *Biomacromolecules* **2009**, *10*, 2515–2532. [CrossRef]
29. Vargün, E.; Usanmaz, A. Degradation of poly (2-hydroxyethyl methacrylate) obtained by radiation in aqueous solution. *J. Macromol. Sci. Part A Pure Appl. Chem.* **2010**, *47*, 882–891. [CrossRef]
30. Liu, T.; Zhang, W.; Wang, J.; Zhang, Y.; Wang, H.; Sun, F.; Cai, L. Improved Dimensional Stability and Mold Resistance of Bamboo via In Situ Growth of Poly (Hydroxyethyl Methacrylate-*N*-Isopropyl Acrylamide). *Polymers* **2020**, *12*, 1584. [CrossRef]
31. Reimschuessel, H. On the glass transition temperature of comblike polymers: Effects of side chain length and backbone chain structure. *J. Polym. Sci. Polym. Chem. Ed.* **1979**, *17*, 2447–2457. [CrossRef]
32. Siddiqui, M.N.; Redhwi, H.H.; Tsagkalias, I.; Softas, C.; Ioannidou, M.D.; Achilias, D.S. Synthesis and characterization of poly (2-hydroxyethyl methacrylate)/silver hydrogel nanocomposites prepared via in situ radical polymerization. *Thermochim. Acta* **2016**, *643*, 53–64. [CrossRef]
33. Castelli, F.; Pitarresi, G.; Giammona, G. Influence of different parameters on drug release from hydrogel systems to a biomembrane model. Evaluation by differential scanning calorimetry technique. *Biomaterials* **2000**, *21*, 821–833. [CrossRef]
34. Martinez, A.W.; Caves, J.M.; Ravi, S.; Li, W.; Chaikof, E.L. Effects of crosslinking on the mechanical properties, drug release and cytocompatibility of protein polymers. *Acta Biomater.* **2014**, *10*, 26–33. [CrossRef]
35. Iwata, M.; Takayama, K.; Takahashi, Y.; Obata, Y.; Machida, Y.; Nagai, T.; Shirotake, S. Effect of temperature on drug release and drug absorption in mixed type diclofenac sodium suppositories. *Yakugaku Zasshi J. Pharm. Soc. Jpn.* **1999**, *119*, 170–177. [CrossRef]
36. Ibezim, E. Effects of dissolution medium, pH and temperature on the in vitro release properties of metronidazole tablets. *J. Pharm. Allied Sci.* **2004**, *2*, 209–213. [CrossRef]
37. Deng, K.; Zhong, H.; Tian, T.; Gou, Y.; Li, Q.; Dong, L. Drug release behavior of a pH/temperature sensitive calcium alginate/poly (*N*-acryloylglycine) bead with core-shelled structure. *Express Polym. Lett.* **2010**, *4*, 773–780. [CrossRef]
38. Kenawy, E.R.; Abdel-Hay, F.; El-Newehy, M.; Ottenbrite, R.M. Effect of pH on the drug release rate from a new polymer–drug conjugate system. *Polym. Int.* **2008**, *57*, 85–91. [CrossRef]
39. Sari, S.C.; Benmouna, M.; Mahlous, M.; Kaci, M. Swelling behavior of poly (2-hydroxyethyl methacrylate) copolymer gels. *MATEC Web Conf.* **2013**, *5*, 04008. [CrossRef]
40. Kost, J.; Horbett, T.A.; Ratner, B.D.; Singh, M. Glucose-sensitive membranes containing glucose oxidase: Activity, swelling, and permeability studies. *J. Biomed. Mater. Res.* **1985**, *19*, 1117–1133. [CrossRef]
41. Chu, K.R.; Lee, E.; Jeong, S.H.; Park, E.-S. Effect of particle size on the dissolution behaviors of poorly water-soluble drugs. *Arch. Pharmacal Res.* **2012**, *35*, 1187–1195. [CrossRef]
42. Sareen, S.; Mathew, G.; Joseph, L. Improvement in solubility of poor water-soluble drugs by solid dispersion. *Int. J. Pharm. Investig.* **2012**, *2*, 12–17. [CrossRef]
43. Bhakay, A.; Rahman, M.; Dave, R.N.; Bilgili, E. Bioavailability enhancement of poorly water-soluble drugs via nanocomposites: Formulation–Processing aspects and challenges. *Pharmaceutics* **2018**, *10*, 86. [CrossRef]
44. Lin, M.; Wang, H.; Meng, S.; Zhong, W.; Li, Z.; Cai, R.; Chen, Z.; Zhou, X.; Du, Q. Structure and release behavior of PMMA/silica composite drug delivery system. *J. Pharm. Sci.* **2007**, *96*, 1518–1526. [CrossRef]
45. Reinhard, C.S.; Radomsky, M.L.; Saltzman, W.M.; Hilton, J.; Brem, H. Polymeric controlled release of dexamethasone in normal rat brain. *J. Control. Release* **1991**, *16*, 331–339. [CrossRef]
46. Cypes, S.H.; Saltzman, W.M.; Giannelis, E.P. Organosilicate-polymer drug delivery systems: Controlled release and enhanced mechanical properties. *J. Control. Release* **2003**, *90*, 163–169. [CrossRef]
47. Frank, A.; Rath, S.K.; Venkatraman, S.S. Controlled release from bioerodible polymers: Effect of drug type and polymer composition. *J. Control. Release* **2005**, *102*, 333–344. [CrossRef]
48. Dilmi, A.; Bartil, T.; Yahia, N.; Benneghmouche, Z. Hydrogels based on 2-hydroxyethylmethacrylate and chitosan: Preparation, swelling behavior, and drug delivery. *Int. J. Polym. Mater. Polym. Biomater.* **2014**, *63*, 502–509. [CrossRef]

49. Wan, L.S.; Heng, P.W.; Wong, L.F. Relationship between swelling and drug release in a hydrophilic matrix. *Drug Dev. Ind. Pharm.* **1993**, *19*, 1201–1210. [CrossRef]
50. Liechty, W.B.; Kryscio, D.R.; Slaughter, B.V.; Peppas, N.A. Polymers for drug delivery systems. *Annu. Rev. Chem. Biomol. Eng.* **2010**, *1*, 149–173. [CrossRef] [PubMed]
51. Hezaveh, H.; Muhamad, I.I.; Noshadi, I.; Shu Fen, L.; Ngadi, N. Swelling behaviour and controlled drug release from cross-linked κ-carrageenan/NaCMC hydrogel by diffusion mechanism. *J. Microencapsul.* **2012**, *29*, 368–379. [CrossRef] [PubMed]
52. Narayanaswamy, R.; Torchilin, V.P. Hydrogels and their applications in targeted drug delivery. *Molecules* **2019**, *24*, 603. [CrossRef] [PubMed]
53. Park, H.-Y.; Choi, C.-R.; Kim, J.-H.; Kim, W.-S. Effect of pH on drug release from polysaccharide tablets. *Drug Deliv.* **1998**, *5*, 13–18. [CrossRef]
54. Ruan, L.L.; Wang, D.X.; Zhang, Y.W.; Zhao, J.X.; Zhang, X.F.; Chen, N.L. Different pH-values of release medium influence the drug release from PTX-PCL microspheres. *Adv. Mater. Res.* **2012**, *482*, 2605–2608. [CrossRef]
55. Cao, Z.; Li, W.; Liu, R.; Li, X.; Li, H.; Liu, L.; Chen, Y.; Lv, C.; Liu, Y. pH-and enzyme-triggered drug release as an important process in the design of anti-tumor drug delivery systems. *Biomed. Pharmacother.* **2019**, *118*, 109340. [CrossRef]
56. Qixing, Z.; Hongbin, W.; Jingyuan, D.; Shipu, L.; Yuhua, Y. The bacterial inhibitory ability andin vivo drug release pattern of a new drug delivery system: Ciprofloxacine/tricalcium phosphate delivery capsule. *J. Tongji Med. Univ.* **1998**, *18*, 172–176. [CrossRef]
57. Ebrahimi, S.; Farhadian, N.; Karimi, M.; Ebrahimi, M. Enhanced bactericidal effect of ceftriaxone drug encapsulated in nanostructured lipid carrier against gram-negative Escherichia coli bacteria: Drug formulation, optimization, and cell culture study. *Antimicrob. Resist. Infect. Control* **2020**, *9*, 28. [CrossRef]
58. Belzer, C.; de Vos, W.M. Microbes inside—From diversity to function: The case of Akkermansia. *ISME J.* **2012**, *6*, 1449–1458. [CrossRef]

polymers

MDPI

Article

The Influence of Hydrophobic Blocks of PEO-Containing Copolymers on Glyceryl Monooleate Lyotropic Liquid Crystalline Nanoparticles for Drug Delivery

Aleksander Forys [1], Maria Chountoulesi [2], Barbara Mendrek [1], Tomasz Konieczny [1], Theodore Sentoukas [1], Marcin Godzierz [1], Aleksandra Kordyka [1], Costas Demetzos [2], Stergios Pispas [3] and Barbara Trzebicka [1,*]

1 Centre of Polymer and Carbon Materials, Polish Academy of Sciences, 34 ul. M. Curie-Skłodowskiej, 41-819 Zabrze, Poland; aforys@cmpw-pan.edu.pl (A.F.); bmendrek@cmpw-pan.edu.pl (B.M.); tkonieczny@cmpw-pan.edu.pl (T.K.); tsentoukas@cmpw-pan.edu.pl (T.S.); mgodzierz@cmpw-pan.edu.pl (M.G.); akordyka@cmpw-pan.edu.pl (A.K.)

2 Section of Pharmaceutical Technology, Department of Pharmacy, School of Health Sciences, National and Kapodistrian University of Athens, Panepistimioupolis Zografou, 15771 Athens, Greece; mchountoules@pharm.uoa.gr (M.C.); demetzos@pharm.uoa.gr (C.D.)

3 Theoretical and Physical Chemistry Institute, National Hellenic Research Foundation, 48 Vassileos Constantinou Avenue, 11635 Athens, Greece; pispas@eie.gr

* Correspondence: btrzebicka@cmpw-pan.edu.pl

Citation: Forys, A.; Chountoulesi, M.; Mendrek, B.; Konieczny, T.; Sentoukas, T.; Godzierz, M.; Kordyka, A.; Demetzos, C.; Pispas, S.; Trzebicka, B. The Influence of Hydrophobic Blocks of PEO-Containing Copolymers on Glyceryl Monooleate Lyotropic Liquid Crystalline Nanoparticles for Drug Delivery. *Polymers* **2021**, *13*, 2607. https://doi.org/10.3390/polym13162607

Academic Editor: Vitaliy Khutoryanskiy

Received: 1 July 2021
Accepted: 2 August 2021
Published: 5 August 2021

Abstract: The investigation of properties of amphiphilic block copolymers as stabilizers for non-lamellar lyotropic liquid crystalline nanoparticles represents a fundamental issue for the formation, stability and upgraded functionality of these nanosystems. The aim of this work is to use amphiphilic block copolymers, not studied before, as stabilizers of glyceryl monooleate 1-(cis-9-octadecenoyl)-*rac*-glycerol (GMO) colloidal dispersions. Nanosystems were prepared with the use of poly(ethylene oxide)-*b*-poly(lactic acid) (PEO-*b*-PLA) and poly(ethylene oxide)-*b*-poly(5-methyl-5-ethyloxycarbonyl-1,3-dioxan-2-one) (PEO-*b*-PMEC) block copolymers. Different GMO:polymer molar ratios lead to formulation of nanoparticles with different size and internal organization, depending on the type of hydrophobic block. Resveratrol was loaded into the nanosystems as a model hydrophobic drug. The physicochemical and morphological characteristics of the prepared nanosystems were investigated by dynamic light scattering (DLS), cryogenic transmission electron microscopy (cryo-TEM), fast Fourier transform (FFT) analysis and X-ray diffraction (XRD). The studies allowed the description of the lyotropic liquid crystalline nanoparticles and evaluation of impact of copolymer composition on these nanosystems. The structures formed in GMO:block copolymer colloidal dispersions were compared with those discussed previously. The investigations broaden the toolbox of polymeric stabilizers for the development of this type of hybrid polymer/lipid nanostructures.

Keywords: amphiphilic block copolymers; lipidic lyotropic liquid crystals; cryogenic transmission electron microscopy; fast Fourier transform; dynamic light scattering; X-ray diffraction; resveratrol; lipid/polymer carriers

1. Introduction

Lipids, as a primary component of liposomal drug delivery systems, have been at the focus of scientific interest owing to their unique self-assembly features. Among the numerous nanostructures formed by lipids, lyotropic liquid crystals (LLC) or liquid crystalline nanoparticles (LCNPs) with highly ordered internal structure are especially important objects for research [1–3]. According to the type of internal organization, LLC can generally be classified into three categories: lamellar phase (L_α), cubic phase (V_2) and hexagonal phase (H_2), while the dispersion of each phase is known as liposomes, cubosomes and hexosomes, respectively [4–6]. Cubic and hexagonal phase have received wide interest due to their physicochemical properties and capability of encapsulating different types of compounds—from hydrophilic, through hydrophobic to amphiphilic. Combined with other properties like biocompatibility, nontoxicity and biodegradability, LLC have become an ideal potential platform for the design of versatile peptide [7] and drug delivery systems [8–15] as well as vaccines [16], imaging [17,18] and theranostics [19–21].

Various types of amphiphilic lipids are reported for the formation of lyotropic liquid crystals, such as phytantriol (PHYT) [16,22], glyceryl monooleate (GMO) [23,24], gadolinium oleate/myverol [17] and algal biomass [25]. To form stable LCNPs, aqueous dispersions of lipids require the addition of stabilizers, such as surfactants or amphiphilic block copolymers, which are one of the main class of polymeric stabilizers for these nanosystems [9,26]. It has been proven that the copolymers play a key role on the resulting formulation and determining the mesophases of final structures, while it also prevents unfavorable interactions between nanoparticles, such as aggregation [27,28]. In this case, hydrophilic block should be on the outer rim in the role of a hydrophilic corona, while the hydrophobic blocks should be anchored inside the lipid bilayer, forming hydrophobic interactions with the hydrophobic lipid tails. A schematic representation of the hybrid lipid/copolymer nanostructure can be found in [29,30]. Up until now, poly(ethylene oxide)-*b*-poly(propylene oxide)-*b*-poly(ethylene oxide) (PEO-PPO-PEO) triblock copolymers have been extensively used as stabilizers in these LLC systems, due to their biocompatibility and "stealth" properties [19,20,31]. However, the polymeric stabilizers can be carefully designed to provide extra properties that are extremely important due to the potential applications of lipid-based LLC for the drug delivery systems, such as stimuli-responsiveness [26,32,33]. In recent years, the finding of alternative amphiphilic block copolymers has been developing very fast, but the selection process of stabilizers is difficult, since it depends on many factors such as chemical structure, molar mass and hydrophobic–hydrophilic ratio of the chosen polymers [26,33,34].

One of the most popular lipids is glyceryl monooleate (GMO) [2,8,26,35] which is a polar, unsaturated monoglyceride. It can self-assemble into different mesophases, depending on water concentration and temperature [8,27,28]. Glyceryl monooleate is a nontoxic, biodegradable and biocompatible material, through it is commonly used for the preparation of liquid crystalline nanoparticles for drug delivery systems [8–10]. The amphiphilic stabilizer is necessary, in order for GMO dispersions to be created, because it provides steric stabilization by having its hydrophobic portion anchored in the lipid bilayers, while its more hydrophilic ends extend into the surrounding solution. In doing so, the inner cubic or hexagonal phase structure is maintained [36,37].

The most often used stabilizers in LLC systems are based on poly(ethylene oxide) (PEO), which is a polyether composed of repeated ethylene glycol units [-$(CH_2CH_2O)_n$]. PEO is hydrophilic, non-ionic, non-toxic, non-immunogenic and biocompatible polymer. It is well investigated and has many potential applications, from industrial manufacturing, through chemistry to medicine and drug delivery [38–41]. The widespread use of PEO owes to the broad range of its possible molar mass and solubility in aqueous media, as well as in many organic solvents. Furthermore, high polarity of PEO, increases hydrophilicity and thus enhances water solubility of PEO containing structures [41–44]. It can be combined with many hydrophobic chains towards the development of block copolymers.

Another promising polymer reported for biomedical applications is poly(lactic acid) (PLA) which has been used extensively since the 1970s. It is comprised of biodegradable, aliphatic polyesters, derived from lactide or 2-hydroxy propionic acid, which is generally obtained by bacterial fermentation of carbohydrates from renewable sources, like agricultural crops such as corn, potato and cassava [45]. PLA is biocompatible and safe [46,47] thermoplastic, high-strength, high-modulus polymer and is widely used in both industry and biomedicine [48,49]. Poly(5-methyl-5-ethyloxycarbonyl-1,3-dioxan-2-one) (PMEC) is biocompatibile, crystalline, degradable and resorbable material being a methylcarboxytrimethylene carbonate derivative. Due to its simple structure, it can be easily prepared from 2,2-bis (methylol)propionic acid (bis-MPA) in high yield and homopolymerized with predictable molar mass and narrow dispersity [50–52].

In the present study, novel potential polymeric stabilizers of glyceryl monooleate 1-(cis-9-octadecenoyl)-*rac*-glycerol lyotropic liquid crystals were investigated. More specifically, amphiphilic block copolymers were synthesized consisting of hydrophilic poly(ethylene oxide) and different hydrophobic blocks of poly(lactic acid) and poly(5-methyl-5-ethyloxycarb onyl-1,3-dioxan-2-one) (PMEC). We also compared its stabilizing ability with previously studied poly(ethylene oxide)-*b*-poly(ε-caprolactone) (PEO-*b*-PCL) block copolymers [28].

The prepared nanosystems were analyzed using a gamut of techniques, such as dynamic light scattering (DLS) for the physicochemical characterization, cryogenic transmission electron microscopy (cryo-TEM), fast Fourier transform (FFT) analysis and X-ray diffraction (XRD) for the morphological evaluation.

Studied LCNPs served as nanocarriers of resveratrol, a model hydrophobic drug that is a well-studied biologically active compound and exhibits various pharmacological properties. Among the many, it can be distinguished as anti-oxidative, anti-inflammatory, neuro-protective, anti-aging and anticancer drug [28,53,54].

The drug was encapsulated in the liquid crystalline nanoparticles, and its influence on the resulting structures was also examined. The resveratrol carriers obtained in this study were compared with these previously prepared with the use of PEO-*b*-PCL as a GMO dispersion stabilizer.

To the best of our knowledge, this is the first report where the above referred block copolymers were used as stabilizers for liquid crystalline nanoparticles.

2. Materials and Methods

2.1. Materials

All liquid crystalline nanosystems were prepared from glyceryl monooleate lipid Monomuls® 90-O18 (1-(cis-9-octadecenoyl)-*rac*-glycerol), (GMO) (BASF, Ludwigshafen, Germany) and used without further purification. All formulations were prepared in HPLC-grade water.

Resveratrol was acquired from Sigma-Aldrich Chemical Co (St. Louis, MO, USA).

Dichloromethane (POCH) was dried over CaH_2 and distilled under reduced pressure prior to use. THF (POCH) was distilled over a sodium-potassium alloy. Dowex 50WX8 (Sigma-Aldrich, Darmstadt, Germany) was washed with dry THF before use. Methoxy ether poly(ethylene oxide) 5000 ($mPEO_{123}$-OH, TCI, M_n(GPC) = 5 200, M_n(NMR) = 5 400) was dried by two azeotropic distillations using anhydrous toluene. L-lactide (LA) (>99.5%, Forusorb) was purified by sublimation two times before use. 1,8-Diazabicyclo[5.4.0]undec-7-ene (DBU) (98%, Sigma-Aldrich) was distilled under reduced pressure over BaO. Triethylamine (TEA, >99%, Sigma-Aldrich) was distilled under reduced pressure over BaO. 2,2-Bis(hydroxymethyl)propionic acid (bis-MPA, >97%, TCI), ethanol (96%, POCH), ethyl chloroformate (97%, Sigma-Aldrich), Amberlyst 15 (Sigma-Aldrich) and magnesium sulphate (Chempur, Piekary Slaskie, Poland) were used as received.

2.2. Methods

2.2.1. MEC Synthesis

Bis-MPA (44.2 g, 0.330 mol, 1.0 eq.) was dissolved in 300 mL of ethanol with 13.6 g of Amberlyst 15. Solution was refluxed for 24 h. Then, the resin was filtered, and unreacted ethanol was stripped off. The residue was dissolved in 400 mL of DCM and the insoluble part was removed via filtration. After removal of DCM, ethyl 2,2-bis(hydroxymethyl)propionate (bis-MPA-Et) was received as a colorless, viscous liquid. Yield: 75%.

In the second step, into the solution of bis-MPA-Et (40.0 g, 0.247 mol, 1 eq.) and ethyl chloroformate (58.7 mL, 0.617 mol, 2.5 eq.) in 400 cm^3 of DCM at 0 °C, TEA (85.9 mL, 0.617 mol, 2.5 eq.) diluted with 100 mL of DCM was added dropwise over a period of 1 h. The reaction mixture was kept in dry nitrogen atmosphere at 0 °C for 2 h and then at room temperature for 24 h. Then, the mixture was filtered, and the filtrate was concentrated under vacuum. Crude product was dissolved in 100 mL of DCM and washed with 50 mL 1

M $HCl_{(aq)}$ twice and 50 mL saturated solution of $NaHCO_3$, 25 mL of brine and 25 mL of distilled water. The organic phase was dried with anhydrous $MgSO_4$. Then, $MgSO_4$ was filtered off and DCM was stripped off. Crude product was then purified by crystallization from ethyl acetate twice to give MEC, as white crystals. Yield: 39%.

2.2.2. Polymerization of the Block Copolymers

The chemical structures and the molecular characteristics of copolymer samples are presented in Figure 1 and in Table 1, respectively.

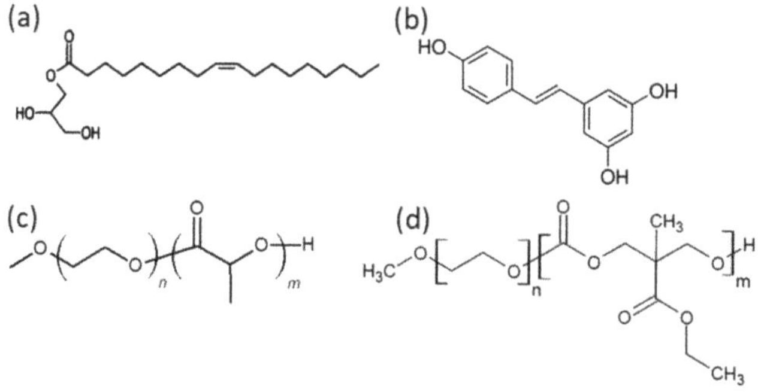

Figure 1. Chemical structures of (**a**) glyceryl monooleate (GMO) lipid, (**b**) resveratrol, (**c**) poly(ethylene oxide)-*b*-poly(lactic acid) (PEO-*b*-PLA), (**d**) poly(ethylene oxide)-*b*-poly(5-methyl-5-ethyloxycarbonyl-1,3-dioxan-2-one) (PEO-*b*-PMEC), employed in this study.

Table 1. Molecular characteristics of copolymers utilized.

Polymer	$DP_{(NMR)}$ of Hydrophobic Block	$M_{n(NMR)}$	$M_{w(GPC)}$	M_w/M_n [a]	%wt [b]
PEO-*b*-PLA	18	6700	6850	1.06	%wt PEO 81
PEO-*b*-PMEC	9	7100	6560	1.05	%wt PEO 76

[a] By SEC in DMF, calculated using PEO standards calibration; [b] by ^{1}H-NMR in $CDCl_3$.

Polymerization of poly(ethylene oxide)-*b*-poly(lactic acid) (PEO-*b*-PLA) was performed in an anhydrous atmosphere (glove-box, $H_2O < 1$ ppm, $O_2 < 3$ ppm). The diblock copolymer was synthesized by ring-opening polymerization (ROP) of LA using $mPEO_{123}$-OH as an initiator and DBU as catalyst in dry CH_2Cl_2 at the monomer concentration equal to 1.000 mmol/mL at room temperature.

Detailed polymerization procedures and chemical characterization of the block copolymers are presented in Supporting Information.

In case of poly(ethylene oxide)-*b*-poly(5-methyl-5-ethyloxycarbonyl-1,3-dioxan-2-one) (PEO-*b*-PMEC), polymerization was performed in an anhydrous atmosphere (glove-box, $H_2O < 1$ ppm, $O_2 < 3$ ppm). The diblock copolymer was synthesized by ring-opening polymerization (ROP) of MEC using $mPEO_{123}$-OH as an initiator and DBU as catalyst in dry CH_2Cl_2 at monomer concentration equal to 1.000 mmol/mL at room temperature.

Detailed polymerization procedures and chemical characterization (Figures S1 and S2) of the block copolymers are summarized in Supporting Information.

2.2.3. Preparation of Liquid Crystalline Nanoparticle Dispersions

PEO-*b*-PLA and PEO-*b*-PMEC copolymers were tested as polymeric stabilizers, each one individually. Two different weight ratios were prepared, namely GMO:polymeric stabilizer 9:1 and 4:1 respectively, which correspond to two different concentrations of the stabilizer, namely 10% and 20% w/w relative to the lipid mass. Copolymers were able to stabilize the lipid in nanoparticles in both lipid:polymer ratios. The lipid concentration was 20 mg/mL in all prepared systems. The temperature used during the preparation process of all the systems was 45 °C.

All liquid crystalline systems were prepared by Top-Down Method (TD). More specifically, GMO was weighted into glass vials and heated to 45 °C, until free flowing. The appropriate volume of HPLC-grade water solution (pH = 6.0), containing the different amounts of the polymeric stabilizers, was added to the vials containing the lipids, in order to achieve a lipid concentration of 20 mg/mL. The mixtures were firstly sonicated using a bath sonicator for 2 min, at 45 °C, followed by two 2-min sonication cycles (amplitude 70, cycle 0.7), interrupted by a 2-min resting period, using a probe sonicator (UW 2070 Bandelin electronic, Berlin, Germany), until a milky dispersion was formed. The resultant dispersions were allowed to anneal for 30 min, then stored at room temperature and measured 5 days after preparation.

In the case of liquid crystalline systems with entrapped resveratrol, the same process as above was followed, by the difference that the appropriate amounts of resveratrol and GMO were fully dissolved in ethanol initially. Ethanol was gently evaporated until a dry film of liquid-resveratrol mixture was achieved. The total resveratrol concentration in the final dispersion was 2 mg/mL.

2.2.4. Physicochemical and Morphological Characterization of the Liquid Crystalline Dispersions

Dynamic and Electrophoretic Light Scattering

Hydrodynamic radius and zeta potential measurements were performed at least in triplicate on a Zetasizer Nano ZS 90 (Malvern Instruments, Malvern, Worcestershire, UK) in disposable cuvettes and processed with Zetasizer Software (Malvern Instruments, Malvern, Worcestershire, UK) version 6.32.

Cryogenic Transmission Electron Microscopy (Cryo-TEM)

Cryogenic Transmission Electron Microscopy (cryo-TEM) images were obtained using a Tecnai F20 X TWIN microscope (FEI Company, Hillsboro, OR, USA) equipped with field emission gun, operating at an acceleration voltage of 200 kV. Images were recorded on the Gatan Rio 16 CMOS 4k camera (Gatan Inc., Pleasanton, CA, USA) and processed with Gatan Microscopy Suite (GMS) software (Gatan Inc., Pleasanton, CA, USA). Specimen preparation was done by vitrification of the aqueous solutions on grids with holey carbon film (Quantifoil R 2/2; Quantifoil Micro Tools GmbH, Großlöbichau, Germany). Prior to use, the grids were activated for 15 s in oxygen plasma using a Femto plasma cleaner (Diener Electronic, Ebhausen, Germany). Cryo-samples were prepared by applying a droplet (3 μL) of the suspension to the grid, blotting with filter paper and immediate freezing in liquid ethane using a fully automated blotting device Vitrobot Mark IV (Thermo Fisher Scientific, Waltham, MA, USA). After preparation, the vitrified specimens were kept under liquid nitrogen until they were inserted into a cryo-TEM-holder Gatan 626 (Gatan Inc., Pleasanton, CA, USA) and analyzed in the TEM at −178 °C.

X-ray Diffraction (XRD)

Samples for X-ray diffraction studies were prepared by evaporation of H_2O at 50 °C, up to a final concentration around 8–10 mg/mL. Samples obtained from dispersions were placed between Capton foil in PMMA holder with additional PMMA 1 mm thick distance, in order to obtain optimal intensity of scattering in water.

XRD was performed using the D8 Advance diffractometer (Bruker, Karlsruhe, Germany) with Cu-Kα cathode (λ = 1.54 nm) working in transmission mode. The scan rate was 0.02°/min with scanning step 0.01° in range of 0.5° to 3° 2Θ. All patterns were acquired at least seven times, then accumulated to obtained higher resolution. Background subtraction was performed using DIFFRAC.EVA program, using air scattering filter. Obtained patterns were smoothed using Fourier smooth filter. Typically, *Im3m* and *Pn3m* cubic structures are reported for this type of liquid crystalline, according to literature [36,55–57]. Model patterns of two cubic, hexoctahedral phases: body centered *Im3m* and primitive *Pn3m* were calculated using DIFFRAC.EVA program, assuming that lattice parameter a = 130 Å [55,57], while exact lattice parameters of fitted phase were calculated using Rietveld refinement in TOPAS 6 program, based on Williamson-Hall theory. The pseudo-Voigt function was used in the description of diffraction line profiles at the Rietveld refinement [58,59]. The R_{wp} (weighted-pattern factor) and GOF (goodness-of-fit) parameters were used as numerical criteria of the quality of the fit of calculated to experimental diffraction data.

2.2.5. Characterization of the Liquid Crystalline Dispersions with Entrapped Resveratrol
Entrapment Efficiency and Drug Loading Determination

The ultrafiltration centrifugal method was utilized for the separation of the free resveratrol from the resveratrol entrapped in the liquid crystalline nanoparticles [15,23]. The liquid crystalline nanoparticle dispersions were centrifuged for 45 min at 8000 rpm inside the centrifugal filter tubes, with a molecular weight cutoff of 10 kDa, at 4 °C. The nanoparticles were separated from the aqueous phase and the supernatant was analyzed via UV-Vis spectroscopy in order to quantitate the free resveratrol concentration. Absorption measurements were performed at a range of 190–600 nm, while the value at 307 nm and a pre-constructed calibration curve were used for the analysis. In addition, plain liquid crystalline nanoparticles were centrifuged, and supernatant was used as blank. The entrapment efficiency (EE)% was calculated using the following equation:

$$(EE)\ \% = (1 - \frac{C_{supernatant}}{C_{total}})\ \%$$

where $C_{supernatant}$ is the resveratrol concentration that was quantified in the supernatant (non-entrapped) and C_{total} is the total concentration of the resveratrol added in the initial hybrid liposomes aqueous solutions (2 mg/mL) [15,23,31].

The drug loading (DL)% was calculated according to the following equation

$$(DL)\ \% = (\frac{C_{total} - C_{supernatant}}{C_{lipid}})\ \%$$

where C_{total} is the total concentration of the resveratrol added in the dispersion (2 mg/mL), $C_{supernatant}$ is the non-entrapped resveratrol concentration and C_{lipid} is the total concentration of the lipid in the aqueous solution (20 mg/mL) [15,31].

In Vitro Resveratrol Release Studies

The drug-release studies were conducted using the dialysis method in the dark, [15,23,31,60,61] based on a previous study [28]. 0.5 mL of the aqueous solution samples were placed into a dialysis sack, with a molecular weight cutoff at 10,000. The dialysis sacks were inserted in a 20 mL PBS (pH = 7.4) shaking water bath set at 37 °C. Aliquots of samples were taken from the external solution at specific time intervals, while the volume was replaced with fresh release medium. Samples were diluted with ethanol and quantified for drug content by an Analytik Jena Specord 200 plus UV-Vis spectrometer (Jena, Thuringia, Germany), in the range of 190–600 nm. The values corresponding at 307 nm of the spectra were used for the analysis, along with a pre-constructed standard curve for resveratrol in PBS/ethanol. In addition, empty hybrid liposomes were used as blanks. The % cumulative mass of resveratrol released versus time was plotted.

3. Results and Discussion

3.1. Physicochemical and Morphological Characteristics of the Liquid Crystalline Nanosystems as Revealed by DLS and Cryo-TEM

The chemical composition of amphiphilic polymeric stabilizers PEO-*b*-PLA and PEO-*b*-PMEC affected the size, as well as the morphology of prepared nanosystems. Both block copolymers used in our study were found to provide stable GMO-based colloidal dispersions, at least at the studied ratios. Samples were measured 5 days after preparation. We assume that the present nanosystems were stable as it was in the case of previously studied GMO:PEO-*b*-PCL [28] and as shown for Pluronic F127 systems [27,62].

The GMO:copolymers hydrodynamic radius and size dispersity (PDI) measured by DLS and average sizes from cryo-TEM images are given in Table 2. Cryo-TEM histograms are presented in Supporting Information (Figures S3 and S4). Size distribution by intensity from DLS are shown in Supporting Information (Figures S5 and S6).

Table 2. Dynamic light scattering, ζ-potential, cryo-TEM histograms, XRD and FFT analysis results.

Sample	Weight Ratio	R_h (nm)	PDI (±SD)	ζ-Pot (mV)	Average Size (nm) Cryo-TEM	Space Group XRD	Space Group FFT
GMO: PEO-*b*-PLA	9:1	88 and 10	0.53 (±0.04)	−13	66	*Pn3m*	*Pn3m*
GMO: PEO-*b*-PLA	4:1	137 and 18	0.62 (±0.01)	−17	88	*Im3m*	*Im3m/Pn3m*
GMO: PEO-*b*-PMEC	9:1	150 and 19	0.63 (±0.03)	−21	85	*Pn3m*	*Pn3m*
GMO: PEO-*b*-PMEC	4:1	141 and 16	0.55 (±0.02)	−16	63	*Pn3m*	*Pn3m*
GMO: PEO-*b*-PLA +resveratrol	9:1	17	0.40 (±0.02)	−26	85	-	-
GMO: PEO-*b*-PLA +resveratrol	4:1	146 and 20	0.55 (±0.04)	−13	78	-	-
GMO: PEO-*b*-PMEC +resveratrol	9:1	179 and 23	0.58 (±0.01)	−27	147	-	*Im3m/Pn3m*
GMO: PEO-*b*-PMEC +resveratrol	4:1	157 and 14	0.85 (±0.04)	−18	57	-	*Im3m*

Starting from the GMO:PEO-*b*-PLA systems, an increase of the total amount of copolymer (from 9:1 to 4:1 w/w) led to an increase of the particle size. Nanosystems with more stabilizer content are more homogenous, exhibiting smaller values of PDI, with negative ζ-potential values for both ratios.

Cryo-TEM revealed morphological variety inside the dispersions (Figure 2a,b). More specifically, in the first case with a smaller amount of PEO-*b*-PLA, multilamellar particles prevailed, with confined, striated, curved inner structure, resembling an onion. This large population was coexisting with a small number of "sponge-like" nanoparticles, consisting of an outer layer built of intersecting lamellas and an apparent dense inner core with highly disordered interior and with absence of long-range order and periodicity [63,64], and particles with a highly ordered internal structure. FFT analysis of images

(Figure 3a) gave a space group close to *Pn3m* (double-diamond type bicontinuous cubic phase). Simple vesicles were not observed by cryo-TEM.

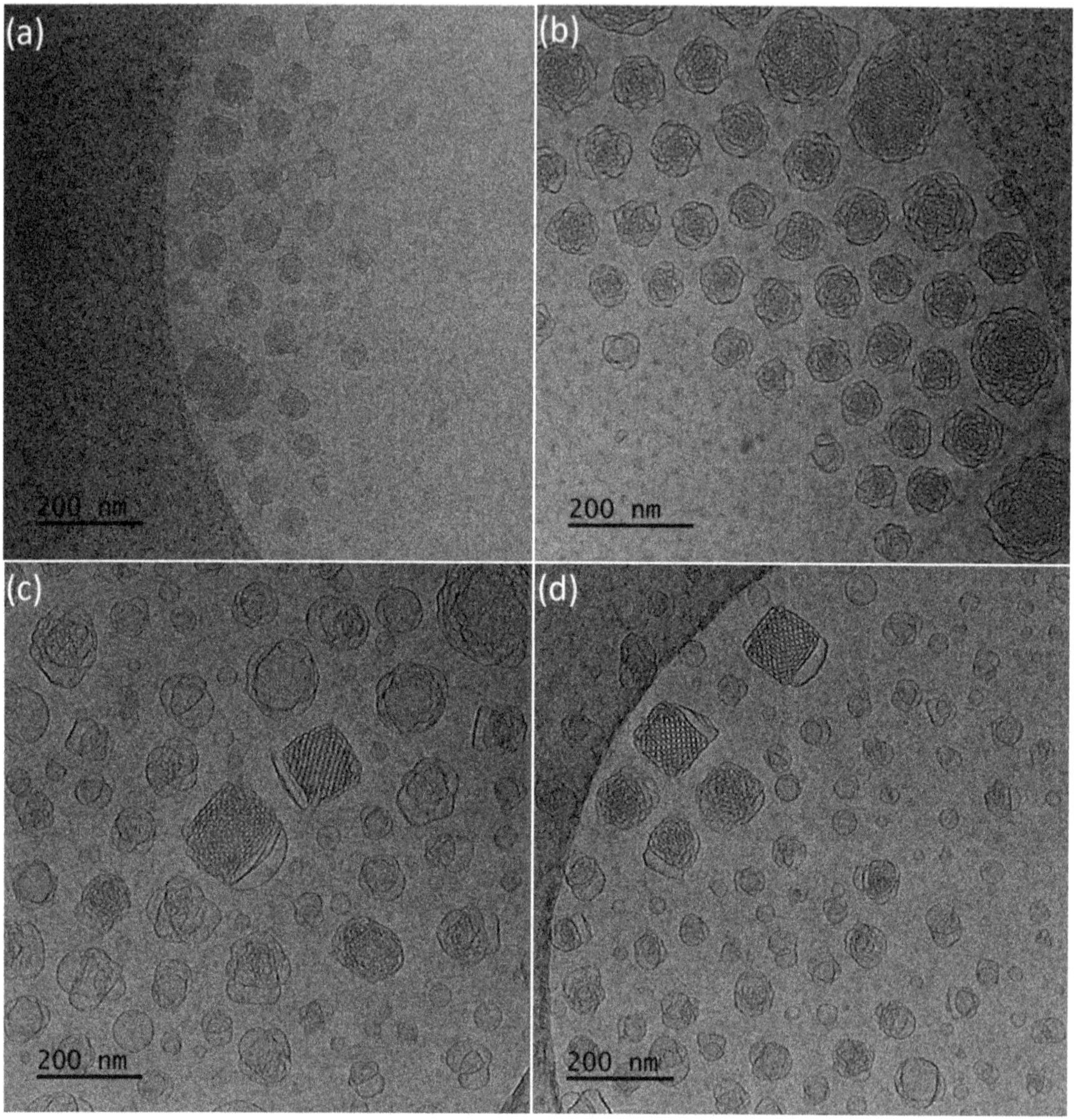

Figure 2. Representative cryo-TEM images of the (**a**) GMO:PEO-*b*-PLA 9:1, (**b**) GMO:PEO-*b*-PLA 4:1, (**c**) GMO:PEO-*b*-PMEC 9:1, (**d**) GMO:PEO-*b*-PMEC 4:1.

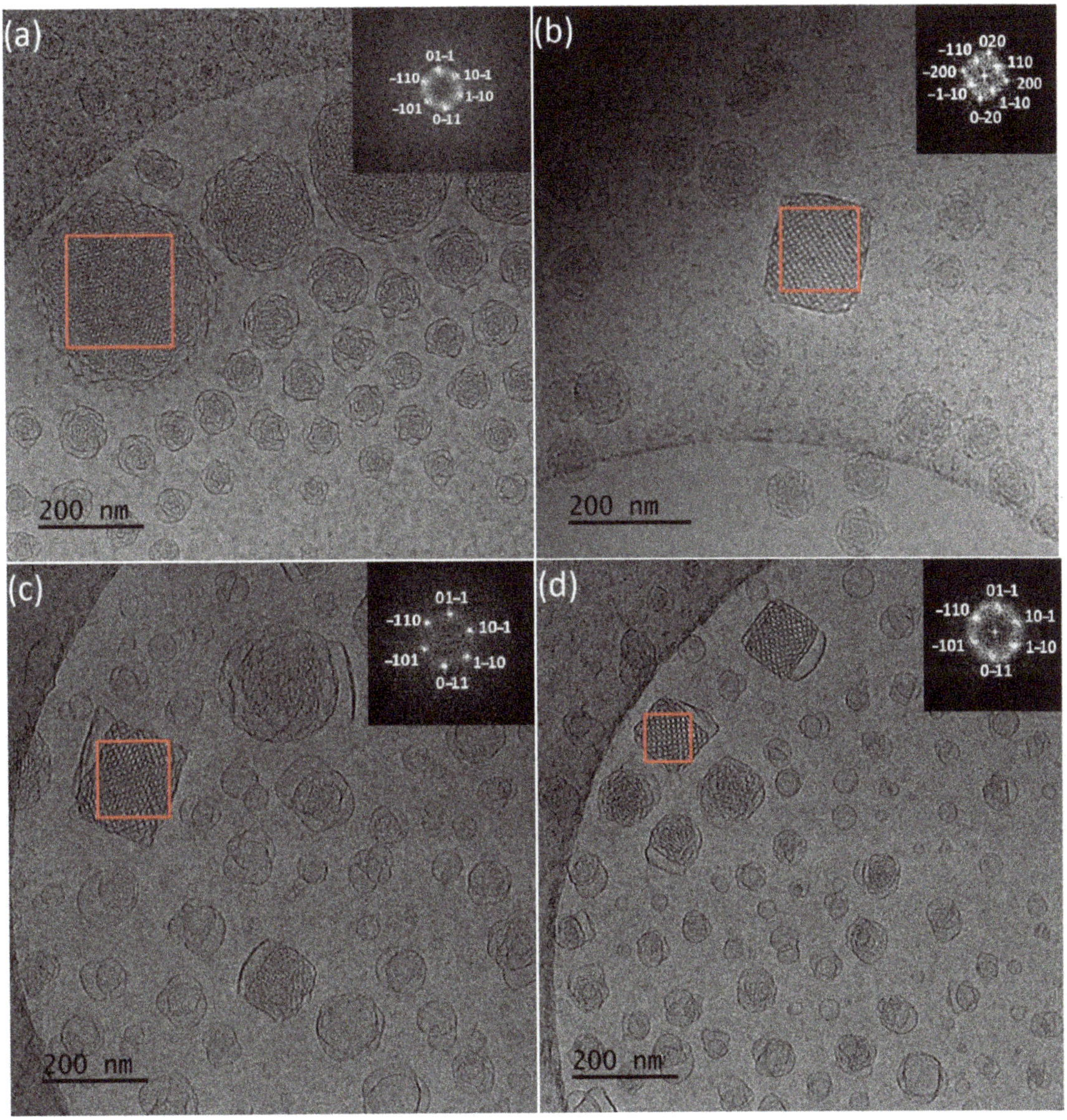

Figure 3. Fast Fourier transform analysis of particles with ordered internal structure for (**a**) GMO:PEO-*b*-PLA 9:1, (**b**) GMO:PEO-*b*-PLA 4:1, (**c**) GMO:PEO-*b*-PMEC 9:1, and (**d**) GMO:PEO-*b*-PMEC 4:1.

In the case of GMO:PEO-*b*-PLA 4:1 ratio, cryo-TEM revealed the predominant spherical, "sponge-like" particles, and particles with a highly ordered internal structure. The fast Fourier transform analysis (Figure 3b) gave a space group symmetry likely to *Im3m* (primitive type bicontinuous cubic phase). In addition, particles with *Pn3m* symmetry were observed (Figure S7). Apart from this category of LCNPs with ordered internal structure, there were also spherical vesicles observed as a minority component.

A small number of large aggregates with irregular shape were observed for both GMO:PEO-*b*-PLA ratios.

The encapsulation of hydrophobic resveratrol affected the physicochemical and morphological characteristics of the prepared nanosystems. The average size of particles is higher in the case of 9:1 ratio and smaller for 4:1, compared to those without the drug (Fig.S3 and S4). They present negative ζ-potential values and smaller size dispersity (Table 2). According to cryo-TEM results, systems with resveratrol (Figure 4a,b) are characterized by significant differences in particles morphology. For 4:1 ratio, a population of spherical vesicles and "sponge-like" particles was observed. Moreover, elongated structures that were not found in drug-free systems were observed. Smaller amount of copolymer results predominant in "sponge-like" particles, coexisting with small number of spherical vesicles. Worm-like structures were in minority.

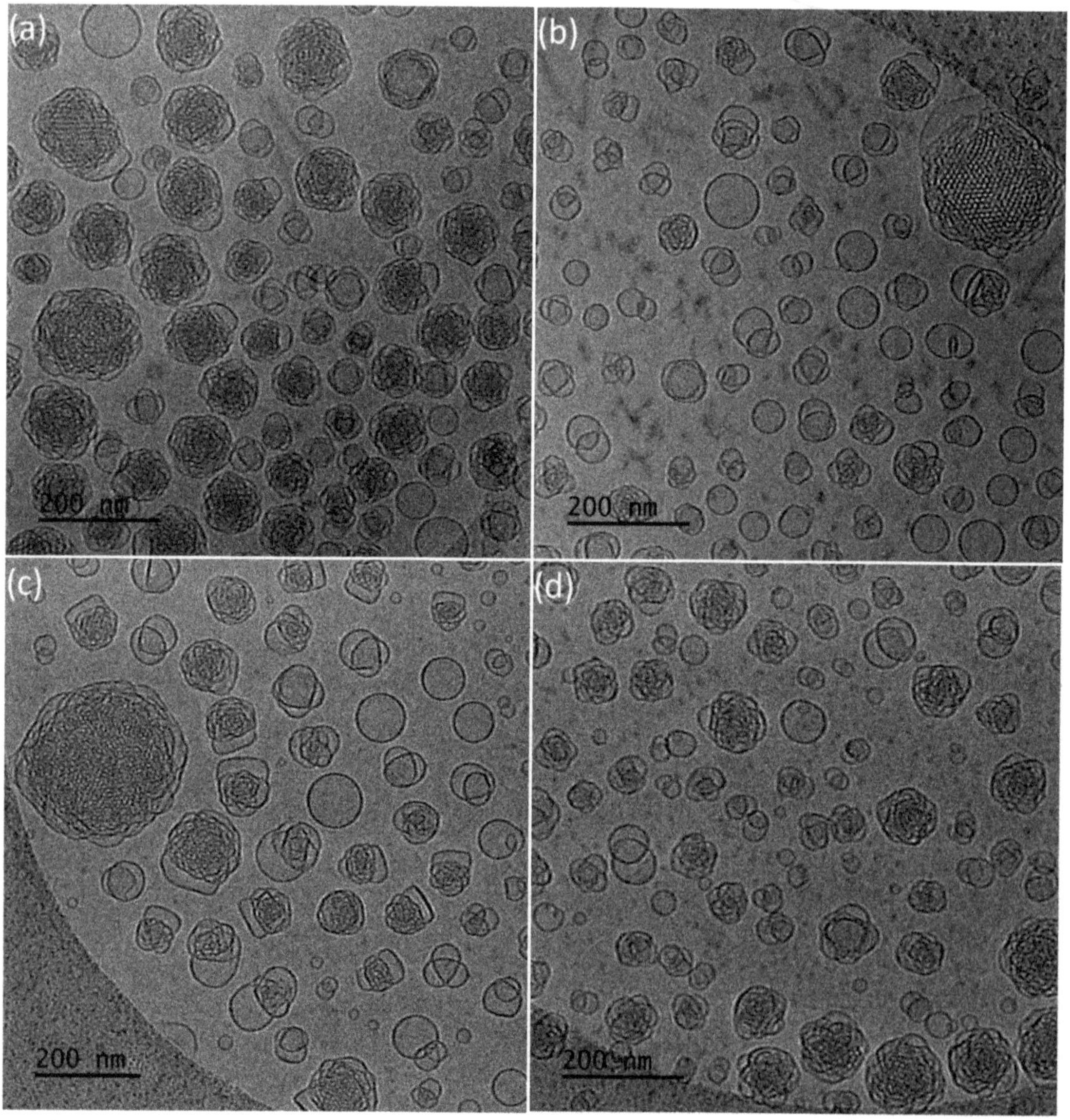

Figure 4. Representative cryo-TEM images of the (**a**) GMO:PEO-*b*-PLA 9:1 + resveratrol, (**b**) GMO:PEO-*b*-PLA 4:1 + resveratrol, (**c**) GMO:PEO-*b*-PMEC 9:1 + resveratrol, (**d**) GMO:PEO-*b*-PMEC 4:1 + resveratrol.

Both GMO:PEO-*b*-PLA ratios systems containing resveratrol resulted in particles of unorganized internal structure, which was assessed with FFT analysis.

Replacement PEO-*b*-PLA with PEO-*b*-PMEC with bigger side group in the hydrophobic blocks, lead to a different, prevailing morphology of nanosystems (Figure 2c,d). Average particles size from DLS and cryo-TEM is higher in case of 9:1, compared to GMO:PEO-*b*-PLA. For 4:1 ratio, average DLS results are similar, but cryo-TEM histograms reveal smaller sizes of particles. A higher amount of this stabilizer caused a decrease of the particle size, compared to a 9:1 ratio, probably because of the larger reduction of interfacial tension between GMO lipid and water phase. Consequently, it led to the formation of a larger surface area and fragmentation of particles towards smaller sizes [32,36,65,66].

Nanosystems were found to present PDI values of 0.55 and 0.63, respectively, with negative ζ-potential values for both ratios (Table 2).

Vesicles, "sponge-like" particles and particles with organized internal structure and cubic shape were observed at both ratios. The fast Fourier transform analysis (Figure 3c,d) gave a space group symmetry likely to *Pn3m*. The increase of the PEO-*b*-PMEC concentration caused decrease of the nanoparticles with regular organization and led to the formation of higher amount of small, simple vesicles exhibiting no internal structure and irregularly shaped particles, which may represent different stages of the fusion processes that are taking place towards the formation of the more organized structures.

This is in accordance with the literature [32,56] which confirms that higher stabilizer concentrations can cause formation of vesicular structures, such as liposomes, reducing the percentage of the existing cubic structures.

The addition of resveratrol results in higher average sizes of particles, compared to pure nanosystems. PDI values are higher for the 4:1 ratio and smaller for the second ratio (Table 2). Cryo-TEM (Figure 4c,d) revealed differences in morphology of the particles for these ratios. Predominantly "sponge-like" structures were observed for higher amount of copolymer with drug present. This large population was coexisting with spherical vesicles. In the case of the 9:1 ratio, the vesicles population was increased compared to "sponge-like" structures.

Apart from these categories of structures, there were also particles with ordered internal structure, observed for both ratios in minority. FFT analysis gave a space group close to *Im3m* (primitive type bicontinuous cubic phase) (Figure 5a,b) and *Pn3m* (double-diamond type bicontinuous cubic phase) (Figure 5a).

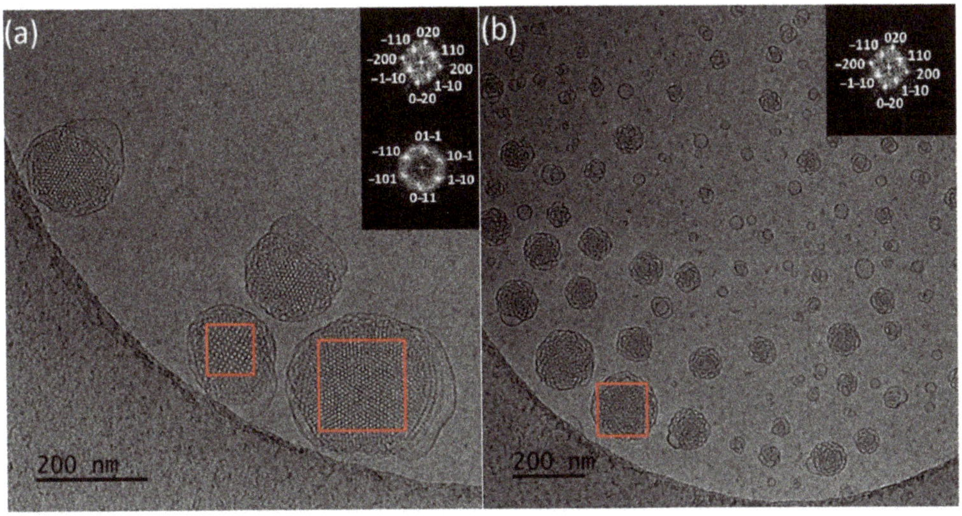

Figure 5. Fast Fourier transform (FFT) analysis of particles with ordered internal structure for (**a**) GMO:PEO-*b*-PMEC 9:1 + resveratrol, and (**b**) GMO:PEO-*b*-PMEC 4:1 + resveratrol.

We have compared the nanosystems of GMO prepared with PEO-*b*-PCL from our previous work [28] with the same PEO block and similar molar mass as the copolymers used here. All PEO-*b*-PCL systems were found to present PDI values ≤ 0.25, with negative ζ-potential values. Higher amount of PEO-*b*-PCL copolymer (4:1 ratio) yielded smaller nanoparticles, similar to GMO:PEO-*b*-PMEC nanosystems. It can be observed that the chemical composition of the used copolymer influenced the size of the nanoparticles. PEO-*b*-PCL caused formation of particles with smaller average dimension than PEO-*b*-PLA and PEO-*b*-PMEC. Furthermore, nanosystems are more homogenous exhibiting smaller values of PDI compared to new ones. All GMO:copolymers systems present negative ζ-potential values.

In addition, significant differences in particles morphology were revealed by cryo-TEM. More specifically, spherical vesicles were coexisting with liquid crystalline nanoparticles with lower grades of internal organization that resemble "sponge-like" structures, for both ratios. Higher concentration of PEO-*b*-PCL results in absence of nanoparticles with ordered internal structure, as confirmed with fast Fourier transform analysis. GMO:PEO-*b*-PCL 9:1 systems exhibited nanoparticles with a highly ordered internal structure, being square or almost squared shaped. FFT analysis patterns give body centered *Im3m* symmetry.

3.2. X-ray Diffraction

The organized internal structure of particles, assessed with fast Fourier transform analysis is likely corresponding to cubic structures of *Pn3m* or *Im3m* symmetry [67]. For the full characterization of the predominant ordered internal structure (Table 2), samples were measured by X-ray diffraction (Figure 6).

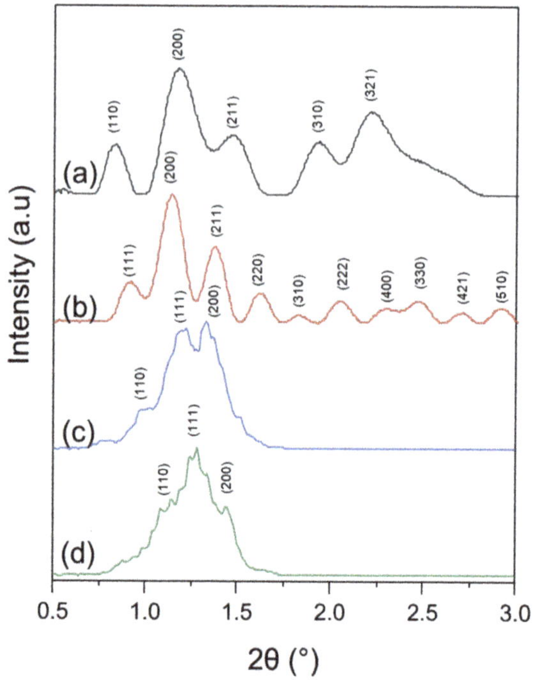

Figure 6. XRD patterns with marked peaks of fitted space groups of (**a**) GMO:PEO-*b*-PLA 4:1, (**b**) GMO:PEO-*b*-PLA 9:1, (**c**) GMO:PEO-*b*-PMEC 4:1, (**d**) GMO:PEO-*b*-PMEC 9:1.

For GMO:PEO-*b*-PLA, the fitting of calculated theoretical structure patterns gives slightly better fit for *Im3m* structure than *Pn3m* in the case of 4:1 stoichiometry. However, considering peak positions and corresponding d-spacings, it should be noticed that for primitive *Pn3m* the (111) peak is missing, which results in worse fit for that structure. Further fitting with Rietveld refinement of the assumed body centered *Im3m* structure shows lattice parameter a = 148.4 ± 0.7 Å. In the case of 9:1 sample, better fit was obtained for primitive *Pn3m* cubic structure with lattice parameter a = 151.1 ± 0.9 Å. Calculated lattice parameters of GMO:PEO-*b*-PLA shows lattice enlargement in comparison to the liquid crystals described in the literature [36,55–57].

In case of GMO:PEO-*b*-PMEC nanosystems, the characteristic (111) peak of *Pn3m* structure was observed (Figure 6). Rietveld refinement confirms presence of *Pn3m* structure for 4:1 and 9:1 nanosystems with lattice parameters 129.6 ± 0.6 and 131.3 ± 0.7 Å, respectively. Moreover, peak shape and intensity are more developed in the case of the 9:1 nanosystem than for the 4:1 ratio.

Comparing all of the examined samples, the best well-developed structure was observed for the GMO:PEO-*b*-PLA 9:1 nanosystem, while low ordered structure was detected for the GMO:PEO-*b*-PMEC 4:1 nanosystem. In both copolymers some regularity was detected—higher amount of GMO lipid (9:1) allows higher order of the obtained structure. In the GMO:PEO-*b*-PLA 4:1 system—a body centered *Im3m* structure was observed instead of primitive *Pn3m*. *Pn3m* structure was reported in literature for PEO-PPO-PEO liquid nanoparticles [36,67] and in our studies for PEO-*b*-PCL cubosomes [28].

3.3. Resveratrol Entrapment and In Vitro Release Studies

In order to investigate the ability of the above lipid:copolymer hybrid nanosystems to encapsulate drug molecules and act as drug nanocarriers, the hydrophobic drug resveratrol was employed as a low molecular weight cargo molecule model.

Resveratrol was incorporated to GMO:PEO-*b*-PLA and GMO:PEO-*b*-PMEC LCNPs to test their drug-loading and release abilities in conditions similar to those inside of a human body. The results of loading and entrapment efficiency are shown in Table 3. The entrapment efficiency was over 99% for resveratrol for all systems under study. The release of the drug from its carriers takes place in three stages as shown in Figure 7. A quick resveratrol release in terms of 10–25% approximately of its total mass can be seen for the first half-hour, probably due to the large surface area to volume ratio of the nanoparticles [23]. A prolonged release is observed for the next three hours which is associated with the strong entrapment of the drug inside the nanosystem [68,69]. Finally, a plateau was observed with no extra drug-release, with the total released resveratrol mass under 40%. This is a typical behavior for such liquid crystalline nanosystems [70,71]. Resveratrol, as a hydrophobic drug can "hide" deeply inside the GMO lipid bilayer, as proved by the high entrapment efficiency percentages. The prolonged release is driven by certain factors, such as the "pore" size [3,72], the organization of the cubic phase network and the entrapped water phase. They are crucial for the diffusion of the drug to the aqueous medium through the "water channels". Thus, highly organized networks lead in tighter nanostructures that result in more drug retention, while looser formations lead to maximum drug release [23,31].

Table 3. Drug loading and entrapment efficiency results.

Sample	Weight Ratio	Entrapment Efficiency (%)	Drug Loading (%)
GMO:PEO-*b*-PLA	9:1	99.91	9.99
GMO:PEO-*b*-PLA	4:1	99.84	9.98
GMO:PEO-*b*-PMEC	9:1	99.84	9.98
GMO:PEO-*b*-PMEC	4:1	99.46	9.94

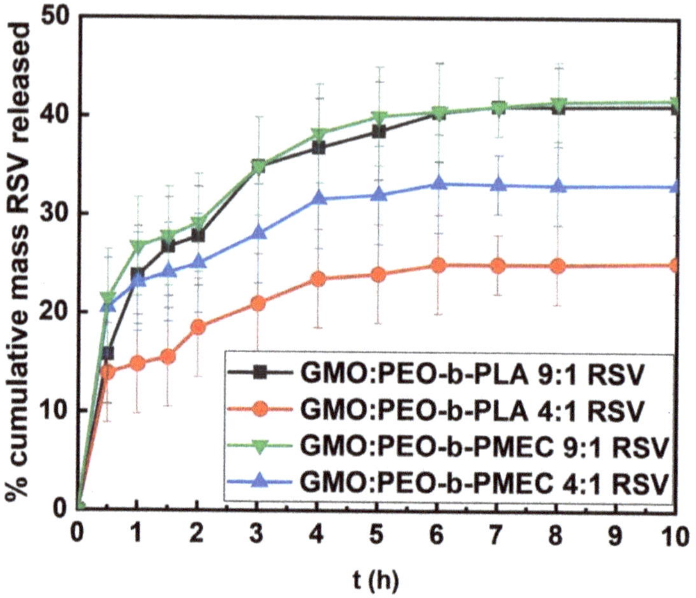

Figure 7. Cumulative mass of resveratrol that is released (%) from the hybrid liquid crystalline nanoparticles (±SD values are given for 5 runs).

The differences shown in the release profile graphs in Figure 7 can be attributed to the block copolymers, which strongly influence the formation of the final nanostructures. The pore size and tortuosity of the water channels of the cubic phase contribute to the sustained release because the entrapped molecule has to be navigated through a nanotubular aqueous network assembled by the continuous lipid bilayer architecture.

All the under-study LCNPs are highly influenced by resveratrol since its encapsulation results in the formation of nanostructures with lower internal order. Vesicles and "sponge-like" nanoparticles prevail when the drug is loaded in comparison to the "empty" ones, as shown in cryo-TEM studies. It is possible that resveratrol, being the most hydrophobic molecule in these systems, antagonizes PMEC and PLA in terms of entrapment inside the GMO phase, altering the final nanostructures.

The GMO:PEO-*b*-PLA 4:1 nanosystem presented a minimum release of resveratrol, depicting a more complex network than the other systems, preventing the drug to "find-its-way" to the external aqueous medium. On the other hand, both GMO:PEO-*b*-PLA and GMO:PEO-*b*-PMEC 9:1 showed the maximum resveratrol release, with an almost identical behavior, probably due to a looser internal structure. Last but not least, drug release from the GMO:PEO-*b*-PMEC 4:1 nanoparticles was somewhere in the middle of all four systems. Such differences in the drug-release behavior of each system are driven by the ratio of the diblock copolymers and the variations in the nature of hydrophobic units. It should be emphasized that the copolymer molar masses, and hydrophobic units in the chains were nearly identical.

Previously studied GMO:poly(ethylene oxide)-*b*-poly(ε-caprolactone) and GMO:poly (2-methyl-2-oxazoline)-*grad*-poly(2-phenyl-2-oxazoline) (MPO_x) LCNPs [28] showed a similar behavior in terms of initial, prolonged and plateau release profile stages. All the systems presented excellent entrapment efficiency, over 97%, which is attributed to the lipophilicity of resveratrol and GMO bilayers formed in chimeric particles. In the case of GMO:PEO-*b*-PCL and GMO:MPOx, the prolonged release lasted more than 3 h. Conditions used in the present study were the same as in a previous studied GMO:PEO-*b*-PCL+resveratrol system (Phosphate Buffer Saline, PBS) [28]. This allowed for the comparison between the systems.

The released amount of resveratrol for the GMO:PEO-*b*-PMEC and GMO:PEO-*b*-PLA LCNPs is smaller in terms of % cumulative mass in respect with the GMO:PEO-*b*-PCL and GMO:MPO_x. The release from hybrid particles can be related to the internal network phase, its complexity and organization. In the studied systems the ceiling release is lower than for GMO:PEO-*b*-PCL which also indicates that in the present case structures are better organized for preventing drug release.

4. Conclusions

Non-lamellar lipid liquid crystalline nanosystems were prepared from GMO and the PEO-*b*-PLA and PEO-*b*-PMEC block copolymers at different GMO:copolymer ratios (9:1 and 4:1). The copolymers were obtained using the same PEO macroinitiator and had hydrophobic blocks of similar weight. All block copolymers were found to act successfully as stabilizers and provide GMO-based colloidal dispersions at the studied ratios. The influence of hydrophobic blocks of the PEO-containing copolymers was investigated in physicochemical and morphological terms. Different GMO and copolymer molar ratios lead to formulation of nanoparticles with different size and internal organization. The hydrophobic block of the copolymers affected the size and morphology of the nanosystems. The nanosystems stabilized by the PEO-*b*-PMEC copolymer exhibited higher grades of internal organization with respect to PEO-*b*-PLA, as revealed by cryo-TEM, fast Fourier transform and X-ray diffraction analysis. They appeared to be better organized than previously studied stabilizer PEO-*b*-PCL.

The loading of hydrophobic resveratrol influenced the morphology and internal organization of the final hybrid lipid-polymer nanostructures with incorporated drug.

These findings support the use of new, alternative polymeric stabilizers for preparation of LCNPs, also with potential applications as the drug delivery systems, broadening the already existing toolbox of polymeric stabilizers and hybrid lipid-copolymer structures.

Supplementary Materials: The following are available online at https://www.mdpi.com/article/10.3390/polym13162607/s1, Figure S1. ^{1}H NMR spectra of the (a) PEO-*b*-PLA, (b) PEO-*b*-PMEC, Figure S2. GPC chromatogram of the (a) PEO-*b*-PLA, (b) PEO-*b*-PMEC, Figure S3. Cryo-TEM histograms of the (a) GMO:PEO-*b*-PLA 9:1, (b) GMO:PEO-*b*-PLA 4:1, (c) GMO:PEO-*b*-PMEC 9:1, (d) GMO:PEO-*b*-PMEC 4:1 nanosystem, Figure S4. Cryo-TEM histograms of the of the (a) GMO:PEO-*b*-PLA 9:1 + resveratrol, (b) GMO:PEO-*b*-PLA 4:1 + resveratrol, (c) GMO:PEO-*b*-MEC 9:1 + resveratrol, (d) GMO:PEO-*b*-MEC 4:1 + resveratrol nanosystems, Figure S5. Size distributions from DLS of the (a) GMO:PEO-*b*-PLA 9:1, (b) GMO:PEO-*b*-PLA 4:1, (c) GMO:PEO-*b*-PMEC 9:1, (d) GMO:PEO-*b*-PMEC 4:1 nanosystems, Figure S6. Size distributions from DLS of the (a) GMO:PEO-*b*-PLA 9:1 + resveratrol, (b) GMO:PEO-*b*-PLA 4:1 + resveratrol, (c) GMO:PEO-*b*-PMEC 9:1 + resveratrol, (d) GMO:PEO-*b*-MEC 4:1 + resveratrol nanosystems, Figure S7. Fast Fourier Transform of particles with ordered internal structure (*Pn3m* symmetry) of GMO:PEO-*b*-PLA 4:1.

Author Contributions: Conceptualization, B.T.; investigation, A.F., M.C., B.M., T.K., T.S., M.G. and A.K.; writing—original draft preparation, A.F.; writing—review and editing, M.C., T.S., S.P. and B.T.; visualization, A.F.; supervision, B.T., C.D. and S.P. All authors have read and agreed to the published version of the manuscript.

Funding: This work was supported by state funds for the Centre of Polymer and Carbon Materials, Polish Academy of Sciences.

Institutional Review Board Statement: Not applicable.

Informed Consent Statement: Not applicable.

Data Availability Statement: The data presented in this study are available on request from the corresponding author.

Conflicts of Interest: The authors declare no conflict of interest.

References

1. Dierking, I.; Neto, A.M.F. Novel trends in lyotropic liquid crystals. *Crystals* **2020**, *10*, 604. [CrossRef]
2. Singhvi, G.; Banerjee, S.; Khosa, A. *Lyotropic Liquid Crystal Nanoparticles*; Elsevier: Amsterdam, The Netherlands, 2018; ISBN 9780128136638.
3. Barriga, H.M.G.; Holme, M.N.; Stevens, M.M. Cubosomes: The next generation of smart lipid nanoparticles? *Angew. Chem. Int. Ed.* **2019**, *58*, 2958–2978. [CrossRef]
4. Nisini, R.; Poerio, N.; Mariotti, S.; De Santis, F.; Fraziano, M. The multirole of liposomes in therapy and prevention of infectious diseases. *Front. Immunol.* **2018**, *9*, 155. [CrossRef] [PubMed]
5. Yaghmur, A.; Mu, H. Recent advances in drug delivery applications of cubosomes, hexosomes, and solid lipid nanoparticles. *Acta Pharm. Sin. B* **2021**, *11*, 871–885. [CrossRef] [PubMed]
6. Barriga, H.M.G.; Ces, O.; Law, R.V.; Seddon, J.M.; Brooks, N.J. Engineering swollen cubosomes using cholesterol and anionic lipids. *Langmuir* **2019**, *35*, 16521–16527. [CrossRef]
7. Boge, L.; Hallstensson, K.; Ringstad, L.; Johansson, J.; Andersson, T.; Davoudi, M.; Larsson, P.T.; Mahlapuu, M.; Håkansson, J.; Andersson, M. Cubosomes for topical delivery of the antimicrobial peptide LL-37. *Eur. J. Pharm. Biopharm.* **2019**, *134*, 60–67. [CrossRef]
8. Karami, Z.; Hamidi, M. Cubosomes: Remarkable drug delivery potential. *Drug Discov. Today* **2016**, *21*, 789–801. [CrossRef]
9. Azmi, I.D.M.; Moghimi, S.M.; Yaghmur, A. Cubosomes and hexosomes as versatile platforms for drug delivery. *Ther. Deliv.* **2015**, *6*, 1347–1364. [CrossRef] [PubMed]
10. Chen, Y.; Ma, P.; Gui, S. Cubic and hexagonal liquid crystals as drug delivery systems. *Biomed. Res. Int.* **2014**, *2014*, 116. [CrossRef] [PubMed]
11. Chountoulesi, M.; Pippa, N.; Pispas, S.; Chrysina, E.D.; Forys, A.; Trzebicka, B.; Demetzos, C. Cubic lyotropic liquid crystals as drug delivery carriers: Physicochemical and morphological studies. *Int. J. Pharm.* **2018**, *550*, 57–70. [CrossRef] [PubMed]
12. Pan, X.; Han, K.; Peng, X.; Yang, Z.; Qin, L.; Zhu, C.; Huang, X.; Shi, X.; Dian, L.; Lu, M.; et al. Nanostructed cubosomes as advanced drug delivery system. *Curr. Pharm. Des.* **2013**, *19*, 6290–6297. [CrossRef]
13. Wu, H.; Li, J.; Zhang, Q.; Yan, X.; Guo, L.; Gao, X.; Qiu, M.; Jiang, X.; Lai, R.; Chen, H. A novel small odorranalectin-bearing cubosomes: Preparation, brain delivery and pharmacodynamic study on amyloid-β 25-35-treated rats following intranasal administration. *Eur. J. Pharm. Biopharm.* **2012**, *80*, 368–378. [CrossRef] [PubMed]
14. Esposito, E.; Mariani, P.; Ravani, L.; Contado, C.; Volta, M.; Bido, S.; Drechsler, M.; Mazzoni, S.; Menegatti, E.; Morari, M.; et al. Nanoparticulate lipid dispersions for bromocriptine delivery: Characterization and in vivo study. *Eur. J. Pharm. Biopharm.* **2012**, *80*, 306–314. [CrossRef] [PubMed]
15. Li, Y.; Angelova, A.; Hu, F.; Garamus, V.M.; Peng, C.; Li, N.; Liu, J.; Liu, D.; Zou, A. PH responsiveness of hexosomes and cubosomes for combined delivery of brucea javanica oil and doxorubicin. *Langmuir* **2019**, *35*, 14532–14542. [CrossRef] [PubMed]
16. Rizwan, S.B.; Assmus, D.; Boehnke, A.; Hanley, T.; Boyd, B.J.; Rades, T.; Hook, S. Preparation of phytantriol cubosomes by solvent precursor dilution for the delivery of protein vaccines. *Eur. J. Pharm. Biopharm.* **2011**, *79*, 15–22. [CrossRef] [PubMed]
17. Liu, G.; Conn, C.E.; Waddington, L.J.; Mudie, S.T.; Drummond, C.J. Colloidal amphiphile self-assembly particles composed of gadolinium oleate and Myverol: Evaluation as contrast agents for magnetic resonance imaging. *Langmuir* **2010**, *26*, 2383–2391. [CrossRef] [PubMed]
18. Tran, N.; Bye, N.; Moffat, B.A.; Wright, D.K.; Cuddihy, A.; Hinton, T.M.; Hawley, A.M.; Reynolds, N.P.; Waddington, L.J.; Mulet, X.; et al. Dual-modality NIRF-MRI cubosomes and hexosomes: High throughput formulation and in vivo biodistribution. *Mater. Sci. Eng. C* **2017**, *71*, 584–593. [CrossRef] [PubMed]
19. Caltagirone, C.; Falchi, A.M.; Lampis, S.; Lippolis, V.; Meli, V.; Monduzzi, M.; Prodi, L.; Schmidt, J.; Sgarzi, M.; Talmon, Y.; et al. Cancer-cell-targeted theranostic cubosomes. *Langmuir* **2014**, *30*, 6228–6236. [CrossRef]
20. Meli, V.; Caltagirone, C.; Falchi, A.M.; Hyde, S.T.; Lippolis, V.; Monduzzi, M.; Obiols-Rabasa, M.; Rosa, A.; Schmidt, J.; Talmon, Y.; et al. Docetaxel-loaded fluorescent liquid-crystalline nanoparticles for cancer theranostics. *Langmuir* **2015**, *31*, 9566–9575. [CrossRef]
21. Zhang, L.; Li, J.; Tian, D.; Sun, L.; Wang, X.; Tian, M. Theranostic combinatorial drug-loaded coated cubosomes for enhanced targeting and efficacy against cancer cells. *Cell Death Dis.* **2020**, *11*, 1–12. [CrossRef]
22. Akhlaghi, S.P.; Loh, W. Interactions and release of two palmitoyl peptides from phytantriol cubosomes. *Eur. J. Pharm. Biopharm.* **2017**, *117*, 60–67. [CrossRef] [PubMed]
23. Mansour, M.; Kamel, A.O.; Mansour, S.; Mortada, N.D. Novel polyglycerol-dioleate based cubosomal dispersion with tailored physical characteristics for controlled delivery of ondansetron. *Colloids Surf. B Biointerfaces* **2017**, *156*, 44–54. [CrossRef] [PubMed]
24. Noor, A.H.; Ghareeb, M.M. Formulation and evaluation of ondansetron HCl nanoparticles for transdermal delivery. *Iraqi J. Pharm. Sci.* **2020**, *29*, 70–79. [CrossRef]
25. Clemente, I.; Bonechi, C.; Rodolfi, L.; Bacia-Verloop, M.; Rossi, C.; Ristori, S. Lipids from algal biomass provide new (nonlamellar) nanovectors with high carrier potentiality for natural antioxidants. *Eur. J. Pharm. Biopharm.* **2021**, *158*, 410–416. [CrossRef] [PubMed]
26. Chong, J.Y.T.T.; Mulet, X.; Postma, A.; Keddie, D.J.; Waddington, L.J.; Boyd, B.J.; Drummond, C.J. Novel RAFT amphiphilic brush copolymer steric stabilisers for cubosomes: Poly(octadecyl acrylate)-block-poly(polyethylene glycol methyl ether acrylate). *Soft Matter* **2014**, *10*, 6666–6676. [CrossRef]

27. Chong, J.Y.T.; Mulet, X.; Boyd, B.J.; Drummond, C.J. *Steric Stabilizers for Cubic Phase Lyotropic Liquid Crystal Nanodispersions (Cubosomes)*, 1st ed.; Elsevier: Amsterdam, The Netherlands, 2015; Volume 21.
28. Chountoulesi, M.; Perinelli, D.R.; Forys, A.; Bonacucina, G.; Trzebicka, B.; Pispas, S.; Demetzos, C. Liquid crystalline nanoparticles for drug delivery: The role of gradient and block copolymers on the morphology, internal organisation and release profile. *Eur. J. Pharm. Biopharm.* **2020**, *158*, 21–34. [CrossRef]
29. Hamada, N.; Gakhar, S.; Longo, M.L. Hybrid lipid/block copolymer vesicles display broad phase coexistence region. *Biochim. Biophys. Acta Biomembr.* **2021**, *1863*, 183552. [CrossRef]
30. Khan, S.; McCabe, J.; Hill, K.; Beales, P.A. Biodegradable hybrid block copolymer—Lipid vesicles as potential drug delivery systems. *J. Colloid Interface Sci.* **2020**, *562*, 418–428. [CrossRef]
31. Kulkarni, C.V.; Vishwapathi, V.K.; Quarshie, A.; Moinuddin, Z.; Page, J.; Kendrekar, P.; Mashele, S.S. Self-assembled lipid cubic phase and cubosomes for the delivery of aspirin as a model drug. *Langmuir* **2017**, *33*, 9907–9915. [CrossRef] [PubMed]
32. Zhai, J.; Hinton, T.M.; Waddington, L.J.; Fong, C.; Tran, N.; Mulet, X.; Drummond, C.J.; Muir, B.W. Lipid-PEG conjugates sterically stabilize and reduce the toxicity of phytantriol-based lyotropic liquid crystalline nanoparticles. *Langmuir* **2015**, *31*, 10871–10880. [CrossRef]
33. Zhai, J.; Suryadinata, R.; Luan, B.; Tran, N.; Hinton, T.M.; Ratcliffe, J.; Hao, X.; Drummond, C.J. Amphiphilic brush polymers produced using the RAFT polymerisation method stabilise and reduce the cell cytotoxicity of lipid lyotropic liquid crystalline nanoparticles. *Faraday Discuss.* **2016**, *191*, 545–563. [CrossRef]
34. Kluzek, M.; Tyler, A.I.I.; Wang, S.; Chen, R.; Marques, C.M.; Thalmann, F.; Seddon, J.M.; Schmutz, M. Influence of a pH-sensitive polymer on the structure of monoolein cubosomes. *Soft Matter* **2017**, *13*, 7571–7577. [CrossRef] [PubMed]
35. Peng, X.; Wen, X.; Pan, X.; Wang, R.; Chen, B.; Wu, C. Design and in vitro evaluation of capsaicin transdermal controlled release cubic phase gels. *AAPS PharmSciTech* **2010**, *11*, 1405–1410. [CrossRef]
36. Gustafsson, J.; Ljusberg-Wahren, H.; Almgren, M.; Larsson, K. Submicron particles of reversed lipid phases in water stabilized by a nonionic amphiphilic polymer. *Langmuir* **1997**, *13*, 6964–6971. [CrossRef]
37. Da Dong, Y.; Larson, I.; Barnes, T.J.; Prestidge, C.A.; Allen, S.; Chen, X.; Roberts, C.J.; Boyd, B.J. Understanding the interfacial properties of nanostructured liquid crystalline materials for surface-specific delivery applications. *Langmuir* **2012**, *28*, 13485–13495. [CrossRef] [PubMed]
38. D'souza, A.A.; Shegokar, R. Polyethylene Glycol (PEG): A versatile polymer for pharmaceutical applications. *Expert Opin. Drug Deliv.* **2016**, *13*, 1257–1275. [CrossRef]
39. Pippa, N.; Kaditi, E.; Pispas, S.; Demetzos, C. PEO-b-PCL-DPPC chimeric nanocarriers: Self-assembly aspects in aqueous and biological media and drug incorporation. *Soft Matter* **2013**, *9*, 4073–4082. [CrossRef]
40. Gou, M.; Wei, X.; Men, K.; Wang, B.; Luo, F.; Zhao, X.; Wei, Y.; Qian, Z. PCL/PEG copolymeric nanoparticles: Potential nanoplatforms for anticancer agent delivery. *Curr. Drug Targets* **2011**, *12*, 1131–1150. [CrossRef] [PubMed]
41. Thomas, A.; Müller, S.S.; Frey, H. Beyond poly(ethylene glycol): Linear polyglycerol as a multifunctional polyether for biomedical and pharmaceutical applications. *Biomacromolecules* **2014**, *15*, 1935–1954. [CrossRef]
42. Abuchowski, A.; McCoy, J.R.; Palczuk, N.C.; van Es, T.; Davis, F.F. Effect of covalent attachment of polyethylene glycol on immunogenicity and circulating life of bovine liver catalase. *J. Biol. Chem.* **1977**, *252*, 3582–3586. [CrossRef]
43. Milton Harris, J.; Chess, R.B. Effect of pegylation on pharmaceuticals. *Nat. Rev. Drug Discov.* **2003**, *2*, 214–221. [CrossRef]
44. Kolate, A.; Baradia, D.; Patil, S.; Vhora, I.; Kore, G.; Misra, A. PEG—A versatile conjugating ligand for drugs and drug delivery systems. *J. Control Release* **2014**, *192*, 67–81. [CrossRef] [PubMed]
45. Södergård, A.; Stolt, M. Industrial Production of High Molecular Weight Poly(lactic acid). In *Poly(Lactic Acid): Synthesis, Structures, Properties, Processing, and Applications*; John Wiley & Sons, Inc.: Hoboken, NJ, USA, 2010; pp. 27–41. [CrossRef]
46. Tyler, B.; Gullotti, D.; Mangraviti, A.; Utsuki, T.; Brem, H. Polylactic Acid (PLA) controlled delivery carriers for biomedical applications. *Adv. Drug Deliv. Rev.* **2016**, *107*, 163–175. [CrossRef]
47. Wang, J.; Li, S.; Han, Y.; Guan, J.; Chung, S.; Wang, C.; Li, D. Poly(ethylene glycol)-polylactide micelles for cancer therapy. *Front. Pharmacol.* **2018**, *9*, 1–15. [CrossRef]
48. Garlotta, D. A literature review of poly (lactic acid). *J. Polym. Environ.* **2019**, *9*, 63–84. [CrossRef]
49. De Queiroz, T.S.; Prado, R.F.; Aparecida, I.; De Brito, W.; De Oliveira, L.D.; Marotta, L.; De Vasconcellos, R.; Camargo, E.A. Cytotoxicity and genotoxicity of PLA and PCL membranes on osteoblasts. *Acta Sci. Dent* **2019**, *3*, 55–59.
50. Fukushima, K.; Pratt, R.C.; Nederberg, F.; Tan, J.P.K.; Yang, Y.Y.; Waymouth, R.M.; Hedrick, J.L. Organocatalytic approach to amphiphillic comb-block copolymers capable of stereocomplexation and self-assembly. *Biomacromolecules* **2008**, *9*, 3051–3056. [CrossRef] [PubMed]
51. Zhu, K.J.; Hendren, R.W.; Jensen, K.; Pitt, C.G. Synthesis, properties, and biodegradation of poly(1,3-trimethylene carbonate). *Macromolecules* **1991**, *24*, 1736–1740. [CrossRef]
52. Liu, Z.L.; Zhou, Y.; Zhuo, R.X. Synthesis and properties of functional aliphatic polycarbonates. *J. Polym. Sci. Part A Polym. Chem.* **2003**, *41*, 4001–4006. [CrossRef]
53. Keylor, M.H.; Matsuura, B.S.; Stephenson, C.R.J. Chemistry and biology of resveratrol-derived natural products. *Chem. Rev.* **2015**, *115*, 8976–9027. [CrossRef] [PubMed]
54. Salehi, B.; Mishra, A.P.; Nigam, M.; Sener, B.; Kilic, M.; Sharifi-Rad, M.; Fokou, P.V.T.; Martins, N.; Sharifi-Rad, J. Resveratrol: A double-edged sword in health benefits. *Biomedicines* **2018**, *6*, 91. [CrossRef]

55. Nakano, M.; Sugita, A.; Matsuoka, H.; Handa, T. Small-angle X-ray scattering and 13C NMR investigation on the internal structure of "cubosomes". *Langmuir* **2001**, *17*, 3917–3922. [CrossRef]
56. Akhlaghi, S.P.; Ribeiro, I.R.; Boyd, B.J.; Loh, W. Impact of preparation method and variables on the internal structure, morphology, and presence of liposomes in phytantriol-Pluronic® F127 cubosomes. *Colloids Surf. B Biointerfaces* **2016**, *145*, 845–853. [CrossRef]
57. Chong, J.Y.T.; Mulet, X.; Waddington, L.J.; Boyd, B.J.; Drummond, C.J. Steric stabilisation of self-assembled cubic lyotropic liquid crystalline nanoparticles: High throughput evaluation of triblock polyethylene oxide-polypropylene oxide-polyethylene oxide copolymers. *Soft Matter* **2011**, *7*, 4768–4777. [CrossRef]
58. Rietveld, H.M. A profile refinement method for nuclear and magnetic structures. *J. Appl. Crystallogr.* **1969**, *2*, 65–71. [CrossRef]
59. Rietveld, H.M. Line profiles of neutron powder-diffraction peaks for structure refinement. *Acta Crystallogr.* **1967**, *22*, 151–152. [CrossRef]
60. Pippa, N.; Merkouraki, M.; Pispas, S.; Demetzos, C. DPPC:MPOx chimeric advanced drug delivery nano systems (chi-aDDnSs): Physicochemical and structural characterization, stability and drug release studies. *Int. J. Pharm.* **2013**, *450*, 225–228. [CrossRef] [PubMed]
61. Pippa, N.; Dokoumetzidis, A.; Pispas, S.; Demetzos, C. The interplay between the rate of release from polymer grafted liposomes and their fractal morphology. *Int. J. Pharm.* **2014**, *465*, 63–69. [CrossRef]
62. Chong, J.Y.T.; Mulet, X.; Waddington, L.J.; Boyd, B.J.; Drummond, C.J. High-throughput discovery of novel steric stabilizers for cubic lyotropic liquid crystal nanoparticle dispersions. *Langmuir* **2012**, *28*, 9223–9232. [CrossRef] [PubMed]
63. Barauskas, J.; Misiunas, A.; Gunnarsson, T.; Tiberg, F.; Johnsson, M. "Sponge" nanoparticle dispersions in aqueous mixtures of diglycerol monooleate, glycerol dioleate, and polysorbate 80. *Langmuir* **2006**, *22*, 6328–6334. [CrossRef]
64. Barauskas, J.; Johnsson, M.; Tiberg, F. Self-assembled lipid superstructures: Beyond vesicles and liposomes. *Nano Lett.* **2005**, *5*, 1615–1619. [CrossRef]
65. Chountoulesi, M.; Perinelli, D.R.; Pippa, N.; Chrysostomou, V.; Forys, A.; Otulakowski, L.; Bonacucina, G.; Trzebicka, B.; Pispas, S.; Demetzos, C. Physicochemical, morphological and thermal evaluation of lyotropic lipidic liquid crystalline nanoparticles: The effect of stimuli-responsive polymeric stabilizer. *Colloids Surf. A Physicochem. Eng. Asp.* **2020**, *595*, 124678. [CrossRef]
66. Tilley, A.J.; Drummond, C.J.; Boyd, B.J. Disposition and association of the steric stabilizer Pluronic® F127 in lyotropic liquid crystalline nanostructured particle dispersions. *J. Colloid Interface Sci.* **2013**, *392*, 288–296. [CrossRef]
67. Sagalowicz, L.; Michel, M.; Adrian, M.; Frossard, P.; Rouvet, M.; Watzke, H.J.; Yaghmur, A.; De Campo, L.; Glatter, O.; Leser, M.E. Crystallography of dispersed liquid crystalline phases studied by cryo-transmission electron microscopy. *J. Microsc.* **2006**, *221*, 110–121. [CrossRef]
68. Badie, H.; Abbas, H. Novel small self-assembled resveratrol-bearing cubosomes and hexosomes: Preparation, charachterization, and ex vivo permeation. *Drug Dev. Ind. Pharm.* **2018**, *44*, 2013–2025. [CrossRef]
69. Elgindy, N.A.; Mehanna, M.M.; Mohyeldin, S.M. Self-assembled nano-architecture liquid crystalline particles as a promising carrier for progesterone transdermal delivery. *Int. J. Pharm.* **2016**, *501*, 167–179. [CrossRef] [PubMed]
70. Boyd, B.J.; Whittaker, D.V.; Khoo, S.M.; Davey, G. Lyotropic liquid crystalline phases formed from glycerate surfactants as sustained release drug delivery systems. *Int. J. Pharm.* **2006**, *309*, 218–226. [CrossRef]
71. Thorn, C.R.; Clulow, A.J.; Boyd, B.J.; Prestidge, C.A.; Thomas, N. Bacterial lipase triggers the release of antibiotics from digestible liquid crystal nanoparticles. *J. Control. Release* **2020**, *319*, 168–182. [CrossRef] [PubMed]
72. Chen, H.; Fan, Y.; Zhang, N.; Trépout, S.; Ptissam, B.; Brûlet, A.; Tang, B.Z.; Li, M.H. Fluorescent polymer cubosomes and hexosomes with aggregation-induced emission. *Chem. Sci.* **2021**, *12*, 5495–5504. [CrossRef] [PubMed]

polymers

MDPI

Article

HPMA-Based Polymer Conjugates for Repurposed Drug Mebendazole and Other Imidazole-Based Therapeutics

Martin Studenovský *, Anna Rumlerová, Libor Kostka and Tomáš Etrych

Institute of Macromolecular Chemistry, Czech Academy of Sciences, Heyrovský Sq. 2, 16206 Prague 6, Czech Republic; annarumlerova@centrum.cz (A.R.); kostka@imc.cas.cz (L.K.); etrych@imc.cas.cz (T.E.)

* Correspondence: studenovsky@imc.cas.cz

Abstract: Recently, the antitumor potential of benzimidazole anthelmintics, such as mebendazole and its analogues, have been reported to have minimal side effects, in addition to their well-known anti-parasitic abilities. However, their administration is strongly limited owing to their extremely poor solubility, which highly depletes their overall bioavailability. This study describes the design, synthesis, and physico-chemical properties of polymer-mebendazole nanomedicines for drug repurposing in cancer therapy. The conjugation of mebendazole to water-soluble and biocompatible polymer carrier was carried out via biodegradable bond, relying on the hydrolytic action of lysosomal hydrolases for mebendazole release inside the tumor cells. Five low-molecular-weight mebendazole derivatives, differing in their inner structure, and two polymer conjugates differing in their linker structure, were synthesized. The overall synthetic strategy was designed to enable the modification and polymer conjugation of most benzimidazole-based anthelmintics, such as albendazole, fenbendazole or albendazole, besides the mebendazole. Furthermore, the described methodology may be suitable for conjugation of other biologically active compounds with a heterocyclic N-H group in their molecules.

Keywords: mebendazole; drug delivery; drug repurposing; polymer; HPMA; controlled release

Citation: Studenovský, M.; Rumlerová, A.; Kostka, L.; Etrych, T. HPMA-Based Polymer Conjugates for Repurposed Drug Mebendazole and Other Imidazole-Based Therapeutics. *Polymers* **2021**, *13*, 2530. https://doi.org/10.3390/polym13152530

Academic Editors: Marek Kowalczuk, Iza Radecka and Barbara Trzebicka

Received: 13 July 2021
Accepted: 28 July 2021
Published: 30 July 2021

1. Introduction

1.1. Benzimidazole Derivatives in Antitumor Therapy

The development of new antitumor drugs has been limited by the high cost of experimental and clinical trials required by the US Food and Administration (FDA) and other health institutions. Drug repurposing, i.e., investigation of existing and already approved drugs with known toxicity profile and pharmacokinetics for novel therapeutic indications [1] should therefore bypass the elevated economic burden of trials and speed up the drug approval process.

Benzimidazoles have long been used as broad-spectrum anthelmintics. However, some benzimidazole derivatives exhibit antitumor potential [2–5]. Namely mebendazole (MBZ) and albendazole (ABZ), already in use in human medicine, and flubendazole (FLZ) and fenbendazole (FBZ), prescribed in veterinary practice, are the most promising anticancer drug candidates (Figure 1). They inhibit tubulin assembly and suppress microtubule formation by binding to active colchicine sites of the β-tubulin subunit leading to mitotic arrest and apoptosis [3,6]. Thus, they can inhibit cell proliferation, angiogenesis, metastatic activity, and show synergism with other chemotherapeutics (e.g., docetaxel) or radiotherapy [6–8]. Since their molecular size is quite small, they are not substrates for P-glycoprotein (P-gp) and multidrug resistance should not occur during their use. Moreover, MBZ has been described as a P-glycoprotein (P-gp) inhibitor [9], thus MBZ and its analogs appear to be promising antitumor chemotherapeutics. However, the persisting problem with benzimidazole derivatives is their extremely poor water solubility, which severely restricts their bioavailability and thus makes it difficult to reach serum concentrations required for antitumor activity.

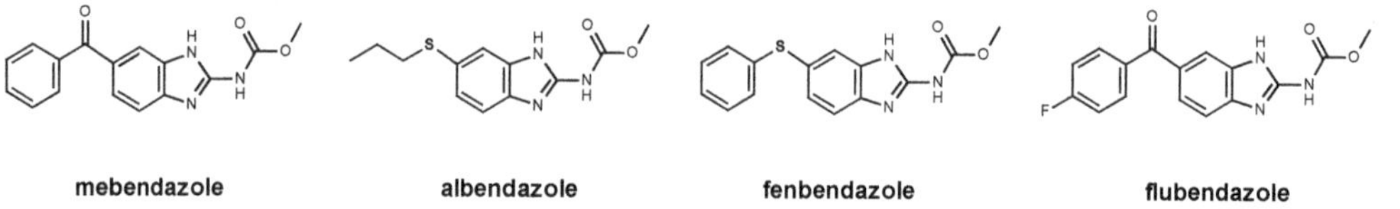

Figure 1. Chemical structures of benzimidazole derivatives.

1.2. Mebendazole Mechanism of Action and Antitumor Potential

Mebendazole is a benzimidazole derivative with extremely low water solubility and thus very poor bioavailability (~17–22% of oral dose reaches the blood circulation with significant inter-individual variation in pharmacokinetics) [10,11]. This is probably the reason why MBZ in per oral treatment does not exert any serious side effects. Clearance of MBZ from organism is mainly via metabolites in bile and feces, with only 2–5% excretion via urine [12].

Mebendazole has been used as an anthelmintic drug since 1974; however, its antitumor activity took almost 30 years to be published [6]. Ever since, various studies have described the cytostatic activity of MBZ in a wide range of tumor types e.g., breast, ovary, and colorectal carcinoma, glioblastoma, gastric cancer, melanoma or lung carcinoma [2,6,13–15]. Either as a single antitumor agent or in combination with other chemotherapeutics or radiotherapy, MBZ treatment inhibits tumor progression together with a decrease of metastatic spread and therefore improved survival. Moreover, it is worth mentioning that MBZ treatment also stimulates antitumor immune response [16,17].

The main mechanism of MBZ cytostatic activity is presumably its ability to interfere with microtubule apparatus of tumor cells resulting in defective cellular structures, intracellular transport, and glucose metabolism [18]. In vitro studies showed inhibition of factors involved in tumor progression upon MBZ treatment, e.g., decrease of matrix metalloproteinase-2 activity, inhibition of antiapoptotic activity of Bcl-2, inhibition of multidrug resistance protein transporters P-gp or MRP1, or inhibition of angiogenesis [9,13,19,20]. However, in vivo studies have not fully confirmed in vitro observations [21,22]. Several studies reported potential side effects of high dose MBZ treatment, including abdominal pain, diarrhea, and in rare cases neutropenia, marrow aplasia, alopecia, and elevation of transaminases levels [23,24].

1.3. Drug Delivery Systems Based on Synthetic Copolymers

Considering current knowledge, developing a general approach of effective cancer therapy would require the design of drugs with strong antitumor activity, concurrently with minimal adverse side effects. Importantly, polymer-based high-molecular-weight (HMW) delivery systems have been recently recognized as nanosystems with potent anticancer efficacies devoid of significant systemic toxicity [25,26]. Thus, the concept of HMW drug delivery systems, including water-soluble polymers such as *N*-(2-(hydroxypropyl)methacrylamide (HPMA) copolymers, has become generally accepted. Preferably, low-molecular-weight (LMW) drugs are covalently bound to a polymer via spacers, which enables the controlled release of the active drug in target tissues or cells. Polymer drug carriers are designed to alter the drug biodistribution and optimize its pharmacokinetics. Binding of the LMW drug to the HMW carrier results in its prolonged blood circulation, enhanced accumulation in solid tumors, controlled drug release, reduction of systemic toxicity and immunogenicity, and possibility in induction of resistance against tumor regrowth upon treatment [27,28]. Moreover, the binding to the polymer carrier can solubilize and stabilize water-insoluble drugs before biotransformation and degradation. The HMW anticancer drug delivery systems rely on passive accumulation in solid tumors based on the enhanced permeability and retention (EPR) effect, which is caused by the morphological and physiological differences between healthy and malignant tissues [29]. Numerous studies have shown higher permeability of tumor blood vessels compared to normal blood vessels [30]. The endothelial lining of the capillaries in the tumor is fenestrated and leaky, and thereby permeable to macromolecules. Moreover, tumor lymphatic drainage is mostly defective or even missing; therefore, macromolecules are retained within tumor tissues. The discovery of the EPR phenomenon [31] led to the development of tumor-selective delivery of polymer conjugates, micellar and liposomal drugs, and gene vectors. In case of HPMA copolymer-based drug conjugates, the EPR effect was observed after administration of conjugates with a molecular weight (M_w) exceeding 20,000 g·mol^{-1} and increased with the higher M_w of the polymer [32].

1.4. Polymer-Drug Linkers Based on a Prodrug Strategy

The linkage between drugs and polymer carriers must be generally sufficiently stable in the bloodstream but capable of releasing the parent drug within the target site. While the structure of the polymer precursor can be tailor-made, the structure of the drug is well established, which determines the probability of chemical modification towards its attachment to the polymer. Thus, for example, active compounds containing hydroxyl can be bound via ester, ether, or acetal linker, while amines can be transformed to amide, urethane or carbamate, carboxyl to ester, and carbonyl to acetal or hydrazone. A great inspiration for a design of polymer-drug conjugates can be found in the prodrug chemistry [33], the broad field of pharmacochemistry. The prodrug strategy is in principle the same as for the polymer-drug conjugate: appropriate modification of the parent drug with the aim to enhance its pharmacokinetics as well as pharmacodymamics. In other words, the polymer-drug conjugates may be a high-molecular-weight prodrug. Nitrogen-containing heterocyclic compounds, like mebendazole and other benzimidazole anthelmintics, can be reversibly modified via the nitrogen atom [34]. In this study, carbamate and acyloxymethyl group, both cleavable by lysosomal hydrolases, were utilized for the preparation of LMW mebendazole derivatives as well as its polymer conjugates. The general formation and reverse degradation of the derivatives is shown in Figure 2.

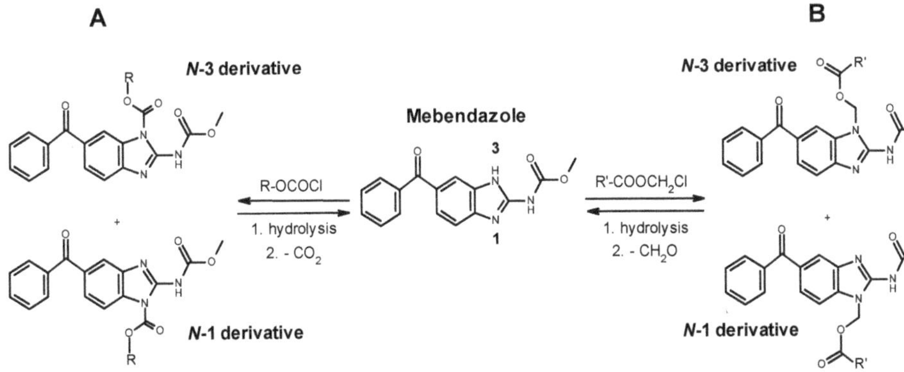

Figure 2. Generation and cleavage of carbamate (**A**) and acyloxymethyl (**B**) derivatives of mebendazole (MBZ). Due to a tautomerism of the MBZ molecule (1,3-H shift), a statistical mixture of *N*-1 and *N*-3 derivatives is formed in all cases. The presence of both isomers was indicated by doubled high-performance liquid chromatography (HPLC) peaks and nuclear magnetic resonance (NMR) signals of carbamate protons. For better clarity, only *N*-3 isomers will be depicted in the next figures.

2. Materials and Methods

2.1. Chemicals

Acetic acid, β-alanine, 1-aminopropan-2-ol, 3-azidopropanol, 2,2′-azobis(2-methylp ropionitrile), 6-(Boc-amino)hexanoic acid, chloromethyl chlorosulfate, chloromethyl pivalate, dibenzocyclooctyne-amine, dicyclohexylcarbodiimide, 4-(dimethylamino)pyridine, ethyl chloroformate, ethyldiisopropylamine, 1-ethyl-3-(3-dimethylaminopropyl)carbodiimide hydrochloride, methacryloyl chloride, mebendazole, 4-nitrobenzyl chloroformate, phosgene solution, silica gel, sodium hydride, tetrabutylammonium hydrogen sulfate, tetramethylurea, thiazolidine-2-thione, and trifluoroacetic acid were purchased from Sigma-Aldrich (Prague, Czech Republic). Acetonitrile, methanol and other common solvents and chemicals were purchased from Merck (Prague, Czech Republic). All chemicals and solvents were of analytical grade.

2.2. Analytical Methods

Analyses were performed on a high-performance liquid chromatograph (HPLC; Shimadzu, Japan) using a reverse-phase column (Chromolith Performance RP-18e 100 × 4.6 mm) with UV detection. A mixture of water and acetonitrile was used as the eluent at a gradient 0–100 vol.% and a flow rate of 2 mL/min. Elemental composition was determined using a Perkin Elmer Elemental Analyzer 2400 CHN (Perkin Elmer, Waltham, MA, USA). Nuclear magnetic resonance (NMR) spectra were measured on a Bruker Avance MSL 300 MHz NMR spectrometer (Bruker Daltonik, Bremen, Germany). Molecular weights M_w and M_n of the polymers were determined by gel permeation chromatography (GPC) using an HPLC Shimadzu system equipped with a GPC column (TSKgel G3000SWxl 300 × 7.8 mm; 5 μm), UV–Vis, refractive index (RI) Optilab®-rEX and multiangle light scattering (MALS) DAWN EOS (Wyatt Technology Co., Goleta, CA, USA) detectors using a methanol:sodium acetate buffer (0.3 M; pH 6.5) mixture (80:20 vol.%; flow rate 0.5 mL/min). Preparative HPLC chromatography was performed on PrepChrom C-700 (Büchi, Flawil, Switzerland). UV–Vis spectra were measured on Specord 205 (Analytik Jena AG, Jena, Germany). Mass spectra (MS) were measured on an ion trap mass spectrometer LCQ Fleet (Thermo Scientific, Waltham, MA, USA). The hydrodynamic diameter (D_h) of the polymer conjugates in PBS buffer pH 7.4 (5 mg/mL, 25 °C) was measured using a Nano-ZS instrument (ZEN3600, Malvern, UK). The intensity of scattered light was detected at angle θ = 173° using a laser with a wavelength of 632.8 nm. Data were evaluated using the DTS (Nano) program. The values were the mean of at least five independent measurements.

2.3. Syntheses

Mebendazole Derivatives **1–5**

A general synthetic procedure for low-molecular-weight of mebendazole is described in Figure 3.

2.4. N-Ethoxycarbonyl Mebendazole (**1**)

Mebendazole (100 mg, 0.34 mmol) was suspended in 3 mL of tetramethylurea and ethyl chloroformate (200 mg, 1.9 mmol) was added. The reaction mixture was stirred until homogenous and then the solvent was removed under vacuum. The residuum was dissolved in dichloromethane and the traces of unreacted mebendazole were removed by triple extraction with 5% HCl. The organic layer was dried under anhydrous magnesium sulfate, filtered, and the solvent was removed under vacuum. The crude product was crystallized from ethyl acetate and dried under vacuum. Yield: 43 mg (34%), 98% pure (HPLC, 220 nm). MS: The corresponding molecular peak M/Z = 390 (M + Na) was detected. Elemental analysis: calcd. C 62.12, H 4.66, N 11.44%, found C 62.05, H 4.7, N 11.32%. ^1H NMR 300 MHz (DMSO-d_6, 297 K): 1.26 t (3H, $\underline{CH_3}CH_2$), 3.63 s (3H, OCH_3), 4.26 q (2H, $CH_3\underline{CH_2}$), 7.56–7.84 m (8H, Ar), 12.38 + 12.61 s (1H, NH).

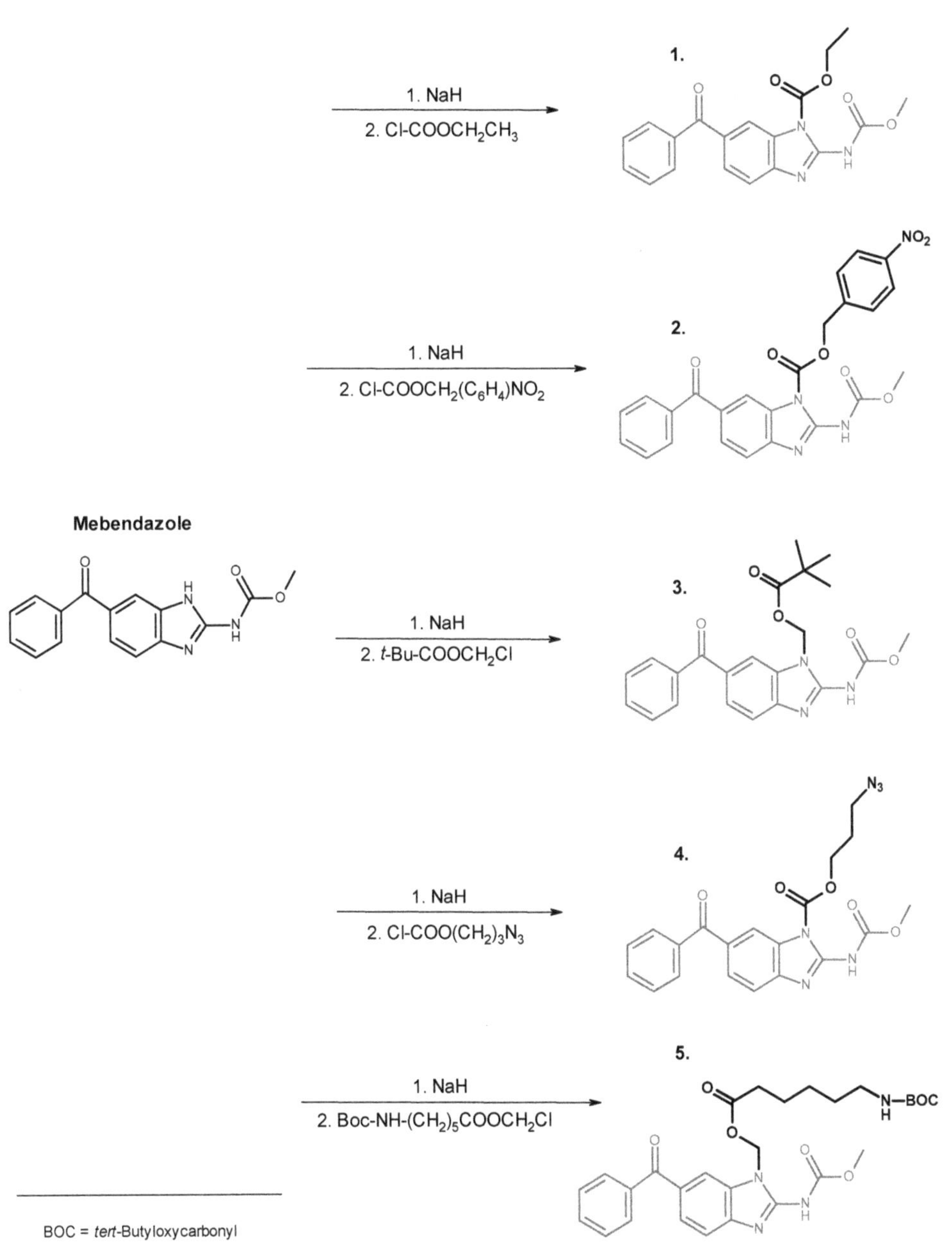

Figure 3. Synthesis of model mebendazole derivatives **1–3** and reactive derivatives **4**, **5** designed for the attachment to a polymer carrier.

2.5. N-(4-Nitrobenzyloxycarbonyl) Mebendazole (**2**)

Mebendazole (100 mg, 0.34 mmol) was suspended in 4 mL of tetramethylurea and 4-nitrobenzyl chloroformate (150 mg, 0.7 mmol) was added. The reaction mixture was stirred until homogenous and then the solvent was removed under vacuum. The residuum was dissolved in dichloromethane, filtered, and any remaining traces of mebendazole were removed by triple extraction with 5% HCl. The organic layer was dried under anhydrous magnesium sulfate, filtered, and the solvent was removed under vacuum. The crude product was crystallized from a mixture of hexane/dichloromethane 1:1 and dried under vacuum. Yield: 34 mg (21%), 97% pure (HPLC, 220 nm). MS: The corresponding molecular peak M/Z = 497 (M + Na) was detected. Elemental analysis: calcd. C 60.76, H 3.82, N 11.81%, found C 60.04, H 3.88, N 11.72%. ^{1}H NMR 300 MHz (DMSO-d_6, 297 K): 3.63 s (3H, OCH_3), 5.18 s (2H, ArC$\underline{H_2}$), 7.51–8.17 m (12H, Ar), 12.35 + 12.59 s (1H, NH).

2.6. *N-Pivaloyloxymethyl Mebendazole* (**3**)

Mebendazole (1 g, 3.4 mmol) and sodium hydride (200 mg of 60% oil suspension, 5 mmol) were suspended in 20 mL of dimethylformamide and sonicated until a clear yellow solution of mebendazole sodium salt was formed. Chloromethyl pivalate (700 mg, 4.7 mmol) was added and the reaction mixture was stirred overnight. The solvent was removed under vacuum, the residuum was dissolved in dichloromethane, filtered, concentrated under vacuum, and separated on the silica gel in ethylacetate/dichloromethane/he xane 5:2:5 with 1% of trifluoroacetic acid. Yield: 378 mg (27%), 99% pure (HPLC, 220 nm). MS: The corresponding molecular peak M/Z = 432 (M + Na) was detected. Elemental analysis: calcd. C 64.54, H 5.66, N 10.26%, found C 64.24, H 5.75, N 10.31%. ^{1}H NMR 300 MHz (DMSO-d_6, 297 K): 1.1 s (9H, *t*-Bu), 3.65 s (3H, OCH_3), 6.08 s (2H, NCH_2), 7.55–7.85 m (8H, Ar), 12.34 + 12.5 s (1H, NH).

2.7. *N-(3-Azidopropyloxycarbonyl) Mebendazole* (**4**)

3-Azidopropanol (211 mg, 2.09 mmol) and sodium carbonate (1 g, 9.4 mmol) were mixed in 3 mL of dichloromethane and 20% phosgene solution in toluene (5 g, 10.1 mmol) was added. The reaction mixture was stirred for 2 h, filtered and the solvents were evaporated under vacuum. The obtained 3-azidopropyl chloroformate was added without further purification to a solution of sodium salt of mebendazole (617 mg mebendazole equiv., 2.09 mmol) in dimethyl formamide (see above) and the solution was stirred for 2 h. The reaction was then quenched with 0.5 mL of trifluoroacetic acid, and the volatiles were evaporated under vacuum. The residuum was dissolved in dichloromethane, filtered and the remaining unreacted mebendazole was removed by triple extraction with 5% HCl. The organic layer was dried under anhydrous magnesium sulfate, filtered, and the solvent was removed under vacuum. The crude product was then purified on silica gel in a mixture of dichloromethane and acetone 20:1. Yield: 240 mg (29%), 97.5% pure (HPLC, 220 nm). MS: The corresponding molecular peak M/Z = 445 (M + Na) was detected. Elemental analysis: calcd. C 56.87, H 4.3, N 19.9%, found C 57.05, H 4.37, N 20.2%. ^{1}H NMR 300 MHz (DMSO-d_6, 297 K): 1.84 qu (2H, C$\underline{H_2}$$CH_2N_3$), 3.31 t (2H, CH_2N_3) 3.63 s (3H, OCH_3), 4.15 t (2H, OCH_2), 7.54–7.84 m (8H, Ar), 12.41 + 12.6 s (1H, NH).

2.8. *N-(6-(Boc-amino)hexanoyloxymethyl) Mebendazole* (**5**)

6-(Boc-amino)hexanoic acid (604 mg, 2.6 mmol), chloromethyl chlorosulfate (350 mg, 2.12 mmol), sodium hydrogen carbonate (870 mg, 10.36 mmol) and tetrabutylammonium hydrogen sulfate (90 mg, 0.27 mmol) were vigorously stirred in a 20 mL of dichoromethane/water emulsion for 24 h. The organic layer was separated, extracted five times with brine, and dried with anhydrous sodium sulfate. The sodium sulfate was filtered off and the solvent was removed from the filtrate under vacuum. The crude chloromethyl 6-(Boc-amino)hexanoate was then added without further purification to a solution of sodium salt of mebendazole (755 mg mebendazole equiv., 2.56 mmol) in dimethyl formamide (see above), and stirred for 24 h. The reaction mixture was concentrated under vacuum, filtered, and separated on a C-18 reversed phase preparative chromatogram in a water/acetonitrile gradient. Yield: 440 mg (32%), 98.5% pure (HPLC, 220 nm). MS: The corresponding molecular peak M/Z = 561 (M + Na) was detected. Elemental analysis: calcd. C 62.44, H 6.36, N 10.4%, found C 62.15, H 6.37, N 10.25%. ^{1}H NMR 300 MHz (DMSO-d_6, 297 K): 1.19 qu (2H, C$\underline{H_2}$$CH_2CH_2N$), 1.31 qu (2H, C$\underline{H_2}$$CH_2CO$), 1.34 s (9H. *t*-Bu), 1.47 qu (2H, C$\underline{H_2}$$CH_2N$),

2.32 t (2H, CH_2CO), 2.83 t (2H, CH_2N), 3.65 s (3H, OCH_3), 6.08 s (2H, NCH_2-O), 6.72 t (1H, N<u>H</u>COOt-Bu), 7.54–7.82 m (8H, Ar), 12.32 + 12.51 s (1H, N<u>H</u>COOMe).

*2.9. Polymer Conjugates of Derivatives **4** (Conjugate **I**) and **5** (Conjugate **II**)*

A general synthetic procedure for polymer conjugates of mebendazole is described in Figure 4.

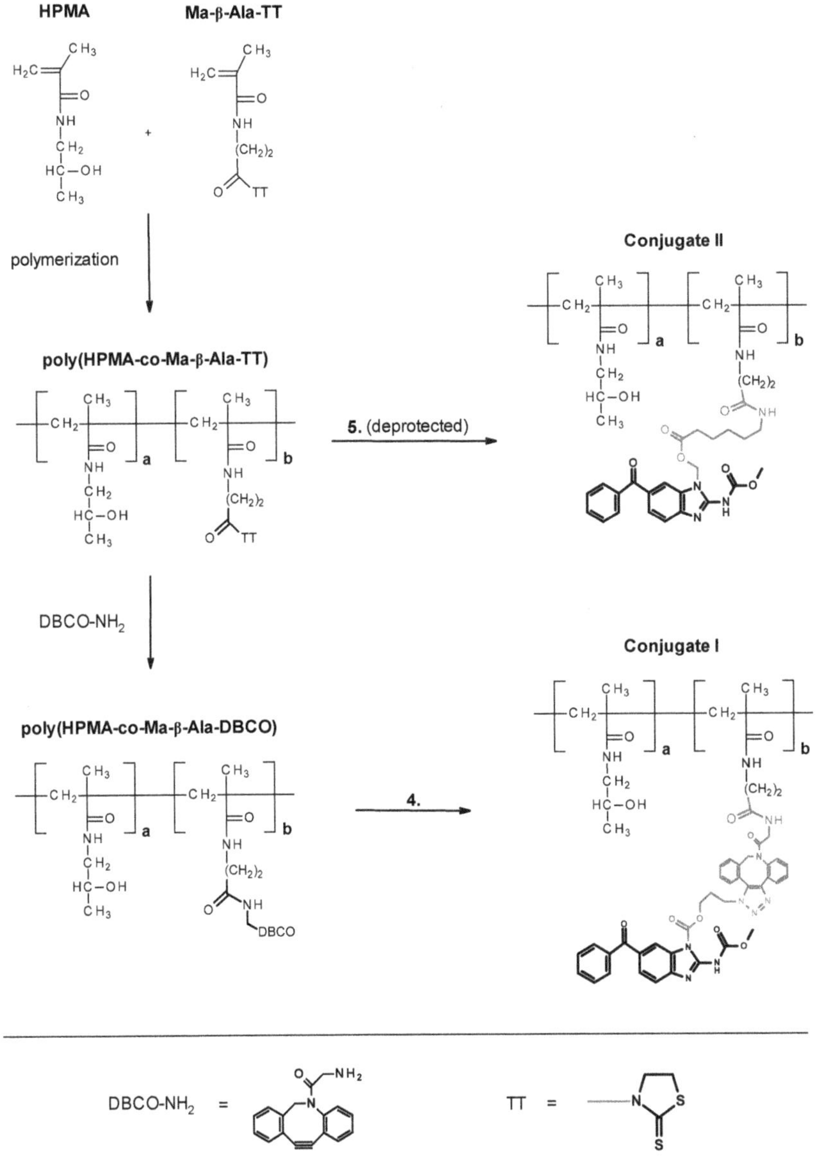

Figure 4. Synthesis of HPMA-based mebendazole polymer conjugates **I** (carbamate linker) and **II** (ester linker).

2.10. N-(2-Hydroxypropyl)methacrylamide (HPMA)

Synthesis of this monomer was carried out according to literature [35] via reaction of methacryloyl chloride with 1-amino propan-2-ol. Yield: 75%, 99% pure (HPLC, 220 nm). Elemental analysis: calcd. C 58.72, H 9.15, N 9.78%, found C 58.90, H 9.10, N 9.88%.

2.11. 3-(3-Methacrylamidopropanoyl)thiazolidine-2-thione (Ma-β-Ala-TT)

Synthesis of this reactive monomer was carried out according to literature [36] using a two-step procedure starting with condensation of methacryloyl chloride with β-alanine with consequent activation to aminoreactive thiazolidine-2-thione amide. Yield: 72%, 99% pure (HPLC, 220 nm). Elemental analysis: calcd. C 59.62, H 7.88, N 9.91%, found C 58.99, H 8.1, N 9.78%.

2.12. poly(HPMA-co-Ma-β-Ala-TT)

This reactive polymer precursor was prepared by controlled radical RAFT copolymerization (reversible addition-fragmentation chain-transfer polymerization) of HPMA and Ma-β-Ala-TT following the procedure described previously [36]. A molar ratio of monomers:CTA:initiator 400:2:1 was used. The molar ratio of HPMA:Ma-β-Ala-TT in the reaction mixture was 92:8. Characteristics: M_w = 35 kDa, M_n = 32 kDa, content of TT groups was determined spectrophotometrically in methanol (ε_{305} = 10,800 Lmol^{-1}cm^{-1}): 0.35 mmol/g.

2.13. poly(HPMA-co-Ma-β-Ala-DBCO)

poly(HPMA-*co*-Ma-β-Ala-TT) (1 g, 0.35 mmol of TT groups) was dissolved in 5 mL of dimethyl formamide and dibenzocyclooctyne-amine (100 mg, 0.38 mmol) was added. The reaction mixture was stirred for 2 h and the modified polymer was then precipitated into diethyl ether, dissolved in small amount of methanol, precipitated again into diethyl ether, filtered off, and dried under vacuum. Yield: 0.95 g, molecular weight (GPC): M_w = 40 kDa, M_n = 34 kDa, content of DBCO groups was determined spectrophotometrically in methanol (ε_{292} = 13,000 Lmol^{-1}cm^{-1}): 0.32 mmol/g.

2.14. Conjugate **I** (Polymer Bound Azido Derivative **4**)

poly(HPMA-*co*-Ma-β-Ala-DBCO) (0.5 g, 0.16 mmol of DBCO groups) was dissolved in 3 mL of dimethyl formamide and *N*-(3-azidopropyloxycarbonyl) mebendazole (**4**) (84.5 mg, 0.2 mmol) was added. The reaction mixture was stirred for 12 h and the polymer conjugate was then precipitated into diethyl ether, dissolved in methanol, precipitated again into diethyl ether, filtered off, and dried under vacuum. Yield: 0.44 g, molecular weight (GPC): M_w = 48 kDa, M_n = 41 kDa, the content of mebendazole was determined by HPLC after treatment of the conjugate with 1% NaOH solution for 30 min: 0.31 mmol/g (8.9 wt.%).

2.15. Conjugate **II** (Polymer Bound Amino Derivative **5**)

N-(6-(Boc-amino)hexanoyloxymethyl) mebendazole (**5**) (950 mg, 1.76 mmol) was dissolved in 28 mL of trifluoroacetic acid and the reaction mixture was shaken for 1 h. After complete deprotection (confirmed by HPLC) the acid was evaporated and the solid was co-distilled twice with 20 mL of acetonitrile. The resulted free *N*-(6-aminohexanoyloxymethyl) mebendazole was dissolved in 28 mL of dimethyl formamide and combined with a solution of poly(HPMA-*co*-Ma-β-Ala-TT) (7.4 g, 2.59 mmol TT) and ethyldiisopropylamine (1 g, 7.7 mmol). After 1 h of stirring, 1-aminopropan-2-ol (350 mg, 4.66 mmol) was added and the mixture was stirred for 15 min to quench the remaining polymer-bound TT groups. The conjugate was then precipitated into diethyl ether, filtered off, re-precipitated twice the same way and dried under vacuum. Yield: 7.8 g, molecular weight (GPC): M_w = 37 kDa, M_n = 33 kDa, content of mebendazole was determined by HPLC after treatment of the conjugate with 1% NaOH solution for 30 min: 0.22 mmol/g (6.45 wt.%).

2.16. Hydrolytic Stability of the Conjugates

A long-term circulation hydrolytic stability under blood-mimicking conditions was evaluated by incubation of the samples (1 mg/mL) in the PBS buffer in pH 7.4 at 37 °C for three days. The concentration of the released free mebendazole was determined by HPLC at 300 nm.

3. Results and Discussion

In this paper, a chemical modification of an anthelmintics mebendazole, a repurposed anticancer drug for parenteral administration, was designed and optimized. We aimed to synthesize polymer conjugates, which will take the benefit from the covalent attachment of mebendazole, and the linkage will be tailored for the tumor-cell specific degradation by the lysosomal hydrolases. We assumed that a such concept will bring a significant increase in the efficacy of the mebendazole, which will be specifically accumulated in the solid tumor via the EPR effect-base polymer accumulation and subsequently released to fully act as an anticancer drug. To reach that goal, three low-molecular-weight derivatives, carbamates **1**, **2** and pivaloyloxymethyl derivative **3**, were prepared to develop and optimize the multistep synthetic procedure prior to synthesis of corresponding polymer conjugates. The heterocyclic nitrogen atom of mebendazole was acylated or alkylated either after previous transformation to a sodium salt with sodium hydride or by an alternative route using tetramethylurea as a base/solvent [37]. The overall low to moderate yields, not exceeding 30%, were mainly due to a need of chromatographic separation. The same synthetic strategy was used for derivatives **4** and **5**, already designed for direct conjugation to a polymer. Compound **4** contained an enzymatic cleavable carbamate group between the mebendazole moiety and the spacer azide-terminated for attaching to the polymer precursor via copper-free click chemistry. Compound **5** comprised a cleavable ester group between the drug and the spacer, which is terminated via protected amine for subsequent aminolytic attachment to a polymer precursor bearing TT groups. As illustrated in Figure 2, all mebendazole derivatives were mixtures of *N-1* and *N-3* isomers due to a 1,3-H shift in the mebendazole molecule. There were no attempts to separate the individual isomers because of their almost identical properties in view of chromatographic separation and chemical reactivity. Figure 5 shows a HPLC chromatogram and ^{1}H NMR spectrum of derivative **3**, where the doubled peak and N–H singlet, respectively, indicate a presence of both isomers.

A B

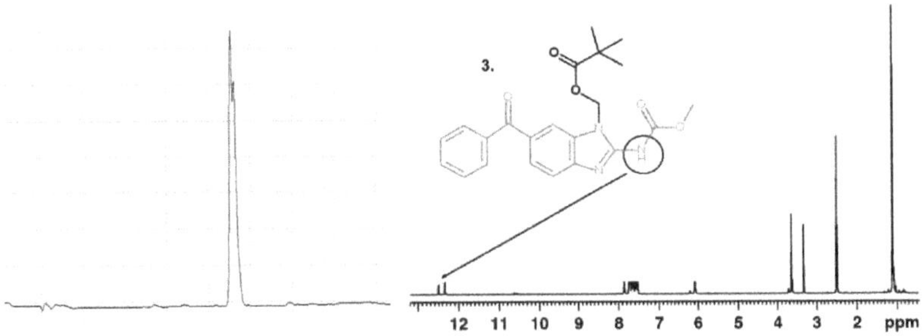

Figure 5. HPLC chromatogram at 300 nm (**A**) and ^{1}H NMR spectrum of derivative **3** (**B**).

Importantly, controlled radical RAFT polymerization [38,39] was employed for the preparation of reactive polymer precursors. This approach enables to obtain polymers with precisely set molecular weight, with narrow distribution of molecular weight and thus minimize the fraction exceeding the renal threshold (~50 kDa). poly(HPMA-*co*-Ma-β-Ala-TT) was synthesized by copolymerization of HPMA and Ma-β-Ala-TT co-monomers where the content of TT groups and the average molecular weight were set by concentration of the monomers and the initiator in a reaction mixture. The typical values are 0.35 mmol TT groups per gram and M_w ~35 kDa (Figure 1). To obtain the conjugate **I**, the poly(HPMA-*co*-Ma-β-Ala-TT) was first modified with DBCO-amine to introduce the reactive DBCO groups. The introduction of DBCO groups did not change the physico-chemical properties of the polymer poly(HPMA-*co*-Ma-β-Ala-DBCO) when compared to its polymer precursor. In the next step, the derivative **4** was clicked in a copper free manner to obtain conjugate **I**. Here, M_w increases to 48 kDa (Figure 6), with the increase of the hydrodynamic size to 16.3 nm. We assume such an increase of the hydrodynamic size of polymer conjugate **I.** to the concurrent presence of the hydrophobic DBCO groups and mebendazole, which led to the formation of the nanoaggregates with hydrophobic core composed of DBCO and mebendazole and hydrophilic shell formed by the water-soluble copolymer. Such behavior would be beneficial in a forthcoming biological evaluation as the increased size of the conjugate **I.** would lead to an enhanced tumor accumulation. Moreover, after the delivery and release of the mebendazole polymer carrier, of which the size will be dropped back to the value close to the limit of renal threshold should be easily removed from the organism via urine excretion. On the contrary, conjugate **II** was prepared by direct aminolytic reaction of poly(HPMA-*co*-Ma-β-Ala-TT) with freshly deprotected mebendazole derivative **5** with subsequent quenching of remaining TT groups with 1-aminopropan-2-ol. Here, no significant increase in both the molecular weight and hydrodynamic size was observed. The conjugate **II** remained water-soluble and no sign of aggregation was observed. The mebendazole content in all conjugates ranged between 6 wt.% and 9 wt.% and was determined by HPLC after alkaline hydrolysis. This procedure was optimized to be effective and quantitative for the precise mebendazole determination without any undesirable side reactions. A treatment in 1% NaOH water solution satisfied the above-mentioned criteria. Characterization of all prepared polymers is summarized in Table 1. The key feature of all the drug delivery systems is based on the sufficient stability during the blood circulation to the site of the action. Any premature release of a carried drug led to the lowering of the drug delivery efficacy, which can cause significant increase in the side effects on healthy organs. Thus, the stability of the conjugates was comprehensively investigated under blood stream mimicking conditions. Figure 7 shows a time dependent release of free mebendazole in PBS buffer in pH 7.4 at 37 °C from conjugates **I** and **II**, mimicking their stability during blood circulation. The data clearly show that the carbamate bond in conjugate **I** is cleaved much faster at blood pH ($t_{50\%}$ ~4 h) in contrast to the ester bond in conjugate **II** ($t_{10\%}$ ~53 h). In other words, the ester linker is more stable under blood-mimicking slightly alkaline conditions and would be therefore more suitable for evaluation in vivo. Importantly, even the formation of nanoaggregates in the case of conjugate I., where the localization of mebendazole in the inner hydrophobic core is expected, did not cause significant reduction of the hydrolytic cleavage of the carbamate bond. Based on these results, the chemical and biological properties of polymer conjugate **II** will be further evaluated in a future study.

Table 1. Properties of polymer precursors and polymer conjugates.

Sample	M_w kDa	M_w/M_n	Size (D_h) nm	Functionality mmolg^{-1}	Group Type
poly(HPMA-*co*-Ma-β-Ala-TT)	35	1.1	8.6	0.35	TT
poly(HPMA-*co*-Ma-β-Ala-DBCO)	40	1.2	10.5	0.32	DBCO
Conjugate **I**	48	1.2	16.3	0.31	MBZ
Conjugate **II**	37	1.1	10.3	0.22	MBZ

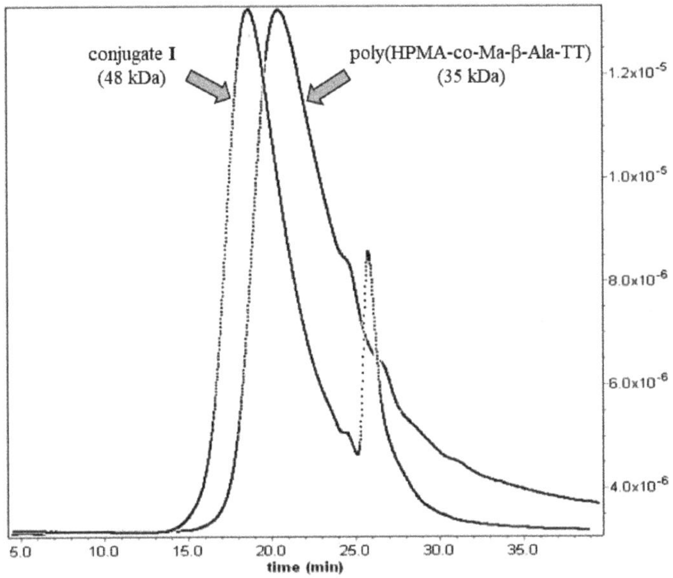

Figure 6. Comparison of GPC curves of poly(HPMA-*co*-Ma-β-Ala-TT) and conjugate **I**.

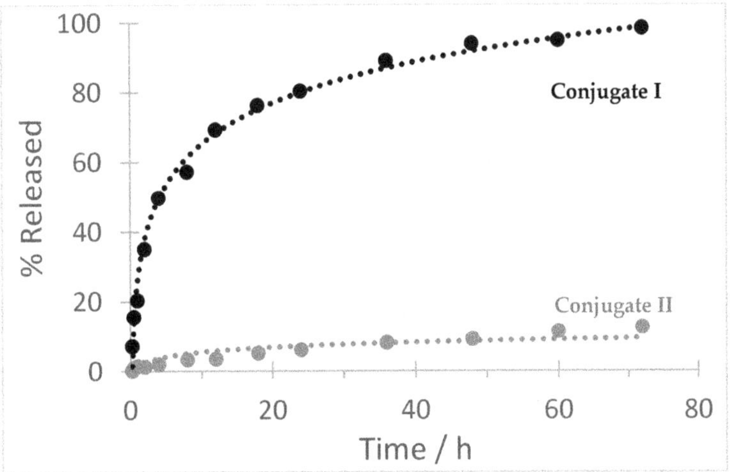

Figure 7. Stability of conjugates **I** and **II** in a buffer pH 7.4.

It is worth noting that the synthetic strategy described in the study can be applied for conjugation in a wide range of other therapies to a water-soluble or amphiphilic biocompatible polymers to expand the family of macromolecular drugs in a broad therapeutic portfolio. In addition to benzimidazole anthelmintics, any active compound with a heterocyclic N-H group in the molecule, can be involved. For example, pyrimidine or purine antimetabolites with a well-established antineoplastic activity (5-Fluorouracil, 6-Mercaptopurine, 6-Thioguanine), indole derivatives (anticancer agents Ellipticine, Vincristine or Panobinostat, antimigrenic drug Sumatriptan, anti-HIV drug Ateviridine) or other compounds, such as antineoplastics Vemurafenib and Axitinib, may be of interest.

4. Conclusions

In summary, two chemically distinct linkers, carbamate and ester, for attachment of the repurposed drug mebendazole to the HPMA-based polymer carrier, were designed and comprehensively studied. Five low-molecular-weight MBZ derivatives were designed and their step-by-step synthesis was optimized. On that basis, carbamate and ester-based polymer conjugates were prepared and their solution behavior and stability in blood-mimicking environment was evaluated. The ester linkage between the MBZ and polymer carrier is, in contrast to the carbamate one, long-term stable and has all the attributes to be selected for attachment of all benzimidazole-based anthelmintics. Further, a combination of hydrophobic and bulky DBCO units with mebendazole residues causes a significant increase in hydrodynamic diameter, which can be beneficial in view of a more pronounced EPR effect. Moreover, we envision that the proposed synthetic strategy may be employed in other biologically active nitrogen heterocycles containing drugs to water soluble polymer carriers. From a clinical point of view, the results of this work allow for use of the above-mentioned anticancer potential of mebendazole by its solubilization, improving the pharmacokinetics and controlled release. The strategy developed can be also utilized for a broad spectrum of other drugs which further increases the clinical relevance of this study.

Author Contributions: Conceptualization, M.S. and T.E.; methodology, M.S.; software, L.K.; validation, T.E.; formal analysis, A.R.; investigation, M.S.; resources, T.E.; data curation, A.R.; writing—original draft preparation, M.S.; writing—review and editing, T.E.; visualization, L.K.; supervision, T.E.; project administration, M.S.; funding acquisition, T.E. All authors have read and agreed to the published version of the manuscript.

Funding: This research was funded by the Czech Science Foundation grant numbers 19-01417S and 19-05649S and Ministry of Education, Youth and Sports of the Czech Republic within the Inter-excellence program grant number LTAUSA18083.

Institutional Review Board Statement: Not applicable.

Informed Consent Statement: Not applicable.

Data Availability Statement: The data presented in this study are available on request from the corresponding author.

Conflicts of Interest: The authors declare no conflict of interest.

References

1. Cha, Y.; Erez, T.; Reynolds, I.J.; Kumar, D.; Ross, J.; Koytiger, G.; Kusko, R.; Zeskind, B.; Risso, S.; Kagan, E.; et al. Drug repurposing from the perspective of pharmaceutical companies. *Br. J. Pharmacol.* **2018**, *175*, 168–180. [CrossRef] [PubMed]
2. Mukhopadhyay, T.; Sasaki, J.; Ramesh, R.; Roth, J.A. Mebendazole elicits a potent antitumor effect on human cancer cell lines both in vitro and in vivo. *Clin. Cancer Res.* **2002**, *8*, 2963–2969. [PubMed]
3. Chu, S.W.; Badar, S.; Morris, D.L.; Pourgholami, M.H. Potent inhibition of tubulin polymerisation and proliferation of paclitaxel-resistant 1A9PTX22 human ovarian cancer cells by albendazole. *Anticancer Res.* **2009**, *29*, 3791–3796.
4. Dogra, N.; Kumar, A.; Mukhopadhyay, T. Fenbendazole acts as a moderate microtubule destabilizing agent and causes cancer cell death by modulating multiple cellular pathways. *Sci. Rep.* **2018**, *8*, 11926. [CrossRef] [PubMed]
5. Hou, Z.J.; Luo, X.; Zhang, W.; Peng, F.; Cui, B.; Wu, S.J.; Zheng, F.M.; Xu, J.; Xu, L.Z.; Long, Z.J.; et al. Flubendazole, FDA-approved anthelmintic, targets breast cancer stem-like cells. *Oncotarget* **2015**, *6*, 6326–6340. [CrossRef]
6. Sasaki, J.; Ramesh, R.; Chada, S.; Gomyo, Y.; Roth, J.A.; Mukhopadhyay, T. The anthelmintic drug mebendazole induces mitotic arrest and apoptosis by depolymerizing tubulin in non-small cell lung cancer cells. *Mol. Cancer Ther.* **2002**, *1*, 1201–1209.
7. Rushworth, L.K.; Hewit, K.; Munnings-Tomes, S.; Somani, S.; James, D.; Shanks, E.; Dufes, C.; Straube, A.; Patel, R.; Leung, H.Y. Repurposing screen identifies mebendazole as a clinical candidate to synergise with docetaxel for prostate cancer treatment. *Br. J. Cancer* **2020**, *122*, 517–527. [CrossRef]
8. Poruchynsky, M.S.; Komlodi-Pasztor, E.; Trostel, S.; Wilkerson, J.; Regairaz, M.; Pommier, Y.; Zhang, X.; Maity, T.K.; Robey, R.; Burotto, M.; et al. Microtubule-targeting agents augment the toxicity of DNA-damaging agents by disrupting intracellular trafficking of DNA repair proteins. *Proc. Natl. Acad. Sci. USA* **2015**, *112*, 1571–1576. [CrossRef]

9. Celestino Pinto, L.; Moreira-Nunes, C.D.F.A.; Moreira Soares, B.; Rodrigues Burbano, R.M.; Rodrigues de Lemos, J.A.; Carvalho Montenegro, R. Mebendazole, an antiparasitic drug, inhibits drug transporters expression in preclinical model of gastric peritoneal carcinomatosis. *Toxicol. In Vitro* **2017**, *43*, 87–91. [CrossRef]
10. Dawson, M.; Braithwaite, P.A.; Roberts, M.S.; Watson, T.R. The pharmacokinetics and bioavailability of a tracer dose of [3H]-mebendazole in man. *Br. J. Clin. Pharmacol.* **1985**, *19*, 79–86. [CrossRef]
11. Dawson, M.; Allan, R.J.; Watson, T.R. The pharmacokinetics and bioavailability of mebendazole in man: A pilot study using [3H]-mebendazole. *Br. J. Clin. Pharmacol.* **1982**, *14*, 453–455. [CrossRef] [PubMed]
12. Dayan, A.D. Albendazole, mebendazole and praziquantel. Review of non-clinical toxicity and pharmacokinetics. *Acta Trop.* **2003**, *86*, 141–159. [CrossRef]
13. Pinto, L.C.; Moreira Soares, B.; Viana Pinheiro, J.J.; Riggins, G.J.; Pimentel Assumpcao, P.; Rodriguez Burbano, R.M.; Carvalho Montenegro, R. The anthelmintic drug mebendazole inhibits growth, migration and invasion in gastric cancer cell model. *Toxicol. In Vitro* **2015**, *29*, 2038–2044. [CrossRef] [PubMed]
14. Bai, R.Y.; Staedke, V.; Aprhys, C.M.; Gallia, G.L.; Riggins, G.J. Antiparasitic mebendazole shows survival benefit in 2 preclinical models of glioblastoma multiforme. *Neuro Oncol.* **2011**, *13*, 974–982. [CrossRef] [PubMed]
15. Simbulan-Rosenthal, C.M.; Dakshanamurthy, S.; Gaur, A.; Chen, Y.S.; Fang, H.B.; Abdussamad, M.; Zhou, H.; Zapas, J.; Calvert, V.; Petricoin, E.F.; et al. The repurposed anthelmintic mebendazole in combination with trametinib suppresses refractory $NRAS^{Q61K}$ melanoma. *Oncotarget* **2017**, *8*, 12576–12595. [CrossRef] [PubMed]
16. Blom, K.; Senkowsky, W.; Jarvius, M.; Berglund, M.; Rubin, J.; Lenhammar, L.; Parrow, V.; Andersson, C.; Loskog, A.; Fryknas, M.; et al. The anticancer effect of mebendazole may be due to M1 monocyte/macrophage activation via ERK1/2 and TLR8-dependent inflammasome activation. *Immunopharmacol. Immunotoxicol.* **2017**, *39*, 199–210. [CrossRef]
17. Blom, K.; Rubin, J.; Berglund, M.; Jarvius, M.; Lenhammar, L.; Parrow, V.; Andersson, C.; Loskog, A.; Fryknas, M.; Nygren, P.; et al. Mebendazole-induced M1 polarisation of THP-1 macrophages may involve DYRK1B inhibition. *BMC Res. Notes* **2019**, *12*, 234. [CrossRef]
18. Jornet, D.; Bosca, F.; Andreu, J.M.; Domingo, L.R.; Tormos, R.; Miranda, M.A. Analysis of mebendazole binding to its target biomolecule by laser flash photolysis. *J. Photochem. Photobiol. B* **2016**, *155*, 1–6. [CrossRef]
19. Doudican, N.; Rodriguez, A.; Osman, I.; Orlow, S.J. Mebendazole induces apoptosis via Bcl-2 inactivation in chemoresistant melanoma cells. *Mol. Cancer Res.* **2008**, *6*, 1308–1315. [CrossRef]
20. Sung, S.J.; Kim, H.K.; Hong, Y.K.; Joe, Y.A. Autophagy is a potential target for enhancing the anti-angiogenic effect of mebendazole in endothelial cells. *Biomol. Ther.* **2019**, *27*, 117–125. [CrossRef]
21. Williamson, T.; Bai, R.Y.; Staedtke, V.; Huso, D.; Riggins, G.J. Mebendazole and a non-steroidal anti-inflammatory combine to reduce tumor initiation in a colon cancer preclinical model. *Oncotarget* **2016**, *7*, 68571–68584. [CrossRef]
22. Zhang, F.; Li, Y.; Zhang, H.; Huang, E.; Gao, L.; Luo, W.; Wei, Q.; Fan, J.; Song, D.; Liao, J.; et al. Anthelmintic mebendazole enhances cisplatin's effect on suppressing cell proliferation and promotes differentiation of head and neck squamous cell carcinoma (HNSCC). *Oncotarget* **2017**, *8*, 12968–12982. [CrossRef]
23. Fernández-Bañares, F.; Gonzalez-Huix, F.; Xiol, X.; Catala, I.; Miro, J.; Lopez, N.; Casais, L. Marrow aplasia during high dose mebendazole treatment. *Am. J. Trop. Med. Hyg.* **1986**, *35*, 350–351. [CrossRef]
24. Colle, I.; Naegels, S.; Hoorens, A.; Hautekeete, M. Granulomatous hepatitis due to mebendazole. *J. Clin. Gastroenterol.* **1999**, *28*, 44–45. [CrossRef]
25. Duncan, R. Development of HPMA copolymer–anticancer conjugates: Clinical experience and lessons learnt. *Adv. Drug Deliv. Rev.* **2009**, *61*, 1131–1148. [CrossRef]
26. Kopeček, J. Polymer-drug conjugates: Origins, progress to date and future directions. *Adv. Drug Deliv. Rev.* **2013**, *65*, 49–59. [CrossRef]
27. Rihova, B.; Kovar, M. Immunogenicity and immunomodulatory properties of HPMA-based polymers. *Adv. Drug Deliv. Rev.* **2010**, *62*, 184–191. [CrossRef]
28. Sirova, M.; Kabesova, M.; Kovar, L.; Etrych, T.; Strohalm, J.; Ulbrich, K.; Rihova, B. HPMA copolymer-bound doxorubicin induces immunogenic tumor cell death. *Curr. Med. Chem.* **2013**, *20*, 4815–4826. [CrossRef]
29. Maeda, H. Tumor-selective delivery of macromolecular drugs via the EPR effect: Background and future prospects. *Bioconjug. Chem.* **2010**, *21*, 797–802. [CrossRef]
30. Taurin, S.; Nehoff, H.; Greish, K. Anticancer nanomedicine and tumor vascular permeability; Where is the missing link? *J. Control. Release* **2012**, *164*, 265–275. [CrossRef]
31. Matsumura, Y.; Maeda, H. A new concept for macromolecular therapeutics in cancer chemotherapy: Mechanism of tumoritropic accumulation of proteins and the antitumor agent smancs. *Cancer Res.* **1986**, *46*, 6387–6392.
32. Seymour, L.W.; Miyamoto, Y.; Maeda, H.; Brereton, M.; Strohalm, J.; Ulbrich, K.; Duncan, R. Influence of molecular weight on passive tumour accumulation of a soluble macromolecular drug carrier. *Eur. J. Cancer* **1995**, *5*, 766–770. [CrossRef]
33. Simplício, A.L.; Clancy, J.M.; Gilmer, J.F. Prodrugs for amines. *Molecules* **2008**, *13*, 519–547. [CrossRef]
34. Zimmermann, S.C.; Tichy, T.; Vavra, J.; Dash, R.P.; Slusher, C.E.; Gadiano, A.J.; Wu, Y.; Jancarik, A.; Tenora, L.; Monincova, L.; et al. N-substituted prodrugs of mebendazole provide improved aqueous solubility and oral bioavailability in mice and dogs. *J. Med. Chem.* **2018**, *61*, 3918–3929. [CrossRef]

35. Studenovsky, M.; Pola, R.; Pechar, M.; Etrych, T.; Ulbrich, K.; Kovar, L.; Kabesova, M.; Rihova, B. Polymer carriers for anticancer drugs targeted to EGF receptor. *Macromol. Biosci.* **2012**, *12*, 1714–1720. [CrossRef]
36. Pola, R.; Janouskova, O.; Etrych, T. The pH-dependent and enzymatic release of cytarabine from hydrophilic polymer conjugates. *Physiol. Res.* **2016**, *65* (Suppl. 2), S225–S232. [CrossRef]
37. Luttringhaus, A.; Dirksen, H.W. Tetramethylurea as a solvent and reagent. *Angew. Chem. Int. Ed.* **1964**, *3*, 260–269. [CrossRef]
38. Graeme, M.; Rizzardo, E.; Thang, S.H. Radical addition–fragmentation chemistry in polymer synthesis. *Polymer* **2008**, *49*, 1079–1131.
39. Kostka, L.; Subr, V.; Laga, R.; Chytil, P.; Ulbrich, K.; Seymour, L.W.; Etrych, T. Nanotherapeutics shielded with a pH responsive polymeric layer. *Physiol. Res.* **2015**, *64*, S29–S44. [CrossRef]

polymers

MDPI

Article

PnBA-b-PNIPAM-b-PDMAEA Thermo-Responsive Triblock Terpolymers and Their Quaternized Analogs as Gene and Drug Delivery Vectors

Athanasios Skandalis, Dimitrios Selianitis and Stergios Pispas *

Theoretical and Physical Chemistry Institute, National Hellenic Research Foundation, 48 Vassileos Constantinou Avenue, 11635 Athens, Greece; thanos.skan@gmail.com (A.S.); dselianitis@eie.gr (D.S.)
* Correspondence: pispas@eie.gr

Abstract: In this work, the ability of thermo-responsive poly [butyl acrylate-b-N-isopropylacrylamide-b-2-(dimethylamino) ethyl acrylate] (PnBA-b-PNIPAM-b-PDMAEA) triblock terpolymer self-assemblies, as well as of their quaternized analogs (PnBA-b-PNIPAM-b-QPDMAEA), to form polyplexes with DNA through electrostatic interactions was examined. Terpolymer/DNA polyplexes were prepared in three different amine over phosphate group ratios (N/P), and linear DNA with a 2000 base pair length was used. In aqueous solutions, the terpolymers formed aggregates of micelles with mixed PNIPAM/(Q)PDMAEA coronas and PnBA cores. The PnBA-b-PNIPAM-b-PDMAEA terpolymers' micellar aggregates were also examined as carriers for the model hydrophobic drug curcumin (CUR). The complexation ability of the terpolymer with DNA was studied by UV–Vis spectroscopy and fluorescence spectroscopy by investigating ethidium bromide quenching. Fluorescence was also used for the determination of the intrinsic fluorescence of the CUR-loaded micellar aggregates. The structural characteristics of the polyplexes and the CUR-loaded aggregates were investigated by dynamic and electrophoretic light scattering techniques. Polyplexes were found to structurally respond to changes in solution temperature and ionic strength, while the intrinsic fluorescence of encapsulated CUR was increased at temperatures above ambient.

Keywords: ABC triblock terpolymers; stimuli-responsive polymers; polyplexes; drug delivery; bioimaging

check for updates

Citation: Skandalis, A.; Selianitis, D.; Pispas, S. PnBA-b-PNIPAM-b-PDMAEA Thermo-Responsive Triblock Terpolymers and Their Quaternized Analogs as Gene and Drug Delivery Vectors. *Polymers* **2021**, *13*, 2361. https://doi.org/10.3390/polym13142361

Academic Editor: Marek Kowalczuk

Received: 30 June 2021
Accepted: 16 July 2021
Published: 19 July 2021

1. Introduction

Gene therapy has gained significant scientific attention for the treatment of diseases that arise from genetic abnormalities [1–3]. The presence of an effective vector that will carry out the efficient delivery of the genetic material through the cell membrane is a prerequisite [1,4,5]. There have been reported two major categories of gene delivery vectors, the viral and the nonviral, each one with its respective advantages and disadvantages [4,6].

Viral-mediated gene delivery systems [2,7] include, among others, adenoviruses and retroviruses and exhibit high transfection efficiency. However, viral gene delivery systems demonstrate limited carrying capacity, immunogenicity toxicity and high cost.

The above-mentioned limitations can be overcome by the utilization of synthetic, non-viral gene delivery vectors as an appealing alternative. Such vectors interact electrostatically with the DNA and must have the ability to condense the DNA, be not immunogenic and toxic, promote cellular uptake and protect DNA from degradation. Cationic polymers [8,9], along with lipids [10–13], have emerged as the most promising nonviral gene delivery vectors. The most used cationic polymers for gene delivery applications are polyethyleneimine (PEI) [3,14], poly(L-lysine) (PLL) [15,16], poly(ethylene glycol) bis (amine) [17,18], poly (2-dimethylamino ethyl methacrylate) (PDMAEMA) [19–21] and poly (2-dimethylamino ethyl acrylate) (PDMAEA) [22,23]. The complexes of cationic polymers with DNA are widely known as polyplexes.

Stimuli-responsive polymers are promising candidates for gene delivery applications, due to their unique ability to adapt to changes in their environment after the application of external stimuli [24,25]. Temperature-responsive polymers demonstrate changes in their conformation/hydration state upon the application of temperature variations. PNIPAM [26–28], probably the most famous and the most-well studied thermo-responsive polymer, exhibits hydrophilic behavior at room temperature and has a lower critical solution temperature (LCST) at approximately 32 °C, above which it becomes less hydrophilic. The biocompatibility, along with its LCST being close to the human body temperature, make PNIPAM an attractive polymer for gene delivery applications [29–33]. Upon heating above the LCST, the PNIPAM chains at the periphery of the polyplexes shrink, leading to increased protection of the DNA against enzymatic degradation. Moreover, the enhanced hydrophobicity without complete dehydration can result in improved transfection efficiency and cellular uptake [34–36].

Amphiphilic block copolymer micelles are widely used as nanocarriers for the encapsulation and delivery of hydrophobic drugs as they seek to avoid possible side effects, decrease drug degradation after administration and enhance the bioavailability of the drug [37–39]. Curcumin (CUR) is a hydrophobic polyphenol compound that has gained significant attention due to its low cytotoxicity, anticancer effects and anti-inflammatory properties [40–43].

The aim of this work was the physicochemical investigation of the capability of PnBA-b-PNIPAM-b-PDMAEA terpolymers and of their quaternized analogs, PnBA-b-PNIPAM-QPDMAEA, to form complexes with DNA. Furthermore, the potential of the initial PnBA-b-PNIPAM-b-PDMAEA terpolymers to be utilized as carriers for the encapsulation of the hydrophobic model drug CUR for potential bioimaging purposes (due to CUR's intrinsic fluorescence) was also examined, making use of the hydrophobic interactions with PnBA blocks. The ABC-type triblocks are composed of one hydrophobic PnBA block, one temperature-responsive PNIPAM block and one weak cationic polyelectrolyte PDMAEA block, which is converted into a strong cationic polyelectrolyte with permanent positive charges after post-polymerization functionalization with methyl iodide. In aqueous solutions, the terpolymers self-assemble in spherical micelles with PnBA cores and mixed PNIPAM/(Q)PDMAEA coronas and exhibit aggregation behavior due to secondary interactions, which are weaker in the case of the quaternized terpolymers. Further aggregation takes place upon heating above the LCST of PNIPAM because of the change in its hydrophobicity and the shrinkage of its chains. The complexation of the terpolymers micellar aggregates with DNA is possible through electrostatic interactions between the positive charges of the (Q)PDMAEA and the negative charges of the DNA.

2. Materials and Methods

2.1. Materials

Deoxyribonucleic acid (DNA) sodium salt from salmon testes (~2000 bp) was obtained from Sigma-Aldrich (Athens, Greece). Curcumin was obtained from Merck (Athens, Greece). Ethidium bromide, sodium chloride and all other reagents were obtained from Sigma-Aldrich and used as received.

2.2. Triblock Terpolymer Synthesis

PnBA-b-PNIPAM-b-PDMAEA triblock terpolymers were synthesized by sequential RAFT polymerization using AIBN as the radical initiator, 2-(dodecylthiocarbonothioylthio)-2-methylpropionic acid as the chain transfer agent and 1,4-dioxane as the polymerization solvent. The conversion of PnBA-b-PNIPAM-b-PDMAEA triblock terpolymers into PnBA-b-PNIPAM-b-QPDMAEA strong cationic polyelectrolytes was achieved by a typical quaternization reaction in THF, using CH_3I as the quaternizing agent.

More details about the synthetic procedure, the quaternization process, molecular characterization and aqueous solution properties of the terpolymers can be found in our previous work [44]. The chemical structures of PnBA-b-PNIPAM-b-PDMAEA and PnBA-b-

PNIPAM-b-QPDMAEA triblock terpolymers are presented in Scheme 1a,b, respectively, and their molecular characteristics can be found in Table 1.

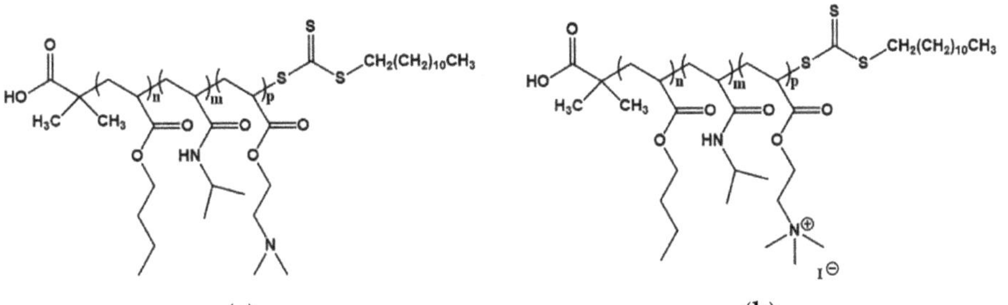

(a) (b)

Scheme 1. Chemical structure of (**a**) PnBA-b-PNIPAM-b-PDMAEA and (**b**) PnBA-b-PNIPAM-b-QPDMAEA triblock terpolymers.

Table 1. Molecular characteristics of PnBA-b-PNIPAM-b-(Q)PDMAEA terpolymers.

Sample	M_w [a] (10^{-4}) (g mol^{-1})	M_w/M_n [a]	Composition [b] (% wt) PnBA/PNIPAM/(Q)PDMAEA
$PnBA_{39}$-*b*-$PNIPAM_{87}$-*b*-$PDMAEA_{17}$	1.82	1.25	28/59/13
$PnBA_{39}$-*b*-$PNIPAM_{87}$-*b*-$PDMAEA_{35}$	2.40	1.27	25/54/21
$PnBA_{39}$-*b*-$PNIPAM_{87}$-*b*-$QPDMAEA_{17}$	2.05 [c]	-	26/51/23
$PnBA_{39}$-*b*-$PNIPAM_{87}$-*b*-$QPDMAEA_{35}$	2.90 [c]	-	21/45/34

[a] Determined by SEC, [b] Determined by ^{1}H-NMR, [c] Calculated for 100% conversion.

2.3. Polyplex Formation

The terpolymer/DNA polyplexes were formed by mixing terpolymer solution (5×10^{-4} g mL^{-1} in 0.01 M NaCl) and DNA solution (3.3×10^{-4} g mL^{-1} in 0.01 M NaCl) under gentle stirring at ambient conditions. The volume of the terpolymer solution was predefined and the corresponding volume of DNA solution was added for all N/P ratios formed. The final volume of the prepared mixed solutions was adjusted at 10 mL. The polyplexes were prepared in three different N/P ratios (0.5, 1.0 and 2.0).

2.4. Preparation of CUR-Loaded Micellar Aggregates

The procedure followed for the encapsulation of CUR into the triblock terpolymer nanostructures is described below. First, two separate stock solutions of the terpolymer and the drug were prepared in THF. The solutions were left to stand overnight to ensure the complete dissolution of the terpolymer and curcumin. Next, the two solutions were mixed in two different ratios and the mixtures were directly injected into distilled water under vigorous stirring until THF evaporation by heating at 65 °C.

2.5. Light Scattering

Dynamic light scattering (DLS) measurements were conducted on an ALV/CGS-3 compact goniometer system (ALVGmbH, Hessen, Germany), equipped with an ALV 5000/EPP multi-τ digital correlator with 288 channels and an ALV/LSE-5003 light scattering electronics unit for stepper motor drive and limit switch control. A JDS Uniphase 22 mW He-Ne laser (λ = 632.8 nm) was used as the light source. The solutions were filtered through 0.45 μm hydrophilic PTFE filters (Millex-LCR from Millipore, Billerica, MA, USA) before measurements. Each solution was measured five times at each angle and temperature and the average was used. The angular and temperature range for the measurements were 30–150° and 25–55 °C, respectively. The cumulants method and CONTIN software were utilized for the analysis of the obtained correlation functions. The size data and figures shown below are from measurements at 90°.

Electrophoretic light scattering (ELS) measurements were performed on a ZetaSizer Nano Series Nano-ZS (Malvern Instruments Ltd., Malvern, UK) equipped with a He-Ne laser beam at a wavelength of 633 nm and a fixed backscattering angle of 173°. Measurements were conducted at 25 °C and 45 °C. Data analysis was performed using the Henry correction of the Smoluchowski equation [45], after equilibration at 25 °C or 45 °C (depending on the measurement temperature). Reported zeta-potential (ζ_p) values are the average of 50 runs.

2.6. *UV–Vis*

UV–V is measurements of the polyplexes were conducted on a Perkin Elmer (Lambda 19) UV–Vis-NIR spectrometer (Waltman, MA, USA) in a wavelength range of 200 to 400 nm. The polyplexes were measured at all three N/P ratios prepared.

2.7. *Fluorescence Spectroscopy*

2.7.1. Ethidium Bromide Quenching Assay

The complexation ability of the terpolymers with DNA was investigated by studying the fluorescence of ethidium bromide (EtBr) in the whole N/P ratio range. A DNA solution (1×10^{-4} g mL^{-1}) was initially prepared in 0.01 M NaCl, followed by the addition of ethidium bromide ([EtBr]/[P] = 4) to it. The EtBr-containing DNA solution was titrated with a concentrated terpolymer solution starting from N/P = 0 and ending at N/P = 8.0. Fluorescence spectroscopy measurements were carried out for the initial DNA/EtBr solution, as well as after each titration with the polymer solution. Fluorescence spectra were recorded on a Fluorolog-3 Jobin Yvon-Spex spectrofluorometer (model GL3–21). The excitation wavelength for the measurements was 535 nm and the emission was monitored at 600 nm.

2.7.2. Fluorescence of CUR-Loaded Aggregates

The same instrument was used for fluorescence measurements of the CUR-loaded terpolymer aggregates and the excitation wavelength was set at 405 nm.

3. Results and Discussion

3.1. *PDMAEMA-b-PNIPAM-b-(Q)PDMAEA/DNA Polyplexes*

The ability of the nanostructures formed by PnBA-*b*-PNIPAM-*b*-PDMAEA and PnBA-*b*-PNIPAM-*b*-QPDMAEA triblock terpolymers in aqueous solutions to complex with DNA was investigated. The complexation is achieved through electrostatic interactions between the positively charged amine groups of the terpolymers (N) and the negatively charged phosphoric groups of the DNA (P). The terpolymer/DNA complex solutions were prepared in three different N/P ratios (N/P = 0.5, 1.0 and 2.0) using DNA with M_w = 2000 bp.

The successful formation of the polyplexes was confirmed by UV–Vis spectroscopy measurements. UV–Vis spectra for (a) PnBA-*b*-PNIPAM-*b*-PDMAEA/DNA and (b) PnBA-*b*-PNIPAM-*b*-QPDMAEA/DNA polyplexes are presented in Figure 1. The characteristic peak that corresponds to free DNA appears at 260 nm, while simultaneously a new peak is observed at 225 nm, which corresponds to the complexed DNA. However, this peak is only visible in the case of the quaternized terpolymers (Figure 1b). More specifically, for N/P = 0.5, both peaks are observed (at 225 and at 260 nm) and this can be attributed to the existence of both complexed and free DNA in the solution at the same time.

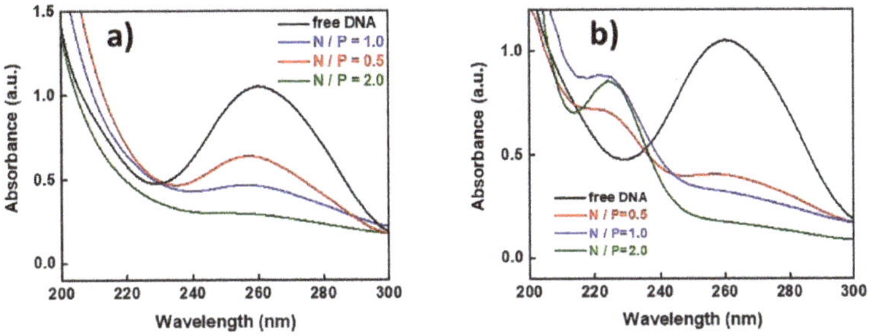

Figure 1. UV–Vis spectra of (**a**) $PnBA_{39}$-*b*-$PNIPAM_{87}$-*b*-$PDMAEA_{35}$/DNA polyplexes and (**b**) $PnBA_{39}$-*b*-$PNIPAM_{87}$-*b*-$QPDMAEA_{35}$/DNA polyplexes (spectrum of free DNA in black).

As the terpolymer concentration in the solution increases, the peak at 260 nm disappears and only the one at 260 nm is visible, meaning that the total amount of DNA participates in the formation of polyplexes. This occurs because the increase in the positive charges in the solution leads to the neutralization of the negative DNA charges.

On the other hand, in the case of PnBA-*b*-PNIPAM-*b*-PDMAEA/DNA polyplexes (Figure 1a), a similar decrease is observed in the height of the free DNA peak as the polymer concentration increases, but the peak at 225 nm that corresponds to the bonded DNA is not visible. This is an indication that the electrostatic interactions between terpolymer and DNA are significantly more intense when the quaternized terpolymers are used, due to the existence of permanent positive charges.

Ethidium bromide fluorescence quenching assays (Figure 2) support the conclusion that the quaternized terpolymers interact more strongly and more efficiently with DNA molecules forming polyplexes. This is evident through the strongest decrease in relative fluorescence for the $PnBA_{39}$-*b*-$PNIPAM_{87}$-*b*-$QPDMAEA_{35}$ polyplexes compared to $PnBA_{39}$-*b*-$PNIPAM_{87}$-*b*-$PDMAEA_{35}$ ones, at N/P ratios greater than 3. However, even at an N/P ratio equal to 8, the relative fluorescence does not drop below 0.7, indicating that the terpolymer micellar aggregates do not interact sufficiently with DNA, possibly due to their aggregate structure, which hinders terpolymer/DNA interactions because of spatial constraints, including low access of amine/quaternary amine groups by DNA chains in the corona of the terpolymer micellar aggregates.

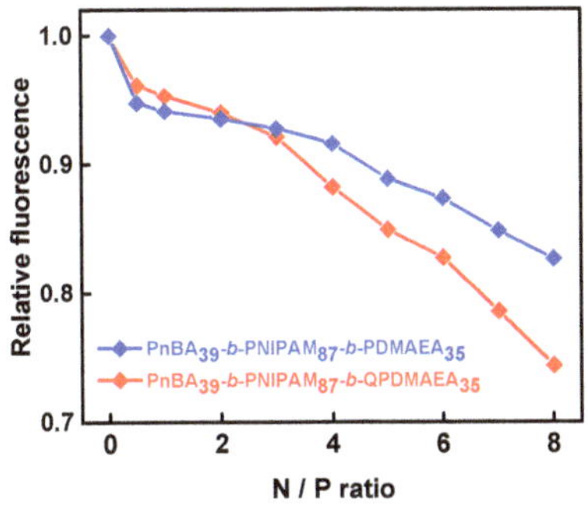

Figure 2. Ethidium bromide fluorescence quenching in $PnBA_{39}$-*b*-$PNIPAM_{87}$-*b*-$PDMAEA_{35}$ (blue line) and $PnBA_{39}$-*b*-$PNIPAM_{87}$-*b*-$QPDMAEA_{35}$ (red line) polyplexes with DNA.

As depicted in Figure 3, the scattering intensity of $PnBA_{39}$-*b*-$PNIPAM_{87}$-*b*-$PDMAEA_{35}$/DNA does not exhibit major changes for N/P = 0.5 and N/P = 1 ratios and decreases significantly for N/P = 2, showing that, in this case, particles with smaller mass are present. On the contrary, as far as the size of the polyplexes is concerned, it is observed that at

N/P = 0.5, there are large aggregates (R_h = 380 nm approximately) and there is a great size decrease at N/P = 1 (R_h = 170 nm approximately). This rapid decrease in the size of the polyplexes is attributed to the formation of the smaller aggregates since the concentration of DNA increases and interactions are more efficient. At N/P = 0.5, the negatively charged phosphate groups are in excess and the formation of stable complexes with the polymeric nanostructures is more difficult. Therefore, the formation of aggregates is favored.

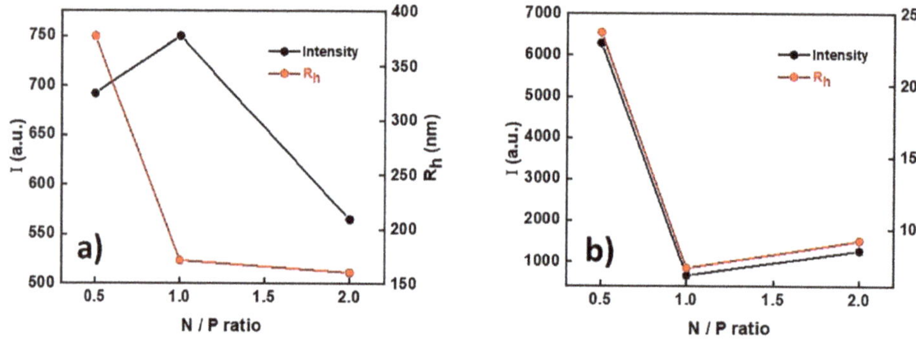

Figure 3. Variations in scattering intensity and R_h with N/P ratio for (**a**) $PnBA_{39}$-*b*-$PNIPAM_{87}$-*b*-$PDMAEA_{35}$/DNA and (**b**) $PnBA_{39}$-*b*-$PNIPAM_{87}$-*b*-$QPDMAEA_{35}$/DNA polyplexes.

At the ratios where the amine groups of the terpolymer are either in equilibrium or in excess (the size of the polyplexes does not change at N/P = 2), it seems that the complexation with DNA is stronger and induces the breaking of the intermicellar aggregates and the formation of more stable structures.

At $PnBA_{39}$-*b*-$PNIPAM_{87}$-*b*-$QPDMAEA_{35}$/DNA polyplexes, the scattering intensity shows a rapid decrease from N/P = 0.5 to N/P = 1 and remains practically unchanged for N/P = 2. As mentioned earlier, at N/P = 0.5, the DNA is in excess and the electrostatic interactions are weaker, leading to the formation of aggregates. The size of the polyplexes also shows a rapid decrease from 250 nm at N/P = 0.5 to 70 nm at N/P = 1. This decrease is attributed to the formation of smaller intermicellar aggregates in the solution.

At N/P = 2, there is a small increase in the size of the polyplexes (R_h = 95 nm). Moreover, it must be highlighted that the electrostatic interactions between the quaternized terpolymers and the DNA are much stronger than the interactions between the non-quaternized ones (amine form) with DNA, due to the existence of more positive charges.

Dynamic and electrophoretic light scattering were also utilized for more detailed investigations of the temperature effect on $PnBA_{39}$-*b*-$PNIPAM_{87}$-*b*-PDMAEA/DNA and PnBA-*b*-$PNIPAM_{87}$-*b*-QPDMAEA/DNA polyplexes, because of the presence of the thermoresponsive PNIPAM block in the structure of the terpolymers.

The size distribution graphs from DLS measurements for $PnBA_{39}$-*b*-$PNIPAM_{87}$-*b*-PDMAEA35/DNA (a, b) and $PnBA_{39}$-*b*-$PNIPAM_{87}$-*b*-$QPDMAEA_{35}$/DNA (c, d) polyplexes at 25 °C (a, c) and at 55 °C (b, d) are presented in Figure 4. Monomodal size distributions are observed in all cases. For $PnBA_{39}$-*b*-$PNIPAM_{87}$-*b*-$PDMAEA_{35}$ /DNA polyplexes at 25 °C (Figure 4a), the size distribution becomes narrower as the polymer concentration in the solution increases. Similar behavior is observed for $PnBA_{39}$-*b*-$PNIPAM_{87}$-*b*-$QPDMAEA_{35}$/DNA polyplexes at 25 °C (Figure 4c).

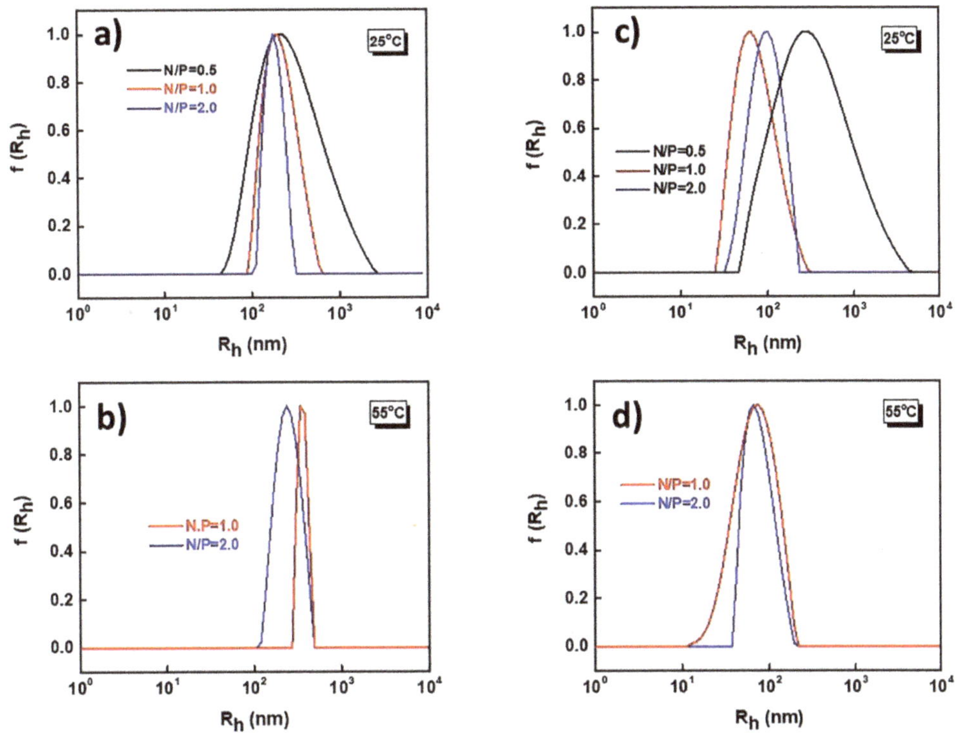

Figure 4. Size distributions of PnBA$_{39}$-*b*-PNIPAM$_{87}$-*b*-PDMAEA$_{35}$/DNA (**a**,**b**) and PnBA$_{39}$-*b*-PNIPAM$_{87}$-*b*-QPDMAEA$_{35}$/DNA (**c**,**d**) polyplexes at 25 °C (**a**,**c**) and at 55 °C (**b**,**d**).

At 55 °C, precipitation phenomena took place for both complexes involving the non-quaternized and the quaternized terpolymers. For the other N/P ratios, the solutions remained stable, and the size distributions remained monomodal. In the case of PnBA$_{39}$-*b*-PNIPAM$_{87}$-*b*-PDMAEA$_{35}$/DNA polyplexes (Figure 4b), it is also observed that the size distributions become narrower as the polymer concentration increases, while for PnBA$_{39}$-*b*-PNIPAM$_{87}$-*b*-QPDMAEA$_{35}$/DNA polyplexes (Figure 4d), there are no significant changes.

Figure 5 shows the dependence of the scattering intensity and hydrodynamic diameter of the terpolymer/DNA polyplexes on temperature. A constant increase in the scattering intensity (increase in the mass of the particles) is seen for PnBA$_{39}$-*b*-PNIPAM$_{87}$-*b*-PDMAEA$_{35}$/DNA polyplexes at N/P = 1 (Figure 5a) and up to 45 °C. In the range of 45–55 °C, a plateau is reached. The R_h follows a similar pattern and increases as the temperature rises up to 35 °C, remains practically unchanged between 35 and 50 °C and increases again from 50 to 55 °C. This behavior shows that as the temperature increases, which means that PNIPAM blocks becomes less hydrophilic and shrink, the polyplexes tend to form larger aggregates. Importantly, the response of the terpolymers to temperature variations is maintained after their complexation with the DNA. Similar behavior is observed for N/P = 2 (Figure 5b).

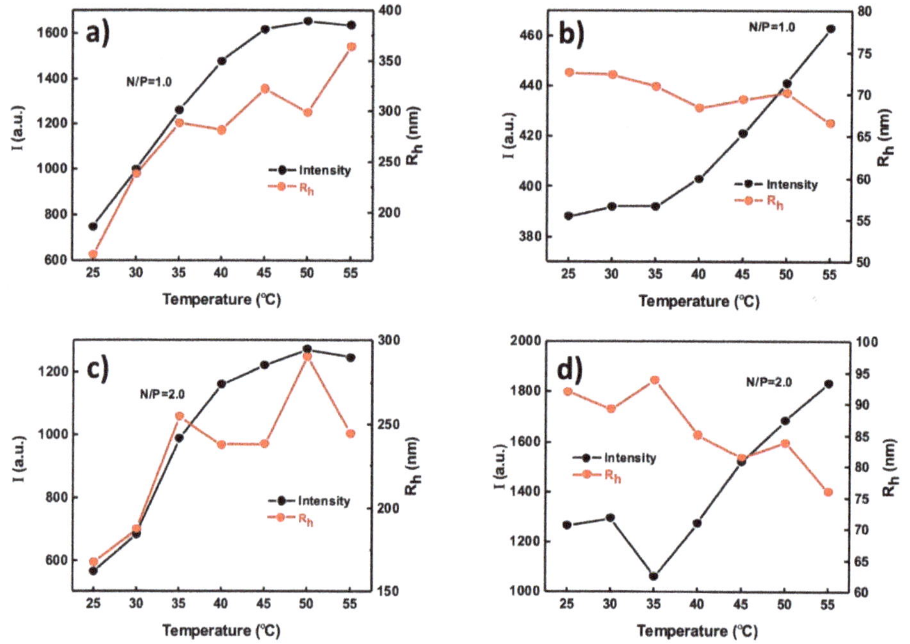

Figure 5. Scattering intensity and R_h as a function of temperature for PnBA$_{39}$-*b*-PNIPAM$_{87}$-*b*-PDMAEA$_{35}$ /DNA polyplexes at (**a**) N/P = 1.0 and (**b**) N/P = 2.0 and PnBA$_{39}$-*b*-PNIPAM$_{87}$-*b*-QPDMAEA$_{35}$/DNA polyplexes at (**c**) N/P = 1.0 and (**d**) N/P = 2.0.

For the quaternized terpolymers/DNA polyplexes at N/P = 1 (Figure 5c), an increase can be observed in the scattering intensity from 35 °C and above, but at a much lower scale than at PnBA$_{39}$-*b*-PNIPAM$_{87}$-*b*-PDMAEA$_{35}$/DNA polyplexes, while the changes in R_h are insignificant. This means that as the temperature increases and the PNIPAM block shrinks, disassociation of the aggregates (if they exist) or shrinkage of the polymeric micelles may take place. Similar results are obtained for N/P = 2, with the changes in the size of the polyplexes being slightly larger, indicating that the polymer/DNA ratio is a very important factor regarding the response of the polyplexes to temperature variations.

The surface charge of the polyplexes, at all N/P ratios, was investigated by electrophoretic light scattering measurements at 25 °C and at 45 °C (Figure 6). For PnBA$_{39}$-*b*-PNIPAM$_{87}$-*b*-PDMAEA$_{35}$/DNA at 25 °C (Figure 6a), the ζ-potential values are negative at all N/P ratios, meaning that there is uncomplexed DNA in the solution. The explanation behind this observation is that the terpolymers are partially protonated and, for this reason, the electrostatic interactions are weak. At 45 °C, a temperature above the LCST of PNIPAM, the largest change is observed at N/P = 0.5, where the solution is not colloidally stable. For the remaining N/P ratios, there were no important changes.

As far as the PnBA$_{39}$-*b*-PNIPAM$_{87}$-*b*-QPDMAEA$_{35}$/DNA polyplexes at 25 °C are concerned (Figure 6b), the ζ-potential values are negative for N/P = 0.5, due to the existence of free DNA in the solution. At N/P = 1 and N/P = 2 ratios, where the quaternized amine groups of the QPDMAEA cationic block are either in equilibrium or in excess in comparison to the negatively charged phosphate groups of DNA, the ζ-potential values are positive. This may indicate either the existence of micelles that have not complexed with the DNA, or that each DNA molecule interacts with more than one micelle, creating aggregates the outer parts of which carry positive charges. The latter seems to be the most probable scenario.

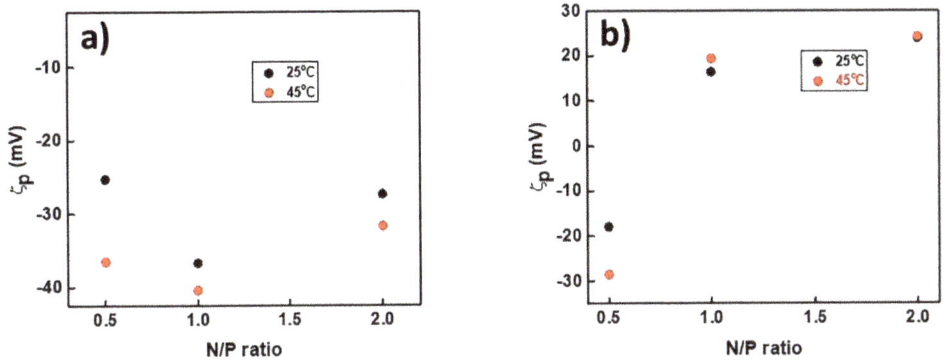

Figure 6. Variation in ζ-potential with N/P ratio for PnBA$_{39}$-*b*-PNIPAM$_{87}$-*b*-PDMAEA$_{35}$/DNA (**a**) and PnBA$_{39}$-*b*-PNIPAM$_{87}$-*b*-QPDMAEA$_{35}$/DNA (**b**) polyplexes at 25 °C and at 45 °C.

The effect of ionic strength in the terpolymer/DNA polyplexes was also studied, as it is a very important parameter for the efficacy of a cationic polymer as nonviral gene delivery system. Figure 7 depicts the changes in the scattering intensity and hydrodynamic diameter of the polyplexes as a function of ionic strength at N/P = 1 and N/P = 2 ratios.

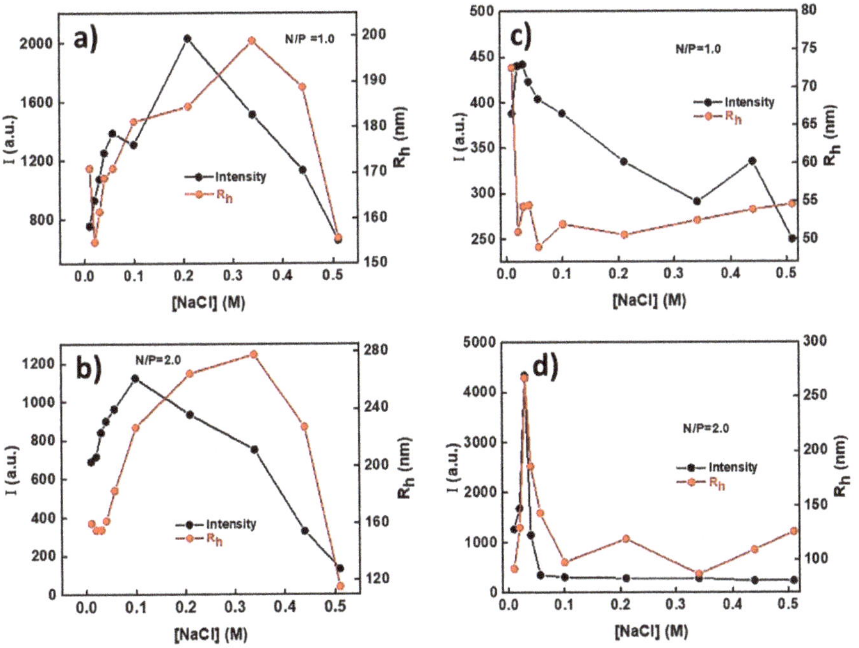

Figure 7. Scattering intensity and R_h as a function of ionic strength for PnBA$_{39}$-*b*-PNIPAM$_{87}$-*b*-PDMAEA$_{35}$/DNA polyplexes at N/P = 1.0 (**a**) and N/P = 2.0 (**b**) and for PnBA$_{39}$-*b*-PNIPAM$_{87}$-*b*-QPDMAEA$_{35}$ at N/P = 1.0 (**c**) and N/P = 2.0 (**d**).

For PnBA$_{39}$-*b*-PNIPAM$_{87}$-*b*-PDMAEA$_{35}$/DNA polyplexes at N/P = 1 (Figure 7a), the scattering intensity increases C_{NaCl} = 0.2 M and decreases for higher salt concentrations. The initial increase is translated into an increase in the mass of the polyplexes, which continues until their disintegration, where the decrease in the scattering intensity begins, showing that the polyplexes are not stable at higher salt concentrations. The R_h of the polyplexes exhibits a similar trend. It increases till C_{NaCl} = 0.32 M and decreases thereafter. A similar pattern is observed for the polyplexes formed at N/P = 2 (Figure 7b).

$PnBA_{39}$-*b*-$PNIPAM_{87}$-*b*-$QPDMAEA_{35}$/DNA polyplexes exhibit different behavior as the salt concentration increases. At the ratio N/P = 1 (Figure 7c), the scattering intensity decreases in all cases, but at a much lower rate than the $PnBA_{39}$-*b*-$PNIPAM_{87}$-*b*-$PDMAEA_{35}$/DNA polyplexes. This is also translated into the disassociation of the complexes and they are not stable upon an ionic strength increase. The R_h decreases from approx. 72 nm to approx. 52 nm with the first addition of salt and remains practically unchanged within the remaining ionic strength range studied.

At N/P = 2 (Figure 7d), an initial increase is observed in the scattering intensity till C_{NaCl} = 0.04 M, followed by a significant decrease till C_{NaCl} = 0.065 M, and the intensity remains constant at higher ionic strength values. The R_h behavior follows the same pattern as the scattering intensity. This shows that dissociation of the polyplexes takes place even at relatively low salt concentrations.

After evaluation of all the results obtained from the physicochemical characterization of the terpolymer/DNA complexes, the complexation process is illustrated in Scheme 2. The effect of temperature is much more visible in the case of PnBA-*b*-PNIPAM-*b*-PDMAEA/DNA complexes, where the electrostatic interactions are weaker.

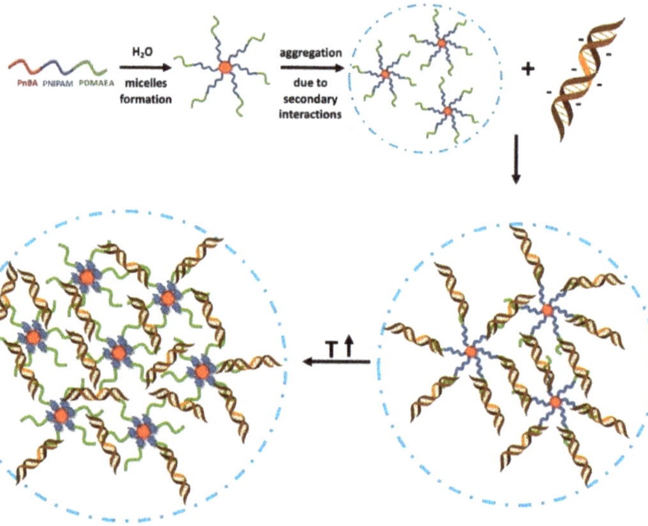

Scheme 2. Schematic illustration of the complexation process of PnBA-b-PNIPAM-b-(Q)PDMAEA triblock terpolymers with DNA.

3.2. *Encapsulation of CUR in the PnBA-b-PNIPAM-b-PDMAEA Polymeric Micellar Aggregates*

The capability of PnBA-b-PNIPAM-b-PDMAEA terpolymers to act as drug delivery carriers was also investigated by encapsulating curcumin (a model hydrophobic drug). Curcumin is considered an anticancer and anti-inflammatory drug with increased hydrophobicity [41,42]. Dynamic light scattering and fluorescence spectroscopy measurements were carried out in order to determine the successful encapsulation of curcumin in the polymeric micelles and to specify the properties that curcumin imparts to the polymeric system. CUR-loaded micellar aggregates with 10% and 20% *w/w* targeted entrapment levels (relative to the hydrophobic PnBA content) for $PnBA_{39}$-*b*-$PNIPAM_{87}$-*b*-$PDMAEA_{35}$ copolymer were prepared according to the protocol described above. UV–Vis determinations, utilizing a calibration curve, showed that the 10% *w/w* formulation actually resulted in 27% drug loading efficiency with a drug loading of 0.7%, while for the 20% *w/w* formulation, the values were 59% and 1.5%, respectively. Despite the rather low CUR entrapment, DLS measurements of the loaded terpolymer nanostructure were recorded at 25 °C, 45 °C and at a 90° measuring angle. The fluorescence spectroscopy study was accomplished in the same temperature range. Figure 8 illustrates comparative plots of the size distributions of CUR-loaded $PnBA_{39}$-b-$PNIPAM_{87}$-b-$PDMAEA_{35}$ nanostructures.

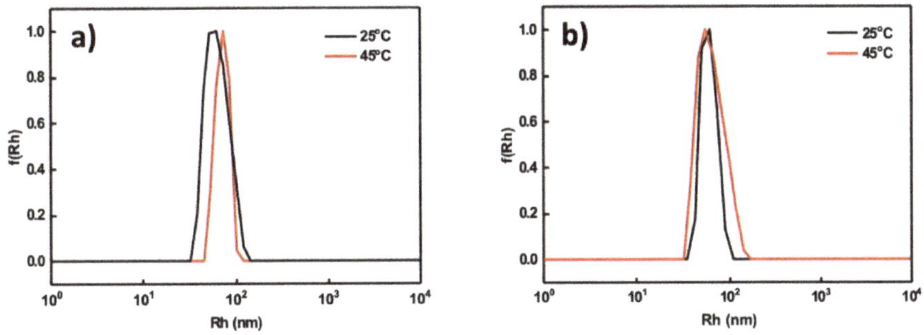

Figure 8. Comparative size distributions graphs from DLS measurements for CUR-loaded $PnBA_{39}$-*b*-$PNIPAM_{87}$-*b*-$PDMAEA_{35}$ micelles as a function of temperature for 10% CUR (**a**) and 20% wt CUR (**b**).

The size distributions from DLS measurements presented in Figure 8 reveal that the addition of curcumin (10 wt% and 20 wt% relative to the hydrophobic PnBA block) causes a slight decrease in the dimensions of the polymeric micelles, which is accompanied by a significant reduction in the polydispersity index. This observation proves that the entrapment of curcumin significantly improves the self-organization of the drug-loaded nanoparticles and may denote that the hydrophobic interactions created between the copolymer and the drug facilitate the effective encapsulation of CUR into the polymeric micelles. From the results shown in Table 2, it is worth noting that when the temperature was increased, a significant increase occurred in the mass of CUR-loaded nanoparticles, which indicates the formation of more compact micellar aggregates when curcumin is present in the nanosystem.

Table 2. DLS data for free and CUR-loaded terpolymer micellar aggregates.

Sample	CUR (%)	Temperature (°C)	Intensity [a] (a.u.) Free/Loaded	R_h [a] (nm) Free/Loaded	PDI [a] Free/Loaded
$PnBA_{39}$-*b*-$PNIPAM_{87}$-*b*-$PDMAEA_{35}$	10	25	5960/14,870	83/71	0.35/0.16
		45	19,000/30,000	54/73	0.22/0.14
	20	25	5960/17,450	83/61	0.35/0.13
		45	19,000/37,000	54/64	0.22/0.15

[a] By DLS at 90° angle.

As discussed in our previous work, in addition to its beneficial properties, curcumin can also be used as an imaging agent for bioimaging applications due to the strong endogenous fluorescence that it presents [46]. Given that, fluorescence spectroscopy measurements were performed on the CUR-loaded micellar aggregates. It has been proved that the solubility of curcumin in water (4.2 μg mL^{-1}) is very low [43]. However, when entrapped in the hydrophobic environment of amphiphilic copolymer self-assemblies, its solubility increases significantly. CUR-loaded terpolymer micellar aggregates displayed an important increase in curcumin solubility, by a factor of 6 (10 wt% CUR, 25 μg mL^{-1}) and 12 times (20 wt% CUR, 50 μg mL^{-1}) compared to the case of water. In Figure 9, it can be observed that the peak of curcumin at 489 nm in THF shifted to 514 nm (10 wt% CUR) and 504 nm (20 wt% CUR), respectively, in aqueous media. This shift is speculated to be due to the hydrophobic interactions that take place between CUR and the hydrophobic PnBA domains of the terpolymer. As far as fluorescence intensity is concerned, an interesting change occurred when temperature was increased at 45 °C. In both cases (i.e., for 10% and 20% wt CUR), a significant increase in fluorescence intensity is observed, indicating that curcumin is arranged within the hydrophobic domains of the macromolecular assemblies according to the thermo-responsive structural reorganization of the copolymer. It is observed that by increasing the solution temperature, aggregates of greater mass are formed. Thus, CUR-loaded nanoparticles become more compact, which changes entrapped CUR spatial arrangements and interactions within the aggregates and ultimately results in higher fluorescence intensity.

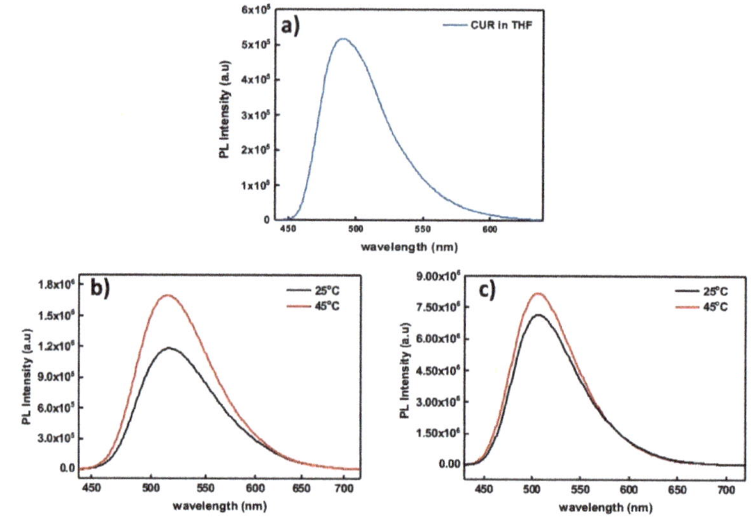

Figure 9. Fluorescence of CUR in THF (c_{CUR} = 100 μg mL^{-1}) (**a**), fluorescence of 10 wt% (**b**) and 20 wt% (**c**) CUR-loaded $PnBA_{39}$-*b*-$PNIPAM_{87}$-*b*-$PDMAEA_{35}$ micellar aggregates.

The encapsulation procedure and the effect of temperature on the CUR-loaded aggregates is depicted in Scheme 3. The observed temperature dependence of fluorescence may be of potential utilization in bioimaging, e.g., in discriminating differences in temperature within cell or tissue environments.

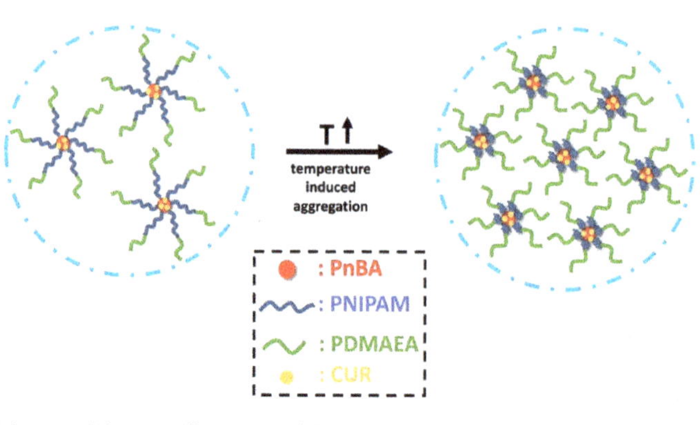

Scheme 3. Schematic illustration of CUR-loaded micellar aggregates and the effect of temperature.

4. Conclusions

The complexation ability of PnBA-*b*-PNIPAM-*b*-PDMAEA and PnBA-*b*-PNIPAM-*b*-QPDMAEA triblock terpolymers with DNA in aqueous solutions was studied. The electrostatic interactions between the quaternized terpolymers and DNA were found to be significantly stronger than the PnBA-*b*-PNIPAM-*b*-PDMAEA/DNA ones, due to the existence of permanent positive charges. The size of the PnBA-*b*-PNIPAM-*b*-PDMAEA/DNA polyplexes is larger (aggregate formation) and they exhibit increased responses to temperature variations because of the presence of the PNIPAM block. On the contrary, the polyplexes formed between PnBA-*b*-PNIPAM-*b*-QPDMAEA and DNA are smaller due to the higher solubility of the particular terpolymers in aqueous media because of the existence of more positive charges. These polyplexes do not present significant changes with temperature variations due to their stronger polyelectrolyte character. The PnBA-*b*-PNIPAM-*b*-PDMAEA terpolymers were also tested as nanocarriers for the encapsulation of the hydrophobic model drug curcumin in the hydrophobic PnBA cores of the formed micelles. The size of the drug-encapsulated nano-assemblies was smaller than the empty ones and this difference became larger at 45 °C, above the LCST of PNIPAM. Moreover, the endogenous fluorescence of curcumin exhibits a significant increase as the temperature increases above the LCST of PNIPAM, showing that this nanosystem can be promising for potential bioimaging applications.

Author Contributions: Conceptualization, S.P.; methodology, A.S. and D.S.; validation, S.P., A.S. and D.S.; data curation, A.S. and D.S.; writing—original draft preparation, A.S. and D.S.; writing—review and editing, S.P., A.S. and D.S.; visualization, A.S.; supervision, S.P. All authors have read and agreed to the published version of the manuscript.

Funding: This research is co-financed by Greece and the European Union (European Social Fund-ESF) through the Operational Program "Human Resources Development, Education and Lifelong Learning" in the context of the project "Strengthening Human Resources Research Potential via Doctorate Research" (MIS-5000432), implemented by the State Scholarships Foundation (IKY).

Institutional Review Board Statement: Not applicable.

Informed Consent Statement: Not applicable.

Data Availability Statement: The data presented in this study are available on request from the corresponding author.

Conflicts of Interest: The authors declare no conflict of interest.

References

1. Pack, D.W.; Hoffman, A.S.; Pun, S.; Stayton, P.S. Design and Development of Polymers for Gene Delivery. *Nat. Rev. Drug Discov.* **2005**, *4*, 581–593. [CrossRef]
2. Sung, Y.K.; Kim, S.W. Recent Advances in the Development of Gene Delivery Systems. *Biomater. Res.* **2019**, *23*, 8. [CrossRef] [PubMed]
3. Thomas, T.J.; Tajmir-Riahi, H.-A.; Pillai, C.K.S. Biodegradable Polymers for Gene Delivery. *Molecules* **2019**, *24*, 3744. [CrossRef] [PubMed]
4. Lai, W.-F.; Wong, W.-T. Design of Polymeric Gene Carriers for Effective Intracellular Delivery. *Trends Biotechnol.* **2018**, *36*, 713–728. [CrossRef] [PubMed]
5. Nelson, C.E.; Gersbach, C.A. Engineering Delivery Vehicles for Genome Editing. *Annu. Rev. Chem. Biomol. Eng.* **2016**, *7*, 637–662. [CrossRef] [PubMed]
6. Guo, X.; Huang, L. Recent Advances in Nonviral Vectors for Gene Delivery. *Acc. Chem. Res.* **2012**, *45*, 971–979. [CrossRef] [PubMed]
7. Nayerossadat, N.; Maedeh, T.; Ali, P.A. Viral and Nonviral Delivery Systems for Gene Delivery. *Adv. Biomed. Res.* **2012**, *1*, 27. [CrossRef]
8. Dinçer, S.; Türk, M.; Pişkin, E. Intelligent Polymers as Nonviral Vectors. *Gene Ther.* **2005**, *12*, S139–S145. [CrossRef]
9. Kundu, P.P.; Sharma, V. Synthetic Polymeric Vectors in Gene Therapy. *Curr. Opin. Solid State Mater. Sci.* **2008**, *12*, 89–102. [CrossRef]
10. Pathak, A.; Patnaik, S.; Gupta, K.C. Recent Trends in Non-Viral Vector-Mediated Gene Delivery. *Biotechnol. J.* **2009**, *4*, 1559–1572. [CrossRef]

11. Yin, H.; Kanasty, R.L.; Eltoukhy, A.A.; Vegas, A.J.; Dorkin, J.R.; Anderson, D.G. Non-Viral Vectors for Gene-Based Therapy. *Nat. Rev. Genet.* **2014**, *15*, 541–555. [CrossRef]
12. Navarro, G.; Pan, J.; Torchilin, V.P. Micelle-like Nanoparticles as Carriers for DNA and SiRNA. *Mol. Pharm.* **2015**, *12*, 301–313. [CrossRef]
13. Zhang, X.-X.; McIntosh, T.J.; Grinstaff, M.W. Functional Lipids and Lipoplexes for Improved Gene Delivery. *Biochimie* **2012**, *94*, 42–58. [CrossRef] [PubMed]
14. Cho, S.K.; Dang, C.; Wang, X.; Ragan, R.; Kwon, Y.J. Mixing-Sequence-Dependent Nucleic Acid Complexation and Gene Transfer Efficiency by Polyethylenimine. *Biomater. Sci.* **2015**, *3*, 1124–1133. [CrossRef]
15. Byrne, M.; Victory, D.; Hibbitts, A.; Lanigan, M.; Heise, A.; Cryan, S.-A. Molecular Weight and Architectural Dependence of Well-Defined Star-Shaped Poly(Lysine) as a Gene Delivery Vector. *Biomater. Sci.* **2013**, *1*, 1223–1234. [CrossRef] [PubMed]
16. Ivanova, E.; Dimitrov, I.; Kozarova, R.; Turmanova, S.; Apostolova, M. Thermally Sensitive Polypeptide-Based Copolymer for DNA Complexation into Stable Nanosized Polyplexes. *J. Nanopart. Res.* **2012**, *15*, 1358. [CrossRef]
17. Garrett, S.W.; Davies, O.R.; Milroy, D.A.; Wood, P.J.; Pouton, C.W.; Threadgill, M.D. Synthesis and Characterisation of Polyamine–Poly(Ethylene Glycol) Constructs for DNA Binding and Gene Delivery. *Bioorganic Med. Chem.* **2000**, *8*, 1779–1797. [CrossRef]
18. Nisha, C.K.; Manorama, S.v.; Ganguli, M.; Maiti, S.; Kizhakkedathu, J.N. Complexes of Poly(Ethylene Glycol)-Based Cationic Random Copolymer and Calf Thymus DNA: A Complete Biophysical Characterization. *Langmuir* **2004**, *20*, 2386–2396. [CrossRef]
19. Crommelin, D.J.A.; Storm, G.; Jiskoot, W.; Stenekes, R.; Mastrobattista, E.; Hennink, W.E. Nanotechnological Approaches for the Delivery of Macromolecules. *J. Control. Release* **2003**, *87*, 81–88. [CrossRef]
20. De Smedt, S.C.; Demeester, J.; Hennink, W.E. Cationic Polymer Based Gene Delivery Systems. *Pharm. Res.* **2000**, *17*, 113–126. [CrossRef] [PubMed]
21. Skandalis, A.; Uchman, M.; Štěpánek, M.; Kereïche, S.; Pispas, S. Complexation of DNA with QPDMAEMA-b-PLMA-b-POEGMA Cationic Triblock Terpolymer Micelles. *Macromolecules* **2020**, *53*, 5747–5755. [CrossRef]
22. Giaouzi, D.; Pispas, S. Complexation Behavior of PNIPAM-b-QPDMAEA Copolymer Aggregates with Linear DNAs of Different Lengths. *Eur. Polym. J.* **2021**, *155*, 110575. [CrossRef]
23. Chroni, A.; Pispas, S. Hydrophilic/Hydrophobic Modifications of a PnBA-b-PDMAEA Copolymer and Complexation Behaviour with Short DNA. *Eur. Polym. J.* **2020**, *129*, 109636. [CrossRef]
24. Shim, M.S.; Kwon, Y.J. Stimuli-Responsive Polymers and Nanomaterials for Gene Delivery and Imaging Applications. *Adv. Drug Deliv. Rev.* **2012**, *64*, 1046–1059. [CrossRef] [PubMed]
25. Wei, M.; Gao, Y.; Li, X.; Serpe, M.J. Stimuli-Responsive Polymers and Their Applications. *Polym. Chem.* **2017**, *8*, 127–143. [CrossRef]
26. Pelton, R. Poly(N-Isopropylacrylamide) (PNIPAM) Is Never Hydrophobic. *J. Colloid Interface Sci.* **2010**, *348*, 673–674. [CrossRef] [PubMed]
27. Halperin, A.; Kröger, M.; Winnik, F.M. Poly(N-Isopropylacrylamide) Phase Diagrams: Fifty Years of Research. *Angew. Chem. Int. Ed.* **2015**, *54*, 15342–15367. [CrossRef]
28. Fliervoet, L.A.L.; van Nostrum, C.F.; Hennink, W.E.; Vermonden, T. Balancing Hydrophobic and Electrostatic Interactions in Thermosensitive Polyplexes for Nucleic Acid Delivery. *Multifunct. Mater.* **2019**, *2*, 024002. [CrossRef]
29. Türk, M.; Dinçer, S.; Yuluğ, I.G.; Pişkin, E. In Vitro Transfection of HeLa Cells with Temperature Sensitive Polycationic Copolymers. *J. Control. Release* **2004**, *96*, 325–340. [CrossRef]
30. Feng, G.; Chen, H.; Li, J.; Huang, Q.; Gupte, M.J.; Liu, H.; Song, Y.; Ge, Z. Gene Therapy for Nucleus Pulposus Regeneration by Heme Oxygenase-1 Plasmid DNA Carried by Mixed Polyplex Micelles with Thermo-Responsive Heterogeneous Coronas. *Biomaterials* **2015**, *52*, 1–13. [CrossRef]
31. Ma, Y.; Hou, S.; Ji, B.; Yao, Y.; Feng, X. A Novel Temperature-Responsive Polymer as a Gene Vector. *Macromol. Biosci.* **2010**, *10*, 202–210. [CrossRef] [PubMed]
32. Türk, M.; Dinçer, S.; Pişkin, E. Smart and Cationic Poly(NIPA)/PEI Block Copolymers as Non-Viral Vectors: In Vitro and in Vivo Transfection Studies. *J. Tissue Eng. Regen. Med.* **2007**, *1*, 377–388. [CrossRef]
33. Mao, Z.; Ma, L.; Yan, J.; Yan, M.; Gao, C.; Shen, J. The Gene Transfection Efficiency of Thermoresponsive N,N,N-Trimethyl Chitosan Chloride-g-Poly(N-Isopropylacrylamide) Copolymer. *Biomaterials* **2007**, *28*, 4488–4500. [CrossRef]
34. Calejo, M.T.; Cardoso, A.M.S.; Kjøniksen, A.-L.; Zhu, K.; Morais, C.M.; Sande, S.A.; Cardoso, A.L.; Lima, M.C.P.d.; Jurado, A.; Nyström, B. Temperature-Responsive Cationic Block Copolymers as Nanocarriers for Gene Delivery. *Int. J. Pharm.* **2013**, *448*, 105–114. [CrossRef]
35. Kanto, R.; Yonenuma, R.; Yamamoto, M.; Furusawa, H.; Yano, S.; Haruki, M.; Mori, H. Mixed Polyplex Micelles with Thermoresponsive and Lysine-Based Zwitterionic Shells Derived from Two Poly(Vinyl Amine)-Based Block Copolymers. *Langmuir* **2021**, *37*, 3001–3014. [CrossRef]
36. Haladjova, E.; Toncheva-Moncheva, N.; Apostolova, M.D.; Trzebicka, B.; Dworak, A.; Petrov, P.; Dimitrov, I.; Rangelov, S.; Tsvetanov, C.B. Polymeric Nanoparticle Engineering: From Temperature-Responsive Polymer Mesoglobules to Gene Delivery Systems. *Biomacromolecules* **2014**, *15*, 4377–4395. [CrossRef] [PubMed]
37. Ahmad, Z.; Shah, A.; Siddiq, M.; Kraatz, H.-B. Polymeric Micelles as Drug Delivery Vehicles. *RSC Adv.* **2014**, *4*, 17028–17038. [CrossRef]

38. Ghezzi, M.; Pescina, S.; Padula, C.; Santi, P.; del Favero, E.; Cantù, L.; Nicoli, S. Polymeric Micelles in Drug Delivery: An Insight of the Techniques for Their Characterization and Assessment in Biorelevant Conditions. *J. Control. Release* **2021**, *332*, 312–336. [CrossRef]
39. Zhang, Y.; Huang, Y.; Li, S. Polymeric Micelles: Nanocarriers for Cancer-Targeted Drug Delivery. *AAPS PharmSciTech* **2014**, *15*, 862–871. [CrossRef] [PubMed]
40. Bisht, S.; Feldmann, G.; Soni, S.; Ravi, R.; Karikar, C.; Maitra, A.; Maitra, A. Polymeric Nanoparticle-Encapsulated Curcumin ("nanocurcumin"): A Novel Strategy for Human Cancer Therapy. *J. Nanobiotechnol.* **2007**, *5*, 3. [CrossRef]
41. Hewlings, S.J.; Kalman, D.S. Curcumin: A Review of Its Effects on Human Health. *Foods* **2017**, *6*, 92. [CrossRef] [PubMed]
42. Tomeh, M.; Hadianamrei, R.; Zhao, X. A Review of Curcumin and Its Derivatives as Anticancer Agents. *Int. J. Mol. Sci.* **2019**, *20*, 1033. [CrossRef] [PubMed]
43. Liu, M.; Teng, C.P.; Win, K.Y.; Chen, Y.; Zhang, X.; Yang, D.-P.; Li, Z.; Ye, E. Polymeric Encapsulation of Turmeric Extract for Bioimaging and Antimicrobial Applications. *Macromol. Rapid Commun.* **2019**, *40*, 1800216. [CrossRef]
44. Skandalis, A.; Pispas, S. PH- and Thermo-Responsive Solution Behavior of Amphiphilic, Linear Triblock Terpolymers. *Polymer* **2018**, *157*, 9–18. [CrossRef]
45. Lowry, G.V.; Hill, R.J.; Harper, S.; Rawle, A.F.; Hendren, C.O.; Klaessig, F.; Nobbmann, U.; Sayre, P.; Rumble, J. Guidance to Improve the Scientific Value of Zeta-Potential Measurements in NanoEHS. *Environ. Sci. Nano* **2016**, *3*, 953–965. [CrossRef]
46. Selianitis, D.; Pispas, S. P(MMA-Co-HPMA)-b-POEGMA Copolymers: Synthesis, Micelle Formation in Aqueous Media and Drug Encapsulation. *Polym. Int.* **2021**. [CrossRef]

polymers

MDPI

Article

Citrate-Coated Magnetic Polyethyleneimine Composites for Plasmid DNA Delivery into Glioblastoma †

Ken Cham-Fai Leung [1,*], Kathy W. Y. Sham [2], Josie M. Y. Lai [2], Yi-Xiang J. Wang [3], Chi-Hin Wong [1] and Christopher H. K. Cheng [2,*]

1 State Key Laboratory of Environmental and Biological Analysis, Department of Chemistry, The Hong Kong Baptist University, Kowloon Tong, KLN, Hong Kong, China; 12467375@life.hkbu.edu.hk
2 School of Biomedical Sciences, The Chinese University of Hong Kong, Shatin, NT, Hong Kong, China; kathysera@cuhk.edu.hk (K.W.Y.S.); josiepig@cuhk.edu.hk (J.M.Y.L.)
3 Department of Imaging and Interventional Radiology, Prince of Wales Hospital, The Chinese University of Hong Kong, Shatin, NT, Hong Kong, China; yixiang_wang@cuhk.edu.hk
* Correspondence: cfleung@hkbu.edu.hk (K.C.-F.L.); chkcheng@cuhk.edu.hk (C.H.K.C.)
† In memory of professor Christopher Hon Ki Cheng.

Abstract: Several ternary composites that are based on branched polyethyleneimine (bPEI 25 kDa, polydispersity 2.5, 0.1 or 0.2 ng), citrate-coated ultrasmall superparamagnetic iron oxide nanoparticles (citrate-NPs, 8–10 nm, 0.1, 1.0, or 2.5 μg), and reporter circular plasmid DNA pEGFP-C1 or pRL-CMV (pDNA 0.5 μg) were studied for optimization of the best composite for transfection into glioblastoma U87MG or U138MG cells. The efficiency in terms of citrate-NP and plasmid DNA gene delivery with the ternary composites could be altered by tuning the bPEI/citrate-NP ratios in the polymer composites, which were characterized by Prussian blue staining, in vitro magnetic resonance imaging as well as green fluorescence protein and luciferase expression. Among the composites prepared, 0.2 ng bPEI/0.5 μg pDNA/1.0 μg citrate-NP ternary composite possessed the best cellular uptake efficiency. Composite comprising 0.1 ng bPEI/0.5 μg pDNA/0.1 μg citrate-NP gave the optimal efficiency for the cellular uptake of the two plasmid DNAs to the nucleus. The best working bPEI concentration range should not exceed 0.2 ng/well to achieve a relatively low cytotoxicity.

Keywords: citrate; gene delivery; magnetic nanoparticle; nanostructure; polyethyleneimine

Citation: Leung, K.C.-F.; Sham, K.W.Y.; Lai, J.M.Y.; Wang, Y.-X.J.; Wong, C.-H.; Cheng, C.H.K. Citrate-Coated Magnetic Polyethyleneimine Composites for Plasmid DNA Delivery into Glioblastoma. *Polymers* **2021**, *13*, 2228. https://doi.org/10.3390/polym13142228

Academic Editor: Marek Kowalczuk

Received: 31 May 2021
Accepted: 2 July 2021
Published: 6 July 2021

1. Introduction

Novel theranostic nanomaterials [1–4] had been demonstrated for their fascinating properties in the co-delivery of genes and drugs. Their intrinsic properties as probes for various imaging techniques had also been developed rapidly for targeted brain cancer diagnosis and treatment [5–7]. Ultrasmall iron oxide nanoparticles (USIO NPs), by way of examples, offered properties including Fenton reactions [8], magnetic resonance imaging (MRI), magnetic targeting, cellular tracking, drug, and gene delivery to specific target site(s) [9–18]. On the other hand, polyethylenimine (PEI) polymers with branched structures had been investigating their properties to deliver genes and drugs with enhanced transfection and targeting efficiencies as well as with minimal cytotoxicities [19–24]. USIO composite materials with branched PEIs with alginate or deferoxamine had been demonstrated [25] with reduced cytotoxicities and the properties of different coatings. USIO-NPs that were coated with negatively charged citrate derivatives were less studied for biomedical purposes [26–28], partly because of the difficulty of transfecting them and drugs to the brain and many other organs without severe agglomeration. It is known that plasmid DNA-based gene therapy by direct delivery is less efficient [29–32]. We aimed to investigate the potential of our theranostic nanomaterials to deliver these kinds of materials with improved and/or promising transfection (delivery) efficiencies. Two expression plasmids encoding two separate reporter genes (EGFP and *Renilla* luciferase) were used to evaluate the transfection efficiency of plasmid DNA into glioblastoma cells. For the purpose of delivering designed plasmid DNAs towards a glioblastoma cell line U87MG transfection with much enhanced cellular uptake efficiency, we report herein the preparation of citrate-coated Fe_3O_4 USIO NPs (citrate-NPs, 8–10 nm), hybridizing with circular plasmid DNAs (pDNAs, pEGFP-C1 and pRL-CMV, 4 kb), and branched PEI (bPEI, MW 25 kDa, polydispersity 2.5) to furnish the ternary composites (Scheme 1) [33–35]. These composites had been for studied for MRI, fluorescence imaging and cytotoxicities. Any reduction or enhancement of the plasmid DNA uptake by the composites can be estimated by using two individual detection methods. Circular plasmid DNA pEGFP-C1 (4.7 kb) encoded with a red-shifted variant of wild-type green fluorescence protein (GFP) in mammalian cells as well as the pRL-CMV (4.0 kb) encoded with a *Renilla* luciferase in various cell types were employed. A variety of cell types and cell lines could be optimized because the reporter genes could be expressed as luminescence or fluorescence intensities. These intensities were directly proportional to the amounts of the luciferase or GPF in the cells. By the strong, enhanced and constitutive expression of the reporter genes, the signals can be easily detected. It was envisaged that after the uptake of the composites into the cells, the NPs in the composites could be cleaved thereby generating the negative (dark) MRI contrast signal. On the other hand, the plasmid DNA of the composites could further be translocated into the nucleus for expressing the reporter genes.

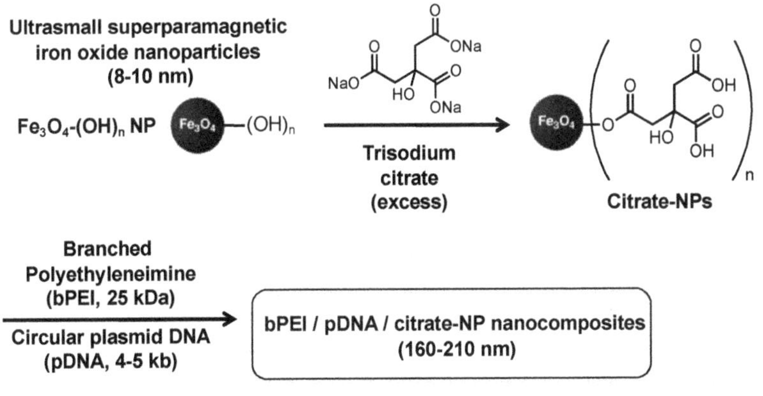

Scheme 1. Schematic illustration of the preparation of the citrate-NPs and the nanocomposites.

2. Materials and Methods

All reactions were carried out under high purity (99.9%) nitrogen atmosphere. Deionized water was obtained from Barnstead RO pure system. All solvents were bubbled with high purity nitrogen for at least 30 min before use. Ultrasmall superparamagnetic iron oxide nanoparticles with average diameter range of 8–10 nm were synthesized according to the literature procedures [36–38], which contain hydroxyl functional group as the periphery, i.e., $Fe_3O_4-(OH)_n$. Citrate-coated Fe_3O_4 USIO NPs (citrate-NPs) were prepared as follows. In particular, $Fe_3O_4-(OH)_n$ (66 mg) NPs were treated with an excess of trisodium citrate (1.01 g, Sigma-Aldrich, St. Louis, MO, USA) in water (25 mL) at 90 °C with mechanical stirring for an hour. The citrate-NPs were magnetically separated and washed repeatedly with water and ethanol. The residue was dried in high vacuum overnight to create the citrate-NPs. FT-IR stretching frequencies: 3443 cm^{-1} (O–H), 1624 cm^{-1} (C = O), 1053 cm^{-1} (C–OH) and 580 cm^{-1} (Fe–O). The size of the ternary complex was determined by the dynamic light scattering (DLS) to be 160–210 nm. Furthermore, stability test of the ternary complex was performed for a week, revealing no obvious size change.

Both U87MG and U138MG glioblastoma cell lines were acquired from the American Type Culture Collection. Cells were cultured with α-MEM (Thermo Fisher Scientific,

Waltham, MA USA) containing 10% fetal bovine serum, 100 μg/mL streptomycin and 100 U/mL penicillin, in a humidified 5% CO_2 atmosphere at 37 °C.

All circular plasmid DNAs (pDNAs) were prepared using the QIAprep Spin Miniprep Kit (QIAGEN) with an A_{260}/A_{280} ratio larger than 1.8. For the synthesis of composites, a stock solution of branched PEI (25 kDa, polydispersity 2.5, Sigma-Aldrich, St. Louis, MO, USA) was prepared with a concentration of 10 ng/μL in water. By serial dilutions, the solutions of branched PEI with different concentrations were added to the culture medium containing the plasmid DNA. After incubation for 30 min, pre-ultrasonicated, citrate-NPs of known particle and iron concentrations (ICP-MS) in water were added to the mixture, gently mixed and incubated for further 30 min to obtain the composites.

To evaluate the cytotoxicities of the magnetic composites, 5000 cells were seeded onto each well of 96-well plates for methylthiazolyldiphenyl-tetrazolium bromide (MTT) assay. On the other hand, 50,000 cells were seeded onto each well of 24-well plates for luciferase assay and fluorescence microscopy. Next day, the culture medium was replaced with the serum-free α-MEM containing different composites. After incubation of the composites for 5 h, the medium was aspirated and refreshed with complete α-MEM. The cells were incubated for further 24 h at 37 °C for subsequent assays (n = 2).

The green fluorescence in glioblastoma cells was visualized by a Nikon TE2000 fluorescence microscope and luminescence was detected by a luminometer (GloMax 20/20 Luminometer, Promega, Madison, WI, USA). Luciferase expression of different composites in glioblastoma cells. *Renilla* luciferase reporter plasmid pRL-CMV was mixed with different amounts of bPEI and citrate-NP to form the composites. The amount of pRL-CMV of all composites was fixed at 0.5 μg/well. These composites were incubated with U87MG and U138MG cells for 5 h. Twenty-four hours after transfection, cells were harvested and lysed. *Renilla* activities (RLU) were then measured and normalized against total cellular protein per well.

The cells were washed with PBS to remove any free composites. Cells were then fixed using paraformaldehyde (4%) for 40 min. Subsequently, cells were washed with PBS and incubated with freshly prepared Perls' reagent (4% potassium ferrocyanide and 12% HCl, 1:1 v/v) for 30 min. Cells were washed with PBS three times, counterstained with neutral red (0.02%), and subsequently observed by an inverted bright-field optical microscope (Nikon TE2000).

The viability of cells incubated with different composites was estimated by MTT assay in glioblastoma cells. Ten microliters of 5 mg/mL MTT solution was added into each well. After incubation for 3 h, the medium was removed, and formazan crystals were dissolved in dimethyl sulfoxide (150 μL) for 10 min on a shaker. A small round disc-like magnet was placed under the plate to the bottom of the well and to attract the magnetic composite-uptaken cells. After that, 100 μL of the supernatant was transferred to another 96-well plate. Absorbance of each well was measured on a microplate reader (Bio-Rad, Model 3550) at a wavelength of 540 nm. The relative cell viability (%) for each sample which is related to the control well, was calculated.

For in vitro MRI, after washing with PBS, the cells were trypsinised and counted. Different numbers (12.5k, 25k, 50k, 100k, 150k, and 300k) of cells were placed in an Eppendorf tube (1.5 mL) separately. After centrifugation at 3000× g for 5 min, the Eppendorf tubes were placed perpendicularly to the main magnetic induction field (B_0) in a $20 \times 12 \times 8$ cm^3 water bath. MRI was performed with a 3.0-T clinical whole-body magnetic resonance unit (Achieva, Philips Medical Systems) using a transmit–receive head coil. The magnetic resonance sequence was a two-dimensional gradient-echo sequence with TR = 400 ms, TE = 48 ms, flip angle = 18°, matrix = 512 × 256, resolution = 0.45 × 0.45 mm, slice thickness = 2 mm, and number of excitations = 2. Sagittal images were obtained through the central section of the bottom tips of the Eppendorf tubes. The areas of signal void at the bottom of the Eppendorf tubes due to the citrate-NP-containing composites uptaken into U87MG cells from which the NPs is the MRI-responsive contrast agent. The T_2 relaxation times were measured by using a standard Carr–Purcell–Meiboom–Gill pulse sequence with the following parameters: repetition time TR = 2000 ms, echo time TE range = 30–960 ms, 32 echoes, field-of-view = 134 × 67 mm^2, matrix = 128 × 64, slice thickness = 5 mm, number of excitations = 3. T_2 relaxation times were calculated by fitting the logarithmic region of interest signal amplitudes versus TE. The T_2 relaxivities (r_2) were determined by a linear fit of the inverse relaxation times as a function of the iron concentrations used.

3. Results

Both USIO nanoparticles with negatively charged citrate coating as well as the negatively charged plasmid DNAs could be noncovalently self-assembled [39] to the positively charged branched PEI polymers to furnish the ternary composites (160–210 nm), stabilized by mainly multiple ionic ($CO_2^{-}\cdots NH^{+}$) charge attractions between the carboxylic acid group of the citrate and the amine group of the PEI. The polymeric nature of the materials provided multivalent, strong interactions to furnish the stable composites. The morphology and surface functional groups of the composites were characterized by transmission electron microscopy (TEM) and Infrared (IR) absorption spectroscopy, which had been reported in the literature [19,30,31]. The as-prepared citrate-NPs had a narrow size distribution (8–10 nm). IR spectra of these NPs reveal their functional group characteristic signals at 580 cm^{-1} for the Fe–O, 1053 cm^{-1} for the C–OH, 1624 cm^{-1} for the carbonyl C=O and 3443 cm^{-1} for the O–H moieties.

The size of the ternary complex was determined by DLS in buffer to be 160–210 nm. Stability test of the ternary complex was performed for a week, revealing no obvious size change (180 ± 20 nm) by DLS.

Composites with varying amounts of citrate-NPs (0.1 to 1.0 μg/well), together with fixed amounts of bPEI (0.2 ng/well) and pDNA (0.5 μg/well) had been separately internalized with the U87MG cells. The cellular uptake efficiencies generally increased when the citrate-NP concentration increased from 0.1 and 1.0 μg/well, as indicated by the Prussian Blue staining of the citrate-NPs (Figure 1A). The typical GFP green, fluorescent images of the U87MG cells which had been separately internalized with four different composites with different amounts of bPEI (0.1 and 0.2 ng/well) and citrate-NPs (0.1 and 1.0 μg/well) are shown in Figure 1B. All four composites showed cellular uptake efficiency with significant green fluorescence observed with the U87MG cells.

U87MG cells that were incubated separately with two different composites of citrate-NP (1.0 and 2.5 μg/well) with fixed bPEI (0.2 ng/well) and pDNA (0.5 μg/well) amounts were analysed by in vitro MRI. Substantial negative (dark) contrast MRI signals with a ballooning effect were observed as shown in Figure 2 with the cells that were centrifuged at the bottom of the Eppendorf tubes. Increasing number of ternary composite-incubated cells gave a larger MRI dark contrast signal. The MRI contrast signal with a cell number of 12.5k was small and yet still observable. The iron concentrations [Fe] of all cell samples were determined by the inductively coupled plasma mass spectrometry (ICP-MS). By measuring the pixels of the dark contrasts related to the iron concentrations, the in vitro T_2 relaxivities (r_2) of the two composites (0.2 ng bPEI/0.5 μg pEGFP-C1 pDNA/1.0 μg citrate-NP and 0.2 ng bPEI/0.5 μg pEGFP-C1 pDNA/2.5 μg citrate-NP) were determined to be 45.9 and 43.9 s^{-1} μM^{-1} Fe, respectively. Comparatively, the composites with 1.0 μg/well citrate-NPs possessed the highest MRI signal intensities than that of the 2.5 μg/well.

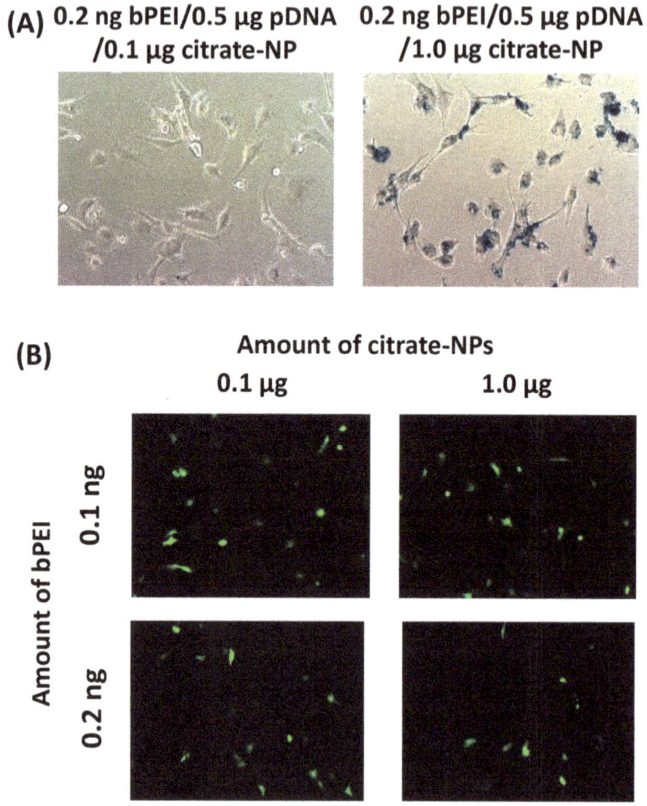

Figure 1. (**A**) Prussian staining images of two different composites 0.2 ng branched bPEI/0.5 μg pEGFP-C1 pDNA/0.1 μg citrate-NP and 0.2 ng branched bPEI/0.5 μg pEGFP-C1 pDNA/1.0 μg citrate-NP 24 h after incubation with each composite for 5 h in U87MG cells. (**B**) GFP fluorescence images of four different composites with a fixed amount of 0.5 μg pEGFP-C1 pDNA 24 h after incubation with each composite for 5 h in U87MG cells.

Transfecting U87MG cells with the pRL-CMV pDNA-containing ternary composites resulted in a range of 10^4 to 10^5 RLU (Figure 3, left). Generally, higher RLU signal was observed with U138MG cells (Figure 3, right) then U87MG cells. Noticeably, the RLU from the composite 0.2 ng bPEI/0.5 μg pRL-CMV pDNA/0.1 μg citrate-NP in U138MG cells was exceptionally low. However, 0.1 ng bPEI/0.5 μg pRL-CMV pDNA/0.1 μg citrate-NP gave the highest luciferase activity in both glioblastoma cell lines.

Ternary magnetic composites and the controls, i.e., medium alone, lipofectamine, branched bPEI alone, and bPEI/pDNA composites, were prepared and that their cytotoxicities in U87MG cells were evaluated by MTT assay (Figure 4). Generally, percentages cell viability decreased with increasing amounts of bPEI from 0.1 ng/well or 0.2 ng (80–90%) to 0.5 ng (40–50%). These results also revealed that the cytotoxicity of using 0.5 ng bPEI was approximately 10% higher than the commercially available transfection agent lipofectamine.

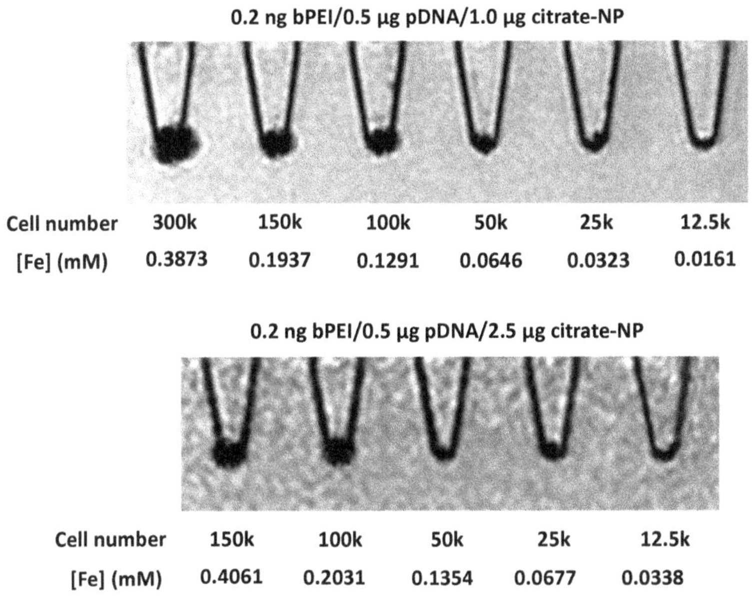

Figure 2. Gradient echo in vitro MRI images of the two composites (0.2 ng branched bPEI/0.5 µg pEGFP-C1 pDNA/1.0 µg citrate-NP and 0.2 ng branched bPEI/0.5 µg pEGFP-C1 pDNA/2.5 µg citrate-NP)-transfected U87MG cells in Eppendorf tubes with culture medium.

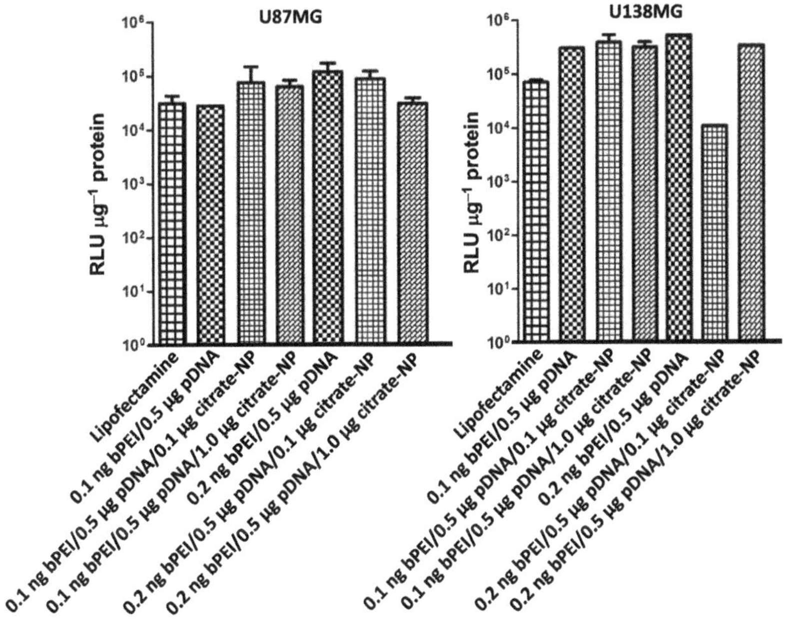

Figure 3. Luciferase expression of different composite transfected to U87MG (left) and U138MG (right) glioblastoma cells for 5 h. The amount of plasmid pRL-CMV pDNA of all composites was fixed at 0.5 µg/well. The amount of lipofectamine was fixed at 2.0 µg/well. *Renilla* activities (RLU) were measured and normalized against total cellular protein per well.

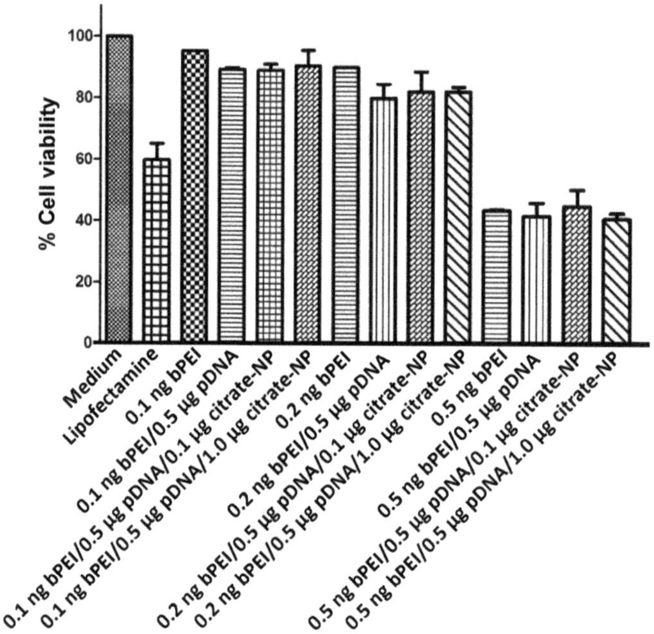

Figure 4. Cell viability of U87MG cells as determined by MTT assay 24 h after incubation with each composite for 5 h. The concentration of plasmid pEGFP-C1 pDNA of all composites was fixed at 0.5 μg/well. The amount of lipofectamine was fixed at 2.0 μg/well.

4. Discussion

The Infrared absorbance of the citrate-NPs demonstrated that the successful surface modification of iron hydroxyl groups to citrate in the presence of the carbonyl absorption at 1624 cm^{-1}. Ternary composites based on different combination amounts of branched bPEI (0.1 or 0.2 ng), citrate-NP (0.1, 1.0, or 2.5 μg), and pEGFP-C1 or pRL-CMV pDNA (0.5 μg) were studied for optimization of the best composite. It is reasonably to consider that the cellular uptake efficiency depends on the surface charge density, stability of the composites and further cleavage of the composite with citrate-NP localized in cytoplasm and pDNA in nucleus. The composites would eventually be dissociated into separate components and after the nanomaterial's uptaken mechanism. Therefore, it is essential to tune the components' ratios to study the uptake of NP and nucleic acids towards U87MG and U138MG glioblastoma cells. For the best MRI observations, the use of 1.0 μg/well citrate-NP would be favourable. However, 0.1 ng bPEI/0.5 μg pDNA/0.1 μg citrate-NP would give an optimal gene delivery into the U87MG cells. pDNA pRL-CMV carried a *Renilia* luciferase gene thus it had to be translocated into the nuclei of U87MG or U138MG cells for the gene transcription process. Cellular uptake efficiencies of plasmid pDNA in the ternary composites towards U138MG cells were generally similar to that of U87MG cells. The amount of pDNA that was successfully transfected, was directly proportional to the luminescent signal generated from luciferase expression. Therefore, a composite with a high luciferase activity revealed higher gene delivery efficiency into the glioblastoma cells. Most of the composites had a RLU value higher than that of the commercially available transfecting agent lipofectamine that acted on a positive control in the present study. The ternary composite 0.1 ng bPEI/0.5 μg pDNA/0.1 μg citrate-NP gave the highest luciferase activity and fluorescence signal in both glioblastoma cell lines. The best working bPEI concentration range should not exceed 0.2 ng/well to achieve a relatively low cytotoxicity.

5. Conclusions

Different ternary composites, based on branched bPEI (0.1 or 0.2 ng), negatively charged citrate-coated, ultrasmall superparamagnetic iron oxide nanoparticles citrate-NPs (0.1, 1.0, or 2.5 μg), and pEGFP-C1 or pRL-CMV circular plasmid pDNA (0.5 μg), were studied for optimization of the best composite for the uptake into U87MG or U138MG glioblastoma cells. The uptake efficiency in terms of citrate-NP and pDNA gene delivery with the ternary composites could be altered by tuning the bPEI/citrate-NP ratios in the composite, thereby characterized by Prussian blue staining, in vitro MRI as well as GFP and luciferase expression. Among the composites prepared, 0.2 ng bPEI/0.5 μg pDNA/1.0 μg citrate-NP ternary composite possessed the best cellular uptake efficiency of NP evident by MRI assessments and Prussian blue staining. Composite comprising 0.1 ng bPEI/0.5 μg pDNA/0.1 μg citrate-NP gave the optimal efficiency for the cellular uptake of the two circular plasmid pDNAs to the nucleus. The cytotoxicity became significant when 0.5 ng/well of bPEI was present in the ternary composites. The best working bPEI concentration range should not exceed 0.2 ng/well to achieve a relatively low cytotoxicity. As a result, as-prepared polymer composites or other novel nanostructured magnetic composites offered potential biomedical applications in simultaneous gene delivery, imaging contrast enhancement, and mechanistic study. To achieve the next generation in vivo nano-theranostic anti-cancer drug delivery systems, more sophisticated, stimuli-responsive polymeric/dendritic nanostructures held by mechanical bonds for bio-evaluation on imaging and active drug release on demand [40–46] would be developed for targeting various brain tumours.

Author Contributions: Conceptualization, K.C.-F.L. and C.H.K.C.; Data curation, K.W.Y.S., J.M.Y.L., Y.-X.J.W. and C.-H.W.; Formal analysis, C.H.K.C.; Funding acquisition, K.C.-F.L.; Methodology, K.C.-F.L. and C.H.K.C.; Project administration, K.C.-F.L.; Supervision, K.C.-F.L. and C.H.K.C. All authors have read and agreed to the published version of the manuscript.

Funding: We acknowledge the financial support partially by The Chinese University of Hong Kong as well as The Hong Kong Baptist University (SKLP_1920_P05, RC-IRCMs/17-18/03 and RC-KRPS-20-21/02).

Institutional Review Board Statement: The study was conducted according to the guidelines approved by The Chinese University of Hong Kong.

Data Availability Statement: The data presented in this study are available on request from the corresponding authors.

Conflicts of Interest: The authors declare no conflict of interest.

References

1. Leung, K.C.F.; Xuan, S.; Zhu, X.; Wang, D.; Chak, C.P.; Lee, S.F.; Ho, W.K.W.; Chung, B.C.T. Gold and iron oxide hybrid nanocomposite materials. *Chem. Soc. Rev.* **2012**, *41*, 1911–1928. [CrossRef]
2. Jo, S.D.; Ku, S.H.; Won, Y.Y.; Kim, S.H.; Kwon, I.C. Targeted nanotheranostics for future personalized medicine: Recent progress in cancer therapy. *Theranostics* **2016**, *6*, 1362–1377. [CrossRef]
3. Smith, B.R.; Gambhir, S.S. Nanomaterials for in vivo imaging. *Chem. Rev.* **2017**, *117*, 901–986. [CrossRef]
4. Cucci, L.M.; Trapani, G.; Hansson, Ö.; La Mendola, D.; Satriano, C. Gold nanoparticles functionalized with angiogenin for wound care application. *Nanomaterials* **2021**, *11*, 201. [CrossRef] [PubMed]
5. Wohlfart, S.; Gelperina, S.; Kreuter, J. Transport of drugs across the blood-brain barrier by nanoparticles. *J. Control. Release* **2012**, *161*, 264–273. [CrossRef] [PubMed]
6. Koffie, R.M.; Farrar, C.T.; Saidi, L.J.; William, C.M.; Hyman, B.T.; Spires-Jones, T.L. Nanoparticles enhance brain delivery of blood-brain barrier-impermeable probes for in vivo optical and magnetic resonance imaging. *Proc. Natl. Acad. Sci. USA* **2011**, *108*, 18837–18842. [CrossRef]
7. Agrawal, P.; Singh, R.P.; Sonali; Kumari, L.; Sharma, G.; Koch, B.; Rajesh, C.V.; Mehata, A.K.; Singh, S.; Pandey, B.L.; et al. TPGS-chitosan cross-linked targeted nanoparticles for effective brain cancer therapy. *Mater. Sci. Eng. C Mater. Biol. Appl.* **2017**, *74*, 167–176. [CrossRef] [PubMed]
8. Shen, Z.Y.; Song, J.B.; Yung, B.C.; Zhou, Z.J.; Wu, A.G.; Chen, X.Y. Emerging strategies of cancer therapy based on ferroptosis. *Adv. Mater.* **2018**, *30*, 1704007. [CrossRef] [PubMed]

9. Xuan, S.H.; Lee, S.F.; Lau, J.T.; Zhu, X.; Wang, Y.X.J.; Wang, F.; Lai, J.M.; Sham, K.W.; Lo, P.C.; Yu, J.C.; et al. Photocytotoxicity and magnetic relaxivity responses of dual-porous γ-Fe2O3@meso-SiO2 microspheres. *ACS Appl. Mater. Interfaces* **2012**, *4*, 2033–2040. [CrossRef]
10. Nair, M.; Guduru, R.; Liang, P.; Hong, J.; Sagar, V.; Khizroev, S. Externally controlled on-demand release of anti-HIV drug using magneto-electric nanoparticles as carriers. *Nat. Commun.* **2013**, *4*, 1707. [CrossRef] [PubMed]
11. Leung, K.C.F.; Lee, S.F.; Wong, C.H.; Chak, C.P.; Lai, J.M.Y.; Zhu, X.M.; Wang, Y.X.J.; Sham, K.W.Y.; Cheng, C.H.K. Nanoparticle-DNA-polymer composites for hepatocellular carcinoma cell labeling, sensing, and magnetic resonance imaging. *Methods* **2013**, *64*, 315–321. [CrossRef] [PubMed]
12. Lee, S.F.; Zhu, X.M.; Wang, Y.X.J.; Xuan, S.H.; You, Q.; Chan, W.H.; Wong, C.H.; Wang, F.; Yu, J.C.; Cheng, C.H.K.; et al. Ultrasound, pH, and magnetically responsive crown-ether-coated core/shell nanoparticles as drug encapsulation and release systems. *ACS Appl. Mater. Interfaces* **2013**, *5*, 1566–1574. [CrossRef] [PubMed]
13. Wang, D.W.; Zhu, X.M.; Lee, S.F.; Chan, H.M.; Li, H.W.; Kong, S.K.; Yu, J.C.; Cheng, C.H.K.; Wang, X.Y.J.; Leung, K.C.F. Folate-conjugated Fe3O4@SiO2@gold nanorods@mesoporous SiO_2 hybrid nanomaterial: A theranostic agent for magnetic resonance imaging and photothermal therapy. *J. Mater. Chem. B* **2013**, *1*, 2934–2942. [CrossRef]
14. Wang, Y.X.J.; Zhu, X.M.; Liang, Q.; Cheng, C.H.K.; Wang, W.; Leung, K.C.F. In vivo chemoembolisation and magnetic resonance imaging of liver tumors by using iron oxide nanoshell/doxorubicin/poly(vinyl alcohol) hybrid composites. *Angew. Chem. Int. Ed.* **2014**, *53*, 4912–4915. [CrossRef]
15. Sun, C.; Lee, J.S.H.; Zhang, M. Magnetic nanoparticles in MR imaging and drug delivery. *Adv. Drug Deliver. Rev.* **2008**, *60*, 1252–1265. [CrossRef] [PubMed]
16. De, M.; Ghosh, P.S.; Rotello, V.M. Applications of nanoparticles in biology. *Adv. Mater.* **2008**, *20*, 4225–4241. [CrossRef]
17. Samamta, A.; Medintz, I.L. Nanoparticles and DNA—A powerful and growing functional combination in bionanotechnology. *Nanoscale* **2016**, *8*, 9037–9095. [CrossRef] [PubMed]
18. Fisher, D.G.; Price, R.J. Recent advances in the use of focused ultrasound for magnetic resonance image-guided therapeutic nanoparticle delivery to the central nervous system. *Front. Pharm.* **2019**, *10*, 1348. [CrossRef]
19. Leung, K.C.F.; Chak, C.P.; Lee, S.F.; Lai, J.M.; Zhu, X.M.; Wang, Y.X.J.; Sham, K.W.Y.; Cheng, C.H.K. Enhanced cellular uptake and gene delivery of glioblastoma with deferoxamine-coated nanoparticle/plasmid DNA/branched polyethylenimine composites. *Chem. Commun.* **2013**, *49*, 549–551. [CrossRef]
20. Kunath, K.; von Harpe, A.; Fischer, D.; Peterson, H.; Bickel, U.; Voigt, K.; Kissel, T. Low-molecular-weight polyethylenimine as a non-viral vector for DNA delivery: Comparison of physicochemical properties, transfection efficiency and in vivo distribution with high-molecular-weight polyethylenimine. *J. Control. Release* **2003**, *89*, 113–125. [CrossRef]
21. Moghimi, S.M.; Symonds, P.; Murray, J.C.; Hunter, A.C.; Debska, G.; Szewczyk, A. A two-stage poly(ethylenimine)-mediated cytotoxicity: Implications for gene transfer/therapy. *Mol. Ther.* **2005**, *11*, 990–995. [CrossRef] [PubMed]
22. Song, J.; Wang, D.; Wang, J.; Shen, Q.; Xie, C.; Lu, W.; Wang, R.; Liu, M. Low molecular weight polyethyleneimine modified by 2-aminoimidazole achieving excellent gene transfection efficiency. *Eur. Polym. J.* **2020**, *140*, 110017. [CrossRef]
23. Li, S.; Lin, L.; Wang, W.; Yan, X.; Chen, B.; Jiang, S.; Liu, S.; Ma, X.; Tian, H.; Yu, X. Aza-crown ether locked on polyethyleneimine: Solving the contradiction between transfection efficiency and safety during in vivo gene delivery. *Chem. Commun.* **2020**, *56*, 5552–5555. [CrossRef]
24. Lu, L.; Chen, H.; Wang, L.; Zhao, L.; Cheng, Y.; Wang, A.; Wang, F.; Zhang, X. A dual receptor targeting- and BBB-penetrating-peptide functionalized polyethyleneimine nanocomplex for secretory endostatin gene delivery to malignant glioma. *Int. J. Nanomed.* **2020**, *15*, 8875–8892. [CrossRef]
25. Leung, K.C.F.; Sham, K.W.Y.; Chak, C.P.; Lai, J.M.Y.; Lee, S.F.; Wang, Y.X.J.; Cheng, C.H.K. Evaluation of biocompatible alginate- and deferoxamine-coated ternary composites for magnetic resonance imaging and gene delivery into glioblastoma cells. *Quant. Imaging Med. Surg.* **2015**, *5*, 382–391.
26. Zhao, F.; Zhao, Y.; Liu, Y.; Chang, X.L.; Chen, C.Y.; Zhao, Y.L. Cellular uptake, intracellular trafficking and cytotoxicity of nanomaterials. *Small* **2011**, *7*, 1322–1337. [CrossRef]
27. Albanese, A.; Tang, P.S.; Chan, W.C.W. The effect of nanoparticle size, shape and surface chemistry on biological systems. *Ann. Rev. Biomed. Eng.* **2012**, *14*, 1–16. [CrossRef] [PubMed]
28. Zhu, X.M.; Wang, Y.X.J.; Leung, K.C.F.; Lee, S.F.; Zhao, F.; Wang, D.W.; Lai, J.M.; Wan, C.; Cheng, C.H.K.; Ahuja, A.T. Enhanced cellular uptake of aminosilane-coated superparamagnetic iron oxide nanoparticles in mammalian cell lines. *Int. J. Nanomed.* **2012**, *7*, 953–964.
29. Heller, L.; Jaroszeski, M.J.; Coppola, D.; Pottinger, C.; Gilbert, R.; Heller, R. Electrically mediated plasmid DNA delivery to hepatocellular carcinomas in vivo. *Gene Ther.* **2000**, *7*, 826–829. [CrossRef] [PubMed]
30. Jiang, Z.; Thayumanavan, S. Noncationic material design for nucleic acid delivery. *Adv. Ther.* **2020**, *3*, 1900206. [CrossRef] [PubMed]
31. Peng, L.; Wagner, E. Polymeric carriers for nucleic acid delivery: Current designs and future directions. *Biomacromolecules* **2019**, *20*, 3613–3626. [CrossRef]
32. Durymanov, M.; Reineke, J. Non-viral delivery of nucleic acids: Insight into mechanisms of overcoming intracellular barriers. *Front. Pharmacol.* **2018**, *9*, 971. [CrossRef]

33. Fan, Y.; Yao, J.; Du, R.H.; Hou, L.; Zhou, J.P.; Lu, Y.; Meng, Q.G.; Zhang, Q. Ternary complexes with core-shell bilayer for double level targeted gene delivery: In vitro and in vivo evaluation. *Pharm. Res.* **2013**, *30*, 1215–1227. [CrossRef] [PubMed]
34. Park, S.J.; Park, W.; Na, K. Tumor intracellular-environment responsive materials shielded nano-complexes for highly efficient light-triggered gene delivery without cargo gene damage. *Adv. Funct. Mater.* **2015**, *25*, 3472–3482. [CrossRef]
35. Saha, N.; Saha, N.; Sáha, T.; Öner, E.T.; Brodnjak, U.V.; Redl, H.; von Byern, J.; Sáha, P. Polymer based bioadhesive biomaterials for medical application—A perspective of redefining healthcare system management. *Polymers* **2020**, *12*, 3015. [CrossRef] [PubMed]
36. Wang, H.H.; Wang, Y.X.; Leung, K.C.F.; Au, D.W.; Xuan, S.; Chak, C.P.; Lee, S.K.; Sheng, H.; Zhang, G.; Qin, L.; et al. Durable mesenchymal stem cell labelling by using polyhedral superparamagnetic iron oxide nanoparticles. *Chem. Eur. J.* **2009**, *15*, 12417–12425. [CrossRef] [PubMed]
37. Wang, Y.X.J.; Leung, K.C.F.; Cheung, W.H.; Wang, H.H.; Shi, L.; Wang, D.F.; Qin, L.; Ahuja, A.T. Low-intensity pulsed ultrasound increases cellular uptake of superparamagnetic iron oxide nanomaterial: Results from human osteosarcoma cell line U2OS. *J. Magn. Reson. Imaging* **2010**, *31*, 1508–1513. [CrossRef]
38. Wang, Y.X.; Quercy-Jouvet, T.; Wang, H.H.; Li, A.K.; Chak, C.P.; Xuan, S.; Shi, L.; Wang, D.F.; Lee, S.F.; Leung, P.C.; et al. Efficacy and durability in direct labeling of mesenchymal stem cells using ultrasmall superparamagnetic iron oxide nanoparticles with organosilica, dextran, and PEG coatings. *Materials* **2011**, *4*, 703–715. [CrossRef]
39. South, C.R.; Leung, K.C.F.; Lanari, D.; Stoddart, J.F.; Weck, M. Noncovalent side-chain functionalization of terpolymers. *Macromolecules* **2006**, *39*, 3738–3744. [CrossRef]
40. Leung, K.C.F.; Lau, K.N. Self-assembly and thermodynamic synthesis of rotaxane dendrimers and related structures. *Polym. Chem.* **2010**, *1*, 988–1000. [CrossRef]
41. Ho, W.K.W.; Lee, S.F.; Wong, C.H.; Zhu, X.M.; Kwan, C.S.; Chak, C.P.; Mendes, P.M.; Cheng, C.H.K.; Leung, K.C.F. Type III-B rotaxane dendrimers. *Chem. Commun.* **2013**, *49*, 10781–10783. [CrossRef] [PubMed]
42. Kwan, C.S.; Zhao, R.; Van Hove, M.A.; Cai, Z.; Leung, K.C.F. Higher-generation type III-B rotaxane dendrimers with controlling particle size in three-dimensional molecular switching. *Nat. Commun.* **2018**, *9*, 497. [CrossRef] [PubMed]
43. Kwan, C.S.; Wang, T.; Li, M.; Chan, A.S.C.; Cai, Z.; Leung, K.C.F. Type III-C rotaxane dendrimers: Synthesis, dual size modulation and in vivo evaluation. *Chem. Commun.* **2019**, *55*, 13426–13429. [CrossRef] [PubMed]
44. Wang, T.; Cai, Z.W.; Chen, Y.Y.; Lee, W.K.; Kwan, C.S.; Li, M.; Chan, A.S.C.; Chen, Z.F.; Cheung, A.K.L.; Leung, K.C.F. MALDI-MS imaging analysis of noninflammatory type III rotaxane dendrimers. *J. Am. Soc. Mass Spectrom.* **2020**, *31*, 2488–2494. [CrossRef] [PubMed]
45. Kwan, C.S.; Leung, K.C.F. Hetero type III-B rotaxane dendrimers. *J. Chin. Chem. Soc.* **2020**, *67*, 1734–1741. [CrossRef]
46. Kwan, C.S.; Leung, K.C.F. Development and advancement of rotaxane dendrimers as switchable macromolecular machines. *Mater. Chem. Front.* **2020**, *4*, 2825–2844. [CrossRef]

MDPI AG
Grosspeteranlage 5
4052 Basel
Switzerland
Tel.: +41 61 683 77 34

Polymers Editorial Office
E-mail: polymers@mdpi.com
www.mdpi.com/journal/polymers

www.ingramcontent.com/pod-product-compliance
Lightning Source LLC
LaVergne TN
LVHW070238100526
838202LV00015B/2148
* 9 7 8 3 7 2 5 8 2 2 1 2 6 *